Third International Conference on Computer Simulation in Physics and Beyond (CSP2018)

Journal of Physics: Conference Series Volume 1163

Moscow, Russia
24-27 September 2018

Editor:

Lev Shchur

ISBN: 978-1-5108-8490-8
ISSN: 1742-6588

Printed from e-media with permission by:

Curran Associates, Inc.
57 Morehouse Lane
Red Hook, NY 12571

Institute of Physics
Dirac House, Temple Back
Bristol BS1 6BE UK

Phone: 44 1 17 929 7481
Fax: 44 1 17 920 0979

techtracking@iop.org

Additional copies of this publication are available from:

Curran Associates, Inc.
57 Morehouse Lane
Red Hook, NY 12571 USA
Phone: 845-758-0400
Fax: 845-758-2633
Email: curran@proceedings.com
Web: www.proceedings.com

TABLE OF CONTENTS

IOP Conf. Series: Journal of Physics: Conf. Series **1163** (2019) 011001 doi:10.1088/1742-6596/1163/1/011001

Preface

Computer simulations are well established third pillar of the modern natural sciences. It is the complement to the experiment and theory, and to the emerging the fourth pillar, the extraction of the new knowledge from the big data sets.

This volume collects the referred papers based on plenary, invited, and oral talks, as well on the posters presented at the Third International Conference on Computer Simulations in Physics and beyond (CSP2018), which took place September 24-27, 2018 in Moscow. The Conference continues the tradition started by an inaugural conference in 2015. It took place on the campus of A.N. Tikhonov Moscow Institute of Electronics and Mathematics in Strogino, was jointly organized by the National Research University Higher School of Economics, the Landau Institute for Theoretical Physics and Science Center in Chernogolovka.

The Conference is a multidisciplinary meeting, with a focus on computational physics and related subjects. Indeed, methods of computational physics prove useful in a broad spectrum of research in multiple branches of natural sciences, and this volume provides a sample.

We hope that this volume will interest readers, and we are already looking forward to the next conference in the series.

Moscow, Russia
November, 2018

CSP2018 Conference Chair
and Volume Editor
Lev Shchur

IOP Conf. Series: Journal of Physics: Conf. Series **1163** (2019) 011002 doi:10.1088/1742-6596/1163/1/011002

Peer review statement

All papers published in this volume of *Journal of Physics: Conference Series* have been peer reviewed through processes administered by the proceedings Editors. Reviews were conducted by expert referees to the professional and scientific standards expected of a proceedings journal published by IOP Publishing.

IOP Conf. Series: Journal of Physics: Conf. Series **1163** (2019) 012001 doi:10.1088/1742-6596/1163/1/012001

Medium-range percolation in two dimensions

Youjin Deng [1], Yunqing Ouyang [1] and Henk W J Blöte [2]

[1] Hefei National Laboratory for Physical Sciences at Microscale, Department of Modern Physics, University of Science and Technology of China, Hefei 230027, China

[2] Lorentz Institute, Leiden University, P.O. Box 9506, 2300 RA Leiden, The Netherlands

E-mail: yjdeng@ustc.edu.cn

Abstract. We investigate equivalent-neighbor percolation models in two dimensions with a variable interaction range, and include the mean-field limit. We employ a Monte Carlo algorithm whose efficiency depends only weakly on the number of equivalent neighbors within the range of the interactions. We consider 2 classes of models, one in which the interacting neighbors fill a circle, and one in which they fill a square. We first determine the critical points, and then determine the critical exponents. These are, for all finite ranges investigated, in agreement with the exactly known exponents of the nearest-neighbor model.

1. Introduction

It is well known that the scaling properties of models with long-range interactions reveal some aspects that are absent in short-range models, in particular the violation of hyperscaling. These aspects were analyzed in detail by Knops et al. [1] for the Ising model, using position-space renormalization. They found that these models can still be described by the renormalization theory, by use of the mechanism of dangerous irrelevant variables, i.e., singular dependence of the scaling function on these irrelevant fields.

In this work we place the percolation model in the wider context of the q-state Potts [2] model, for which a general theoretical context is available, as well as numerical results for $q = 2$, 3 and 4. The crossover between the mean-field (MF) Ising model ($q = 2$) and the short-range Ising model was numerically investigated by Luijten et al. [3]. It was found that, in two dimensions, all equivalent-neighbor Ising models with a finite range belong to the short-range universality class. Only the limit of infinitely many neighbors is mean-field-like. This result fits well with the principle that the universal properties of a phase transition are typically independent on the microscopic details of the Hamiltonian. Nevertheless, there are instances of models disobeying this principle. For instance, a seemingly innocent increase of the range of interactions to $\sqrt{5}$ times the nearest-neighbor distance turns the critical transition of the two-dimensional, 4-state Potts model into a discontinuous one [4]. Also the phase transition of the three-state Potts model becomes discontinuous for a sufficiently long range. The critical and first-order ranges are separated by a tricritical point located near the model with 80 equivalent neighbors [4].

These existing results for the $q = 2$ (Ising), and the $q = 3$ and 4 Potts models still have to be supplemented with a numerical analysis for the $q \to 1$ limit of the random-cluster model, i.e., the percolation model. In the present work, we investigate the role of the range of interactions in the two-dimensional percolation model. More specifically we consider equivalent-neighbor models, in which each site equally interacts with a local environment, i.e., a set of neighbors

IOP Conf. Series: Journal of Physics: Conf. Series **1163** (2019) 012001 doi:10.1088/1742-6596/1163/1/012001

filling a region around that site. We distinguish two cases of such local environments: first, we define the type-1 models, with local interactions between a site and all neighboring sites within a circle of radius r; and second, we define type-2 models with interactions including all neighboring sites within a $2r \times 2r$ square. More precisely, in type-1 models a site i with coordinates x_i, y_i couples to all sites j satisfying $\sqrt{(x_i - x_j)^2 + (y_i - y_j)^2} \leq r$, and in type-2 models to all sites j simultaneously satisfying $|x_i - x_j| \leq r$ and $|y_i - y_j| \leq r$.

We formulate this percolation problem using the language of the q-state Potts model and the equivalent random-cluster model [5]. The Potts variables s_k can, for all sites k assume the values 1,2,..., q. The Potts Hamiltonian is

$$\mathcal{H} = -K \sum_{<i,j>} \delta(s_i, s_j) - H \sum_{k=1}^{N} \delta(s_k, 1)\,, \tag{1}$$

where the sum on $< i, j >$ is as defined above for cases 1 or 2. We restrict ourselves to the ferromagnetic case $K > 0$. The Kasteleyn-Fortuin mapping [5] involves the definition of bond variables with values $b_{ij} = 0$ or 1 between all site pairs i, j fitting within the interaction range r. In the absence of the external field H, summation of the Potts partition sum on the Potts variables then leads to the random-cluster partition sum

$$Z_{\mathrm{RC}} = \sum_{\mathcal{G}} (e^K - 1)^{N_{\mathrm{b}}} q^{N_{\mathrm{c}}}\,, \tag{2}$$

where the graph $\mathcal{G}$ is the set of bond variables with $b_{ij} = 1$, N_{b} is the sum over all b_{ij}, and N_{c} the number of percolating clusters formed by the graph $\mathcal{G}$. The connection with the percolation problem is obtained by setting $q = 1$ and dividing out a factor $e^{-KN_{\mathrm{e}}}$ where N_{e} is the total number of bond variables:

$$Z_{\mathrm{perc}} = e^{-KN_{\mathrm{e}}} Z_{\mathrm{RC}} \sum_{\mathcal{G}} p^{N_{\mathrm{b}}} (1-p)^{N_{\mathrm{e}} - N_{\mathrm{b}}} q^{N_{\mathrm{c}}}\,, \tag{3}$$

with $p = 1 - e^{-K}$. This sum is trivial for $q = 1$, but it may serve as the generating function of percolation configurations. Differentiation to q at $q = 1$ leads to the expectation value of the number of clusters N_{c}. The relation between the Potts magnetic properties and the cluster-size distribution of the percolation model also requires the $q \to 1$ limit [6] of the random-cluster model.

The physics of the medium-range percolation model was recently investigated by Ouyang et al. [7], using a Monte Carlo method, and subsequent analysis of the Binder [8] ratio and the wrapping probabilities. The present work includes several new models, and, in addition explores the scaling behavior of the wrapping probabilities as defined by Scullard and Jacobsen [9]. These wrapping probabilities display very favorable scaling properties for planar lattices. Furthermore, it has been shown [10] that non-planar bonds introduce only rather weak corrections to scaling. Therefore, it seems worthwhile to analyze the Scullard-Jacobsen wrapping probabilities in the present non-planar medium-range models.

2. Computational aspects

2.1. Simulation method

All simulations were performed on square lattices with periodic boundary conditions. The Monte Carlo generation of bond percolation configurations involves the placing of bond variables with probability p. This can be done by computing a uniformly distributed random number X, for

IOP Conf. Series: Journal of Physics: Conf. Series **1163** (2019) 012001 doi:10.1088/1742-6596/1163/1/012001

instance in the range $0 < X \leq 1$, and placing a bond if $X < p$. Typical values of p are $p \approx 1/z$. For models with a large number z of interacting neighbors, for instance $z = 4016$, one thus would typically have to compute thousands of random numbers before a bond can be placed. A more efficient approach is to analytically compute the probability $P(j) = (1-p)^{j-1}p$ that the jth interacting neighbor is the first one to be connected by the bond selection process [11]. The cumulative distribution of the first selected neighbor is then

$$C(k) = \sum_{j=1}^{k} P(k) = 1 - (1-p)^k \tag{4}$$

The number k can thus be obtained from X by the condition

$$C(k-1) \leq X \leq C(k)\,, \tag{5}$$

which requires a number of operations of order 1 instead of z.

2.2. Sampled quantities

After completion of the random process of defining the bond variables b_{ij}, the graph $\mathcal{G}$ is decomposed in disconnected clusters. For each graph $\mathcal{G}$ thus generated, we sampled the density c_1 of the largest cluster, the sum $\sum_i c_i^2$ of the squared cluster densities, and $\sum_i c_i^4$ of the fourth powers. The cluster decomposition allows the sampling of magnetic quantities in the random-cluster model. In the Ising case $q = 2$, the expectation values of the second and fourth moments of the magnetization density satisfies

$$\langle m^2 \rangle = \langle \sum_i c_i^2 \rangle, \quad \langle m^4 \rangle = \left\langle 3\left(\sum_i c_i^2\right)^2 - 2\sum_i c_i^4 \right\rangle, \tag{6}$$

and thus the Binder ratio of an $L \times L$ system, at a distance t to the critical point, can be expressed as

$$Q(t,L) = \langle m^2 \rangle^2 / \langle m^4 \rangle = \langle \sum_i c_i^2 \rangle^2 / \left\langle 3\left(\sum_i c_i^2\right)^2 - 2\sum_i c_i^4 \right\rangle. \tag{7}$$

This definition can also be used for $q \neq 2$ Potts and random-cluster models, including the percolation model. In the scaling limit, it converges to a q-dependent universal constant. For the percolation case $q = 1$ its value has been numerically determined as $Q = 0.87057(1)$ [7].

For each bond percolation configuration, we test whether there exists closed loops wrapping around the torus defined by the periodic boundary conditions. We thus sample three wrapping probabilities R_{b}, R_{e} and R_1. R_1 is the probability that there is a cluster connection to itself spanning the x direction but not the y direction. R_b is the probability that the configuration percolates in the x and the y direction. This includes the case of percolation in a diagonal direction. R_{e} is the probability that the configuration wraps at least once around the torus in any direction. The universal values of these quantities are exactly known in the scaling limit as $R_{\mathrm{b}} = 0.351642855$, $R_{\mathrm{e}} = 0.690473725$ and $R_1 = 0.169415435$ [14].

Scullard and Jacobsen [9] used instead wrapping probabilities R_0 (no wrapping) and R_2 (wrapping in two independent directions) which are dual in self-dual lattices, so that the difference $R_{\mathrm{d}} \equiv R_2 - R_0$ vanishes at the critical point. For non-self-dual planar lattices, R_{d} still displays rapid finite-size convergence to zero at the critical point [9].

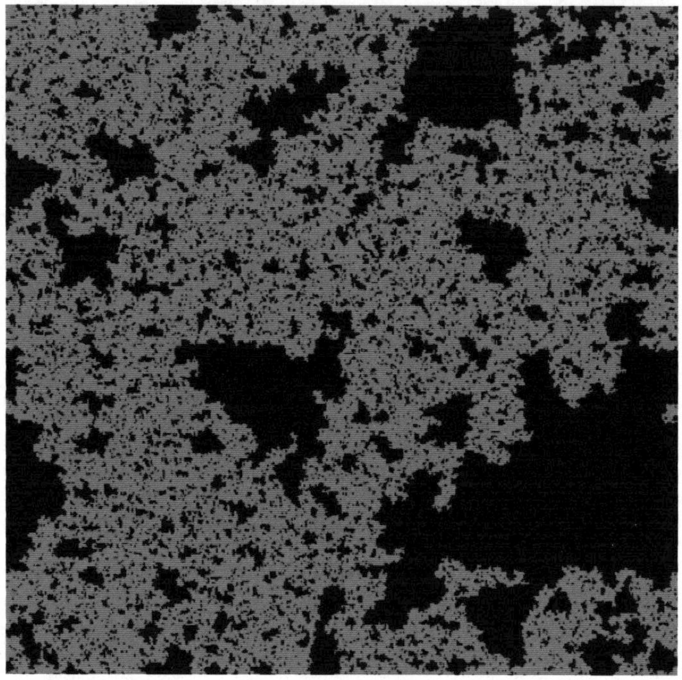

Figure 1. Typical critical configuration of the largest cluster in a 480×480 percolation model of type 1 with $z = 4$ ($r = 1$) interacting neighbors.

2.3. Finite-size scaling
According to finite-size scaling, the Binder ratio behaves as

$$Q(t,L) = Q + \sum_{j=1}^{\infty} q_j t^j L^{j y_t} + b_{\mathrm{i}} L^{y_{\mathrm{i}}} + b_{\mathrm{l}} \ln(L) L^{y_{\mathrm{i}}} + b_3 L^{d-2y_h} + b_4 L^{y_t - 2y_h} + \cdots \tag{8}$$

which may serve as a basis for least-squares fits to the Monte Carlo results for Q. To this purpose, we compute a set of $Q(t,L)$ data for a number of finite sizes L near the critical point. The fit may yield estimates of the critical point ($t \simeq K_{\mathrm{c}} - K$), Q and y_t. It is helpful that the critical exponents are exactly known [12, 13]. Somewhat better results for the critical point may be obtained from similar fits to the results for the wrapping probabilities, because the terms with exponents $d - 2y_h$ and $y_t - 2y_h$ are absent, and the universal wrapping probabilities in the scaling limit are exactly known.

3. Results

3.1. Remarks on the nature of percolation configurations
Typical configurations of the largest cluster for lattices with size 480×480 are shown in figures 1, 2, 3 and 4, for type-1 systems with z = 4, 60, 4016, and the mean-field limit respectively. These figures illustrate the crossover between the uniformly distributed clusters of the mean-field model and the more sharply defined clusters of the short-range model. The density of the largest cluster decreases considerably with increasing range of interactions.

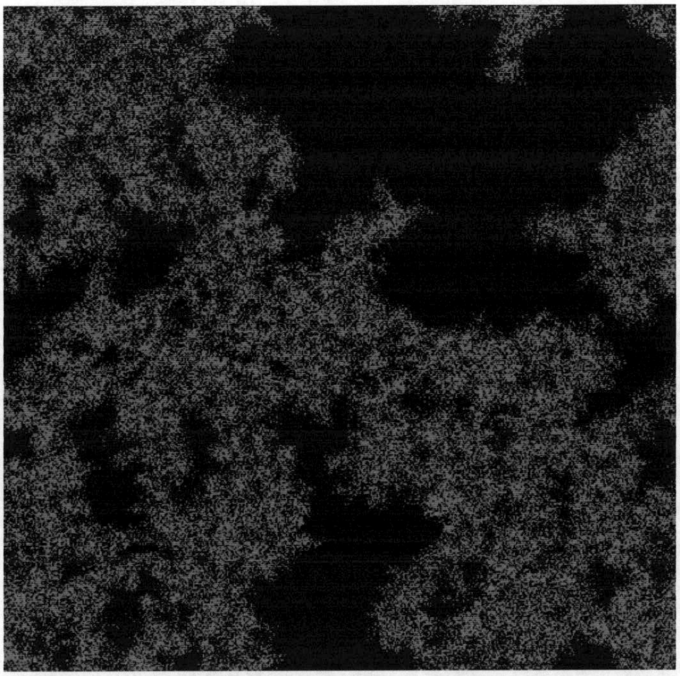

Figure 2. Typical critical configuration of the largest cluster in a 480×480 percolation model of type 1 with $z = 60$ ($r = 4.3$) interacting neighbors.

Figure 3. Typical critical configuration of the largest cluster in a 480×480 percolation model of type 1 with $z = 4016$ ($r = 35.8$) interacting neighbors.

Figure 4. Typical critical configuration of the largest cluster in a 480×480 percolation model on the complete graph, i.e., the mean-field limit $r \rightarrow \infty$.

3.2. Numerical analysis of the type-1 models

The procedure used to determine the critical points, and its results for type-1 models were described in Ref. [7]. That work also determined the critical exponents and other universal quantities, and reported that all investigated models with finite z belong to the short-range percolation universality class. In this work, we extend the series of finite-range models with several additional values of the coordination number z, making use of the wrapping probability R_d defined in section 1. The results are listed in table 1.

Table 1. Results for the percolation threshold zp_c of type-1 models with several coordination numbers z. Statistical uncertainties in the last decimal place given are shown between parentheses.

z	r	zp_c
148	7	1.234 704(2)
708	15	1.102 812(3)
1652	23	1.066 297(7)
3000	31	1.048 803(8)
4016	35.8	1.042 043(5)
6920	47	1.031 871(16)
12452	63	1.023 640(20)
50616	127	1.011 655(20)

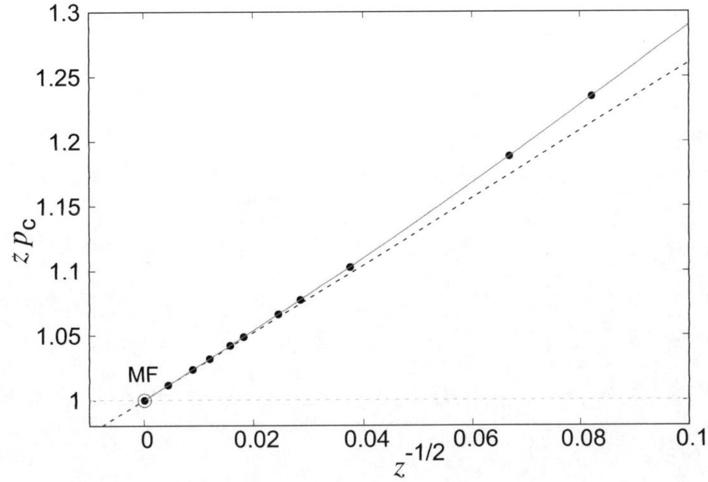

Figure 5. Percolation threshold $p_c(z)$ versus $1/\sqrt{z}$ for type-1 models. The dashed line represents a straight line with slope 2.608, representing the first term in a fit according to $zp_c - 1 = a_1 z^{-1/2} + a_2 z^{-1} + a_3 z^{-3/2}$. The magenta curve includes all these terms.

The new data for zp_c allow a more sensitive test of its linearity as a function of $z^{-1/2}$ in comparison with Ref. [7]. We plot these critical points in figure 5 accordingly. No deviation from linearity is observable in the neighborhood of the MF point.

As a more quantitative test, we fitted the data by the expression

$$zp_c - 1 \approx (a_1 + a_2 z^{-1/2} + a_3 z^{-1})(z^{-1/2})^{1/\phi} \tag{9}$$

which did adequately describe the data for $z \geq 224$, and yielded $1/\phi = 1.002(2)$.

3.3. Numerical analysis of the type-2 models

Table 2. Results for the percolation threshold zp_c of type-2 models with several coordination numbers z. Statistical uncertainties in the last decimal place given are shown between parentheses.

z	r	zp_c
120	5	1.257 695(7)
224	7	1.184 443(5)
960	15	1.085 839(5)
2208	23	1.055 830(10)
3968	31	1.041 349(7)
9024	47	1.027 217(15)
16128	63	1.020 270(15)
65024	127	1.010 050(30)

Like for the type-1 models, the critical exponents appear to be in accurate agreement with the short-range values for all z investigated. In figure 6, the critical points of the type-2 model

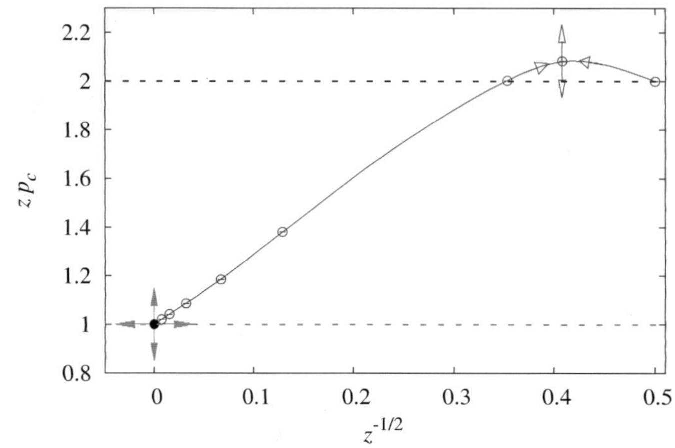

Figure 6. Critical points zp_c for type-2 equivalent-neighbor percolation models in two dimensions, and illustration of the location of the fixed points. The stable fixed point is set at the critical point $p_c = 2\sin(\pi/18)$ for the triangular-lattice bond percolation model. These results are analogous to those for type-1 models as given in Ref. [7].

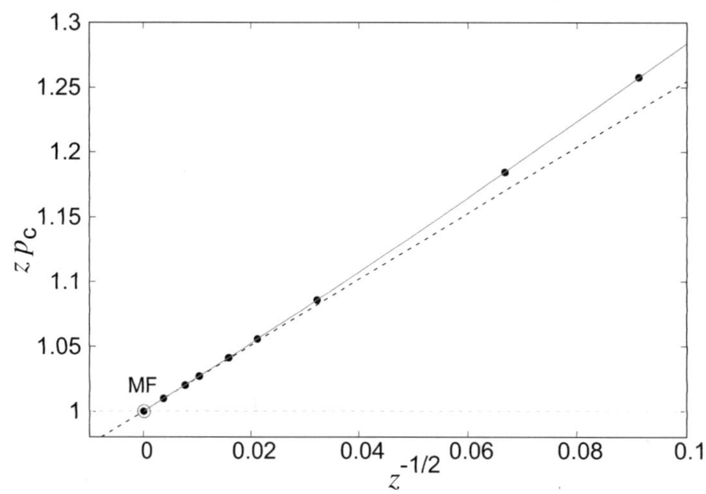

Figure 7. Percolation threshold $p_c(z)$ versus $1/\sqrt{z}$ for type-2 models.. The dashed line represents a straight line with slope 2.547, representing the first term in a fit according to $zp_c - 1 = a_1 z^{-1/2} + a_2 z^{-1} + a_3 z^{-3/2}$. The magenta curve includes all these terms.

zp_c are shown as a function of $1/\sqrt{z}$. Again, the critical points near the MF point in this figure lie approximately on a straight line. An enlarged picture is shown in figure 7. We have also fitted the type-2 critical points on the basis of Eq. (9), which yielded, after truncation, the same result $1/\phi = 1.002(2)$ as for the type-1 models.

Finally, we have constructed a data collapse of the Binder ratio Q versus L results at the critical points of the type-2 models, using a z-dependent rescaling of L. These data, together

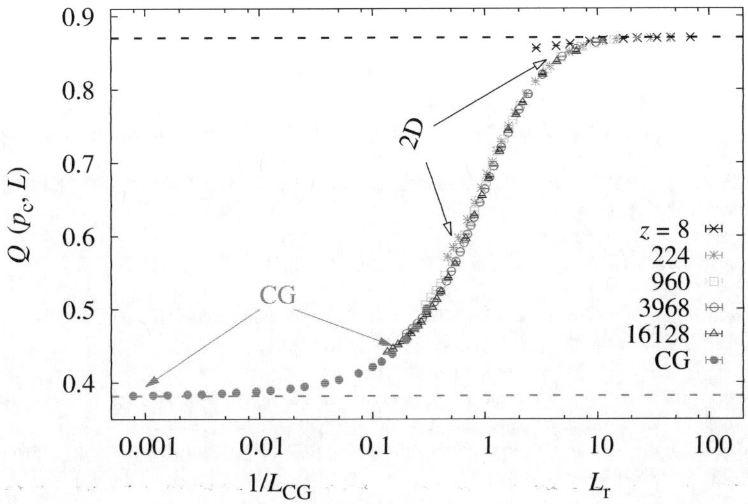

Figure 8. $Q(p_c, L)$ vs. rescaled size L_r for $z < \infty$. The rescaling factor $b(z)$ in $L_r = b(z)L$ is determined such as to collapse the finite-z data with $z > 8$ in this figure. The figure includes data for the complete graph (CG), using the inverse finite size $1/L_{CG}$ as the finite-size parameter. Dashed lines in the figure indicate the limiting values of Q for $L \to \infty$. These are 0.87057 for short-range models, and 0.38126 for the MF limit.

with finite- size results for the complete graph, cover the entire range of Q between the MF and 2D short-range fixed points. The procedure is entirely similar to that for the type-1 models analyzed in Ref. [7]. The results are also similar, but the overlap with the finite-size regime of the MF limit, while accidental and dependent on the rescaling procedure, is much larger because of the larger coordination numbers used in the present work. These results are shown in figure 8 and clearly demonstrate the crossover from the MF to the short-range behavior.

4. Conclusion

The "square" type-2 models closely follow the crossover behavior between the mean-field limit and the short-range fixed point as already found for the "almost circular" type-1 models. We confirm, with much improved accuracy, that the crossover exponent ϕ at the mean-field fixed point is equal to 1, corresponding with an exponent $y_r = 2/3$ describing the scaling of the inverse range parameter $z^{-1/2}$ at the MF fixed point [20]. We find that for the non-planar cases ($z \geq 8$), finite-size corrections in the wrapping probability R_d are, although not absent, still so small that R_d appears to be a useful quantity to determine critical points of non-planar models on the basis of Monte Carlo simulations and finite-size scaling.

Acknowledgments

Y. D. thanks the Ministry of Science and Technology of China for Grant No. 2016YFA0301604 and the National Natural Science Foundation of China for Grant No. 11625522. H. B. acknowledges the hospitality of the University of Science and Technology of China where this work was started.

IOP Conf. Series: Journal of Physics: Conf. Series **1163** (2019) 012001 doi:10.1088/1742-6596/1163/1/012001

References

[1] Knops H J F, van Leeuwen J M J and Hemmer P C *J. Stat. Phys.* **17** 197 (1977)
[2] Potts R B *Proc. Cambridge Philos. Soc.* **48** 106 (1952)
[3] Luijten E, Blöte H W J and Binder K *Phys. Rev.* E **54** 4626 (1996); *Phys. Rev.* E **56** 6540 (1997)
[4] Qian X F, Deng Y and Blöte H W J *Phys. Rev.* E **72** 056132 (2005)
[5] Kasteleyn P W and Fortuin C M *J. Phys. Soc. Jpn.* **26** (Suppl.) 11 (1969); Fortuin C M and Kasteleyn P W *Physica* (Amsterdam) **57** 536 (1972)
[6] Wu F Y *Rev. Mod. Phys.* **54** 235 (1982)
[7] Ouyang Y, Deng Y, and Blöte H W J submitted *Phys. Rev.* E (2018)
[8] Binder K *Z. Phys.* B **43** 119 (1981)
[9] Scullard C R and Jacobsen J L *J. Phys.* A **49** 125003 (2016), and references therein
[10] Guo W-A, Deng Y, and Blöte H W J *Phys. Rev.* E **79** 061118 (2009).
[11] Luijten E and Blöte H W J *Int. J. Mod. Phys.* C **6** 359 (1995)
[12] Nienhuis B in *Phase Transitions and Critical Phenomena* vol 11, ed C Domb and J L Lebowitz (London: Academic 1987). In Eq. (4.26) of this work, +1 has to be replaced by +2
[13] Cardy J L in *Phase Transitions and Critical Phenomena* vol 11, ed C Domb and J L Lebowitz (London: Academic 1987) p. 55, and references therein
[14] Ziff R M, Lorenz C D, and Kleban P *Physica* A **266** 17 (1999)
[15] Fisher M E, Ma S-K, and Nickel B G *Phys. Rev. Lett.* **29** 917 (1972)
[16] Sak J *Phys. Rev.* B **8** 281 (1973)
[17] Luijten E and Blöte H W J *Phys. Rev.* B **56** 8945 (1997); *Phys. Rev. Lett.* **89** 025703 (2002)
[18] Qian X F, Deng Y, Liu Y, Guo W-A, and Blöte H W J *Phys. Rev.* E **94** 052103 (2016)
[19] Nienhuis B, Berker A N, Riedel E K and Schick M *Phys. Rev. Lett.* **43** 737 (1979)
[20] Huang W, Hou P, Wang J, Ziff R M and Deng Y *Phys. Rev.* E **97** 022107 (2018)

IOP Conf. Series: Journal of Physics: Conf. Series **1163** (2019) 012002 doi:10.1088/1742-6596/1163/1/012002

Coarsening in the long-range Ising model: Metropolis versus Glauber criterion

Wolfhard Janke, Henrik Christiansen and Suman Majumder
Institut für Theoretische Physik, Universität Leipzig, Postfach 100 920, 04009 Leipzig, Germany

E-mail:
wolfhard.janke@itp.uni-leipzig.de
henrik.christiansen@itp.uni-leipzig.de
suman.majumder@itp.uni-leipzig.de

Abstract. Coarsening kinetics of systems with long-range interactions has, for a long time, only been attempted by truncating the potential using a cut-off distance. In such simulations of the long-range Ising model, one finds effectively short-range like behavior. This contradicts a longstanding theoretical prediction for the growth of the characteristic length scale for this model. We recently performed simulations of this system without the use of a cut-off and, for the first time, confirmed the analytic prediction numerically. Here, we investigate the properties of the used algorithm in more detail and present a comparison of coarsening using the Metropolis and the Glauber criteria.

1. Introduction

Coarsening is a very general phenomenon occurring in many physical situations. At high temperature, systems are generally in a disordered state with high symmetry (e.g., paramagnetic), whereas at low temperature the symmetries are broken and the system is in an ordered state (e.g., ferromagnetic). When one now subjects such a disordered system to external conditions that would correspond to an ordered state, one can observe the growth of ordered regions or domains with time t. The purpose of coarsening analysis is to quantify this nonequilibrium process. This is usually done by calculating a characteristic length scale $\ell(t)$, which quantifies the size of the ordered structures. For the short-range interacting Ising model with non-conserved order parameter, the growth follows the Lifshitz-Cahn-Allen law $\ell(t) \sim t^{1/2}$ [1, 2]. There exist several simulation studies which numerically confirm this prediction for the short-range Ising model [3, 4]. Long-range interactions are omnipresent in nature and other sciences, ranging from electrostatic forces over neuroscience to economical phenomena [5, 6, 7, 8, 9]. Thus to correctly describe such physical systems, one needs to take those long-range interactions into account. Here, we study the long-range interacting Ising model with Hamiltonian

$$\mathcal{H} = -\sum_{i}\sum_{i<j} J(r_{ij}) s_i s_j, \tag{1}$$

where $s_i = \pm 1$ are spins and the long-range interactions are chosen to decay algebraically with

$$J(r_{ij}) = \frac{1}{r_{ij}^{d+\sigma}}. \tag{2}$$

Here $r_{ij} = |\vec{r}_{ij}| = |\vec{r}_i - \vec{r}_j|$ is the distance between two spins and d is the dimension of the system. How long-range the potential is can be tuned by σ, where $\sigma = \infty$ corresponds to the short-range Ising model. Note, that the summation is performed over the whole lattice, i.e., every spin interacts with every other spin. For this model, there exists a prediction for the growth of $\ell(t)$ in dependence of σ, given by [10, 11, 12]

$$\ell(t) = \begin{cases} t^{1/2} & \sigma > 1 \\ [t\log(t)]^{1/2} & \sigma = 1 \\ t^{1/(1+\sigma)} & \sigma < 1 \end{cases}, \tag{3}$$

implying that the short-range scaling behavior already sets in for $\sigma > 1$. Here, the crossover value $\sigma_\times = 1$ is expected to be dimension independent. This is in contrast to the crossover values $\sigma_\times$ in equilibrium, which are still disputed and dependent on the spatial dimension [13, 14, 15, 16, 17, 18]. There have been previous numerical studies using Monte Carlo simulations for the coarsening in the long-range Ising model, in which a cut-off of the potential was used to make the simulation tractable [19, 9]. Ref. [9] reported that the coarsening dynamics was effectively short-range, i.e., $\ell(t) \sim t^{1/2}$ for all σ.

The treatment of the full range of long-range interactions in simulations often leads to problems that are no longer computationally tractable. In some cases, e.g., the long-range interacting Ising model, there exist specialized cluster algorithms based on the Swendsen-Wang algorithm [20] that enable efficient simulations of such systems [21, 22, 23]. However, those algorithms are only meaningful in equilibrium simulations where one is interested in sampling the equilibrium distribution, i.e., the moves do not need to have a physical equivalence. However, when studying dynamics, one requires a physical evolution in time and thus is restricted to single spin or particle dynamics. We recently performed simulations by utilizing the trick of storing the effective field for each spin.[1] With this method, when applied to the coarsening of the two-dimensional Ising model, we obtained a significant speedup by a factor of the order of 1000 [25]. Regarding the coarsening dynamics, we were able to confirm, for the first time, the above mentioned long-standing analytical prediction (3) [10, 11, 12]. In this work, we will present some further details of the method as well as the algorithm and additionally compare Metropolis and Glauber update dynamics. We end with a detailed analysis of the growth exponent by performing a finite-size scaling analysis.

2. Model, Method, and Algorithm

2.1. *Algorithm*

In usual canonical, i.e., fixed temperature, Monte Carlo simulations of the short-range Ising model with Hamiltonian

$$\mathcal{H} = -J \sum_{\langle ij \rangle} s_i s_j, \tag{4}$$

one calculates the change in energy ΔE by considering the local energy change by essentially summing over the nearest neighbors. This is done at every spin-flip attempt. The spin flip is then most commonly accepted by the Metropolis criterion [26] for a flip of spin s_i,

$$p_{\rm M}(s_i) = \min[1, \exp(-\Delta E_i/T)]. \tag{5}$$

[1] We thank Alfred Hucht for pointing out to us at this conference, that such a method has already been briefly mentioned in Ref. [24] dealing with simulations of the Heisenberg model with dipole interactions.

Generally, and also in our simulation, all physical constants, such as the Boltzmann constant k_B, are set to unity. Alternatively, one could also use the acceptance criterion of Glauber [27],

$$p_{\mathrm{G}}(s_i) = 0.5[1 - \tanh(0.5\Delta E_i/T)], \tag{6}$$

which is considered to be more physical. Both p_{M} and p_{G} fulfill the criterion of detailed balance in equilibrium. In an equilibrium simulation the Glauber update generally possesses a smaller autocorrelation time. For the long-range interacting Ising model, the calculation of ΔE is computationally much more demanding, as not only the nearest neighbors but also all other spins have to be considered. Thus one tries to limit the number of times one has to calculate this sum. The idea of storing the effective field for each spin arises from the very trivial observation, that during a single Monte Carlo sweep typically never all spins flip (unless when simulating at $T = \infty$). Thus it would be beneficial to have a method which is significantly faster if a spin does not flip; reminiscent of the kinetic Monte Carlo approach [28, 29]. The kinetic Monte Carlo approach is rejection free and one chooses which spin will flip according to its proposed energy change. In succession one then updates the time according to how long this spin flip would have taken in a Metropolis Monte Carlo simulation. We, however, have taken another approach which will be explained in the context of the Ising model with the archetypical Hamiltonian of (1). In our approach, we store the effective field h_j for every spin s_j, where

$$h_j = \sum_i J(r_{i,j}) s_i. \tag{7}$$

This enables us to calculate ΔE_i, the change in energy if spin s_i would flip, in a very efficient way:

$$\Delta E_i = 2 s_i h_i. \tag{8}$$

Using the proposed formulation, one only would need to recalculate the h_j for all $j \neq i$ if a spin flips. However, there is a much more efficient way to do this. We can instead iterate over the lattice and update all $h_j \rightarrow h_j + 2s_i J(r_{ij})$, making this operation comparable in speed to a single calculation of ΔE_i. For the nearest-neighbor Ising model, one would thus only update the effective field of the surrounding spins (4 in a two-dimensional, 6 in a three-dimensional simple-cubic lattice). This leads to a speedup of only $\approx 10\% - 20\%$ when simulating below the critical temperature T_c, because the computational effort of calculating $\Delta E_i = 2s_i J(1) \sum_{\langle ij \rangle} s_j$ is roughly comparable to calculating $\Delta E_i = 2s_i h_i$ (here $\langle ij \rangle$ refers to all j with $r_{ij} \equiv 1$). However, for long-range interactions this difference is very big, thus leading to significant speedups. In Ref. [25], we found a speedup of $\approx 10^3$ for the Metropolis simulation of coarsening in the two-dimensional Ising model.

2.2. Ewald summation

When dealing with long-range interactions, the finite-size effects become very strong. In this work, we chose the power-law like decaying potential presented in (2). For long-range interacting systems, one often uses periodic boundary conditions via the minimum-image convention. Another way to imitate periodic boundary conditions and to minimize the effect of a finite system size is the Ewald summation [30, 18]. The basic idea is to envision infinitely many copies or identical image systems in all directions. One then can identify the distance of each replica by $\vec{v} = (m_x, m_y)$, where $m_i = 0, \pm L, \pm 2L, \ldots$ for $i = x, y$ and L is the linear size of the system. The interaction strength between two spins can then be more accurately calculated as

$$J^{\star}(s_i, s_j) = \sum_{m_x} \sum_{m_y} \frac{1}{|\vec{v} + \vec{r}_{ij}|^{d+\sigma}}. \tag{9}$$

IOP Conf. Series: Journal of Physics: Conf. Series **1163** (2019) 012002 doi:10.1088/1742-6596/1163/1/012002

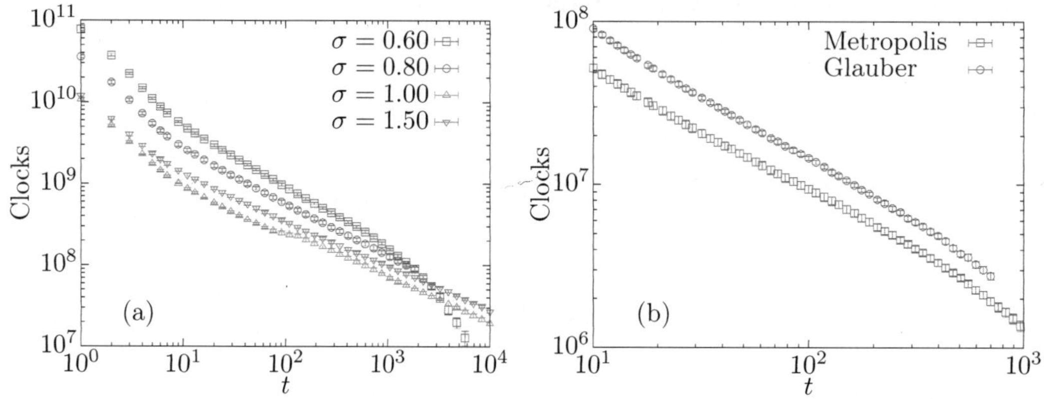

Figure 1. The number of clocks needed per sweep vs. time t for (a) different values of σ with $L = 2048$ using the Metropolis criterion and (b) for $\sigma = 0.6$ with $L = 1024$ using the Metropolis and the Glauber criterion. The system was quenched to $T = 0.1T_c$ from an initially random configuration.

However, in principle this sum has to be performed over infinitely many mirror systems due to slow convergence. Luckily, one way to significantly speedup the convergence is to split the sum in a sum in real space and a sum in reciprocal space:

$$J^\star = J^\star_s + J^\star_l, \tag{10}$$

where $J^\star_s$ is the short-range term and $J^\star_l$ is the long-range term. Those sums can be expressed as

$$J^\star_s = \sum_{\vec{v}} \frac{1}{\Gamma\left(\frac{d+\sigma}{2}\right)} \frac{1}{|\vec{v} + \vec{r}_{ij}|^{d+\sigma}} \Gamma\left(\frac{d+\sigma}{2}, (\kappa|\vec{v} + \vec{r}_{ij}|)^2\right), \tag{11}$$

$$J^\star_l = \frac{2\pi^{\frac{d}{2}}}{\Gamma\left(\frac{d+\sigma}{2}\right) V} \sum_{\vec{k}} \cos(2\pi\vec{k}\vec{r}_{ij}) \frac{1}{2} (\pi k)^\sigma \Gamma\left(-\frac{\sigma}{2}, \left(\frac{\pi k}{\kappa}\right)^2\right). \tag{12}$$

Here $\Gamma(x) = \int_0^\infty t^{x-1} e^{-t} dt$ is the gamma function and $\Gamma(s, x) = \int_x^\infty t^{x-1} e^{-t} dt$ is the incomplete gamma function. Both gamma functions were evaluated numerically by the GNU scientific library (GSL). The value of κ is important for the speed of convergence and was chosen to $\kappa = 2/L$ in accordance with Ref. [18]. The approach of using Ewald summation to determine a more accurate representation of the interaction potential is widely used, e.g., in long-range interacting all-atom simulations of peptides [31]. There, the positions of the particles are not fixed and this approach is used "on-the-fly", whereas in our simulation the positions of the spins are fixed and $J^\star$ is calculated only once at the beginning of the simulation for every pair of σ and L.

3. Results

We now want to use the above introduced methodology for simulating the coarsening in the two-dimensional Ising model with long-range interactions for different values of σ and for the mentioned acceptance criteria according to Metropolis and Glauber.

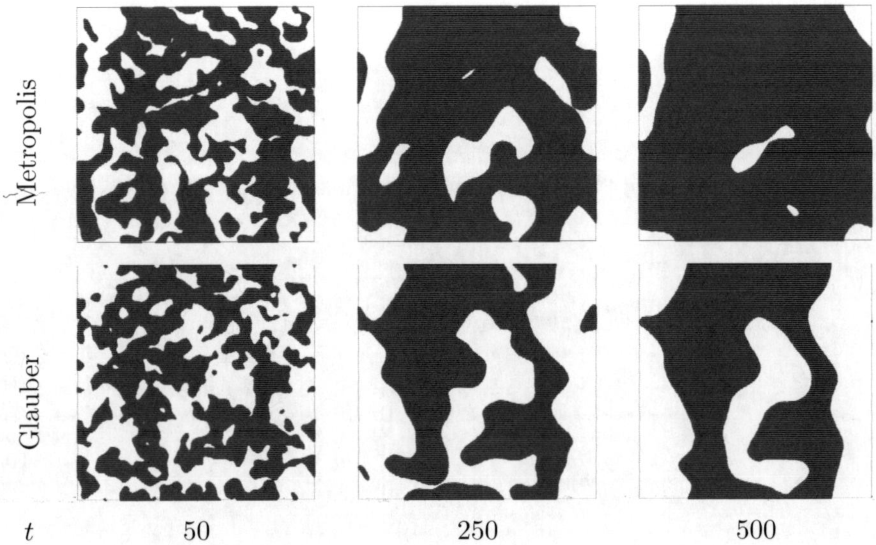

Figure 2. Snapshots at time t of the evolution for a single seed using the Metropolis (upper panels) and Glauber (lower panels) update criterion for $\sigma = 0.6$ and $L = 1024$ when the system is quenched from a random start configuration to $T = 0.1T_c$.

3.1. Runtime analysis

In Figs. 1(a) and (b) we show the dependence of the clocks needed per sweep versus the time t for (a) different σ using the Metropolis criterion and for (b) $\sigma = 0.6$ using both the Metropolis and Glauber criterion. It becomes clear from this figure, that the runtime per sweep dramatically decreases as the systems becomes more ordered, indicating that less spin-flip attempts are successful. In the scaling region, one can observe a power-law like decay with varying exponent, depending on σ. If one would not store the effective field, the number of clocks per sweep would be roughly constant for all times and always greater than the presented data. One important factor, somehow neglected in the above graph, is the additional time needed to calculate the initial h_j. In principle, this overhead could be so big, that this approach is not advantageous anymore. However, one can reformulate the algorithm in such a way, that the runtime is always strictly below the runtime of the traditional algorithm (details will be presented in Ref. [32]). The comparison presented in Fig. 1(b) shows that for $\sigma = 0.6$ and $L = 1024$, the Glauber criterion generally has a bigger number of accepted spin flips in such a quench, reflected in the higher number of clocks needed per sweep for every t. It is clear, that the approach of storing the effective field per spin and subsequently updating those values significantly decreases the amount of needed computational resources. In Ref. [25], we presented such a comparison for the Metropolis criterion and found an improvement factor of $\approx 10^3$.

3.2. Kinetics of growth

In Fig. 2 we present snapshots for the coarsening of a single run for $\sigma = 0.6$ and $L = 1024$ for both the Metropolis and Glauber criterion. The system was in this work quenched to $T = 0.1T_c$, where the values of T_c, which for the long-range Ising model depend on the decay exponent σ, were extracted by a power-law fit of the form $T_c(\sigma) = T_c(\infty) + a\sigma^b$ to the data presented in Ref. [18]. For nonequilibrium studies, knowing the exact value of $T_c(\sigma)$ is not important and this rather crude estimation is sufficient. Phenomenologically, the evolution appears to be driven in a similar way as for the short-range model by reduction of phase boundaries. However, it

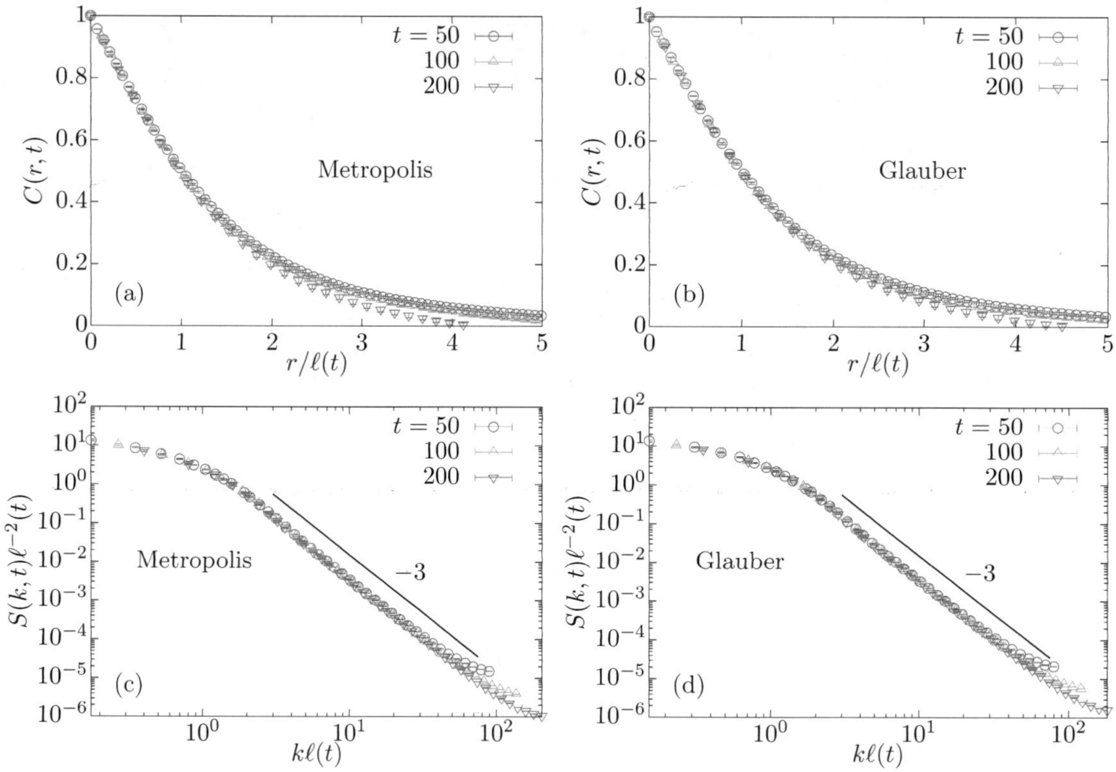

Figure 3. In (a) for the Metropolis criterion and in (b) for the Glauber criterion, the scaling of the correlation function $C(r,t)$ is shown by plotting it against $r/\ell(t)$. In (c) and (d) we present the scaling of the structure factor $S(k,t)$ for the Metropolis respectively Glauber update criterion. The solid lines with slope -3 indicate Porod's law. All presented data were obtained for systems with $\sigma = 0.6$ and $L = 1024$ with a quench temperature of $T = 0.1T_c$.

appears that the dynamics of the Metropolis update is faster than for the Glauber update, which is of course already reflected in the update criteria p_{M} and p_{G} and the analysis of the run time in Fig. 1. The starting configuration was identical for both criteria. Importantly, the phase boundaries are smooth rather than rugged.

We calculate the equal-time spin-spin correlation function

$$C(r,t) = \langle s_i s_j \rangle - \langle s_i \rangle \langle s_j \rangle \tag{13}$$

to quantify the coarsening. With increasing time after the quench, the correlation is expected to become stronger, i.e., $C(r,t)$ decays slower. From this fact, one can then extract a length scale $\ell(t)$ by taking the intersection of the correlation function with a constant value $c \in (0,1)$. We chose $C(r,t) = \text{const.} = 0.5$ as one often does in simulations of coarsening. In Figs. 3(a) and (b) the data collapse of $C(r,t)$ if plotted against $r/\ell(t)$ for (a) the Metropolis criterion and (b) the Glauber criterion. This scaling is predicted by theory. For later times t and larger r, one sees some deviation of the collapse, which we attribute to the very strong finite-size effects. However, overall the collapse is very good. Next we analyse the scaling of the structure factor $S(\vec{k},t) = \int d\vec{r} C(\vec{r},t) e^{i\vec{k}\vec{r}}$, the Fourier transform of the correlation function $C(r,t)$ in Fig. 3(c) for the Metropolis criterion and in (d) for the Glauber criterion. The data collapse is very good, until at high $k\ell(t)$ the finite-size effects come into play. The decay follows a power law with

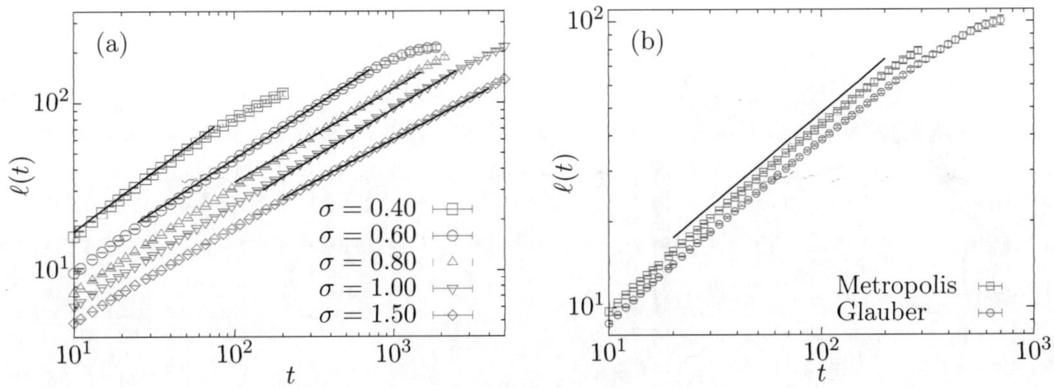

Figure 4. (a) Growth of length scale $\ell(t)$ with time t using the Metropolis update for different σ for $L = 2048$. In (b) the growth of $\ell(t)$ is shown for both the Metropolis and Glauber criterion for $\sigma = 0.6$ and $L = 1024$. The solid lines in (a) and (b) correspond to the prediction (3).

exponent -3, in accordance with Porod's law [33]. This fact is used in the derivation of (3) and thus is important. With this necessary condition of the coarsening being a scaling phenomenon out of the way, we now aim to quantify the growth of $\ell(t)$.

To quantify the differences in the growth of $\ell(t)$ for different σ, we simulated the coarsening for $L = 2048$ and the Metropolis criterion. In Fig. 4(a), the characteristic length $\ell(t)$ for these simulations is shown. The solid lines here correspond to the prediction (3) with an appropriate prefactor to make the solid lines fall on top of the data. We now want to compare the Metropolis criterion to the Glauber criterion. For this, we present the corresponding plot in Fig. 4(b) for $\sigma = 0.6$ and $L = 1024$. As expected, both the Metropolis and Glauber simulations show agreement with the prediction (3) (given by the solid line in the same figure). However, the simulation using the Glauber criterion has a smaller amplitude and overall runs longer. This is in agreement with the higher number of clocks/spin flips per sweep observed in Fig. 1(b).

3.3. Exponent analysis

In Fig. 4 we can observe a good agreement with the prediction (3). However, it is always important to obtain some independent quantitative estimation of observables to mitigate the effect of observation bias. We therefore fit a power law of the form

$$\ell(t) = At^{\frac{1}{1+\sigma_f}} \tag{14}$$

to the growth of $\ell(t)$ for the Metropolis simulation with $\sigma = 0.6$ and $L = 2048$. When doing such fits, it is always very difficult to find proper ranges of minimal and maximal values of the independent variable. We therefore vary t and record the reduced chi-square χ_r^2 value and the values of the fit parameters A and σ_f. In Fig. 5(a) we present a heatmap plot, where t_{min} is the lower bound of t and t_{max} the upper bound of t. The colormap encodes the value of σ_f. All fits with $\chi_r^2 > 1.5$ and $\chi_r^2 < 0.5$ were discarded. One can observe that $\sigma_f = 1.0$, the value expected from such a quench in the short-range Ising model, is not present at all. One finds a rather big plateau around $\sigma_f = 0.6$, indicating this to be the correct value. If one moves towards higher values of t_{min}, one of course observes an increase of σ_f due to finite-size effects. There is, however, no longer any plateau visible, a strong indication that $\sigma_f = 0.6$ is indeed the correct value, further validating the found results.

IOP Conf. Series: Journal of Physics: Conf. Series **1163** (2019) 012002 doi:10.1088/1742-6596/1163/1/012002

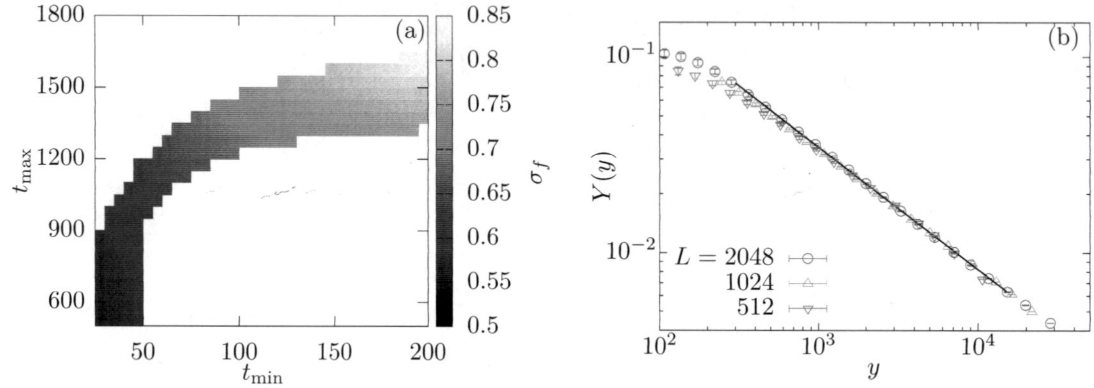

Figure 5. (a) Heatmap of the fit parameter σ_f for different ranges of the independent variable $t_{min} \leq t \leq t_{max}$ for a fit of $\ell(t) = At^{1/(1+\sigma_f)}$ to $\ell(t)$ for $\sigma = 0.6$ and $L = 2048$. The colormap here encodes the value of σ_f found. Only fits with $0.5 \leq \chi_r^2 \leq 1.5$ are shown. (b) Finite-size scaling analysis for $\ell(t)$ with $\sigma = 0.6$ and $L = 2048, 1024, 512$. The solid line corresponds to the predicted behavior.

To further strengthen the claim that indeed the analytic prediction is correct, we finally perform a finite-size scaling analysis [34, 35] by allowing an off-set in the asymptotic scaling law for $\ell(t)$,

$$\ell(t) = \ell_0 + At^{\frac{1}{1+\sigma}}, \tag{15}$$

where ℓ_0 is some initial length, which proved to be significant for the finite-time scaling behavior of $\ell(t)$. Then one writes down the finite-size scaling function as

$$Y(y) = \frac{\ell(t) - \ell_0}{L - \ell_0} \tag{16}$$

with appropriate choice of the argument,

$$y = \frac{(L - \ell_0)^{1+\sigma}}{t - t_0}. \tag{17}$$

In the finite-size unaffected regime, i.e., in the scaling regime it can be shown that

$$Y \sim y^{-\frac{1}{1+\sigma}}. \tag{18}$$

Thus, when plotting $Y(y)$ for different L, one expects data collapse with a slope corresponding to $y^{-\frac{1}{1+\sigma}}$. A similar analysis has previously successfully been applied in various systems [34, 35, 36, 37, 38]. In Fig. 5(b) we show such a data collapse exercise. For large values of y the data for different L collapse well and are in very good agreement with the solid line showing the power-law behavior with exponent $-1.0/1.6$ expected for $\sigma = 0.6$. For small values of y, however, one can observe a flat behavior due to the influence of finite-size effects. From the point of the deviation of the data from the asymptotic $Y \sim y^{-\frac{1}{1+\sigma}}$ behavior to the flat regime one can estimate the onset point of finite-size effects, which in our case occurs at $\ell/\ell_{max} \approx 0.78$, where ℓ_{max} is the maximal length scale $\ell(t)$ can attain for a finite L. Here we obtained ℓ_{max} from the asymptotic value of $\ell(t)$ observed in Fig. 4(a). A larger value of L would manifest in a longer period of smaller values of y following the slope.

4. Conclusion

We have compared the coarsening in the two-dimensional Ising model with long-range interactions using both the Metropolis and Glauber criterion. It was confirmed that both update criteria produce results that are in agreement with the analytical prediction. We further have analyzed the runtime per sweep during a quench for both criteria. Apart from the obvious adaptation of this approach to simulate the dynamics of long-range interacting systems to other related models and the investigation of other nonequilibrium quenching properties (such as, e.g., aging or persistence [39, 32]), we would like to point out that one could also effectively investigate the Kibble-Zurek mechanism in Ising models with long-range interactions during a nonequilibrium quench [40, 41, 42, 43]. This would hopefully lead to significant new physical insight in the process of slow quenches.

Acknowledgments

This project was funded by the Deutsche Forschungsgemeinschaft (DFG) under Grant Nos. JA 483/33-1 and SFB/TRR 102 (project B04), and further supported by the Deutsch-Französische Hochschule (DFH-UFA) through the Doctoral College "$\mathbb{L}^4$" under Grant No. CDFA-02-07, the EU Marie Curie IRSES network DIONICOS under Grant No. PIRSES-GA-2013-612707, and the Leipzig Graduate School of Natural Sciences "BuildMoNa".

References

[1] Lifshitz I M 1962 *J. Exp. Theor. Phys.* **15** 939
[2] Allen S M and Cahn J W 1979 *Acta Metall.* **27** 1085
[3] Bray A 2002 *Adv. Phys.* **51** 481
[4] Puri S and Wadhawan V (eds) 2009 *Kinetics of Phase Transitions* (CRC Press, Boca Raton)
[5] Bak P 1997 *How Nature Works* (Oxford University Press, Oxford)
[6] Lux T and Marchesi M 1999 *Nature* **397** 498
[7] Beggs J and Plenz D 2003 *J. Neurosci.* **24** 11167
[8] Petrs O and Neelin D 2006 *Nat. Phys.* **2** 393
[9] Gundh J, Singh A and Singh R 2015 *PloS one* **10** e0141463
[10] Bray A J 1993 *Phys. Rev. E* **47** 3191
[11] Bray A and Rutenberg A 1994 *Phys. Rev. E* **49** R27
[12] Rutenberg A and Bray A 1994 *Phys. Rev. E* **50** 1900
[13] Stell G 1970 *Phys. Rev. B* **1** 2265
[14] Fisher M, Ma S and Nickel B 1972 *Phys. Rev. Lett.* **29** 917
[15] Sak J 1973 *Phys. Rev. B* **8** 281
[16] Luijten E and Blöte H 1997 *Phys. Rev. B* **56** 8945
[17] Luijten E and Blöte H 2002 *Phys. Rev. Lett.* **89** 025703
[18] Horita T, Suwa H and Todo S 2017 *Phys. Rev. E* **95** 012143
[19] Singh A, Ahmad S, Puri S and Singh S 2014 *Eur. Phys. J. E* **37** 2
[20] Swendsen R and Wang J 1987 *Phys. Rev. Lett.* **58** 86
[21] Luijten E and Blöte H 1995 *Int. J. Mod. Phys. C* **6** 359
[22] Fukui K and Todo S 2009 *J. Comput. Phys.* **228** 2629
[23] Flores-Sola E, Weigel M, Kenna R and Berche B 2017 *Eur. Phys. J. Spec. Top.* **226** 581
[24] Hucht A, Moschel A and Usadel K 1995 *J. Magn. Magn. Mater.* **148** 32
[25] Christiansen H, Majumder S and Janke W 2018 *preprint arXiv:1808.10426*
[26] Metropolis N, Rosenbluth A W, Rosenbluth M N, Teller A H and Teller E 1953 *J. Chem. Phys.* **21** 1087
[27] Glauber R J 1963 *J. Math. Phys.* **4** 294
[28] Newman M and Barkema G 1999 *Monte Carlo Methods in Statistical Physics* (Oxford University Press, Oxford)
[29] Voter A F 2007 *Introduction to the kinetic Monte Carlo method* in *Radiation Effects in Solids* ed Sickafus K E, Kotomin E A and Uberuaga B P (Springer, Dordrecht) pp 1–23
[30] Ewald P 1921 *Ann. Phys.* **369** 253
[31] Frenkel D and Smit B 2001 *Understanding Molecular Simulation: From Algorithms to Applications* (Academic Press, Cambridge)
[32] Christiansen H, Majumder S and Janke W to be published

[33] Porod G 1982 *General theory* in *Small Angle X-ray Scattering* ed Glatter O and Kratky O (Academic Press, London) p 17
[34] Majumder S and Das S 2010 *Phys. Rev. E* **81** 050102
[35] Das S, Roy S, Majumder S and Ahmad S 2012 *Europhys. Lett.* **97** 66006
[36] Majumder S and Janke W 2015 *Europhys. Lett.* **110** 58001
[37] Majumder S, Zierenberg J and Janke W 2017 *Soft Matter* **13** 1276
[38] Christiansen H, Majumder S and Janke W 2017 *J. Chem. Phys.* **147** 094902
[39] Henkel M and Pleimling M 2010 *Non-Equilibrium Phase Transitions, Vol. 2: Ageing and Dynamical Scaling far from Equilibrium* (Springer, Heidelberg)
[40] Kibble T W B 1976 *J. Phys. A* **9** 1387
[41] Zurek W H 1985 *Nature* **317** 505
[42] Liu C W, Polkovnikov A and Sandvik A W 2014 *Phys. Rev. B* **89** 054307
[43] Puebla R, Nigmatullin R, Mehlstäubler T E and Plenio M B 2017 *Phys. Rev. B* **95** 134104

IOP Conf. Series: Journal of Physics: Conf. Series **1163** (2019) 012003 doi:10.1088/1742-6596/1163/1/012003

Studies on the Rabi Model

H. Q. Lin[1], M. X. Liu[1,2], Stefano Chesi[1], H. G. Luo[1,3]

[1] Beijing Computational Science Research Center, Beijing 100193
[2] State Key Laboratory of Information Photonics and Optical Communications, Beijing University of Posts and Telecommunications, Beijing 100876
[3] Center for Interdisciplinary Studies and Key Laboratory for Magnetism and Magnetic Materials of the MoE, Lanzhou University, Lanzhou 730000

Abstract. We present our recent studies on the quantum Rabi model (QRM). Firstly, by using a variational wave function, which facilitates to extract physics in entire parameter regime with high accuracy, we unveil a ground-state phase diagram of the QRM and argue that the main constituents are polaron and anti-polaron. Secondly, introducing an anisotropy into the QRM, in which the rotating- and counter-rotating terms are allowed to have different coupling strength, we explore that how the criticality of the model interpolates between two known limits with distinct universal properties. Through a combination of analytic and numerical approaches we compute phase diagram, scaling functions and critical exponents, and establish that the universality class at the finite anisotropy is the same as that of the isotropic limit.

1. Introduction

In this section we review our efforts in studying the Rabi model [1, 2]. A quantum version of the Rabi model could be obtained by quantizing the cavity, with the Hamiltonian

$$H = \omega a^+ a + \frac{\Omega}{2}\sigma_x + g\sigma_z(a^+ + a), \tag{1}$$

where ω is the oscillator frequency, Ω the energy splitting of the qubit and g is the coupling constant. The model has $Z2$ symmetry with well-defined parity. It might be one of the simplest models of quantum light interacting with matter, and plays a fundamental role in quantum optics (cavity/circuit QED), quantum information, and condensed matter physics (magnetic resonance, molecular physics), etc. It can be experimentally realized in optical cavity, superconductor circuit, trapped ions, quantum dot with spin-orbital coupling, and so on. The model has attracted great attention in recent years due to the fact that experimentally one can achieve strong or even ultra-strong coupling.

By combing various analytical approaches and numerical methods, we carried systematic studies on the quantum Rabi model. Many of our results were published already [3]. We present two important results in this talk. Firstly we proposed a variational wave function based on the polaron and anti-polaron picture. The wave function has 5 minimum variational parameters, accounting for the shift of well minimum on the left and right, renormalization of oscillator frequencies, and the normalization factor. The proposed variational wave function works outstanding for the ground state and excellent for low lying excited states. We also obtained the ground state phase diagram. Secondly, we study the role of counter-rotating term on the phase diagram and relevant quantum phase transitions.

IOP Conf. Series: Journal of Physics: Conf. Series **1163** (2019) 012003 doi:10.1088/1742-6596/1163/1/012003

$$H = \omega a^+ a + \frac{\Omega}{2}\sigma_x + g[(\sigma^+ a + \sigma^- a^+) + \lambda(\sigma^+ a^+ + \sigma^- a)]. \tag{2}$$

The anisotropy λ varies from negative value to positive, including $\lambda = 0$ (the Jaynes-Cummings model [4]) and $\lambda = 1$ (the Rabi model), representing the strength of counter-rotating term. In the limit of $\eta = \Omega/\omega \to \infty$, we performed Schrieffer-Wolff transformation to obtain an effective Hamiltonian and did scaling analysis. A phase diagram was obtained and we found that all quantum phase transitions along coupling direction are second-order. Along the anisotropy direction, we find a first order phase transition at $\lambda = 0$ (JC line) if coupling reaches the superradiant phase. It is reveals that the transitions are universal for all $\lambda > 0$. There exists symmetry between x-type super-radiant phase (in $\lambda > 0$) and p-type super-radiant phase (in $\lambda < 0$). Moreover our results also apply to the Dicke model, i.e., to the system contains arbitrary N spins as well.

2. Entanglement of the oscillator and the qubit

When one takes the frequency ratio limit $\eta = \Omega/\omega \to \infty$, the model undergoes a quantum phase transition at $g_c = \sqrt{\omega\Omega}/2$. For a finite η, the scaling form turns out. To address the criticality issue, we take the critical behavior of the entanglement entropy as an example. The entanglement of the oscillator and the qubit is quantified by the entropy S of the qubit state. It is defined by the trace that

$$S = -Tr\{\rho_q \log_2 \rho_q\}, \tag{3}$$

where ρ_q is the reduced density matrix of the qubit, which satisfies

$$\rho_q = -Tr\{|G\rangle\langle G|\}, \tag{4}$$

$|G\rangle$ is the ground state.

At the critical regime, there should be the scaling form for the entanglement entropy that (we take Ω as the unity)

$$S = \omega^{-\beta_s/\nu_s}\tilde{S}(Z), \tag{5}$$

where $Z = \frac{g - g_c}{g_c}\omega^{-1/\nu_s}$ is the scaling variable, $\tilde{S}$ is the universal scaling function, β_s and ν_s are two critical exponents. Fig. 1 shows the scaling function of the entanglement entropy calculated by the numerical way. Curves for different scales of ω collapse to a single one, to which related the scaling function $\tilde{S}$. This numerical result confirms the scaling relation in Eq. 4, and thus supports the existence of the critical behavior for the entanglement. The critical exponent ν_s is of the value about 0.667, which agrees with 2/3 predicted by the mean-field-like analysis. Interestingly, β_s is about 0.5. It has not been understood by the mean-field-like framework, and requires further exploration. The above-unpublished results are new and will not be published elsewhere.

In summary, we proposed a variational wave function constructed from polaron and antipolaron induced by spin-bosonic mode coupling to study the ground state properties of the quantum Rabi model. This variational wave function provides high accuracy for all coupling strengths. We also obtained a phase diagram as function of coupling strength and oscillator frequency. Moreover, our proposed wave function can also be used to extract excited state information of the quantum Rabi model. We would like to mention that preliminary results suggest that similar wave function could be applied to study the spin-boson model. Adding an anisotropy to the counter-rotating term of the quantum Rabi model, which interpolates between the two known limits with distinct universal properties, we have characterized quantum phase

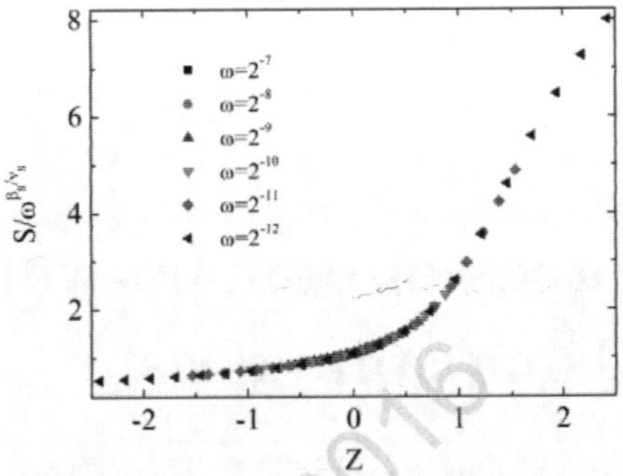

Figure 1. The scaling function of the entanglement entropy.

transitions of the quantum Rabi model and demonstrated its universality. Scaling behaviors were analyzed and we found that except at zero anisotropy, i.e., the JC model, all other anisotropies belong to the same universality class. Our results emphasize the critical role played by counter-rotating terms, and it also applies to multi spin cases. Our findings are relevant to a variety of systems that are able to realize strong coupling between matter and light.

Acknowledgments
The NSAF U1530401 supports this work and we thank CSRC for providing computation resources.

References
[1] I. I. Rabi, On the Process of Space Quantization. Phys. Rev. **49**, 324-328 (1936).
[2] I. I. Rabi, Space Quantization in a Gyrating Magnetic Field. Phys. Rev. **51**, 652-654 (1937).
[3] Z.-J. Ying, M. X. Liu, H.-G. Luo, H.-Q. Lin and J.-Q.You, Phys. Rev. A **92**, 053823 (2015); M.X. Liu, Z.-J. Ying, J.-H. An, H.-G. Luo and H.-Q. Lin, J. Phys. A **50**, 084003 (2017); L. Cong, X.-M. Sun, M.X. Liu, Z.-J. Ying, and H.-G. Luo Phys. Rev. A **95**, 063803 (2017); M. X. Liu, Stefano Chesi, Z.-J. Ying, X. S. Chen, H.-G. Luo, and H.-Q. Lin, Phys. Rev. Lett. **119**, 220601 (2017); and references therein.
[4] E. T. Jaynes and F. W. Cummings, Proceedings of the IEEE 51, **89** (1963).

IOP Conf. Series: Journal of Physics: Conf. Series **1163** (2019) 012004 doi:10.1088/1742-6596/1163/1/012004

High-precision studies of domain-wall properties in the 2D Gaussian Ising spin glass

Martin Weigel[1] **and Hamid Khoshbakht**[1,2]

[1] Applied Mathematics Research Centre, Coventry University, Coventry CV1 5FB, England
[2] Institut für Physik, Johannes Gutenberg-Universität Mainz, Staudinger Weg 7, D-55099 Mainz, Germany

E-mail: `martin.weigel@complexity-coventry.org`

Abstract.
In two dimensions, short-range spin glasses order only at zero temperature, where efficient combinatorial optimization techniques can be used to study these systems with high precision. The use of large system sizes and high statistics in disorder averages allows for reliable finite-size extrapolations to the thermodynamic limit. Here, we use a recently introduced mapping of the Ising spin-glass ground-state problem to a minimum-weight perfect matching problem on a *sparse* auxiliary graph to study square-lattice samples of up to $10\,000 \times 10\,000$ spins. We propose a windowing technique that allows to extend this method, that is formally restricted to planar graphs, to the case of systems with fully periodic boundary conditions. These methods enable highly accurate estimates of the spin-stiffness exponent and domain-wall fractal dimension of the 2D Edwards-Anderson spin-glass with Gaussian couplings. Studying the compatibility of domain walls in this system with traces of stochastic Loewner evolution (SLE), we find a strong dependence on boundary conditions and compatibility with SLE only for one out of several setups.

1. Introduction

Spin glasses are by now a mature subject with a multitude of spin-off directions including applications in neural networks and machine learning, error correcting codes and gene regulation [1, 2]. After more than half a century of research, some of the fundamental questions are still unsolved, including the nature of the ordering in the spin-glass phase. Although many more complicated models including systems with continuous spins [3–7], long-range interactions [8] or samples subject to external fields [9] have been studied, the most fundamental questions are mostly studied for the paradigmatic example of the short-range Ising spin glass.

A spin glass phase has not been observed in a two-dimensional system with short-range interactions, and in such systems spin glass is confined to zero temperature which is hence the critical temperature of this system. Nevertheless, the physics of the 2D Ising spin glass is rather rich and many features are still under active debate [10–15]. One of its most convenient features is that in contrast to systems in more than two dimensions ground states can be found in polynomial time using efficient algorithms [16–18]. Here we report on advances in the machinery of such combinatorial optimization algorithms that allow us to determine the critical exponents characterizing the behavior of the Gaussian Ising spin glass in two dimensions for large system sizes and thus arrive at estimates of these critical parameters that are of unprecedented accuracy.

IOP Conf. Series: Journal of Physics: Conf. Series **1163** (2019) 012004 doi:10.1088/1742-6596/1163/1/012004

2. Model and methods

We consider the standard Edwards-Anderson Ising spin glass model with Hamiltonian [19]

$$\mathcal{H} = -\sum_{\langle i,j \rangle} J_{ij} s_i s_j. \tag{1}$$

For the purposes of this paper, the spins are located at the vertices of a square lattice, where interactions are only between pairs of nearest neighbors. We consider a number of different boundary conditions: if periodic boundaries are chosen in one direction only while free boundaries are assumed in the other (PFBC), the graph remains planar, which is in contrast to the case of periodic-periodic boundaries (PPBC), which correspond to a torus topology. In both cases, the common periodic direction is switched to antiperiodic for the defect-energy calculations reported below. In the present work we focus on the Gaussian distribution of exchange couplings, i.e.,

$$P(J_{ij}) = \frac{1}{\sqrt{2\pi}} \exp\left[-J_{ij}^2/2\right].$$

While the ground-state problem for Ising systems in three dimensions and above has been shown to be *NP* hard [20], this is not the case in two dimensions, where ground states [16, 21] and even partition functions [22] can be computed in polynomial time. The ground-state algorithm of Ref. [16] is based on the concept of defect lines connecting frustrated plaquettes. As was initially realized by Toulouse [23], the elementary closed loops of a lattice carry the basic information about the frustration state of a spin configuration, and one calls a plaquette frustrated if it has an odd number of antiferromagnetic bonds, such that it is impossible to satisfy all the interactions with a spin configuration around the loop. As a consequence, the ground state of an Ising system corresponds to a complete pairing of frustrated plaquettes along paths on the dual lattice that traverse a minimal sum of $|J_{ij}|$, a problem known as minimum-weight perfect matching [24]. This is a polynomial problem solved by the blossom algorithm introduced by Edmonds [25] and further improved in a number of generations up to the most recent Blossom V version of the code that is used here [26].

While this mapping to a matching problem transforms the Ising spin-glass ground-state problem into a practically tractable task, there remain a number of problems: Firstly, the matching of plaquettes formally requires a calculation on the complete graph of frustrated plaquettes, thus turning a problem on an originally sparse graph into an auxiliary calculation on a dense graph with the resulting overheads in memory and time to solution. Secondly, as a consequence of this situation, it is a well-established practice to thin out this dense graph to only include matching edges up to a certain maximum weight as very distant frustrated plaquettes are very unlikely to be part of the matching of overall minimum weight [16, 27, 28]. As a result, strictly speaking this approach does no longer guarantee to yield exact ground states as there always remains a certain small, but non-zero probability for rare events to occur. Thirdly, the approach of Ref. [16] only applies to samples on planar graphs. On non-planar graphs the solution of the auxiliary matching problem is not guaranteed to correspond to a realizable spin configuration, such that a ground-state cannot always be found.

These problems are addressed by the approach used here as first discussed in the more comprehensive report of Ref. [29]. It relies on a different representation of the spin system as a combinatorial optimization problem that was first proposed in Refs. [17, 18]. In this model, the dual graph of the original spin lattice is expanded into an auxiliary graph where each plaquette of the spin lattice corresponds to a complete graph of four nodes also known as a Kasteleyn city. Once the Kasteleyn cities are contracted back to individual vertices, a perfect matching on this auxiliary graph is seen to correspond to a set of closed loops on the dual of the spin lattice separating up from down spins. Writing the task to find a spin-glass ground state as a

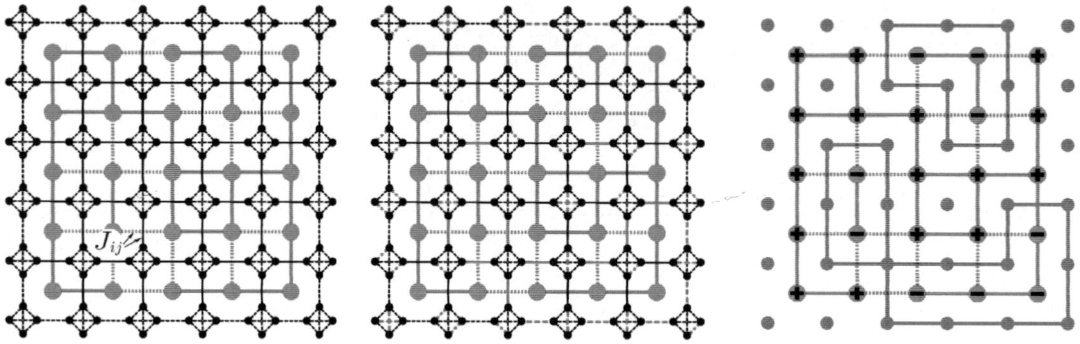

Figure 1. Illustration of the transformation linking the Ising spin-glass ground-state calculation to a minimum-weight perfect matching problem on an auxiliary graph. Left: Expansion of the Ising graph (light grey) to an auxiliary graph of Kasteleyn cities for each plaquette of the original lattice. The edge weights are chosen as J_{ij} for links corresponding to the original lattice bonds and 0 otherwise. Middle: A minimum-weight perfect matching on the auxiliary graph. Right: On contracting back the Kasteleyn cities to individual vertices, the matching corresponds to a set of closed contours on the dual of the original lattice. In the ground-state spin configuration these contours separate up from down spins.

minimum-cut problem, it is seen that in this setup a spin-glass ground state corresponds to a minimum-weight perfect matching on the auxiliary graph if the weights of physical edges are chosen to be identical to the couplings J_{ij} and the weights of auxiliary edges are assumed to be zero [18]. The whole transformation is illustrated in Fig. 1. The matching problem solved in this case remains to be on a sparse graph, hence allowing to treat much larger system sizes than with the model proposed by Bieche *et al.* [16]. In combination with an efficient implementation of the matching algorithm [26], the new mapping allowed us to treat systems up to $10\,000 \times 10\,000$ spins

While the new transformation leads to much more efficient ground-state calculations for planar graphs, the ambiguity and resulting non-applicability for non-planar lattices still remains. For a system with periodic boundary conditions, for instance, some of the loops identified by the matching (cf. the right panel of Fig. 1) will wrap around the lattice in one of the two inequivalent directions and, as a consequence, on traversing across the system in one of the periodic directions one will cross an odd number of such loop lines, and a spin configuration consistently changing sign on crossing such lines is then impossible to find. As a workaround, in Ref. [29] we proposed a windowing technique that allows to compute ground states with high reliability also for systems with periodic boundaries. It proceeds by computing an exact ground state using the minimum-weight perfect matching (MWPM) approach on a square sub-system or "window" of size $L' \times L'$ while keeping all spins outside of the window fixed. This can be achieved by choosing sufficiently strong bonds between the spins just outside of the window that keep them in the chosen relative orientation. The setup is illustrated in the left panel of Fig. 2. In practice, we find that a maximal window size, namely $L' = L - 2$ yields the best results. A single such computation does not always yield a ground-state, but repeated calculations on windows that are randomly shifted over the lattice yields a ground state with a probability that weakly depends on the number of iterations (with about 80% probability for 30 iterations) and that is practically independent of system size. The success probability can be arbitrarily increased by additional repetitions of the whole procedure as the total success probability of m

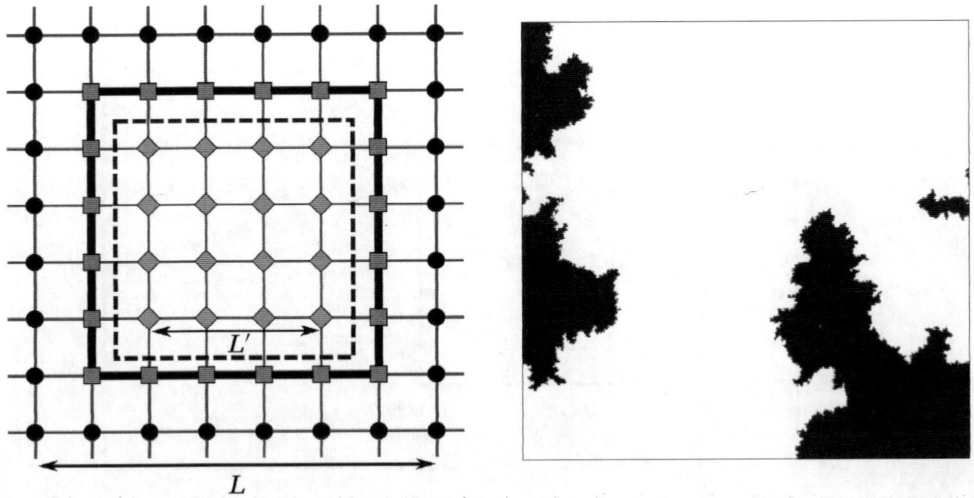

Figure 2. Left: Setup used for the windowing technique to compute ground states for samples with fully periodic boundary conditions.. Right: The domain wall separating regions of equal and opposite ground-state configurations for a specific disorder sample considered with periodic and with antiperiodic boundaries.

independent runs of this type is [30, 29]

$$P_s(\{J_{ij}\}) = 1 - [1 - P_n(\{J_{ij}\})]^m. \tag{2}$$

Studying these dependencies systematically [29], we find that 8 independent runs with 10 minimization steps with randomly chosen window locations yield a total success probability of $P_s = 0.999$, independent of system size. If desired, this can be further increased performing additional independent runs. Through the independence of the success probability from the system size, this setup results in an algorithm that produces exact ground states with a prescribed success probability with a constant-factor overhead over the underlying MWPM algorithm which is known to be polynomial [20].

3. Defect energies

The behavior of defects in the ordered state is among the fundamental differences in the various models that have been proposed to describe the spin-glass state [2]. Banavar and Cieplak [31] were the first to propose a study of defect energies as a means to understand the nature of the spin-glass phase. In particular, it was argued to consider the energy difference between the ground states of systems with periodic and antiperiodic boundary conditions, for which a scaling according to

$$E_{\mathrm{def}} \propto L^{\theta}, \tag{3}$$

was proposed [32]. Generalizing Peierls' argument for the stability of the ordered phase, it was argued that a spin-stiffness exponent $\theta < 0$ corresponds to a situation were droplet excitations are unstable against thermal fluctuations and hence the spin-glass phase is not stable at non-zero temperatures. For the opposite case $\theta > 0$, on the other hand, spin glass should be able to exist for $0 < T \leq T_g$.

We studied defect energies by changing the boundary conditions for individual samples from periodic to antiperiodic and considering the resulting difference in ground-state energies, i.e,

$$E_{\mathrm{def}} = |E_{\mathrm{P}} - E_{\mathrm{AP}}|.$$

IOP Conf. Series: Journal of Physics: Conf. Series **1163** (2019) 012004 doi:10.1088/1742-6596/1163/1/012004

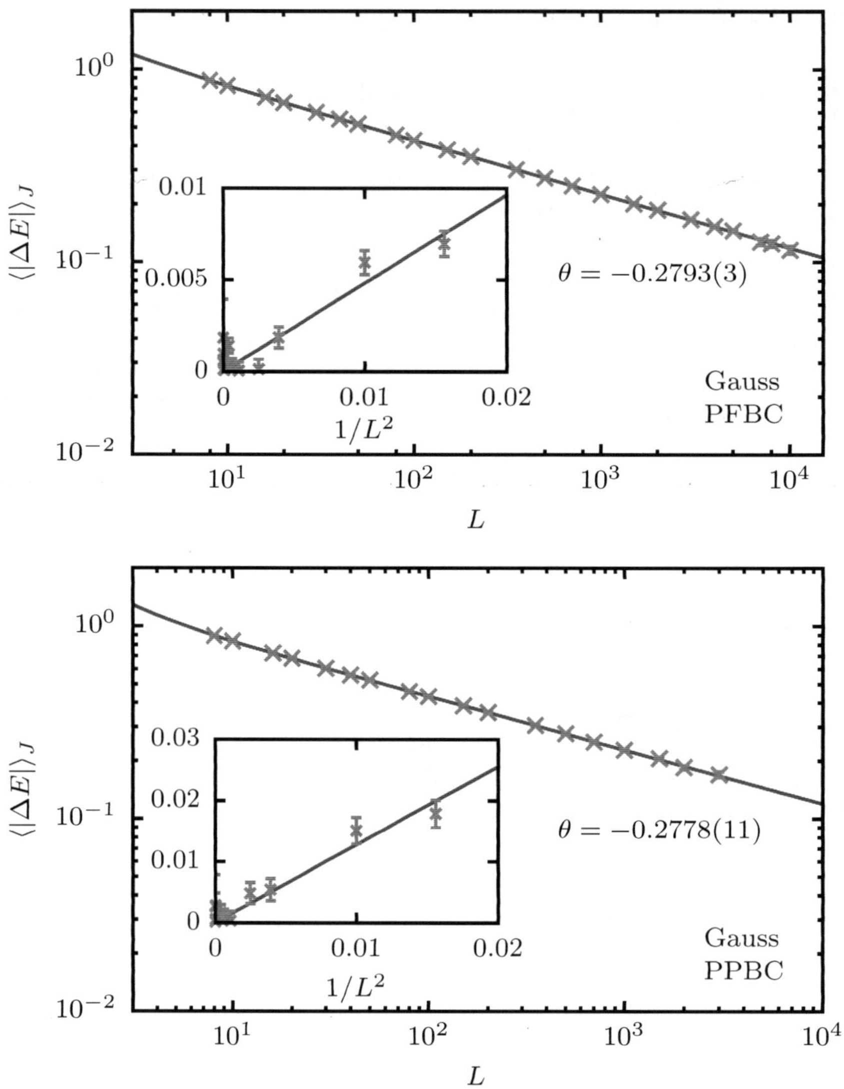

Figure 3. Average defect energies $\langle|\Delta E|\rangle_J$ for the square-lattice Ising spin glass with Gaussian interactions as calculated from the difference in ground-state energies between periodic and antiperiodic boundary conditions in the x direction, and with free boundaries in y direction (PFBC, top) and periodic boundaries in y direction (PPBC, bottom), respectively. The lines show fits of the functional form (4) to the data. The insets show the correction $\langle|\Delta E|\rangle_J(L)-A_\theta L^\theta$ plotted against $1/L^2$ illustrating that this single term describes the corrections very well.

In the other lattice direction we either chose free boundaries, leading to periodic-free boundary conditions (PFBC), or also periodic boundaries (PPBC), where in the latter case we made use of the newly developed algorithm for fully periodic lattices based on the windowing technique [29]. The resulting defect energies, averaged over up to 10^6 disorder samples for the smallest system sizes, and for systems of up to $10\,000 \times 10\,000$ and 3000×3000 spins for the cases of PFBC and PPBC, respectively, are shown in Fig. 3. It is clear that the data follow the power-law implied

IOP Conf. Series: Journal of Physics: Conf. Series **1163** (2019) 012004 doi:10.1088/1742-6596/1163/1/012004

by Eq. (3) rather clearly, but there are some small corrections to scaling that, as the insets of Fig. 3 reveal, are well described by a single $1/L^2$ correction. We hence performed fits of the functional form

$$\langle|\Delta E|\rangle_J(L) = A_\theta L^\theta + C_\theta/L^2 \tag{4}$$

to the data, finding very good fit qualities [33] of $Q = 0.16$ for PFBC and $Q = 0.92$ for PPBC, including systems with $L \geq 8$ for PFBC and $L \geq 10$ for PPBC. The resulting estimates of the stiffness exponent, $\theta = -0.2793(3)$ (PFBC) and $\theta = -0.2778(11)$ (PPBC), respectively, are in excellent agreement with each other and of significantly higher accuracy than previous estimates of the stiffness exponent of this model, see, e.g., Ref. [34].

4. Domain wall fractal dimension

The change of boundary conditions considered above induces a change in the ground-state spin configuration in some part of each sample. If one draws lattice sites with unchanged ground-state spin configurations in black and lattice sites with overturned spin configurations in white, the typical configurations of such overlaps look like what is depicted in the right panel of Fig. 3. The boundary between the two domains forms a fractal curve with a Hausdorff dimension $1 < d_\mathrm{f} < 2$ [35]. The data for the domain-wall lengths for PFBC and PPBC boundary conditions are shown in the upper and lower panel of Fig. 4, respectively. We find the lengths to follow a power law up to very high accuracy, and as the insets of Fig. 4 show, it is hardly possible to resolve any corrections to this leading dependence. We hence performed fits to the pure power-law behavior,

$$\langle\ell\rangle_J = A_\ell L^{d_\mathrm{f}}, \tag{5}$$

to the data for $L \geq L_\mathrm{min} = 40$. The fractal dimensions extracted from these fits are $d_\mathrm{f} = 1.273\,19(9)$ for PFBC and $d_\mathrm{f} = 1.2732(5)$ for PPBC, respectively. As is seen from the insets of Fig. 4, scaling corrections for the PPFC geometry are even smaller than those for PFBC, an observation that is in line with the general expectation of reduced scaling corrections for fully periodic systems. The two values extracted from PFBC and PPBC are well compatible with each and, again, significantly more accurate than previously published estimates, see, e.g., Ref. [36].

5. Domain walls as stochastic Loewner evolution

The fact that the continuum limit of spin systems at a critical point is scale invariant is at the heart of our understanding of critical phenomena and forms the basis for the renormalization group [37]. The fact that there is also rotational symmetry and, more generally, *local* scale invariance leads to the notion of conformal symmetry, that is a powerful mathematical instrument for the description of critical systems [38]. While there are results in three dimensions [39–42], in two dimensions, specifically, the conformal group is infinite-dimensional since any holomorphic map is conformal, and hence conformal invariance alone is able to fully determine the form of certain correlation functions and in many cases gives exact results for all critical exponents. While this has enabled a comprehensive characterization of the operator content of all standard universality classes in two dimensions, this approach runs into difficulties for systems with quenched disorder that, prior to performing the disorder average, do not have conformal or, in fact, even translational invariance. Conformal field theories describing spin glasses, random-field problems or even systems with weak disorder such as disordered ferromagnets in two dimensions have hence not been found [38].

A complementary approach to critical lattice systems focused on geometrical aspects is given by stochastic Loewner evolution (SLE). It allows to connect the geometry of lattice paths such as domain boundaries of critical systems to the statistics of one-dimensional Brownian motion [43], thus resulting in a wide range of exact results for geometrical properties of such paths in

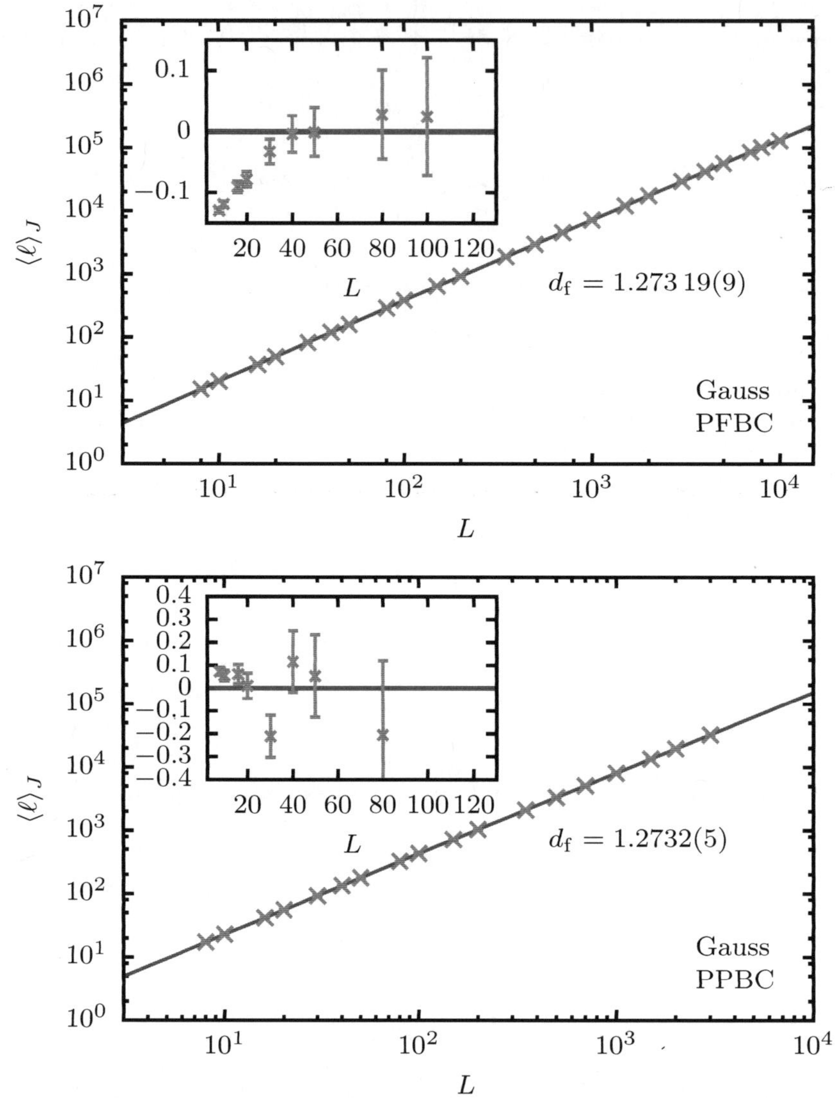

Figure 4. Average length of the domain wall in the 2D Edwards-Anderson Ising spin glass with Gaussian couplings induced by a change of boundary conditions from periodic to antiperiodic. The lines show fits of the functional form (5) to the data. The insets indicate the deviations $L^{d_f} - \langle \ell \rangle_J$.

the scaling limit. Domain boundaries in percolation and the Ising model, for example, have been shown rigorously to be described by SLE [44, 45]. Domain walls in disordered systems have also been investigated regarding compatibility or not with SLE [46–49], and in some cases suprising consistencies have been found, nourishing hopes of the construction of conformal field theories for such systems.

One well-known prediction of SLE regards the probability of an SLE trace to pass to the left of a given point in the domain under consideration [43]. While SLE is normally defined in the upper half plane, the fact that all holomorphic maps in two dimensions are conformal means that

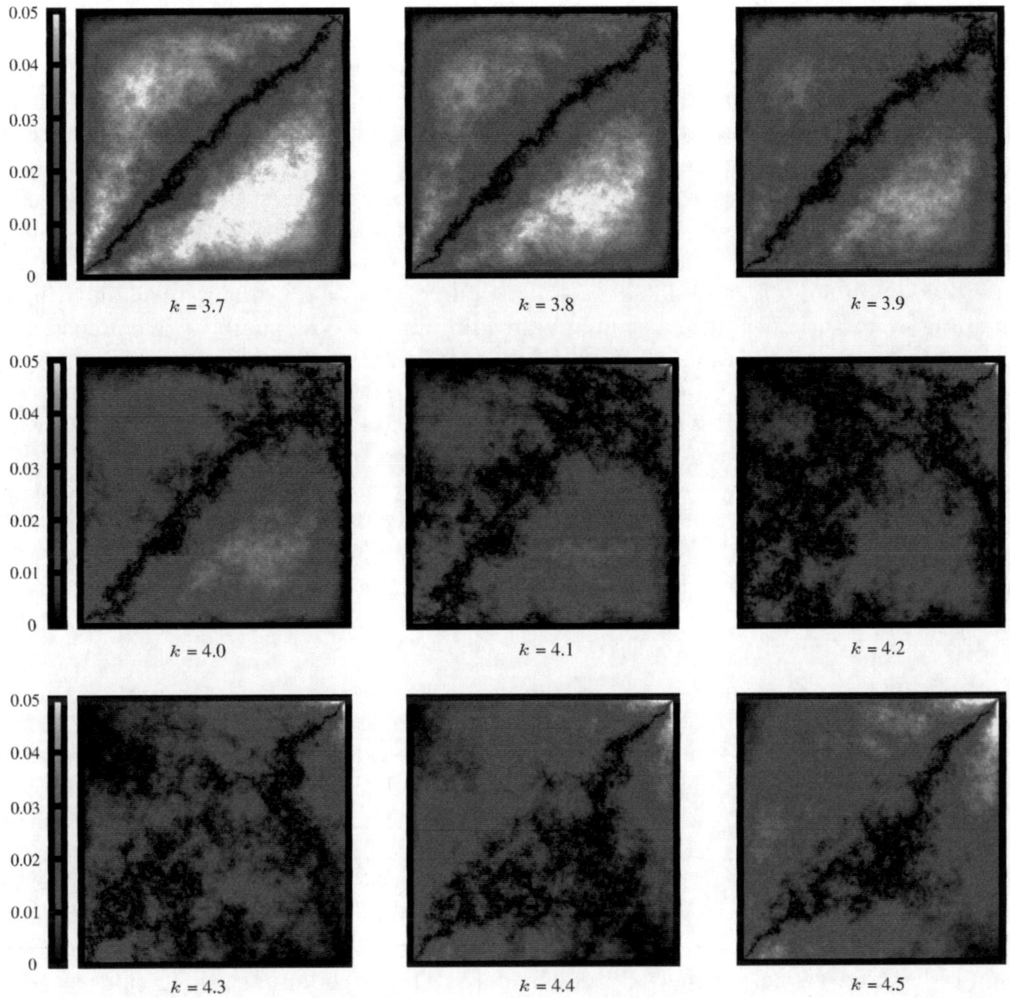

Figure 5. Average deviations of the observed left-passage probabilities of domain walls in the Gaussian 2D Ising spin glass from the exact results for SLE traces derived from the expression (6) for the upper half plane for different values of κ. The data are for systems with 100×100 spins.

one can also consider the problem in any other geometry that can be conformally connected to the upper half plane, including for instance a system of square shape as it is normally considered in lattice systems. The left-passage probability of an SLE trace of parameter κ in the upper half plane is given by [50]

$$P_{\mathrm{LP}}^{\kappa}(x,y) = \frac{1}{2} + \frac{\Gamma(4/\kappa)}{\sqrt{\pi}\,\Gamma\left(\frac{8-\kappa}{2\kappa}\right)}\frac{x}{y}\,{}_2F_1\left(\frac{1}{2},\frac{4}{\kappa};\frac{3}{2};-\left(\frac{x}{y}\right)^2\right), \tag{6}$$

and the corresponding expression for the domain at hand can be calculated from the conformal transformation connecting the two geometries.

While a general consistency of domain walls in the two-dimensional spin glass with SLE had already been observed in Refs. [46, 47], here we study the dependence of these observations on

IOP Conf. Series: Journal of Physics: Conf. Series **1163** (2019) 012004 doi:10.1088/1742-6596/1163/1/012004

the boundary conditions used to force a domain wall into the system. Under the conformal map connecting the upper half plane and the square geometry considered for the spin model, the origin of the upper half plane is mapped onto the lower left corner of the square, while the infinite point of the plane is mapped onto the top right corner. The domain wall should hence be pinned to run exactly across diagonally opposing corners. This can be achieved by introducing strong bounds that make it impossible for the domain wall to break any couplings along the circumference of the square apart from the bonds at the two corner points of the system. In Fig. 5 we show the deviations of measured left-passage probabilities from the ones calculated from Eq. (6) under the assumption of different values of the diffusion constant κ. It is seen that a relative minimum of these deviations is attained for $\kappa \approx 4.3$, but it turns out that there are remaining systematic deviations even at that minimum. Additionally, κ is connected to the fractal dimension of domain walls as $d_f = 1 + \kappa/8$ [43] and $\kappa \approx 4.3$ hence implies $d_f \approx 1.54$, which is not compatible with the accurate estimates of d_f reported above. It hence appears clear that domain walls in the 2D Ising spin glass with fixed boundaries are not compatible with SLE, and only for effectively free boundaries as the ones studied in Refs. [46, 47] compatibility with SLE is possible. Sensitivity to boundary conditions is not unknown in the study of SLE as for example the loop-erased random-walk is compatible with SLE for certain boundary conditions but not for others [51]. Still, the observations reported here require additional work to come to a definite understanding of the relation of domain-wall boundaries in the 2D Ising spin glass to SLE traces.

6. Conclusions

Using a rarely used mapping of the Ising spin-glass ground-state problem on planar lattices to a minimum-weight perfect matching problem on a sparse graph we study ground-states of the system for a wide range of system sizes ranging up to $10\,000 \times 10\,000$ spins. With the help of a windowing algorithm, it is also possible to determine ground-states of systems with non-planar, fully periodic samples with just a constant-factor overhead over the polynomial run-time of the matching algorithm. Employing these techniques, we determine defect energies for the system and estimate a stiffness exponent $\theta = -0.2793(3)$ and a domain-wall fractal dimension $d_f = 1.273\,19(9)$. Testing domain walls in the system for their compatibility with SLE traces, it is found that the result strongly depends on the chosen boundary conditions, and while SLE was found to apply for effectively free boundaries, it does not hold for fixed boundaries. Compatibility with SLE is hence a less universal property than, for example, a critical exponent.

Acknowledgments

The authors acknowledge funding from the DFG in the Emmy Noether Programme (WE4425/1-1) and from the European Commission through the IRSES network DIONICOS (PIRSES-GA-2013-612707).

References

[1] Mézard M, Parisi G and Virasoro M A 1987 *Spin Glass Theory and Beyond* (Singapore: World Scientific)
[2] Kawashima N and Rieger H 2005 Recent progress in spin glasses *Frustrated Spin Systems* ed Diep H T (Singapore: World Scientific) chap 9, p 491
[3] Weigel M and Gingras M J P 2006 *Phys. Rev. Lett.* **96** 097206
[4] Lee L W and Young A P 2003 *Phys. Rev. Lett.* **90** 227203
[5] Pixley J H and Young A P 2008 *Phys. Rev. B* **78** 014419
[6] Fernández L A, Martín-Mayor V, Perez-Gaviro S, Tarancón A and Young A P 2009 *Phys. Rev. B* **80** 024422
[7] Beyer F, Weigel M and Moore M A 2012 *Phys. Rev. B* **86** 014431
[8] Katzgraber H G and Young A P 2003 *Phys. Rev. B* **67** 134410
[9] Katzgraber H G, Larson D and Young A P 2009 *Phys. Rev. Lett.* **102** 177205
[10] Thomas C K, Huse D A and Middleton A A 2011 *Phys. Rev. Lett.* **107** 047203

[11] Parisen Toldin F, Pelissetto A and Vicari E 2011 *Phys. Rev. E* **84** 051116
[12] Jörg T and Krzakala F 2012 *JSTAT* **2012** L01001
[13] Jinuntuya N and Poulter J 2012 *JSTAT* **2012** P01010
[14] Fernández L A, Marinari E, Martín-Mayor V, Parisi G and Ruiz-Lorenzo J J 2016 *Phys. Rev. B* **94** 024402
[15] Lundow P H and Campbell I A 2016 *Phys. Rev. E* **93** 022119
[16] Bieche I, Maynard R, Rammal R and Uhry J P 1980 *J. Phys. A* **13** 2553
[17] Thomas C K and Middleton A A 2007 *Phys. Rev. B* **76** 220406
[18] Pardella G and Liers F 2008 *Phys. Rev. E* **78** 056705
[19] Edwards S F and Anderson P W 1975 *J. Phys. F* **5** 965
[20] Barahona F 1982 *J. Phys. A* **15** 3241
[21] Barahona F, Maynard R, Rammal R and Uhry J P 1982 *J. Phys. A* **15** 673
[22] Galluccio A, Loebl M and Vondrák J 2000 *Phys. Rev. Lett.* **84** 5924
[23] Toulouse G 1977 *Commun. Phys.* **2** 115
[24] Cormen T H, Leiserson C E, Rivest R L and Stein C 2009 *Introduction to Algorithms* 3rd ed (Cambridge, MA: MIT Press)
[25] Edmonds J 1965 *J. Res. Natl. Bur. Stand. B* **69** 125
[26] Kolmogorov V 2009 *Math. Prof. Comp.* **1** 43
[27] Palmer R G and Adler J 1999 *Int. J. Mod. Phys. C* **10** 667
[28] Hartmann A K and Rieger H 2002 *Optimization Algorithms in Physics* (Berlin: Wiley)
[29] Khoshbakht H and Weigel M 2018 *Phys. Rev. B* **97** 064410 (*Preprint* 1710.01670)
[30] Weigel M 2007 *Phys. Rev. E* **76** 066706
[31] Banavar J R and Cieplak M 1982 *Phys. Rev. Lett.* **48** 832
[32] Bray A J and Moore M A 1984 *J. Phys. C* **17** L463
[33] Press W H, Teukolsky S A, Vetterling W T and Flannery B P 2007 *Numerical Recipes: The Art of Scientific Computing* 3rd ed (Cambridge: Cambridge University Press)
[34] Hartmann A K and Young A P 2001 *Phys. Rev. B* **64** 180404
[35] Fisher D S and Huse D A 1988 *Phys. Rev. B* **38** 386
[36] Weigel M and Johnston D A 2007 *Phys. Rev. B* **76** 054408
[37] Zinn-Justin J 2002 *Quantum Field Theory and Critical Phenomena* 4th ed (Oxford: Oxford University Press)
[38] Henkel M 1999 *Conformal Invariance and Critical Phenomena* (Berlin/Heidelberg/New York: Springer)
[39] Weigel M and Janke W 1999 *Phys. Rev. Lett.* **82** 2318
[40] Weigel M and Janke W 2000 *Phys. Rev. B* **62** 6343
[41] Weigel M and Janke W 2000 *Europhys. Lett.* **51** 578
[42] El-Showk S, Paulos M F, Poland D, Rychkov S, Simmons-Duffin D and Vichi A 2012 *Phys. Rev. D* **86** 025022
[43] Cardy J 2005 *Ann. of Phys.* **318** 81–118 special Issue
[44] Smirnov S and Werner W 2001 *Math. Res. Lett.* **8** 729–744
[45] Chelkak D, Duminil-Copin H, Hongler C, Kemppainen A and Smirnov S 2013 Convergence of Ising interfaces to Schramm's SLE curves (*Preprint* 1312.0533)
[46] Amoruso C, Hartmann A K, Hastings M B and Moore M A 2006 *Phys. Rev. Lett.* **97** 267202
[47] Bernard D, Le Doussal P and Middleton A A 2007 *Phys. Rev. B* **76** 020403
[48] Jacobsen J L, Le Doussal P, Picco M, Santachiara R and Wiese K J 2009 *Phys. Rev. Lett.* **102** 070601
[49] Stevenson J D and Weigel M 2011 *Comput. Phys. Commun.* **182** 1879–1882
[50] Schramm O 2001 *Electron. Commun. Probab.* **6** 115–120
[51] Bauer M and Bernard D 2006 *Phys. Rep.* **432** 115–221

Influence of initial states and structure defects on non-equilibrium critical behavior of 3D and 2D Ising models

V V Prudnikov[†], P V Prudnikov[‡], and P N Malyarenko[§]
Omsk State University, Pr. Mira 55A, Omsk 644077, Russia
E-mail: [†]prudnikv@univer.omsk.su, [‡]prudnikp@univer.omsk.su,
[§]petr.malyarenko@yandex.ru

Abstract. Results of Monte Carlo description of features of non-equilibrium critical behavior in three- and two-dimensional Ising models conditioned by presence of structural quenched disorder are presented. It was revealed that the increase in the concentration of defects is accompanied by the strengthening of the aging effects manifested as the slowing down of correlation and relaxation processes in structurally disordered systems as compared to the pure system. Non-equilibrium initial states begin to increasingly more strongly influence peculiar features and characteristics of the system's evolution. It was shown that limiting fluctuation-dissipation ratio values X^{∞} characterizing the degree of system departure from equilibrium satisfy inequality $X^{\infty} < 1$ and depend on the universality class of non-equilibrium critical behavior to which they belong: one of these classes corresponds to the high-temperature, and the other to the low-temperature initial state of the system. Values of these non-equilibrium characteristics in 3D Ising model demonstrate belonging to universality classes of critical behavior of weakly and strongly diluted systems, but non-equilibrium characteristics in 2D Ising model depend continuously on concentration of defects owing to the crossover effects in the percolation behavior.

1. Basic concepts and model representations of the non-equilibrium behavior

One of the peculiarities arisen in describing the critical behavior of systems is the critical slowing down effect. It is associated with an anomalous increase in the system relaxation time t_{rel} when approaching the second-order phase transition temperature T_c. As a result, the system being at the critical point appears in no condition to reach equilibrium throughout the entire relaxation process. Therefore, at times $t \ll t_{\text{rel}}$, extraordinary non-equilibrium phenomena characteristic of systems with slow dynamics, such as aging, fluctuation-dissipation theorem (FDT) violation, and the effect of various initial non-equilibrium system states arise in the behavior of systems [1].

The aging effects appearing in the non-equilibrium relaxation stage of the system with slow dynamics are characterized by the existence of two-time dependencies for such functions as the autocorrelation function $C(t, t_w)$ and the response functions $R(t, t_w)$ on the waiting time t_w and observation time $t - t_w$ determined for the spin system with spin density $S(x)$ by the relations

$$\begin{aligned} C(t, t_w) &= \frac{1}{V} \int d^d x \left[\langle S(x,t) S(0,t_w) \rangle - \langle S(x,t) \rangle \langle S(0,t_w) \rangle \right], \\ R(t, t_w) &= \frac{1}{V} \int d^d x \left. \frac{\delta \langle S(x,t) \rangle}{\delta h(x,t_w)} \right|_{h=0}, \end{aligned} \tag{1}$$

where $\langle ... \rangle$ characterizes the statistical averaging over different realizations of initial spin configurations, h is a weak magnetic field applied to a system at the time point t_w. The waiting time t_w is defined by time between a sample preparation and the beginning of measurement of its characteristics. During $t, t_w \ll t_{rel}$, the effect of initial states of the system and aging effects manifests itself in the temporal system behavior, which is characterized by both translation symmetry breaking in time and slowing down relaxation and correlation processes with increasing the system "age" t_w.

It is supposed that the non-equilibrium behavior of a system is realized via its transition at the starting instant t_0 from the initial state at temperature T_0 to the state with temperature T_s differing from T_0. The accompanying equilibration process is characterized by relaxation time $t_{\rm rel}(T_s)$, and equilibrium corresponding to temperature T_s is reached in times $t \gg t_{\rm rel}(T_s)$, while the system dynamics prove stationary and invariant with respect to time reversal. However, in times $t \ll t_{\rm rel}(T_s)$, the evolution of the system depends on its initial state. In this connection, the non-equilibrium behavior of the system depends on whether it evolves from a high-temperature $T_0 > T_s$ or a low-temperature $T_0 < T_s$ initial state.

According to general concepts about non-equilibrium processes, it is expected that for times $t > tw \gg t_{\rm rel}(T_s)$ $C(t, t_w) = C^{\rm eq}(t - t_w)$ and $R(t, t_w) = R^{\rm eq}(t - t_w)$, where $C^{\rm eq}$ and $R^{\rm eq}$ are corresponding equilibrium values related by the fluctuation-dissipation theorem (FDT)

$$R^{\rm eq}(t - t_w) = \frac{1}{T_s} \frac{dC^{\rm eq}(t - t_w)}{dt_w}. \tag{2}$$

The consequences of the FDT provide theoretical grounds of various experimental methods for the measurement of radiation scattering and absorption by matter.

In the case of non-equilibrium behavior of systems for $t, t_w \ll t_{\rm rel}(T_s)$, the generalized FDT assumes the form

$$R(t, t_w) = \frac{X(t, t_w)}{T_s} \frac{\partial C(t, t_w)}{\partial t_w}, \tag{3}$$

with introduction of the quantity $X(t, t_w)$ which is called as the fluctuation-dissipation ratio (FDR):

$$X(t, t_w) = \frac{T_s R(t, t_w)}{\partial_{t_w} C(t, t_w)}. \tag{4}$$

For times with $t > t_w \gg t_{\rm rel}$, the FDT establishes that $X(t, t_w) = 1$. However, in the general case, $X(t, t_w) \neq 1$ for times with $t, t_w \ll t_{\rm rel}$. The asymptotic FDR value introduced as

$$X^\infty = \lim_{t_w \to \infty} \lim_{t \to \infty} X(t, t_w) \tag{5}$$

is an important universal characteristic of non-equilibrium processes in the various systems.

Let us consider the general properties of X^∞ and its dependence on the system's quenching temperature T_s. For systems demonstrating dissipative slow dynamics from the high-temperature initial state with initial magnetization $m_0 = 0$ ($m_0 \ll 1$), the free propagator $R_0(q, t, t_w)$ and correlator $C_0(q, t, t_w)$ as the bare response function and correlation function are characterized by following expressions [1, 2]:

$$R_0(q, t - t_w) = \theta(t - t_w) \exp\left[-\lambda(q^2 + \tau)(t - t_w)\right], \tag{6}$$

$$C_0^{(D)}(q, t, t_w) = C_0^{(e)}(q, t - t_w) + C_0^{(i)}(q, t + t_w), \tag{7}$$

where

$$C_0^{(e)}(q, t - t_w) = \frac{1}{q^2 + \tau} e^{-\lambda(q^2+\tau)|t - t_w|}, \tag{8}$$

is the so-called equilibrium correlator and

$$C_0^{(i)}(q, t+t_w) = -\frac{1}{q^2+\tau}\, e^{-\lambda(q^2+\tau)(t+t_w)}, \tag{9}$$

is the correlator characterizing the influence of the starting conditions. In (6-9), parameters $\mathbf{q}$ and $\tau \sim (T - T_c)/T_c$ are wave vectors and the reduced temperature, correspondingly.

The FDR in momentum space is given by the expression

$$X(q, t, t_w) = \frac{T_s R(q, t, t_w)}{\partial_{t_w} C(q, t, t_w)}, \tag{10}$$

and we obtain in the Gaussian approximation

$$X_0(q, \tau, t, t_w) = \left(1 + e^{-2\lambda(q^2+\tau)t_w}\right)^{-1}. \tag{11}$$

The asymptotic FDR in the Gaussian approximation is characterized by the following values

$$X_0^\infty = \lim_{t_w\to\infty} \lim_{t\to\infty} X_0(q, \tau, t, t_w) = \begin{cases} 1, & q \neq 0, \tau \neq 0, \\ 1, & q = 0, \tau > 0, \\ 0, & q = 0, \tau < 0, \\ 1/2, & q = 0, \tau = 0. \end{cases} \tag{12}$$

This means that only the zero order-parameter mode with $q = 0$ is characterized by aging effects at the critical point; in other words, it does not relax to equilibrium state and the FDT for this mode is violated.

Non-equilibrium dynamics of systems evolving from a low-temperature initial state with initial magnetization $m_0 \neq 0$ is characterized by following expressions for the bare response function and correlation function [1,3]:

$$R_0(q, t, t_w, t_m) = \theta(t - t_w) \left(\frac{t_w + t_m}{t + t_m}\right)^{3/2} e^{-\lambda(q^2+\tau)(t-t_w)}, \tag{13}$$

$$C_0(q, t, t_w, t_m) = \frac{2\lambda e^{-\lambda(q^2+\tau)(t+t_w)}}{[(t+t_m)(t_w+t_m)]^{3/2}} \int\limits_0^{t_w} dt'(t'+t_m)^3 e^{2\lambda(q^2+\tau)t'}, \tag{14}$$

where $t_m = B_m m_0^{-k}$ is a new time-scale with $k = 2$ and $B_m = \sqrt{2/3}$ in the Gaussian approximation.

At the critical point $\tau = 0$ the FDR can be presented in the Gaussian approximation as

$$X_0(q, t, t_w) = X_0(t_w/t_m, q^2 t_w) \equiv X_0(x, y). \tag{15}$$

with $x = t_w/t_m = 2t_w m_0^2/3$. The FDR in (15) is unrelated to t, i.e., $X_0(q, t, t_w) \equiv X_0(t_w/t_m, q^2 t_w)$, which is a distinctive feature of the FDR in the Gaussian approximation. $X_0(x = 0, y)$ corresponds to case with evolution from the high-temperature initial state.

For the zero mode with $q = 0$ $(y = 0)$, the FDR as a function of variable x is characterized by the relation

$$X_0(x, y = 0) = \frac{4}{5}\left[1 + \frac{3}{5}\frac{1}{(1+x)^4}\right]^{-1}. \tag{16}$$

Its value grows monotonically from $X_0(x=0, y=0) = 1/2$ at $x = 0$ to $X_0(x \to \infty, y=0) = 4/5$ as $x \to \infty$. Thus, the FDR for the global order parameter (complete magnetization) in the long-time limit ($x \to \infty$) reaches the value of $X_0^\infty = 4/5$ if initial magnetization $m_0 \neq 0$ in the case of evolution from a magnetized low-temperature state, and $X_0^\infty = 1/2$ if initial magnetization $m_0 = 0$ in the case of evolution from a high-temperature state. The value of $X_0^\infty = 4/5$ invariably remains independent of a concrete $m_0 \neq 0$ value.

Strong fluctuation effects accompanying second-order phase transitions result in fluctuation corrections to values of these characteristics of non-equilibrium critical behavior considered in the Gaussian approximation.

Consequently, we can generalize that for states of the system with temperature $T_s > T_c$, $X^\infty(T_s > T_c) = 1$. For a low- temperature ordered phase with $T_s < T_c$, $X^\infty(T_s < T_c) = 0$. These results are unrelated to specific properties of individual systems. In the case of $T_s = T_c$, there are no general arguments defining the value of $X^\infty(T_s = T_c)$, which necessitates its derivation for each individual statistical model, and $X^\infty(T_s = T_c)$ is a universal quantity associated with the universality class of the model's critical dynamics.

2. Influence of defects on critical behavior

The renormalization-group [4, 5], numerical [6–9], and experimental [10] investigations of the critical dynamics of structurally disordered systems have clearly established that the presence of uncorrelated structural defects and defects with effects of long-range correlations in the studied systems leads to new types of critical behavior and to a significant enhancement of critical slowing down effects in comparison with pure systems. In this respect, specific features of a non-equilibrium behavior, such as aging effects, should undoubtedly find a more striking manifestation in structurally disordered systems with new universal values of the fluctuationdissipation ratio.

Well-known that phenomenological approach to description of an universal critical behavior of systems with disorder is determined by the Hamiltonian

$$H = \int d^d x \left\{ \frac{1}{2} \sum_i^n \left[(\nabla \phi_i)^2 + \tau_0 \phi_i^2 + V(x) \phi_i^2 \right] + \frac{g_0}{4!} \left(\sum_i^n \phi_i^2 \right)^2 \right\}, \tag{17}$$

where $\phi(x)$ is the n-component order parameter, $V(x)$ characterizes a potential of random field of defects. Under the assumption that spatial distribution of the potential $V(x)$ is specified by a Gaussian distribution, statistical properties of $V(x)$ are determined by only first moment with $\langle\langle V(x) \rangle\rangle = 0$ and second moment $\langle\langle V(x)V(y) \rangle\rangle \sim u(\boldsymbol{x} - \boldsymbol{y})$. In the case of uncorrelated point-like defects [11],

$$u(\boldsymbol{x} - \boldsymbol{y}) = v\delta^D(\boldsymbol{x} - \boldsymbol{y}), \tag{18}$$

where $v \sim V^2 c_{imp}$, c_{imp} is a concentration of defects. The model with ε_D-extended defects parallel oriented relative to one another and random distributed in sample perpendicularly to defect direction [12] is characterized by

$$u(\boldsymbol{x} - \boldsymbol{y}) = v\delta^{D-\varepsilon_D}(\boldsymbol{x}_\perp - \boldsymbol{y}_\perp). \tag{19}$$

The model of long-range correlated defects with random orientations was proposed in the work of Weinrib and Halperin [13] is determine by

$$u(\boldsymbol{x} - \boldsymbol{y}) \sim |\boldsymbol{x} - \boldsymbol{y}|^{-a}. \tag{20}$$

The free energy of disordered system is given by relation

$$\langle\langle F/kT \rangle\rangle = (F_0/kT) - \frac{1}{8} \int \mathrm{d}^d x \, \mathrm{d}^d y \, \langle\langle V(\boldsymbol{x}) V(\boldsymbol{y}) \rangle\rangle \left\langle \phi^2(\boldsymbol{x}) \phi^2(\boldsymbol{y}) \right\rangle_0 + \ldots, \tag{21}$$

where $\langle\langle\ldots\rangle\rangle$ indicates an averaging over disorder configurations, $\langle\ldots\rangle_0$ means statistical averaging with the Hamiltonian of pure system. The specific heat of pure system can be introduced as

$$C_0 = \frac{d^2(F_0/kT)}{dT^2} = \int d^dx\, d^dy \left\langle\phi^2(x)\phi^2(y)\right\rangle_0 \sim \tau^{-\alpha_0}, \tag{22}$$

where α_0 is critical exponent of the specific heat for pure system. With the use of scaling analysis, we can determine the free energy $F_0 \sim \tau^{2-\alpha_0}$ and the correlator $\left\langle\phi^2(\boldsymbol{x})\phi^2(\boldsymbol{y})\right\rangle_0 = \tau^{-\alpha_0+D\nu_0}G\left(\frac{|\boldsymbol{x}-\boldsymbol{y}|}{\xi}\right)$. Then asymptotic behavior of the free energy of disordered system on temperature can be presented in the following form

$$\langle\langle F/kT\rangle\rangle = \tau^{2-\alpha_0}A + \frac{\tau^{-\alpha_0+d\nu_0}}{8}\int \mathrm{d}^dx\; u(\boldsymbol{x})G\left(\frac{x}{\xi}\right) = \tau^{2-\alpha_0}(A + B\tau^{-\varphi}), \tag{23}$$

where φ is crossover critical exponent, which characterizes influence of defects on critical behavior.

As it is seen from (23), influence of defects on critical behavior for $\varphi > 0$ is relevant and leads to a new type of critical behavior with new values of critical exponents, while for $\varphi < 0$ influence of defects on critical behavior is irrelevant and critical behavior of disordered system is characterized by values of critical exponents for pure system.

For uncorrelated point-like defects, $\int \mathrm{d}^dx\; u(\boldsymbol{x})G(x/\xi) \sim$ const that leads to $\varphi = 2 - D\nu_0 = \alpha_0$. Therefore, the presence of point-like defects in sample is relevant for critical behavior if critical exponent of the specific heat for pure system α_0 is positive. This statement forms a content of well-known Harris criterion [14].

This paper devotes to study of influence of quenched point-like defects on properties of non-equilibrium critical behavior and we must note that only three-dimensional Ising-like systems are characterized by positive value of the specific heat critical exponent α_0: 0.1091(24) (RG calculations), 0.1109(15) (MC calculations), 0.1099(7) (experiment). The 2D Ising model is a marginal case with the specific heat $C_v \sim \ln|\tau|$ and suitable exponent $\alpha_0 = 0$.

In case of systems with ε_D-extended or isotropic long-range correlated defects, the crossover exponent $\varphi = \alpha_0 + \varepsilon_D\nu_0$ and $\varphi = \alpha_0 + (D - a)\nu_0 = 2 - a\nu_0$, respectively, and defects affect on critical behavior of more wide class of systems than Ising-like systems. We recommend for familiarity with question of influence of ε_D-extended and isotropic long-range correlated defects on critical behavior the articles [15, 16].

Investigations into the critical behavior of diluted Ising type magnets with the use of RG methods, MC numerical simulations, and experimental techniques confirm the existence of a new universality class of critical behavior, but the dependence of asymptotic values of critical exponents on the degree of their dilution and the influence of crossover effects in weakly and strongly disordered systems await elucidation and remain subjects of ardent discussion. It is worthwhile to note that analytical RG methods applied to study the critical behavior of impurity systems are suitable only for weakly diluted magnetic materials with concentration of defects $c_{\mathrm{imp}} = 1 - p \ll 1$. As a system becomes increasingly more diluted with nonmagnetic impurity atoms at spin concentrations $p_c^{(s)} < p < p_c^{(\mathrm{imp})} = 1 - p_c^{(s)}$, where $p_c^{(s)}$ and $p_c^{(\mathrm{imp})}$ are the spin and impurity percolation thresholds, respectively (for simple cubic lattices with nearest neighbor interactions one has ps $p_c^{(s)} \simeq 0.3117$, $p_c^{(\mathrm{imp})} \simeq 0.6883$), the impurities aggregate into a binding cluster coexisting for $T < T_c$ with the spin binding cluster at spin concentrations up to $p_c^{(s)}$ and form a fractal-like structure with effective long- range spatial correlation in impurity distribution [16].

A change in effects of order parameter fluctuation scattering from impurity atoms must give rise to new fixed points for the vertices of interaction between order parameter fluctuations.

Therefore, the region with $p_c^{(s)} < p < p_c^{(\mathrm{imp})}$ is characterized by a new type of critical behavior of three- dimensional Ising models, corresponding to the region of the strong structural disorder.

Presence of point-like defects in the 2D Ising model does not changes static critical exponents but logarithmic corrections may become important in the vicinity of the critical point. It has been found analytically within bond disorder model for the specific heat $C_v \sim \ln|\ln|\tau|$ and correlation length $\xi \sim |\tau|^{-1}(1 + g\ln(1/|\tau|))^{1/2}$ with $g \simeq 4.8(1-p)/p$ [17] as well as for the magnetic susceptibility $\chi \sim |\tau|^{-7/4} ln^{7/8}|\tau|$ [18]. This prediction for weak bond disorder was confirmed numerically in a number of papers [19–21]. In this paper we study the 2D site-diluted Ising model, for which the phase diagram contains two crucial points, the pure Ising fixed point and the percolation fixed point at $p_c^{(s)} \simeq 0.5927$. This model was investigated numerically in a number of papers [22–25]. Some of authors claim in [22,23] that the static critical exponents β, γ vary with defect concentration, but the ratios β/ν, γ/ν do not change. Others [24,25] concluded that defects lead only to logarithmic corrections as for weak bond dilution. At the same time, the studies of the critical dynamics of the model reported in [26,27] demonstrated that, near the spin percolation threshold (for the systems with the spin concentrations $p \leq 0.85$), the dynamic critical exponent z, determining the temperature dependence of the relaxation time, exhibits the dependence on the concentration of defects with the violation of the standard form of dynamic scaling and takes the form

$$z = A\,|\ln(p - p_c)| + B. \tag{24}$$

This suggests a rather nontrivial effect of defects on the characteristics of the non-equilibrium behavior of the 2D Ising model.

3. Non-equilibrium relaxation of the magnetization in 3D and 2D Ising models

In this part, we present the results of numerical Monte Carlo study of the effects of initial states and structural defects on relaxation properties of the magnetization $M(t)$ in the 3D and 2D Ising models. The Hamiltonian of a site-diluted Ising model with external magnetic field h is given by the expression

$$H = -J\sum_{\langle i,j\rangle} p_i p_j S_i S_j - h\sum_i p_i S_i, \tag{25}$$

where $J > 0$ is the short-range exchange interaction between spins $S_i = \pm 1$ fixed at the sites of a lattice, p_i are the random occupation numbers indicating the presence of quenched structurally uncorrelated disorder in the system: $p_i = 1$ for the i site with a spin, and $p_i = 0$ for the site with a nonmagnetic impurity atom. Structural defects are canonically distributed in the system according to the distribution function $P(p_i) = (1-p)\delta(p_i) + p\delta(p_i)$, where $p = \langle p_i \rangle$ defines the spin density in the system. The position of structural defects is fixed for a particular impurity configuration.

The simulation was performed on the simple cubic lattice for the 3D Ising model and on the planar square lattice for the 2D Ising model with imposed periodic boundary conditions. $N_s = pL^D$ characterizes the number of spins in the lattice with linear size L.

To calculate characteristics of the non-equilibrium critical behavior, the dynamic process of one-spin flips was numerically simulated within the Monte Carlo statistical method. The dynamic process of one-spin flips was implemented using the Metropolis algorithm and the heat bath algorithm [28]. For an Ising model with two possible spin states $S_j = \pm 1$, the probability of a spin flip with the use of the heat bath algorithm can be represented in the form

$$W_{\mathrm{hb}}(S_i \to S_i^{'}) = \frac{e^{-H(S_i^{'})/T}}{e^{H(S_i)/T} + e^{-H(S_i)/T}}, \tag{26}$$

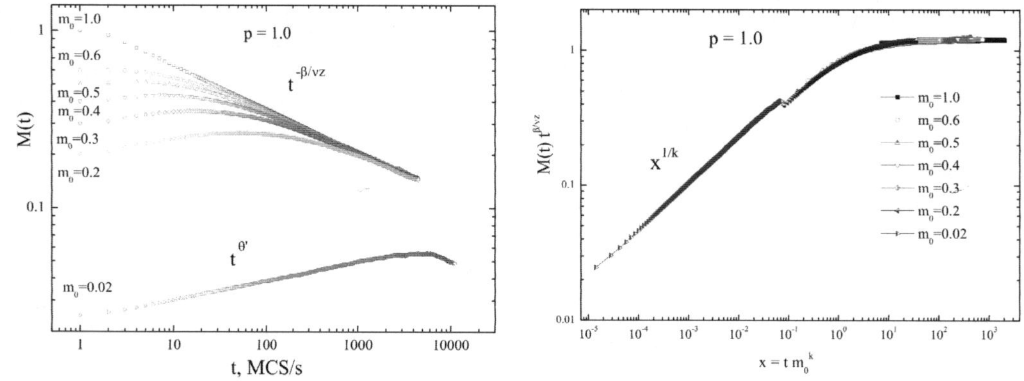

Figure 1. Time dependence of magnetization $M(t)$ (left) and its scaling function $F_M(t/t_m) = M(t)t^{\beta/z\nu}$ (right) for various initial states of m_0 in pure 3D Ising model.

with realization of the so-called Glauber dynamics. As a time unit of the dynamic process, the Monte Carlo step per spin (MCS/s) which defines N_s sequential spin flips at lattice sites was chosen.

The simulation was performed on a 3D cubic lattice with linear size $L = 128$ at the spin densities $p = 1.0$, 0.95, 0.8, 0.6, and 0.5 and corresponding critical temperatures $T_c(p)$: $T_c(1.0) = 4.5114(1)$ [29], $T_c(0.95) = 4.26267(4)$, $T_c(0.8) = 3.4995(2)$, $T_c(0.6) = 2.4241(1)$, and $T_c(0.5) = 1.84509(6)$ [6]. In the 2D case, the simulation was performed on a square lattice with linear size $L = 1024$ at the spin densities $p = 1.0$, 0.95, 0.9, 0.85, 0.8, 0.75 and 0.7 and corresponding critical temperatures $T_c(p)$: $T_c(1.0) = 2.26918$, $T_c(0.95) = 2.08989(8)$, $T_c(0.9) = 1.9032(5)$, $T_c(0.85) = 1.7098(4)$, $T_c(0.8) = 1.5103(4)$, $T_c(0.75) = 1.2980(10)$ [30], and $T_c(0.7) = 1.0729(10)$ [31].

Non-equilibrium behaviour of the magnetization relaxed from an initial state with magnetization m_0 is characterized at the critical point by the following scaling dependence

$$M(t, t_m) = A_M t^{-\beta/z\nu} F_M(t/t_m) \tag{27}$$

with time-scale $t_m \sim m_0^{-k}$, exponent $k = 1/(\theta' + \beta/z\nu)$, and scaling function $F_M(t/t_m) = F_M(tm_0^k)$. The time dependence of the magnetization $M(t, t_m)$ is determined by increasing power-law dependence $M(t) \simeq m_0 t^{\theta'}$ for small initial magnetization $m_0 \ll 1$ (the high-temperature initial state) and decreasing power-law dependence $M(t) \sim t^{-\beta/z\nu}$ for $m_0 = 1$ (the completely ordered initial state). Results of simulation of non-equilibrium behaviour of the magnetization relaxed from various initial states are presented for the 3D Ising model in Figs.1-3 and for the 2D Ising model in Figs.4-5 with different spin concentrations p.

The curves $M(t)$ clearly show the significant both qualitative and quantitative differences in magnetization relaxation from a high-temperature initial state with $m_0 \ll 1$, a low-temperature completely ordered state with $m_0 = 1$, and intermediate states with $0.1 \leq m_0 \leq 0.6$. We can see that the relaxation curves for systems started from initial states $m_0 \neq 1$ asymptote to the relaxation curve from a low-temperature initial state with $m_0 = 1$. In this case, for systems with $m_0 \ll 1$, a characteristic increase in the magnetization is observed at the non-equilibrium evolution stage, which is described by the power law $M(t) \sim t^{\theta'}$ with $\theta' = 0.111(4)$ for $p = 1.0$, $\theta' = 0.127(16)$ for $p = 0.8$, and $\theta' = 0.167(18)$ for $p = 0.6$. At times $t > t_{cr} \equiv t_m \sim m_0^{-k}$, this evolution stage is followed by the mode characterized by the power-law time dependence of the magnetization $M(t) \sim t^{-\beta/z\nu}$. In the case of system evolution from the initial ordered state with $m_0 = 1$, the time dependence of the magnetization at the critical point is immediately defined by

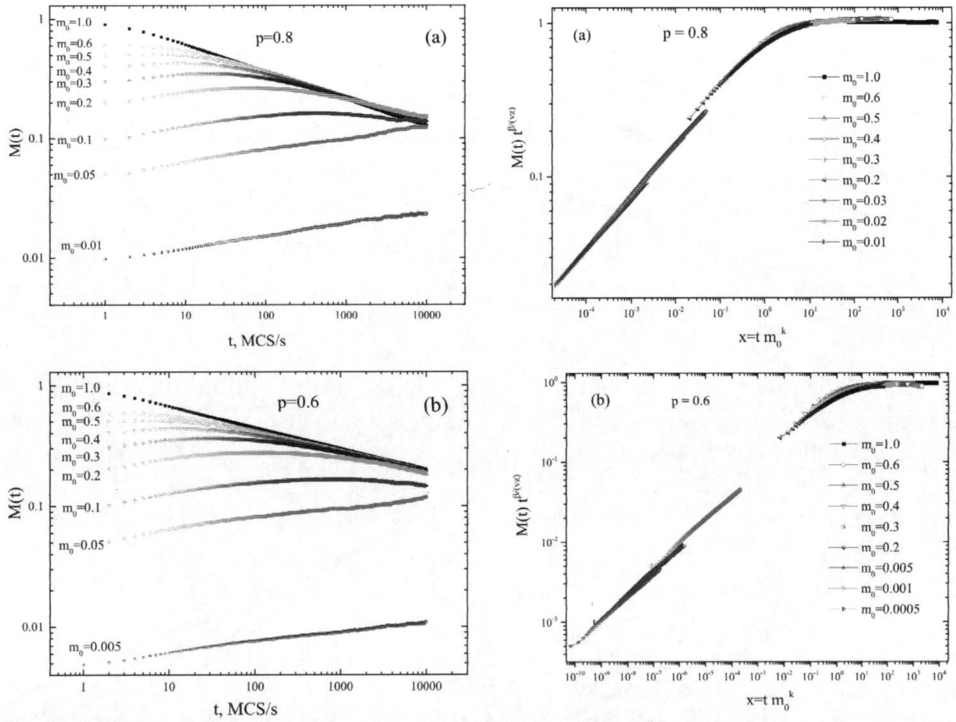

Figure 2. Time dependence of magnetization $M(t)$ (left) and its scaling function $F_M(t/t_m) = M(t)t^{\beta/z\nu}$ (right) for various initial states of m_0 in site-diluted 3D Ising model with spin concentrations $p = 0.8$ (a) and $p = 0.6$ (b).

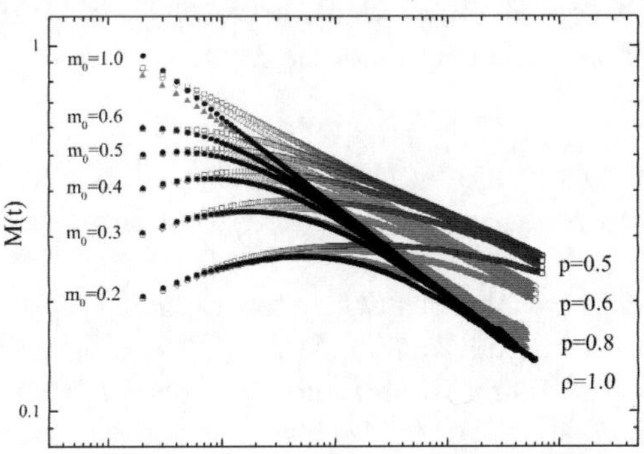

Figure 3. Time dependence of magnetization $M(t)$ for various initial states of m_0 in site-diluted 3D Ising model with spin concentrations $p = 1.0$, 0.8, 0.6 and 0.5.

the power-law dependence $M(t) \sim t^{-\beta/z\nu}$ with exponents $\beta/z\nu$ depending on the spin density p. It was determined that $\beta/z\nu = 0.243(6)$ for $p = 1.0$, $\beta/z\nu = 0.224(10)$ for $p = 0.8$, and $\beta/z\nu = 0.176(13)$ for $p = 0.6$. Intermediate states with $0.1 \leq m_0 \leq 0.6$ are characterized by a short magnetization increase stage by the law $M(t) \sim t^{\theta'}$ followed by the transition to a longer relaxation stage according to the law $M(t) \sim t^{-\beta/z\nu}$.

Summary Fig.3 shows the curves of the critical magnetization relaxation $M(t)$ for systems with various m_0 and spin densities $p = 1.0, 0.8, 0.6$, and 0.5 generalizing the effect of structural

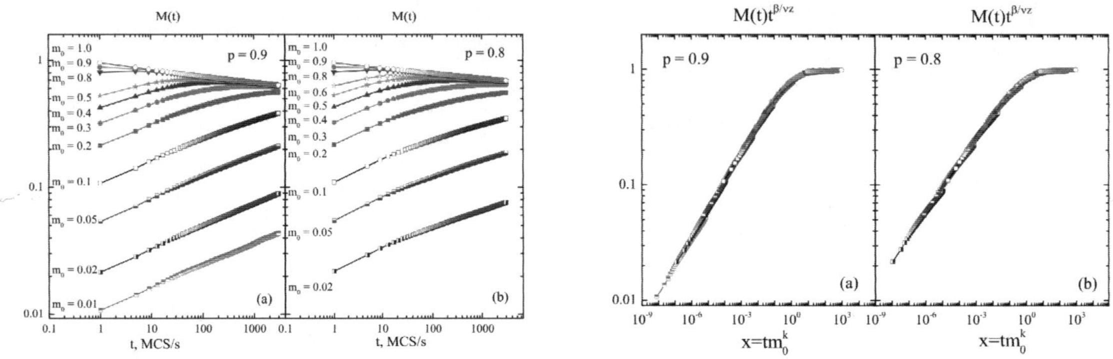

Figure 4. Time dependence of magnetization $M(t)$ (left) and its scaling function $F_M(t/t_m) = M(t)t^{\beta/z\nu}$ (right) for various initial states of m_0 in site-diluted 2D Ising model with spin concentrations $p = 0.9$ (a) and $p = 0.8$ (b).

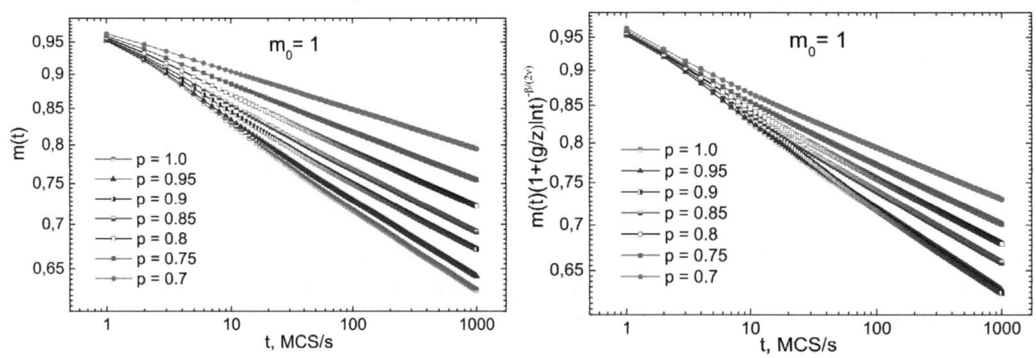

Figure 5. Time dependence of the magnetization $m(t)$ (left) and $m(t)\,(1 + (g/z)\ln t)^{-\beta/2\nu}$ (right) for $m_0 = 1$ subject to the logarithmic corrections.

defects on relaxation of the 3D Ising model. We can see that the presence of structural defects slows down the system relaxation, leading to an increase in the relaxation time with the defect density.

In Figs.1-2 we also demonstrate the results of numerical tests of predictions of the time scaling dependence (27) for the magnetization $M(t, t_m)$ as a function of initial values of the magnetization m_0 for the 3D pure Ising model and systems with $p = 0.8$ and $p = 0.6$. For the curves of the scaling function $F_M(x) = t^{\beta/z\nu} M(t, t_m)$ of the variable $x = tm_0^k$, a "collapse" of the data obtained for various m_0 is observed in the unified universal curve whose shape depends on the system spin density p. The universal curves for the scaling function $F_M(x)$ in the log-log plot are characterized by a linear initial part corresponding to the power-law dependence $F_M(x)\ x^{1/k}$. By the slope of these linear parts, the following exponents k for the 3D Ising model were calculated: $k = 2.77(2)$ for $p = 1.0$, $k = 2.83(3)$ for $p = 0.8$, and $k = 3.08(7)$ for $p = 0.6$.

The similar behavior are demonstrated in Fig.4 by the curves of the critical magnetization relaxation $M(t)$ and its scaling function $F_M(t/t_m) = M(t)t^{\beta/z\nu}$ for the 2D site-diluted Ising model. Differences against the 3D Ising model are determined by smaller values of exponent $\beta/z\nu$ which is equal, for example, 0.046(1) for $p = 0.8$ while $\beta/z\nu = 0.224(10)$ for the 3D Ising model with the same dilution $p = 0.8$. It leads to increase of the critical slowing-down effects in the 2D Ising model and gives possibility better to study non-equilibrium behavior of this model numerically.

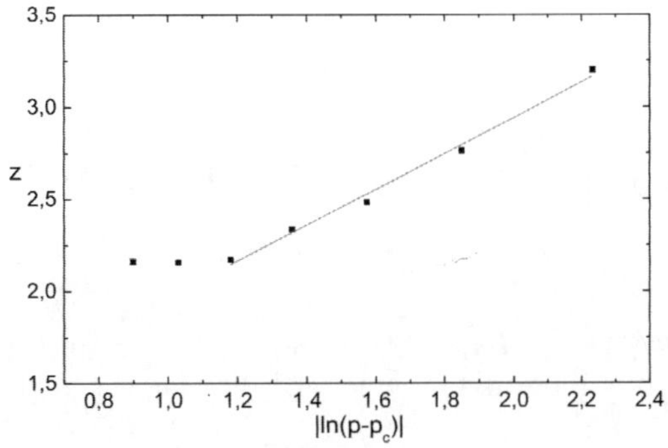

Figure 6. Dependence of the exponent z on spin density p at logarithmic scale $|\ln(p - p_c^s)|$.

Particularly, we show in Fig.5 the procedure of taking into consideration of the logarithmic corrections for description of the time dependence of the magnetization in the case of evolution from the low-temperature initial state with $m_0 = 1$ using these corrections in the form:

$$M(t) \sim t^{-\beta/(\nu z)} \left(1 + (g/z) \ln t\right)^{-\beta/(2\nu)}, \tag{28}$$

where $g = \frac{4}{\pi} g_0 = \frac{4}{\pi} \frac{8}{(1+\sqrt{2}/\pi)^2} \frac{1-p}{p} \approx 4.843 \frac{1-p}{p}$ [17, 30]. In result, we calculated influence of the logarithmic corrections on values of the exponents $\beta/(\nu z)$ and z which demonstrate logarithmic dependence of the exponent z on the spin concentrations (Fig.6). The relation (24) is confirmed in the case of the spin concentrations $p \leq 0.85$. However, the values of the exponents $\beta/(\nu z)$ and z for systems with $p = 0.95$ and $p = 0.9$ coincide within the statistical errors with the values for the pure system $\beta/(\nu z) = 0.05784(30)$ and $z = 2.161(11)$. This result confirms that the values of the dynamic critical exponent z for weakly disordered systems with $p \geq 0.9$ remain constant and the critical dynamics of the 2D site-diluted Ising model belongs to the universality class of the pure model [26, 27]. The value of z for the pure system is in a good agreement with the value $z = 2.1667(5)$ from the work [32]. We must note that the logarithmic corrections provide a marked contribution to the values of z.

Analysis of the scaling functions $F_M(tm_0^k)$ for the 2D Ising model gives to calculate the exponent $1/k$ which has the values $1/k \simeq 0.245(11)$ for spin concentrations $p = 1.0 \div 0.9$ and $1/k \simeq 0.228(4) \div 0.160(5)$ for $p = 0.85 \div 0.7$ with dependence on p by a logarithmic law $|\ln(p - p_c^s)|$.

4. Aging in two-time behavior of the autocorrelation and the response functions. Calculation of the FDR

Aging effects during the non-equilibrium process demonstrate such two-time dependent quantities as the autocorrelation function $C(t, t_w)$, and the linear response function $R(t, t_w)$ to a weak field applied at time t_w. The functions $C(t, t_w)$ and $R(t, t_w)$ can be defined as

$$C(t, t_w) = \frac{1}{V} \int d^d x \left[\langle S(x,t) S(x,t_w) \rangle - \langle S(x,t) \rangle \langle S(x,t_w) \rangle \right], \tag{29}$$

$$R(t, t_w) = \frac{1}{V} \int d^d x \left. \frac{\delta \left[\langle S(x,t) \rangle \right]}{\delta h(x, t_w)} \right|_{h=0}, \tag{30}$$

where the angle brackets stand for averaging over the different realizations of initial state and the square brackets are for averaging over the different impurity configurations.

In simulation of system's dynamics with the use of the thermal bath algorithm, the response function in (30) is calculated using the following relations [35]

$$R(t,t_w) = \frac{1}{TN_s}\sum_{i=1}^{N_s}\Big[\Big\langle p_i S_i(t)\Big(S_i(t_w+1) - S_i^W(t_w+1)\Big)\Big\rangle\Big], \tag{31}$$

where $S_i^W(s) = \tanh(J\sum_{j\neq i} p_j S_j/T)$.

The linear response function $R(t,t_w)$ corresponding to definition (30), cannot be directly measured experimentally. The generalized susceptibility in the form of the integral response function (thermostatic susceptibility)

$$\chi(t,t_w) = \int_0^{t_w} dt' R(t,t') = \frac{1}{T_c N}\sum_i [\langle p_i S_i(t)\Delta S_i(t_w)\rangle] \tag{32}$$

can be measured experimentally and calculated numerically. The quantity $\Delta S_i(t_w) = \sum_{s=0}^{t_w}[S_i(s) - S_i^W(s)]$ in (32) is calculated in the simulation process from initial state at $t=0$ until the time t_w, where $S_i^W(s)$ is the same as in (31).

Currently, it is known that, in the case of evolution from an initial state with $m_0 \neq 0$, the non-equilibrium critical behavior of the autocorrelation function, the response function, and dynamic susceptibility is characterized by the following scaling dependencies [3,33,34]:

$$\begin{aligned} C(t,t_w,t_m) =& A_C(t-t_w)^{a+1-d/z}\,(t/t_w)^{\theta-1}\,F_C(t_w/t,t/t_m),\\ R(t,t_w,t_m) =& A_R(t-t_w)^{a-d/z}\,(t/t_w)^{\theta}\,F_R(t_w/t,t/t_m),\\ \chi(t,t_w,t_m) =& A_\chi(t-t_w)^{a+1-d/z}(t/t_w)^{\theta-1}F_\chi(t_w/t,t/t_m), \end{aligned} \tag{33}$$

where the exponents θ and a are related to the critical exponents of the system under consideration: $a = (2-\eta-z)/z$, $\theta = \theta' - (2-z-\eta)/z$. The scaling functions $F_C(t_w/t,t/t_m)$, $F_R(t_w/t,t/t_m)$, and $F_\chi(t_w/t,t/t_m)$ are finite at $t_w \to 0$ and $t/t_m \to 0$, A_C, A_R, and A_χ are the non-universal amplitudes whose values are fixed by the conditions $F_C(0,0) = 1$, $F_R(0,0) = 1$, $F_\chi(0,0) = 1$.

The quantities in (33) are generally homogeneous functions of three time scales tt_w, t_w, and t_m. When the following their correspondence $t_w < t \ll t_m$ is satisfied, which is realized in the case of evolutions from a high-temperature initial state with $m_0 = 0$, dependencies (33) transform to the relations corresponding to this case [2,36].

In the opposite case with times $t_m \ll t_w < t$, which is realized in the case of evolution from a low-temperature completely ordered initial state with $m_0 = 1$, the scaling dependencies (33) take the form

$$\begin{aligned} C(t,t_w) =& \bar{A}_C(t-t_w)^{a+1-d/z}\,(t/t_w)^{\bar\theta-1}\,\bar{F}_C(t_w/t),\\ R(t,t_w) =& \bar{A}_R(t-t_w)^{a-d/z}\,(t/t_w)^{\bar\theta}\,\bar{F}_R(t_w/t),\\ \chi(t,t_w) =& \bar{A}_\chi(t-t_w)^{a+1-d/z}\,(t/t_w)^{\bar\theta-1}\,\bar{F}_\chi(t_w/t), \end{aligned} \tag{34}$$

where the new exponent $\bar\theta = -\beta\delta/(\nu z) = -(1+a+\beta/(\nu z))$ is introduced, and $\bar{F}_{C,R,\chi}$ are universal scaling functions obtained from $F_{C,R,\chi}(t_w/t,t/t_m)$ in (33) at $t/t_m \to \infty$.

In the aging regime realized at observation times $t - t_w \sim t_w \gg t_m$, in which the two-time dependence of functions $C(t,t_w)$, $R(t,t_w)$, and $\chi(t,t_w)$ is most pronounced, the scaling

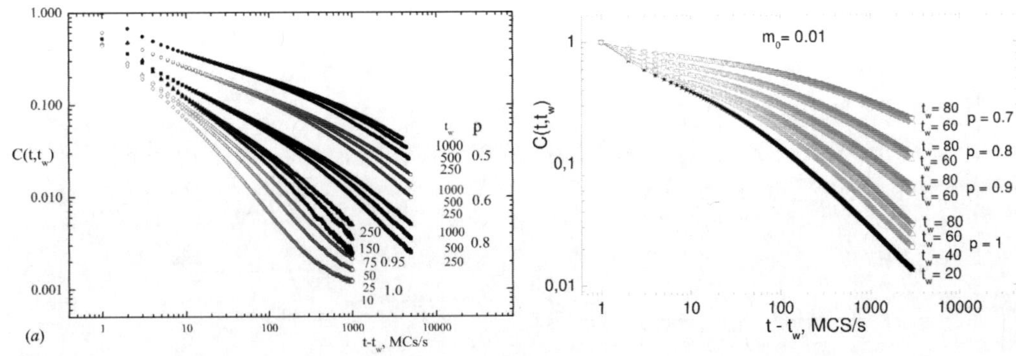

Figure 7. Dependencies of the autocorrelation function $C(t, t_w)$ on observation times $t - t_w$ for various spin concentrations p and waiting times t_w with evolution from the high-temperature initial state with $m_0 \ll 1$ (left) for the 3D Ising model and (right) the 2D Ising model.

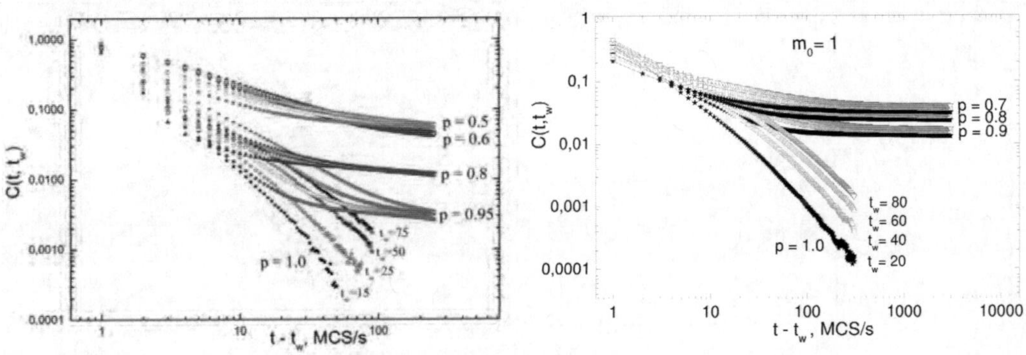

Figure 8. Dependencies of the autocorrelation function $C(t, t_w)$ on observation times $t - t_w$ for various spin concentrations p and waiting times t_w with evolution from the low-temperature initial state with $m_0 = 1$ (left) for the 3D Ising model and (right) the 2D Ising model.

dependence of these functions on waiting times t_w and t is given by relations

$$\begin{aligned} C(t, t_w) &\sim t_w^{-2\beta/(\nu z)} \tilde{F}_C(t/t_w), \\ R(t, t_w) &\sim t_w^{-2\beta/(\nu z)-1} \tilde{F}_R(t/t_w), \\ \chi(t, t_w) &\sim t_w^{-2\beta/z\nu} \tilde{F}_\chi(t/t_w), \end{aligned} \tag{35}$$

with scaling functions $\tilde{F}_{C,R,\chi}(t/t_w)$, which decrease at the long-term stage of their variation with $t - t_w \gg t_w \gg t_m$ according to the power law

$$\tilde{F}_{C,R,\chi}(t/t_w) \sim (t/t_w)^{-\phi}, \tag{36}$$

where the exponent $\phi = d/z - a + \beta\delta/z\nu$ [3]. For the high-temperature initial state with $m_0 = 0$,

$$\tilde{F}_{C,R,\chi}(t/t_w) \sim (t/t_w)^{-c_{A,R,\chi}} \tag{37}$$

with $c_A = c_R = c_\chi = d/z - \theta' < \phi$.

Results of calculations of two-time dependencies of the autocorrelation function $C(t, t_w)$ and the dynamic susceptibility $\chi(t, t_w)$ on observation times $t - t_w$ for various spin concentrations

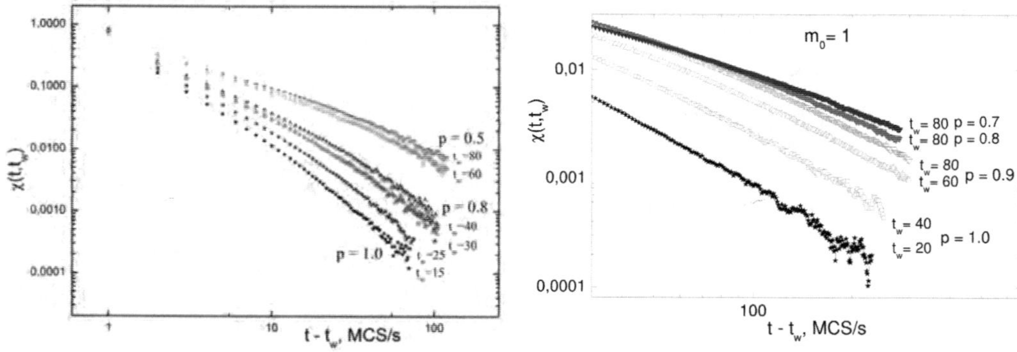

Figure 9. Dependencies of the dynamic susceptibility $\chi(t,t_w)$ on observation times $t - t_w$ for various spin concentrations p and waiting times t_w with evolution from the low-temperature initial state with $m_0 = 1$ (left) for the 3D Ising model and (right) the 2D Ising model.

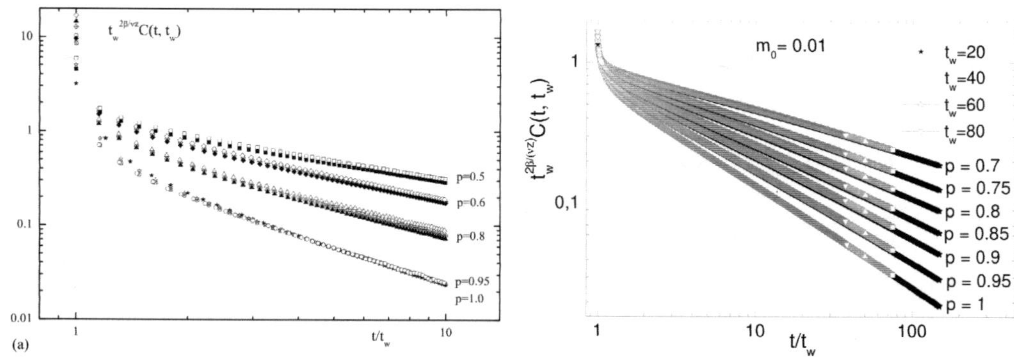

Figure 10. Scaling functions $F_C(t/t_w) = t_w^{2\beta/z\nu} C(t,t_w)$ for the autocorrelation function with initial state $m_0 \ll 1$ demonstrating the data collapse for different t_w into single curve with fixed spin concentration p (left) for the 3D Ising model and (right) the 2D Ising model.

p and waiting times t_w with evolution from the high-temperature and low-temperature initial states are presented in Figs.7-9. The aging effects clearly manifest themselves at the observation times $t - t_w \sim t_w$ and are characterized by the slowing down of these functions with increase of the waiting time t_w. The aging effects increase in the structurally disordered systems in comparison to those in the pure system.

Comparison of the autocorrelation function plots in Figs.7-8 with different initial states shows that the $C(t,t_w)$ are decreased more slowly for the case of evolution from high-temperature initial state with $m_0 \ll 1$ than from the low-temperature initial state with $m_0 = 1$. The role of defects is the most clearly pronounced in the strong slowing down of the correlation effects in structurally disordered systems as compared to the pure system for the case of evolution from the low-temperature initial state (Fig. 8). We attribute these pronounced changes in the behavior of the autocorrelation function to the pinning of domain walls by structural defects. The pinning occurs in the course of non-equilibrium changes in the domain structure of the system at its transition from the single-domain state at $T_0 = 0$ to a multidomain fluctuation-induced structure arising at the critical temperature T_c.

To confirm the scaling dependence for autocorrelation function $C(t,t_w)$ and susceptibility $\chi(t,t_w)$ at the aging regime with $t - t_w \sim t_w$ (33-35), we draw the plots of $t_w^{2\beta/z\nu} C(t,t_w)$ and

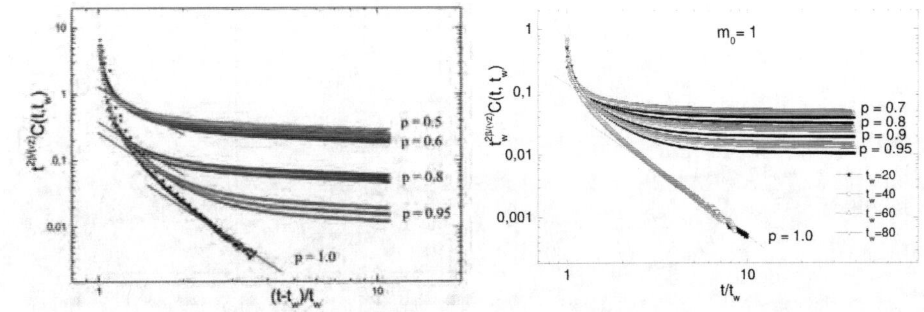

Figure 11. Scaling functions $F_C(t/t_w) = t_w^{2\beta/z\nu} C(t, t_w)$ for the autocorrelation function with initial state $m_0 = 1$ (left) for the 3D Ising model and (right) the 2D Ising model.

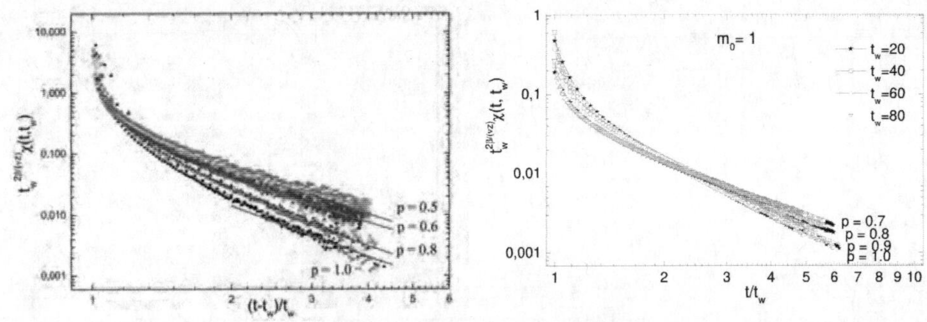

Figure 12. Scaling functions $F_\chi(t/t_w) = t_w^{2\beta/z\nu} \chi(t, t_w)$ for the dynamic susceptibility with initial state $m_0 = 1$ (left) for the 3D Ising model and (right) the 2D Ising model.

$t_w^{2\beta/z\nu} \chi(t, t_w)$ versus $(t - t_w)/t_w$ in Figs.10-12 for cases with different initial states with the use of exponent $\beta/z\nu$ which values were determined from study of relaxation properties of the magnetization. These plots demonstrate the collapse of the data obtained for different values of t_w on the universal curves corresponding to different spin densities p and characterized by the scaling functions $F_C(t/t_w)$ and $F_\chi(t/t_w)$ for case with evolution from the high-temperature initial state with $m_0 \ll 1$. However, the strong slowing down of the correlation effects in structurally disordered systems as compared to the pure system which was revealed for the case of evolution from the low-temperature initial state leads to violation of usual scaling behavior of function $F_C(t/t_w)$ on t/t_w with non-availability of collapse of the data obtained for different values of t_w on the universal curves. Putting into operation the scaling function $F_C(t/t_w^\mu)$ on t/t_w^μ with exponent $\mu > 1$ restores the collapse of the data obtained for different values of t_w on the universal curves as it is demonstrated in Fig.13. This phenomenon is classified as superaging and characterized by the exponent $\mu = 2.30(6)$ for weakly disordered 3D Ising systems and by $\mu = 2.80(7)$ for the 3D Ising systems with strong disorder. We revealed that the exponent μ for the 2D Ising model demonstrates the dependence on spin density p: $\mu(0.95) = 6.15(5)$, $\mu(0.9) = 6.25(5)$, $\mu(0.85) = 6.50(5)$, $\mu(0.8) = 6.75$, $\mu(0.75) = 7.00(5)$, and $\mu(0.7) = 7.50(5)$. For scaling function $F_\chi(t/t_w^\mu)$, the canonical aging with $\mu = 1$ was revealed for both pure and diluted 3D and 2D Ising systems.

At the next stage of our studies, we calculated the fluctuation-dissipation ratio according to the relation [37, 38]:

$$X(t, t_w) = \lim_{C \to 0} T_c \frac{\partial \chi(t, t_w)}{\partial C(t, t_w)}. \quad (38)$$

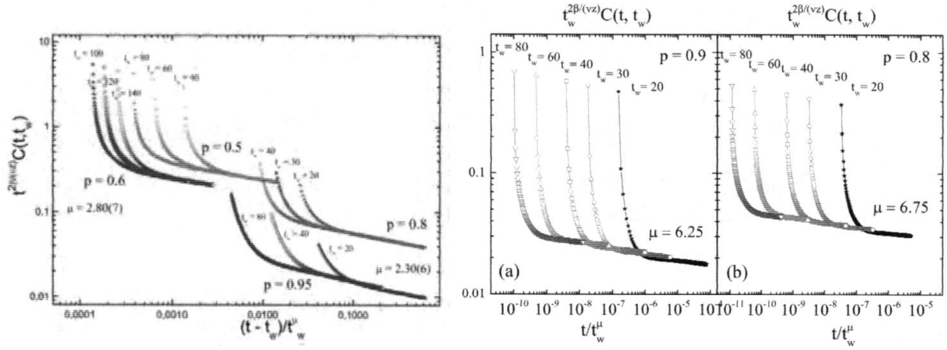

Figure 13. Superaging effect in the scaling behavior of the autocorrelation function $t_w^{2\beta/z\nu}C(t,t_w)$ versus t/t_w^μ with initial state $m_0 = 1$ (left) for the 3D Ising model and (right (a), (b)) the 2D Ising model.

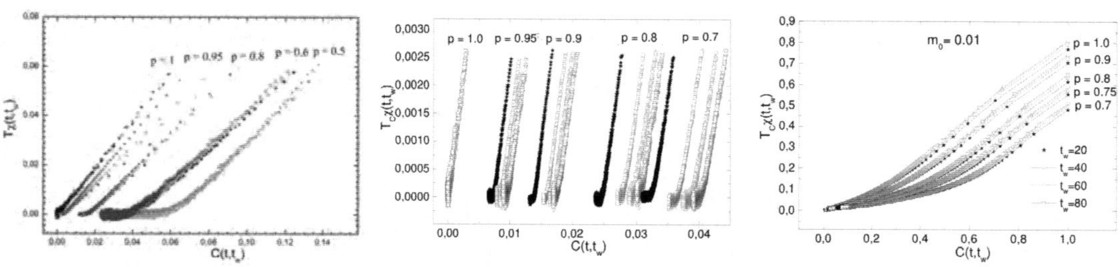

Figure 14. Dependencies of susceptibility on the autocorrelation function determining the FDR for various spin concentrations p with initial state $m_0 = 1$ (left) for the 3D Ising model and (center) the 2D Ising model and (right) with initial state $m_0 \ll 1$ for the 2D Ising model.

Table 1. The asymptotic values of the FDR X_{HT}^{∞} for case of system evolution from the high-temperature initial state and X_{LT}^{∞} the low-temperature initial state for 3D and 2D Ising models

p	X_{HT}^{∞} 3DIM	X_{HT}^{∞} 2DIM	X_{LT}^{∞} 3DIM	$X_{LT}^{(t-t_w\sim t_w)}$ 3DIM	X_{LT}^{∞} 2DIM	$X_{LT}^{(t-t_w\sim t_w)}$ 2DIM
1.0	0.380(13)	0.339(19)	0.784(5)	0.784(5)	0.751(24)	0.751(24)
0.95	0.413(7)	0.329(20)	0	0.740(3)	0	0.750(25)
0.9	-	0.317(21)	-	-	0	0.748(25)
0.85	-	0.265(21)	-	-	0	0.726(26)
0.8	0.413(11)	0.256(23)	0	0.736(6)	0	0.714(26)
0.75	-	0.215(24)	-	-	0	0.682(28)
0.7	-	0.178(24)	-	-	0	0.669(31)
0.6	0.446(8)	-	0	0.726(5)	-	-
0.5	0.441(13)	-	0	0.726(1)	-	-

Dependencies $T\chi$ on C are presented in Fig.14 for the evolution of systems from different initial states. We calculated the FDR for each value of the waiting time t_w, and then the asymptotic value of the FDR X^∞ was determined by extrapolation $X(t_w \to \infty)$. The asymptotic values of the FDR are presented in Table 1 for case of system evolution from the high-temperature X_{HT}^{∞} and the low-temperature X_{LT}^{∞} initial states for the 3D and 2D Ising models.

For the 3D Ising model similarly to exponents, calculated values of $X_{HT}^{\infty}(p)$ could consider as universal characteristics of three different classes of non-equilibrium critical behavior for pure, weakly and strongly diluted Ising systems [6, 7]. The obtained values X^∞ for the 3D weakly diluted Ising models with $p = 0.95, 0.8$ are in very good agreement with result $X^\infty_{\rm 3dRIM} \simeq 0.416$ from low-order renormalization-group computations by the $\sqrt{\varepsilon}$-expansion method [39], but the values X^∞ for the 3D strongly diluted Ising models with $p = 0.6, 0.5$ do not agree with above denoted value in the limits of errors. Also, we must mention that the obtained value $X^\infty(p = 1) = 0.380(13)$ for the 3D pure Ising model does not agree with $X^\infty_{\rm 3dIs} = 0.429(6)$ which has been calculated with the use of ε-expansion method in two-loop approximation for pure systems with a non-conserved order parameter dynamics in [40], but it is characterized by close agreement with $X^\infty(p = 1) \approx 0.40$ noted in [41] as result of preliminary simulations on the three-dimensional pure Ising model.

For the 2D Ising model, the values of $X^\infty_{HT}(p)$ for spin densities $p \geq 0.9$ demonstrate belonging to universal class of the pure 2D Ising model within the statistical errors while $X^\infty_{HT}(p)$ are characterized by dependence on $p \leq 0.85$ with decreasing values with growth of defects concentration owing to the crossover effects in the percolation behavior [27]. The value $X^\infty_{HT} = 0.339(19)$ for the pure system is in very good agreement with the value $X^\infty_{HT} = 0.33(1)$ determined in [3] by the MC simulation method.

For the pure 3D and 2D Ising models relaxing from a low-temperature initial state with $m_0 = 1$, the plot of $T\chi$ versus C is linear for the time range $t - t_w \geq t_w$ of variation of the autocorrelation function and characterized by the asymptotic values of the FDR $X^\infty_{LT}(3DIM) = 0.784(5)$ and $X^\infty_{LT}(2DIM) = 0.751(24)$. These values are in very good agreement with the field-theoretical values $X^\infty_{LT}(3DIM) = 0.78$ and $X^\infty_{LT}(2DIM) = 0.75$ determined in [3]. However, the dependencies of $T\chi$ versus C for the structurally disordered systems relaxing from the initial ordered state with $m_0 = 1$ exhibit two linear segments because of the found slowing down of the correlation effects at times $t - t_w \geq t_w$ owing to the pinning of domain walls by the defects. The first segment corresponds to the changes in the autocorrelation function $C(t, t_w)$ at times $t - t_w \sim t_w$, whereas the second segment corresponds to the long-term stage of the evolution with $t - t_w \gg t_w$ (Fig.14). Second segments lead to the asymptotic FDR values $X^\infty_{LT} = 0$ for all spin concentrations $p < 1$ under study. At the same time, the analysis of the dependences of $T\chi$ versus C in the first segments using (38) without considering the limit $C \to 0$ demonstrates that the extrapolation $X_{LT}(t_w \to \infty)$ applied to certain $X_{LT}(t_w)$ gives the values $X^{(t-t_w\sim t_w)}_{LT}$ presented in the Table 1. These values of the FDR for the 2D Ising model with $p = 0.95$ and $p = 0.9$ are close to the asymptotic FDR value for the pure Ising model.

5. Conclusions

To conclude, we note that our numerical studies have revealed an appreciable role of defects in the non-equilibrium critical dynamics of both the 3D and 2D Ising models. Specifically, a rise in defect concentration results in the strengthening of aging effects manifested as the slowing down of correlation and relaxation processes in structurally disordered systems, in contrast to pure systems. Non-equilibrium initial states begin to increasingly more strongly influence peculiar features and characteristics of the system's evolution. For example, in the case of evolution from a high-temperature initial state, the influence of defects is manifested in quantitative changes to universal characteristics of non-equilibrium critical behavior, such as critical exponents and the asymptotic FDR $X^\infty_{HT}(p)$.

It was shown for the 3D Ising model that the presence of defects results in setting new $X^\infty_{HT}(p)$ values, with $X^\infty_{HT}(p)$. for strongly diluted systems being higher than for weakly diluted systems, which, in turn, is higher than $X^\infty_{HT}(p = 1)$ for a pure model.

For evolution from a low-temperature initial state for both the 3D and 2D Ising models, the autocorrelation function for times $t - t_w \gg t_w$ decreases as a power-law of critical magnetization relaxation due to domain wall pinning on structural defects, while asymptotic FDR values determined by the domain dynamics in the long-time regime become equal to zero. In this case,

IOP Conf. Series: Journal of Physics: Conf. Series **1163** (2019) 012005 doi:10.1088/1742-6596/1163/1/012005

the two-time scaling dependence of the autocorrelation function was found to obey relations of the superaging theory suggesting enhanced influence of the system's age (the time of onset of t_w measurement) determined by the power-law dependence t_w^{μ} with exponent $\mu > 1$. The values of this exponent were shown to differ for weakly and strongly diluted 3D Ising models while exponent μ for the 2D Ising model demonstrates dependence on p.

Acknowledgments

This work was supported by the Russian Foundation for Basic Research, project nos. 17-02-00279 and 18-42-550003, and by the Council of the President of the Russian Federation, project no.MD-6868.2018.2. For calculations, we used the resources provided by the Shared Facility Data Center FEB RAS (Khabarovsk).

References

[1] Prudnikov V V, Prudnikov P V, and Mamonova M V 2017 *Phys. Usp.* **60** 762
[2] Calabrese P, Gambassi A 2005 *J. Phys. A: Math. Gen.* **38** R133
[3] Calabrese P, Gambassi A, and Krzakala F 2006 *J. Stat. Mech.* **6** 2
[4] Prudnikov V V, Prudnikov P V, Kalashnikov I A, and Tsirkin S S 2008 *JETP* **106** 1095
[5] Prudnikov V V, Prudnikov P V, Kalashnikov I A, and Rychkov M V 2010 *JETP* **110** 253
[6] Prudnikov V V, Prudnikov P V, Vakilov A N, and Krinitsyn A S 2007 *JETP* **105** 371
[7] Prudnikov V V, Prudnikov P V, Vakilov A N, et al. 2010 *Phys. Rev. E* **81** 011130
[8] Prudnikov V V, Prudnikov P V, Zheng B, et al. 2007 *Prog. Theor. Phys.* **117** 973
[9] Prudnikov P V and Medvedeva M A 2012 *Prog. Theor. Phys.* **127** 369
[10] Rosov N, Hohenemser C, and Eibschutz M 1992 *Phys. Rev. B* **46** 3452
[11] Khmelnitskii D E 1975 *Sov. Phys. JETP* **41** 981
[12] Dorogovtsev S N 1981 *Sov. Phys. JETP* **53** 1070
[13] Weinrib A and Halperin B I 1983 *Phys. Rev. B* **27** 413
[14] Harris A B 1974 *J. Phys. C* **7** 1671
[15] Lawrie I D and Prudnikov V V 1984 *J. Phys. C* **17** 1655
[16] Prudnikov V V, Prudnikov P V, and Fedorenko A A 2000 *Phys. Rev. B* **62** 8777
[17] Dotsenko Vik S and Dotsenko Vl S 1981 *JETP Lett.* **33** 37
[18] Shalaev B N 1984 *Sov. Phys. Solid State* **26** 1089; 1994 *Phys. Rep.* **237** 129
[19] Wang J S, Selke W, Dotsenko Vl S, and Andreichenko V B 1990 *Physica A* **164** 221
[20] Selke W, Shchur L N, and Talapov A L 1995 *Annual Reviews of Computational Physics* **1** 17
[21] Roder A, Adler J, and Janke W 1999 *Physica A* **265** 28
[22] Kim J K and Patrascioiu A 1994 *Phys. Rev. Lett.* **72** 2785
[23] de Queiroz S L A and Stinchcombe R B 1994 *Phys. Rev. B* **50** 9976
[24] Ballesteros H G et al. 1997 *J. Phys. A* **30** 8379
[25] Selke W, Shchur L N, and Vasilyev O A, 1998 *Physica A* **259** 388
[26] Markov O N and Prudnikov V V 1994 *JETP Lett.* **60** 23
[27] Prudnikov V V and Markov O N 1995 *Europhys. Lett.* **29** 245
[28] Janke W 2008 *Lect. Notes Phys.* **739** 79
[29] Ferrenberg A M and Landau D P 1991 *Phys. Rev. B* **44** 5081
[30] Shchur L N and O A Vasilyev 2001 *Phys. Rev.* E **65** 016107
[31] Martins P H L and Plascak J A 2007 *Phys. Rev.* E **76** 012102
[32] Nightingale M P and Blote H W J 1996 *Phys. Rev. Lett.* **76** 4548
[33] Prudnikov V V, Prudnikov P V, Pospelov E A, and Malyarenko P N 2015 *JETP Lett.* **102** 167
[34] Prudnikov P V, Prudnikov V V, Malyarenko P N, et al. 2015 *Prog. Theor. Exp. Phys.* **2015** 053A01
[35] Prudnikov P V, Prudnikov V V, and Pospelov E A 2013 *JETP Lett.* **98** 619
[36] Prudnikov V V, Prudnikov P V, Pospelov E A, and Vakilov A N 2015 *Phys. Lett. A* **379** 774
[37] Prudnikov V V, Prudnikov P V, and Pospelov E A 2016 *J. Stat. Mech.* 043303
[38] Prudnikov V V, Prudnikov P V, and Malyarenko P N 2017 *JETP* **152** 1102
[39] Calabrese P and Gambassi A 2002 *Phys.Rev. B* **66** 212407
[40] Calabrese P and Gambassi A 2002 *Phys.Rev. E* **66** 066101
[41] Godréche C and Luck J M 2000 *J. Phys. A: Math. Gen.* **33** 9141

Effect of dispersity of particle length on electrical conductivity of two-dimensional systems

Yuri Yu Tarasevich[1], Andrei V Eserkepov[1], Irina V Vodolazskaya[1], Petr G Selin[1], Valentina V Chirkova[1], Nikolai I Lebovka[2,3]

E-mail: tarasevich@asu.edu.ru
[1]Astrakhan State University, Astrakhan, 414056, Russia
[2]Department of Physical Chemistry of Disperse Minerals, F. D. Ovcharenko Institute of Biocolloidal Chemistry, NAS of Ukraine, Kiev, Ukraine, 03142
[3]Department of Physics, Taras Shevchenko Kiev National University, Kiev, Ukraine, 01033

Abstract. By means of computer simulation, we examined the effect of dispersity of filler length on electrical conductivity of two-dimensional (2D) composites with rod-like fillers. A continuous approach was used. Highly conductive zero-width rod-like particles were deposited uniformly with given anisotropy onto a poorly conductive substrate. Length of particles varied according to the lognormal distribution. Our simulation evidenced that slightly disordered systems may have both high figure of merit, i.e., both high electrical conductivity and optical transmission simultaneously, and high electrical anisotropy.

1. Introduction

Transparent and flexible electrodes are important components of modern optoelectronic devices (see, e.g., [1] for review). Such electrodes are transparent, poorly conductive films containing highly conductive particles. Recently, simulations of electrical conductivity ,of two-dimensional systems with rod-like fillers of equal length have been performed both in lattice [2, 3, 4] and in continuous [5, 6] approaches. By contrast, fillers in real systems of conductive nanocomposites have different lengths following a lognormal distribution [7]. The goal of the present conference paper is to exam the effect of length dispersity on electrical conductivity of two-dimensional systems with rod-like fillers.

2. Methods

A continuous approach was used to prepare a 2D sample. Highly conductive zero-width rod-like particles were deposited uniformly with given anisotropy onto a poorly conductive substrate of size $L \times L$ subjected to periodic boundary conditions, i.e., onto a torus. The rods were allowed to intersect each other. The electrical conductivities of the particles and the substrate were σ_p and σ_m, correspondingly. High electrical contrast $\Delta = \sigma_p/\sigma_m$ was assumed. Length of the particles, l, varied according to the lognormal distribution. We performed simulations for the fixed value of the mean length $\langle l \rangle = 1$ and different values of the standard deviation, $\mathrm{SD} = 0, 0.1, 0.5, 1.0$.

The simulations were performed for different values of the order parameter defined as

$$s = N^{-1} \sum_{i=1}^{N} \cos 2\theta_i, \tag{1}$$

where θ_i is the angle between the axis of the i-th rod and the horizontal axis x, and N is the total number of rods in the system. In our simulations, values of the order parameter were $s = 0, 0.5, 0.8, 1$.

To calculate the electrical conductivity of the film, the torus was unwrapped and transformed into a random resistor network (RRN) by means of discretization. A square mesh of $m \times m$ was applied to the film. When a cell of the mesh contained any part of a rod, it was assumed to be conductive and opaque. Otherwise, it was assumed to be insulating and transparent. The fraction of conductive (opaque) cells was denoted as p. The Frank–Lobb algorithm [8] was utilized to calculate the conductivity of the RRN. The percolation threshold, p_c, was found using the Hoshen–Kopelman algorithm [9].

A detailed description of our approach can be found in Ref. [10].

To characterize the electrical anisotropy, we used the quantity [2, 11]

$$\delta = \frac{\log_{10}\left(\sigma_{\|}/\sigma_{\perp}\right)}{\log_{10}\Delta}, \tag{2}$$

where $\sigma_{\|}$ and $\sigma_{\perp}$ are the electrical conductivity along and perpendicular to the alignment of the fillers, respectively.

The figure of merit [12]

$$\Phi_{\mathrm{TC}} = T^{10}/R_s \tag{3}$$

was calculated to characterize the films. Here, T is the optical transmission and R_s is the sheet resistance. In our computations, the optical transmission was taken as $T = 1 - p$. The sheet resistance R_s was defined by

$$R_s = 1/\sigma t, \tag{4}$$

where σ is the electrical conductivity and t is the thickness of the film. Since the thickness of the film was constant in all our simulations, it was assumed to be 1 arbitrary unit in calculations of the figure of merit. Hence, the figure of merit was measured in arbitrary units.

To characterize the electrical properties of films at low concentrations of conductive rods, the "intrinsic conductivity"

$$[\sigma] = \left.\frac{\mathrm{d}\ln\sigma}{\mathrm{d}p}\right|_{p\to 0} \tag{5}$$

was calculated. This quantity was introduced [13, 14] by means the following virial expansion

$$\frac{\sigma}{\sigma_m} = 1 + [\sigma]p + \mathrm{O}\left(p^2\right). \tag{6}$$

Throughout the text, the error bars in the figures correspond to the standard deviation of the means. When not shown explicitly, they are of the order of the marker size.

3. Results

Figure 1 demonstrates the effect of filler length dispersity on the figure of merit for two values of the order parameter. Both in the case of an isotropic system ($s = 0$) and in the case of a strictly anisotropic system, i.e., a system with completely aligned fillers ($s = 1$), an increase in the dispersity led to both an increase of the maximal value of the figure of merit and decrease of the number of fillers per area when this maximum occurred. Nevertheless, the figure of merit was significantly smaller (about 5 times) for the anisotropic system in compare with the isotropic case.

The results presented below were obtained for lattices with $L = 32$. Figure 2 shows, as an example, the results of a numerical calculation of the dependence of the logarithm of the

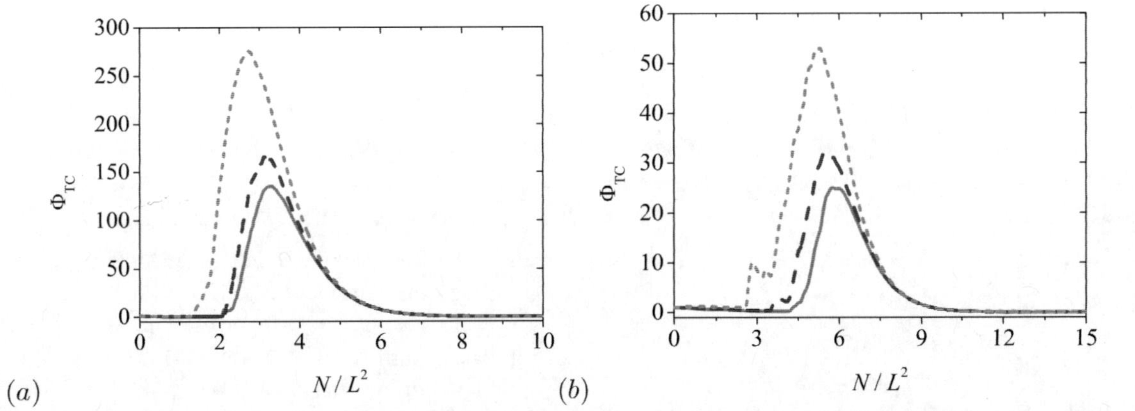

Figure 1. Examples of the figure of merit, Φ_{TC}, vs number of particles per unite area, $m = 256$. —— SD $= 0$, - - - SD $= 0.5$, - - - - - SD $= 1$. (a) $s = 0$; (b) $s = 1$, the electrical conductivity was computed along the direction of fillers alignment.

averaged effective electrical conductivity, σ, on the concentration, p, along the direction of alignment of the rods (a) and in the perpendicular direction (b) at SD $= 0, 0.1, 0.5, 1.0$ and $s = 0.5, m = 512$. The curves had the shape of a sigmoid; the jump in electrical conductivity indicated an insulator–metal transition. The concentration at which this transition occurred was reduced when the length dispersity increases in both perpendicular directions at partial ordering of the rods ($s = 0.5$). The number of long rods increases when the dispersity increased, which improved the connectivity of the system in both directions.

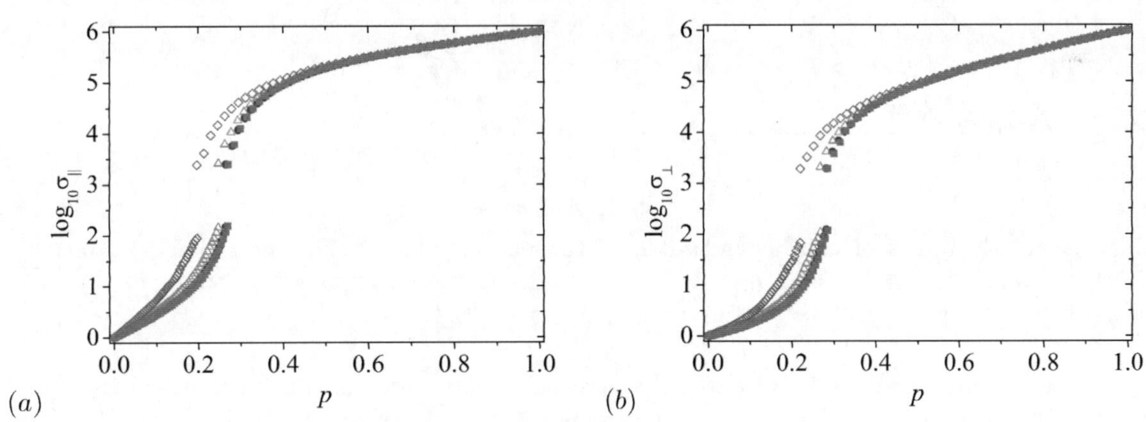

Figure 2. Examples of the dependencies of the logarithm of the averaged effective electrical conductivity, σ, on the concentration, p, at ■ SD $= 0$, ● SD $= 0.1$, △ SD $= 0.5$, ◇ SD $= 1.0$ and $s = 0.5$, $m = 512$, $L = 32$ (a) along the direction of alignment of the rods, (b) in the perpendicular direction.

For an anisotropic system, the concentrations at which the transition occurred in perpendicular directions must be different, and figure 3 demonstrated this fact. Examples of the dependencies of the averaged effective electrical conductivity, σ, on the concentration, p, along the direction of alignment of the rods and in the perpendicular direction are presented in this figure at SD $= 0.1$, $s = 1$, $m = 1024$. The transition will occur at lower concentrations along the alignment direction of the rods, since the appearance of long rods will affect the connectivity of the system primarily along the alignment direction.

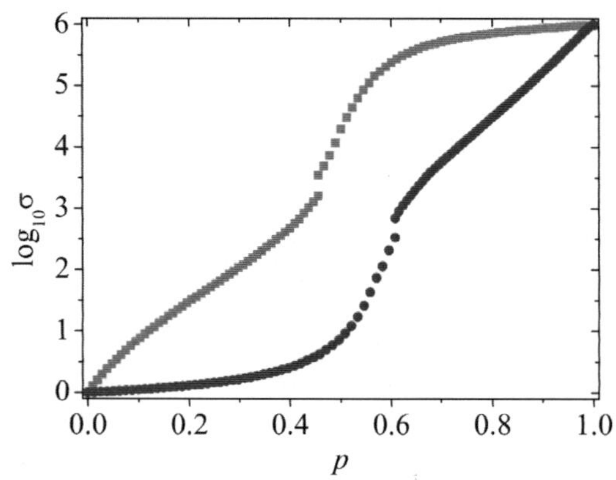

Figure 3. Examples of the dependencies of the effective electrical conductivity, σ, on the concentration, p, SD = 0.1, s = 1, m = 1024, L = 32 along the direction of alignment of the rods (■) and in the perpendicular direction (●).

The more orderly the system was, the stronger this difference became (compare figure 4(a) for $s = 1$ and figure 4(b) for $s = 0.8$) at SD = 1, $s = 1$, $m = 1024$.

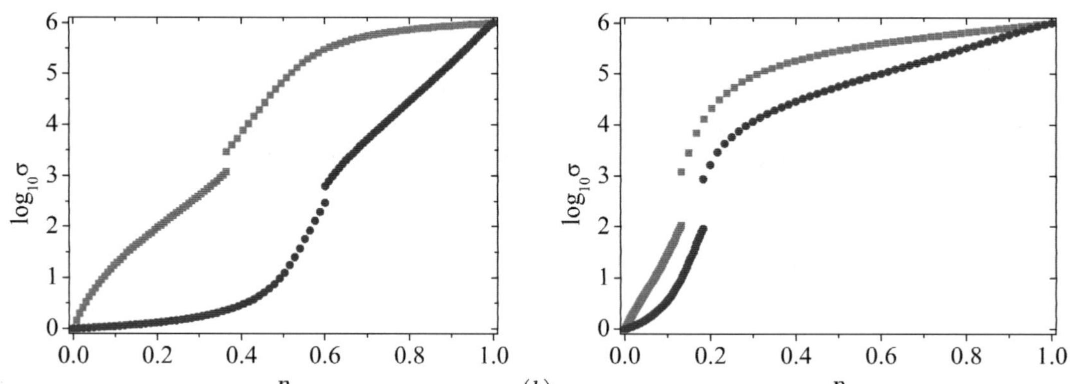

Figure 4. Examples of the dependencies of the effective electrical conductivity, σ, on the concentration, p, SD = 1, $s = 1$, $m = 1024$, $L = 32$ along the direction of alignment of the rods (■) and in the perpendicular direction (●): (a) $s = 1$; (b) $s = 0.8$.

The results obtained for the values of the percolation threshold, p_c, for different values of the length dispersity and the order parameter (s = 0.5, 0.8, 1.0) are summarized in figure 5. Here the value of the percolation threshold was normalized to ones at SD = 0, the results were given both for along the direction of alignment of the rods (closed symbols) and in the perpendicular direction (open symbols) and $m = 512$. The previously obtained conclusions were retained for all the investigated parameters s and SD, except for the case when all the rods were aligned in one direction ($s = 1$) and the percolation threshold was determined in the direction perpendicular to the alignment direction. In this case, the value of the percolation threshold does not depend on the dispersity of the length of the rods, since the connectivity of the system in the direction perpendicular to the alignment did not depend on the length of the rods.

One of the important characteristics of the system is intrinsic conductivity; the results of its calculation are shown in figure 6. Here, the intrinsic conductivity along the direction of alignment of the rods, $[\sigma_{||}]$, versus the aspect ratio of the rods, k, and the intrinsic conductivity in the perpendicular direction, $[\sigma_\perp]$ versus $1/k$ for the case of completely aligned rods ($s = 1$) are given at SD=0, 0.1, 0.5, 1. The aspect ratio was calculated as m/L for three values m=256, 512, 1024. The lines are a linear approximation.

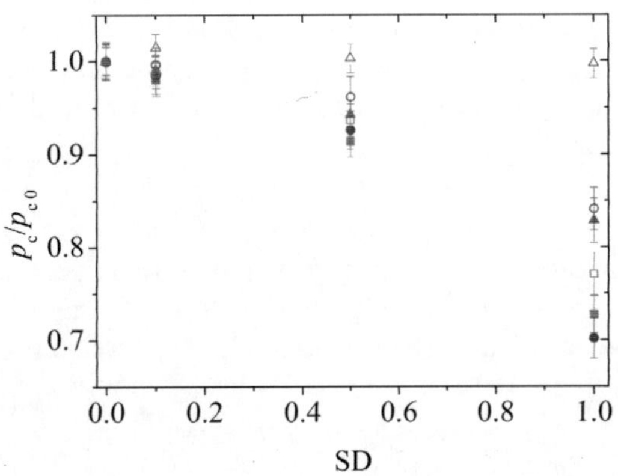

Figure 5. Examples of the normalized percolation threshold, p_c/p_{c0}, versus SD for three values of the standard deviation SD at $s = 0.5$ (□, ■), 0.8 (○, ●), 1.0 (△, ▲) (closed and open symbols correspond to the direction along alignment of the rods and in the perpendicular direction, correspondingly) and $m = 256$. Here, p_{c0} corresponds to SD = 0.

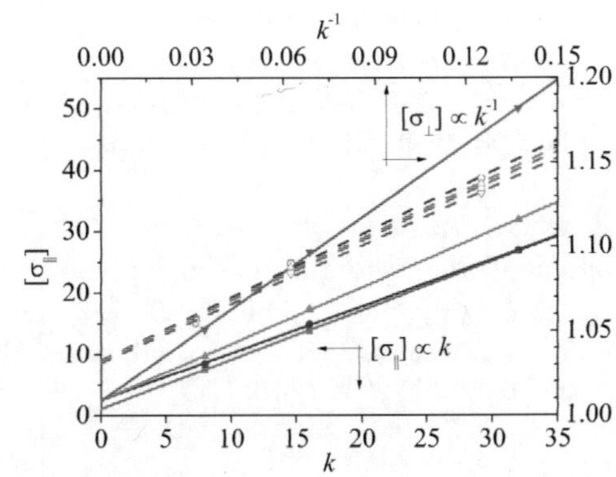

Figure 6. The intrinsic conductivity along the direction of alignment of the rods, $[\sigma_\parallel]$, vs the aspect ratio, k, of the rods (closed symbols, solid lines) and the intrinsic conductivity in the perpendicular direction, $[\sigma_\perp]$ vs $1/k$ (open symbols, dotted lines) at SD=0 (□, ■), 0.1 (○, ●), 0.5 (△, ▲), 1 (▽, ▼) for the case of completely aligned rods ($s = 1$).

Figure 7 presents examples of the electrical anisotropy ratio, δ, versus the concentration, p, for SD $= 0.1, 0.5, 1$, $m = 512$: (a) $s = 1$, (b) $s = 0.8$. In the case $s = 0.8$, partial arrangement of the rods (figure 7(b)), the maximum of the electrical anisotropy ratio was shifted to the region of smaller values p and the interval of p for large electrical anisotropy was smaller compared with completely aligned fillers ($s = 1$) (figure 7(a)).

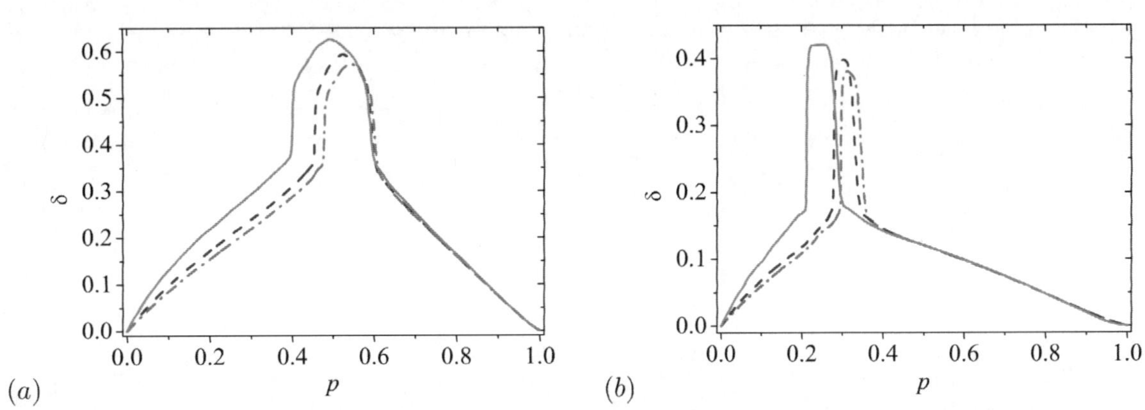

Figure 7. Example of the dependencies of the electrical anisotropy, δ, vs the concentration, p, for different values of the length dispersity SD = 0.1— · —, 0.5- - - -, 1——, $m = 512$ and $L = 32$: (a) completely aligned rods $s = 1$; (b) $s = 0.8$.

4. Conclusion

Our computer simulation suggests that the dispersity of the filler length had a significant impact only at the electrical conductivity of films along the direction of the filler alignment; the electrical conductivity in the direction perpendicular to the filler alignment was insensitive to the dispersity of the filler length. This dispersity had a positive impact on the figure of merit of films. It decreased the percolation threshold and, as a result, the critical concentration of fillers when the insulator–conductor phase transition occurred. In the vicinity of the phase transition, the electrical conductivity of the film exhibited high anisotropy.

Acknowledgments

We acknowledge the funding from the Ministry of Education and Science of the Russian Federation, Project No. 3.959.2017/4.6(Y.Y.T., A.V.E., I.V.V., P.G.S., V.V.C.) and from the National Academy of Sciences of Ukraine, Projects No. 2.16.1.6. (0117U004046), 43/19-H, and 15F (0117U006352) (N.I.L.).

References

[1] Hecht D S, Hu L and Irvin G 2011 *Adv. Mater.* **23** 1482–1513 ISSN 1521-4095
[2] Tarasevich Y Y, Goltseva V A, Laptev V V and Lebovka N I 2016 *Phys. Rev. E* **94**(4) 042112 ISSN 2470-0045
[3] Tarasevich Y Y, Laptev V V, Goltseva V A and Lebovka N I 2017 *Physica A* **477** 195–203 ISSN 0378-4371
[4] Tarasevich Y Y, Laptev V V, Burmistrov A S and Lebovka N I 2018 *J. Phys.: Conf. Ser.* **955** 012006 ISSN 1742-6588
[5] Lebovka N I, Tarasevich Y Y, Vygornitskii N V, Eserkepov A V and Akhunzhanov R K 2018 *Phys. Rev. E* **98**(1) 012104 ISSN 2470-0045
[6] Tarasevich Y Y, Lebovka N I, Vodolazskaya I V, Eserkepov A V, Goltseva V A and Chirkova V V 2018 *Phys. Rev. E* **98**(1) 012105 ISSN 2470-0045
[7] Forró L, Gaal R, Grimaldi C, Mionić M, Ribič P R, Smajda R and Magrez A 2013 *AIP Advances* **3** 092117 ISSN 2158-3226
[8] Frank D J and Lobb C J 1988 *Phys. Rev. B* **37**(1) 302–307 ISSN 2469-9950
[9] Hoshen J and Kopelman R 1976 *Phys. Rev. B* **14**(8) 3438–3445 ISSN 2469-9950
[10] Tarasevich Y Y, Vodolazskaya I V, Eserkepov A V, Goltseva V A, Selin P G and Lebovka N I 2018 *arXiv:1808.02217, submitted to J. Appl. Phys.*
[11] Tarasevich Y Y, Laptev V V, Goltseva V A and Lebovka N I 2017 *Physica A* **477** 195–203 ISSN 0378-4371
[12] Haacke G 1976 *J. Appl. Phys.* **47** 4086–4089 ISSN 0021-8979
[13] Douglas J F and Garboczi E J 1995 Intrinsic viscosity and the polarizability of particles having a wide range of shapes *Advances in Chemical Physics* vol XCI ed Prigogine I and Rice S A (John Wiley & Sons Inc) pp 85–154 ISBN 0-471-12002-2
[14] Garboczi E J and Douglas J F 1996 *Phys. Rev. E* **53**(6) 6169–6180 ISSN 2470-0045

Monte Carlo simulation of entropy-driven pattern formation in a two-dimensional system of rectangular particles

Yuri Yu Tarasevich, Andrei V Eserkepov, Valentina V Chirkova, Valeria A Goltseva

E-mail: tarasevich@asu.edu.ru
Astrakhan State University, Astrakhan, 414056, Russia

Abstract. We simulated a random walk of rectangular particles on a square lattice with periodic boundary conditions. Two kinds of particles were investigated, viz., so-called "blind" and "myopic" particles. We found that steady state patterns occurred only for some values of the ratio L_x/L_y, where L_x and Ly are the linear sizes of the system, and only for "needles", i.e., the particles of size $1 \times k$. Different patterns were observed for "blind" and "myopic" particles.

1. Introduction

Recently, diffusion-driven pattern formation in a 2D discrete system has been studied by means of Monte Carlo (MC) simulations [1, 2, 3]. The observed behaviour resembled pattern formations in thin layers that are composed of elongated granules subjected to vibrations [4, 5, 6].

The present conference paper is devoted to a detailed analysis of how different kinetics of particles ("blind" and "myopic" particles) affects pattern formation in a two-species diffusion system. Moreover, we have examined the effect of shape of particles and shape of the substrate on formation of such a pattern.

The rest of the paper is organised as follows. In section 2, the technical details of the simulations are described and all necessary quantities are introduced. Section 3 presents our principal findings. Section 4 summarises the main results.

2. Methods

We used square lattices of size $L_x \times L_y$ with periodic boundary conditions. We considered rectangular particles of size $k_x \times k_y$ (2×2, 4×4, 8×8, 2×12, 2×16). Special attention was paid to the case $1 \times k$ ("needles" or linear k-mers). The initial state of the system under consideration corresponded to jamming. The jammed state was generated using the random sequential adsorption [7]. No additional particle can be added to the jammed system due to the absence of an appropriate empty space.

Hard-core interaction between the particles was assumed. Two kinds of particle movement were examined. The first kind of the particles corresponds to "blind" particles, i.e., a particle chooses one of the four possible directions and attempts to shift in this direction [8]. If the attempt is unsuccessful, other particle attempts to move. The second kind corresponds to "myopic" particles [9], such a particle is "smart", viz., it chooses one of the four possible directions and attempts to shift in this direction. If the attempt is unsuccessful, the particle chooses other direction until the first successful attempt or when all four directions would examined.

In the present study, we utilize four main quantities, i.e., the normalized number of clusters, n, the degrees of freedom (DoF) per particle, f, the order parameter, s, and the relative number of contacts, n^*_{xy}.

In percolation theory, a cluster is a group of neighbouring occupied sites [10]. In our particular case, a cluster is a group of particles of the same orientation connected with one another. Two particles are considered to be connected when there is at least one pair of neighbouring sites belonging to different particles. The clusters built of horizontal particles and the ones built of vertical particles were counted separately and then their numbers were averaged. We used the Hoshen–Kopelman algorithm [11] to count the clusters. The normalised number of clusters, n, is the current number of clusters divided by the value at the initial state, $t_{\mathrm{MC}} = 0$.

We define the DoF as the number of possible unit movements of a particle from its current location.

For elongated particles, we exploited the nematic order parameter (see, e.g., [12]). Adopted to two-dimensional lattice case, this quantity is defined as

$$s = \frac{|N_y - N_x|}{N}, \tag{1}$$

where N_x and N_y are the numbers of sites belonging to the particles oriented along the x and y directions, respectively, and $N = N_y + N_x$ is the total number of the occupied sites. This order parameter, s, was calculated within a sliding window of $w \times w$ sites and then was averaged over the entire set of windows, i.e., over $L_x \times L_y$ windows except for the empty ones. We utilized $w = 2^n$, where $n = 1, 2, \dots, 7$. In all cases, we considered only isotropic systems, i.e., vertically and horizontally oriented particles were equiprobable in their deposition, hence, the order parameter calculated in the window of size $L_x \times L_y$ equals to zero, $s = 0$.

The relative number of interspecific contacts $n^*_{xy} = n_{xy}/(n_{xy} + n_x + n_y)$, where n_{xy} is the number of interspecific contacts between the different sorts of particles (i.e., horizontal–vertical), n_x and n_y are the numbers of intraspecific contacts between particles of the same kind (i.e., horizontal–horizontal and vertical–vertical, respectively).

3. Results

We investigated the random walk of particles with different aspect ratios k_x/k_y. Since the pattern formation was observed only for "needles", i.e., the particles of size $1 \times k$, the text below is devoted only to this particular particle shape.

3.1. Comparison of "blind" and "myopic" kinetics

This subsection is devoted to "needles", i.e., the particles of size $1 \times k$, on the lattices of size $L_x = L_y = L$ with the fixed ratio $L/k = 32$ unless otherwise is not explicitly indicated.

Figure 1 demonstrates examples of the local order parameter, s, vs the ratio w/k for some particular values of k at $t_{\mathrm{MC}} = 0$. Here, w is the size of the window in which the local order parameter was calculated. The order parameter evidenced that, in the initial jammed state, particles formed stacks of the typical size $k \times k$. This stack structure was unstable, i.e., the random walk of needles destroyed this structure and brought the system to a new state. The transitions to a new state were different for "blind" and "myopic" particles. For both kinds of particles, the normalized number of clusters increased up to $t_{\mathrm{MC}} \approx 10^2$ and then decreased up to $t_{\mathrm{MC}} \approx 10^6$ ("myopic" particles) or $t_{\mathrm{MC}} \approx 10^7$ ("blind" particles) (figure 2). Nevertheless, for "myopic" particles, the maximum of the normalized number of clusters was reached earlier. For this kind of particles, the curve $n\,(t_{\mathrm{MC}})$ had a step when $10^5 \lesssim t_{\mathrm{MC}} \lesssim 10^6$.

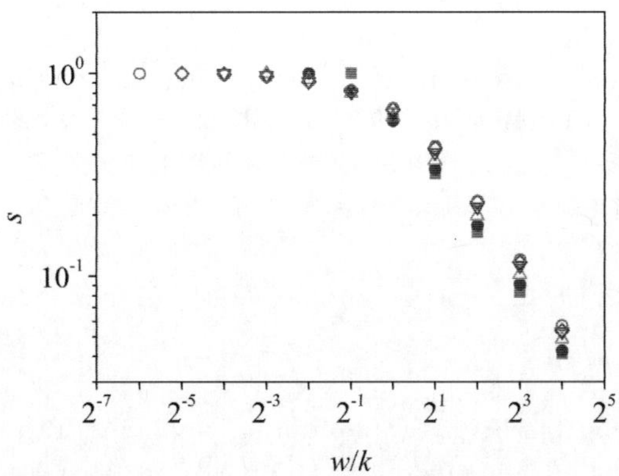

Figure 1. Examples of dependencies of the local order parameter, s, at the initial jammed state ($t_{\mathrm{MC}} = 0$) vs the ratio w/k, where w is the size of the window in which the local order parameter was calculated, k is the "needle" length. ■ $k = 2$, ● $k = 4$, △ $k = 8$, ▽ $k = 16$, ◇ $k = 32$, ○ $k = 64$.

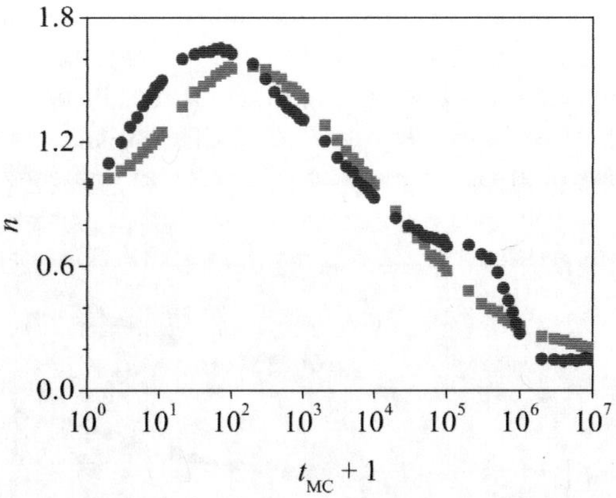

Figure 2. Examples of dependencies of the normalized number of clusters, n, vs MC steps, t_{MC}. $k = 8$. ● "myopic" particles, ■ "blind" particles.

Figure 3 demonstrates temporal dynamics of the relative number of the contacts, n^*_{xy}. For both kinds of particles, the system tended to a state with a small number of contacts between particles of different orientations, i.e., segregation of particles with different orientations occurred. This state was reached at $t_{\mathrm{MC}} \approx 10^6$ in the case of "myopic" particles and at $t_{\mathrm{MC}} \gtrsim 10^7$ in the case of "blind" particles. Figure 4 shows that the system evolved from a short-range order (ordered regions or stacks with the typical size $k \times k$) to a long-range order (ordered regions with the typical size $L/4$).

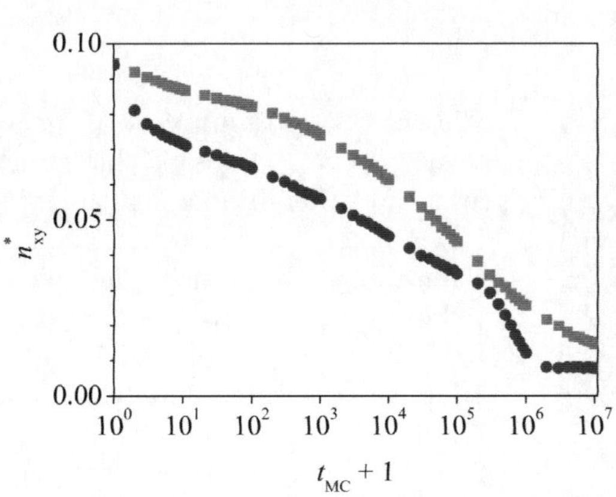

Figure 3. Examples of dependencies of the relative number of contacts, n^*_{xy}, vs MC steps, t_{MC}. $k = 8$. ● "myopic" particles, ■ "blind" particles.

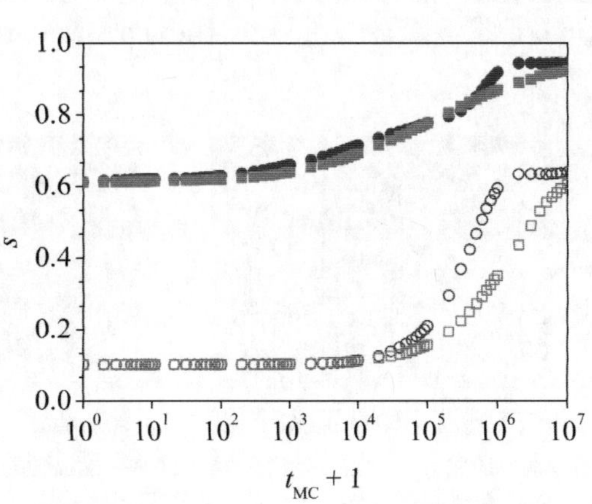

Figure 4. Examples of dependencies of the local order parameter, s, calculated in windows of size 8×8 (closed symbols) and 64×64 (open symbols) sites vs MC steps, t_{MC}. $k = 8$. ●, ○ "myopic" particles, ■, □ "blind" particles.

For both kinds of particles, the random walk led to an increase of the freedom of particles (figure 5). Due to hard-core interaction between particles, pattern formation should be treated as entropy-driven. Formation of the macroscopic order was accompanied by an increase of the microscopic disorder, i.e., in the initial jammed state, a particle had less possibilities to change its location in comparison with any succeeding state.

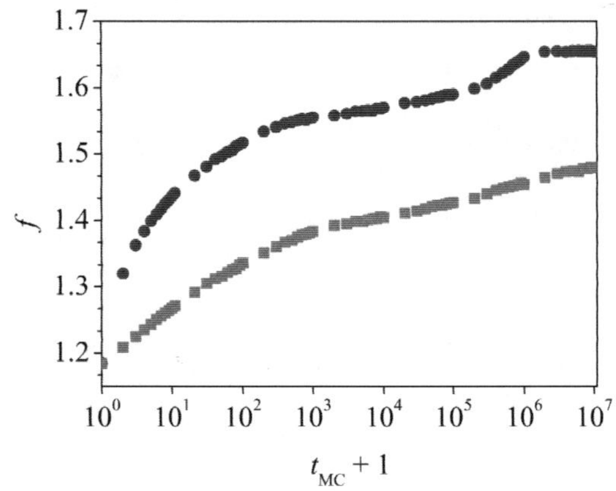

Figure 5. Examples of dependencies of the degrees of freedom, f, vs MC steps, t_{MC}. $k = 8$. ● "myopic" particles, ■ "blind" particles.

3.2. Effect of the substrate shape

Special attention has been paid to the effect of the lattice shape on pattern formation in the systems of the "myopic" particles. We examined the lattices with different ratio L_x/L_y.

When $L_x/L_y \in \mathbb{N}$, stable stripe domains were observed at $t_{\mathrm{MC}} \sim 10^7$. The width of a stripe is determined by the length of the narrower side, i.e., by the value of L_y. Widths of the stripes were the same when the lattices were $L_y \times L_y$ and $L_x \times L_y$. Figure 6 demonstrates an example of pattern formation in an elongate rectangular region ($L_x/L_y = 16$). Final patterns in a square region with the same area are shown in figure 6(a) for comparison. Presence of numerous linear defects evidenced that the relaxation time exceeds 10^7; hence, the relaxation time is also determined by the narrower side of the region.

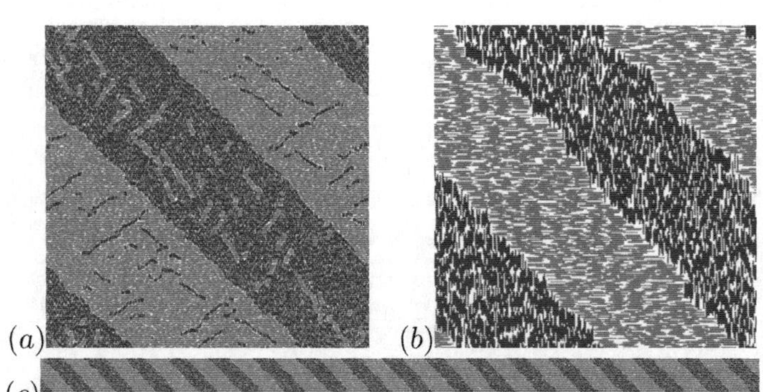

Figure 6. Examples of patterns in two regions of the same area but different shapes; the initial jammed state at isotropic distribution, PBCs, "myopic" particles, $t_{\mathrm{MC}} = 10^7$, $k = 12$. (a) Square region of size 1024×1024. (b) Fragment 256×256 of a rectangular region of size 4096×256. (c) The entire rectangular region of size 4096×256.

Quite a different situation was observed when L_x and L_y are incommensurate. No steady state was observed even until $t_{\mathrm{MC}} = 10^8$. By contrast, the local order parameter was still changing at this time. Different patterns were observed at this time with different frequencies. For instance, narrow stripes were mainly observed when $L_x/L_y = 1.27734375\ldots$ (figure 7).

Wide stripes were mainly observed when $L_x/L_y = 1.23828125\ldots$ (figure 8).

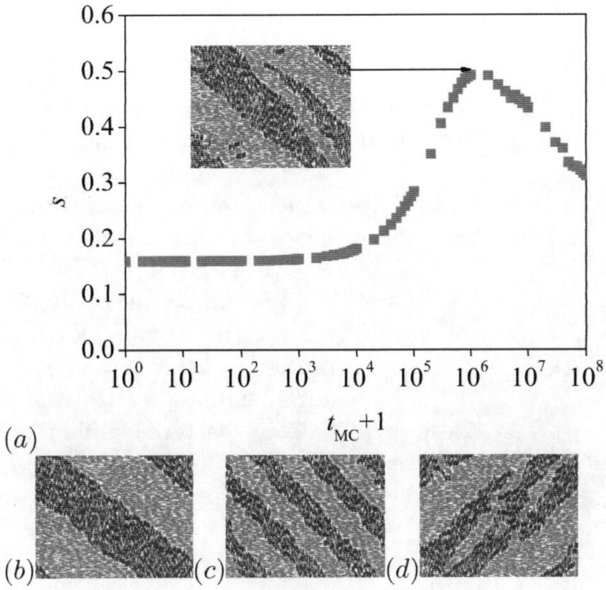

Figure 7. Example of pattern evolution vs MC steps, t_{MC}, for $k = 12$. The lattice size is 327×256. 100 independent statistical runs. (a) dependence of the local order parameter, s, calculated in windows of size 64×64 sites; inset: an example of a typical pattern at $t_{\mathrm{MC}} = 10^6$, (b, c, d) examples of patterns at $t_{\mathrm{MC}} = 10^8$. Patterns similar to (b) were observed 3 times, patterns similar to (c) were observed 84 times, patterns similar to (d) were observed 13 times.

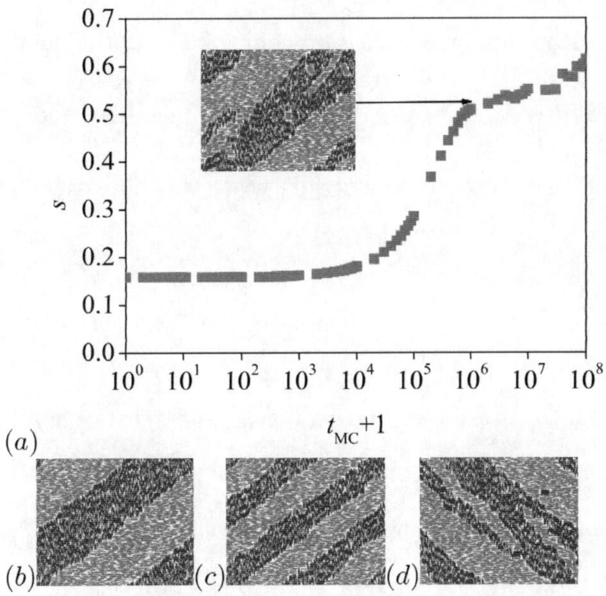

Figure 8. Example of pattern evolution vs MC steps, t_{MC}, for $k = 12$. The lattice size is 317×256. 100 independent statistical runs. (a) dependence of the local order parameter, s, calculated in windows of size 64×64 sites; inset: an example of a typical pattern at $t_{\mathrm{MC}} = 10^6$, (b, c, d) examples of patterns at $t_{\mathrm{MC}} = 10^8$. Patterns similar to (b) were observed 58 times, patterns similar to (c) were observed 4 times, patterns similar to (d) were observed 38 times.

Distorted stripes were mainly observed when $L_x/L_y = 1.2578125\ldots$ (figure 9). We may suggest that no steady state is achievable for systems with incommensurate aspect ratio, highly likely, such a system is highly likely to constantly rebuild its patterns.

4. Conclusion

Using Monte Carlo simulation and the lattice approach, we found out that

- pattern formation was observed only for particles $1 \times k$; no patterns were observed for particles with both sides greater than 1;
- when a pattern formation occurred, macroscopic regions with nematic order formed,

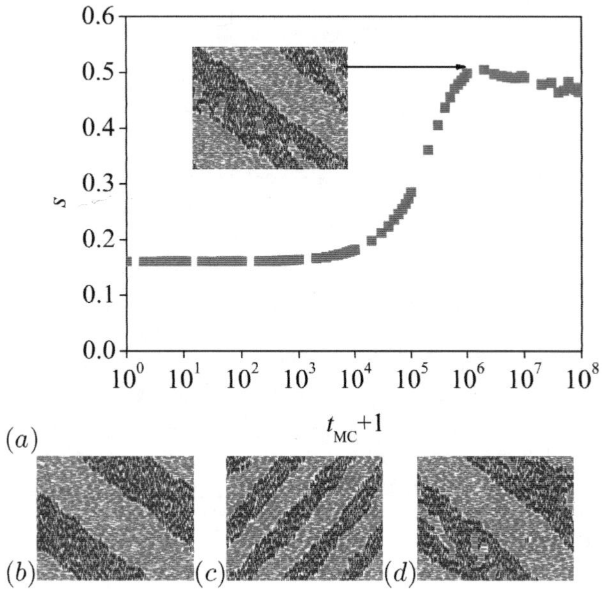

Figure 9. Example of pattern evolution vs MC steps, t_{MC}, for $k = 12$. The lattice size is 322×256. 100 independent statistical runs. (a) dependence of the local order parameter, s, calculated in windows of size 64×64 sites; inset: an example of a typical pattern at $t_{\mathrm{MC}} = 10^6$, (b, c, d) examples of patterns at $t_{\mathrm{MC}} = 10^8$. Patterns similar to (b) were observed 22 times, patterns similar to (c) were observed 33 times, patterns similar to (d) were observed 45 times.

- "blind" and "myopic" particles demonstrated similar behaviour; nevertheless, the final patterns are different;
- the random walk of particles led the system to a state in which the particles have more possibilities to move than in the initial jammed state;
- steady states were observed only for the lattices $L_x/L_y \in \mathbb{N}$; a steady state was not reached even until $t_{\mathrm{MC}} = 10^8$ for incommensurable aspect ratio.

Acknowledgments
The reported study was funded by RFBR according to the research project No. 18-07-00343.

References

[1] Lebovka N I, Tarasevich Y Y, Gigiberiya V A and Vygornitskii N V 2017 *Phys. Rev. E* **95**(5) 052130 ISSN 2470-0045
[2] Tarasevich Y Y, Laptev V V, Burmistrov A S and Lebovka N I 2017 *J. Stat. Mech.* **2017** 093203 ISSN 1742-5468
[3] Patra S, Das D, Rajesh R and Mitra M K 2018 *Phys. Rev. E* **97**(2) 022108 ISSN 2470-0045
[4] Börzsönyi T and Stannarius R 2013 *Soft Matter* **9** 7401–7418 ISSN 1744-6848
[5] Müller T, de las Heras D, Rehberg I and Huang K 2015 *Phys. Rev. E* **91**(6) 062207 ISSN 2470-0045
[6] González-Pinto M, Borondo F, Martinez-Ratón Y and Velasco E 2017 *Soft Matter* **13**(14) 2571–2582 ISSN 1744-6848
[7] Evans J W 1993 *Rev. Mod. Phys.* **65**(4) 1281–1329 ISSN 0034-6861
[8] de Gennes P G 1976 *La Recherche* **7** 919–927
[9] Mitescu C D and Roussenq J 1983 Diffusion on percolation clusters *Percolation Processes and Structures* (*Ann. Israel Phys. Soc.* vol 5) ed Deutscher G, Zallen R and Adler J (Bristol: Adam Hilger) pp 81–100 ISBN 978-0852744772
[10] Stauffer D and Aharony A 1992 *Introduction to Percolation Theory* (London: Taylor & Francis) ISBN 0748400273
[11] Hoshen J and Kopelman R 1976 *Phys. Rev. B* **14**(8) 3438–3445 ISSN 2469-9950
[12] Andrienko D 2018 *Journal of Molecular Liquids* **267** 520–541 ISSN 0167-7322 special Issue Dedicated to the Memory of Professor Y. Reznikov

IOP Conf. Series: Journal of Physics: Conf. Series **1163** (2019) 012008 doi:10.1088/1742-6596/1163/1/012008

Numerical studies of solutions for kinetic equations with many-particle collisions

S A Matveev[1], D A Stefonishin[1,2], A P Smirnov[2], A A Sorokin[3] and E E Tyrtyshnikov[4]

[1] Skoltech Center for Computational and Data-Intensive Science and Engineering, Skolkovo Institute for Science and Technology, Moscow 121205, Russia
[2] Faculty of Computational Mathematics and Cybernetics, Lomonosov Moscow State University, Moscow 119992, Russia
[3] Nuclear Safety Institute, Russian Academy of Sciences, Moscow 115191, Russia
[4] Marchuk Institute of Numerical Mathematics, Russian Academy of Sciences, Moscow 119333, Russia

E-mail: stefonishin@gmail.com

Abstract. In this work we present a method of low-rank tensor decompositions to accelerate evaluation of the right-hand side of systems of kinetic equations with many-particle collisional terms. These equations can be interpreted as a generalization of classical Smoluchowski aggregation equations allowing one to consider not only binary collisions of particles but also triple collisions. Straight-forward evaluation of the right-hand side for such system of N equations with $k = 1, 2, \dots, N$ requires $O(N^3)$ numerical operations and we find such complexity too high for practical investigations. However, under assumptions that kinetic coefficients can be represented with either canonical polyadic (CP) or tensor train decomposition (TT) with the rank $R \ll N$ we can propose algorithms evaluating the right-hand side with much lower complexities: $O(NR \log N)$ and $O(NR^2 \log N)$ for CP and TT respectively. We compare numerical solutions with different triple-collision rates and obtain a significant influence of accounting triple collisional effects.

1. Introduction

Smoluchowski-type kinetic equations can be considered as a traditional approach for description of the aggregation process in a complex spatially homogeneous physical system. Such ordinary differential equations correspond to a discrete model, where particles of the system collide inelastically in consequence of chaotic movement [1]. Each particle (cluster) consists of some minimal size particles (monomers). This model is applicable to various natural and industrial phenomena from the polymer growth and ultrafine metal clusters catalysts to formation of stars [2,3]. However in most studies only binary collisions of particles are taken into consideration [2,4]. That can be reasonably explained, because many-particle collisions lead to a growth of the computational cost of the solution and are often supposed to be rare events. Nevertheless some authors made a progress in studying such class of problems [5–7].

In this work we assume the presence of triple particle collisions along with binary ones including both the collisions between monomers as well as between clusters. Thus, the k-th kinetic equation defines the evolution of the concentration $n_k(t)$ per volume unit of k-size particles and can be written as $\frac{dn_k}{dt} = S_k(n_1, n_2, \ldots)$ where

$$S_k = \frac{1}{2}\sum_{i_1+i_2=k} C^{(2)}_{i_1,i_2} n_{i_1} n_{i_2} - n_k \sum_{i_1\geq 1} C^{(2)}_{i_1,k} n_{i_1} + \frac{1}{6}\sum_{i_1+i_2+i_3=k} C^{(3)}_{i_1,i_2,i_3} n_{i_1} n_{i_2} n_{i_3} - \frac{n_k}{2}\sum_{i_1,i_2\geq 1} C^{(3)}_{i_1,i_2,k} n_{i_1} n_{i_2}.$$

Here, the first and the second terms correspond respectively to concentration increase and decrease rates of k-size particles due to binary collisions. The third and fourth terms analogously define effects consequent to ternary collisions. Kinetic coefficients $C^{(2)}_{i_1,i_2}$ and $C^{(3)}_{i_1,i_2,i_3}$ are binary and ternary collision rates of particles of the sizes i_1, i_2, i_3. Straight-forward evaluation of the right-hand side term S_k for all sizes k is too heavy for practical investigations, because it requires $O(N^3)$ numerical operations for system of N equations with $k = 1, 2, \ldots, N$.

A Cauchy problem for such system of considered above kinetic equations can be posed by defining initial condition $n_k(0) = n_{k,0}$ for each size k. The analytical solution for this problem is known only for kinetic coefficients of simple form (constant, additive and multiplicative) [5,6,8]. In the present work we provide a fast finite difference method for solving the Cauchy problem for mentioned kinetic equations. With the use of tensor decompositions and fast methods of numerical linear algebra we reach high level of accuracy of numerical solutions in really modest computational times. We also describe the obtained numerical results for a realistic model of aggregation process accounting for triple collisional effects.

2. Model of many-particle aggregation

We consider the following kinetic coefficients for binary and ternary collisions

$$C^{(2)}_{i_1,i_2} = \frac{\sqrt{2}}{4}\cdot\left(\sqrt[3]{i_1}+\sqrt[3]{i_2}\right)^2\cdot\sqrt{1/i_1+1/i_2}, \tag{1}$$

$$C^{(3)}_{i_1,i_2,i_3} = \frac{\alpha_{3/2}\sqrt{6}}{324}\cdot\frac{\left(\sqrt[3]{i_1}+\sqrt[3]{i_2}+\sqrt[3]{i_3}\right)^2}{\sqrt{i_1\, i_2\, i_3/(i_1+i_2+i_3)}}\cdot\left[\frac{\left(\sqrt[3]{i_1}+\sqrt[3]{i_2}\right)^3}{\sqrt{1/i_1+1/i_2}}+\frac{\left(\sqrt[3]{i_1}+\sqrt[3]{i_3}\right)^3}{\sqrt{1/i_1+1/i_3}}+\frac{\left(\sqrt[3]{i_2}+\sqrt[3]{i_3}\right)^3}{\sqrt{1/i_2+1/i_3}}\right]. \tag{2}$$

Here each particle is represented as a sphere. Model is based on the assumption, that any result of ternary collision is in fact the result of stabilization of intermediate nonstable formation of two particles by collision with the third one. We form the ternary collisional kinetic coefficients empirically by analogy with the classical binary ballistic coefficients. Moreover their symmetry and positiveness follow from the properties of studied aggregation process. The parameter $\alpha_{3/2}$ defines the relative rate of ternary collisional events as compared to binary ones (i. e. the ratio of characteristic times of these processes). We use the monodisperse initial conditions $n_{k,0} = \delta_{1,k}$ for all sizes k where $\delta_{i,j}$ is the Kronecker symbol.

3. Numerical method

An approximate solution of the Cauchy problem for the system of Smoluchowski-type kinetic equations can be evaluated by means of finite-difference schemes. For example one can apply the second order Runge-Kutta scheme with a time step Δt for sizes $k = 1, 2, \ldots, N$ of the form

$$\frac{n_k^{s+1/2} - n_k^s}{\Delta t/2} = S_k(n_1^s, n_2^s, \ldots, n_N^s), \qquad \frac{n_k^{s+1} - n_k^s}{\Delta t} = S_k\left(n_1^{s+1/2}, n_2^{s+1/2}, \ldots, n_N^{s+1/2}\right).$$

Here the value n_k^s approximates $n_k(s\cdot\Delta t)$. The scheme has an approximation error $O\left((\Delta t)^2\right)$ of the analytical solution. However each step of the scheme requires $O(N^3)$ numerical operations with the naive calculations as was said in the introduction. Yet it can be accelerated by the use of tensor decompositions and fast methods of numerical linear algebra. This fact can be illustrated by ternary collision kinetic coefficients of the rank-one form $C^{(3)}_{i_1,i_2,i_3} = u_{i_1} \cdot v_{i_2} \cdot w_{i_3}$ (the same holds for rank-one binary collision coefficients). Thus if we are to compute the following values of the third term in S_k for all sizes k

$$P_k = \sum_{i_1+i_2+i_3=k} C^{(3)}_{i_1,i_2,i_3} n_{i_1} n_{i_2} n_{i_3} = \sum_{i_1+i_2+i_3=k} (u_{i_1} n_{i_1}) \cdot (v_{i_2} n_{i_2}) \cdot (w_{i_3} n_{i_3}), \qquad k = 1, 2, \ldots, N,$$

then we just have to succesively calculate the results of two shifted discrete convolutions

$$P_k = \sum_{i=1}^{k-2} (u_i n_i) \cdot z_{k-i}, \qquad z_k = \sum_{j=1}^{k-1} (v_j n_j) \cdot (w_{k-j} n_{k-j}), \qquad k = 1, 2, \ldots, N.$$

These calculations can be done via fast Fourier transform in $O(N \log N)$ numerical operations [9]. One can generalize the mentioned procedure with kinetic coefficients of more complex form in order to evaluate the operators S_k for all k. Suppose the array $C^{(2)} = \left[C^{(2)}_{i_1,i_2}\right]$ is approximately represented as a skeleton decomposition and the array $C^{(3)} = \left[C^{(3)}_{i_1,i_2,i_3}\right]$ has one of the following two forms

$$C^{(3)}_{i_1,i_2,i_3} \approx \sum_{\rho=1}^{R} u_{i_1,\rho} \cdot v_{i_2,\rho} \cdot w_{i_3,\rho}; \qquad C^{(3)}_{i_1,i_2,i_3} \approx \sum_{\rho_1,\rho_2=1}^{R} u_{i_1,\rho_1} \cdot v_{\rho_1,i_2,\rho_2} \cdot w_{\rho_2,i_3}.$$

The first decomposition is a canonical polyadic (CP) and the second one is a tensor train (TT). Then the total computational cost of values S_k for all sizes $k = 1, 2, \ldots, N$ together is $O(NR \log N)$ and $O(NR^2 \log N)$ for CP and TT-representations respectively. Note that there exists a robust method of tensor approximation by a TT-decomposition with complexity $O(NR^3)$ numerical operations [10].

There is also a sequence of related results in terms of computational complexity and use of low-rank matrix and tensor decompositions. They are presented in recent studies of the physical systems allowing only binary collisions: in application to continuous coagulation models [11, 12], multicomponent coagulation processes [13, 14] and discrete equations of aggregation and fragmentation processes in both time-dependent [15, 16] and stationary cases [17].

4. Numerical results

In the present section we decribe the numerical results concerning the influence of triple particle collisions. Here we do not show the results of tests concerning perfomance and accuracy of our method. For such results we refer the readers to our recent article [18]. Note, that in the mentioned work a good agreement was obtained between a numerical solution and the known analytical solution in special case of constant kinetic coefficients and monodisperse initial conditions.

In the numerical experiments we used the monodisperse initial conditions $n_{k,0} = \delta_{1,k}$ and the kinetic coefficients of forms (1) and (2) for binary and ternary collisions respectively. From the results of the experiments we conclude that there is a significant influence of triple-particle collisions on the cluster distribution. One can see that on figure 1 illustrating the numerical solutions for Cauchy problem with and without accounting ternary collisional events. There the concentrations of large particles become nonzero two times faster with a quite small (twenty percent) relative rate of ternary collisions.

We also compared the behavior of the full cluster concentration $n(t)$ and the relative geometric size R_1 and the volume size R_3 of clusters depending on the parameter $\alpha_{3/2}$. These values are

IOP Conf. Series: Journal of Physics: Conf. Series **1163** (2019) 012008 doi:10.1088/1742-6596/1163/1/012008

defined as follows

$$n(t) = \sum_{k=1}^{\infty} n_k(t), \qquad R_\lambda(t) = \sum_{k=1}^{\infty} k^\lambda \cdot n_k(t)/n(t).$$

The values of the governing parameter were chosen as $\alpha_{3/2} = 0.05, 0.1, 0.2$. We also compared the obtained numerical results with ones in case of only binary collisions model (i. e. when $C^{(3)} = 0$).

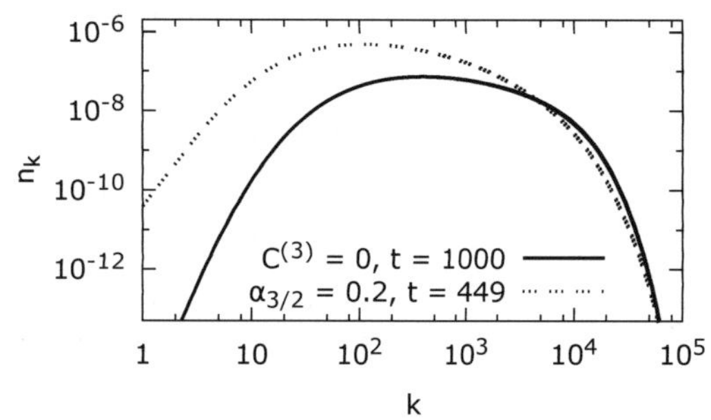

Figure 1. Cluster distribution with and without accounting ternary collisions. Logarithmic scale on both axes.

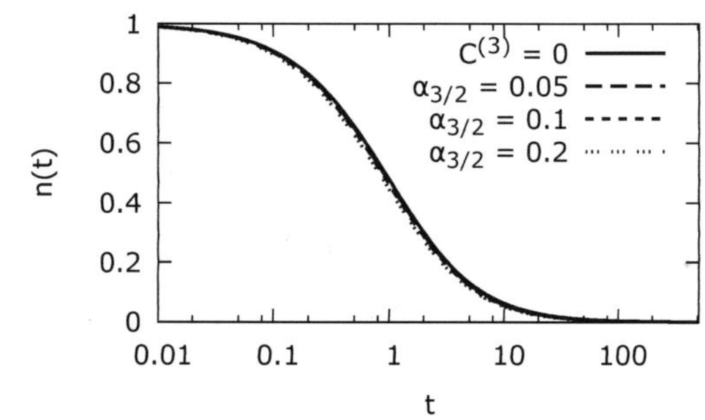

Figure 2. Evolution of the full cluster concentration $n(t)$. Logarithmic scale on t-axis.

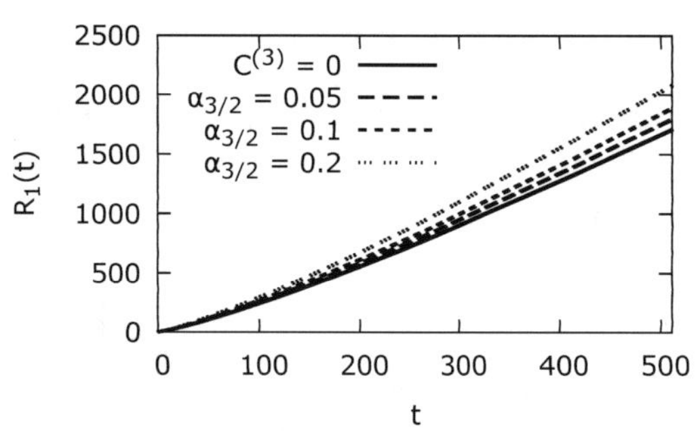

Figure 3. Evolution of the relative geometric size R_1 of clusters.

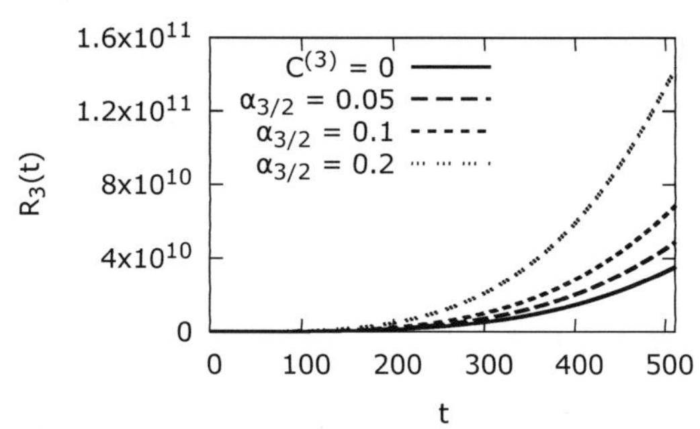

Figure 4. Evolution of the relative volume R_3 of clusters.

Figure 2 shows that with the growth of the ternary collision rate the full concentration of particles changes insignificantly. At the same time the mass of the system redistributes faster in particles of large sizes with the growth of the parameter $\alpha_{3/2}$. The latter fact is shown on figures 3 and 4. It is clear, that the presence of ternary collisions has a great influence on the growth of relative moments R_λ of the solution of higher orders λ. This also means that the total concentration of particles is dominated by small particles, but their size is defined by larger particles. The last result shows that only the ten percents of ternary collisions define approximately the double growth in the mean particle volume size.

5. Conclusions

In this work we recall a fast numerical method for solving the Cauchy problem for system of Smoluchowski-type kinetic equations of aggregation with triple-particle collisional terms.

The method is based on applications of low-rank tensor decompositions such as CP or TT. Such approach can be used with different physical models accounting ternary collisions.

With the use of the method we were able to compare numerical solutions corresponding to different relative ternary collision rates that manifest a significant influence of accounting triple collisional effects.

Acknowledgments

We would like to thank group of prof. A Shapeev for sharing of their computational resources during preparation of this work.

This work is supported by the Program of the Presidium of the Russian Academy of Sciences no. 01 "Fundamental Mathematics and its Applications" under grant PRAS-18-01.

References

[1] Smoluchowski M V 1917 *Z. Phys. Chem.* **92** 129–68
[2] Galkin V 2001 *The Smoluchowski Equation* (Moscow: Fizmatlit)
[3] Leyvraz F 2003 *Phys. Reports* **383** 95–212
[4] Melzak Z A 1957 *Transactions of the American Mathematical Society* **85** 547–60
[5] Krapivsky P L 1991 *J. Phys.* A **24** 4697–703
[6] Krapivsky P L, Redner S and Ben-Naim E 2010 *A Kinetic View of Statistical Physics* (Cambridge: Cambridge University Press)
[7] Brener A M 2014 *Chem. Eng. Trans.* **38** 145–50
[8] Jiang Y and Gang H 1989 *Phys. Rev.* B **39** 4659–65
[9] Tyrtyshnikov E E 2012 *A Brief Introduction to Numerical Analysis* (Berlin: Springer Science & Business Media)
[10] Oseledets I and Tyrtyshnikov E 2008 *Linear Algebra and Its Appl.* **432** 70–88
[11] Matveev S A, Smirnov A P and Tyrtyshnikov E E 2015 *J. of Comp. Phys.* **282** 23–32
[12] Hackbusch W 2007 *Numer. Math.* **106** 627–57
[13] Matveev S A, Zheltkov D A, Tyrtyshnikov E E and Smirnov A P 2016 *J. of Comp. Phys.* **316** 164–79
[14] Smirnov A, Matveev S, Zheltkov D and Tyrtyshnikov E 2016 *Procedia Comput. Sci.* **80** 2141–6
[15] Matveev S A, Krapivsky P L, Smirnov A P, Tyrtyshnikov E E and Brilliantov N V 2017 *Phys. Rev. Lett.* **119** 260601
[16] Brilliantov N V, Otieno W, Matveev S A, Smirnov A P, Tyrtyshnikov E E and Krapivsky P L 2018 *Phys. Rev. E* **98** 012109
[17] Matveev S A, Ampilogova N V, Stadnichuk V I, Tyrtyshnikov E E, Smirnov A P and Brilliantov N V 2018 *Comp. Phys. Communications* **224** 154–63
[18] Stefonishin D A, Matveev S A, Smirnov A P and Tyrtyshnikov E 2018 *Num. Methods and Programming* **19** 261–9

IOP Conf. Series: Journal of Physics: Conf. Series **1163** (2019) 012009 doi:10.1088/1742-6596/1163/1/012009

Simulation of delay reaction-drift-diffusion system applied to charging effects in electron-irradiated dielectrics

A G Maslovskaya and A V Pavelchuk

Mathematics and Computer Science Department, Amur State University, 21 Ignatyevskoe Shosse, Blagoveshchensk, 675000, Russia

E-mail: maslovskayaag@mail.ru

Abstract. The paper is devoted to reaction-convection-diffusion system modeling with time delay and its application for simulation of charging process in dielectrics irradiated by electron bunches. The model of dielectric charging is expressed by mixed initial-boundary value problem for locally instantaneous Poisson equation and time-dependent transport equation as reaction-convection-diffusion equation with time delay. An advanced finite difference scheme was proposed to numerically solve the delay parabolic partial differential equation. Monte-Carlo simulation of electron transport in electron irradiated target was also used to estimate a source function. The special program application was designed using Matlab. The computing experiments were performed to estimate the charge density distribution, potential distribution and electron beam-induced field intensity in electron irradiated ferroelectrics.

1. Introduction

Nowadays "reaction-convection-diffusion" systems have become increasingly relevant for various physical applications. Among numerous models one can specify a class of the systems formalized by partial differential equations (PDE) with time-delay or hereditary effect [1-3]. In particular, functional differential equations can be also referred to delay PDE. When solving applied problems physical sense of delay is often attributed to the finite velocity of distribution of perturbations or the inertial nature of the systems. These systems can not generate a response instantly under external exposure and some time lag can be found. The results of mathematical modelling dynamic systems with time delay have been discussed elsewhere [1-6]. Note also that analytical solution can be obtained for rather limited class of parabolic PDE with delay [4]. Specifically previous studies have been reported the analysis of solution properties such as stability, asymptomaticity, frequency, oscillations et al. In addition, in order to study the model object using computational experiment we need to construct flexible and effective calculation scheme. For this reason numerical methods (especially finite difference methods) play a major role in practice to simulate dynamic "reaction-convection-diffusion" system with time delay [5-6].

The model of dynamic charging process in dielectrics under nonequilibrium conditions can be expressed with use of drift-diffusion approach, which allows one formulates this model mathematically as the system of Poisson equation and time-dependent transport equation [7-10]. The series of the our previous works [9-10] has been devoted to development of mathematical models describing stationary and evolutionary charging processes of ferroelectrics exposed to electron beam irradiation in scanning electron microscope (SEM). Ferroelectrics as a subclass of polar dielectrics are one of the most promising research objects used in practical applications. SEM enables creating new modes of image formation and modification of the ferroelectric properties. In this case one of the relevant effect of external impact is the charging of ferroelectrics by electrons injected into the sample surface. The great importance of charging effect is due to practical problems of using SEM techniques for formation of controlled submicro- and nanodimensional domain structures in ferroelectrics [11].

Development of mathematical models permits us to simulate the maps of attendant effects, analyze the observed injection, thermal and polarizing processes in an irradiated samples and also predict the conditions of the control polarization switching and domain structure formation. The modification of dynamic charging model has been proposed in view of importance of intrinsic radiation-stimulated conductivity of a sample and also with use of preliminary Monte-Carlo simulation of electron transport in irradiated target [12]. The mathematical model of dynamic charging has been defined as an initial-boundary problem for the system of the PDE, where the basic equation represents by nonlinear reaction-drift-diffusion equation. Considering physical mechanisms which underline diffusive character of charging process we may assume that the delay effect as a reaction to external exposure can be introduced in the corresponding equation of mathematical problem definition. Therefore the current study was designed to develop mathematical model of "reaction-convection (drift)-diffusion" system with time delay and its application for numerical simulation of time-dependent charging processes in polar dielectrics irradiated with SEM.

2. Problem formulation and governing equations

In general case, mathematical model of charging process in dielectrics is defined by locally instantaneous Poisson equation as well as continuity equation. Since high-voltage SEM mode is considered we can use only two independent spatial coordinates: the radius r and the height z. In addition the internal source has semispherical or semiellipsoid approximation. Therefore we assume that problem is characterized by cylindrical symmetry as it is shown in Fig. 1.

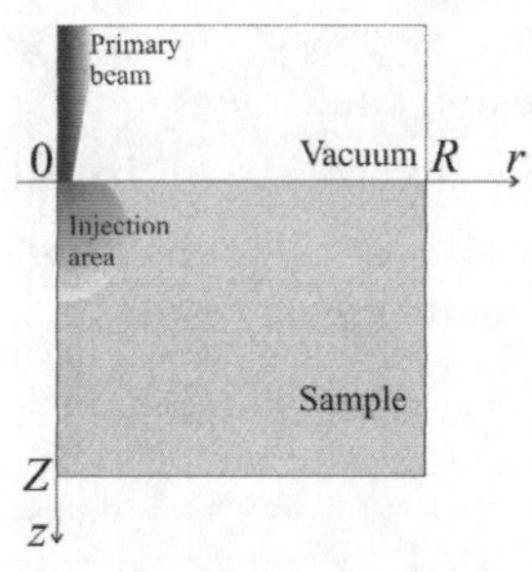

Figure 1. Geometrical scheme of the irradiated sample.

Further, suppose inductively that the action of focused electron beam starts at the time moment t=0 perpendicularly to a plane of a sample surface z=0. After some transformations [9-10] we can express the nonstationary model of charging process taking into account intrinsic radiation-stimulated conductivity of a sample as initial-boundary value problem for the following system of equations:

$$\begin{cases} \dfrac{\partial \rho}{\partial t} = D\Delta\rho - \mu_n\left(E_r\dfrac{\partial \rho}{\partial r} + E_z\dfrac{\partial \rho}{\partial z}\right) - \dfrac{\mu_n}{\varepsilon\varepsilon_0}\rho^2 + G(r,z), \\ \Delta\varphi = -\dfrac{\rho}{\varepsilon\varepsilon_0}, \qquad \mathbf{E} = -\operatorname{grad}\varphi, \end{cases} \tag{1}$$

where $\rho(r,z,t)$ is charge density distribution, C/m^3; $0 \le r \le R$, $0 \le z \le Z$, $0 < t \le \overline{t}$; R, Z are geometrical parameters of sample, m; **E** is field intensity, V/m; φ is potential, V; ε is dielectric permittivity; ε_0 is dielectric constant, F/m; $G(r,z)$ is term related to internal source, C/(m^3·s); $D = \mu_n kT / e$ is diffusion coefficient, m^2/s; k is Boltzmann constant, J/K; T is temperature, K; μ_n is drift mobility of electrons, $m^2/(V \cdot s)$; e is electron charge, C; $\Delta = \begin{cases} \frac{\partial^2}{\partial r^2} + \frac{1}{r}\frac{\partial}{\partial r} + \frac{\partial^2}{\partial z^2}, & \text{if } r \neq 0 \\ 2\frac{\partial^2}{\partial r^2} + \frac{\partial^2}{\partial z^2}, & \text{if } r = 0 \end{cases}$ is Laplace operator.

The reaction-drift-diffusion equation and Poisson equation need to be complemented by initial conditions at time moment $t_0 = 0$ for $0 \le r \le R,\ 0 \le z \le Z$:

$$\rho(r,z,t_0) = \rho^0(r,z),\quad E(r,z,t_0) = E^0(r,z),\quad \varphi(r,z,t_0) = \varphi^0(r,z) \tag{2}$$

and boundary conditions corresponding to physical sense for $t \in (0,\bar{T}]$:

$$\left.\frac{\partial\rho}{\partial r}\right|_{\substack{r=0,\\ 0\le z\le Z}} = 0,\quad \left.\frac{\partial\rho}{\partial z}\right|_{\substack{z=0,\\ 0\le r\le R}} = 0,\quad \left.\frac{\partial\varphi}{\partial r}\right|_{\substack{r=0,\\ 0\le z\le Z}} = 0,\quad \left.\frac{\partial\varphi}{\partial z}\right|_{\substack{z=0,\\ 0\le r\le R}} = 0, \tag{3}$$

$$\rho\big|_{\substack{r=R,\\ 0\le z\le Z}} = 0,\quad \rho\big|_{\substack{z=Z,\\ 0\le r\le R}} = 0,\quad \varphi\big|_{\substack{r=R,\\ 0\le z\le Z}} = 0,\quad \varphi\big|_{\substack{z=Z,\\ 0\le r\le R}} = 0. \tag{4}$$

Furthermore, let us suppose that changes in gradient of the injected charges density lead to following effect. The system responds by changes in diffusion current component not instantly, but with some time lag t^*. Using the concept of "single-phase-lag" model [13-14] the expression of diffusion current density $\mathbf{j}^{dif}$ can be modified taking into account time delay effect: $\mathbf{j}^{dif}(r,z,t+t^*) = -D \cdot \operatorname{grad}\rho(r,z,t)$. Here the latter is transformed to generalized relation of diffusion current density with respect to time moment t. Whence we specify the contribution of diffusion component into current density change at instant time moment as well as at time moment "in the past":

$$\mathbf{j}^{dif}(r,z,t) = -D \cdot \operatorname{grad}\rho(r,z,t) - \bar{D} \cdot \operatorname{grad}\rho(r,z,t-t^*), \tag{5}$$

where $\bar{D} = \gamma \cdot D$ is corresponding diffusion coefficient and $0 \le \gamma \le 1$ is constant of delay.

This allow us to write the expression for divergence of diffusion current density in consideration of the time delay t^*. The continuity equation is given by following equation as superposition of diffusion current and drift current contributions respectively:

$$\begin{aligned} \frac{\partial\rho(r,z,t)}{\partial t} &= D\Delta\rho(r,z,t) + \bar{D}\Delta\rho(r,z,t-t^*) - \frac{\mu_n}{\varepsilon\varepsilon_0}\cdot\rho^2(r,z,t) - \\ &-\mu_n\left(E_r\frac{\partial\rho(r,z,t)}{\partial r} + E_z\frac{\partial\rho(r,z,t)}{\partial z}\right) + G(r,z). \end{aligned} \tag{6}$$

Here initial condition (2) is modified using prehistory of process:

$$\rho(r,z,t) = \rho^0(r,z,t) \text{ for } -t^* \le t \le 0,\ 0 \le r \le R,\ 0 \le z \le Z. \tag{7}$$

This implies that problem definition is given by continuity equation (6), the Poisson equation, the expression for field intensity in system (1), prehistory (7) and boundary conditions (3)-(4).

3. Numerical method

The equation (6) represents the so-called "reaction-convection-diffusion" equation. This equation is defined as parabolic PDE with constant time delay and it can be solved by finite difference method [5-6]. Note that the relevant issue of numerical solving such equation is approximation of convection term. In order to solve the reaction-convection-diffusion equation we proposed the modification of computational splitting scheme based on alternating direction method of Peacman-Rachford [15] for approximation of "diffusion" part of equation. Also the Roberts-Weiss scheme [16] was used for approximation of "convection" part. In order to approximate the "reaction" part of PDE the "frozen coefficient" method was also applied. Let us consider space-time grid with steps h_1, h_2, τ related to variables r, z, t respectively: $\omega^{\tau}_{h_1h_2} = \{r_i = ih_1, i = \overline{1,N}, z_j = jh_2, j = \overline{1,M}, t^k = k\tau, k = 1,2,...\}$. The computational subshemes on $k+1/2$ and $k+1$ time layers can be written as:

$$\frac{\rho_{ij}^{k+1/2} - \rho_{ij}^{k}}{\tau} = \frac{1}{2} G_{ij}^{k+1/2} - \frac{\mu n}{2\varepsilon\varepsilon_0} \cdot \left(\rho_{ij}^{k}\right)^2 - \frac{\mu_n}{2}\left(E_r \frac{\rho_{ij}^{k+1/2} - \rho_{i-1j}^{k+1/2} + \rho_{i+1j}^{k} - \rho_{ij}^{k}}{2h_1} + E_z \frac{\rho_{ij+1}^{k} - \rho_{ij-1}^{k}}{2h_2} \right) +$$
$$+ \frac{D}{2}\left(\frac{\rho_{i+1j}^{k+1/2} - 2\rho_{ij}^{k+1/2} + \rho_{i-1j}^{k+1/2}}{h_1^2} + \frac{1}{r_i}\frac{\rho_{i+1j}^{k+1/2} - \rho_{i-1j}^{k+1/2}}{2h_1} + \frac{\rho_{ij+1}^{k} - 2\rho_{ij}^{k} + \rho_{ij-1}^{k}}{h_2^2} \right), \quad (8)$$

$$\frac{\rho_{ij}^{k+1} - \rho_{ij}^{k+1/2}}{\tau} = \frac{1}{2} G_{ij}^{k+1/2} - \frac{\mu_n}{2\varepsilon\varepsilon_0} \cdot \rho_{ij}^{k} \cdot \rho_{ij}^{k+1} - \frac{\mu_n}{2}\left(E_r \frac{\rho_{i+1j}^{k+1/2} - \rho_{i-1j}^{k+1/2}}{2h_1} + E_z \frac{\rho_{ij}^{k+1} - \rho_{ij-1}^{k+1/2} + \rho_{ij+1}^{k+1/2} - \rho_{ij}^{k+1/2}}{2h_2} \right) +$$
$$+ \frac{D}{2}\left(\frac{\rho_{i+1j}^{k+1/2} - 2\rho_{ij}^{k+1/2} + \rho_{i-1j}^{k+1/2}}{h_1^2} + \frac{1}{r_i}\frac{\rho_{i+1j}^{k+1/2} - \rho_{i-1j}^{k+1/2}}{2h_1} + \frac{\rho_{ij+1}^{k+1} - 2\rho_{ij}^{k+1} + \rho_{ij-1}^{k+1}}{h_2^2} \right). \quad (9)$$

Subschemes (8) and (9) are completed by approximations of Derichlet boundary conditions: $\rho_{N,j}^{k+1/2} = 0$, $\rho_{i,M}^{k+1/2} = 0$ and $\rho_{N,j}^{k+1} = 0$, $\rho_{i,M}^{k+1} = 0$. Neumann boundary conditions are automatically included in the modified differential equations with use of additional layers on coordinates with fictitious points and approximations: $\frac{\rho_{0,j}^{k+1/2} - \rho_{2,j}^{k+1/2}}{2h_1} = 0$ and $\frac{\rho_{i,0}^{k+1} - \rho_{i,2}^{k+1}}{2h_2} = 0$. The constructed numerical scheme was modified taking into account delay effect. According to approximation for diffusion component with time delay the subschemes (8) and (9) are complemented by expression: $\frac{\bar{D}}{2}\left(\frac{\rho_{i+1j}^{k-q} - 2\rho_{ij}^{k-q} + \rho_{i-1j}^{k-q}}{h_1^2} + \frac{1}{r_i}\frac{\rho_{i+1j}^{k-q} - \rho_{i-1j}^{k-q}}{2h_1} + \frac{\rho_{ij+1}^{k-q} - 2\rho_{ij}^{k-q} + \rho_{ij-1}^{k-q}}{h_2^2} \right)$, where $q = t^* / \tau$ is time lag index. Note that numerical simulation of delay system requires controlling memory-stored structure of "window of delay" (values of variables on each time layer included to delay period). This aspect leads to resource-intensive computational process. The obtained systems are solved by sweep method on each time layer. The Poisson equation on each time layer is solved using alternating direction method (relaxation scheme). The procedure of approximation analysis results in second order accuracy for space and time variables $O\left(\tau^2 + h_1^2 + h_2^2\right)$. The stability analysis by means of maximum principle suggests the absolute stability of the scheme. Also, the constructed scheme is monotone.

The absolute value of field intensity is evaluated using field components (E_r, E_z). In addition, Monte-Carlo simulation of electron trajectories in irradiated sample allows us to specify the initial charge distribution $\rho^0(r,z,t)$ (for the model of relaxition charging process) and the sorce funtion $G(r,z)$ (for the problem of dynamic charge accumulation).

4. Simulation results and discussion

The special program application developed with Matlab package was designed to simulate dynamic electron beam-stimulated charging processes in dielectrics. Three modes of charging process were implemented in the program: the stationary mode, the dynamic mode of charge relaxation and the dynamic mode of charge accumulation. The relation of deposits of "drift" (convection) and "diffusion" parts in the model for dynamic modes can be evaluated due to analysis of value of Peclet number $\mathrm{Pe} = v_{dr} \cdot L/D$, where $v_{dr} = \mu_n E$ is the drift velocity in m/s and L is characteristic scale of distance, for instance, linear size of a gradient zone in m. Note that $\mathrm{Pe} << 1$ and diffusion dominates in the mode of charge accumulation. Whereas $\mathrm{Pe} >> 1$ and drift dominates in the mode of charge relaxation. Using the analog of Fourier number $\mathrm{Fo} = D \cdot t/L^2$ enables the scales ratio of time and distance to be specified. The computing experiments were performed by the example of promising ferroelectric materials lithium niobate ($LiNbO_3$) and lithium tantalate ($LiTaO_3$).

The mode of charge relaxation. The Fig. 2 demonstrates the simulation results for spatial distribution of potential and field intensity distribution induced by the injected charges in $LiTaO_3$ in the mode of charge relaxation at time of observation t=1 ps (calculation parameters are h_1=0.25 μm, $\tau=0.5 \cdot h_1$ μs at start beam energy E_0=25 keV and surface charge density σ_{surf}=10 C/m^2). The estimation of Peclet number gives $5 \cdot 10^5$ (more than unity), which means drift dominates and we have weak diffusion. The computing experiment allows us to observe charge relaxation dynamics and identify a relaxation time in this mode. Here the relaxation time is going to be more than 100 ps.

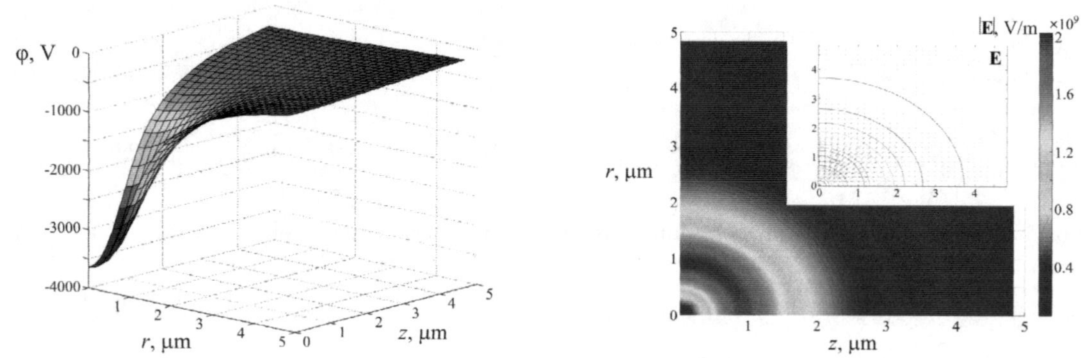

Figure 2. The simulation results: potential distribution – (a);
distribution of field intensity absolute value and its vector field of (in the insert figure) – (b).

Mode of charge accumulation dynamics. The Fig. 3 shows the compuatational results for charging characteristics of $LiNbO_3$ taking into account delay effect.

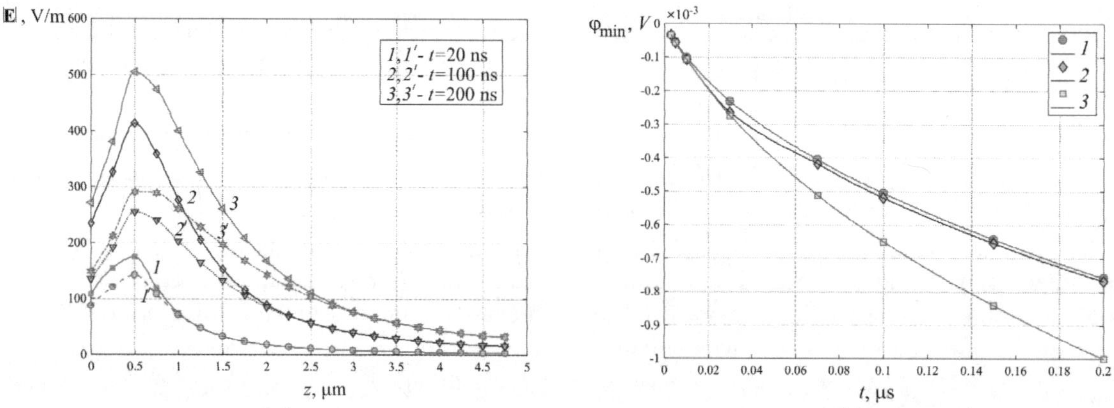

Figure 3. The absolute value profile of field intensity at the fixed time moments (curves correspond to the modes: 1, 2, 3 – without delay, 1′, 2′, 3′ – with delay, time lag is t^*=4 ns) – (a); changes in potential minimum value at various time lag: 1 – t^*=20 ns, 2 – t^*=2 ns, 3 – without delay – (b).

The computing parameters are h_1=0.25 μm, τ=1 ns, γ=1. Calculated data are displayed in comparison with simulation data provided nonregistering delay at start beam energy E_0=10 keV, irradiation time 10 s and surface charge density σ_{surf}=20 C/m^2. In this mode Peclet number equals to 0.04 (less than unity). This means that diffusion process dominates. The computing experiment indicates significant difference in values of estimated characteristics both of maximum value level and value at a point of source impact in a gradient zone.

The mode of stationary charging. The estimation of characteristic value of dynamic time by Fourier criterion yields 50 μs. The exposure time is approximately changes in the range from 1 s to 10 s when SEM is used to control polarization switching in ferroelectrics. This allows us to conclude that stationary mode is realized. The analysis of simulation data suggests that field intensity in nonirradiated part of a sample exceeds the coercive field ($2.1\cdot10^7$ V/m for $LiTaO_3$). This can result to polarization switching in oppositely orientated domains. Provided start beam energy E_0 equals to 15 keV and surface charge density σ_{surf} equals to 5 C/m^2 the maximum value of field intensity is $2.1\cdot10^9$ B/м at the distance of 0.05 μm from the source and it corresponds to $3\cdot10^7$ V/m at the distance of 5 μm. At the same time the use of σ_{surf}=1 C/m^2 does not allow switching process to be stimulated in nonirradiated part of crystal at different values of start beam energy. Also the use of reduced accelerating voltages (~5 kV) can not create the condition for polarization switching when the surface charge density varies over the range 5-20 C/m^2. This confirms that surface charge density and accelerating voltages require to be optimally controlled.

5. Conclusions

The reaction-drift-diffusion model with time delay was proposed to simulate dynamic charging process in ferroelectric electron beam-irradiated with the SEM. An advanced splitting scheme was constructed to solve the delay parabolic PDE numerically. The analysis indicates the second order accuracy for space and time variables, monotony and absolute stability of the scheme. The main peculiarity of delay system numerical simulation is necessity of memory-stored control for delay "window" structure. This aspect leads to resource-intensive computational process. The relation of "drift" and "diffusion" deposits for dynamic modes of the model was also discussed using analysis of value of Peclet number. The computing experiments by the example of $LiNbO_3$ and $LiTaO_3$ ferroelectrics were performed to estimate the charging characteristics in the stationary mode, the dynamic mode of charge relaxation and the dynamic mode of charge accumulation.

References

[1] Wu J 1996 *Theory and applications of partial functional differential equations* (New York: Springer) p 368
[2] Kolmanovskii V B and Myshkis A D 1999 *Introduction to the theory and applications of functional differential equations* (Kluver: Dordrecht) p 648
[3] Kyrychko Y N and Hogan S J 2010 *Journal of vibration and control* **16** 943
[4] Polyanin A D and Zhurov A I 2013 *International Journal of Non-Linear Mechanics* **54** 115
[5] Jin Y F, Jiang J, Hou C M and Guan D H 2012 *Journal of Information & Computational Science* **18** 5579
[6] Garcia P, Castro M A, Martin J A and Sirvent A 2009 *Mathematical and Computer Modelling* **50** 860
[7] Kotera M, Yamaguchi K and Suga H 1999 *Jpn. J. Appl. Phys* **38** 7176
[8] Raftari B, Budko N V and Vuik C 2015 *J. Appl. Phys.* **118** 204101
[9] Maslovskaya A and Pavelchuk A 2015 *Ferroelectrics* **476** 157
[10] Pavelchuk A, Barabash T and Maslovskaya A 2016 *IOP Conf. Series: Materials Science and Engineering* **110** 012080
[11] He J, Tang S H, Qin Y Q, Dong P, Zhang H Z, Kang C H, Sun W X, Shen Z X 2003 *J. Appl. Phys.* **93** 9943
[12] Joy D C 1995 *Monte-Carlo modeling for electron microscopy and microanalysis* (New York: Oxford University Press)
[13] Tzou D Y 1995 *International Journal of Heat and Mass Transfer* **38** 3231
[14] Hong B-S and Chou C-Y 2014 *Entropy* **16** 1101
[15] Peacman D W and Rachford H H 1995 *J.Soc. Indust. Appl. Math.* **3** 28
[16] Roberts K V and Weiss N O 1966 *Convective Difference Schemes* (Abingdon, Berks: U.K.A.E.A., Culham Laboratory) pp 272-279

IOP Conf. Series: Journal of Physics: Conf. Series **1163** (2019) 012010 doi:10.1088/1742-6596/1163/1/012010

Simulation of the Super-scalar Processor Core Operation

A V Vishnekov[1] and E M Ivanova[1]

[1]National Research University Higher School of Economics, 34, Tallinskaya St., Moscow, 123458, Russia

emivanova@hse.ru

Abstract. The article considers the features of super-scalar processors, their way of performing several operations on several pairs of operands simultaneously. The research focuses on the organization of processor pipeline execution operation of several machine instructions in one processor core. The simulating kit was developed for better understanding of a processor core microarchitecture. It includes two parts: program and methodical recommendations with multiple task options. The simulating kit demonstrates the pipeline architecture consisting of two clusters: front-end and back-end and the principle of translating complex multi-cycle CISC-like instructions into simpler RISC-like micro-operations. The main types of machine instructions are considered: data transfer between registers and memory cells (four variations), data processing of couple of operands from registers and memory cells (four variations), conditional jump to the specified address. The program-simulator makes it possible to conduct a more detailed simulation of one of the three mechanisms for calculations accelerating in the processor core: multi-functional (super-scalar) processing, out-of-order processing, speculative instructions execution after the branch prediction. The simulating kit is used in educational process when training masters of Higher School of Economics National Research University.

1. Introduction

Modern super-scalar processors [1-4] can perform several operations on several pairs of operands simultaneously. This is achieved by a variety of different technological and microarchitectural solutions which parallelize calculations. The main method is the pipeline processing of machine instructions [5-8]. Each processor may contain one or several cores. Each core contains one or more instruction pipelines. In fact, the pipeline is the main part of the processor. Understanding its execution mechanism gives an understanding of computing principles in modern computers. A huge number of scientific and practical works are devoted to this issue, but the faster and more visual way to study the principle of pipeline processing is the launch and study of the processor core simulation model. The complexity of choosing the degree of modeling detail depends on the problem statement. Electronic device developers create similar cycle accurate simulator in languages supported by CAD tools (such as SystemVerilog, SystemC) to verify the specific device. These models are not freely available. In our case there is an educational task. Free simulators of superscalar processing (which can be found for example here [9]) are either simplified models or simulators of individual technological solutions. Neither professional simulators, nor partial solutions are not suitable for use in the educational process. Therefore, it was decided to develop our own simulator. The peculiarity and advantage of the developed simulator is following. Students get an idea about the microarchitecture features of the modern CPU cores by performing a number of simple operations. The structure and type of these operations in combination with the features of the developed simulator provide study efficiency and simplicity.

The processor is very complex device, so the model cannot display every aspect of its functioning. We focus only on some of them. We have built and investigated such a model that allows us to trace the process of machine instructions executing by the processor core pipeline. The modeling is based on universal features of Intel processor cores, such microarchitecture as Nehalem [2], Sandy Bridge [5], Haswell [1, 10], Skylake [1, 11]. The analyzed mechanisms are inherent to other modern processors as well, for example AMD Zen family microarchitecture [12]. The most interesting mechanisms, which are the most complicated for understanding, are the following: multi-functional (super-scalar) processing [13], out-of-order (OOO) processing [1, 10, 14], branch prediction [15, 16], speculative instructions execution [17], pipeline reset after misprediction [2], the accelerated instructions fetch from L1i cache [18, 19], origins of pipeline hazards and bubbles [5, 11, 14], principles of pipeline division into two clusters: front-end and back-end [3, 4, 20, 21], exception of data re-reading from L1d cache [1, 14], etc. Many scientific works and articles describe Intel CPU core microarchitecture. So students will easy find any additional information on all theoretical issues.

2. The front-end and back-end clusters of pipeline

Two parts of the pipeline, front-end and back-end, allow to integrate advantages of two processors architecture: convenience of the CISC program model and RISC high speed execution [21].

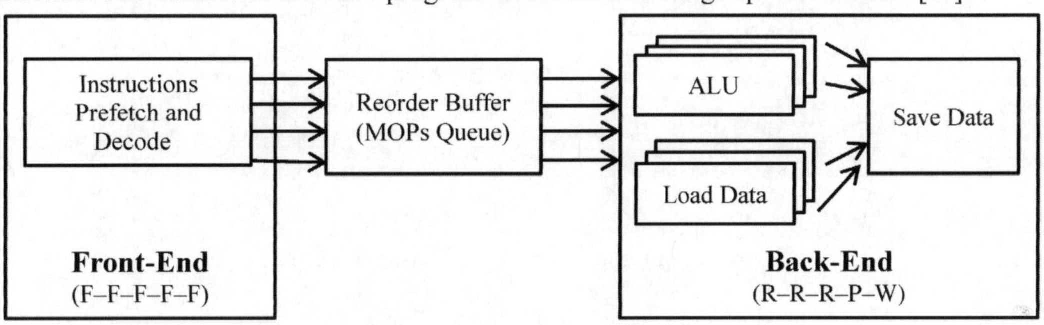

Figure 1. The simulated pipeline

The convenience is provided by the difficult instructions capable to process data from registers and memory. This is implemented via the Front-end CPU cluster. High speed is provided by simple and short operations. CISC-instructions are translated in simple RISC-like microoperations (MOPs) [18] which are processed in the Back-end CPU cluster by a set of executive devices (Figure 1): ALU (Processing), Load (Read data), Save (Write data). There is an instruction phase/ MOPs designation: F–F–F–F–F (for five Front-end phases)–R–R–R–P–W (for Back-end phases) under the scheme.

3. The simulating kit description

Simulating kit includes the program and the methodical recommendations [22] – a manual with multiple task options. Simulating kit demonstrates the pipeline architecture, which consists of two clusters: Front-end and Back-end, and the principle of translating complex multi-cycle CISC-like instructions into simpler RISC-like micro operations (MOPs). In addition, it is possible to conduct a more detailed simulation of one of the three accelerating mechanisms in the processor core:

- L.R. №1 – multi-functional processing,
- L.R. №2 – out-of-order processing,
- L.R. №3 – speculative instructions execution after the branch prediction.

The simulating program includes four windows of the model, which reflect the input parameters and simulation results. The computer simulator considers a code fragment written in a simplified

Assembler as an initial data. The simulation program automatically generates a new version of the code fragment at startup or at the user's request. The "Task" window looks as shown in Figure 2. To study the pipeline executing principles, the following parameters of the pipeline and of the program code are additionally supposed to specify: the number of store/load devices, the number of ALU (Executive devices), and the percentage of operations with memory cells (Figure 2).

Program provides the following options as a result of the simulation: to explore the principle of translation of instructions in micro-operation ("MOPs" window), shown in Figure 3, to study the employment cycles of main back-end pipeline cluster when executing the given instructions ("Pipeline" window), shown in Figure 4, to examine the time diagram of instructions execution sequence from a given code fragment ("Diagram" window), shown in Figure 5.

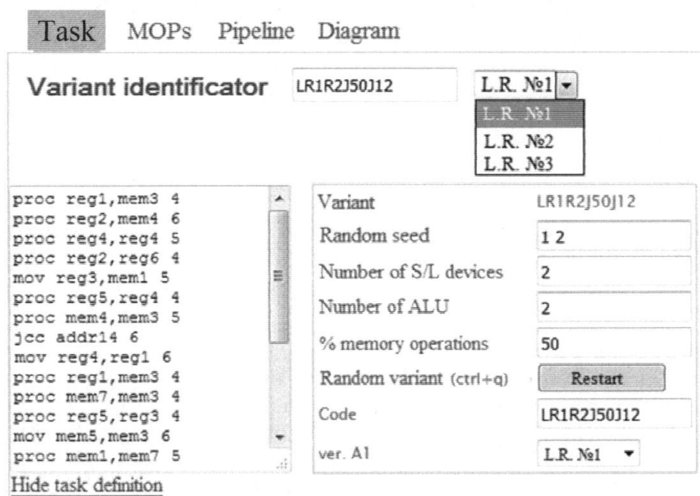

Figure 2. Simulator interface, "Task" window

The code fragment consists of about twenty instructions. Three main types of machine instructions [23] are considered: data transfer between registers and memory cells (four variations), data processing from registers and memory cells (four variations) and conditional jump to the specified address. These instructions differ from each other in a set of the MOPs which are carried out after decoding (Figure 3).

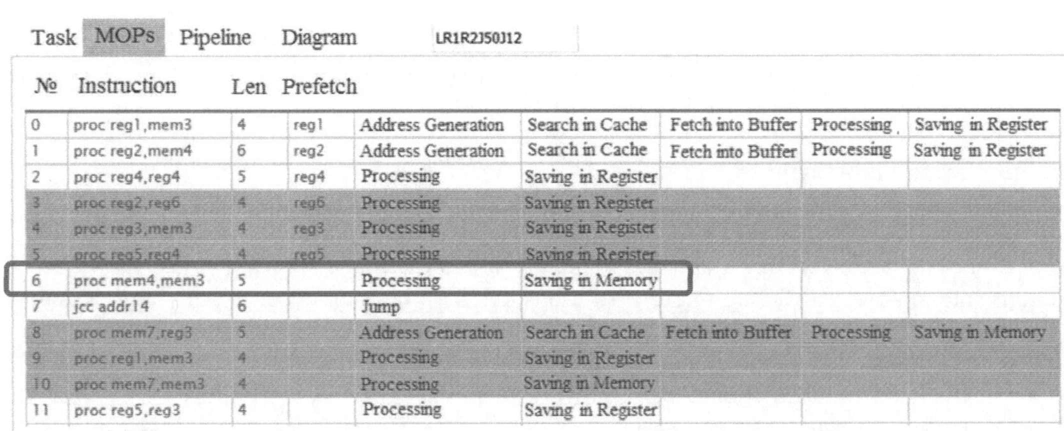

Task MOPs Pipeline Diagram LR1R2J50J12

№	Instruction	Len	Prefetch					
0	proc reg1,mem3	4	reg1	Address Generation	Search in Cache	Fetch into Buffer	Processing	Saving in Register
1	proc reg2,mem4	6	reg2	Address Generation	Search in Cache	Fetch into Buffer	Processing	Saving in Register
2	proc reg4,reg4	5	reg4	Processing	Saving in Register			
3	proc reg2,reg6	4	reg6	Processing	Saving in Register			
4	proc reg3,mem3	4	reg3	Processing	Saving in Register			
5	proc reg5,reg4	4	reg5	Processing	Saving in Register			
6	proc mem4,mem3	5		Processing	Saving in Memory			
7	jcc addr14	6		Jump				
8	proc mem7,reg3	5		Address Generation	Search in Cache	Fetch into Buffer	Processing	Saving in Memory
9	proc reg1,mem3	4		Processing	Saving in Register			
10	proc mem7,mem3	4		Processing	Saving in Memory			
11	proc reg5,reg3	4		Processing	Saving in Register			

. . .

Figure 3. Simulator interface, "MOPs" window

IOP Conf. Series: Journal of Physics: Conf. Series **1163** (2019) 012010 doi:10.1088/1742-6596/1163/1/012010

The 16-byte block of program code (which includes several instructions with a total length up to 16 bytes) is read from the L1i cache to Prefetch Buffer during each CPU clock cycle. Each window displays the executable code fragment. "MOPs" window shows relative instruction number (first column), instruction mnemonics (second column), instruction length (third column), prefetching register names (fourth column), and the columns 5-9 show us the executable MOPs in CPU Back-end cluster devices (Address Generation – GA, Searching data in Cache L1d – SC, Fetching data from cache into Buffer – FB, data Processing – P, Saving data in Register – SR or Saving data in Memory – SM). Sequentially fetching blocks of program code are depicted by alternating light and dark lines. Here one can explore the exception of data re-reading from L1d cache.

For example, instruction 6 must decode into eight MOPs (Table 1). But it's both operands (mem3, mem4) were already read by previous instruction0 (mem3) and instruction1 (mem4). Six MOPs are cancelled. CPU performs fewer operations. Program code executes faster.

Table 1. Theoretically possible and actually executed MOPs for instruction 6 (*proc mem4,mem3*).

Theoretically possible MOPs (*with re-reading*)	Removed MOPs	Actually executed MOPs (*without re-reading*)
1) mem4 address generation (6GA),	1) 6GA,	7) data processing (6P),
2) searching mem4 in cache (6SC),	2) 6SC,	8) saving result in memory (6SM)
3) load mem4 from cache into buffer (6FB),	3) 6FB,	
4) mem3 address generation (6GA),	4) 6GA,	
5) searching mem3 in cache (6SC),	5) 6SC,	
6) load mem3 from cache into buffer (6FB),	6) 6FB	
7) data processing (6P),		
8) saving result in memory (6SM)		

Figure 4 displays the "Pipeline" window: relative number of CPU clock cycles (first column), block of instructions (second column) and the columns 3-9 show us the the employment cycles of named CPU Back-end cluster devices (Read0…2, ALU0...1, SM, SR). Empty cells mean the halt of specified device in specified cycle. Each MOP has a number of the appropriate instruction.

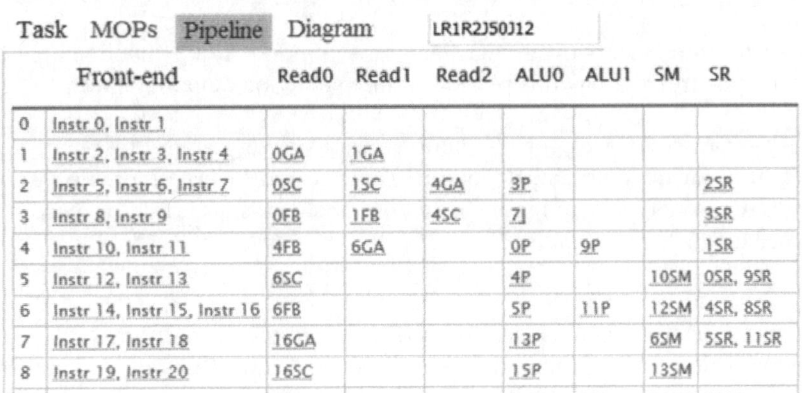

Figure 4. Simulator interface, "Pipeline" window

One can study the parallel cycle-by-cycle MOPs execution. Out-of-order MOPs processing can be tracked according to the number of the executed MOPs on each clock cycle. For example, five MOPs are executed simultaneously on a cycle number 4: loading data from cache L1d in buffer for instruction 4 (4FB), address generation for instruction 6 (6GA), data processing for instruction 0 (0P), data processing for instruction 9 (9P), saving data in register for instruction 1 (1SR). MOP 9P is processed out of order.

"Diagram" window allows one to examine the sequence and duration of instructions execution from given code fragment. Figure 5 (big picture) illustrates the diagram of the twenty instructions execution sequence: a relative instruction number (first column), instruction mnemonic (second column) and the CPU clock cycles from 0 to 16 (3-19 columns) show us the MOPs of instruction execution phase (F-fetch, R-read from memory, P-processing, W-result writing/saving). In the same Figure (small picture) one can see instructions execution sequence for a case of pipeline reset after misprediction for both front-end (F-F-F-F-F) and back-end (R-R-R-P-W) cluster. The situation of pipeline reset is represented by crossing out of the cancelled phases/MOPs (red colour letters).

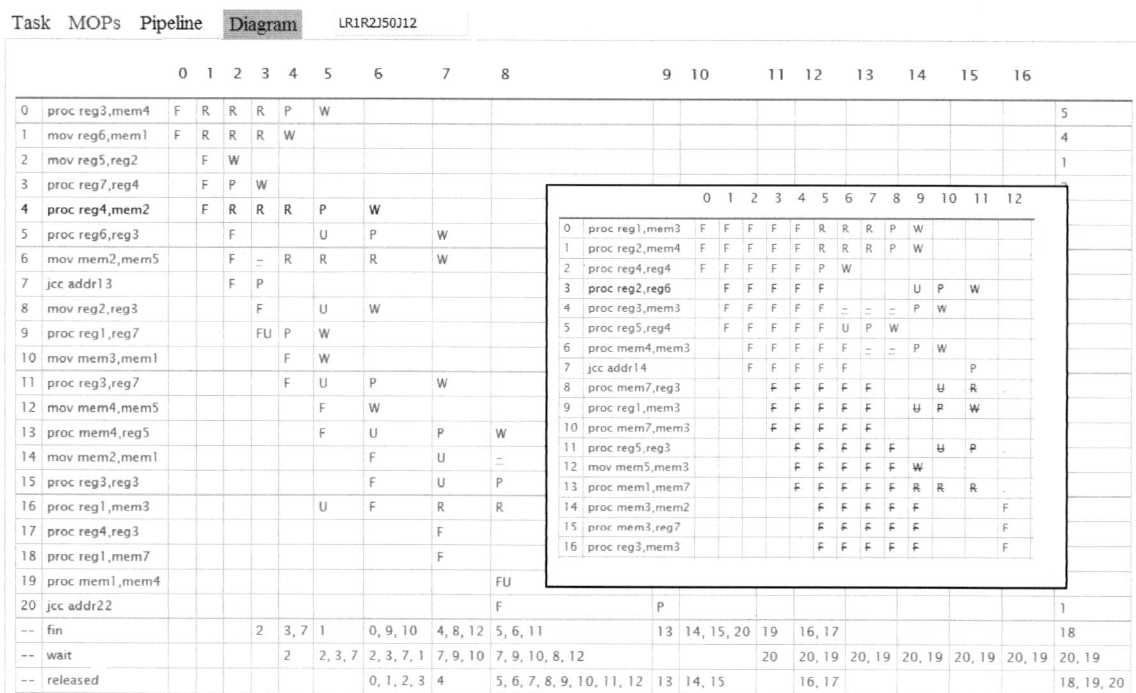

Figure 5. Simulator interface, two variants "Diagram" window: multi-functional processing (big picture) and pipeline reset after misprediction (small picture)

The sequence in-order recovering for instructions results saving after OOO execution is shown in two lines in the bottom of the diagram. Recovering is executed in two steps: fin (wait in the queue) and release. Diagram makes it easy to realize how common data and devices are used by instructions. One can easily trace the instruction halt.

4. Conclusion

The offered simulation of the super-scalar processor core allows to study the CPU core execution in dynamics cycle-by-cycle. This Simulating kit is useful for studying computer architecture. Understanding the features of the processor is useful for both system developers and programmers. Improving the style of writing programs will let the computer speed up their execution. The Simulating kit is used when teaching the discipline "Computing systems" in Higher School of Economics.

5. Acknowledgments

Authors thank the HSE student Lazutov A.N. for help in creation of the kit program part.

6. References

[1] Intel® 64 and IA-32 Architectures Optimization Reference Manual. 2018. https://docslide.net/download/link/intel-64-and-ia-32-architectures-optimization-reference-64-and-ia-32-architectures
[2] Korogod Sergey New in the Silvermont architecture of Intel Atom Z3xxx Processors. 2013. http://www.ixbt.com/portopc/atom-clover5.shtml
[3] Swinburne Richard. Feature - Intel Core i7 - Nehalem Architecture Dive. bit-tech. 2008. http://www.bit-tech.net/reviews/tech/cpus/intel-core-i7-nehalem-architecture-dive/5/
[4] Ozerov Sergei. CPU architecture. Computerra. 2005, 37(609). http://www.kinnet.ru/cterra/609/233266.html
[5] Kanter David. Intel's Sandy Bridge Microarchitecture. Website Real World Tech. 2010. https://www.realworldtech.com/sandy-bridge/10/
[6] Patterson David A., Hennessy John L. Computer Architecture: A Quantitative Approach, 5th Edition. — Morgan Kaufmann, 2011.
[7] Tanenbaum A.S., Austin T. Structured Computer Organization. Prentice Hall, 2012.
[8] Tullsen D.M., Eggers S.J., Levy H.M. "Simultaneous multithreading: Maximizing on-chip parallelism". 22nd Annual International Symposium on Computer Architecture. IEEE. 1995: 392–403.
[9] The Complete Open-Source Software Platform. https://sourceforge.net/
[10] Rappoport Lihu. Intel Core Microarchitecture. https://moodle.technion.ac.il/enrol/index.php?id=5916
[11] Skylake (client) - Microarchitectures – Intel. https://en.wikichip.org/wiki/intel/microarchitectures/skylake_(client)
[12] Zen - Microarchitectures – AMD. https://en.wikichip.org/wiki/amd/microarchitectures/zen
[13] Abeydeera Maleen, Subramanian Suvinay, Jeffrey Mark C., Emer Joel, Sanchez Daniel, "SAM: Optimizing Multithreaded Cores for Speculative Parallelism", Parallel Architectures and Compilation Techniques (PACT) 2017 26th International Conference. 2017: 64-78.
[14] Intel Sandy Bridge microarchitecture overview. Out-of-Order Cluster. https://hw-lab.com/intel-sandy-bridge-cpu-microarchitecture.html/4
[15] Coursera. Branch Prediction Introduction https://www.coursera.org/lecture/comparch/branch-prediction-introduction-BF58C
[16] Branch Prediction Review https://courses.cs.washington.edu/courses/csep548/06au/lectures/branchPred.pdf
[17] Intel Analysis of Speculative Execution Side Channels. White Paper. 2018. https://www.intel.com/content/www/us/en/architecture-and-technology/intel-analysis-of-speculative-execution-side-channels-paper.html?wapkw=speculative+execution
[18] Polyvyalov Alexander. In anticipation of the Willamette — the history of the IA-32 architecture and the way the P6 family processors work. 2000. https://www.ixbt.com/cpu/pentium4-1.html
[19] Garmatyuk Stanislav. The "new old" Core i7 architecture: what's more-similarities or differences? Website ixbt.com/PC Platform. 2008. https://www.ixbt.com/cpu/intel-ci7-theory.shtml
[20] Nechai Oleg. Selection of author's articles http://www.libma.ru/kompyutery_i_internet/cifrovoi_zhurnal_kompyuterra_74/p2.php
[21] Hinton Glenn, Sager Dave, Upton Mike, etc. The Microarchitecture of the Pentium4 Processor. Intel Technology Journal Q1, 2001 http://www.ecs.umass.edu/ece/koren/ece568/papers/Pentium4.pdf
[22] Ivanova E. M. Core superscalar CPU simulation. Methodical recommendations on "Computer systems" for the direction 09.04.01. "Computer science and engineering" master's degree. M.: HSE, 2018. https://publications.hse.ru/en/books/214154058
[23] Intel 64 and IA-32 Architectures Software Developer's Manual. http://www.intel.ie/content/dam/www/public/us/en/documents/manuals/64-ia-32-architectures-software-developer-manual-325462.pdf

IOP Conf. Series: Journal of Physics: Conf. Series **1163** (2019) 012011 doi:10.1088/1742-6596/1163/1/012011

New type of centrifugal instability in a thin rotating spherical layer

D Zhilenko[1], O Krivonosova[1] and M Gritsevich[2,3]

[1] Moscow State University, GSP-1, Leninskie Gory, 119991,Moscow, Russia
[2] Department of Physics, University of Helsinki, Gustaf Hällströmin katu 2a, FI-00014 Helsinki, Finland
[3] Institute of Physics and Technology, Ural Federal University,Mira str. 19, 620002, Ekaterinburg, Russia

E-mail: jilenko@imec.msu.ru

Abstract. We numerically model the instability of viscous incompressible fluid flows caused by torsional oscillations of the inner sphere in a thin spherical layer with respect to the state of rest. We show that an increase in the frequency of torsional oscillations leads to a change in the mode of the instability, with a transition from secondary flows in the form of Taylor vortices to the structures, which were not previously observed. The revealed instability is found in the frequency range from 0.61 to 2.45 Hz or, if the wavelengths are taken relative to the layer thickness, from 0.67 to 1.33.

1. Introduction

Flows caused by torsional oscillations can be used to determine rheological properties of fluids [1] and for intensifying filtration [2]. Periodic perturbations in the velocity may affect the transition to turbulence [3, 4]. Therefore it is important to have information on the stability limit and on the structure of secondary flows [1, 5].

A possible dependence of the secondary flow structures and the types of instability of the oscillation frequency is of special interest. Such dependence was revealed based on the measurements of velocity fields in a cylindrical thin layer $\beta = (r_2 - r_1)/r_1 = 0.087$, where r_1 and r_2 are the inner and outer radii [1, 5]. The structures resembling Taylor and Görtler vortices were observed in the low and high frequency limits, respectively. Subsequently, there was a knowledge gap about the structures in the intermediate frequency range. In a spherical layer with $\beta = 4.3$ [6] at torsional oscillations of the inner sphere the structure of secondary flows in two-dimensional calculations remains constant over a wide range of frequency variations. At other values of β, the dependence of the instability type on the frequency in spherical layers has not been previously studied.

The aim of this work is a numerical investigation of how the form of the secondary flow structures and stability limit depend on the frequency at torsional oscillations of the inner sphere with respect to the state of rest. To enable qualitative comparison of our results with [1, 5], we model a thin layer $\beta = 0.19$ in which Taylor vortices appear at a constant rotational speed of the inner sphere at the stability limit [7].

2. Methodology

IOP Conf. Series: Journal of Physics: Conf. Series **1163** (2019) 012011 doi:10.1088/1742-6596/1163/1/012011

An isothermal flow of a viscous incompressible fluid is described by the Navier–Stokes and continuity equations:

$$\frac{\partial \boldsymbol{U}}{\partial t} = \boldsymbol{U} \times \mathrm{rot}\boldsymbol{U} - \mathrm{grad}\left(\frac{p}{\rho} + \frac{\boldsymbol{U}^2}{2}\right) - \nu\mathrm{rotrot}\boldsymbol{U},\ \mathrm{div}\boldsymbol{U} = 0.$$

We use a spherical coordinate system with radial r, polar θ, and azimuthal φ directions. The no-slip and non percolation conditions at the boundaries have the form: $u_\varphi(r = r_{1,2}) = \Omega_{1,2}(t)r_{1,2}\sin\theta$, $u_r(r = r_{1,2}) = 0$, $u_\theta(r = r_{1,2}) = 0$, where 1 denotes conditions for the inner sphere and 2 – for the outer one. Here, U, p, and ρ are the velocity, pressure, and density of the fluid; u_φ, u_r, and u_θ are the azimuthal, radial, and polar components of the velocity, respectively. Ω_k is the angular rotational speed of the corresponding sphere and ν is the kinematic viscosity of the fluid in the layer. The rotational speed of the inner sphere varies periodically: $\Omega_1(t) = A\sin(2\pi ft)$, where A and f are the modulation amplitude and the frequency. The outer sphere is immobile: $\Omega_2 = 0$. We have used the algorithm and numerical solution program [8] based on a finite-difference spatial discretization scheme for the Navier–Stokes equations and a semi-implicit Runge–Kutta scheme for integration with respect to time. The spatial discretization was performed with decreasing cell size near the boundaries (with respect to r) and the equatorial plane (with respect to θ). The ratio of the maximum to minimum cell size varied from 2 to 5 with a total number of nodes up to $4.6\text{x}10^5$. The computations were performed at the following dimension parameters: $\nu = 5 \times 10^{-5}$ m^2/s, $r_1 = 0.126$ m, $r_2 = 0.15$ m, and $f = 0.0682$–4.363 Hz. In this case, thickness of the dynamic boundary layer is $\delta = (\nu/(\pi f))^{1/2} = (1.91-15.3)10^{-3}\,\mathrm{m}$ and minimum size δ corresponds to the seven nodes of the computational grid. The character of the flow depends on the quantities A, f, r_1,r_2 and ν and is determined by the three similarity parameters. These similarity parameters are relative layer thickness β, dimensionless wavelength $\mathrm{L} = \lambda/(r_2 - r_1)$, where $\lambda = 2\pi\delta$, and the Reynolds number $\mathrm{Re} = (Ar_1^2/\nu)(\delta/r_1)$ [6].

3. Results and Discussion

Oscillations of the inner sphere cause circulation in the meridional plane of the flow [6, 9]. The circulation is similar to that at a constant rotational speed [7], but periodically varies with a double modulation frequency. The evolution of such flow in time before it loses its stability has already been well studied (see e.g. [6]). The flow structure was determined by the form of the azimuthal vorticity component [10] in the meridional plane ω_φ:

$$\omega_\varphi = \frac{1}{rr'}\frac{\partial ru_\theta}{\partial r} - \frac{1}{r\theta'}\frac{\partial u_r}{\partial \theta}$$

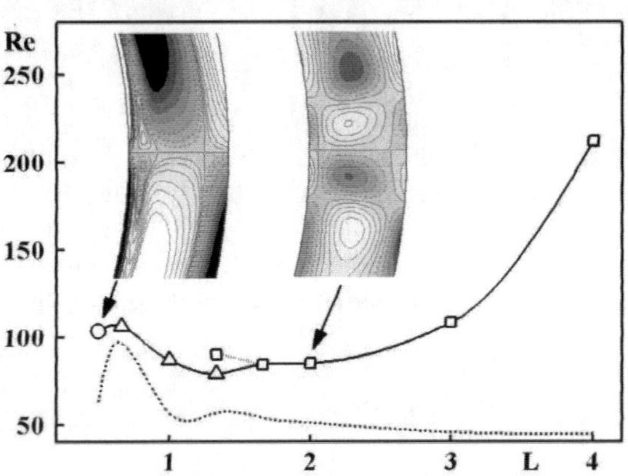

Figure 1. Stability limit (solid line) and difference of oscillation phases E_φ and E_ψ (dotted line). Shown here are levels ω_φ [s^{-1}] ($-10 < \omega_\varphi < 10$). On the right: Taylor vortices at L = 2, Re = 84.7, and $\Delta\omega_\varphi = 2$ (squares).On the left: Görtler vortices at L = 0.5, Re = 105.9, and $\Delta\omega_\varphi = 0.5$ (circle). The dotted lines for the levels ω_φ correspond to the counterclockwise direction. The triangles show the instability at intermediate wavelengths.

The attenuation decrement of the azimuthal velocity in the radial direction increases with increasing f until loss of stability [9]. This leads to a decrease in M_{out} (i.e., in the amplitude of the friction torque transferred to the outer sphere). Near the stability limit (both before and after it), the flow is symmetric with respect to the rotation axis and equatorial plane. All flow parameters vary with frequencies f or $2f$. Following [8], supercritical values of Re were used as initial conditions. The stability limit was determined by decreasing Re at f = const according to the form of the flow structure, magnitude of M_{out}, and ratio E_ψ/E_φ (where $E_\varphi = \int u_\varphi^2$ and $E_\psi = \int (u_r^2 + u_\theta^2)$ are the azimuthal and meridional components of the kinetic energy of the flow determined by integration over the whole volume of the spherical layer).

Our results demonstrate that, after the loss of stability, unsteady toroidal structures are formed in the flow near the equatorial plane. Their characteristic size decreases with a decrease in L and the meridional circulation (MC) is pushed back to the poles. In the case of long waves ($1.66 \leq L \leq 4$), the thoroughly studied Taylor vortices are formed at the stability limit. Their direction of rotation is opposite of the MC direction (Figure 1). The formation and decay of the vortices occur with hysteresis. At $L \leq 2$, the vortices exist during the whole oscillation cycle. At L = 4, the vortices are observed during almost the whole cycle, with the exception of two small intervals when $\Omega_1(t)$ approaches A ($0.075 < tf < 0.177$ and $0.581 < tf < 0.676$, where t is the time from the beginning of the oscillation cycle, the left boundaries of the intervals correspond to the decay of vortices, and the right boundaries correspond to the beginning of the formation). At L = 3, the vortices observed in the region of the maximum of $\Omega_1(t)$ ($0.12 < tf < 0.3$ and $0.6 < tf < 0.8$). The Taylor vortices change their configuration during the oscillation period: during the formation, they are expanded in the radial direction; by the instant of decay, the dimensions in r and θ are close. The stability limit for Taylor vortices has a minimum at $1.66 \leq L \leq 2$ (Figure 1). Taylor vortices are also observed at L = 1.33. In this case, however, they are formed above the stability limit: the leftmost square in Figure 1 (Re = 91.1) and squares in Figure 2. At the same time, the flow is stable at Re ≤ 80.9.

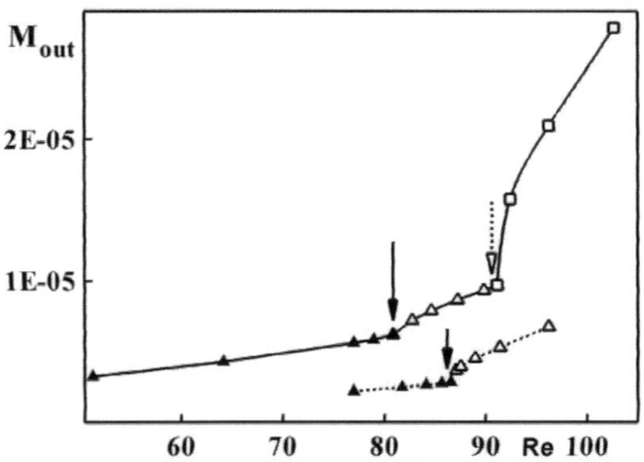

Figure 2. Quantity M_{out} normalized by the liquid density as a function of the Re number at L = 1.33 (solid line) and 1.0 (dotted line). The dark symbols correspond to the stable flow and the light ones - to the unstable flow. The first and second bifurcations are shown by the arrows with a solid and dotted lines, respectively.

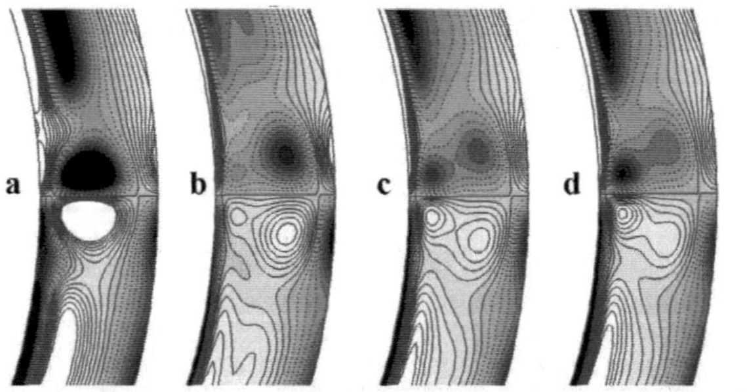

Figure 3. Distribution of the levels ω_φ [s^{-1}] in the meridional plane of the flow at L = 1.33 and Re = 80.8, $-8 < \omega_\varphi < 8$, and $\Delta\omega_\varphi = 1$. tf = (a) 0, 0.5; (b) 0.25, 0.75; (c) 0.28, 0.78; and (d) 0.3, 0.8. t is the time from the beginning of the oscillation cycle.

Thus, at L = 1.33, the transition to Taylor vortices is preceded by another type of instability. This type of instability is observed in the interval of intermediate wavelengths $0.67 \leq L \leq 1.33$ (triangles in Figures 1, 2).The structure of the secondary flow in this case is shown in Figure 3. The direction of the vortex rotation near the equatorial plane coincides with the MC direction and the characteristic scale does not exceed half of the layer thickness. Let us consider the evolution of these structures in time. With an increase in $\Omega_1(t)$ from 0 (Figure 3a) to A (Figure 3b), individual vortices are pushed back from the inner sphere and from the equatorial plane. Two vortices are formed. The vortex with the maximum ω_φ is closer to the outer boundary. Later, with a decrease in $\Omega_1(t)$ (Figure 3c), a rapid redistribution of ω_φ occurs: the maximum is shifted to the inner sphere, then this maximum enhances (Figure 3d), and the flow returns to the initial state (Figure 3a). The minimum of the stability limit is observed near L = 1.33. At L < 1.33, an increase in Re is accompanied not by the transition to Taylor vortices but by the transition to a flow without the axial and equatorial symmetries. At the same time, the direction of vortex rotation in the secondary flow remains the same.

In the case of short waves (L = 0.5), at the stability limit after passing the extreme values ($0.28 < tf < 0.31$ and $0.78 < tf < 0.81$), vortices with a characteristic scale that is much less than the layer thickness are formed near the inner sphere and equatorial plane (Figure 1). The typical size, rotation direction (opposite to the MC direction), and sharp change in the quantity ω_φ near the inner sphere may imply that Görtler vortices caused by the instability of the dynamic boundary layer are observed in this case. Probably, at L = 0.67, the instability in the form of Görtler vortices precedes the instability at intermediate wavelengths (Figure 3), just as the instability at intermediate wavelengths precedes Taylor vortices at L = 1.33 (Figure 1).

4. Summary

At torsional oscillations of the inner sphere in a thin layer $\beta = 0.19$, three types of instability are observed, and each type is associated with its own variation range of dimensionless wavelength L. The boundaries of the ranges can be determined by the position of maxima of both, the stability limit and the phase difference between E_φ and E_ψ depending on L (Figure 1). The toroidal structures forming at the stability limit near the equatorial plane in the case of short (L ≤ 0.5) and long (L ≥ 1.67) waves qualitatively correspond to the results of experiments [1, 5]. The rotation direction for them is opposite to the MC direction. At the instability revealed at intermediate wavelengths (0.67 ≤ L ≤ 1.33 and $2.45 \geq f \geq 0.61$ Hz), rotation directions of the vortices and MC coincide. The results of our calculations imply that further experimental verification of the revealed instability relies on measuring not only velocity fields (as it was done, e.g., in [1, 5]), but also friction torques.

Acknowledgments

This work was supported by the Russian Foundation for Basic Research project No. 16-05-00004 and 18-08-00074. MG acknowledges, in part, support from the ERC Advanced Grant No. 320773. Research at the Ural Federal University is supported by the Act 211 of the Government of the Russian Federation, agreement No 02.A03.21.0006. We thank Dr. Laura K. Zschaechner (University of Helsinki, Finnish Center for Astronomy with ESO) for her help with language corrections.

References

[1] Fardin M, Perge C, Casanellas L, Hollis T, Taberlet N, Ortin J, Lerouge S and Manneville S 2014 *Rheol. Acta* **53** 885
[2] Jaffrin M 2012 *Ann. Rev. Fluid Mech.* **44** 77
[3] Zhilenko D Yu and Krivonosova O E 2015 *Tech. Phys. Lett.* **41** 5
[4] Bountin D A and Maslov A A 2017 *Tech. Phys. Lett.* **43** 623
[5] Fardin M, Perge C, Taberlet N and Manneville S 2014 *Phys. Rev.E* **89** 011001
[6] Hollerbach R, Wiener R, Sullivan I, Donnelly R, and Barenghi C 2002 *Phys. Fluids* **14** 4192
[7] Belyaev Yu N and Yavorskaya I M 1980 *Itogi Nauki Tekh. Ser. Mekh. Zhidk. Gaza* **15** 3
[8] Nikitin N 2006 *J. Comp. Phys.* **217** 759
[9] Zhilenko D Yu and Krivonosova O E 2016 *JETP Lett.* **104** 527
[10] Landau L D and Lifshitz E M 1986 *Course of Theoretical Physics*, *Vol. 6: Fluid Mechanics* Pergamon, New York.

IOP Conf. Series: Journal of Physics: Conf. Series **1163** (2019) 012012 doi:10.1088/1742-6596/1163/1/012012

On the Possibility of a Numerical Solution of the Heat and Mass Transfer Problem with the Combined Matrix&Generalized Powers of Bers Method

V V Kalmanovich[1], E V Seregina[2], M A Stepovich[1]

[1] Tsiolkovsky Kaluga State University, 26 Stepan Razin Street, Kaluga, 248023 Russia
[2] Bauman Moscow State Technical University (National Research University), Kaluga Branch, 2 Bazhenov Street, Kaluga, 248000 Russia

[1]v572264@yandex.ru, m.stepovich@rambler.ru,
[2]evfs@yandex.ru

Abstract. Some possibilities for the combined use of the matrix method and the method of generalized powers of Bers for the numerical solution of the heat and mass transfer problem is considered. Initially, the matrix method under consideration was used by us for the analytical solution of the problem of heat and mass transfer in a homogeneous or multilayered medium with translation, axial or central symmetry for an arbitrary number of layers. Simulation is reduced to the sequential multiplication of second-order functional matrices whose components at each point are determined by the physical and geometric parameters of the current layer. The main advantage of the matrix method is that it can be used for any number of layers. It is this property of the method that makes it possible to consider of its application as a numerical one if the medium is artificially splitted into many layers.
For modelling, the problem of one-dimensional diffusion of minority charge carriers generated by a wide electron beam in a semiconductor material was taken. The relative error in the computations was estimated by a uniform norm in comparison with the analytical solution. The errors made were from 0.8% to 2.9%.

1. Introduction

In this paper, we consider some possibilities for combining the matrix method and the method of generalized powers of Bers ([1], [2]) for the numerical solution of the heat and mass transfer problem by solving the problem of diffusion of minority charge carriers (MCC) generated by a wide electron beam in a semiconductor target. The idea of the matrix method for solving heat transfer problems in multilayer medium is found in [3]. Initially, the matrix method under consideration was used by us for the analytical solution of the problem of heat and mass transfer in a homogeneous or multilayered medium with translation, axial or central symmetry for an arbitrary number of layers [4]. Such a generality of the method is achieved due to the application of the apparatus of generalized powers of Bers, which makes it possible to obtain a solution for all the above types of symmetries of the medium in a single analytical form. The main advantage of the matrix method is that it can be used for any number of layers. It is this property of the method that makes it possible to consider of its application as a numerical one if we artificially split the medium into many layers.

2. The method of calculation generalized powers of Bers

Let the operators

$$D_1 = a_1(x)\frac{\mathrm{d}}{\mathrm{d}x},\ D_2 = a_2(x)\frac{\mathrm{d}}{\mathrm{d}x},$$

be given in a linear function space of functions $f(x)$ defined, continuous, and differentiable in the interval (a, b). Here, $a_1(x)$ and $a_2(x)$ are the positive functions in the interval. There are the right inverse operators I_1 and I_2, in what follows, these are operators of the simplest form

$$I_1 = \int_{x_1}^{x} \frac{\mathrm{d}\xi}{a_1(\xi)}\cdots,\ I_2 = \int_{x_1}^{x} \frac{\mathrm{d}\xi}{a_2(\xi)}\cdots.$$

The generalized powers of Bers with the zero point x_1 are a sequence of functions defined by the expressions

$$X^{(0)}(x,x_1) = \tilde{X}^{(0)}(x,x_1) = 1$$

$$X^{(n)}(x,x_1) = n!\begin{cases}(I_1 I_2)^p \cdot 1, \text{ at } n = 2p\\ I_1(I_2 I_1)^p \cdot 1, \text{ at } n = 2p+1.\end{cases} \qquad \tilde{X}^{(n)}(x,x_1) = n!\begin{cases}(I_2 I_1)^p \cdot 1, \text{ at } n = 2p\\ I_2(I_1 I_2)^p \cdot 1, \text{ at } n = 2p+1.\end{cases}$$

The rules of differentiation are as follows

$$D_1 X^{(n)}(x,x_1) = n\tilde{X}^{(n-1)}(x,x_1),\ D_2\tilde{X}^{(n)}(x,x_1) = nX^{(n-1)}(x,x_1).$$

The power series $f(x) = \sum_{i=0}^{\infty} c_i (x - x_1)^i$ is associated with a series of generalized powers $f(x) = \sum_{i=0}^{\infty} c_i X^{(i)}(x,x_1)$. In particular, series $\cosh\lambda x$, $\sinh\lambda x$, where $\lambda = \text{const}$, are associated with a series of generalized powers of the form

$$\cosh\lambda X(x,x_1) = \sum_{i=0}^{\infty} \frac{\lambda^{2i}}{(2i)!} X^{(2i)}(x,x_1),\ \sinh\lambda X(x,x_1) = \sum_{i=0}^{\infty} \frac{\lambda^{2i+1}}{(2i+1)!} X^{(2i+1)}(x,x_1).$$

The functions $\cosh\lambda X(x,x_1)$ and $\sinh\lambda X(x,x_1)$ have differential properties:

$$D_1\cosh\lambda X(x,x_1) = \lambda\sinh\lambda\tilde{X}(x,x_1),\quad D_1\sinh\lambda X(x,x_1) = \lambda\cosh\lambda\tilde{X}(x,x_1),$$
$$D_2\cosh\lambda\tilde{X}(x,x_1) = \lambda\sinh\lambda X(x,x_1),\quad D_2\sinh\lambda\tilde{X}(x,x_1) = \lambda\cosh\lambda X(x,x_1),$$
$$D_2 D_1\cosh\lambda X(x,x_1) = \lambda^2\cosh\lambda X(x,x_1),\quad D_2 D_1\sinh\lambda X(x,x_1) = \lambda^2\sinh\lambda X(x,x_1).$$

In the case when the functions $a_1(x)$, $a_2(x)$ are constant, the following equalities are valid:

$$\cosh mX(x,x_1) = \cosh m\tilde{X}(x,x_1) = \cosh\frac{m(x - x_1)}{(a_1 a_2)^{1/2}},$$

$$\sinh mX(x,x_1) = \left(\frac{a_2}{a_1}\right)^{1/2}\sinh\frac{m(x - x_1)}{(a_1 a_2)^{1/2}},\quad \sinh m\tilde{X}(x,x_1) = \left(\frac{a_1}{a_2}\right)^{1/2}\sinh\frac{m(x - x_1)}{(a_1 a_2)^{1/2}}.$$

3. Statement of the problem and the method of calculation

We consider the one-dimensional stationary heat and mass transfer process in a multilayer medium, determined by the equation

$$D_2^{(i)} D_1^{(i)} \Phi^{(i)}(x) - m_i^2 \Phi^{(i)}(x) = \mu^{(i)}(x),\ i = \overline{1,n}, \tag{1}$$

where the potential $\Phi^{(i)}(x)$ is the unknown function,

$$D_1^{(i)} = a_1^{(i)}(x)\frac{\mathrm{d}}{\mathrm{d}x},\ D_2^{(i)} = a_2^{(i)}(x)\frac{\mathrm{d}}{\mathrm{d}x},$$

$a_1^{(i)}(x)$, $a_2^{(i)}(x)$ are positive functions determined by the physical and geometric parameters of the layers, $\mu^{(i)}(x)$ is a quantity proportional to the volume power density of the source distribution, and i is the layer number. The flow is defined by the formula $J^{(i)}(x) = -D_1^{(i)}\Phi^{(i)}(x)$. At the point of contact between layers, continuity condition for the potential and of the flow are given

$$\Phi^{(i)}(x_{i+1}) = \Phi^{(i+1)}(x_{i+1}),\ J^{(i)}(x_{i+1}) = J^{(i+1)}(x_{i+1}).$$

Let the values of the potential $\Phi^{(1)}(x_1)$ and flow $J^{(1)}(x_1)$ be specified, i.e., let the Cauchy problem be posed for Eq. (1).

We introduce the column vectors V and W and the matrix K

$$V^{(i)}(x) = \begin{pmatrix} \Phi^{(i)}(x) \\ J^{(i)}(x) \end{pmatrix},\ W^{(i)}(x) = \begin{pmatrix} w^{(i)}(x) \\ -D_1^{(i)} w^{(i)}(x) \end{pmatrix},\ K^{(i)}(x,x_i) = \begin{pmatrix} \cosh mX_i(x,x_i) & -\frac{1}{m}\sinh mX_i(x,x_i) \\ -m\sinh m\tilde{X}_i(x,x_i) & \cosh m\tilde{X}_i(x,x_i) \end{pmatrix}.$$

Here $w^{(i)}(x)$ is a particular solution of Eq. (1) for the ith layer and $X_i(x,x_i)$, $\tilde{X}_i(x,x_i)$ are the generalized power and the adjoint generalized power in the segment $[x_i, x_{i+1}]$, respectively. Then the solution the Cauchy problem in fist layer is

$$V^{(1)}(x) = K^{(1)}(x,x_1)\left(V^{(1)}(x_1) - W^{(1)}(x_1)\right) + W^{(1)}(x).$$

Then in second layer, considering the conditions of continuity at a point x_2, we obtain

$$V^{(2)}(x) = K^{(2)}(x,x_2)\left(V^{(2)}(x_2) - W^{(2)}(x_2)\right) + W^{(2)}(x) =$$
$$= K^{(2)}(x,x_2)\left(K^{(1)}(x_2,x_1)\left(V^{(1)}(x_1) - W^{(1)}(x_1)\right) + W^{(1)}(x_2) - W^{(2)}(x_2)\right) + W^{(2)}(x) =$$
$$= K^{(2)}(x,x_2)K^{(1)}(x_2,x_1)\left(V^{(1)}(x_1) - W^{(1)}(x_1)\right) + K^{(2)}(x,x_2)\left(W^{(1)}(x_2) - W^{(2)}(x_2)\right) + W^{(2)}(x).$$

Moving successively from layer to layer and performing similar substitutions at the points of contact of the layers, we obtain

$$V^{(i)}(x) = L^{(i,1)}(x,x_1)V^{(1)}(x_1) + \sum_{j=1}^{i} L^{(i,j)}(x,x_j)\left(W^{(j-1)}(x_j) - W^{(j)}(x_j)\right) + W^{(i)}(x), \tag{2}$$

where $W^{(0)}(x_1) = 0$, $L^{(i,j)}(x,x_j) = K^{(i)}(x,x_i)K^{(i-1)}(x_i,x_{i-1}) \ldots K^{(j)}(x_{j+1},x_j)$, $i \geq j$.

Let $K^{(u)}(x_{n+1}) = L^{(n,1)}(x_{n+1}, x_1)$ denote the general K-matrix of the whole system of layers and $W^{(u)}(x_{n+1}) = \sum_{j=1}^{n} L^{(n,j)}(x_{n+1}, x_j)\left(W^{(j-1)}(x_j) - W^{(j)}(x_j)\right) + W^{(n)}(x_{n+1})$ be the vector of additional flows. Then, at the endpoint of the system of layers, we obtain

$$V(x_{n+1}) = K^{(u)}(x_{n+1})V^{(1)}(x_1) + W^{(u)}(x_{n+1}). \qquad (3)$$

Formula (3) yields an analytical solution of the Cauchy problem for Eq. (1) for an arbitrary number of layers and allows one to find two unknown value of $\Phi^{(1)}(x_1)$, $\Phi^{(n)}(x_{n+1})$, $J^{(1)}(x_1)$, $J^{(n)}(x_{n+1})$ for given others two value.

In numerical calculation by this method medium is divided at the grid points of the selected number of layers, the calculations are carried out by the formulas (2) and (3). The particular solution, included in the vector W, can be found analytically, or may be constructed to function approximating the right side of equation (1).

4. Results of calculations and their discussion

We considered the problem of diffusion of MCC generated by a wide electron beam in a semiconductor target, which can be described by the model of collective motion of charge carriers. According to this model, electrons (or holes) from one microzone of the material affect the diffusion of nonequilibrium MCC generated from any microvolume of the semiconductor material by external energy. Then, for one-dimensional diffusion in a semiconductor, the distribution of minority charge carriers in depth $\Delta p(z)$ is found as a solution of the differential equation

$$D\frac{d^2\Delta p(z)}{dz^2} - \frac{\Delta p(z)}{\tau} = -\rho(z) \qquad (4)$$

with boundary conditions

$$D\left.\frac{d\Delta p(z)}{dz}\right|_{z=0} = v_s \Delta p(0), \quad \Delta p(l) = 0. \qquad (5)$$

Here D, τ and v_s are the electrophysical parameters of the semiconductor target: the diffusion coefficient, the lifetime and the surface recombination rate of the MCC, respectively, and $\rho(z)$ is the number of MCC generated per unit time in a thin target layer at a depth z. The values $\rho(z)$ can be determined from the relationship for the energy density $\rho^*(z)$ released in this target layer per unit time, by dividing $\rho^*(z)$ by the formation energy of the electron-hole pair. Semiempirical formula [5] is valid for a wide electron beam

$$\rho^*(z) = \frac{1.085(1-\eta)P_0}{\sqrt{\pi}\left(1-\eta+\eta z_{ss}/z_{ms}\right)}\left(\exp\left[-\left(\frac{z-z_{ms}}{z_{ms}}\right)^2\right] + \frac{\eta}{\eta-1}\exp\left[-\left(\frac{z-z_{ss}}{z_{ss}}\right)^2\right]\right), \qquad (6)$$

where η is the backscattering coefficient of the beam electrons, z_{ms} is the depth of the maximum energy loss of the primary electrons that have experienced small-angle scattering and are absorbed by the target, z_{ss} is the depth of the maximum energy losses by backscattered electrons, and P_0 is the power of the beam electrons.

Applying the described method for the numerical solution of equation (4) – (5), we obtain that $a_1^{(i)}(z)$ is equal to the diffusion coefficient $D^{(i)}$ in the ith segment of partition, and $a_2^{(i)}(z)$ is equal to the lifetime of the MCC in the ith segment of partition, $m = 1$. Since the coefficients of the equation in each segment are constant the matrix K takes the form under substitution $a_1^{(i)}(z)$ and $a_2^{(i)}(z)$ in its elements:

$$K^{(i)}(z_{i+1}, z_i) = \begin{pmatrix} \cosh\dfrac{z_{i+1}-z_i}{\left(D^{(i)}\tau^{(i)}\right)^{1/2}} & -\left(\dfrac{\tau^{(i)}}{D^{(i)}}\right)^{1/2}\sinh\dfrac{z_{i+1}-z_i}{\left(D^{(i)}\tau^{(i)}\right)^{1/2}} \\ -\left(\dfrac{D^{(i)}}{\tau^{(i)}}\right)^{1/2}\sinh\dfrac{z_{i+1}-z_i}{\left(D^{(i)}\tau^{(i)}\right)^{1/2}} & \cosh\dfrac{z_{i+1}-z_i}{\left(D^{(i)}\tau^{(i)}\right)^{1/2}} \end{pmatrix}.$$

To evaluate the possibilities of the proposed method, we considered the solution of equation (4) – (5) for a homogeneous semiconductor under the action by a wide electron beam with energies of 10 keV, 20 keV, and 30 keV. The calculations were performed with typical electrophysical parameters for the chosen target type (silicon) [6]: the lifetime of the MCC $\tau = 10^{-8}$ s, diffusion coefficient $D = 10^8$ μm^2/s, diffusion length $L = \sqrt{D\tau} = 1$ μm, surface recombination rate $v_s = 10^{10}$ μm/s.

Conditions of the third type, according to formula (3), irrespective of the number of mesh nodes, lead to the solution of one linear equation with one unknown – the concentration $\Delta p(0)$ of the MCC on the surface of the semiconductor target. After solving it, we can calculate by formula (2) at all mesh nodes.

The function $\rho^*(z)$ (see formula (6)) was approximated on each interval (z_i, z_{i+1}) by a constant function equal to the value of the function $\rho^*(z)$ at the midpoint between neighboring mesh node, i.e. in a point $0,5(z_i + z_{i+1})$.

Calculations were carried out in the mathematical package Maple 18.00 (Widows 7, Service Pack 1, processor Intel® Core™ i3-370M, 2.40GHz, 4B DDR3 Memory). To compare the computational errors, the length l was divided into 25, 50, 100, 150 and 200 knots. The error was estimated by the formula

$$\delta = \frac{\max\limits_{[0,l]}\left|\Delta p(z_i) - \Delta\tilde{p}(z_i)\right|}{\max\limits_{[0,l]}\left|\Delta p(z_i)\right|}\cdot 100\%,$$

where $\Delta p(z_i)$ – the values of the analytical solution at the grid nodes, $\Delta\tilde{p}(z_i)$ – the values at the mesh nodes obtained by the numerical method.

In calculations for beam energies of 10 keV, 20 keV, and 30 keV, the values of 7 μm, 10 μm and 13 μm, respectively, were taken as the length l for zero concentration. The results of the simulation at $n = 100$ are shown in the figure 1.

For the beam energy of 10 keV, 20 keV, and 30 keV, the relative error was, respectively, 2.847%, 1.220%, 0.835% for $n = 25$; 2.739%, 1.211%, 0.825% for $n = 50$; 2.732%, 1.206%, 0.824% for $n = 100$; 2.731%, 1.206%, 0.823% for $n = 150$, and 2.730%, 1.205%, 0.823% for $n = 200$.

Thus, in comparison with the analytical solution, approximate calculations have an insignificant error even with a small number of mesh nodes. The greatest value of the relative error of the results is for $z = 0$ μm, i.e. on the surface of the semiconductor. With an increase in the number of mesh nodes, the relative error goes down a little and for $n \geq 100$ practically does not change.

The computer time expenditure for $n = 150$ was approximately 28 s. The computer time spent on calculating the same problem by the projection least square method was approximately 4 s and by the projection methods of Galerkin and Ritz was approximately 8 min [7].

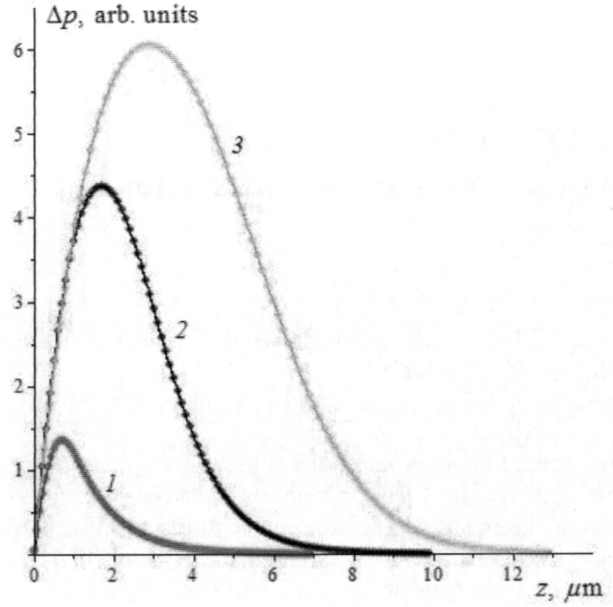

Figure 1. Distribution of MCC generated in silicon by a wide electron beam with energies: curve 1 – 10 keV, curve 2 – 20 keV, curve 3 – 30 keV. The solid lines represent graphs of analytical curves, the selected points are the results of numerical calculations for $n = 100$ layers of discretization.

5. The conclusion

The main advantage of the combined use of the matrix method and the method of generalized powers of Bers for heat and mass transfer modeling lies in the possibility of applying it to an arbitrary number of layers of the medium, which makes it possible to use this approach as a numerical method. The results of the calculations show the prospects of further research and development of the proposed approach for modeling the transport process in semiconductor materials, including for numerical modeling, both in the case of a homogeneous and inhomogeneous medium.

Acknowledgments

This work was supported by the Russian Foundation for Basic Research, project no. 16-03-00515, and the Russian Foundation for Basic Research and the Government of Kaluga Oblast, project no. 18-41-400001.

References

[1] Bers L and Gelbart A 1944 *Transactions of the American Mathematical Society* **56** 67
[2] Gladyshev Y A 2011 *Method of Generalized Powers of Bers and Its Applications in Mathematical Physics* (Kaluga: Tsiolkovsky Kaluga State University Press), p 204
[3] Carslaw H S and Jaeger J C 1959 *Conducion of Heat in Solids* (Oxford: Oxford University Press) p 517
[4] Gladyshev Y A, Kalmanovich V V and Stepovich M A 2017 *Journal of Surface Investigation: X-Ray, Synchrotron and Neutron Techniques* **11** 1096
[5] Mikheev N N, Petrov V I and Stepovich M A 1991 *Bulletin of the Academy of Sciences of the USSR. Physical Series* **55** 1
[6] Seregina E V, Makarenkov A M and Stepovich M A 2011 *Surface Investigation. X-ray, Synchrotron and Neutron Techniques* **5** 746
[7] Seregina E V, Makarenkov A M, Stepovich M A and Filippov M N 2016 *Journal of Surface Investigation: X-Ray, Synchrotron and Neutron Techniques* **10** 445

IOP Conf. Series: Journal of Physics: Conf. Series **1163** (2019) 012013 doi:10.1088/1742-6596/1163/1/012013

Modeling of heating in the epitaxial structure $Cd_xHg_{1-x}Te/CdTe$ with the projection least squares method

E V Seregina[1], M A Stepovich[2], V V Kalmanovich[2]

[1] Bauman Moscow State Technical University (National Research University), Kaluga Branch, 2 Bazhenov Street, Kaluga, 248000 Russia
[2] Tsiolkovsky Kaluga State University, 26 Stepan Razin Street, Kaluga, 248023 Russia

Abstract. The work contains the results of using the modified projective method of the least squares for modeling of unsteady heat conduction in epitaxial structure of infrared techniques, exposed a short laser pulse. The solution is in the form of partial sums of double Fourier series on Jacobi-Laguerre polynomials. The ordinal error estimation of the projection scheme of the least squares method used for the solution the heat equation with discontinuous coefficients.

1. Introduction

Unsteady heat conduction differential equations with discontinuous coefficients describe different time-dependent physical processes, e.g., temperature distribution of non-stationary source in heterogeneous or multilayer structure. The authors of [1, 2] performed the substantiation of the proposed modified scheme of the least squares method for simulating the one-dimensional minority charge carriers distribution in semiconductor materials. In article [3] the Galerkin projection method was used to solve stationary diffusion equation of modified Laguerre functions. The present work develops such investigations: the estimate of the projection scheme of the least squares method is applied to find the temperature field in epitaxial structure $\mathrm{Cd_xHg_{1-x}Te/CdTe}$.

2. Problem statement

In this paper, we used a model of unsteady heat conduction in epitaxial structure of infrared techniques, exposed a short laser pulse.

Let us consider the differential equation of heat in a semi-infinite region

$$\rho c\frac{\partial \Delta T}{\partial t}-\frac{\partial}{\partial z}\left(\lambda(z,t)\frac{\partial \Delta T}{\partial z}\right)=\alpha(z)(1-A)I_0(t)\exp(-\alpha(z)z) \tag{2.1}$$

with boundary conditions

$$\Delta T(z,0)=0,\ \left.\frac{\partial \Delta T}{\partial z}\right|_{z=0}=0,\ \Delta T(+\infty,t)=0,\ \Delta T(z,+\infty)=0,\ 0\le z<\infty, 0\le t<\infty. \tag{2.2}$$

Here $\Delta T(z,t)=T(z,t)-T_0$, where T is the sample temperature at point (z,t), T_0 is the temperature after the establishment of a non-stationary regime under the influence of a short laser pulse; A is the reflection coefficient, $\alpha(z)$ is the absorption coefficient; $I_0(t)=\left(W/(6\tau_m)\right)\left(4t/\tau_m\right)^4\exp\left(-4t/\tau_m\right)$ is the density of a stream of energy [4], where W is the density of energy of a laser pulse, τ_m is the duration of forward front of an pulse. At the point $z=z_0$ the function $\lambda(z,t)$ has a gap of the first kind. Then at this point the conditions

$$[\Delta T]=\Delta T(z_0+0,t)-\Delta T(z_0-0,t)=0, \tag{2.3}$$

$$\left[\lambda\frac{\partial\Delta T}{\partial z}\right]=0 \text{ when } z=z_0. \tag{2.4}$$

3. Projection approximation of the initial model based on using the least squares method

Following the classical least_squares method, we introduce the functional

$$J(\Delta T(z,t))=\left\|L_1^{(\Delta T)}(\Delta T_1(z,t))-f_1(z,t)\right\|_{L_2}^2+\left\|L_1^{(\Delta T)}(\Delta T_2(z,t))-f_2(z,t)\right\|_{L_2}^2+\left\|\Delta T_1(z,0)\right\|_{L_2}^2+$$

$$+\left\|\Delta T_2(z,0)\right\|_{L_2}^2+\left\|\frac{\partial\Delta T_1}{\partial z}\bigg|_{z=0}\right\|_{L_2}^2+\left\|\Delta T_1(z_0,t)-\Delta T_2(z_0,t)\right\|_{L_2}^2+$$

$$+\left\|\left(\lambda_1\frac{\partial\Delta T_1}{\partial z}-\lambda_2\frac{\partial\Delta T_2}{\partial z}\right)\bigg|_{z=z_0}\right\|_{L_2}^2+\left(\Delta T_1(z,+\infty)\right)^2+\left(\Delta T_2(+\infty,t)\right)^2$$

where $L_1^{(\Delta T)}$ is the operator on the left-hand sides of the system of equations (2.1) and the solution

$$\Delta T(z,t)=\begin{cases}\Delta T_1(z,t),\ 0\le z\le z_0, 0\le t<\infty,\\ \Delta T_2(z,t),\ z_0\le z<\infty, 0\le t<\infty\end{cases}$$

is sought from the requirement that it provides the minimum of the functional $J(\Delta T(z,t))$. Here $f_1(z,t)=\alpha_1(1-A)I_0(t)\exp(-\alpha_1 z)/(\rho_1 c_1)$ and $f_2(z,t)=\alpha_2(1-A)I_0(t)\exp(-\alpha_2 z)/(\rho_2 c_2)$. We introduce basis of Jacobi polynomials $P_i(z;\alpha,\beta)$ [5] on variable $z\in[0,z_0]$ with the parameters α and β, modified Laguerre functions on variable $z\in[z_0,\infty)$:

$$\varphi_i^{z_0}(z)=\exp\left(-\gamma_1(z-z_0)/2\right)L_i\left(\gamma_1(z-z_0)\right)$$

and on variable $t\in[0,\infty)$:

$$\varphi_i(t)=\exp\left(-\gamma_2 t/2\right)L_i\left(\gamma_2 t\right),$$

which are defined in terms of the Chebyshev-Laguerre polynomials [5] on variable z - $L_i\left(\gamma_1(z-z_0)\right)$ and in terms of polynomials on variable t - $L_i\left(\gamma_2 t\right)$, $i=0,1,2,\ldots$. Here $\gamma_1>0$, $\gamma_2>0$ are used to optimize the computing scheme.

Then each of the functions of two variables z and t in the system (2.1)-(2.4) is approximated by the partial (m,m)-order sum of its double Fourier series using the system of Jacobi polynomials and modified Laguerre functions.

We let $C^{\Delta T_1}_{mm}$, $C^{\Delta T_2}_{mm}$ and $C^{f_1}_{mm}$, $C^{f_2}_{mm}$ denote the columns (extended in matrix columns) of the coefficients of expansion of the unknown functions $\Delta T_1(z,t)$, $\Delta T_2(z,t)$ and the known functions $f_1(z,t)$, $f_2(z,t)$ in terms of the chosen basis, respectively.

We introduce the matrix of differentiation with respect to the variable z - $D^{z_1} = E \otimes D_1$, $D^{z_2} = E \otimes D_2$ and with respect to the variable t - $D^t = D_3 \otimes E$, where D_1 is the differentiation matrix in the one-dimensional basis of Jacobi polynomials and D_2, D_3 are the differentiation matrixes in the one-dimensional basis of modified Laguerre functions, whose elements are found using elementary algebraic operations [3]:

$$d^{2,3}_{ij} = \begin{cases} -\gamma_{1,2}/2, \text{ if } i = j, \\ -\gamma_{1,2}, \text{ if } i < j, \\ 0, \text{ if } i > j. \end{cases}$$

Matrix of differentiation elements D_1 are found by the formula

$$d^1_{ij} = \begin{cases} \left((\alpha+\beta+j)\Gamma(\alpha+\beta+2)/\left(z_0\Gamma(\alpha+\beta+j+1)\right)\right)\times \\ \times\left(\Gamma(\beta+j)/\Gamma(\beta+1)+(-1)^{j-2}\Gamma(\alpha+j)/\Gamma(\alpha+1)\right), \text{ if } i=1, \\ \left((\alpha+\beta+j)(\alpha+\beta+2i-1)\Gamma(\alpha+\beta+i)/\left(z_0\Gamma(\alpha+\beta+j+1)\right)\right)\times \\ \times\left(\Gamma(\beta+j)/\Gamma(\beta+i)+(-1)^{j-i-1}\Gamma(\alpha+j)/\Gamma(\alpha+i)\right), \text{ if } i \le j, \end{cases}$$

which are determined using the following formula for the Jacobi polynomials:

$$\begin{gathered} P_i(z;\alpha+1,\beta) - P_i(z;\alpha,\beta+1) = \left(\Gamma(\alpha+\beta+2)/\left(\Gamma(\alpha+\beta+i+2)\right)\right)\times \\ \times\left(\Gamma(\beta+i+1)/\Gamma(\beta+1)+(-1)^{i-1}\Gamma(\alpha+i+1)/\Gamma(\alpha+1)\right)+ \\ +\sum_{k=1}^{i-1} P_k(z;\alpha,\beta)\left((\alpha+\beta+2k+1)\Gamma(\alpha+\beta+k+1)/\left(\Gamma(\alpha+\beta+i+2)\right)\right)\times \\ \times\left(\Gamma(\beta+i+1)/\Gamma(\beta+k+1)+(-1)^{i-k-1}\Gamma(\alpha+i+1)/\Gamma(\alpha+k+1)\right) \end{gathered}$$

the last relations are easily derived from the formulas for the Jacobi polynomials [5]:

$$\begin{gathered} P_i'(z;\alpha,\beta) = \frac{1}{2}(\alpha+\beta+i+1)P_{i-1}(z;\alpha+1,\beta+1), \\ P_{i-1}(z;\alpha+1,\beta+1) = P_i(z;\alpha+1,\beta) - P_i(z;\alpha,\beta+1), \\ (i+\alpha+\beta+1)P_i(z;\alpha+1,\beta) - (i+\beta)P_{i-1}(z;\alpha+1,\beta) = (2i+\alpha+\beta+1)P_i(z;\alpha,\beta), \\ (i+\alpha+\beta+1)P_i(z;\alpha,\beta+1) + (i+\alpha)P_{i-1}(z;\alpha,\beta+1) = (2i+\alpha+\beta+1)P_i(z;\alpha,\beta). \end{gathered}$$

Then we pass to the algebraic vector-matrix system of equations:

$$AC^{\Delta T} = G, \tag{3.1}$$

where A is the matrix of the redefined system (3.1), which has the form

$$A=\begin{bmatrix} D^t-\frac{\lambda_1}{\rho_1 c_1}\left(D^{z_1}\right)^2 & O_{m^2\times m^2} \\ O_{m^2\times m^2} & D^t-\frac{\lambda_2}{\rho_2 c_2}\left(D^{z_2}\right)^2 \\ \left(\varphi_m(0)\right)^T\otimes E & O_{m\times 1} \\ O_{m\times 1} & \left(\varphi_m(0)\right)^T\otimes E \\ E\otimes\left(P_m^{\alpha,\beta}(0)\right)^T D_1 & O_{m\times 1} \\ E\otimes\left(P_m^{\alpha,\beta}(z_0)\right)^T & -E\otimes\left(\varphi_m^{z_0}(z_0)\right)^T \\ \lambda_1\left(E\otimes\left(P_m^{\alpha,\beta}(z_0)\right)^T D_1\right) & \lambda_2\left(E\otimes\left(\varphi_m^{z_0}(z_0)\right)^T D_2\right) \end{bmatrix}.$$

The column on the right-hand side of the system of equations (3.1) is defined as

$$G=\left[C_{mm}^{f_1}\ C_{mm}^{f_2}\ O_{m\times 1}\ O_{m\times 1}\ O_{m\times 1}\ O_{m\times 1}\ O_{m\times 1}\right]^T.$$

Then we find the column $C^{\Delta T}=\left[C^{\Delta T_1}\quad C^{\Delta T_2}\right]^T$, which minimizes the sum of squared residuals of all equations contained in system (3.1):

$$J\left(C^{\Delta T}\right)=\left(AC^{\Delta T}-G\right)^T\left(AC^{\Delta T}-G\right)\to \inf;$$

i.e., we solve it using the least squares method. Then the normal pseudosolution of the redefined system of equations (3.1) can be found using the pseudoinverse matrix [6]:

$$C^{\Delta T^+}=\left(A^T A\right)^{-1}A^T G. \tag{3.2}$$

If the normal pseudosolution $C^{\Delta T^+}$ is obtained, then the approximate solution for the initial problem (2.1)-(2.4) can be recovered using the formula

$$\Delta T_1(z,t)\approx \Delta T_1^m(z,t)=\left(\varphi_m(t)\right)^T C^{\Delta T_1^+}P_m^{\alpha,\beta}(z),\quad z\in[0,z_0],\ t\in[0,\infty),$$
$$\Delta T_2(z,t)\approx \Delta T_2^m(z,t)=\left(\varphi_m(t)\right)^T C^{\Delta T_2^+}\varphi_m^{z_0}(z),\quad z\in[z_0,\infty),\ t\in[0,\infty).$$

We pass to substantiation of the method.

Following [7], we introduce the notation:

$$D_1=\left(z-z_0\right)\frac{d^2}{dz^2}+\left(1-z+z_0\right)\frac{d}{dz}+t\frac{d^2}{dt^2}+\left(1-t\right)\frac{d}{dt};$$

$$D_2=\left(1-z^2\right)\frac{d^2}{dz^2}+\left(2z\right)\frac{d}{dz}+t\frac{d^2}{dt^2}+\left(1-t\right)\frac{d}{dt}.$$

Here $L_2^n(D_1)$ $(n=0,1,...)$ is the class of functions f such that the functions $\tilde f(z,t)=f(z,t)\exp\left(\gamma_1(z-z_0)/2\right)\exp\left(\gamma_2 t/2\right)\in L_2$ and have generalized partial derivatives in the

Levy sense $\partial^k \tilde{f}(z,t)/\partial z^i \partial t^j$, $i+j=k$, $k=0,1,\ldots$, belonging to the space L_2, for which $D_1^n \tilde{f} \in L_2$, $n=0,1,\ldots$, where $D_1^0 \tilde{f} = \tilde{f}$, $D_1^n \tilde{f} = D_1\left(D_1^{n-1}\tilde{f}\right)$, $n=1,2,\ldots$, $L_2^0(D_1)=L_2$.

$L_2^n(D_2)$ $(n=0,1,\ldots)$ is the class of functions f such that the functions $f \in L_2$ and have generalized partial derivatives in the Levy sense $\partial^k f(z,t)/\partial z^i \partial t^j$, $i+j=k$, $k=0,1,\ldots$, belonging to the space L_2, for which $D_2^n f \in L_2$, $n=0,1,\ldots$, where $D_2^0 f = f$, $D_2^n f = D_2\left(D_2^{n-1} f\right)$, $n=1,2,\ldots$, $L_2^0(D_2)=L_2$.

Based on the results in [3, 6, 7], it is possible to determine the estimate of the residual for the functional $J\left(C_{mm}^{\Delta T^+}\right)$.

Let the function $f_1(z,t) \in L_2^n(D_1)$, $f_2(z,t) \in L_2^n(D_2)$ and have continuous derivatives up to the order of $n>1$. Then the sequence of normal pseudosolutions $\left\{C_{mm}^{\Delta T^+}\right\}$ for $m \geq 4$ and $\alpha=\beta=0$ is minimizing for the functional $J\left(C_{mm}^{\Delta T}\right)$, and the following estimate is valid:

$$\sqrt{J\left(C_{mm}^{\Delta T^+}\right)} < \left(6+\frac{\lambda_1}{\rho_1 c_1}+\frac{3\lambda_2}{\rho_2 c_2}\right)\left[1-\left(1-m^{-1}\right)^m\right]^{-k}(m)^{-n}\,\Omega\left(m^{-1}\right)+$$
$$+O\left(m^{-n+\frac{11}{4}}\right)\omega\left(m^{-1}\right) \quad m\to\infty,$$

where $\Omega\left(m^{-1}\right)$ is the majorant of the generalized moduli of continuity, and $\omega\left(m^{-1}\right)$ is the given majorant of the continuity moduli.

4. Results of computations

Figures 1-2 shows the results of computations carried out using the Matlab (MathWorks, Inc.) package for target parameters that are characteristic for $Cd_xHg_{1-x}Te$ with $x=0.2$: specific heat $c_1=0.19\,\mathrm{J/(gm\,K)}$ [8], thermal conductivity coefficient $\lambda_1=0.013\,\mathrm{W/(cm\,K)}$ [9], absorption coefficient $\alpha_1=200\,\mathrm{cm^{-1}}$ [9]; for CdTe: specific heat $c_2=0.205\,\mathrm{J/(gm\,K)}$, thermal conductivity coefficient $\lambda_2=0.057\,\mathrm{W/(cm\,K)}$, absorption coefficient $\alpha_2=2\cdot10^{-3}\,\mathrm{cm^{-1}}$ [10] at room temperature. The density of energy of a laser pulse was assumed $W=0.5\,\mathrm{J/(cm^2)}$, and duration of forward front of an pulse $\tau_m=1\,\mathrm{ns}$, wavelength – 10 μm, $z_0=1\,\mu\mathrm{m}$.

The estimate of the relative error with respect to the norm of the space L_2 between the temperature distribution $\Delta T^9(z,t)$ and $\Delta T^8(z,t)$ calculated using the least squares method is

$$\Delta_1\left(\delta T^9,\delta T^8\right)=\frac{\left\|\delta T^9(z,t)-\delta T^8(z,t)\right\|_{L_2}}{\left\|\delta T^9(z,t)\right\|_{L_2}}\times100\%<1\%.$$

The values of the modified parameters are $\alpha=\beta=0$, $\gamma_1=2\cdot10^{-7}$, $\gamma_2=3$. Some results of solution of the differential equation of heat (2.1)-(2.4) are shown in the figure 1a and figure 1b presents the results of mathematical modeling for temperature field in epitaxial structure $Cd_{0.2}Hg_{0.8}Te/CdTe$.

IOP Conf. Series: Journal of Physics: Conf. Series **1163** (2019) 012013 doi:10.1088/1742-6596/1163/1/012013

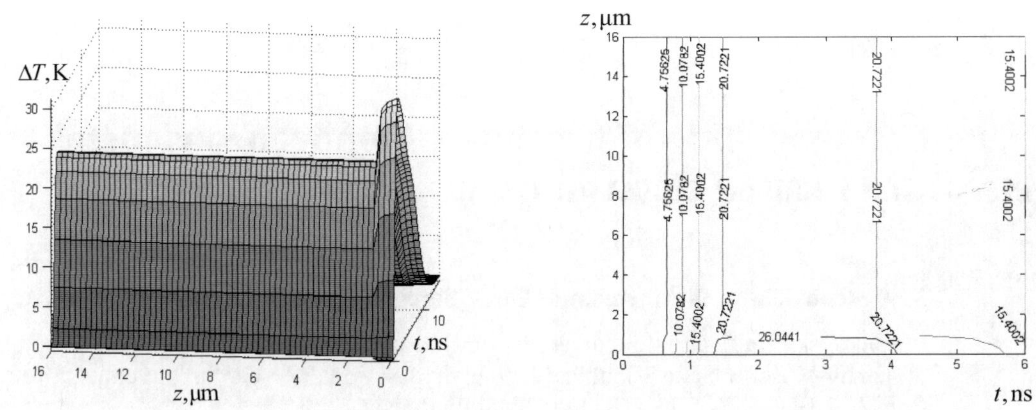

a **b**

Figure 1. a – Temperature distribution in $Cd_{0.2}Hg_{0.8}Te/CdTe$; b – level lines of the temperature distribution in $Cd_{0.2}Hg_{0.8}Te/CdTe$; the level lines were constructed with a step of h=5.3220K. Calculations carried out in 9th approximation.

5. Conclusions

The possibility of using the least squares method for mathematical modelling of temperature distribution in $Cd_{0.2}Hg_{0.8}Te/CdTe$ is considered and the order estimate of the accuracy of the proposed projection scheme of the method is given.

Acknowledgments
This work was supported by the Russian Foundation for Basic Research and the Government of Kaluga District, project no. 18-41-400001.

References

[1] Seregina E V, Stepovich M A, Makarenkov A M 2013 *Journal of Surface Investigation. X-ray, Synchrotron and Neutron Techniques* **7** 1077
[2] Seregina E V, Stepovich M A, Makarenkov A M 2018 *Journal of Surface Investigation. X-ray, Synchrotron and Neutron Techniques* **12** 80
[3] Makarenkov A M, Seregina E V, and Stepovich M A 2017 *Comput. Math. Math. Phys.* **57** 802
[4] Khaneft A V, Dolgachev V A, Duginov E V, Ivanov G A 2013 *Bulletin of the Kemerovo State University* **3** 31 [in Russian]
[5] Suetin P K 2005 *Classical Orthogonal Polynomials* (Fizmatlit, Moscow) 480 pp. [in Russian]
[6] Lapin S V and Egupov N D 1997 *Theory of Matrix Operators and its Application to Problems on Automatic Control* (Bauman Moscow State Technical Univ., Moscow) 496 pp. [in Russian]
[7] Abilov V A, Abilov M V and Kerimov M K 2015 *Comput. Math. Math. Phys.* **55** 1094
[8] Nikiforov V N, Sredin V G 2006 *Russian Physics Journal* **49** 337
[9] Ponomarenko V P 2018 *Quantum Fotosensorika* (AO "NPO "Orion", Moscow) 648 pp. [in Russian]
[10] Shaldin Y V, Varhulska I, Ivanov Y M 2004 *Physics and Technology of Semiconductors* **38** 172 [in Russian]

IOP Conf. Series: Journal of Physics: Conf. Series **1163** (2019) 012014 doi:10.1088/1742-6596/1163/1/012014

Mathematical modelling of heating of homogeneous metal targets with a focused electron beam

M A Stepovich[1], A N Amrastanov[1], E V Seregina[2] and M N Filippov[3]

[1] Tsiolkovsky Kaluga State University, 26 Stepan Razin Street, Kaluga, 248023 Russia
[2] Bauman Moscow State Technical University (National Research University), Kaluga Branch, 2 Bazhenov Street, Kaluga, 248000 Russia
[3] Kurnakov Institute of General and Inorganic Chemistry RAS, 31 Leninsky Prospect Street, Moscow, 119991 Russia

E-mail: m.stepovich@rambler.ru

Abstract. The problem of heat distribution in metal materials irradiated with sharply focused electron beams in the absence of heat exchange between the target and the external medium is considered with mathematical modeling methods. For a quantitative description of energy losses by probe electrons a model based on a separate description of the contributions of absorbed in the target and backscattered electrons is used. This method is applicable to a wide class of solids and a range of primary electron energies. Some modeling results are illustrated for different metal materials.

1. Introduction

Some methods for local analysis of solids are based on the excitation of an informative signal with a beam of accelerated electrons (electron microscopy, spectroscopy of characteristic losses of electron energy, X-ray spectral microanalysis, etc.). However, when the electrons of low (1-8 keV) and medium (8-50 keV) energies decelerate in the targets, only a small part of energy goes to the formation of informative signals, the most part of the energy goes to heating the sample [1, 2]. Target heating can be especially important when using modern electron-probe technology, since the effective diameter of an electron beam interacting with a semiconductor target can be as high as 1-2 nm. When this happens, the density of the energy emitted by the electron beam in a target can range from 10 to 1000 W/cm^3 [2], and the characteristic size (diameter) of the area of electron beam energy loss is no more than 1 or 1.5 micrometers. So, when the target is irradiated with a sharply focused electron beam, the heating of the sample can be significant, which can lead to an increase in the local temperature and this may lead to the necessity to take this into account on carrying out the quantitative measurements. In [3-5] we considered the influence of electron beam on semiconductor targets. Modeling the heating of metal targets with an electron beam is the subject of this paper.

2. Simulation of electron beam energy losses

For computation of energy losses of an electron beam in the condensed substance, we use a model based on the possibility of a separate quantitative description of the contribution of the energy of the electrons absorbed in the target and backscattered electrons [6, 7]. This model can be successfully used to perform quantitative calculations for a wide class of materials in a wide range of primary-electron energies. According to this model, the energy loss density of the beam electrons has the form:

$$P(x,y,z)=\frac{1.085(1-\eta)P_0}{\pi^{3/2}a_1^2 z_{ms}\left(1-\eta+\eta\, z_{ss}/z_{ms}\right)}\left(\exp\left(-\left[\frac{x^2+y^2}{a_1^2}+\left(\frac{z-z_{ms}}{z_{ms}}\right)^2\right]\right)+\right.$$
$$\left.+\frac{\eta a_1^2}{(1-\eta)a_2^2}\exp\left(-\left[\frac{x^2+y^2}{a_2^2}+\left(\frac{z-z_{ss}}{z_{ss}}\right)^2\right]\right)\right). \qquad (1)$$

Here the origin coincides with the point of incidence of the electron beam on the sample; P_0 is the primary beam power; z_{ms} is the depth of maximum energy loss with primary electrons, that have experienced small-angle scattering and absorbed in the target, and z_{ss} is the depth of the maximum energy loss with backscattered electrons, $z_{ss}=Z^{-1/3}z_{ms}$, Z is the effective target material number; η is backscattering coefficient of electrons. Parameters a_1 and a_2 are determined from $a_1^2=z_{ms}^2+0.72d_b^2$ and $a_2^2=0.25z_{ss}^2+0.72d_b^2$, respectively, where d_b is diameter of the probe.

For probe electrons with energy E_0 [keV] incident perpendicular to a target surface with density ρ_0 [g/cm^3], z_{ms} can be expressed in terms of total beam electron path R in a solid (see diffusion model [8]):

$$z_{ms}[\mu\mathrm{m}]=\frac{R}{2}\left[1-\left(\frac{C\gamma}{1+\gamma}\right)^2\right],\quad R[\mu\mathrm{m}]=\frac{2.76\cdot10^{-2}AE_0^{5/3}}{\rho_0 Z^{8/9}}\cdot\frac{\left(1+0.978\cdot10^{-3}E_0\right)^{5/3}}{\left(1+1.957\cdot10^{-3}E_0\right)^{1/3}}.$$

Here A is the effective target material atomic weight, $\gamma=0.187Z$ and $C\approx1.1$.

The distributions of energy loss densities calculated according to (1) for kilovolt electrons in Al (material with low effective target material atomic weight – light target), Cu (medium target), and Au (heavy target) normalized by the amount of energy released per target per unit time are illustrated in figure 1.

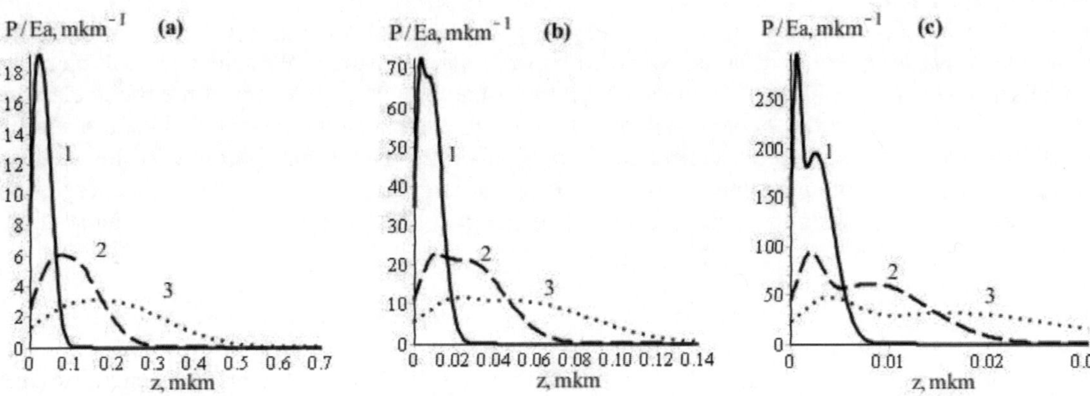

Figure 1. Distributions of the energy loss densities of beam of electrons in Al (a), Cu (b) and Au (c) normalized by the amount of power E_a released in these targets and calculated for the electron beam energy E_0 = 2 (curves 1), 4 (2) and 6 (3) keV.

IOP Conf. Series: Journal of Physics: Conf. Series **1163** (2019) 012014 doi:10.1088/1742-6596/1163/1/012014

Note that in metal targets with medium and heavy effective targets material atomic weights the energy loss density of the beam electrons is significantly higher than in the corresponding semiconductor targets at the same energies [3-5].

The possibilities of using this model for estimating the heat distribution in various metal targets are the subject of this work: the simulation is carried out for the case of the absence of heat transfer of the target with the external medium (unlike [3-5]), which is usually realized in electron-probe technology.

3. Modeling of heat distribution

The time to establish a steady-state temperature regime of the analyzed microvolume is usually much shorter than the data set time during which the electron probe is positioned at a given point on the surface of the sample [6]. This makes it possible to look for a temperature distribution in the region of interaction of the probe electrons with the sample and in the adjacent regions of the sample on the basis of the solution of the stationary heat equation:

$$\left(\frac{\partial^2 \Delta T(x,y,z)}{\partial x^2}+\frac{\partial^2 \Delta T(x,y,z)}{\partial y^2}+\frac{\partial^2 \Delta T(x,y,z)}{\partial z^2}\right)=-\frac{P(x,y,z)}{k}. \qquad (2)$$

Here $\Delta T(x,y,z)=T(x,y,z)-T_0$ is the value of target heating under the action of the electron beam, where T is the sample temperature at point (x,y,z) after the establishment of a stationary regime under the influence of an electron beam, T_0 is the temperature before the influence of an electron beam, and $P(x, y, z)$ is function describing energy losses with an electron beam in a target. $\Delta T(x,y,z)$ must satisfy the following boundary conditions: $\lim\limits_{x\to-\infty}\Delta T=0$, $\lim\limits_{x\to+\infty}\Delta T=0$, $\lim\limits_{y\to-\infty}\Delta T=0$, $\lim\limits_{y\to+\infty}\Delta T=0$, $\lim\limits_{z\to+\infty}\Delta T=0$. Since in electron-probe technology the target irradiation by an electron beam is mainly carried out under conditions close to vacuum, the following boundary condition meaning the absence of heat exchange with the external medium: $\partial\Delta T(x,\, y,\, z=0)/\partial z=0$.

Equation (2) were solved in a standard way using the Green's function [3-5].

4. Results of calculations

The simulation results showed that the highest heating of all samples is observed at the lowest (from the considered range) energies what is explained by the following: the size of the deceleration region of the electron beam is approximately inversely proportional to the fourth power of the initial electrons energy, therefore, as the initial electrons energy of the beam decreases, there is a significant increase in the bulk density of thermal energy generated in the sample [2]. Therefore, the range of low electron energies of the probe is the most interesting for estimating target heating. At fixed beam energies, the smallest heating is observed for a light target (Al), and the heaviest sample (Au) is heated most of all – see figure 2 and figure 3.

The simulation results show that heating of heavy target (Au) at the used beam current is approximately an order of magnitude greater than heating of light target (Al). Moreover, for light targets, heating is essential for low electron energies of the beam, and for medium energies (from 10 keV and above) in many cases the heating of the light metal can be ignored, especially when using beam currents characteristic for scanning electron microscopes. For heavy metals with beam currents characteristic for X-ray microanalyzers, for low beam energies, heating can be quite significant: from ten degrees to several tens of degrees.

IOP Conf. Series: Journal of Physics: Conf. Series **1163** (2019) 012014 doi:10.1088/1742-6596/1163/1/012014

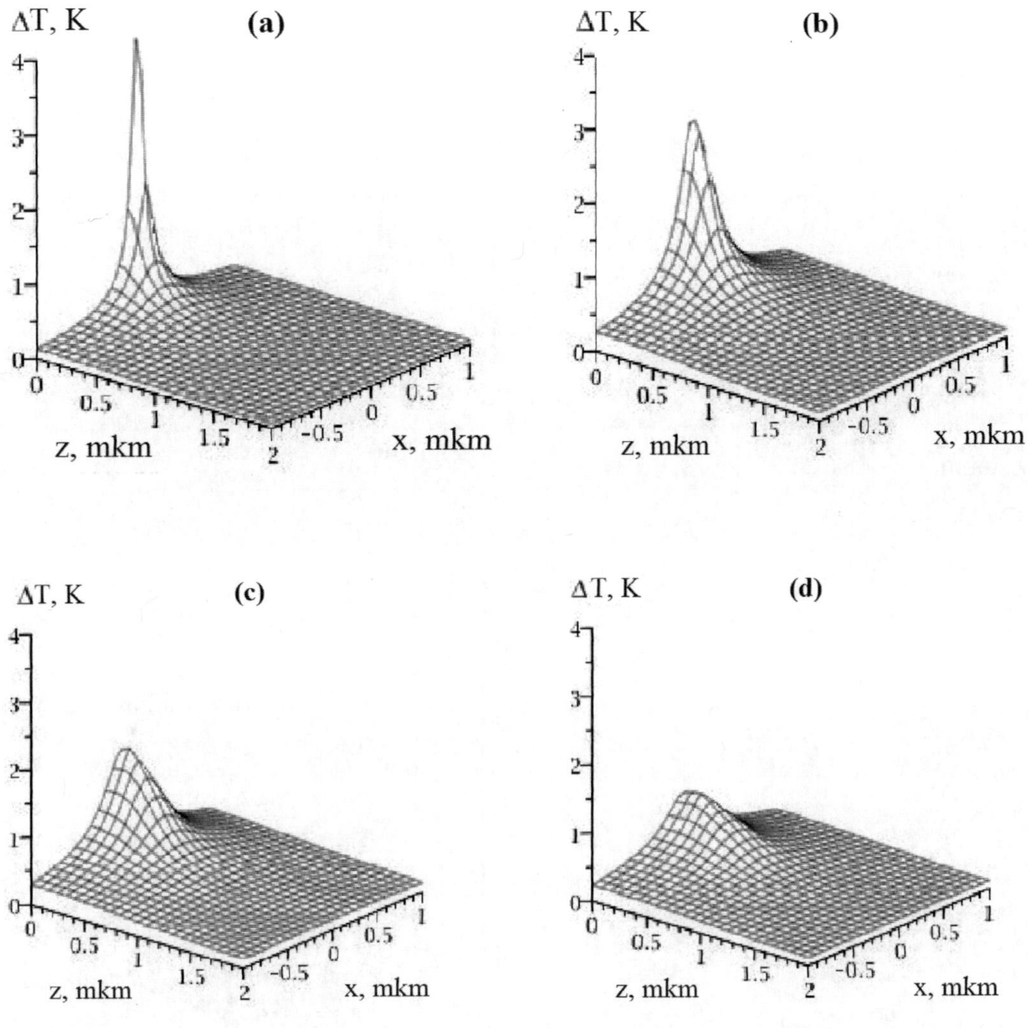

Figure 2. Temperature distribution in Al for coordinates x and z. The calculations were carried out at $y = 0$, for primary electron energies $E_0 = 2$ (a), 4 (b), 6 (c) and 8 (d) keV, at probe current 10^{-7} A.

Figure 4 shows the dependences of the of maximum temperature values of target heating under the action of the electron beam in the Al and Au targets on the energy of the primary electrons E_0; the contribution of the energy scattered in the target with the absorbed and backscattered (reflected) electrons is shown separately. The calculations are performed for x and y values equal to zero.

For light metals the contribution of electrons absorbed in the target to the total energy losses of the electron probe in the target and, correspondingly, to its heating, is determinant at a probe energy of less than 7-8 keV and the energy losses of these electrons become practically zero at an energy of about 10-11 keV. For heavy metals, the energy losses of these electrons become practically zero at an energy of about 8 keV. For light samples and samples with average ordinal numbers, the energy losses absorbed in the target and backscattered electrons become commensurate at an electron energy of the probe of about 6-8 keV. As for backscattered electrons in these metals, they are characterized by the presence of a maximum of about 3-5 keV, and then the curves decrease monotonically.

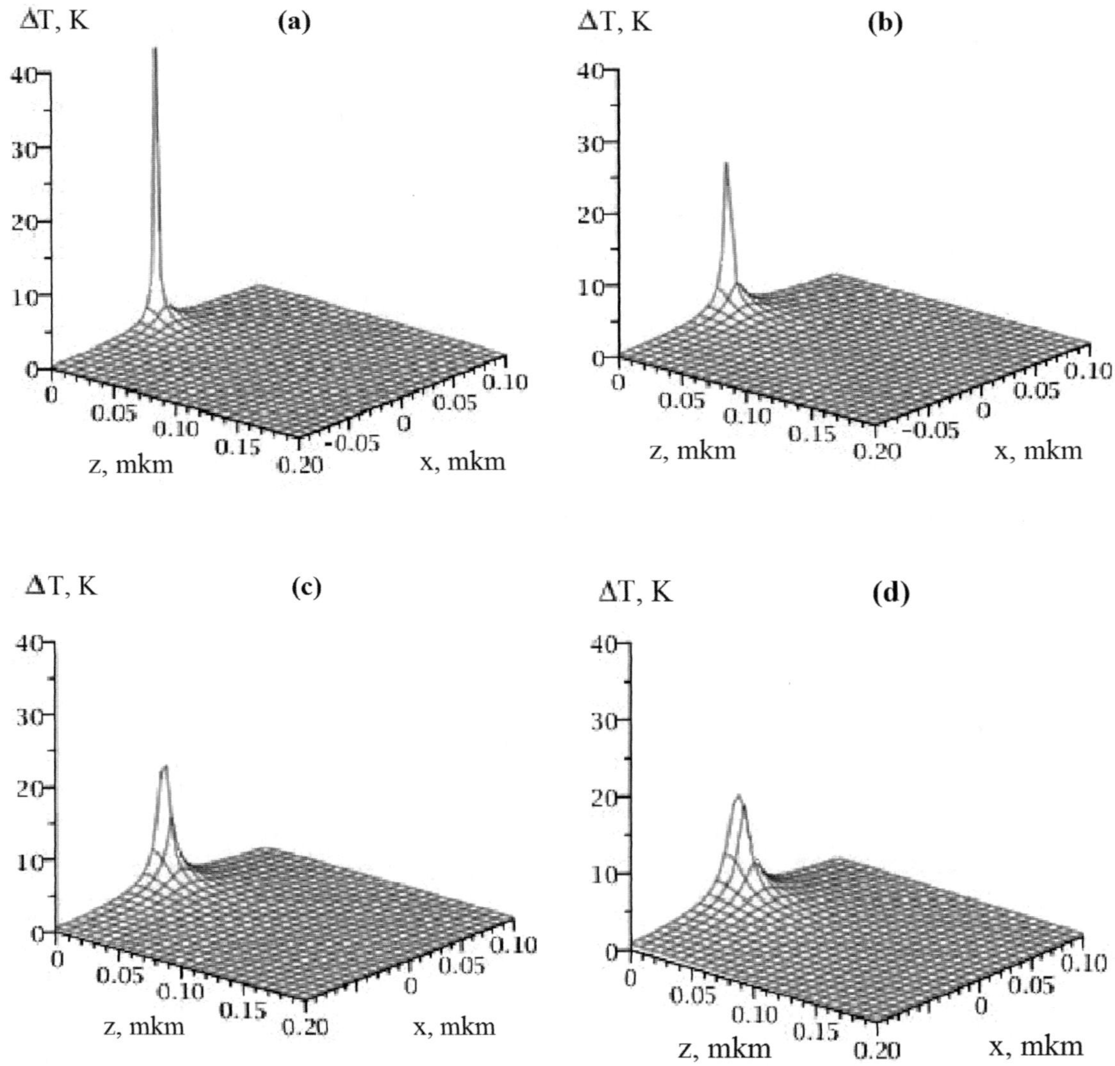

Figure 3. Temperature distribution in Au by coordinates x and z. The calculations were carried out at $y = 0$, for primary electron energies $E_0 = 2$ (a), 4 (b), 6 (c) and 8 (d) keV, at probe current 10^{-7} A.

The presence of a rather sharp increase in the contribution of backscattered electrons to the total curve $\Delta T(E_0)$ leads to the formation of a "step" (for medium metals) or local maximum (for heavy metals) on the dependences of the maximum temperature on the energy of the primary electrons. We also note that the backscattered electrons have the least effect on the value of the maximum heating for light samples; this effect is observed most in heavy targets. The presence of such effect is explained with the difference of the densities of energy loss in these metals: for example, in figure 1 in these distributions there is a “step” for medium metals, and a local maximum for heavy metals. The presence of these features leads to a difference in the nature of the dependences $\Delta T(E_0)$ of the groups of metals under consideration.

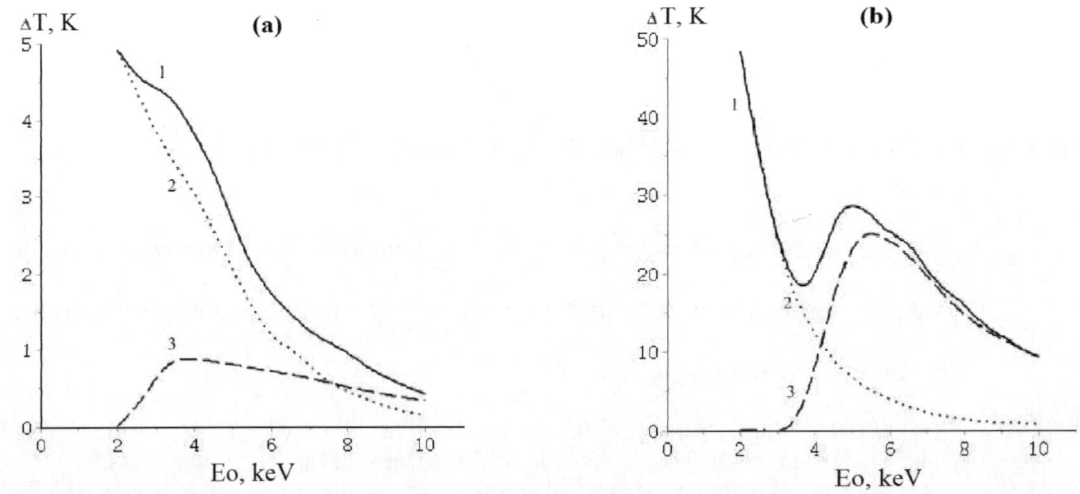

Figure 4. The results of modeling the heating of the metal targets for parameters characteristic of Al (a) and Au (b): the simulation of the distribution of the dependences of the maximum temperature on the energy of the primary electrons (curves 1) also the contribution to the curves 1 of the energy scattered by the electrons absorbed in the target (curves 2) and reflected (curves 3) were shown. Calculations were carried out at the probe current 10^{-7} A.

5. Conclusions

The methods of mathematical modeling were used to study the heating of metal targets using an electron beam of medium and low energies. The greatest heating of all samples by medium-energy electrons is observed at the lowest (from the considered range) energies. At fixed beam energies, the smallest heating is observed for light targets. The obtained heating estimates make it possible to select the optimal mode of action of the beam electrons on a metallic target when planning the experiment.

Acknowledgments

This work was supported by the Russian Foundation for Basic Research, project no. 16-03-00515, and by the Russian Foundation for Basic Research and the Government of Kaluga District, project no. 18-41-400001.

References

[1] Rau E I, Ditsman S A, Zaitsev S V, Lermontov N V, Lukyanov A E and Kupreenko S Yu 2013 *Bulletin of the Russian Academy of Sciences: Physics* **77** 951

[2] Amrastanov A N, Kuzin A Yu, Mityukhlyaev V B, Seregina E V, Stepovich M A, Todua P A and Filippov M N 2017 *Measurement Techniques* **60** 534

[3] Stepovich M A, Amrastanov A N, Seregina E V and Filippov M N 2018 *Journal of Physics: Conf. Series* **955** 012040

[4] Amrastanov A.N., Seregina E V and Stepovich M A 2018 *Bulletin of the Russian Academy of Sciences: Physics* **82** 1187

[5] Amrastanov A N, Seregina E V, Stepovich M A and Filippov M N 2018 *Journal of Surface Investigation. X-ray, Synchrotron and Neutron Techniques* **12** 778

[6] Mikheev N N, Petrov V I and Stepovich M A 1991 *Bulletin of the Academy of Sciences of the USSR. Physical Series* **55** 1

[7] Mikheev N N and Stepovich M A 1996 *Industrial Laboratory* **62** 221

[8] Kanaya K and Okayama S 1972 *J. Phys. D: Appl. Phys.* **5** 43

Development of an Educational Kit for Learning IoT

Ilia Lyzhin, Sergey Efremov, Alexey Rolich, Leonid Voskov, Dmitrii Abrameshin

National Research University Higher School of Economics, Moscow 123458, Russia

E-mail: lyzhin.iliya@gmail.com

Abstract. The primary purpose of this paper is to provide an overview of existing education solutions for IoT and develop proposals for their improvement. The study draws analysis of current conditions of the educational IoT sphere, a comparative analysis of educational products used for teaching of undergraduate students. With that the article describes the architecture of our own software and hardware platform for learning IOT. Moreover, this paper reviews methods and technical instruments employed to design software and hardware appliances.

1. Introduction

Today more and more electronic devices are designed with a built-in access to a global network, this concept is widely known as the Internet of Things (IoT). With many isolated infrastructures of such devices, there is a growing need in integration of these smart things into a single network, development of rules for their interconnection and determination of interaction scenarios [1].

The educational side of the Internet of Things is still very much in development. With existing courses at the university level related to electrical, electronic engineering, programming, microcontroller architectures, robotics, data storage and analysis and others it is very much an open question of whether teaching IoT requires something in addition. Having analyzed the related works we have concluded that the general approach to teaching IoT is to design extensible educational constructors for carrying out multidisciplinary practice sessions. Hence the main goal of this paper is to provide an overview of the work that has been done in this direction so far and to present our own proposals.

This paper is based on the results of a project carried out by our research group, which aims at designing a hardware-software appliance (IoT constructor) that would provide an improved way of teaching embedded solutions and Internet of Things.

The following tasks are solved sequentially to achieve the stated goal:

- Systematic review and comparison of existing technical solutions
- Development of a physical architecture for the hardware-software appliance
- Design of hardware components
- Development of software components

The key novelty of the study is a complex approach to the already created educational constructors and development of a device, being both optimal in economic and technical terms and capable of direct interactions with a network as compared to those emulating this process.

2. Related works

The problem of choosing efficient instruments for IoT education has already been extensively studied by different authors. Specifically, it has been addressed by researchers from the University of Novi Sad, Serbia [2] who suggested a package created for university education. This pack is based upon

Arduino/Genuino UNO Rev3 and Ethernet Shield, in addition, a set of breadboards and jumper wires are used for attachment of sensors. In accordance with the information provided by the authors, the cost of the kit without sensors varies from 11.89 to 57.5 euros. The prospects of further advancement are seen in integration of Wi-Fi, ZigBee, RS-232 modules and other communication technologies.

At the same time, staff members of the College of Engineering and Technology, Hubballi, Karnataka, India have created a multidisciplinary IoT course for their educational establishment [3]. From the technical point of view, the course is based on the Intel Galileo board, Arduino-Uno, Raspberry Pi-2. For data transmission GSM, Wi-fi and Ethernet are used. During the course students are suggested to solve cases using the backend tools and equipment enumerated above. The possible versions of tasks are the creation of prototype of IoT device or the development of user interface including Django framework and integration with outside cloud-hosted services.

Research results show that during the course 75% of student groups have applied cloud services, while 25% have used notification function for mobile devices. Authors point out that during the studies mentioned learners have attained basic knowledge of developing systems in IoT sphere.

In addition to that, the educational programme contains both hardware and software learning practices including the integration of devices into an existing cloud infrastructure. In particular, students designed the following applications: smart lighting, smart irrigation, weather reporting bot and other projects in the IoT sphere.

This example illustrates the fact that, taking part in multidisciplinary course in specialized secondary education institution may contribute to the creation of notable and practically oriented projects in case the course programme is thoroughly designed and suitable technical means of learning are available.

Another notable study has been conducted by P. Plaza et al [4]. Their work includes the comparative analysis of PCBs, such as Arduino WiFi Shield, Arduino Yún Shield, Arduino MKR1000, Node MCU ESP8266 and Onion Omega. The following features have been revealed:

- The higher amount of memory (both volatile and non-volatile) is associated with the Genuino Yún Shield and the Onion Omega, while lower amount is incorporated in the Genuino MKR1000.

- The majority of programming languages are supported by the Node MCU ESP8266 and the Onion Omega.

- The least expensive board is the NodeMCU ESP8266, while the costliest one is the Arduino WiFi Shield.

Researchers have concluded that any of enumerated devices are appropriate in the context of IoT education. Specifically, it should be noted that the NodeMCU ESP8266 is 2.25 times less expensive than the Onion Omega and 5 times less expensive than the Genuino Yún Shield.

The Maze game [5] has been created in the context of a student project called «Connection systems» and may be useful for education in both IoT and robotic systems spheres. The basic idea of the activity is a competition of different robotic devices in finding an exit from a labyrinth. The programmers aimed to make the system able to work with ESP-12E, ESPDuino, D1, NodeMCU, ESP8266-01, Wee, Arduino, C.H.I.P.

Authors claim that their product may be employed for educational purposes. The equipment used in project is of low cost and can be easily modified; moreover, there are open libraries of functions. The noteworthy aspect is that 5 out of 8 applied devices are based on the ESP8266 microcontroller.

Overall, these studies consistently indicate that the popularity of technologies application in education in IoT sphere is widespread. Furthermore, the reviewed works suggest that the development of educational projects and learning kits is mainly done with a help of Arduino or PCBs based on the ESP8266. Single-board computers (such as, Raspberry Pi-2 and Onion Omega) are employed a bit less frequently.

As a result of the analysis of current technical solutions and their strengths and weaknesses, the requirement for considered product has been elaborated. The major point is that our device is going to be different from the current analogues in a way that it will have wireless connection to the Internet, larger embedded memory and a built-in support for cloud integration. A detailed table of comparative analysis can be found in A.1. [6]

3. Proposal of a new IoT constructor

3.1. Hardware

The literature review has revealed that in the context of IoT DIY constructors the most widely used boards are Arduino or PCBs based on ESP8266. We plan to extend this list by using the Wemos D1 to develop a new constructor as it supports the Arduino IDE, sensors and partly supports libraries for Arduino projects.

At the same time, the Wemos D1 provides a Wi-Fi connection and costs less than a similar Arduino PCB. The more in-depth results of boards comparison are given in Appendices, Table 2. Consequently, Wemos D1 having both cost-effectiveness and optimal technical characteristics was considered as appropriate for the new product.

The constructor would include the Wemos D1 PCB, a board for module's fastening, a PCB case, specially created magnetic contacts, sensors (collecting data on lighting, temperature and vibration), actuating devices (ordinary and addressable RGB LEDs, relay for switching high-voltage appliances).

The capacity of our university's Laboratory for 3D modelling and visualization is planned to be used for development of terminals and bodies of device.

3.2. Software

Data collected by sensors will be transmitted through the wireless IEEE 802.11 network to the cloud platform. The platform will accumulate data and send back commands for actuators based on the programmable interconnection scenarios.

For the development of the cloud platform the Separate template engine [7] has been chosen as it has the following advantages: relatively insignificant size, concise syntax, provided with extensive documentation and produced under of free license of MIT. Twitter Bootstrap has been selected for the visual web design because of its wide recognition and compatibility with all existing browsers.

As for data storage we plan to use two database types – a relational database for storing device descriptors and user data and a NoSQL database for storing unstructured sensor data. We have chosen MySQL as a relational database as it is supported by the majority of hosting services and it has proven itself efficient in the web development sphere. The unstructured data storage is still at the stage of comparison of existing solutions and testing.

It is further anticipated that the new platform will involve a preconfigured server and web-application with user interface. Moreover, there would be functions of registration and authorization, addition and configuration of devices, device sharing, scenario editor and interactive tutorials provided.

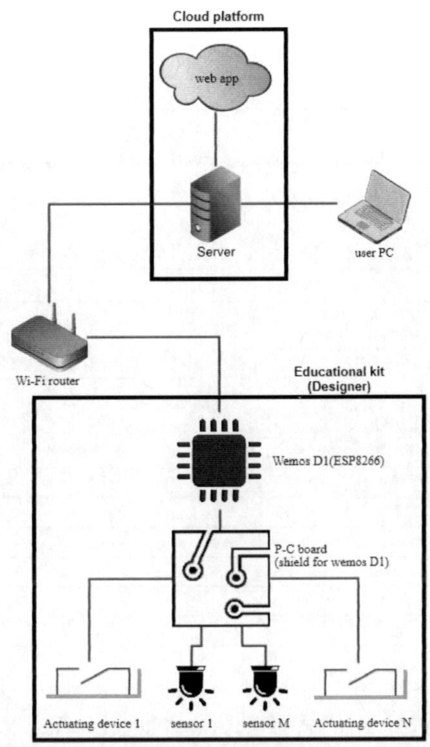

Pic. 1. Architecture of the software and hardware appliance

4. Conclusions

In this paper we have made an attempt to bring together existing approaches to developing IoT solutions that can be used for educational purposes. Motivated by the greatly increasing demand from the educational sphere related to teaching IoT, we have reviewed the work done so far and presented an architecture of a novel IoT constructor and built its first prototype.

Acknowledgments. The article was prepared within the framework of the Academic Fund Program at the National Research University Higher School of Economics (HSE) in 2017 — 2018 (grant № 17-05-0017) and by the Russian Academic Excellence Project «5-100».

5. Appendices

Table 1. Comparison of the existing educational kits

Criteria\Product	Skart-LAB "Umnyj dom"[8]	Skretchduino.Laboratorija[9]	Jodo[10]	LAZY SMART LS Monitoring[11]	IoT Smart Agriculture[12]	Obrazovatel'naja laboratorija JS "Internet veshhej"[13]
Controller	ATmega2560	ATmega328	Cortex-M4	ATmega1280	ATmega328	Atmega
Internet connection	-	-	-	+(GSM/GPRS)	+(Ethernet)	-
Number of sensors	9	2	4	1	4	9
Number of supported programming languages	1	4	1	1	1	1

Availability of cloud platform	-	-	-	+	-	-
Possibility to connect actuating devices	+	-	+	+	+	+
Configurator manager	-	-	-	+	-	-
Price (EUR)	274	124	65	47	72	871

Table 2. Results of PCB comparison

Criteria\P-c board	Arduino Uno	Arduino Leonardo ETH	Arduino M0	Iskra JS	PSoC 4200	IFLAT-32 v2.0	WeMos D1
Controller	ATmega328p	ATmega32U4	Cortex M0	Cortex-M4	Cortex M0	ARM 7	ESP8266EX
Frequency	16 MHz	16 MHz	48 MHz	168 MHz	48 MHz	72 MHz	80 Hz
RAM	2 kB	2.5 kB	32 kB	192 kB	4 kB	96 kB	80 kB
Memory	32 kB	32 kB	256 kB	1 mB	32 kB	12 kB	4 mB
Digital pins	14	8	14	14	36	54	11
Analog pins	6	12	6	12	0	6	1
I2C	+	+	+	+	+	+	+
Ethernet	-	+	-	-	-	-	-
Wi-fi	-	-	-	-	-	-	+
Price (EUR)	13	48	26	22	19	30	6,50

6. References

[1] Efremov S., Pilipenko N., Voskov L. An integrated approach to common problems in the Internet of Things, in: Procedia Engineering Vol. 100: 25th DAAAM International Symposium on Intelligent Manufacturing and Automation, 2014. Switzerland : Elsevier, 2015. P. 1215-1223.

[2] Dobrilovic D., Zeljko S. Design of open-source platform for introducing Internet of Things in university curriculum. *C. IEEE 11th Int. Symp. on SACI* Timisoara, Romania, 2016, p. 273-6

[3] Raikar M. M., Desai P., Naragund J. G.. Active Learning Explored in Open Elective Course: Internet of Things (IoT). *C. IEEE T4E* Mumbai, India, 2016, pp. 15-18

[4] Plaza, Pedro, et al. Wireless Development Boards to Connect the World. Online Engineering & Internet of Things. *J. Lec. Notes in Net. and Sys., Springer Int. Pub.* 2018, vol. **22**, pp.19-27

[5] Dobrilovic D. et al. Building ESP8266 Wi-Fi module network based on open-source hardware and single-board computers. *C MECHEDU* 2017. Serbia, 2017, pp.101-6

[6] Lyzhin I.G. Obrazovatel'nye konstruktory v sfere internet veshhej (IoT). *C Innovacionnoe razvitie rossijskoj jekonomiki* Moscow, 2017, vol.**5**, pp. 397-9

[7] Separate is a modern open-source Template Engine for PHP Available at: http://separate.esud.info/ (accessed 28 February 2018)

[8] SCART Available at: http://skartshop.ru/konstruktory-skart/konstruktory-skart/product/view/42/220 (accessed 28 February 2018)

[9] Skretch Laboratoriya Available at: http://www.linuxcenter.ru/shop/ops_hard/scratch/scratch_lab / (accessed 28 February 2018)

[10] Amperka. Jodo. Available at: http://amperka.ru/product/yodo / (accessed 28 February 2018)

[11] Lazy Smart: LS Monitoring Available at: http://lazysmart.ru/LSMonitor (accessed 28 February 2018)

[12] Mgbot. IoT Smart Agriculture Available at: https://mgbot.ru/catalog/iot/bazovyj-uchebnyj-komplekt-iot-smart-agriculture (accessed 28 February 2018)

[13] SCART Available at: http://skartshop.ru/6-10-let/product/view/47/247 (accessed 28 February 2018)

IOP Conf. Series: Journal of Physics: Conf. Series **1163** (2019) 012016 doi:10.1088/1742-6596/1163/1/012016

Peculiarities of non-equilibrium critical behavior of site-diluted 2D Ising model

P N Malyarenko[†], **V V Prudnikov**[‡], **and P V Prudnikov**[§]
Omsk State University, Pr. Mira 55A, Omsk 644077, Russia
E-mail: [†]petr.malyarenko@yandex.ru, [‡]prudnikv@univer.omsk.su, [§]prudnikp@univer.omsk.su

Abstract. The results of a numerical Monte Carlo study of features of nonequilibrium critical behavior in a two-dimensional pure and structurally disordered Ising models are presented with realization of evolution from different initial states within the wide range of spin concentrations p = 1.0, 0.95, 0.9, 0.85, 0.8, 0.75, and 0.7. Analysis of the two-time dependences of the autocorrelation function and dynamic susceptibility revealed a substantial influence of the initial states on the aging effects. The violation of the fluctuation-dissipation theorem is studied, and the asymptotic values of the fluctuation-dissipation ratio are calculated.

At the present time systems with slow dynamics attract considerable interest of researchers. This is associated with the aging properties and violations of the fluctuation-dissipation theorem [1]. These features of the nonequilibrium behavior can be observed in systems undergoing the second order phase transition [2], because their critical dynamics is characterized by anomalously long relaxation times t_{rel}.

The effect of critical slowing-down is an one of the features appearing in the critical behaviour of various systems. This phenomenon is associated with an abnormal increase in the relaxation time of a system $t_{\rm rel}$ when the temperature tends to the critical temperature T_c of the second order phase transition: $t_{\rm rel} \sim |T - T_c|^{-z\nu}$, where z and ν are the dynamic critical exponent and the critical exponent for the correlation length, respectively. As a result of this, a system in the critical point turns out be unable to come to equilibrium state during the relaxation process. The aging effects manifest themselves in the nonequilibrium behavior just at times $t \ll t_{\rm rel}$ and are characterized by the two-time dependence of the correlation and response functions on the observation time t and waiting time t_w. The waiting time characterizes the time interval from the preparation of the sample until starting measurements of its characteristics. The basically important manifestation of the slow dynamics is the violation of the fluctuation-dissipation theorem [1, 2], which suggests relation between autocorrelation function $C(t, t_w)$ and response function $R(t, t_w)$ on some external field as

$$R(t, t_w) = \frac{X(t, t_w)}{T} \frac{\partial C(t, t_w)}{\partial t_w}, \tag{1}$$

where $X(t, t_w)$ is an additional parameter referred to as the fluctuation-dissipation ratio (FDR). For times $t > t_w \gg t_{\rm rel}$, the fluctuation-dissipation theorem gives $X(t, t_w) = 1$. However, in the general case, for times $t, t_w \ll t_{\rm rel}$, we have $X(t, t_w) \neq 1$. The asymptotic FDR is introduced as

$$X^{\infty} = \lim_{t_w \to \infty} \lim_{t \to \infty} X(t, t_w) \tag{2}$$

and can be used as an universal quantity of nonequilibrium processes in various systems.

In this work, we report the numerical Monte Carlo study of the effects of initial states and structural defects on the aging phenomena and the asymptotic values of the FDR in the two-dimensional Ising model. The Hamiltonian for the ferromagnetic Ising model diluted by nonmagnetic impurities with external magnetic field h is given by the expression

$$H = -J \sum_{\langle i,j \rangle} p_i p_j S_i S_j - h \sum_i p_i S_i, \tag{3}$$

where $J > 0$ is the short-range exchange interaction between spins $S_i = \pm 1$ fixed at the sites of a planar square lattice. The occupation numbers p_i assume the value 0 or 1, and p_i equals 1 if site contains spin and 0 otherwise if spin is absent (a magnetic atom is replaced by a nonmagnetic impurity atom). The structural defects are distributed in the system in a canonical manner according to the distribution function $P(p_i) = (1-p)\delta(p_i) + p\delta(p_i)$, where $p = \langle p_i \rangle$ specifies the spin density in the system.

The nonequilibrium process is characterized by the magnetization

$$M(t) = \frac{1}{V}\int d^d x [\langle S(x,t)\rangle] = \left[\left\langle \frac{1}{N_s}\sum_{i=1}^{N_s} p_i S_i(t) \right\rangle\right], \tag{4}$$

two-time correlation function $C(t, t_w)$, and linear response function $R(t, t_w)$ to a weak field applied at time t_w. The functions $C(t, t_w)$ and $R(t, t_w)$ can be defined as

$$C(t,t_w) = \frac{1}{V}\int d^d x \left[\langle S(x,t)S(x,t_w)\rangle - \langle S(x,t)\rangle \langle S(x,t_w)\rangle\right], \tag{5}$$

$$R(t,t_w) = \frac{1}{V}\int d^d x \left.\frac{\delta\left[\langle S(x,t)\rangle\right]}{\delta h(x,t_w)}\right|_{h=0}, \tag{6}$$

where the angle brackets stand for averaging over the different realizations of initial state and the square brackets are for averaging over the different impurity configurations. $N_s = pL^2$ in (4) characterizes the number of spins in the lattice with the linear size L.

In this work, we use by analogy with [3,4] the technique allowing us to calculate the response function at zero applied magnetic field by calculating the generalized susceptibility in the form of the integral response function (thermostatic susceptibility):

$$\chi(t,t_w) = \int_0^{t_w} dt' R(t,t') = \frac{1}{T_c N}\sum_i \left[\langle p_i S_i(t)\Delta S_i(t_w)\rangle\right] \tag{7}$$

with the response function specified by (6) and $\Delta S_i(t_w) = \sum_{s=0}^{t_w}[S_i(s) - S_i^W(s)]$ calculated in the simulation process from initial state at $t = 0$ until the time t_w, where $S_i^W(s) = \tanh(J\sum_{j\neq i} p_j S_j/T)$.

We have simulated systems with different spin concentrations p for the square lattice with the linear size $L = 1024$ at the corresponding critical temperatures shown in Table 1. We used for simulation the thermal bath algorithm with single-spin overturns. At the early stage of the evolution of the system, the correlation length is still sufficiently small and the finiteness of the system used in the simulations is insignificant. Therefore, the use of the lattice with the large enough linear size $L = 1024$ allows us to neglect the finite size effects in comparison to their manifestations in the simulations of the equilibrium critical phenomena. We prepared initial states of the system with values of the magnetization $m_0 = 0.01 \ll 1$ and $m_0 = 1$. The behavior

Table 1. Values of the critical temperature T_c, critical exponents θ', $\beta/(\nu z)$, z, and the fluctuation-dissipation ratio for a low-temperature X_{LT} and a high-temperature X_{HT} initial states at different spin concentrations p. Values of T_c were taken from [5,6].

p	T_c	θ'	$\beta/(\nu z)$	z	$X_{LT}^{(t-t_w \sim t_w)}$	X_{LT}^{∞}	X_{HT}^{∞}
1.0	2.2692	0.192(4)	0.05784(30)	2.161(11)	0.751(24)	0.751(24)	0.339(19)
0.95	2.08989(8)	0.191(3)	0.05794(32)	2.157(12)	0.750(25)	0	0.329(20)
0.9	1.9032(5)	0.190(3)	0.05798(33)	2.172(13)	0.748(25)	0	0.317(21)
0.85	1.7098(4)	0.180(2)	0.05352(33)	2.336(15)	0.726(26)	0	0.265(21)
0.8	1.5103(4)	0.166(1)	0.05034(34)	2.483(17)	0.714(26)	0	0.256(23)
0.75	1.2980(10)	0.155(1)	0.04522(35)	2.764(21)	0.682(28)	0	0.215(24)
0.7	1.0729(10)	0.128(1)	0.03906(33)	3.197(25)	0.669(31)	0	0.178(24)

of autocorrelation function and dynamic susceptibility was studied for different values of waiting time. The positions of defects are fixed for a specified impurity configuration. The behavior of the system was studied at times up to 10000 Monte Carlo steps. At the simulations of the pure system with $p = 1$, we performed the statistical averaging over 15000 runs. At the simulations of the structurally disordered Ising model, the averaging of the calculated parameters was performed over 2000 impurity configurations and over 15 runs for each impurity configuration.

For the high-temperature initial states with $m_0 \ll 1$, the stage of nonequilibrium evolution exhibits the characteristic growth of magnetization described by a power law $m(t) \sim t^{\theta'}$, where θ' is the independent dynamic critical exponent [2] with calculated values presented in Table 1. This magnetization growth is of purely fluctuation nature. Indeed, if the fluctuations are not taken into account, we have $\theta' = 0$. At times $t > t_m \sim m_0^{-k}$ with $k = 1/(\theta' + \beta/\nu z)$, this stage of evolution is changed by the regime characterized by the relation $m(t) \sim t^{-\beta/(\nu z)}$. At the evolution from the initial state with $m_0 = 1$, the relaxation of magnetization is determined from the very beginning by the power law dependence $m(t) \sim t^{-\beta/(\nu z)}$ [7].

In [8,9], it was shown that the existence of uncorrelated structural defects is insignificant for the statical critical behavior of the two-dimensional Ising model. Indeed, the critical exponents retain the same values as those for the pure model, namely, $\beta = 0.125$ and $\nu = 1$, and the defects lead only to logarithmic corrections to the thermodynamic and correlation characteristics. This prediction was confirmed numerically in a number of papers [5,10,11]. At the same time, the studies of the critical dynamics of the model reported in [12,13] demonstrated that, near the spin percolation threshold (for the systems with the spin concentrations $p \leq 0.85$), the dynamic critical exponent z, determining the temperature dependence of the relaxation time, exhibits the dependence on the concentration of defects with the violation of the standard form of dynamic scaling and takes the form

$$z = A\,|\ln(p - p_c)| + B. \tag{8}$$

Taking into consideration the logarithmic corrections, we assume the time dependence of the magnetization for low-temperature initial state in the form:

$$m(t) \sim t^{-\beta/(\nu z)} \left(1 + (g/z)\ln t\right)^{-\beta/(2\nu)}, \tag{9}$$

where $g = \frac{4}{\pi} g_0 = \frac{4}{\pi} \frac{8}{(1+\sqrt{2}/\pi)^2} \frac{1-p}{p} \approx 4.843 \frac{1-p}{p}$ [5,8]. For the pure system it corresponds to the conventional power-law time dependence of the magnetization $m(t) \sim t^{-\beta/(\nu z)}$. To check (8), we calculated the values of the exponents $\beta/(\nu z)$ and z according to (9) and considered logarithmic dependence of the exponents on the spin concentrations (Fig.1). The relation (8) is confirmed in

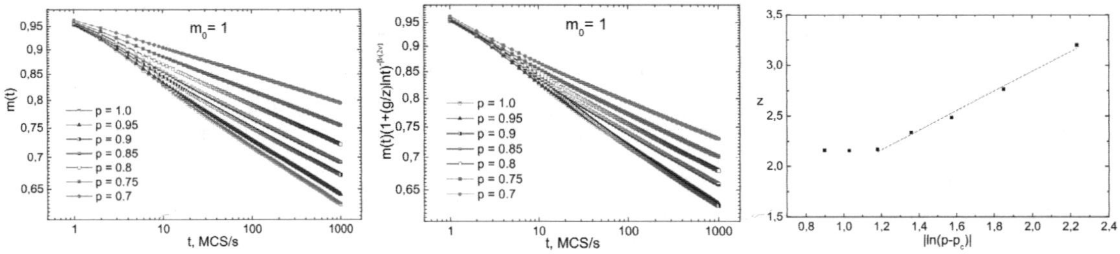

Figure 1. Time dependence of the magnetization $m(t)$ for $m_0 = 1$ (left) and $m(t)\,(1 + (g/z)\ln t)^{-\beta/2\nu}$ (center), determining the dynamical critical exponent z for different spin concentrations p. The right graph illustrates the logarithmic dependence of the exponent z on the spin concentration.

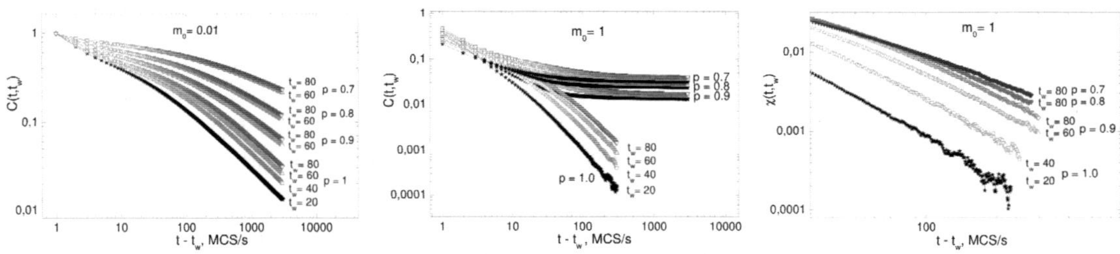

Figure 2. Nonequilibrium dependence of the autocorrelation function $C(t, t_w)$ for $m_0 = 0.01 \ll 1$ (left) and $m_0 = 1$ (center) and the dynamic susceptibility $\chi(t, t_w)$ for $m_0 = 1$ (right) on the observation time $t - t_w$ at different spin concentrations p and waiting times t_w.

the case of the spin concentrations $p \leq 0.85$. However, the values of the exponents $\beta/(\nu z)$ and z for systems with $p = 0.95$ and $p = 0.9$ coincide within the statistical errors with the values for the pure system $\beta/(\nu z) = 0.05784(30)$ and $z = 2.161(11)$. This result confirms that the values of the dynamic critical exponent z for weakly disordered systems with $p \geq 0.9$ remain constant and the critical dynamics of the structurally disordered Ising model belongs to the universality class of the pure model [13, 14]. The value of z for the pure system is in a good agreement with the value $z = 2.1667(5)$ from the work [15]. The exponent $z'(p)$ calculated for structurally disordered systems as $m(t) \sim t^{-\beta/(\nu z')}$ is characterized by the values $z'(0.9) = 2.646(13)$, $z'(0.8) = 3.085(19)$, $z'(0.7) = 4.004(35)$, which are much more in comparison to values of $z(p)$ computed subject to the logarithmic corrections (Table 1). Therefore, logarithmic corrections provide a marked contribution to the values of z. As for the values of the exponent θ', they coincide for systems with $p = 0.95$ and $p = 0.9$ within the error bars with the value $\theta' = 0.192(4)$ for the pure system.

We also calculated the two-time dependence of the autocorrelation function $C(t, t_w)$ and the susceptibility $\chi(t, t_w)$ (Fig.2). The aging effects clearly manifest themselves at the observation times $t - t_w \sim t_w$ and are characterized by the slowing down of these functions with increase of the waiting time t_w in compliance with the generalized scaling forms [2]

$$C(t, t_w) \sim t_w^{-2\beta/z\nu} F_C(t/t_w^{\mu}), \tag{10}$$

$$\chi(t, t_w) \sim t_w^{-2\beta/z\nu} F_\chi(t/t_w^{\mu}). \tag{11}$$

The aging effects increase in the structurally disordered systems in comparison to those in the pure system. Analysis of the scaling functions presented in Fig. 3 revealed that the canonical aging with exponent $\mu = 1$ is observed for evolution from the high-temperature initial state with $m_0 \ll 1$. The scaling function $F_C(t/t_w^{\mu})$ is characterized for case of evolution from

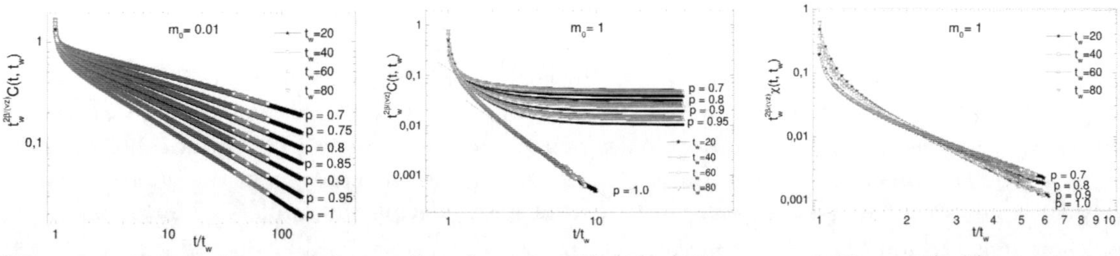

Figure 3. Dependence of the scaling function $F_C(t/t_w)$ for $m_0 = 0.01 \ll 1$ (left) and $m_0 = 1$ (center) and the scaling function $F_\chi(t/t_w)$ for $m_0 = 1$ (right) on the variable $x = t/t_w$ at different spin concentrations p.

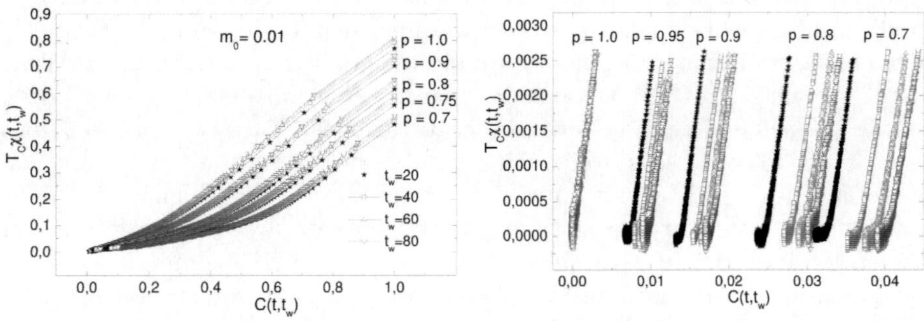

Figure 4. Dependences of susceptibility on the autocorrelation function, determining the fluctuation-dissipation ratio for $m_0 = 0.01$ (left) and $m_0 = 1$ (right) at different spin concentrations p.

the low-temperature initial state by $\mu = 1$ only for the pure system, whereas in structurally disordered systems the correlation effects are strongly slowed down in comparison to the pure system. It leads to superaging effects with values of the exponent $\mu(p) > 1$: $\mu(0.95) = 6.15(5)$, $\mu(0.9) = 6.25(5)$, $\mu(0.85) = 6.50(5)$, $\mu(0.8) = 6.75$, $\mu(0.75) = 7.00(5)$, and $\mu(0.7) = 7.50(5)$. For $F_\chi(t/t_w^\mu)$, the canonical aging is observed for both pure and structurally disordered systems.

At the next stage of our studies, we calculated the fluctuation-dissipation ratio according to the relation [16, 17]:

$$X(t, t_w) = \lim_{C \to 0} T_c \frac{\partial \chi(t, t_w)}{\partial C(t, t_w)}. \tag{12}$$

Dependences $T\chi$ on C are presented in Fig. 4 for the evolution of system from initial states with $m_0 = 0.01$ and $m_0 = 1$. We calculated the FDR for each value of the waiting time t_w, and then the asymptotic value of the FDR X^∞ was determined by extrapolation $X(t_w \to \infty)$. The sought asymptotic values of the FDR are presented in Table 1. The value $X^\infty_{HT} = 0.339(19)$ for the pure system relaxing from the high-temperature initial state is in very good agreement with the value $X^\infty_{HT} = 0.33(1)$ determined in [3] by the MC simulation method. The asymptotic values of the FDR determined for a weakly disordered systems with $p = 0.95$ and $p = 0.9$ coincide within the statistical errors with the value for the pure system. Therefore, these systems belong to the universality class of the pure system. The asymptotic values of the FDR for a strongly disordered systems are appreciably dependent on the concentration of defects owing to the crossover effects in the percolation behavior [12].

For the pure system relaxing from a low-temperature initial state with $m_0 = 1$, the plot of $T\chi$ versus C is linear for the time range $t - t_w \geq t_w$ of variation of the autocorrelation function and characterized by the asymptotic value of the FDR $X^\infty_{LT} = 0.751(24)$. This value is in very good agreement with the field-theoretical value $X^\infty_{LT} = 0.75$ determined in [3]. However, the dependences of $T\chi$ versus C for the structurally disordered systems relaxing from the initial ordered state with $m_0 = 1$ exhibit two linear segments because of the found slowing down of the correlation effects at times $t - t_w \geq t_w$ owing to the pinning of domain walls by the defects. The first segment corresponds to the changes in the autocorrelation function $C(t, t_w)$ at times $t - t_w \sim t_w$, whereas the second segment corresponds to the long-term stage of the evolution with $t - t_w \gg t_w$ (Fig. 4). Second segments lead to the asymptotic FDR values $X_{LT}^{\infty} = 0$ for all spin concentrations $p < 1$ under study. At the same time, the analysis of the dependences of $T\chi$ versus C in the first segments using (12) without considering the limit $C \to 0$ demonstrates that the extrapolation $X_{LT}(t_w \to \infty)$ applied to certain $X_{LT}(t_w)$ gives the values $X_{LT}^{(t-t_w \sim t_w)}$ presented in the Table 1. These values of the FDR for systems with $p = 0.95$ and $p = 0.9$ are close to the asymptotic FDR value for the pure Ising model.

To conclude, we note that our numerical studies have revealed an appreciable role of defects in the nonequilibrium critical dynamics of the two-dimensional Ising model. It was shown that for diluted systems logarithmic corrections provide a significant contribution to values of the dynamic critical exponent z. The realization of two types of universal behavior corresponding to influence of high-temperature and low-temperature initial states is revealed. It has been shown that the weakly disordered systems with $p \geq 0.9$ belong to the universality class of the pure system and are described by the same characteristics of critical behavior. In the case of structurally disordered systems relaxing from the low-temperature initial state, the significant slowing down of the correlation is revealed due to the pinning of domain walls on defects. As a result, superaging effects occur in time behavior of the autocorrelation function and the asymptotic FDR values determined by the domain dynamics in the long-term become equal to zero. For the high-temperature initial state, the asymptotic values of the FDR weakly disordered systems with $p \geq 0.9$ are coincide within the statistical errors with the value for the pure two-dimensional Ising model $X^{\infty} = 0.339(19)$, whereas for strongly disordered systems with $p \leq 0.85$ the asymptotic values of FDR demonstrate a significant decrease. This indicates that nonequilibrium critical behavior of systems with $p \geq 0.9$ belongs to the universality class of the pure system, and the nonequilibrium critical behavior of systems with $p \leq 0.85$ depends on the concentration of defects with the violation of the standard form of dynamic scaling due to the influence of crossover effects of percolation behavior [12, 13].

Acknowledgments

This work was supported by the Russian Foundation for Basic Research, project nos. 17-02-00279 and 18-42-550003, and by the Council of the President of the Russian Federation, project no.MD-6868.2018.2. For calculations, we used the resources provided by the Shared Facility Data Center FEB RAS (Khabarovsk).

References

[1] Vincent E, Hammann J, Ocio M, Bouchaud J P, and Cugliandolo L F 1997 Lect. Notes Phys. **492** 184
[2] Prudnikov V V, Prudnikov P V, and Mamonova M V 2017 *Phys. Usp.* **60** 762
[3] Calabrese P, Gambassi A, and Krzakala F 2006 *J. Stat. Mech.* **6** 2
[4] Prudnikov P V, Prudnikov V V, and Pospelov E A 2013 *JETP Lett.* **98** 619
[5] Shchur L N and O A Vasilyev 2001 *Phys. Rev.* E **65** 016107
[6] Martins P H L and Plascak J A 2007 *Phys. Rev.* E **76** 012102
[7] Prudnikov V V, Prudnikov P V, Pospelov E A, and Malyarenko P N 2018 *JETP Lett.* **107** 569
[8] Dotsenko Vik S and Dotsenko Vl S 1981 *JETP Lett.* **33** 37
[9] Shalaev B N 1984 *Sov. Phys. Solid State* **26** 1089; 1994 *Phys. Rep.* **237** 129
[10] Ballesteros H G et al. 1997 *J. Phys. A* **30** 8379
[11] Selke W, Shchur L N, and Vasilyev O A, 1998 *Physica A* **259** 388
[12] Markov O N and Prudnikov V V 1994 *JETP Lett.* **60** 23
[13] Prudnikov V V and Markov O N 1995 *Europhys. Lett.* **29** 245
[14] Markov O N and Prudnikov V V 1994 *JETP Lett.* **60** 23
[15] Nightingale M P and Blote H W J 1996 *Phys. Rev. Lett.* **76** 4548
[16] Prudnikov V V, Prudnikov P V, and Pospelov E A 2016 *J. Stat. Mech.* 043303
[17] Prudnikov V V, Prudnikov P V, and Malyarenko P N 2017 *JETP* **152** 1102

IOP Conf. Series: Journal of Physics: Conf. Series **1163** (2019) 012017 doi:10.1088/1742-6596/1163/1/012017

The influence of difference in the surface properties on the axisymmetric vibrations of an oblate drop in an AC field

M A Kashina and A A Alabuzhev
Institute of Continuous Media Mechanics UB RAS, Perm 614013, Russia
Perm State University, Perm 614990, Russia

E-mail: alabuzhev@mail.ru

Abstract. The forced oscillations of an incompressible fluid drop in the non-uniform electric field are considered. The external electric field acts as an external force that causes motion of the contact line. In order to describe this contact line motion the modified Hocking boundary condition is applied: the velocity of the contact line is proportional to the deviation of the contact angle and the rate of the fast relaxation processes, whose frequency is proportional to twice the frequency of the electric field. The equilibrium drop has the form of a cylinder bounded by axially parallel solid planes. These plates have different surface (wetting etc.) properties. The solution of the problem is represented as a Fourier series in eigenfunctions of the Laplace operator. The resulting system of heterogeneous equations for unknown amplitudes was solved numerically. The amplitude-frequency characteristics and the evolution of the drop shape are plotted for different values of the problem parameters.

1. Introduction

Control of inclusions (drops, bubbles, particles) in fluid systems is an important research topic. One of the promising methods is changing of wetting properties of a substrate by a liquid, for example, due to an external electric field (electrowetting-on-dielectric – EWOD) [1,2]. Now EWOD has found wide application in various fields, such as electronic display technology [3,4], variable-focus liquid lenses [5,6], digital (droplet) microfluidic devices for bioanalysis (lab-on-a-chip) [7,8], etc.

The Young–Lippmann equation is used in most articles for theoretical description of the contact angle ϑ at EWOD (fig. 1) [1,9]. Nevertheless, the experimental results show a considerable departure from the theoretical predictions of the Young–Lippmann equation: although a zero contact angle is obtained after reaching some critical voltage angle (complete wetting and the contact angle does not change), real experiments invariably yield a finite value of the angle [1,2,10].The mechanism of the contact angle saturation has not yet been elucidated and is the object of discussion [1].

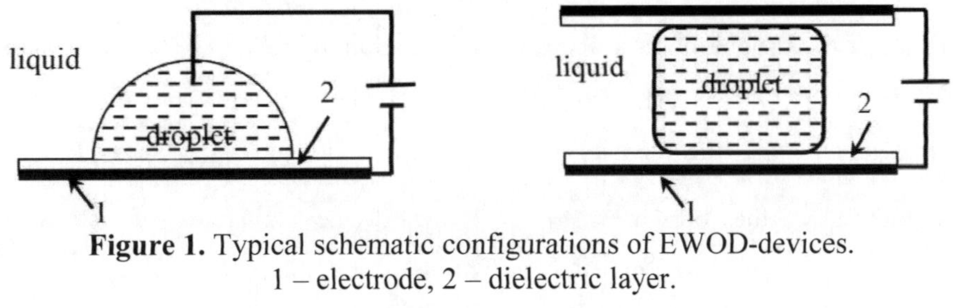

Figure 1. Typical schematic configurations of EWOD-devices.
1 – electrode, 2 – dielectric layer.

As an alternative, another effective boundary condition is proposed in [11] taking as a premise the Hocking equation [12] for a cylindrical drop:

$$\frac{\partial \zeta^*}{\partial t^*} = \pm\Lambda^* \left(\frac{\partial \zeta^*}{\partial z^*} + A^* \cos\left(2\omega^* t^*\right) \right), \tag{1}$$

where ζ^* is the deviation of the drop interface from the equilibrium position, z^* is the axial coordinate, Λ^* is a phenomenological constant (the so-called wetting parameter or Hocking parameter), having the dimension of velocity, A^* is the effective AC amplitude, ω^* is the AC frequency. We also investigated the effect of a non-uniform electric field [11,13] and heterogeneous substrates [13,14]. In the opinion of the authors of [15], a non-uniform field or heterogeneity of substrate can cause the observed effects. In the general case, the Hocking equation [12] is used for a theoretical description of the contact line dynamics under free and forced oscillations [16-22].

This paper is intended as an extension of work [11]. In contrast to all other works [11,13-14,19-22], we consider the case of different plates, i.e. the plates have different surface (wetting etc.) properties. In order to describe the motion of the contact line, the modified boundary condition (1) is used: the velocity of the contact line is proportional to the deviation of the contact angle and the rate of fast relaxation processes, whose frequency is proportional to twice the electric field frequency.

2. Problem formulation

The problem formulation largely coincides with that developed in articles [11,13,14]. An incompressible liquid drop of density ρ_i^* is surrounded by another liquid of density ρ_e^* (see figure 2). In what follows, the quantities with subscript i refer to the drop, and those with subscript e – to the surrounding liquid. Both liquids are bounded by two parallel solid surfaces at a distance h^* from one another. The equilibrium contact angle ϑ_0 between the lateral surface of the drop and the solid surface is equal to $\pi/2$. The external non-uniform alternating electric field acts as an external force that causes the contact line motion.

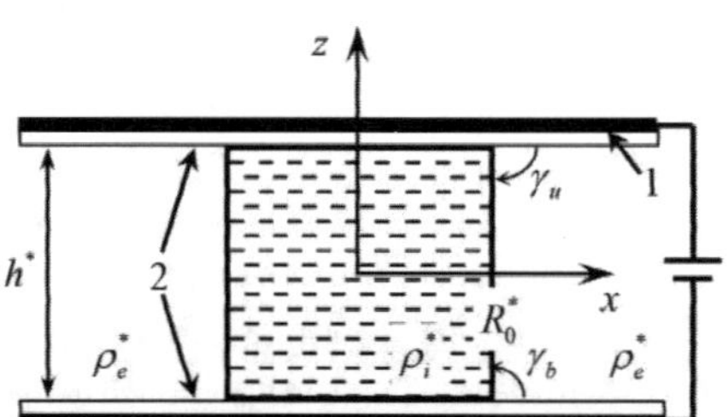

Figure 2. Problem geometry (1 – electrode, 2 – dielectric layer).

Let the surface of the droplet be described by the equation $r^* = R_0^* + \zeta^*\left(\alpha, z^*, t^*\right)$ according to the cylindrical coordinates r^*, α, z^*. The azimuthal angle α is reckoned from the x-axis. Taking the length R_0^*, the height h^*, the density $\rho_e^* + \rho_i^*$, the time $\sigma^{-1/2}\sqrt{\left(\rho_e^* + \rho_i^*\right)R_0^{*3}}$, the velocity $A^*\sqrt{\sigma}\left(\left(\rho_e^* + \rho_i^*\right)R_0^{*3}\right)^{-1/2}$, the pressure $A^*\sigma\left(R_0^*\right)^{-2}$, and the deviation of the surface A^* as characteristic quantities, we pass to dimensionless variables and obtain the following linear problem:

$$p_j = -\rho_j \varphi_{jt}, \ \Delta\varphi_j = 0, \ j = i, e, \tag{2}$$

$$\Delta = \frac{1}{r}\frac{\partial}{\partial r}\left(r \frac{\partial}{\partial r} \right) + \frac{1}{r^2}\frac{\partial^2}{\partial \alpha^2} + b^2 \frac{\partial^2}{\partial z^2},$$

$$r=1:\ [\varphi_r]=0,\ \zeta_t=\varphi_r,\ [p]=\zeta+\zeta_{\alpha\alpha}+b^2\zeta_{zz},\tag{3}$$

$$z=\pm 0.5:\ \varphi_z=0,\tag{4}$$

$$r=1,\ z=\pm 0.5:\ \zeta_t=\mp\lambda_{b,u}\left(\zeta_z+af(\alpha)\cos(2\omega t)\right),\tag{5}$$

where p is the fluid pressure, φ is the velocity potential, $f(\alpha)$ is the function of the non-uniform electric field, the square brackets denote the jump in the quantity at the interface between the external liquid and the drop, λ_u and λ_b are the Hocking parameter of the "top" $(z=0.5)$ and "bottom" $(z=-0.5)$ substrate, respectively. The subscripts (t, r, α or z) at the unknown functions denote differentiation with respect to the corresponding variables. The boundary-value problem (2)–(5) involves six parameters: the aspect ratio, the dimensionless density, the wetting parameter, the AC frequency and amplitude

$$b=R_0^*h^{*-1},\ \rho_i=\rho_i^*\left(\rho_e^*+\rho_i^*\right)^{-1},\ \rho_e=\rho_e^*\left(\rho_e^*+\rho_i^*\right)^{-1},\ \lambda=\Lambda^*\sigma^{-1/2}b\sqrt{\left(\rho_e^*+\rho_i^*\right)R_0^*},$$

$$\omega=\omega^*\sigma^{-1/2}\sqrt{\left(\rho_e^*+\rho_i^*\right)R_0^{*3}},\ a=0.5A^*C\sigma^{-3/2}\sqrt{\left(\rho_e^*+\rho_i^*\right)R_0^{*3}},$$

where $C=\varepsilon_0\varepsilon_d d^{-1}$ is capacitance per unit area, ε_0 and ε_d are the vacuum and the dielectric layer permittivity, respectively, d is the thickness of the dielectric film.

3. Forced oscillations

The function $f(\alpha)$ is expanded in the Fourier series in terms of the eigenfunctions of the Laplace operator. Let us consider a particular case of the non-uniform electric field $f(\alpha)=\left|\sin\left(k_a\cos(\alpha)\right)\right|$, where k_λ are the wavenumber. Note, that we used this function $f(\alpha)$ in [11]. The solutions for the velocity potential and the surface deviation are written as

$$\varphi_i(r,\alpha,z,t)=\mathrm{Re}\left(i2\omega\sum_{m=0}^{\infty}\sum_{k=0}^{\infty}\left(a_{1mk}R_{1mk}^i(r)\sin((2k+1)\pi z)+a_{2mk}R_{2mk}^i(r)\cos(2\pi kz)\right)\cos(2m\alpha)e^{i2\omega t}\right),\tag{6}$$

$$\varphi_e(r,\alpha,z,t)=\mathrm{Re}\left(i2\omega\sum_{m=0}^{\infty}\sum_{k=0}^{\infty}\left(b_{1mk}R_{1mk}^e(r)\sin((2k+1)\pi z)+b_{2mk}R_{2mk}^e(r)\cos(2k\pi z)\right)\cos(2m\alpha)e^{i2\omega t}\right),\tag{7}$$

$$\begin{aligned}\zeta(z,t)=\mathrm{Re}\Bigg(\Bigg(&\sum_{m=0}^{\infty}\sum_{k=0}^{\infty}\left(c_{1mk}\sin((2k+1)\pi z)+c_{2mk}\cos(2\pi kz)\right)\cos(2m\alpha)\\&+d_{10}\sin\left(\frac{z}{b}\right)+d_{20}\cos\left(\frac{z}{b}\right)+\sum_{m=1}^{\infty}\left(d_{1m}\,\mathrm{sh}\left(\frac{\sqrt{4m^2-1}}{b}z\right)+d_{2m}\,\mathrm{ch}\left(\frac{\sqrt{4m^2-1}}{b}z\right)\right)\cos(2m\alpha)\Bigg)e^{i2\omega t}\Bigg),\end{aligned}\tag{8}$$

where I_m and K_m are the modified Bessel functions of the m-th order. Substituting solutions (6)–(8) into (2)–(5), we obtain the expressions for the unknown amplitudes a_{jmk}, b_{jmk}, c_{jmk}, d_{j0} and d_{jm}. These expressions are equivalent to the similar solutions obtained in [11,13,14] for $\lambda_u=\lambda_b=\lambda$.

For convenience, as a maximum deviation of the drop surface from the equilibrium position, we prescribe on the "upper" plate $z=0.5$ – $\zeta_u=\max(\zeta(0,0.5,0))$, on the "bottom" plate $z=-0.5$ – $\zeta_u=\max(\zeta(0,-0.5,0))$, in the center of the layer $z=0$ – $\zeta_0=\max(\zeta(0,0,0))$ and a "quarter" position $z=0.25$ – $\zeta_q=\max(\zeta(0,0.25,0))$; the values of the internal contact angle γ on the "upper" plate are γ_u, at the "bottom" plate – γ_b; the deviation from the equilibrium contact angle on the "upper" plate is $\delta_u=\max(\gamma_u-0.5\pi)$ and on the "bottom" plate – $\delta_b=\max(\gamma_b-0.5\pi)$.

The main contribution to a change in the drop dynamics is made by the Hocking parameter λ [16-22]. Figure 3 shows the oscillation amplitude of the drop surface and the deviation of the contact angle as a function of the frequency of the uniform electric field for several values of the Hocking parameters λ_u and λ_b. The amplitudes of the surface oscillations and the contact angle reach maximum values in a linear resonance. It is also seen from the graphs that the values of the resonant frequencies decrease with an increase of λ_u or λ_b. Despite weak dissipation at small values of the parameter λ, the amplitude of surface oscillations is finite (fig. 3a) and the amplitude of oscillations of the contact line is small (fig. 3c). The contact angle varies in a wide range (fig. 3e, f). It is important to note that if both parameters λ_u and λ_b are small but not equal to each other, the amplitude of the surface oscillations is always finite. However, if these parameters are small but equal, the amplitude of oscillations increases indefinitely.

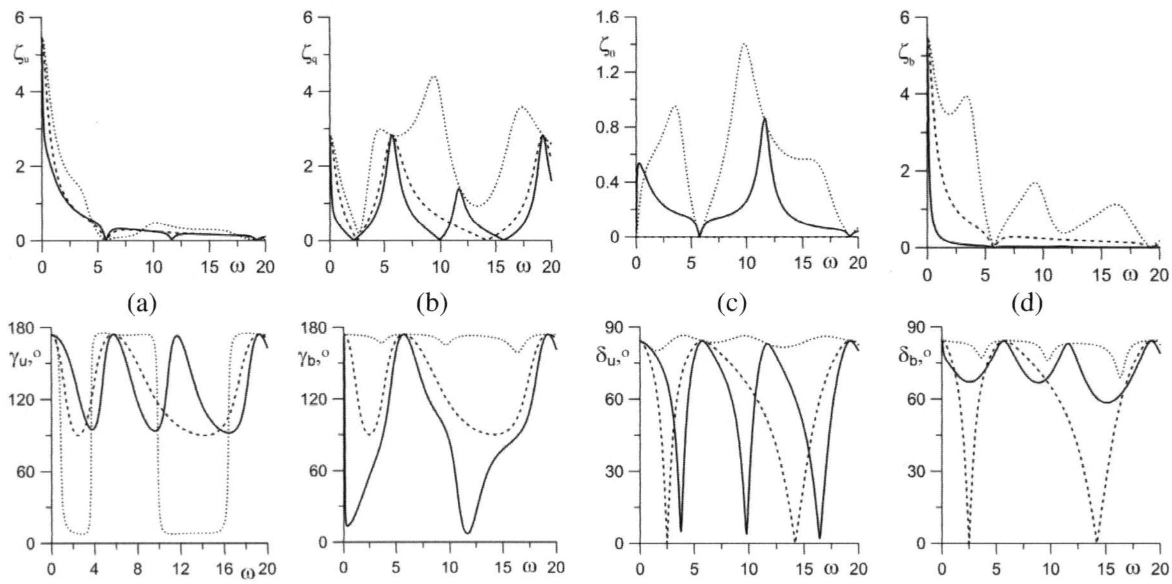

(a) (b) (c) (d)

(e) (f) (g) (h)

Figure 3. The accuracy of representation of the drop surface (a-d) and the contact angle (d-f) vs the frequency ω

($a=10$, $b=1$, $\rho_i=0.7$, $\lambda_u=1$, $f(\alpha)=1$), $\lambda_b=0.1$ – solid line, $\lambda_b=1$ – dashed, $\lambda_b=10$ – dotted.

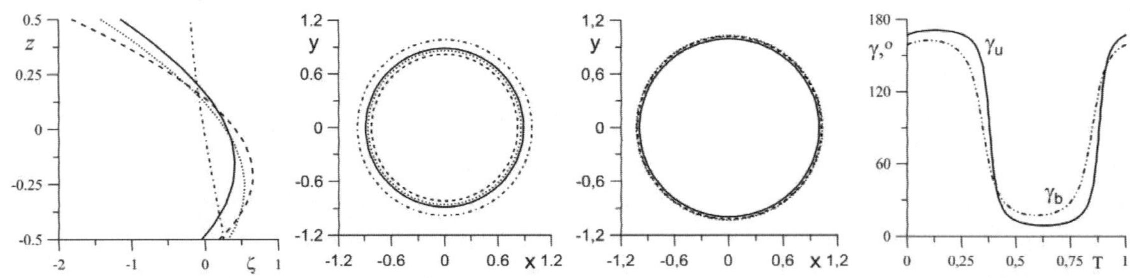

(a) (b) (c) (d)

Figure 4. Evolution of the drop surface shape (a), the shape of the contact line (b,c) and the contact angles (d). $T=2\pi\omega^{-1}$ is the oscillation period.

($b=1$, $\rho_i=0.7$, $a=10$, $\lambda_u=1$, $\lambda_b=0.1$, $\omega=1$, $\varepsilon=0.1$, $f(\alpha)=1$),

(a-c) $t=0$ – solid line, $t=0.125T$ – dashed, $t=0.25T$ – dotted, $t=0.375T$ – dash-dotted.

For clarity, fig. 3 shows the case of equality of the Hocking parameters $\lambda_u = \lambda_b = 1$. In this situation, the external force excites only odd spatial modes, so that there is no deviation of the drop surface in the center of the layer (fig. 3 a). Recall that for finite values of the parameter λ, a dissipation is maximum during the movement of the contact line, and therefore, the plots do not display pronounced resonance peaks (fig. 3 b, c). The motion of a drop does not depend on the parameter λ at certain frequencies ω: the contact line is like a fixed contact line for any values of λ (fig. 3 b, c). The values of such "anti-resonant" frequencies are determined from the solution (6)–(8). Figure 4 shows the profile of the lateral surface (fig. 4 a) and the contact line (fig. 4 b, c) and changes in the internal contact angle (fig. 4 d) at different moments of the oscillation period. The shape of the drop surface depends on the frequency of the electric field. For example, in fig. 4, most of the vibration energy at a given frequency $\omega = 1$ (see fig. 3 a) is transmitted to the lowest spatial mode.

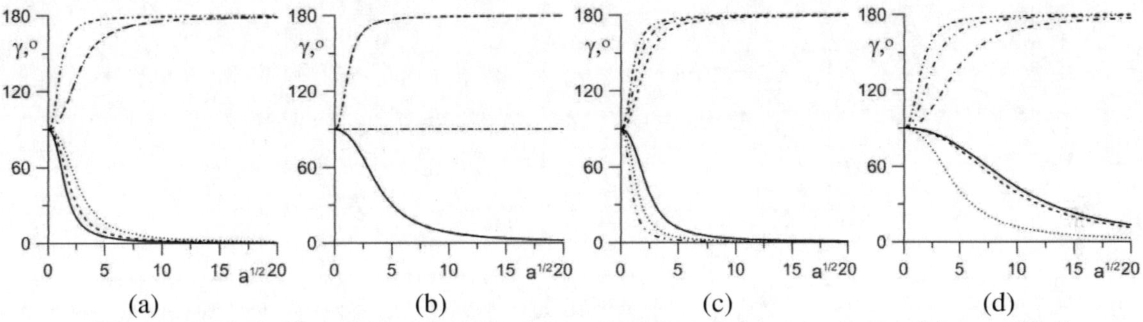

Figure 5. Deviation of the contact angle as a function of the square root of the amplitude ($b = 1$, $\rho_i = 0.7$, $a = 10$, $\lambda_u = 1$, $f(\alpha) = 1$), (a) $\omega = 1$, (b) $\omega = 2.5$, (c) $\omega = 5$, (d) $\omega = 10$,

γ_u: $\lambda_b = 0.1$ – solid line, $\lambda_b = 1$ – dotted, $\lambda_b = 10$ – dash-2-dotted;

γ_b: $\lambda_b = 0.1$ – dashed line, $\lambda_b = 1$ – dash-dotted, $\lambda_b = 10$ – 2-dash-2-dotted

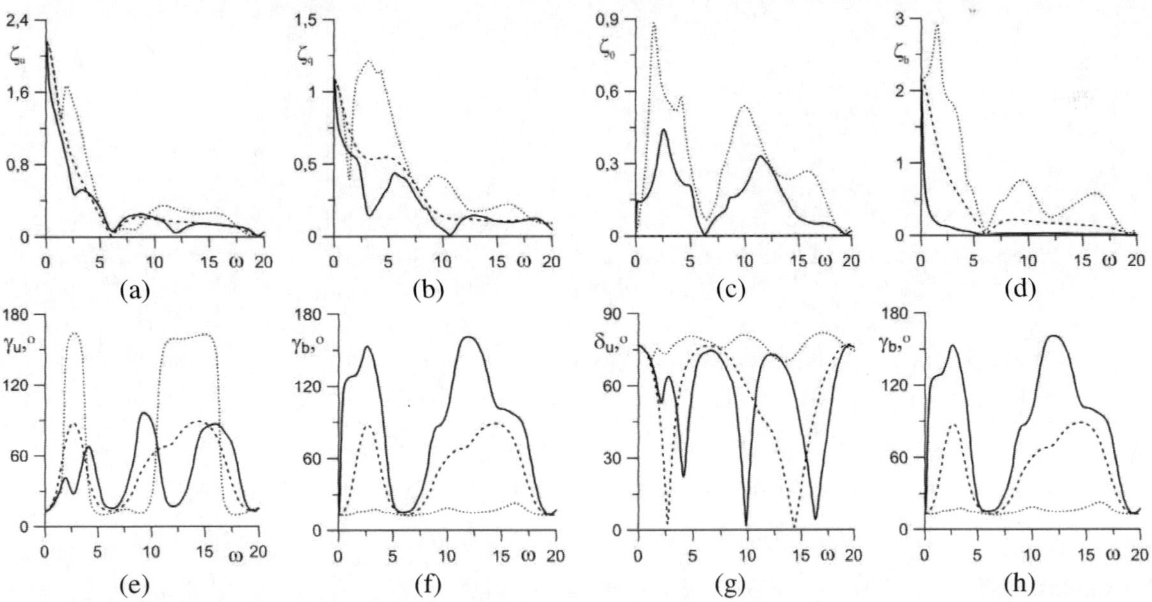

Figure 6. The accuracy of representation of the drop surface (a-d) and the contact angle (d-f) vs frequency ω

($a = 5$, $b = 1$, $\rho_i = 0.7$, $\lambda_u = 1$, $k_a = 1$), $\lambda_b = 0.1$ – solid line, $\lambda_b = 1$ – dashed, $\lambda_b = 10$ – dotted.

The deviation of the contact angle as a function of the square root of the amplitude a (i.e. proportional to AC potential V) is given in fig. 5 for different values of the Hocking parameter λ and AC frequency ω. The responses obtained are in qualitative agreement with the experimental data. The maximum deviation of the contact angle tends to π/2, i.e. $\vartheta \to 0$ or $\vartheta \to \pi$ whereas in experiments the contact angle is finite.

The non-uniform electric field $f(\alpha)=\left|\sin\left(k_a \cos(\alpha)\right)\right|$ excites axisymmetric and azimuthal modes. Consequently the dynamics of the drop differs significantly from its behavior in the uniform field. Similar dependences are shown in fig. 6-8. The amplitude of the lateral surface oscillations has identical local maximum unlike the situation with the uniform electric field. Additional resonance amplitude peaks are associated with the excitation of azimuth modes. A quadrupole mode has a significant effect (fig. 7b): the drop is compressed along field inhomogeneity.

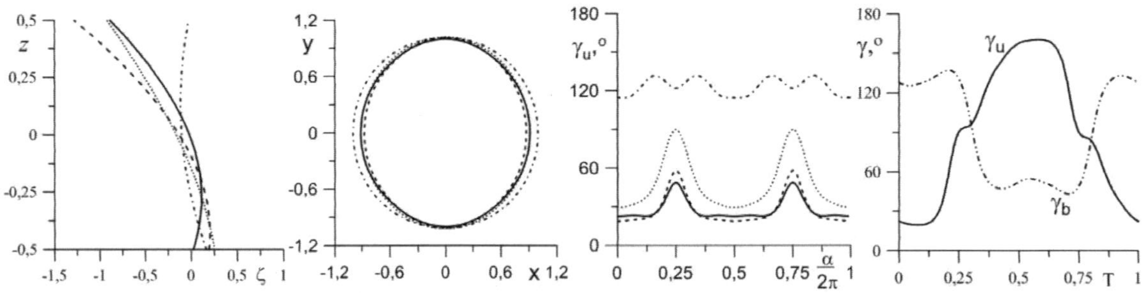

(a) (b) (c) (d)

Figure 7. Evolution of the drop surface shape (a), the shape of the "upper" contact line (b) and the contact angles (c,d). $T = 2\pi\omega^{-1}$ is the oscillation period.
($b=1$, $\rho_i=0.7$, $a=10$, $\lambda_u=1$, $\lambda_b=0.1$, $\omega=1$, $\varepsilon=0.1$, $f(\alpha)=1$),
(a-c) $t=0$ – solid line, $t=0.125T$ – dashed, $t=0.25T$ – dotted, $t=0.375T$ – dash-dotted.

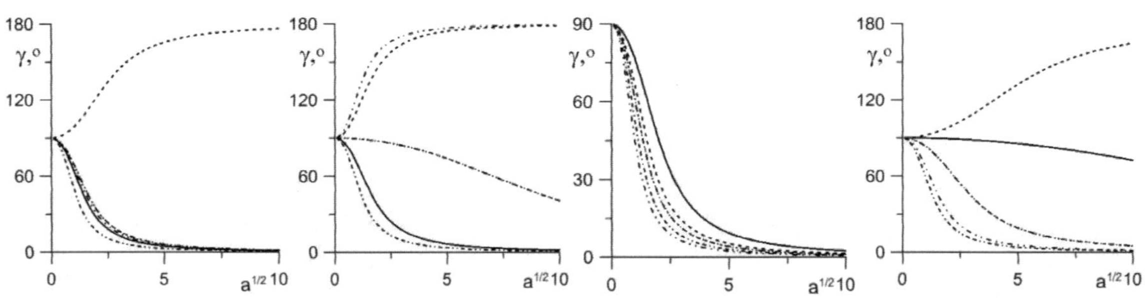

(a) (b) (c) (d)

Figure 8. Deviation of the contact angle as a function of the square root of the amplitude
($b=1$, $\rho_i=0.7$, $a=10$, $\lambda_u=1$, $f(\alpha)=1$), (a) $\omega=1$, (b) $\omega=2.5$, (c) $\omega=5$, (d) $\omega=10$,
γ_u: $\lambda_b=0.1$ – solid line, $\lambda_b=1$ – dotted, $\lambda_b=10$ – dash-2-dotted;
γ_b: $\lambda_b=0.1$ – dashed line, $\lambda_b=1$ – dash-dotted, $\lambda_b=10$ – 2-dash-2-dotted

4. Conclusions

In this paper we studied the behavior of the cylindrical drop between solid plates under the action of the non-uniform electric field $\left|\sin(k_a x)\right|$ taking into account the dynamics of the contact angle. The solid plates had different surfaces.

The study made allowed us to estimate the deviation of frequency and surface characteristics as a function of the Hocking parameter, frequency and amplitude of the external non-uniform electric field and geometric parameters of the system. It was found that a non-uniform field can excite azimuthal modes as was observed, for example, in experiments [15]. The maximum deviation of the contact angle tends to π/2, i.e. $\vartheta \to 0$ or $\vartheta \to \pi$ whereas in experiments the contact angle is always finite. However, the estimates of dimensionless values indicate that the dimensionless amplitude $\sqrt{a}$ is achieved in experiments $\sqrt{a} \le 10$. At this amplitude the contact angle is finite in a large range of the problem parameters. This allows us to expect good agreement between the theoretical predictions and experimental data in the case of a more detailed comparison. The main problem is to measure the value of the Hocking parameter λ at this stage.

Acknowledgments

This work was supported by the Russian Science Foundation (project 14-21-00090).

5. References

[1] Mugele F and Baret J-C 2005 *J. Phys.: Condens. Matter.* **17**
[2] Chen L and Bonaccurso E 2014 *Adv. Colloid Interface Sci.* **210**
[3] Hayes R A and Feenstra B J 2003 *Nature* **425** 383–5
[4] Roques-Carmes T, Hayes R A, Feenstra B J and Schlangen L J M 2004 *J. Appl. Phys* **95** 4389–96
[5] Kuiper S and Hendriks B H W 2004 *Appl. Phys. Lett.* **85**, 1128–30
[6] Li C and Jiang H 2014 *Micromachines* **5** 432-41
[7] Hua Z, Rouse J L, Eckhardt A E, Srinivasan V, Pamula V K, Schell W A, Benton J L, Mitchell T G and Pollack M G 2010 *Anal Chem* **82** 2310-16
[8] Li J, Wang Y, Chen H and Wan J 2014 *Lab Chip* **14** 4334-37
[9] Berge B 1993 *Comptes Rendus Acad. Sci. II* **317**
[10] Chevalliot S, Kuiper S and Heikenfeld J 2012 *J. of Adhesion Sci. Tech.* **26** 1909–30
[11] Alabuzhev A A and Kashina M A 2016 *J. Phys.: Conf. Ser.* **681** 012042
[12] Hocking L M 1987 *J. Fluid Mech.* **179** 253-66
[13] Kashina M A and Alabuzhev A A 2018 *Microgravity Sci. Technol.* **30** (1–2) 11–17
[14] Kashina M A and Alabuzhev A A 2018 *J. Phys.: Conf. Ser.* **955** 012016
[15] Mampallil D, Eral H B, Staicu A, Mugele F and van den Ende D 2013 *Phys. Rev. E* **88** 053015
[16] Shklyaev S and Straube A V 2008 *Phys. Fluids* **20** 052102
[17] Fayzrakhmanova I S and Straube A V 2009 *Phys. Fluids* **21** 072104
[18] Fayzrakhmanova I S, Straube A V and Shklyaev S 2011 *Phys. Fluids* **23** 102105
[19] Borkar A and Tsamopoulus J 1991 *Phys. Fluids A* **3** 2866–74
[20] Perlin M, Schultz W W and Liu Z 2004 *Wave Motion* **40** 41–56
[21] Alabuzhev A A 2016 *Appl. Mech. Tech. Phys.* **53** 9–19
[22] Alabuzhev A A 2018 *Microgravity Sci. Technol.* **30** (1–2) 25–32

IOP Conf. Series: Journal of Physics: Conf. Series **1163** (2019) 012018 doi:10.1088/1742-6596/1163/1/012018

The study of the unstably-stratified marine atmospheric boundary layer by direct numerical simulation

O Druzhinin[1] Yu Troitskaya[1] and S Zilitinkevich[2]

[1]Institute of Applied Physics RAS, Nizhny Novgorod, Russia

[2]Finnish Meteorological Institute, Helsinki, Finland

E-mail: droujmail@mail.ru

Abstract. Prediction of the properties of the atmospheric flows over sea is important for local and regional weather forecast. This prediction typically relies on the performance of large-scale meteorological models based on bulk formulae for air velocity, temperature and humidity. The bulk formulas relate turbulent fluxes of momentum, heat and vapor to bulk air velocity, humidity and air-sea temperature difference. The most widely used parameterizations used to compute the coefficients of proportionality in the bulk formulae are formulated on the basis of the Monin-Obukhov similarity theory (MOST) for different types of air stratification (stable, neutral and unstable). In our previous studies we showed that MOST quite accurately predicts the properties of the air-flow over waved water surface under neutral and stable stratification conditions provided that stratification effects are relatively weak and flow is in a statistically stationary state. In the present work, we perform direct numerical simulation and study the air flow over a waved water surface under unstable stratification conditions where the sea surface temperature is larger than the bulk air temperature. Such situation occurs in the tropical cyclone conditions as well as in polar lows at high latitudes. Our results show that in this case, the air flow dynamics is dominated by the development of large-scale cylindrical coherent vortex structures elongated in the direction of the mean wind. These structures cause a notable deviations from the MOST predictions for the air velocity and temperature profiles.

1. Introduction

Detailed knowledge of the interaction of turbulent wind with surface water waves is necessary for correct parameterization of turbulent exchange at the air-sea interface in prognostic models. This interaction may crucially depend on the temperature difference between air and water surface. Under the conditions of relatively small (about several degrees of K) air-water temperature difference and sufficiently high wind speed (typically, about several m/s) the air flow in the boundary layer (BL) is weakly stratified and turbulent, and its properties are well predicted by the Monin-Obukhov similarity theory (MOST) [1], regardless of the sign of the air-water temperature difference (i.e. for both stable and unstable stratification of the air). If

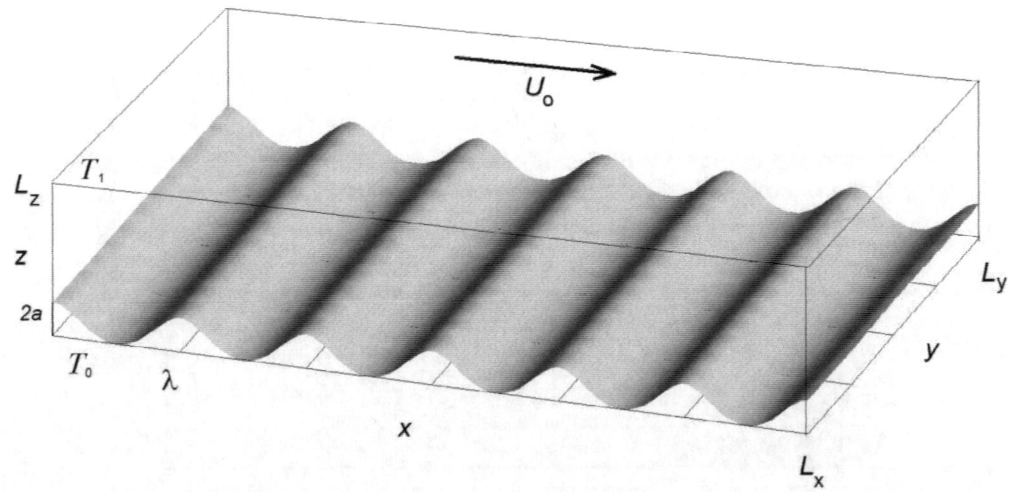

Fig. 1. Schematic of numerical experiment: L_x, L_y, L_z are the domain sizes in the horizontal (x), spanwise (y), and vertical (z) directions; a and λ are the surface water wave amplitude and length; T_0 and T_1 are the bulk temperatures of water and air; U_0 the bulk velocity of the air-flow. The acceleration due to gravity is directed downwards.

this temperature difference increases to tens of Ks, with fixed bulk air velocity, stratification effects may become important.

The case of a sufficiently strong stable stratification (i.e. when the water is cold and the air is warm) was studied recently by performing direct numerical simulation [2]. The results show that strong stratification effectively suppresses turbulence and the flow becomes laminar if the surface wave amplitude is negligible. However, finite air fluctuations in the near-surface layer still survive for sufficiently large wave amplitudes. This regime, where turbulence is induced by surface waves, is poorly described by MOST.

A different situation is realized when the water surface is warm and the air is cold. In this case, stratification is unstable and buoyancy forces induce convective air motions. It is well known that in the absence of shear flow, turbulence can be supported by sufficiently strong (as compared to molecular viscosity) buoyancy forces [1]. Moreover, large-scale convective motions develop into coherent structures forming cells and/or rollers depending on the geometry of the physical domain in the laboratory experiment (or, by topography and planetary boundary layer height in field conditions). These coherent vortical structures are not taken into account by known classical closures. However, recent numerical and laboratory studies indicate that coherent structure significantly influence turbulence statistics and production and dissipation balance [4,5].

In the present paper, we investigate how coherent structures influence the mean velocity and temperature fields in a turbulent Couette flow over a waved water surface. This flow is widely used in direct numerical simulations of atmospheric boundary layer since it is characterized by a constant total momentum flux since it is periodic in the streamwise direction and driven by viscous shear stress at the upper boundary, not by pressure gradient [3]. Section 2 below presents the governing equations and numerical method. Numerical results are discussed in Section 3, and final conclusions and discussion are provided in Section 4.

IOP Conf. Series: Journal of Physics: Conf. Series **1163** (2019) 012018 doi:10.1088/1742-6596/1163/1/012018

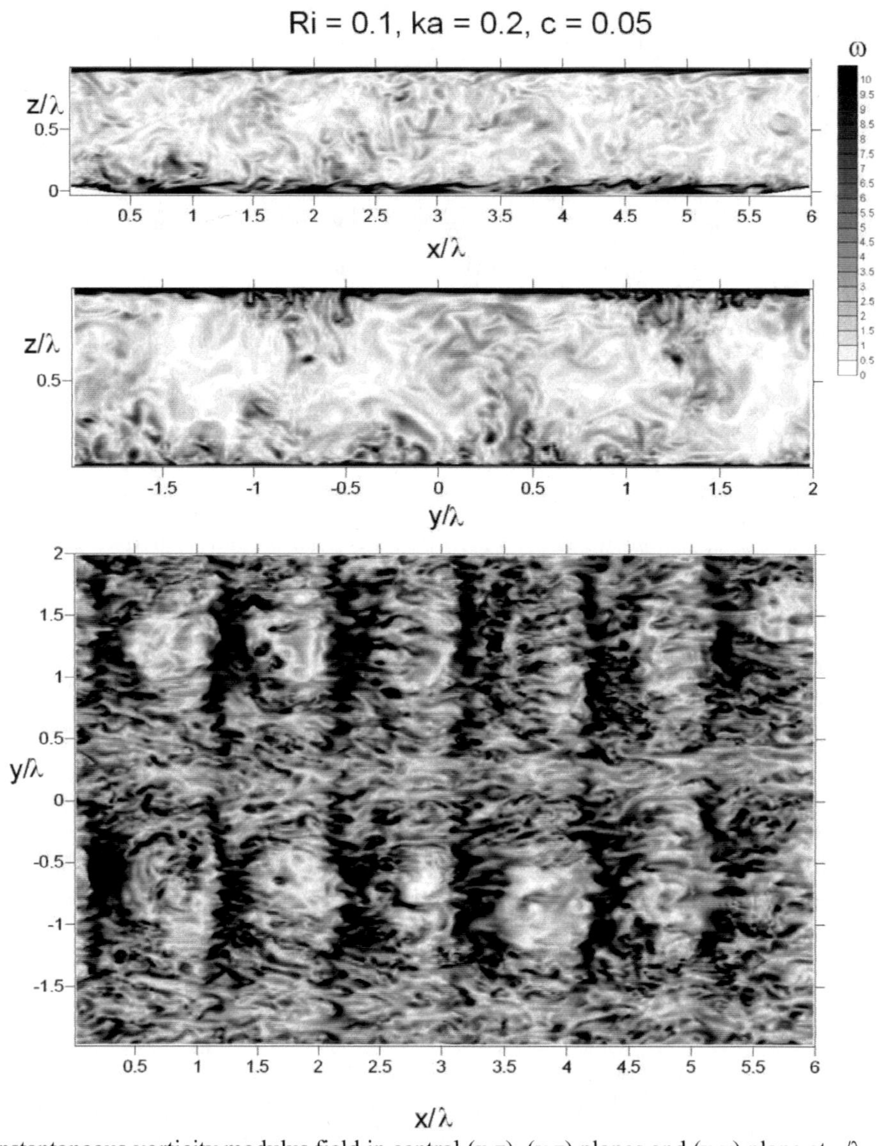

Fig. 2. Instantaneous vorticity modulus field in central (x,z), (y,z) planes and (x,y) plane at z/λ = 0.04.

2. Numerical method and results

We perform direct numerical simulation of turbulent, droplet-laden Couette flow above waved water surface. The schematic of the numerical experiment is similar to that used in our previous DNS study of turbulent air-flow over waved surface [2,3] (Fig. 1). DNS is performed in a reference frame moving with the wave phase velocity, *c*, so that the lower boundary, representing the wave surface, is stationary in the moving reference frame. The no-slip boundary condition is prescribed at the lower boundary, so that the wind velocity at the boundary coincides with the velocity in the water wave. The no-slip boundary condition is also prescribed at the upper horizontal plane moving in *x*-direction with bulk velocity U_0. This condition provides an external source of momentum due to the viscous shear stress, which compensates viscous dissipation in the boundary layer and makes the flow statistically

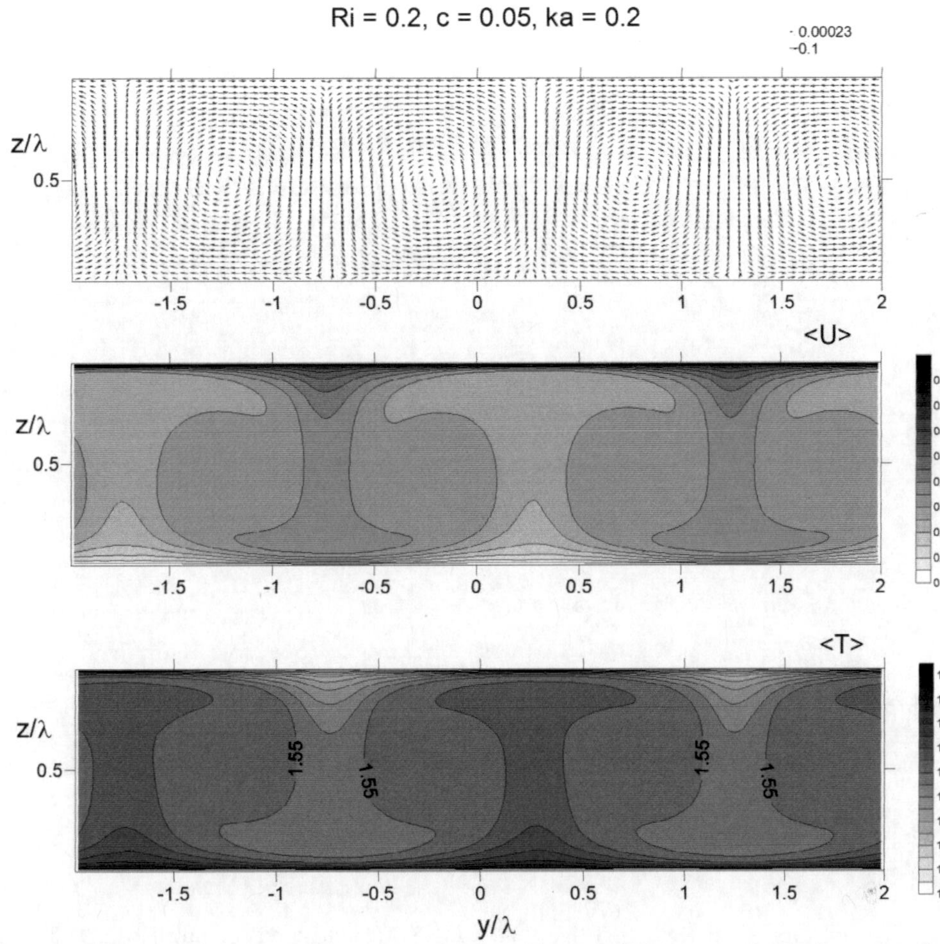

Fig. 3. Ensemble-averaged velocity and temperature fields. Top panel shows mean velocity vector field in (x,z)-plane, and middle and bottom panels present mean x-velocity component and temperature in (y,z) plane, respectively.

stationary. Air and water surface temperatures is prescribed at the top and bottom boundary planes, respectively, as T_0 at $z = 0$ and T_1 at $z = L_z$; where $T_0 > T_1$.

The numerical algorithm is based on the integration of full, 3D Navier-Stokes equations for the carrier air-flow velocity and temperature written under the Boussinesq approximation [3]. The air-flow bulk Reynolds and Richardson numbers in DNS are defined as:

$$\mathrm{Re} = \frac{U_0\lambda}{\nu_a}, \ \mathrm{Ri} = g(T_0 - T_1)\lambda/(T_0U_0^2) \tag{1}$$

and set equal to Re = 15000 and Ri = 0.02 and 0.2 for weak and strong stratification cases. Two different wave-slopes, $ka = 0.1$ and $ka = 0.2$, are considered. We also prescribe the wave celerity to be sufficiently small, $c/U_0 = 0.05$, which corresponds to "slow" waves and strong wind-forcing conditions [3].

Figures 2 and 3 present instantaneous and ensemble-averaged flow fields obtained in DNS. The figures show that the flow field is dominated by large-scale coherent cylindrical counter-rotating rollers elongated in the streamwise direction. Figures 4 and 5 present mean velocity and temperature profiles obtained in DNS and compare them with the MOST prediction [4]:

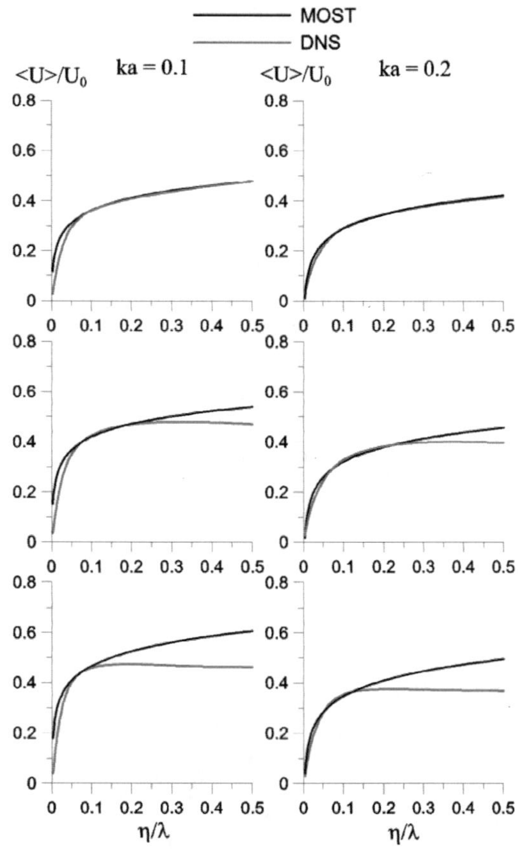

Fig. 4. Mean velocity profiles in DNS (red line) and MOST prediction (black line) obtained for Re = 15000, and Ri = 0, 0.02 and 0.2 (top, middle and bottom panels, respectively) for wave slope ka = 0.1 (left) and ka = 0.2 (right). Wave celerity c/U_0 = 0.05.

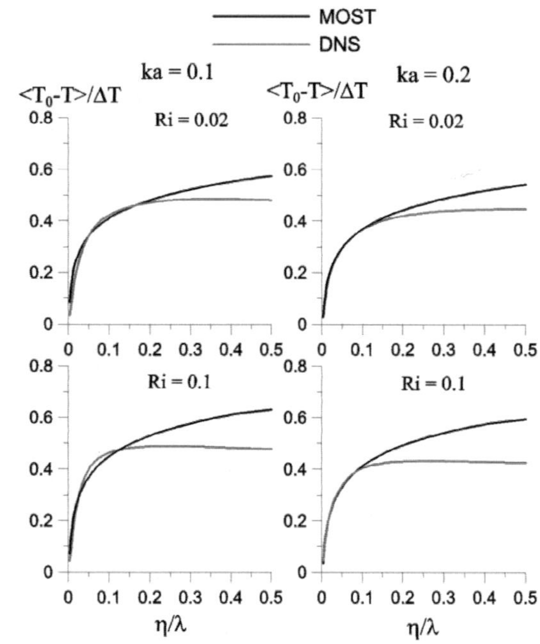

Fig.5. Mean temperature profiles in DNS (red line) and MOST prediction (black line) obtained for for Re = 15000, and Ri = 0.02 and 0.2 (top and bottom panels, respectively) for wave slope ka = 0.1 (left) and ka = 0.2 (right). Wave celerity c/U_0 = 0.05

$$\frac{<U_x>}{u_*}=\frac{1}{\kappa}\left[\ln\frac{z}{z_{0U}}-\psi_U\left(\frac{z}{L}\right)\right],\quad \frac{z}{L}<\frac{z_m}{L}=1.574 \tag{2}$$

$$\frac{<U_x>}{u_*}=\frac{1}{\kappa}\left\{\left[\ln\frac{z_m}{z_{0U}}-\psi_U\left(\frac{z_m}{L}\right)\right]+1.14\left[\left(\frac{z}{L}\right)^{1/3}-\left(\frac{z_m}{L}\right)^{1/3}\right]\right\},\quad z>z_m \tag{3}$$

$$\frac{<T_0-T>}{T_*}=\frac{1}{\kappa}\left[\ln\frac{z}{z_{0T}}-\psi_T\left(\frac{z}{L}\right)\right],\quad \frac{z}{L}<\frac{z_T}{L}=0.465 \tag{4}$$

$$\frac{<T_0-T>}{T_*}=\frac{1}{\kappa}\left\{\left[\ln\frac{z_T}{z_{0T}}-\psi_T\left(\frac{z_T}{L}\right)\right]+0.8\left[\left(\frac{z_T}{L}\right)^{-1/3}-\left(\frac{z}{L}\right)^{-1/3}\right]\right\},\quad z>z_T \tag{5}$$

where $L = \dfrac{u_*^2}{RiT_*}$ is the Obukhov length scale, defined by the fricition velocity and temperature, $z_{0U,T}$ are velocity and temperature roughness lengths and

$$\psi_U = 2\ln\left(\frac{1+\chi}{2}\right) + \ln\left(\frac{1+\chi^2}{2}\right) - 2\tan^{-1}\chi + \frac{\pi}{2},\ \psi_T = 2\ln\left(\frac{1+\chi^2}{2}\right),\ \chi = \left(1+16\frac{z}{L}\right)^{1/4}. \quad (6)$$

The figures show that MOST prediction, Eqs. (1-6), works well throughout the domain in the non-stratified case and in close vicinity of the water surface for stratified cases. However, in the stratified cases, the similarity theory significantly overestimates both mean velocity and temperature as compared to DNS results. The reason for this can be gleaned from the structure of coherent rollers and their effects on ensemble-averaged velocity and temperature fields in Fig. 3. This figure shows that rotating rollers transport slow-moving warmer air upwards and replace it with relatively colder and faster moving air from the upper layers. This transport due to coherent rollers promotes an enhanced mixing which affects mean velocity and temperature profiles, thus causing the deviation of these profiles from MOST prediction which does not take into account the presence of coherent rollers.

4. Conclusions

We have performed direct numerical simulation (DNS) of a turbulent atmospheric boundary layer over waved water surface and investigated the effects of unstable stratification of the air (i.e., the situation where the relatively cold air flows over warm water). The DNS results reveal the presence of large-scale counter-rotating coherent rollers elongated in the streamwise direction. Under the influence of these rollers, an enhanced mixing of the temperature and velocity fields occurs. This additional mixing causes substantial deviations of the mean velocity and temperature profiles from the MOST prediction. Thus, common closure procedure needs improvement under sufficiently strong unstable stratification in order to account for the effects of these coherent vortical structures. This is a subject for future research.

Acknowledgement: This work is supported by RFBR (Nos. 17-05-00703, 18-05-00265, 18-55-50005) and by the Russian Science Foundation (Nos. 14-17-00667, 15-17-20009). SZ additionally acknowledges support from the Academy of Finland project ClimEco No. 314 798/799.

References

[1] Monin A, Yaglom A 1971 *Statistical Fluid Mechanics: Mechanics of Turbulence* **1** (Cambridge).

[2] Druzhinin O A, Troitskaya Yu I, Zilitinkevich S S 2016 Q. J. R. Meteorol. Soc.. **142**. 773-780. doi:10.1002/qj.2678.

[3] Druzhinin O Troitskaya Yu Zilitinkevish S 2017 J. Geophys. Res. Oceans **122** 1789-1807. doi: 10.1002/2016JCO12134.

[4] Zilitinkevich S S, Hunt J C R, Esau I N, Grachev A A, Lalas D P, Akylas S E, Tombrou M, Fairall C W, Fernando H J S, Baklanov A A, Joffre S M 2006 The influence of large convective eddies on the surface-layer turbulence Q. J. R. Meteorol. Soc. 132 1423–1456 doi: 10.1256/qj.05.79

[5] Bukai M, Eidelman A, Elperin T, Kleeorin N, Rogachevskii I., Sapir-Katiraie I 2009 Effect of large-scale coherent structures on turbulent convection Phys. Rev. E **79,** 066302, doi: 10.1103/PhysRevE.79.066302

[6] Zeng X, Zhao M, Dickinson R E 1998 J. of Climate 11, 2628-2644.

Alternative approximations using the similarity theory for turbulent moments of the atmospheric surface layer

A N Vulfson[1,2], **P V Nikolaev**[3]

[1]Institute of Oil and Gas Problems, Russian Academy of Sciences, ul. Gubkina 3, 119333 Moscow, Russian Federation
[2]National Research University Higher School of Economics, ul. Myasnitskaya 20, 101000 Moscow, Russian Federation
[3]National University of Science and Technology MISIS, Leninsky pr. 4, 119991 Moscow, Russian Federation

vulfson@ipng.ru

Abstract. Two limiting cases can be discerned within the atmospheric surface layer by the Monin-Obukhov similarity theory: the dynamic limit with a logarithmic profile of wind velocity and a zero buoyancy flux at the underlying surface; and the free-convection limit with zero wind velocity and the positive buoyancy flux. In this article, a two sublayer model of the convective surface layer is presented. It identifies the lower dynamic sublayer and the upper forced-convection sublayer. The turbulent moments can be determined separately for each of these sublayers by the linear approximations, which are proposed for the first and second turbulent moments of vertical velocity and potential temperature variance. The free-convection limits of the Monin-Obukhov theory corresponds to the first-order expansion terms of the developed approximations. The second-order expansion terms describe profiles of the turbulent moments under conditions with moderate wind. A comparison between the experimental data sets and the proposed approximations shows that the accepted approximations are correct within a forced convection sublayer.

1. Introduction

The classical similarity theory of the atmospheric surface layer was first formulated in [1–3] for the approximation of the first-order turbulent moments.

The Monin-Obukhov similarity theory for the convective surface layer distinguishes two limiting cases: a dynamic limit and a free-convection limit.

The dynamic limit of the convective surface layer is defined as a flow with a logarithmic profile of wind and a zero buoyancy flux at the underlying surface. In the dynamic limit $z/|L_*|=0$, where z is the height above the surface and $|L_*|=\infty$ is the Monin-Obukhov length parameter.

The free-convection limit of the surface layer characterizes a flow with a zero wind speed and a positive buoyancy flux at the underlying surface. In the free convection limit $z/|L_*|=\infty$, where z is the height above the surface and $|L_*|=0$ is the Monin-Obukhov length parameter.

For the limiting cases, the generalized Monin-Obukhov similarity theory makes it possible to calculate higher order turbulent moments as well.

A direct generalization of Monin-Obukhov similarity theory for the approximation of the second order turbulent moments in the dynamic regime was suggested in [4,5]. For the free-convection regime, the Monin-Obukhov similarity theory was first generalized in [6,7].

In this paper, it is supposed that the convective surface layer consists of two sublayers: the lower dynamic sublayer adjacent to the surface and the upper forced-convection sublayer. Therefore, independent approximations can be used for describing turbulent moments of each sublayer.

The limiting regimes of the convective surface layer can be used as first-order approach in constructing linear approximations of the turbulent moments of each of the sublayers.

Second-order linear approximations of turbulent moments in the lower dynamic sublayer, which use the dynamic limits as a first-order approximation, are described in [3,4].

This paper examines the possibility of using second-order linear approximations of turbulent moments of vertical velocity and buoyancy in the upper forced-convection sublayer. The first-order expansion terms of the linear approximations correspond to the free-convection limits of Monin-Obukhov similarity theory under no-wind conditions. The second-order expansion terms of the linear approximations describe the profiles of turbulent moments under wind conditions. A presented comparison with experimental data strongly suggests that linear approximations are correct within the range of $2\cdot 10^{-2} < z/|L_*| < \infty$.

2. Local parameters of friction velocity and buoyancy flux

Let t be time and x, y, z be a Cartesian coordinate system placed at the underlying surface $z=0$ so that the z axis is opposite to gravitational acceleration g.

Let us assume that $u=u(x,y,z,t)$ and $w=w(x,y,z,t)$ are the velocity vector components along the horizontal and vertical axes; $\bar{u}=\bar{u}(z)$ is the mean value of the horizontal wind velocity vector along the x axis; and $u'(x,y,z,t)=u(x,y,z,t)-\bar{u}$ is perturbations of the horizontal velocity.

It is supposed that $\Theta=\Theta(x,y,z,t)$ is the potential air temperature and Θ_0 is a constant mean value of the potential temperature at the upper boundary of the surface layer [8]. Following [8], we introduce a potential temperature fluctuation $\Theta'(x,y,z,t)=\Theta(x,y,z,t)-\Theta_0$ and a dimensionless potential-temperature fluctuation $\theta(x,y,z,t)=\Theta'(x,y,z,t)/\Theta_0$. The quantity $g\theta(x,y,z,t)$ will denote local buoyancy [9].

Let us assume that

$$U_* = \lim_{z\to 0}\left(-\overline{u'w'}\right)^{1/2} > 0, \quad gS_\theta = \lim_{z\to 0}\overline{g\theta w} > 0 \tag{1}$$

Here, U_* and gS_θ have dimensions $[U_*]=\mathrm{m/s}$ and $[gS_\theta]=\mathrm{m^2/s^3}$.

To implement the similarity theory in the convective surface layer, three key parameters: z, gS_θ and U_* are used. The parameter of height z is a variable; the buoyancy flux gS_θ and friction velocity U_* parameters are constants.

The parameters $U_*>0$ and $gS_\theta>0$ make it possible to introduce constant parameters of length $L_*<0$ and buoyancy $g\Theta_*>0$, so that

$$L_* = -\frac{U_*^3}{k_v gS_\theta}, \quad g\Theta_* = \frac{gS_\theta}{U_*} \tag{2}$$

where $k_v=0.4$ is the von Kármán constant.

Let us assume that H is the mean heat flux entering the atmosphere from the underlying surface, ρ_0 is the mean air density at the underlying surface, c_P is the specific heat capacity of dry air, and T_* is the temperature parameter of the Monin-Obukhov theory.

With (2) taken into account, we then obtain

$$gS_0 = \left(\frac{g}{\Theta_0}\right)\frac{H}{c_P\rho_0}, \quad g\Theta_* = -\left(\frac{g}{\Theta_0}\right)T_*, \quad T_* = -\Theta_*\Theta_0 \tag{3}$$

Equalities (3) suggest the proportionality of the key parameters gS_θ and $g\Theta_*$ to the conventional Monin-Obukhov similarity theory parameters H and T_*.

3. Monin-Obukhov similarity theory and free-convection limits of the surface layer

Let us consider a convective surface layer under the free convection conditions $U_* = 0$. In this case, there are only two key parameters z and gS_θ. In accordance with [3], the first moments of the vertical velocity and buoyancy have the form

$$\overline{w} = 0, \quad \frac{\partial}{\partial z}\overline{g\theta} = -\lambda_\theta z^{-4/3}(gS_\theta)^{2/3} \tag{4}$$

where $\lambda_\theta = 1$ from [6].

According to [6], for the second moments of vertical velocity and buoyancy have the form,

$$\overline{w^2} = \lambda_{ww}(gS_\theta)^{2/3} z^{2/3}, \quad \overline{g\theta w} = gS_\theta, \quad \overline{(g\theta)^2} = \lambda_{\theta\theta}(gS_\theta)^{4/3} z^{-2/3} \tag{5}$$

where $\lambda_{ww} = \lambda_{\theta\theta} = 1.8$ from [10].

In accordance with [11–13], the equations (5) are formed by a stochastic ensemble of convective jets.

Similar relationships can also be deduced for the third moments (see [14]).

Let us consider the convective surface layer in forced convection with low winds $0 \neq U_* << 1$. In this case, all three key parameters z, gS_θ and U_* are finite.

To construct the Monin-Obukhov free-convection limit, let us assume that the moments of vertical velocity and buoyancy in low wind $0 \neq U_* << 1$ are the same as those when there is no wind $U_* = 0$.

The transformation of the first buoyancy moment (5) by taking (2) and (4) into account leads to the equality

$$-\frac{k_v z}{g\Theta_*}\frac{\partial}{\partial z}\overline{g\theta} = -\frac{k_v z}{\Theta_*}\frac{\partial}{\partial z}\overline{\theta} = -\frac{k_v z}{|T_*|}\frac{\partial}{\partial z}\overline{\Theta} = \alpha_\theta\left(z/|L_*|\right)^{-1/3} \tag{6}$$

where $\alpha_\theta = k_v^{4/3}\lambda_\theta \approx 0.3$ is a positive constant (see [5,15]).

The transformation of the second moments (6), with (2) and (4) taken into account, leads to equalities

$$\frac{\overline{w^2}}{U_*^2} = \alpha_{ww}^2\left(z/|L_*|\right)^{2/3}, \quad \frac{\overline{(g\theta)^2}}{(g\Theta_*)^2} = \frac{\overline{\theta^2}}{\Theta_*^2} = \frac{\overline{(\Theta')^2}}{|T_*|^2} = \alpha_{\theta\theta}^2\left(z/|L_*|\right)^{-2/3} \tag{7}$$

where $\alpha_{ww}^2 = k_v^{-2/3}\lambda_{ww} \approx 3.3$ and $\alpha_{\theta\theta}^2 = k_v^{2/3}\lambda_{\theta\theta} \approx 1$ are constant coefficients (see [7,15,16]).

4. Linear approximations of the Monin-Obukhov universal similarity functions, the first and second moments of the surface layer

Following [17], let us assume that the convective surface layer consists of two sublayers: the lower dynamic sublayer adjacent to the surface and the upper sublayer of forced convection. A narrow transition area forms a diffuse boundary between these sublayers. This area is not considered as a separate sublayer.

An alternative model of the convective surface layer, that includes three-sublayer structure, was developed in [5,18].

Obviously, the turbulent moments of vertical velocity and buoyancy can be approximated uniformly in the entire convective layer. In the two-sublayer model of the convective surface layer, the turbulent moments can be approximated separately for each sublayer. This approach is quite justifiable, because turbulence in the dynamic sublayer and turbulence in the forced-convection sublayer are of different hydrodynamic natures.

Length parameter $L_* \le 0$ is used to construct the dimensionless height $z/|L_*|$. Let us suppose that region $0 < z/|L_*| \le 2\cdot 10^{-2}$, adjacent to the underlying surface, corresponds to the dynamic sublayer and the transition sublayer. The region $2\cdot 10^{-2} < z/|L_*| < \infty$ is denoted as a forced-convection sublayer with moderate winds. In the forced-convection sublayer, let us also distinguish a free-convection sublayer $1 << z/|L_*|$.

Let us construct linear approximations of the moments of vertical velocity and potential temperature fluctuations in the forced-convection sublayer. The first-order expansion terms of the linear approximations is chosen in accordance with the free-convection limits of the Monin-Obukhov similarity theory in no-wind conditions $U_* = 0$. The second-order expansion terms of linear approximations will then describe profiles of turbulent moments under wind condition $U_* \ne 0$.

It can be assumed, with no restriction of generality, that the dimensionless form of the equation for first buoyancy moment (6), and the forms of the equations for the second moments of vertical velocity and buoyancy (7) in the forced-convection sublayer with $2\cdot 10^{-2} < z/|L_*| < \infty$ can be written as

$$-\frac{k_v z}{|T_*|}\frac{\partial}{\partial z}\overline{\Theta} = \alpha_\theta \left(z/|L_*|\right)^{-1/3} F_h\left(z/|L_*|\right), \quad \lim_{z/|L_*|\to\infty} F_h\left(z/|L_*|\right) = 1 \tag{8}$$

$$\frac{\overline{w^2}}{U_*^2} = \alpha_{ww}^2 \left(z/|L_*|\right)^{2/3} F_{ww}^2\left(z/|L_*|\right), \quad \lim_{z/|L_*|\to\infty} F_{ww}^2\left(z/|L_*|\right) = 1 \tag{9}$$

$$\frac{\overline{(\Theta')^2}}{|T_*|^2} = \alpha_{\theta\theta}^2 \left(z/|L_*|\right)^{-2/3} F_{\theta\theta}^2\left(z/|L_*|\right), \quad \lim_{z/|L_*|\to\infty} F_{\theta\theta}^2\left(z/|L_*|\right) = 1 \tag{10}$$

Let us note that the coefficient $\alpha_\theta = k_v^{4/3}\lambda_\theta \approx 0.3$ from the data of [5,6,15]. According to [16,19], the coefficient $\alpha_{ww}^2 \approx 3.25$, and $\alpha_{\theta\theta}^2 \approx 0.95$ from the data of [20–22].

Limiting ourselves to a linear Taylor expansion of $F_h(z/|L_*|)$, $F_{ww}\left(z/|L_*|\right)$ and $F_{\theta\theta}\left(z/|L_*|\right)$ in parameter $\left(z/|L_*|\right)^{-2/3}$, we obtain

$$-\frac{k_v z}{|T_*|}\frac{\partial}{\partial z}\overline{\Theta} = \alpha_\theta \left(z/|L_*|\right)^{-1/3}\left\{1 - \beta_h \left(z/|L_*|\right)^{-2/3}\right\} \tag{11}$$

$$\frac{\overline{w^2}}{U_*^2} = \frac{\sigma_w^2}{U_*^2} = \alpha_{ww}^2 \left(z/|L_*|\right)^{2/3}\left\{1 + \beta_{ww}\left(z/|L_*|\right)^{-2/3}\right\} \tag{12}$$

$$\frac{\overline{(\Theta')^2}}{|T_*|^2} = \frac{\sigma_\theta^2}{|T_*|^2} = \alpha_{\theta\theta}^2 \left(z/|L_*|\right)^{-2/3}\left\{1 - \beta_{\theta\theta}\left(z/|L_*|\right)^{-2/3}\right\} \tag{13}$$

where σ_w, σ_θ are variances of vertical velocity and potential-temperature fluctuation; β_h, β_{ww}, $\beta_{\theta\theta}$ are constant coefficients.

Fig. 1 presents a comparison of the linear approximation of Monin-Obukhov similarity theory (11) at $\alpha_\theta = 0.3$, $\beta_h = 2\cdot 10^{-2}$ with the field data [23]. It shows that the approximation (11) slightly overestimates this field data within the free-convection sublayer.

A comparison between the field data [24] of the dimensionless vertical velocity variance σ_w / U_* and its linear approximation in the form of (12) with $\alpha_{ww}^2 = 3.25$ and $\beta_{ww} = 0.3$ is shown in Fig. 2. A comparison between the field data [24] of the dimensionless potential temperature variance $\sigma_\theta / |T_*|$ and its linear approximation in the form of (13) with $\alpha_{\theta\theta}^2 = 0.95$, $\beta_{\theta\theta} = 0.06$ is shown in Fig. 3.

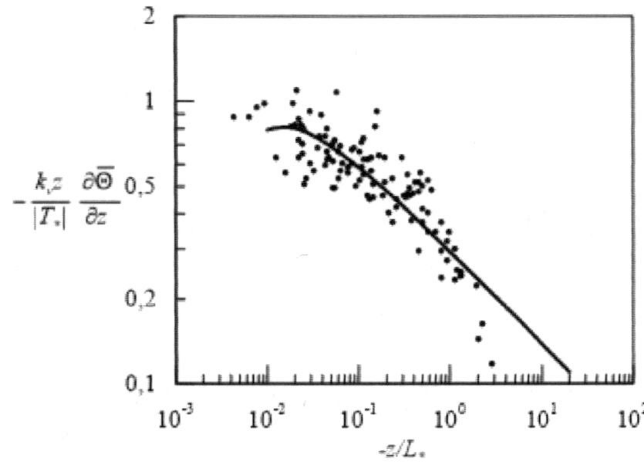

Figure 1. Comparison of linear approximation of similarity theory (11) with empirical values of the function $-k_v z|T_*|^{-1}\partial\bar{\Theta}/\partial z$ from [23]. The solid line corresponds to approximation (11) with coefficients $\alpha_0 = 0.3$ and $\beta_h = 2\cdot 10^{-2}$.

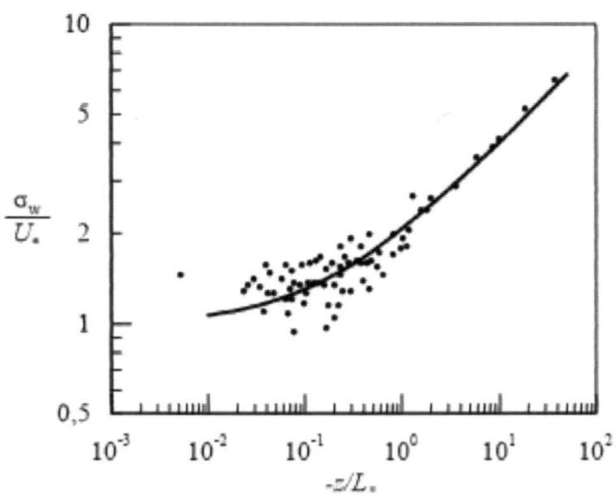

Figure 2. Comparison of linear approximation of similarity theory (12) with empirical values of the velocity variance σ_w / U_* from [24]. The solid line corresponds to approximation (12) with coefficients $\alpha_{ww}^2 = 3.25$ and $\beta_{ww} = 0.3$.

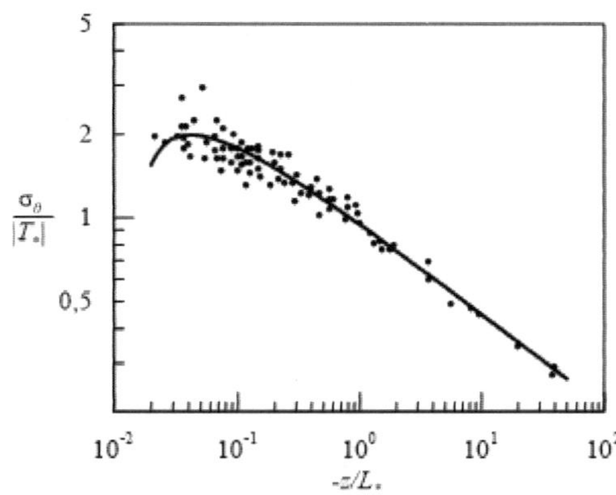

Figure 3. Comparison of linear approximation of similarity theory (13) with empirical values of the buoyancy variance $\sigma_\theta / |T_*|$ from [24]. The solid line corresponds to approximation (13) with coefficients $\alpha_{\theta\theta}^2 = 0.95$ and $\beta_{\theta\theta} = 0.06$.

The results of comparisons in Fig. 1-3 suggest that linear approximations (11) - (13) are correct within the range of $2\cdot 10^{-2} < z/|L_*| < \infty$.

Another expansion, which is similar to (11) - (13), and its comparison with experimental data sets are shown in [25].

Alternative approaches that are using the local similarity theory are presented in [26–29].

Modern comprehensive field data of atmospheric surface layer parameters and techniques of its processing are described in [30,31].

5. Conclusion

The results show that natural approximations of the turbulent moments in the convective surface layer have various forms and are not confined to the classical form of the Monin-Obukhov universal functions.

In the two-sublayer model of the convective surface layer, the linear approximations of turbulent moments are quite efficient in describing the observed data in a forced-convection region.

The proposed version of the linear approximations of the universal functions of similarity theory considers the free-convection limits of Monin-Obukhov theory only as a first-order approach. The second order expansion terms of the linear approximations take into account the existence of wind in the convective surface layer.

Acknowledgments
This study was supported by the Russian Foundation for Basic Research, grant 17-05-01116.

References

[1] Obukhov A M 1946. *Tr. Inst. Teor. Geofiz. Akad. Nauk SSSR* **1** 95–115
[2] Monin A S and Obukhov A M 1953 *Dokl. Akad. Nauk SSSR* **93** 257-60
[3] Monin A S and Obukhov A M F 1954 *Tr. Geofiz. Inst. Akad. Nauk SSSR* **151** 167-87
[4] Monin A S and Yaglom A M 1975 *Statistical Fluid Mechanics, Vol. 1: Mechanics o Turbulence* (MIT Cambridge)
[5] Kader B A and Yaglom A M 1990 *J. Fluid Mech.* **212** 637–62
[6] Priestley C H B 1959 *Turbulent transfer in the lower atmosphere* (University of Chicago Press Chicago)
[7] Wyngaard J C, Cote O R and Izumi Y 1971 *J. Atmos. Sci.* **28** 1171–82
[8] Vulfson A N 1981 *Izv. Atmos. Ocean. Phys.* **17** 646–9
[9] Turner J S 2009 *Buoyancy Effects in Fluids* (Cambridge University Press)
[10] Kaimal J C, Wyngaard J C, Haugen D A, Coté O R, Izumi Y, Caughey S J and Readings C J 1976 *J. Atmos. Sci.* **33** 2152–69
[11] Vulfson A N and Borodin O O 2009 *Russ. Meteorol. Hydrol.* **34** 491
[12] Vulfson A, Borodin O and Nikolaev P 2017 Convective Jets: Volcanic Activity and Turbulent Mixing in the Boundary Layers of the Atmosphere and Ocean *Physical and Mathematical Modeling of Earth and Environment Processes* (Springer) pp 71–83
[13] Vulfson A N and Nikolaev P V 2018 *J. Phys. Conf. Ser.* **955** 012013
[14] Zilitinkevich S S 1973 *Boundary-Layer Meteorol.* **3** 416–23
[15] Li D, Katul G G and Bou-Zeid E 2012 *Phys. Fluids* **24** 105105
[16] Panofsky H A, Tennekes H, Lenschow D H D H and Wyngaard J C 1977 *Boundary-Layer Meteorol.* **11** 355–61
[17] Andreas E L, Hill R J, Gosz J R, Moore D I, Otto W D and Sarma A D 1998 *Boundary-Layer Meteorol.* **86** 379–408
[18] Albertson J D, Parlange M B, Kiely G and Eichinger W E 1997 *J. Geophys. Res. Atmos.* **102** 13423–32
[19] Kaimal J C and Finnigan J J 1994 *Atmospheric boundary layer flows: their structure and measurement* (Oxford university press)
[20] Katul G G and Hsieh C-I I 1999 *Boundary-Layer Meteorol.* **90** 327–38
[21] Asanuma J and Brutsaert W 1999 *Water Resour. Res.* **35** 515–21
[22] Choi T, Hong J, Kim J, Lee H, Asanuma J, Ishikawa H, Tsukamoto O, Zhiqui G, Ma Y, Ueno K, Wang J, Koike T and Yasunari T 2004 *J. Geophys. Res. D Atmos.* **109** 1–12
[23] Wilson D . K 2001 *Boundary-layer Meteorol.* **99** 151–8
[24] Liu X, Tsukamoto O, Oikawa T and Ohtaki E 1998 *Boundary-Layer Meteorol.* **87** 499–508
[25] Vulfson A N and Nikolaev P V 2018 *Izv. Atmos. Ocean. Phys.* **54** 472–9
[26] Vulfson A N, Volodin I A and Borodin O O 2004 *Russ. Meteorol. Hydrol.* **10** 1–10
[27] Shao Y and Hacker J M 1990 *Boundary-Layer Meteorol.* **52** 17–40
[28] Wood C R, Lacser A, Barlow J F, Padhra A, Belcher S E, Nemitz E, Helfter C, Famulari D and Grimmond C S B 2010 *Boundary-Layer Meteorol.* **137** 77–96
[29] Nadeau D F, Pardyjak E R, Higgins C W and Parlange M B 2013 *Boundary-layer Meteorol.* **147** 401–19
[30] Liu Y, Mamtimin A, Huo W, Yang X, Liu X, Yang F and He Q 2016 *Adv. Meteorol.* **2016** 1–8
[31] Kang D and Wang Q 2016 Optimized *Atmosphere (Basel).* **7** 14

Non-equilibrium critical behavior of the 3D classical Heisenberg model

E A Pospelov[†], **V V Prudnikov**[‡], **P V Prudnikov**[§], **and A S Lyakh**
Omsk State University, Pr. Mira 55A, Omsk 644077, Russia
E-mail: [†]posevg@yandex.ru, [‡]prudnikv@univer.omsk.su, [§]prudnikp@univer.omsk.su

Abstract. We have investigated the non-equilibrium critical behavior of the three-dimensional classical Heisenberg model with evolution from various initial states. Initial states with magnetization $m_0 \neq 0$ lead to a new time scale $t_m \sim m_0^{-k}$ with the exponent $k > 0$ which produces complex universal scaling forms on the temporal behavior of the magnetization, the autocorrelation function, and the response function. Aging effects have been revealed in two-time behavior of the autocorrelation function with evolution from various initial states. We show that the slowing down of the autocorrelation function during the evolution from high-temperature initial state is considerably more slow than for case of the evolution from low-temperature initial state. Using scaling forms of the magnetization and the autocorrelation function, the values of critical exponents $z = 2.035(4)$, $\theta' = 0.490(1)$, and $k = 1/(\theta' + \beta/\nu z) = 1.340(4)$ were calculated.

The 3D classical ferromagnetic Heisenberg model is the one of the traditional statistical model, which is used to describe the second order phase transition in the different spin systems, including the such materials as Fe, Co, Ni and its alloys.The static critical behavior of the systems described by this model was investigated with the use of different techniques and approximations. For example, these are the experimental methods, the summation method of the series of the high-temperature expansion, the renormalization-group methods, and the simulations by the Monte Carlo methods. The values of the critical temperature and static critical exponents obtained by these methods are in good agreement within the statistical errors in spite of the results have some differences (see the review [1]).

A much smaller attention is focused on to the study of the critical dynamics and there are less works on the non-equilibrium critical behavior of the model. At early works with the renormalization-group description of different dynamic models [2], the critical dynamics of the Heisenberg isotropic model was classified as model J with conserved order parameter (the magnetization) and non-conserved energy. The dynamic equation of the magnetization near the critical point although contains the relaxation term, but the long-wave and low-frequency dynamics of the magnetization is determined by the hydrodynamic term in the equation which describes the slow precessional rotational motion of the magnetization. The dimensional analysis of the magnetization dynamics of the Heisenberg isotropic model allows one to obtain the expression for the critical dynamic exponent $z = (d+2-\eta_H)/2$, where d is the spatial dimension of the system, η is the Fisher critical exponent that determines the power-law decay of the correlation function with the distance near the critical point. The value $z \simeq 5/2$ is predicted for the 3D Heisenberg model, since value of $\eta_H \simeq 0.021$ (with the use of two-loop approximation ε^2 within the ε-expansion method) is negligible compared to $5/2$. However, the dipole interaction always presents in the magnetics and it is especially significant near the critical point for the isotropic magnets due to the long-range nature of the dipole interaction. The influence of the dipole interaction on the static properties of Heisenberg model was studied by the ϵ-expansion method in the work [3] and on it's dynamic properties in the work [4]. The important property of the influence of the dipole interaction on the critical dynamics is noted in [4]. It leads to non-conservation of the order parameter and relaxation of the magnetization with dynamic critical exponent $z_d = 2+c_d\eta_d \simeq 2.022$, where $c_d = (27/4)\ln(4/3)-1 \simeq 0.942$ according to work [4], and the Fisher's exponent η_d has the value $\eta_d \simeq 0.023$ under the influence of dipole interaction(in the two-loop approximation within ε-expansion method [3]). Although the numerical values of static critical exponents change slightly under influence of dipole interaction, however the influence of the dipole interaction on the values of the dynamic critical exponent turns out to be significant due to the changes in dynamics of the magnetization.

Taking into account influence of dipole interaction, the Heisenberg model dynamics expressed by the relaxation dynamic of the model A in Hohenberg-Halperin classification (see review [2]), in which both the order parameter and the energy are not conserved. The model has dynamic critical exponent $z = 2 + c\eta$ with $c = 6\ln(4/3) - 1 \simeq 0.726$ and one could find the value $z_H = 2 + c\eta_H \simeq 2.015$ for the Heisenberg model using the value $\eta_H \simeq 0.021$, which also is close enough to z_d for the Heisenberg model with influence of the dipole interaction. In the paper [5, 6] the dynamic critical exponents $z = 1.975(10)$ and $z = 1.976(9)$ were found by the short-time dynamics method of simulations. The values are consistent within statistical errors with the value $z = 1.96(6)$ that is earlier calculated in the work [7] using the equilibrium technique simulation of the autocorrelation function. However, the values of z obtained in [5, 6] are less than 2 and it provokes an alertness by reason of these values should lead to the divergence of the kinetic coefficients. It is conflicted with fundamentals of the renormalization-group theory for the relaxation model A and never observed in the experiments.

The behavior of the systems characterized by the anomalously slow dynamics is of the great current interest. This is related to the aging properties of non-equilibrium characteristics and the violations of the fluctuation-dissipation theorem predicted and observed at the slow evolution of the systems from the non-equilibrium initial state [8]. Such features of the non-equilibrium behavior are also characteristic of the systems exhibiting the second order phase transitions [9], because the critical dynamics of such systems is characterized by anomalously large relaxation times.

Near the second order phase transition critical temperature T_c the relaxation time of the system is divergent value $t_{rel} \sim |T - T_c|^{-z\nu}$, where z, ν are critical exponents. Thus, the systems at the critical point never reach equilibrium during the whole relaxation process. The aging in the non-equilibrium behavior of systems appears at times $t \ll t_{rel}$. It is expressed in the existence of the two-time dependencies for the correlation and response functions and in dependence of its evolution on the initial states. It is distinguished a high-temperature initial state created at the temperatures $T \gg T_c$ and a low-temperature initial state created at $T < T_c$. The high-temperature state is characterized by the initial magnetization $m_0 \ll 1$, when the low-temperature one is characterized by the initial magnetization $m_0 = 1$ (this state also called as completely ordered initial state). In the paper we have investigated the influence of various initial states with magnetization in the range $0 < m_0 \leq 1$ on critical relaxation of the magnetization and the autocorrelation function for the 3D Heisenberg model.

The Hamiltonian of the ferromagnetic isotropic Heisenberg model is given by the expression

$$H = -J \sum_{<i,j>} \vec{S}_i \vec{S}_j, \tag{1}$$

where $J > 0$ is the short-range exchange interaction between spins $\vec{S}_i$ fixed at the sites of the simple cubic lattice. Spin $\vec{S}_i = (S_i^x, S_i^y, S_i^z)$ is determined as the classical unit vector. Summation over j affects only nearest neighbors of the i-th spin. The simulations were carried out on lattice with sizes $L \times L \times L$ and with applied periodic boundary conditions. To calculate characteristics of the non-equilibrium critical behavior, the dynamic process of one-spin flips was numerically simulated within the Monte Carlo statistical method. The dynamic process of one-spin flips was implemented using the Metropolis algorithm. Simulations were carried out at the critical temperature $T_c = 1.44292(8)$ calculated in work [10].

To investigate the non-equilibrium evolution, we computed such characteristics as the magnetization $\vec{M}(t)$ and the two-time dependent autocorrelation function $C(t, t_w)$, which are defined by relations

$$\vec{M}(t) = \frac{1}{V}\int d^d x \left\langle \vec{S}(x,t) \right\rangle = \left\langle \frac{1}{L^3}\sum_{i=1}^{L^3} \vec{S}_i(t) \right\rangle, \tag{2}$$

$$C(t,t_w) = \frac{1}{V}\int d^d x \left(\left\langle \vec{S}(x,t)\vec{S}(x,t_w) \right\rangle - \vec{M}(t)\vec{M}(t_w) \right) = \left\langle \frac{1}{L^3}\sum_{i=1}^{L^3} \vec{S}_i(t)\vec{S}_i(t_w) \right\rangle - \vec{M}(t)\vec{M}(t_w), \tag{3}$$

where the angle brackets stand for averaging over the different realizations of initial state.

Near the critical point the magnetization and the correlation function are the generalized homogeneous functions with respect to the observation time t, the waiting time t_w and the new time scale $t_m \sim m_0^{-k}$. The $m_0 = |\vec{M}(0)|$ is the initial magnetization. The exponent $k = 1/(\theta' + \beta/z\nu) > 0$ is expressed on static critical exponents β and ν and dynamic ones θ' and z. At the present time, it become established that the time dependence for the magnetization and the autocorrelation function for systems starting from an initial state with $m_0 \neq 0$ satisfies the following scaling forms [11]:

$$\begin{aligned} M(t,t_m) &\sim t^{-\beta/\nu z} F_m(t/t_m) \\ C(t,t_w,t_m) &\sim t^{-2\beta/\nu z} F_C(t/t_w, t/t_m), \end{aligned} \tag{4}$$

where F_m and F_C are the generalized homogeneous functions with respect to its parameters.

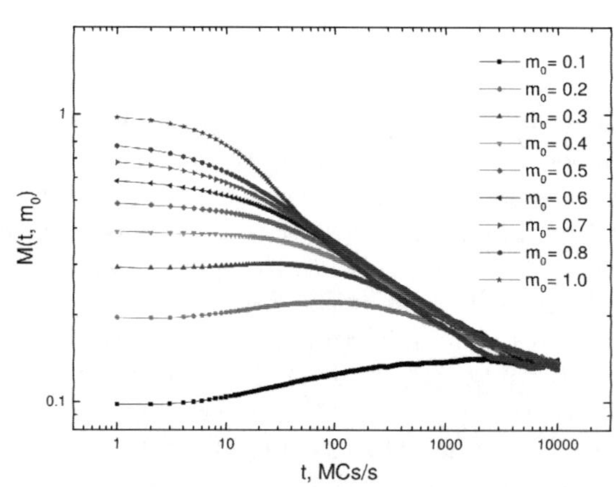

Figure 1. Relaxation of the magnetization $M(t)$ for various values of the initial magnetization m_0.

To check the scaling form (6) for the 3D Heisenberg model, the magnetization was computed for various initial states with magnetization $m_0 \leq 1$. The obtained data are plotted in Fig.1.

As shown in Fig. 1, the relaxation curves for the initial magnetization $m_0 \neq 1$ clearly tends to the curve characterized by the completely ordered initial state with $m_0 = 1$. The non-equilibrium power-law time dependence of the magnetization $M(t) \sim m_0 t^{\theta'}$ is well displayed for the systems with initial states $m_0 \ll 1$ with the exponent $\theta' > 0$ [12, 13]. The increasing magnetization of such configurations is changed to power-law decay of the magnetization $M(t) \sim t^{-\beta/\nu z}$ at times $t > t_{cr} \sim m_0^{-k}$. The critical relaxation from the completely ordered initial state with $m_0 = 1$ is immediately characterized by power-law decay of the magnetization $M(t) \sim t^{-\beta/\nu z}$.

We have considered the systems with the initial states $m_0 = 0.001$, 0.003 and 0.005 in order to compute the exponent θ'. We have used about 1000 realizations for each initial system state with given. Using the critical exponents $\theta'(m_0)$, we obtained the approximated value $\theta(m_0 \to 0) = 0.490(1)$. The exponent computing data are presented in Fig. 2 **a**). By analyzing the magnetization curve for the system with evolution from $m_0 = 1$, the exponent $\beta/\nu z$ was computed with final value $\beta/\nu z = 0.256(1)$. Using the value $\beta/\nu = 0.521(1)$ from work [14], we found the dynamic critical exponent $z = 2.035(4)$.

We have checked the scaling form for magnetization (5). To do that we plotted $t^{\beta/\nu z} M(t, t_m)$ versus t/t_m with $t_m = m_0^{-k}$. It is shown in Fig. 2 **b**. The curves for various initial states are well collapsed at one universal curve that defines the scaling function $F_m(x = t/t_m)$. It has a power-law dependence $F_m(x) \sim x^{1/k}$ at initial evolution stage. Using obtained data, we have found the value of exponent $k = 1.340(4)$.

In addition to the magnetization, we have analyzed realization of aging for the two-time connected correlation function $C(t, t_w)$. We used the waiting times $t_w = 20$, 40, 80, and 160 MCS/s for simulations of evolution from the low-temperature initial state with $m_0 = 1$ and the high-temperature state with $m_0 = 0.001$. The correlation functions versus the observation time are plotted in Fig. 3. The both systems well display the aging in the its time dependencies. Comparing the plots one could see more "strong" aging for evolution from the low-temperature initial state opposite to the high-temperature one. At the same time, the $C(t, t_w)$ is decreased more slowly for the $m_0 \ll 1$ than for case with $m_0 = 1$.

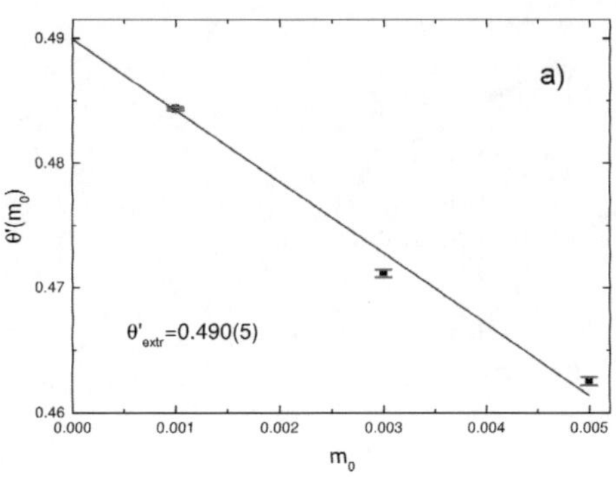

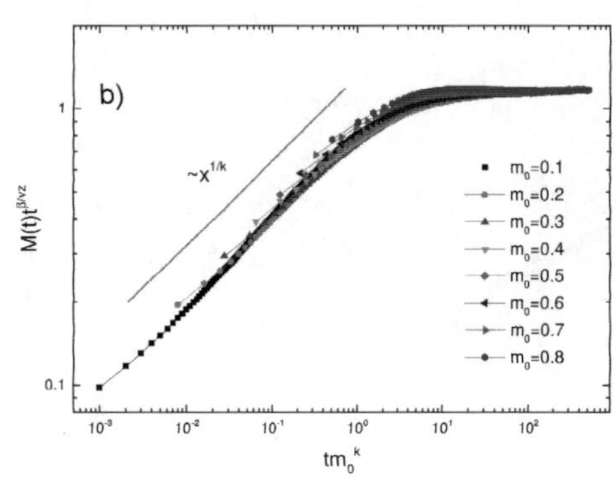

Figure 2. **a)** Extrapolation of exponent values $\theta'(m_0 \to 0)$ computed for $m_0 = 0.001; 0.003; 0.005$. **b)** Dependence of the magnetization scaling function $F_m(tm_0^k) = t^{\beta/\nu z} M(t, m_0)$ on $x = tm_0^k$.

At aging regime with $t - t_w \sim t_w \gg 1$, one expects the following scaling form for correlation function [9, 15]

$$C(t, t_w, t_m) \sim t_w^{-2\beta/\nu z} F_C(t/t_w) \quad (5)$$

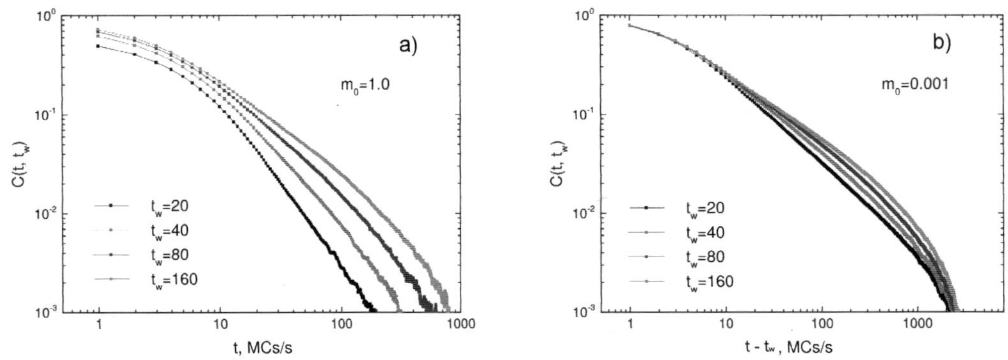

Figure 3. Time dependence of the autocorrelation function $C(t, t_w)$ for various waiting times t_w with evolution from **a)** the completely initial state with $m_0 = 1$ and **b)** the low-temperature initial state with $m_0 = 0.001$.

with the scaling function $F_C(t/t_w)$ that has next power-law at times $t - t_w \gg t_w \gg 1$

$$F_C(t/t_w) \sim \left(\frac{t}{t_w}\right)^{-c_a}. \tag{6}$$

The exponent c_a is equal to $c_a^{LT} = 1 + \frac{\beta(\delta+2)}{z\nu}$ for the completely ordered initial state $m_0 = 1$ [11, 16, 17] and $c_a^{HT} = \frac{d}{z} - \theta'$ for the high-temperature initial state $m_0 \ll 1$ [18, 19]. In order to check the scaling form (5) we have plotted $t_w^{2\beta/\nu z} C(t, t_w)$ versus t/t_w using our result $\beta/\nu z = 0.256(1)$. The data are shown in Fig. 4 for the both initial states.

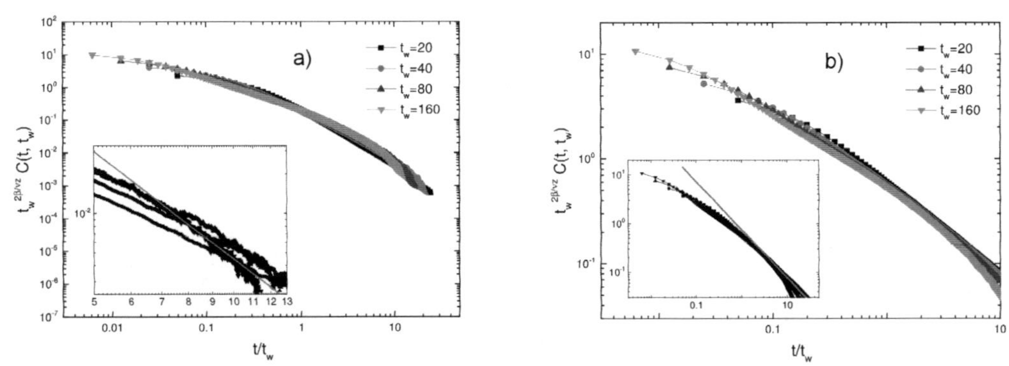

Figure 4. Dependence of the autocorrelation scaling function $t_w^{2\beta/\nu z} C(t, t_w)$ on t/t_w for evolution from **a)** the low-temperature initial state and **b)** the high-temperature initial state.

The plots show well collapse for the different waiting times t_w. The universal curve defines the scaling function $F_C(t/t_w)$. We have found the value of exponent $c_a^{LT} = 2.734(7)$ for the completely ordered initial state with $m_0 = 1$ and $c_a^{HT} = 0.979(6)$ for the high-temperature initial state with $m_0 = 0.001$ at time stage $t/t_w \gg 1$. The values have good match with the theory predictions giving $c_a^{LT} \approx 2.730$ (for the Heisenberg model $\delta = 4.758(11)$) and $c_a^{HT} = \frac{3}{z} - \theta' = 0.984$.

We have investigated the non-equilibrium behavior of the 3D ferromagnetic Heisenberg models for the wide range of the initial states. We have revealed the ageing in the behavior of the two-time connected correlation function. The scaling forms are confirmed for the magnetization and the correlation function.

We have obtained the dynamical critical exponents $\theta' = 0.490(1)$ and $z = 2.035(4)$. The exponent θ' is well matched with the one $\theta' = 0.482(3)$ that was found at [5, 6] using short-time dynamics method. Additionally we note that we used the lattice with more large linear sizes ($L = 60, 100$ compared to $L = 40$) than in works [5, 6] and more small values of the initial magnetization $m_0 \ll 1$. It allows us to state the value $\theta' = 0.490(1)$ more precise.

As for the exponent z, our result with the value $z = 2.035(4)$ has good agreement with $z_H = 2 + c\eta_H \simeq 2.015$ obtained in two-loop approximation within ε-expansion method for the model A [2]. Moreover using the value $\eta_H = 0.0413(16)$ [14] and the value $c = 6\ln(4/3) - 1 \simeq 0.726$, one could find $z = 2.0315(11)$. The method of short-time dynamics gives $z = 1.976(9)$ [5, 6], but this result has strong disagreement with the expected value of $z > 2$ for the relaxation model A with non-conserved order parameter and was never confirmed experimentally.

We have analyzed the magnetization relaxation for wide range of the initial states. It is shown the power-law $M(t) \sim t^{-beta/\nu z}$ for the perfect ordered initial state. The data for the $m_0 \neq 1$ are clearly displayed the asymptotic tends to that law at some stage of the relaxation. For the initial states $m_0 \ll 1$ and the regime $t < t_{cr} \sim m_0^{-k}$ the curves are shown another power-law $M(t) \sim m_0 t^{\theta'}$.

For the two-time correlation function, the aging and the scaling are checked and proved. The dependence of the correlation function on the waiting times is more strong for the case of evolution from the low-temperature initial state with $m_0 = 1$ than from the high-temperature initial state with $m_0 = 0.001$.

Acknowledgments

The reported study was supported by the Russian Foundation for Basic Research, project nos. 17-02-00279 and 18-42-550003, and by the Council of the President of the Russian Federation, project no.MD-6868.2018.2. For calculations, we used the resources provided by the Shared Facility Data Center FEB RAS (Khabarovsk), the Supercomputing Center of Lomonosov Moscow State University, and the St. Petersburg Supercomputer Center of the RAS.

References

[1] Pelissetto A and Vicari E 2002 *Physics Reports* **368** No 6 549-727
[2] Hohenberg P C and Halperin B I 1977 *Rev. Mod. Phys* **49** No. 3 436-479
[3] Fisher M E and Aharony A 1973 *Phys. Rev. Lett* **30** 559; 1973 Phys. Rev. B **8** 3323; 1973 Phys. Rev. B **8** 3342
[4] Teitelbaum G B 1975 *JETP Letters* **21** issue 6 339-341
[5] Fernandes H A, Drugowich de Felcio J R and Caparica A A 2005 *Phys. Rev.* B **72** 054434
[6] Fernandes H A, da Silva R and Drugowich de Felcio J R 2006 *J. Stat. Mech* **6** P10002
[7] Peczak P and Landau D P 1993 *Phys. Rev.* B **47** P14260
[8] Vincent E, Hammann J, Ocio M, Bouchaud J P and Cugliandolo L F 1997 *Lect. Notes Phys* **492** P184-219
[9] Prudnikov V V, Prudnikov P V and Mamonova M V 2017 *Uspekhi Fizicheskikh Nauk* **60** 762797
[10] Chen K, Ferrenberg A M and Landau D P 1993 *Phys. Rev.* B. **48** P3249
[11] Calabrese P, Gambassi A and Krzakala F 2006 *J. Stat. Mech* **6** P06016
[12] Prudnikov V V, Prudnikov P V, Krinitsyn A S, Vakilov A N, Pospelov , Rychkov M V 2010 *Phys. Rev.* E. **81** 011130
[13] Prudnikov V V, Prudnikov P V, Kalashnikov I A, Rychkov M V 2010 *JETP* **137** No 2 pp 287300
[14] Ballesteros H G, Fernandez L A, Martn-Mayor V and Munoz Sudupe A 1996 *Phys. Lett.* B **387** P125
[15] Prudnikov P V, Prudnikov V V, Pospelov E A, Malyarenko P N, Vakilov A N 2015 *Prog. Theor. Exp. Phys* 053A01
[16] Prudnikov P V, Prudnikov V V, Pospelov E A, Malyarenko P N 2015 *JETP Letters* **102** No 3 pp 192201
[17] Prudnikov V V, Prudnikov P V and Pospelov E A 2016 *J. Stat. Mech* **2016** P043303
[18] Prudnikov P V, Prudnikov V V and Pospelov E A 2014 *JETP* **118** No 3 P401409
[19] Prudnikov P V, Prudnikov V V, Pospelov E A, Vakilov A N 2015 *Phys. Lett. A* **379** P774778

Computation of dielectric properties of nano-probe local surroundings by solvatochromic spectral shift measurement of organic dye molecule

A. S. Stepko[1], P. V. Lebedev-Stepanov[1,2]

[1]Photochemistry Center of FSRC Crystallography & Photonics RAS Novatorov St., 7a/1, Moscow, 119421, Russia
[2]National Research Nuclear University MEPhI (Moscow Engineering Physics Institute), 31 Kashirskoe shosse, 115409, Moscow, Russia

E-mail: tousen@bk.ru

Abstract. Study of dielectric properties of the macrocyclic molecules cavity that can form inclusion complexes with dye molecules is an important problem. Changes in absorption spectra of chromophore inserted into macrocyclic cavity with the formation of the supramolecular complex can be considered as solvatochromic shift. This shift caused by changes of dielectric properties of the environment around the chromophore when it moves from water solution to macrocyclic cavity. Theoretical model that can describe the effective dielectric permittivity of the cavity allows predicting solvatochromic spectral shift of dye molecules due to non-covalent supramolecular complexation. Such a dye molecule could apply as a nano-probe for measurement of the effective dielectric constant of any local surrounding media in which its chromophore can be inserted.

1. Introduction

Theoretical model allows to work with a wide range of dielectric permittivity therefore there are no constraints as in the experimental study. The method needs only set of values for parameterization of the Onsager–Liptay equation. In the present work we have been using this approach for determination of dielectric properties of cavitand (cucurbit [7] April (CB [7])) which involved in complexation with styryl dye molecules.

Experimental and theoretical study of the dependence of optical spectra on the position of dyes chromophore (1-(2-ammonioethinyll)-4-[(E)-2-(3,4-dimethoxyphenyl)ethinyll]pyridinium diperchlorate (D1), 1-(3-ammoniopropyl)-4-[(E)-2-(3,4-dimethoxyphenyl)ethinyll]pyridinium diperchlorate (D2), and 1-(5-ammoniopentyl)-4-[(E)-2-(3,4-dimethoxyphenyl)ethinyll]pyridinium diperchlorate (D3), containing two, three, and five methinyllene groups, respectively) were carried out in previous work [1].

Our calculations showed that in the D1@CB [7] and D2@CB [7] inclusion complexes (where @ is the conventional symbol for supramolecular complex), the pyridinium residue of the dye is located inside the CB [7] cavity, which is less polar than water. This leads to a bathochromic (red) shift of absorption bands of the D1@CB[7] and D2@CB[7] complexes as compared with the spectra of these dyes in water. On the contrary, in the spectrum of D3@CB [7] the chromophore moiety of which remains in an aqueous environment, a weak hypsochromic (blue) shift of the long wavelength absorption band is observed (fig. 1).

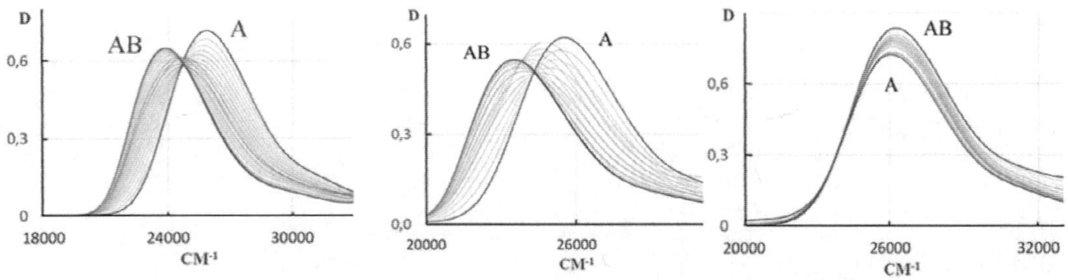

Δν = -1983 cм$^{-1}$ Δν = -2385 cм$^{-1}$ Δν = 274 cм$^{-1}$

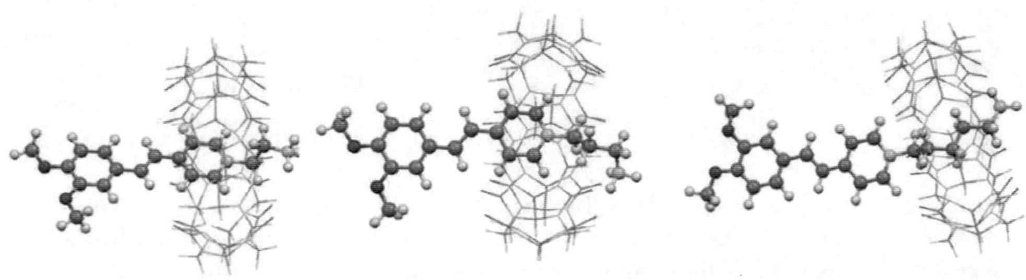

D1@CB[7]. L = 0,4 Å D2@CB[7]. L = 1,1 Å D3@CB[7]. L = 4 Å

Figure 1. Upper panel: absorption spectra of the dyes and their complexes (A is a dye, and AB is a complex). Lower panel: optimized structures of the D1@CB [7], D2@CB[7], and D3@CB[7] complexes (L is the distance between the centers of masses of SD and CB[7] in complexes.

The changes in the spectra of D2 in different solvents and by complexation with CB [7] can be understood on the basis of the ideas of the Onsager theory [2] about the dipole–dipole (universal) interaction of styryl dye (SD) with water molecules and with the CB [7] cavity. Water is one of the most polar solvents. It is known that the more polar the solvent, the larger the short-wavelength shift of their absorption bands of pyridine derivatives. In our experiments, on the contrary, the addition of CB[7] cavitand with a nonpolar cavity to the water leads to a transition of the dye from the polar medium to the nonpolar one, which should be accompanied by a red shift of the long-wavelength absorption band. This is actually observed for D1 and D2 since their pyridine residues are located in the CB[7] cavity.

2. Model

In the present work, we have been using Onsager–Liptay [2-4] theory to calculate solvatochromic shift. Our experimental study of solvatochromic effect was carried out for D2 with a set of different solvents: methanol, ethanol, *1*-propanol, *n*-butanol and water with known dielectric and optical properties (fig. 2).

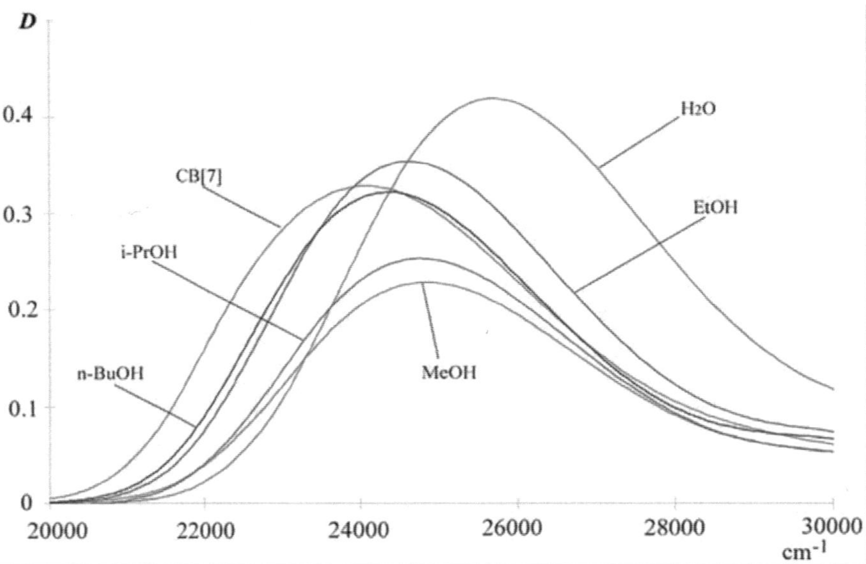

Figure 2. Shifts of D2 absorption spectra depending on permittivity of the solvent.

According to Onsager theory, the value of the absorption spectra shift of a molecule in solution relative to the same molecule in vacuum can be represented as the sum of contributions of dispersion, dynamic polarization and orientation-induction interactions.

Essence of the effect is that the orientation polarization contribution of the dipole moments of ground and excited state changes under the action of the solvent reaction field. The difference in the energies of the ground and excited states also changes, so that is the basis for calculating the shift of the spectral line. Thus, the Onsager model gives a quite rough approximation due to the difference in shape of the studying molecules from the spherical one for which the theory was developed. This model allows to describe solvatochromic processes for different systems as it was shown in many works [4,5]. Nevertheless, this model is an estimate and simple approach that accurately conveys the general physics of the effect.

According to the theory, the dye molecule can be represented as a sphere with radius a and dipole moment placed in the center. In this approximation, the energy shifts of electronic states caused by nonspecific interactions, as follows from the theory of the Onsager reaction field, described by the equations:

$$\boldsymbol{\mu} = \boldsymbol{\mu}_0 + \alpha \mathbf{E}_R, \mathbf{E}_R = (f_{ind} + f_{or})\boldsymbol{\mu},$$

$$f_{or} = \frac{2}{a^3}\left(\frac{\varepsilon - 1}{2\varepsilon + 1} - \frac{n^2 - 1}{2n^2 + 1}\right), \; f_{ind} = \frac{2}{a^3}\left(\frac{n^2 - 1}{2n^2 + 1}\right), \tag{1}$$

where $\mathbf{E}_R$ is Onsager reaction field strength; $\boldsymbol{\mu}$ is the dipole moment of the dissolved molecule; f_{ind} and f_{or} air induction and orientation factors; $\boldsymbol{\mu}_0$ and α are dipole moment and polarizability of the molecule in the vacuum; a is the effective size of the molecule; ε and n are dielectric constant and refractive index of solvent respectively. In the case of charged molecule (for example, dication), $\boldsymbol{\mu}$ and $\boldsymbol{\mu}_0$ are the orientation-induction contribution it the total dipole moment of the dye molecule in solution and in vacuum respectively. This contribution is invariant relative to coordinate transformation. Then, according to the Onsager-Liptay theory [2-4], the change of the wave number $hc\Delta\nu$ is determined by the equation:

$$hc\Delta\nu = -\frac{1}{2}\left[\frac{2}{a^3}\left(\frac{n^2-1}{2n^2+1}\right)\right]\left(\mu_e-\mu_g\right)^2\left[1-\frac{2}{a^3}\left(\frac{n^2-1}{2n^2+1}\right)\alpha\right]^{-1} -$$

$$-\frac{2\left(\mu_e-\mu_g\right)}{a^3}\left(\frac{\varepsilon-1}{2\varepsilon+1}\right)\left[1-\frac{2}{a^3}\left(\frac{\varepsilon-1}{2\varepsilon+1}\right)\alpha\right]^{-1}\mu_g, \quad (2)$$

where h is Planck constant, c is a light velocity, μ_g and μ_e are dipole moments of the ground and excited states of dye molecule respectively, $\Delta\nu$ is a solvatochromic frequency shift, α is polarizability of the molecule.

3. Calculations

Quantum chemical method TDDFT with functional CAM-B3LYP and basis 6-311G (d,p) implemented in Gamess-US software package was used to calculate molecular structure, dipole moments of the ground (39.7 D) and excited (34.3 D) states, and polarizability (609.3 a.u.) of the dye molecule.

Formula (2) allows identifying spectral shift for different values of ε and n. As a result of the calculations, absorption spectrum shifts of D2 for set of solvents with different dielectric permeability relative to H_2O were obtained. These results (fig. 2) are presented in Table 1 in comparison with our experiments. The extrapolation calibration graph for determination of dielectric permeability by experimental solvatochromic spectral shift is presented in figure 3. This graph was obtained by the formula (2) with using a single fitting parameter, the effective size of the molecule a. The fitted value of a is 5.6Å.

On the other hand, an effective size of the molecule can be estimated by calculation of its volume, taking into account van der Waals radii of atoms and their mutual overlap (fig.4). Van der Waals radii of the atoms are 1.2 Å (H), 1.7Å (C), 1.52 Å (O), and 1.55 Å (N).

Algorithm of size estimation includes he next items.

1. A parallelepiped is constructed which completely includes the space region occupied by the given molecule (fig. 4).

2. This parallelepiped is divided into many small identical parallelepipeds with the volume of $\Delta V=\Delta x\Delta y\Delta z$ with the sizes of the sides Δx, Δy, Δz which are much smaller than van der Waals radii of the atoms.

3. The total volume of the molecule is calculated by the formula:

$$V = \Delta V \sum_i \sum_j \sum_k g(i,j,k), \quad (3)$$

where $g(i,j.k)$=1 if a small parallelepiped with coordinates (i, j, k) located inside of at least one of Van der Waals spheres of the molecule; in the opposite case, this value is given by $g(i,j.k)$=0.

The estimated volume of the dye molecule (fig.4) is 296 Å^3. This volume corresponds to a sphere with a radius of 4.2 Å. It is quite close to fitted value of characteristic size of the molecule. We found that solvatochromic shift of CB[7] cavity corresponds to effective dielectric constant 13.6 (fig. 3).

According to experimental data [5], the dielectric constant of the cucurbit[7]uril cavity is comparable to octanol (ε = 10.3). This result is quite close to our estimation.

Table1. Calculated and experimental absorption spectrum shifts of D2 for set of solvents with different dielectric permeability relative to H_2O

Solvent	E	N	Calculated shift relative to H_2O, cm^{-1}	Experimental shift relative to H_2O, cm^{-1}
H_2O	80.1	1.333	0	0
MeOH	32.6	1.327	605	825
EtOH	24.3	1.361	859	1070
i-PrOH	20.8	1.378	1068	887
n-BuOH	17.8	1.399	1291	1251

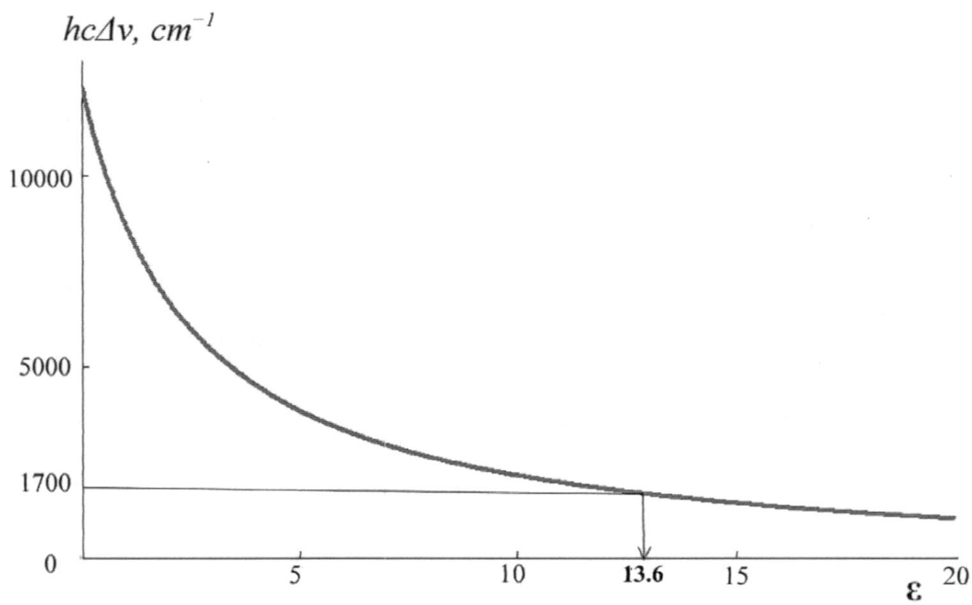

Figure 3. Extrapolation calibration curve for determination of the effective permittivity of the cavity of CB [7] based on the theoretical dependence of the solvatochromic shift on ε (formula (1)).

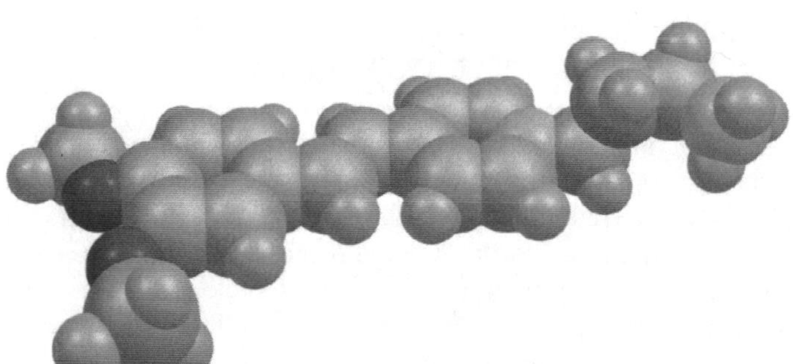

Figure 4. The space region occupied by the molecule 1-(3-ammoniopropyl)-4-[(E) -2-(3,4-dimethoxyphenyl)ethinyll]pyridinium diperchlorate (D2).

4. Conclusion

Our approach gives the value of the effective dielectric constant of the local environment of the dye molecule which works like nano-probe, by experimental determination of the solvatochromic shift of nano-probe inserted in such environment (fig. 4). The method allows working with wide intervals of dielectric permittivity by extrapolation of small sets of experimental beyond the initial interval of ε by formula (2). The use of an extrapolation calibration graph can significantly reduce the number of experiments. Input data for extrapolation includes only set of the calculating values for Onsager-Liptay equation (2). The characteristic size of the dye is a fitting parameter, the value of which is in good comparison with exact calculation (3).

The innovative potential of the method is in development of nano-probe devices for measuring the dielectric constant of small objects and nano-regions into bulk fluid volume.

Acknowledgements
This work was supported by the Ministry of Science and Higher Education within the State assignment FSRC «Crystallography and Photonics» RAS in part of quantum chemistry calculation and experimental study of solvatochromic effect. In addition, this work was supported by the "Improving of the competitiveness" program of the National Research Nuclear University MEPhI (Moscow Engineering Physics Institute) in part of molecule volume and characteristic size determination.

References

[1] Avakyan V.G., Stepko A.S., Lobova N.A., Grushnikova E.Yu., Sazonov S.K., Gromov S.P., Plotnikov V.G., Lebedev-Stepanov P.V. 2017 *Doklady Physical Chemistry* **476** pp 406-409

[2] Liptay W.Sinanoglu O. (Ed.) Modern Quantum Chemistry Part II: Interactions 1965 *Academic Press New York*p 282(Chapter5)

[3] Komarov P.V., Plotnikov V.G. 2012 *International Journal of Quantum Chemistry* **112** pp 3039-3045

[4] Komarova K.G., Sakipov S.N., Plotnikov V.G., Alfimov M.V. 2015 *Journal of Luminescence* **164** pp 57-63

[5] Khaleel I. A., Werner M. N. 2015 *Chemical Society Reviews* **44** pp 394-418

IOP Conf. Series: Journal of Physics: Conf. Series **1163** (2019) 012022 doi:10.1088/1742-6596/1163/1/012022

Development of radial turbines in low-power Gas-turbine engines

A S Pugachuk[1,2], V G Pribylov[1,3], J A Gavrilova[2] and V N Konoplev[3]

[1] Joint Institute for High Temperatures, Russian Academy of Sciences, 13 Bd.2, Izhorskaya st., 125412, Moscow, Russia
[2] Bauman Moscow State Technical University, 5 2-ya Baumanskaya st., 105005, Moscow, Russia
[3] RUDN University, 6 Miklukho-Maklaya st., 117198, Moscow, Russia

E-mail: pugachukalexandr@mail.ru

Abstract. The article deals with the problem of providing required gasdynamic and strength parameters of turbines in low-power Gas-turbine engines. A method of turbine geometrical characteristics correction is offered. It includes blade profile evaluation and modification, turbine load complex modeling and CFD-modeling of gasdynamic processes. This method is applied to the turbine impeller prototype. The most loaded places of the impeller are revealed. Hazardous locations are places of blades joints with the disk in circumference where safety factor is close to the minimum acceptable value. One of the most effective methods of strength enhancement is use of parabolic joints between the wheel and blades. These joints not only can increment safety factor up to 1.3 but also preserves the gasdynamic efficiency of the turbine. This presented method which is based on CFD-modeling and strength analysis, allows to minimize fabrication time and costs of new micro-turbines' developing process.

1. Introduction

Gas turbine engines are used widely in the electric power industry, heating supply systems and industrial facilities. The main component of such unit is a turbine: steam, gas or water powered [1]. One-stage centripetal axial-radial gas turbines are used in low power gas turbine engines (micro-turbines) because of their small dimensions (in comparison with axial turbines) and relatively high efficiency at the low gas flow rate [2]. A high-speed radial gas turbine engine, assembled on a single shaft with a high-frequency generator, composes the main part of a micro-turbine. In the market, fewer low power gas turbine engines are offered compared to high power engines [3].

Cost-effectiveness, reliability of development and enhancement of modern stationary gas-turbine engine construction depend on the capability of gas turbine engine flow range and flow process perfection. Micro-turbine designers achieve the task of using in designed micro-machines geometrical characteristics of existing prototype impellers with proven characteristics (e.g., aircraft impellers). It is necessary to demand higher standards of impeller durability while preserving required gasdynamic parameters of flow range, in order to provide the required frequency speed of micro-turbines. Thus, one of the most important tasks of micro-turbine design is radial turbine impeller gasdynamic and durability [4] optimization.

2. Method of optimization

In order to solve these tasks, a method of turbine geometrical characteristics correction is offered. The method makes possible to lower the stress that occurs in high loaded parts of impellers while preserving gasdynamic characteristics of prototype turbine flow range. Gas turbine engine turbine impeller optimization was conducted in several stages:

- cross-section estimation and prototype blade modification with profile input and output angle preservation;
- turbine impeller model creation according to the geometry of prototype flow range;
- turbine impeller strength calculation and comparison with prototype parameters;
- turbine impeller modification with consideration of greater demands in high-loaded parts;
- turbine gasdynamic calculation and comparison with prototype parameters.

3. Strength enhancement of turbine impeller

As prototype turbine impeller a typical turbine impeller with cooled blades was taken. Using Ansys CFX [5], the stress loads for 96000 rpm frequency speed and 700°C of the prototype turbine impeller were acquired. Prototype turbine impeller FEM strength calculation data is displayed in figure 1. Maximum loads appear at blade and disc connection areas (980 MPa) and on impeller hub (878 MPa). The ultimate stress of the material used ChS88U-VI (ЧС88У-ВИ) at working temperature is equal to 960 MPa. Minimal safety factor is equal to 0.98 while allowed safety factor for continuous work is equal to 1.25 [6]. Prototype turbine impeller does not meet the strength requirement for micro-turbine rotor with nominal frequency speed.

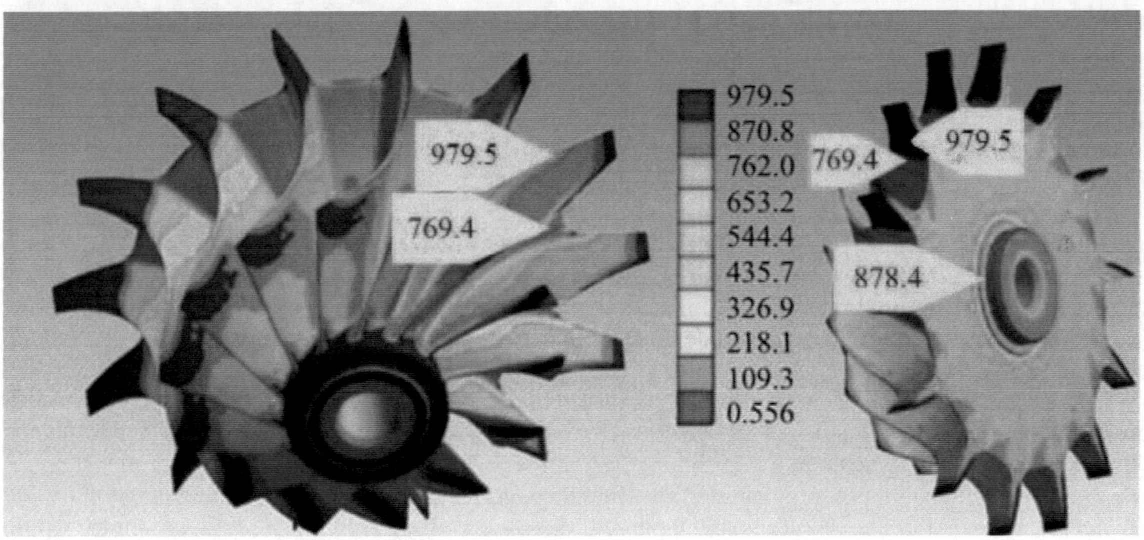

Figure1. Stress distribution in the prototype of turbine wheel

High-stress loads in impeller hub are caused by substantial peripheral blade thickness, therefore, higher values of rotating mass and corresponding stress loads are present. High values of stress loads in blade and disc connection areas are caused by blade flexural strain that comes out of centrifugal forces.

On impeller periphery, blades have a slope to the direction of rotation causing high-stress loads and deformation in disc and blade connection areas at high frequency speeds. On the first stage, to lower the centrifugal forces influence, the prototype turbine impeller blade profile is estimated with blade cross-section (perpendicular to rotation axis) analysis. The main task of the analysis is to define α_{av} — an average angle of blades cross-section center-line deviation from a radial direction. Next, basic

IOP Conf. Series: Journal of Physics: Conf. Series **1163** (2019) 012022 doi:10.1088/1742-6596/1163/1/012022

cylindrical cross-sections of prototype blade are separated layer by layer (figure 2(a)), rotated by rotation axis according to α_{av} in a way that the farthest from axis cross-section is rotated by α_{av} and all other are rotated proportionally to selected cross-section radius (figure 2(b)).

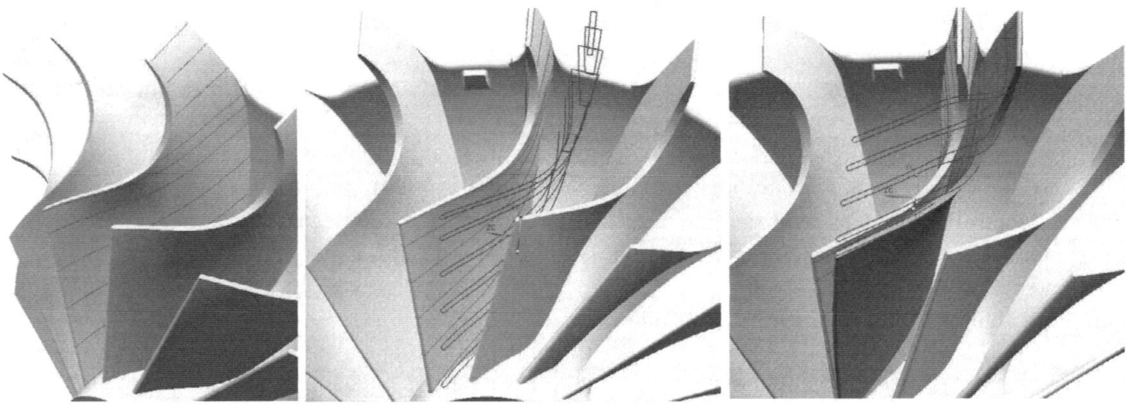

(a) **(b)** **(c)**

Figure 2. Modification of the geometric model of the turbine blade prototype: the allocation of the meridional cross-sections (**a**), rotate the meridional cross-sections (**b**), completion of sections and construction of the modified blade (**c**).

To reduce stresses in the impeller hub, decrease of a blade thickness is carried out by means of reducing thickness of profiles on the back part of a wheel (figure 2(c)). Model of a blade is designed considering turbine flow section and based on obtained basic profiles (figure 2(c)).

On the second stage, the wheel of initial impeller is combined with obtained blades. Coupling of elements is realized by means of a fillet with a radius of $1{,}4 \cdot 10^{-3}$ m. Then comparison of cross sections in initial blade and modified one is carried out. Profile analysis showed that they are partially deviated from radial direction in both sides. It can positively affect on centrifugal forces loads, because they will compensate each other.

Strength analysis in conditions of a nominal mode is conducted in designed turbine impeller to control the effectiveness of a suggested method. Minimal safety factor of a modified turbine impeller increased to 1.18. In this case the most loaded place is a blade joint with a wheel on a circumference. That's why it's offered to investigate and chose the most effective variant considering gasdynamic and strength parameters.

In this research practical measures are offered to reduce equivalent stresses in the most loaded areas: variant 1 with a radius of the fillet increased to $5 \cdot 10^{-3}$ m in joint of a blade and a wheel (figure 3(a)), variant 2 with a triangle support (figure 3(b)), variant 3 with a parabolic support (figure 3(d)). The foundation of suggested configuration variants is strengthening of a blade joint with a wheel in a circumference by means of supports of different configurations. The first variant has good strength parameters, but increasing of a fillet diameter leads to decrease of gasdynamic parameters of flow and efficiency more than 1.5% because of geometric impediment for gas fluid during its movement in a turbine flow section. The second variant allows to avoid changes in the flow section, but, due to fact that thickness of a wheel in circumference is limited, blade support in stress raiser has a small thickness and low value of safety factor.

To enhance blade support in the third variant, back blade surface was changed without any changes in configuration of a flow section in turbine. Radius of connection conic and flat back surfaces of blade are increased. In figure 3c a turbine blade is shown in a meridian section view of the impeller and also an area of strengthened material in the blade. Interface diameter from flat surface to conic one was increased without changes in an outlet blade edge.

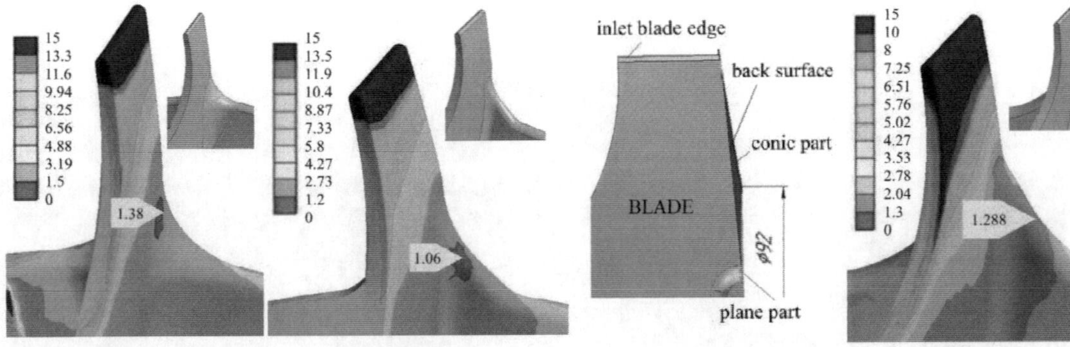

(a) **(b)** **(c)** **(d)**

Figure 3. Blade modifications and stress distribution in the modifications of turbine impeller: variant 1 with a radius of the fillet increased to $5 \cdot 10^{-3}$ m in joint of a blade and a wheel (**a**), variant 2 with a triangular support (**b**), enhancement of a blade back surface (**c**), variant 3 with a parabolic support (**d**).

Variant of a wheel joint modification, which is based on parabolic support, satisfies strength requirements and doesn't affect on a turbine flow section.

4. Evaluation of gasdynamic efficiency in turbine impeller

On the last stage it's necessary to evaluate an efficiency of a modified prototype impeller. To achieve that, using Ansys CFX (licence number 339001) research of gasdynamic parameters (gas flow rate, capability, efficiency) in a flow section considered in initial and modified impellers and comparison of obtained results are carried out.

Computational domain (figure 4) involves: inlet section 1, nozzle assembly 2, turbine computational domain 3, outlet section (to the recuperator) 4, recuperator section 5, the arrows show circular interfaces. Each domain consists of segment, which is closed in circumferential direction. Turbine computational domain doesn't consider radial split between a blade and a stator, configuration of a blade joint with a wheel and a split from the side of the wheel.

The mathematical model for calculating the parameters of the stationary gas flow in the selected region is based on the following dependencies:

$$\nabla \cdot (\rho \mathbf{u} \times \mathbf{u}) = -\nabla p + \mu \nabla^2 \mathbf{u} + \frac{1}{3} \mu \nabla (\nabla \cdot \mathbf{u}) + \rho \mathbf{g} \tag{1}$$

$$\nabla \cdot (\rho \mathbf{u}) = 0 \tag{2}$$

$$\mathrm{div}(\rho c_p T \mathbf{u}) = \mathrm{div}(\lambda \,\mathrm{grad} T) + f \tag{3}$$

$$p = \rho R T \tag{4}$$

This is the Navier–Stokes equations (1), the continuity equation (2), the energy equation (3) and equation of state (4). They contain the value of $\mathbf{u}$ is the vector velocity of the working medium, μ is dynamic viscosity, ρ – density, p – pressure, $\mathbf{g}$ is vector of gravitational acceleration, c_p is heat capacity at constant pressure, T – temperature, λ – thermal conductivity coefficient, R is the gas constant, f is a term of heat source. The mathematical model is supplemented by the equations of SST turbulence model.

Boundary conditions of this analysis are: inlet full pressure (inlet surface) is 350.8 kPa, full temperature is 1000K. On the outlet surface averaged static pressure is defined: 101.3 kPa. Computational turbine (domain 3) rotation speed is 96000 rpm.

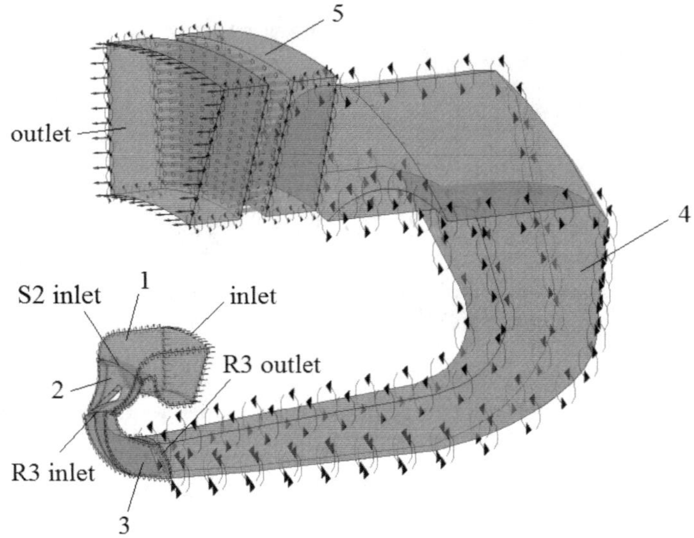

Figure 4. Computational domain for gasdynamic analysis of turbine impeller.

High demands are imposed on the accuracy of the calculation, in particular, on the density of the finite element grid. A regular hexagonal finite element grid is used to solve this problem. The number of elements in the domains is chosen with a minimum loss of accuracy (less than 1% compared to the smaller grid). For the domain 3 no more than 1 million cells are enough.

The solution of the developed model was carried out using Ansys CFX. The convergence criterion is the minimum change of the total temperature on the output surface. As a result, distributions of pressure, velocities and temperature in the computational domain were achieved. Gasdynamic parameters in the flow section are presented in table 1 and also comparison of results of gasdynamic analysis of flow sections in the initial prototype and modified turbine impeller.

Table 1. Comparison of gasdynamic parameters of flow sections in initial turbine and modified one.

Parameter	Initial turbine		Modified turbine	
	S2 inlet	R3 outlet	S2 inlet	R3 outlet
Gas flow rate (kg s^{-1})	0.294	0.294	0.294	0.294
Full temperature (K)	1100	839	1100	839
Power (kW)	91.8		91.7	
Ratio of expansion	3.33		3.34	
Adiabatic efficiency in turbine impeller	0.943		0.941	
Adiabatic efficiency in turbine stage	0.923		0.922	

Analysis of obtained parameters showed, that modified turbine has quite close parameters comparing with initial one's. Because of a modification and a turn of blade inlet profile, there is an blow and jet separation on the inlet edge of blade. That's why adiabatic efficiency slightly reduces (0,1-0,2%). But designed impeller provides initial parameters of gas flow rate, ratio of expansion and power and also satisfies strength requirements.

5. Validation of results

To verify and confirm the obtained calculated data on the basis of JIHT RAS, a stand of physical gasdynamic experiment was developed and installed [7]. The creation of such a gasdynamic stand allows for the implementation of a new technology for the study of impeller machines. This technology includes tests of impeller machines in the form of models which can be manufactured quickly of cheap materials. This set of equipment will allow testing models of turbomachines of full-scale geometric dimensions and obtaining experimental data to verify all the calculated data and validation of the methods used.

6. Conclusion

Method of reducing stresses in a turbine impeller is developed. This method is tested by means of a prototype turbine impeller. Configuration of the prototype is modified in presented method: blade profile is estimated in case of deviation from radial profiles, a turn of blade profiles and development a new blade, design of a turbine impeller considering enhanced requirements related to high loaded places and strength analysis of prototype turbine impeller and modified one are carried out. In contrast to initial prototype, modified turbine impeller satisfies strength requirements in working mode of a micro-turbine. Results of gasdynamic analysis are: the main gasdynamic parameters and efficiency of modified turbine impeller have inconsiderable differences (less than 1%) from appropriate parameters of initial prototype. Due to modification, safety factor of turbine impeller increased in 30% without sacrificing of gasdynamic parameters. This method can be used not only in turbine impellers but in compressor axial-radial ones.

References

[1] Sherstyuk A N, Zaryankin A E 1976 *Radial-axial turbines of low power* (Moscow: Mechanical engineering) p 208

[2] Kosoi A S, Popel' O S, Zeigarnik Y A, Sinkevich M V and Beschastnykh V N 2017 Small Gas-Turbine Units For The Power Industry: Ways For Improving The Efficiency And The Scale Of Implementation *Thermal Engineering* **64** (10) pp 723–28

[3] Sobolev A G 2014 Radial gas and steam turbines *Turbines and diesel engines* **2** pp 62–66

[4] Kostyuk A G 2007 *Dynamics and durability of turbomachines* (Moscow: MPEI publishing house) p 476

[5] Bruyaka V A 2010 *The engineering analysis in ANSYS Workbench* (Samara: Samara State University) p 271

[6] Chernousenko O Yu 2008 The enrichment and analysis of the calculated investigation of individual resource of corps, rotors, catch valves of the CHP and CMP K-200-130 block of 200 MVt *Proc. Conf. Energy and heat engineering processes and equipment (Kharkiv)* vol 6 (*Kharkiv*: NTU "KhPI") pp 107–11

[7] Kosoi A S, Monin S V, Sinkevich M V 2018 Contemporary approaches to research supporting the development of microturbine power generation systems *Vestnik Koncerna VKO «Almaz – Antey»* **1** pp 72–79

Development of a system generator for Marsohod development boards

P Dmitriy[1], I Alexey[1], A Romanov[1,2,3]

[1] National Research University Higher School of Economics, 20 Myasnitskaya Ulitsa, 101000, Moscow, Russia
[2] e-mail: a.romanov@hse.ru
[3] Author to whom any correspondence should be addressed

Abstract. This article describes a technical solution of the system generator of configuration files of development boards “Marsohod 2”, “Marsohod 2bis”, “Marsohod 3”, “Marsohod 3bis” for Quartus Prime software. The solution includes a web interface for the system generator, generation of configuration files, introduction of additional modules into the generated project, such as a frequency divider, Uart8 (RS-232), a module for preventing of contact bounce, and several types of simplest MIPS processor cores. This technical solution improves the convenience and speed of FPGA development, as well as reduces its entry threshold which can be significant for starting developers.

Keywords: FPGA, System Generator, Quartus, MIPS, Marsohod, Marsohod 2, Marsohod 2bis, Marsohod 3, Marsohod 3bis, Uart, RS232.

Introduction

When working with development boards based on FPGA and system-on-chip design, it is necessary to set up the project within the framework of the tasks assigned. The automation of this task contributes not only to reduction in time for commencement of work on a project, but also to decrease in the number of potential problems and mistakes by eliminating the need to go into peculiarities of working and specifics of realization of development boards.

Many manufacturers create their own tools for making custom preconfigured projects for their boards. For example, Terasic Company with each of their development boards provides special programs (system generators) on its website. They make it possible to choose which modules should be integrated into the project. Each board requires its own generator, which must meet the requirements of software portability and be supported by different operating systems. In addition, existing system generators are limited by the existing periphery and functionality of development boards only; they cannot integrate additional modules or common add-ins into the project.

Many boards of less advanced manufacturers do not have system generators at all. This work is aimed at decreasing the entry threshold and reducing the time required to create projects using Marsohod boards by automation of initial configuration process, which will also make it easier to study FPGA design. The developed solution is suitable for executing projects which do not require a specific configuration and allows obtaining configuration files and preconfigured projects for development boards of Marsohod family.

1. Description of technologies used in the project

Currently, there are two main types of programmable logic devices (PLD): CPLD (Complex Programmable Logic Device) and FPGA (Field-Programmable Gate Array). The main difference between them is firmware saving after switching off the power supply of the board (such properties are possessed by devices like CPLD). Among Marsohod development board family, it is the first Marsohod that has CPLD chip. MAX II chip is installed on the first Marsohod, and Marsohod3, Marsohod3bis boards have MAX 10 chips [15]. Marsohod2 and Marsohod2bis are equipped with FPGA Cyclone III [13], and Cyclone IV [14] chips, respectively. All considered boards use PLDs from Intel FPGA company.

For each board from Terasic company of DE1-SoC type [12], based on Cyclone V, there is a specialized software – System Generator through which .qpf, .qsf, .v, .sdc files can be generated, and it is also possible to disable / enable some integrated peripherals on PLD and on the board for further operation in Quartus Prime project [3]. In addition, in Quartus Prime, there is a Template mechanism, so that a project for a specific board can be generated in Project Wizard.

The main differences of the developed technical solution for "Marsohod" boards are:

- the way of user interaction through web interface instead of a separate application;
- the possibility of introducing additional modules (in Verilog language) to the project;
- orientation to "Marsohod" development boards;

2. Description of development drift As a result of analyzing of design principles of projects for PLD with following prototyping, as well as of analyzing of structure of the configuration files, the following main configuration files were identified:

- .qpf file containing the project description for Quartus Prime;
- .qsf file containing the information about the type of board, the used resources, and the assignments of the crystal pins;
- the top level file of the hierarchy (file in Verilog language) containing the description of the top level module in hardware description language [6];
- .sdc file containing information about the time delays of the board;

The main documents for working with "Marsohod" development boards are their technical specifications [2] which are a declarative description of the characteristics of the board, the available resources, and the ways to connect them. To present the board characteristics in a convenient form, and to further generate the configuration files out of them, the following set of formats was chosen: YAML [7] (used by default), JSON [5], and binary serialization Python (pickle). To support these formats, the developed software solution contains a special module that supports the conversion of documents of different formats. The choice of a format is due to the following considerations (in descending order of importance): convenience of reading and editing by a human; document size; prevalence. YAML is chosen as a compromise variant, because it has better readability, a satisfactory file size, and is common enough.

The developed software has a flexible modular structure that allows its parts being used for various variants of their implementation and consists of the following:

- The system core ("thumb") is a module that encapsulates the generation of configuration files and performs related actions, such as writing the configurations into files on the disk, packaging configurations into an archive, loading templates for generation, event logging, etc;
- External interfaces – a set of modules that exploit the system core to provide a friendly interface to the end user. These modules include web service that allows generating configuration files;
- Configuration module of the high-level parameters for individual parts of the system representing a centralized repository of settings that define the system behavior without the necessity to change the entire logic of the program;
- Unit tests and functional tests for individual parts of the system;

- Documentation for the developed system;

The generator of configuration files is included into the system "core" and provides its main functionality. Each type of development boards is presented in the form of a separate class that provides loading corresponding document which describes the board through reflection method [8]. Each class of development boards is a child object of a generic class that implements the file generation in accordance with the loaded configuration, archiving and writing created files into the disk.

To generate the files, Jinja2 template library [4] is used which is an interface that allows performing substitutions with special internal language DSL (Domain-Specific Language) [9]. The use of this solution is due to the following reasons:

- The presence of a large number of different templates for generation. This makes it possible to avoid own tool development leading to more time consumption to perform such design stages as architecture development, implementation and debugging of the system, fine tuning, etc;
- The necessity to ensure the possibility to make patches and additions into the templates by the third party which provides extension of the proposed solution onto new development boards not limited to "Marsohod" boards only;
- The ability to highlight the common parts in the templates of the files. This Jinja2 feature allows introducing the additional information at the beginning and at the end of each generated file;

Basically, the generating process of the configuration files consists of the following sequence of actions:

- Determining the type of development board;
- Downloading relevant documents with board descriptions;
- Performing the substitutions in accordance with the requirements of the user;
- Saving the resulting configurations into the specified directory;

During the software behavior, performed actions and occurring errors logging are maintained. The log is kept in accordance with the required level of detail (debugging information, information about major events, information about errors, etc.) and includes information about the date of occurrence, level, place of event occurrence, and the additional message about the event. Event logging at the files generator level allows analyzing possible user problems occurred during the operation of the entire system.

To provide a cross-platform character of the developed solution, the variant of creating a web service was chosen which allows generating the required configurations for development boards. This solution has the following advantages:

- Availability on any device under the control of any OS including the mobile one;
- Centralized control of the entered changes which ensures that there is no need to make changes to application clients for each OS;
- Distribution control (providing additional content to a limited number of registered users depending on their access rights, for example, providing special setup options for students of certain institutions of higher education);
- Collecting of statistics of the application operation (the user registration mechanism allows receiving statistics of system usage including the addition of a feedback mechanism);

The necessity of Internet connection availability for functioning of the service is a limiting factor, but in modern realities, it is not so important.

The developed service provides the following functions:

- Generation of configuration files for development boards;
- Logging in by using an email address to gain an access to additional functionality;
- Viewing of articles associated with the system-on-chip development (implemented for authorized users);

IOP Conf. Series: Journal of Physics: Conf. Series **1163** (2019) 012023 doi:10.1088/1742-6596/1163/1/012023

- Downloading additional materials (for example, to perform laboratory work on Marsohod boards) for authorized users;

The service distribution is provided by message mailing to specified emails with a special link that contains a characters sequence (token) in JSON Web Tokens (JWS) format inside which the link expiration time and the user id in the database are encoded. The data is encrypted according to HS256 algorithm. This format is also used to generate the links for the materials distributed by the service which allows hiding the real structure of the files location and protecting the service from configuration files leakage and executing of arbitrary code. To manipulate the database, an object-relational model is used [10]. The developed software has a mobile version of the site, as well as Russian and English interface versions.

3. Additional modules for the system generator

In addition to the main modules, the system generator is capable to add various add-ons that allow additional modules to be integrated into the final generated projects (analogue of IP cores in Quartus Prime).

One of the analogues is Platform Designer (previously, QSYS). Platform Designer is a system for device designing at the level of system components; it allows combining IP cores while creating its own device logic based on the composition of these cores. Platform Designer also supports the ability to integrate own IP cores [17]. The second analog is MegaWizard Plugin [18]. This tool allows integrating Megafunctions into the project. Megafunctions are IP blocks parameterized and optimized for Intel FPGA (Altera) device architectures. Intel FPGA is a mega-function library including Library of Parameterized Module (LPM) functions and other parameterized functions that offer more efficient logic synthesis and implementation of digital devices [19].

The difference between the proposed implementation and the analogs, considered above, is the support for additional modules that are absent in MegaWizard Plugin and Platform Designer, such as the contact bounce module and UART module. Also, a distinctive feature is the support for inclusion of the simplest processor core in the project. Additional modules:

- *Debounce module* – the problem of contact bounce is relevant in the implementation of the processing systems of signals applied as a result of mechanical action (for example, by pressing the mechanical button). The essence of the problem includes in the fact that the contacts closure of mechanical element can catch several cycles of device operation, and, therefore, switching will occur several times. The debounce module provides the required delay before switching. The size of the delay can be adjusted by the parameters changing of CLOCK_RATE and MILLIS_DELAY. The first parameter indicates the clock frequency of the internal generator of the board and is set automatically while the second parameter, indicating the delay amount in milliseconds, is set by the user from the system generator interface (Fig. 1).

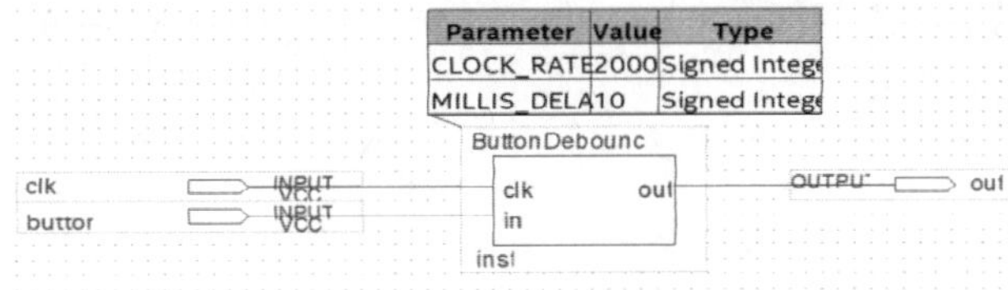

Figure 1. Block diagram of the contact bounce eliminating module connection.

- *Frequency divider module* – the limited number of frequency generators with a fixed frequency can lead to problems when working with peripherals, since the devices can operate at completely different frequencies. The implemented frequency divider module reduces the frequency of the board internal generator to the set one. CLOCK_FREQ parameter is

IOP Conf. Series: Journal of Physics: Conf. Series **1163** (2019) 012023 doi:10.1088/1742-6596/1163/1/012023

responsible for the incoming frequency (cycles per second); OUT_FREQ parameter is responsible for output frequency (cycles per second).

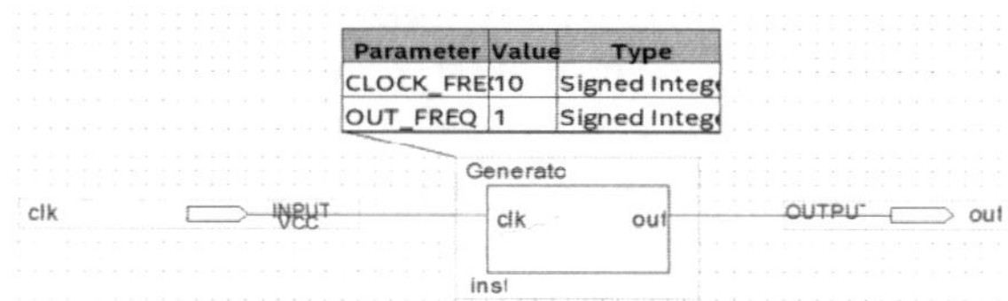

Figure 2. Block diagram of the frequency divider module connection.

- *UART (RS-232) module* – Universal Asynchronous Receiver Transmitter (UART) is a serial universal asynchronous interface that is responsible for transfer data between digital devices [11]. This implementation makes it possible to separately control the operation of the receiver and transmitter (rxEn, txEn) using RS-232 protocol and provides for the installation of 2 parameters for flexible adjustment of the device operation. CLOCK_RATE parameter is responsible for the internal frequency of the development board; BAUD_RATE parameter is responsible for the frequency of the receiving / transfer bus.

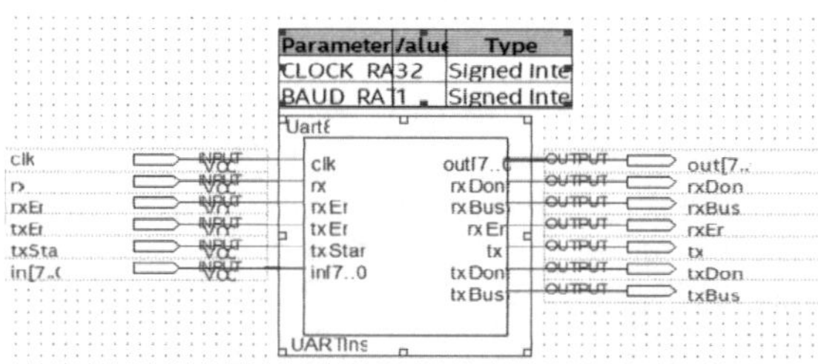

Figure 3. UART module Block diagram.

- *schoolMIPS* – a small MIPS processor core is originally based on Sarah L MiraS processor from Sarah L. Harris[1], and its first version was developed for Young Russian Chip Architects Summer School [16]. The core has several versions. All implementations (simple, mmio, irq, pipelined, pipeline_irq, pipeline_ahb) were taken from the github repository and integrated into the current solution.

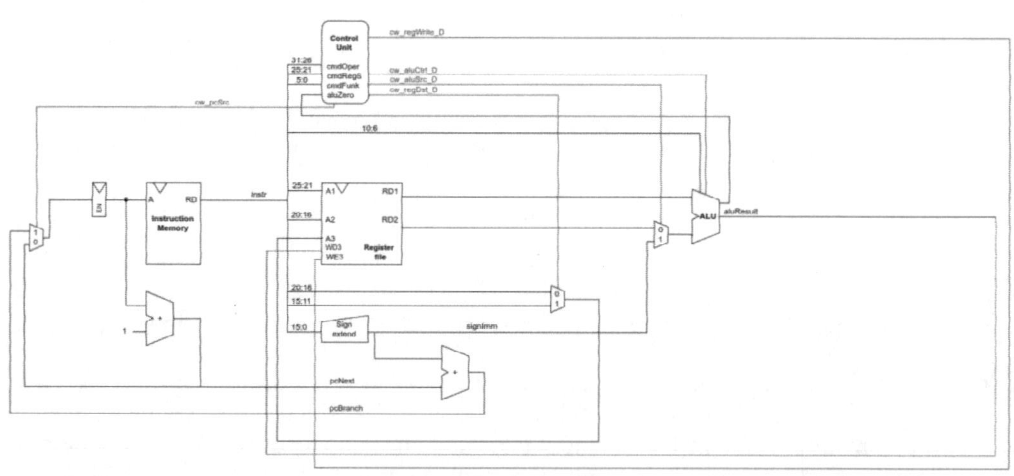

Figure 4. Simple MIPS.

4. Testing

Testing of the developed system was carried out using Google Chrome 68 (x64) browser, on Marsohod3 and Marsohod2 platforms; for CAD Quartus Prime 17 – Marsohod3, and for Quartus II 9.0 – Marsohod2, on Windows 10 (x64) operating system.

Estimated calculation of consumed resources by the additional modules:

- Module for eliminating of contact bounce with parameters CLOCK_RATE = 100 MHz, MILLIS_DELAY = 100 Hz – 31 LUT, 23 REG.
- Module of the frequency divider with the parameters CLOCK_FREQ = 100 MHz, OUT_FREQ = 1 MHz – 12 LUT, 7 REG.
- Control module by seven-segment indicator based on a counter at 16 - 12 LUT, 4 REG.
- Module UART8 with parameters CLOCK_RATE = 100 MHz, BAUD_RATE = 9600 Hz, – 65 LUT, 38REG.

Estimated calculation of consumed resources for school MIPS processor:

- simple – 164 LUT, 65 REG.
- mmio – 159 LUT, 65 REG.
- irq – 171 LUT, 66 REG.
- pipeline – 1012 LUT, 403 REG, 2048 memory bits.
- pipeline_irq – 1004 LUT, 403 REG, 2048 memory bits.
- pipeline_ahb – 1177 LUT, 403 REG, 2048 memory bits.

Testing of the server side:

- Server startup time ~ 5 s.
- Resources consumed: 100 MB.
- Response time ~ 1 ms.
- Simultaneous finding of users on the site – more than 500.

5. Conclusion

As a result of technical solution implementation, the following results were obtained:

- The configuration files generator was implemented;
- Additional functional modules were developed in Verilog language;
- The implementing possibility of the developed functional modules in the project were realized;
- The possibility of integration of a simple processor core schoolMIPS into the project was realized;
- web interface for user interaction was implemented;
- User authorization mechanism was created to provide selective access to the required sections of the proposed technical solution;
- Documentation for rapid deployment of a developed technical solution.

All the above made it possible to significantly improve the operation convenience with the development boards Marsohod 2, Marsohod 3, Marsohod 2bis, Marsohod 3bis and to increase the speed of projects development on these boards. The entry threshold for development on these boards was also reduced which may increase their attractiveness for training in FPGA development. The possibility to implement modules in the projects, which are not realized in such analogues as MegaWizard Plugin, or Platform Designer, allows quick solving of a number of additional tasks.

Acknowledgment

The publication was prepared within the framework of the Academic Fund Program at the National Research University Higher School of Economics (HSE) in 2018-2019 (grant № 18-01-0074) and by the Russian Academic Excellence Project “5-100”.

References

[1] Harris D and Harris S 2014 *Digital Design and Computer Architecture*

[2] Strogonov A and Bystritsky A 2013 *Designing systolic FIR filters in the FPGA basis using the ModelSim-Altera modeling system* (Components and technologies) chapter 9 pp 58-62

[3] Gupta P 2016 *Accelerating datacenter workloads* (26th International Conference on Field Programmable Logic and Applications)

[4] Ronacher and Armin 2008 *Jinja2 Documentation*

[5] Bradley J and Sakimura N 2015 *Json web token* (RFC 7519)

[6] Polyakov A 2018 *VHDL and VERILOG languages in the design of digital equipment* (Litres)

[7] Sinha V Doucet F Siska C Gupta R Liao S and Ghosh A 2000 *YAML: a tool for hardware design visualization and capture* (Proceedings of the 13th international symposium on System synthesis. IEEE Computer Society)

[8] Malenfant J Jacques M and Demers F 1996 *A tutorial on behavioral reflection and its implementation* (Proceedings of the Reflection) p 1-20

[9] Mernik M Heering J and Sloane A 2005 *When and how to develop domain-specific languages* (ACM computing surveys) p 316-344

[10] Schnicariol K Schnicariol M and Schnicariol M 2009 *Object-relational mapping* (Pro JPA 2. Apress) p 69-106

[11] Wei-qing, and Wei-Zhou 2009 *Design of USB-RS232 converter module based on FT2232H [J]* (Electronic Design Engineering) chapter 7

[12] Kashani-Akhavan Sahand and Beuchat R *SoC-FPGA Design Guide DE1-SoC Edition*

[13] Corperation Altera 2012 *Cyclone III device handbook*

[14] Corperation Altera 2016 *Cyclone IV, Device Handbook*

[15] Membrey P and Pang A 2017 *Lock and Load* (Beginning FPGA: Programming Metal. Apress, Berkeley) p 23-45

[16] Ryazanova A Romanova I 2018 *Analysis of options for the implementation of the interaction subsystem for a multiprocessor-on-chip system based on schoolmips*

[17] Edwards S 2015 *Embedded System Design Lab: Peripherals and Device Drivers* (CSSE)

[18] Bochem A Deschenes J Williams J and Kent B 2014 *An Overview of Embedded Systems and Prosthesis Technology* (Faculty of Computer Science University of New Brunswick Frederiction, NB, Canada)

[19] Jose S 2009 *Megafunction overview user guide*

IOP Conf. Series: Journal of Physics: Conf. Series **1163** (2019) 012024 doi:10.1088/1742-6596/1163/1/012024

The application of the Lattice Boltzmann method to the one-dimensional modeling of blood flow in elastic vessels

Oleg Ilyin

Dorodnicyn Computing Center of Federal Research Center Computer Science and Control of Russian Academy of Sciences, Vavilova - 40, Moscow, 119333, Russia

E-mail: oilyin@gmail.com

Abstract. We consider mass conservation and momentum equations for inviscid blood flow in elastic vessels. It is shown that the Lattice Boltzmann(LB) models for non-ideal gas in the limit of low Knudsen numbers are asymptotically equivalent to the blood flow equations. The equation of state for non-ideal gas becomes vessel's pressure-area response when LB models are applied for blood motion modeling. Two test case problems are considered: a propagation of a sole nonlinear wave and pulse wave propagation in a stented vessel.

1. Introduction

The models of blood motion in cardiovascular system range from the 0-D lumped models, 1-D pulse propagation equations to 3D viscous flow equations [1]. In many cases the 3D approach based on the solution of the Navier-Stokes equations is too detailed while 0D lumped models are oversimplified and applicable only for the distal vasculature. In the case of the 1D models it is assumed that the radial velocity is negligible [2]. Then, integrating the Navier-Stokes equations over the radial variable the 1D nonlinear system of equations (depending only on one spatial variable, axial coordinate) for the luminal area change and the axial blood velocity is derived. The most popular method of solving of these equations is locally conservative Galerkin (LCG) method.

I will present an alternative way to model 1D blood dynamics based on the kinetic equations, namely, using the Lattice Boltzmann approach [3]. This method describes the motion of particles on Cartesian spatial lattice (advection part), the collision of the particles in spatial nodes is modeled by assuming that the velocity distribution of the particles tends to a local equilibrium state (trend to Gaussian distribution). The Lattice Bolzmann (LB) method correctly reproduces low-Mach incompressible flows like blood motion and can be used in the modeling of the cardiovascular network. The details about 3D modeling with the LB approach can be found in several papers [4, 5, 6, 7, 8].

We start with the one-dimensional model D1Q3 and show that there exists an interesting analogy between the hydrodynamics obtained from LB model at the limit of small Knudsen numbers (small times between collisions) and the blood flow equations in elastic vessels. More precisely, the LB hydrodynamics for D1Q3 model is equivalent to the 1D blood motion equations if one changes in the LB model the lattice sound velocity by the pulse propagation velocity and

IOP Conf. Series: Journal of Physics: Conf. Series **1163** (2019) 012024 doi:10.1088/1742-6596/1163/1/012024

the density by the luminal area. In this case the LB model describes a particular case of constant pulse propagation velocity though this case is nonlinear. The method can be generalized using the LB methods for non-ideal gases: the addition of fictitious force allows to model arbitrary area-pressure vessel responses.

The presented method is simpler than the finite-difference methods for the nonlinear 1D blood equations [9]. The method correctly describes the change in shape of the initial pulse wave and moreover can be applied for the assessment of the forward and the backward blood pressure-velocity wave (reflected wave) superposition. The blood velocity waveforms can be obtained for the different elastic area-to-pressure responses of the vessel wall, the wall viscoelastic effects can be also potentially implemented.

2. 1D blood motion equations and the Lattice Boltzmann method

Consider a long elastic vessel filled with incompressible fluid (blood). The conservation equations of mass and momentum in the elastic vessel are obtained from the NavierStokes equations by the integration over the cross-sectional spatial coordinate (radial variable). We assume that the radial component of the blood velocity is close to zero, and the velocity profile is flat. Then the following equations are obtained [9]

$$\frac{\partial A}{\partial t} + \frac{\partial Au}{\partial x} = 0, \quad \frac{\partial u}{\partial t} + \frac{\partial u^2/2}{\partial x} = -\frac{1}{\rho}\frac{\partial p}{\partial x}, \tag{1}$$

where ρ is the constant blood density (incompressible fluid), A, u is the luminal area, blood velocity respectively. Here x is the axial coordinate and p is the blood pressure. We mention that viscous forces are not included in Eqs (1).

It is obvious that one has three unknowns u, A, p and only the two governing equations. To close the system (1) one needs to introduce a *pressure-area relation*

$$p = f(A), \tag{2}$$

where f is a some function, its form should be defined from the elastic properties of the considered vessel. For realistic vessels this dependence can be complicated and saturation effects are observed [10]. In practice the Laplace law is popular [1, 2].

From the pressure-area relation one can find the pulse-wave velocity using the formula

$$c_{pulse}^2 = \frac{A}{\rho}\frac{\partial p}{\partial A}.$$

Using the pulse wave velocity definition Eqs (1)-(2) can be rewritten in the following equivalent form

$$\frac{\partial A}{\partial t} + \frac{\partial Au}{\partial x} = 0, \quad \frac{\partial u}{\partial t} + \frac{\partial u^2/2}{\partial x} = -\frac{c_{pulse}^2(A)}{A}\frac{\partial A}{\partial x}. \tag{3}$$

We will model these equations using the Lattice Boltzmann approach [3]. A simplest three-velocity model termed as $D1Q3$ is considered below. This model describes three populations of particles traveling on a 1D lattice with equal spacing c between the lattice nodes. During an each discrete time step Δt the particles at a node x can hop to adjacent nodes $x \pm c$ or stay at the node x. We define the corresponding concentrations of the particles as $f_{\pm 1}(t,x)$ and $f_0(t,x)$ respectively. The dynamics of $f_{\pm 1}, f_0$ obeys the following discrete equations in space and time

$$f_{-1}(t+\Delta t, x - c\Delta t) - f_{-1}(t,x) = \frac{\Delta t}{\tau + \frac{1}{2}}\left(f_{-1}^{eq}(t,x) - f_{-1}(t,x)\right),$$

$$f_0(t+\Delta t, x) - f_0(t,x) = \frac{\Delta t}{\tau + \frac{1}{2}} \left(f_0^{eq}(t,x) - f_0(t,x)\right),$$

$$f_{+1}(t+\Delta t, x + c\Delta t) - f_{+1}(t,x) = \frac{\Delta t}{\tau + \frac{1}{2}} \left(f_{+1}^{eq}(t,x) - f_{+1}(t,x)\right),$$

where τ is the relaxation time which can be considered as a free parameter and $f_{\pm 1}^{eq}, f_0^{eq}$ are the equilibrium states (analogs of the Maxwell distribution)

$$f_{-1}^{eq}(t,x) = \frac{\rho(t,x)}{6}\left(1 - 3\frac{u(t,x)}{c} + 3\frac{u(t,x)^2}{c^2}\right),$$

$$f_0^{eq}(t,x) = \frac{4\rho(t,x)}{6}\left(1 - 3\frac{u(t,x)^2}{2c^2}\right),$$

$$f_{+1}^{eq}(t,x) = \frac{\rho(t,x)}{6}\left(1 + 3\frac{u(t,x)}{c} + 3\frac{u(t,x)^2}{c^2}\right),$$

where ρ, u are the density and flow velocity

$$\rho(t,x) = f_{-1}(t,x) + f_0(t,x) + f_{+1}(t,x),$$

$$\rho(t,x)u(t,x) = -f_{-1}(t,x)c + f_{+1}(t,x)c,$$

moreover, one can define the full energy of the lattice gas in every node by the formula

$$\rho(t,x)(u(t,x)^2 + c_s^2) = (f_{-1}(t,x) + f_{+1}(t,x))c$$

and here c_s is the sound velocity of the lattice gas defined by

$$c_s = \sqrt{\frac{1}{3}}c.$$

At the limit of small values of τ the lattice gas can be considered as a continuous media since the time between collisions tends to zero (τ as the relaxation time is proportional to the expected time between the collisions). In the continuous limit the considered LB model describes some hydrodynamics which can be obtained using the Chapman-Enskog expansion [11] on a small parameter (τ). For the $D1Q3$ the following equations are obtained

$$\frac{\partial \rho}{\partial t} + \frac{\partial \rho u}{\partial x} = 0, \quad \frac{\partial u}{\partial t} + \frac{\partial u^2/2}{\partial x} = -\frac{c_s^2}{\rho}\frac{\partial \rho}{\partial x} + \frac{\nu}{\rho}\frac{\partial}{\partial x}\left(\rho \frac{\partial u}{\partial x}\right) + O(Ma^3) + O(\Delta t^2), \tag{4}$$

where $O(Ma^3)$ is error term which appears due to the polynomial form of the local equilibrium on the bulk velocity (here Ma is Mach number), $O(\Delta t^2)$ is space-time discretization error. If the flow is slow ($Ma \sim 0$) and the time step is small then we can state that the LB model describes 1D hydrodynamics of the compressible gas.

Now if one changes the density ρ by the luminal area A and the sound velocity c_s by the pulse wave velocity c_{pulse} then Eqs (4) turn into Eqs (3) if the flow is slow, discretization error is negligible and viscous term can be omitted from consideration. These condions are realistic due to a) A typical blood flow is slow, the blood velocity is over $0.4m/s$, the pulse wave speed is over $4m/s$, then $Ma \sim 0.1$ b) the viscous terms can be neglected in large vessels since the typical Reynolds number in aorta is close to 1000 or even greater c) the discretization error can be controlled by the choice of Δt.

Note that the equations describe only the particular case of constant pulse velocity since c_s is constant in the LB model. Indeed, the $D1Q3$ model describes isothermal flows with the equation of state $p = \rho c_s^2$ and if one changes ρ by A and c_s by c_{pulse} then we get area-pressure relation $p = Ac_{pulse}^2$.

The generalization of the presented method for the arbitrary area-pressure relation can be performed using the extension of the Lattice Boltzmann method for non-ideal gases and arbitrary equations of state which results in general area-pressure relations in the our case [12, 13, 14].

One of the possible approaches to generalize the equation of state is an addition of a fictitious force term in the LB model [12]

$$f_{-1}(t+\Delta t, x - c\Delta t) - f_{-1}(t,x) = \frac{1}{\tau + \frac{1}{2}} \left(f_{-1}^{eq}(t,x) - f_{-1}(t,x) \right) + F_{-1},$$

$$f_0(t+\Delta t, x) - f_0(t,x) = \frac{1}{\tau + \frac{1}{2}} \left(f_0^{eq}(t,x) - f_0(t,x) \right) + F_0,$$

$$f_{+1}(t+\Delta t, x + c\Delta t) - f_{+1}(t,x) = \frac{1}{\tau + \frac{1}{2}} \left(f_{+1}^{eq}(t,x) - f_{+1}(t,x) \right) + F_{+1},$$

where

$$F_0 = 0, \quad F_{\pm 1} = \pm 2 \frac{c}{6} \Phi \frac{\partial}{\partial x} \Phi, \quad \Phi \equiv \sqrt{A c_{pulse}^2 - P(A)}$$

and $P(A)$ is the desired area-pressure relation. The stability conditions for the presented method [13] are the follows

$$\left(\frac{\Delta x}{\Delta t} \right)^2 \geq \max \left\{ 3 \frac{P(A)}{A}, 1 + \frac{\partial P(A)}{\partial A} \right\}.$$

3. Test problems: nonlinear forward sole wave propagation, pulse wave propagation in a stented vessel

Now we consider two examples.

The first problem is nonlinear sole forward pulse wave propagation [15]. We again consider the equations (1)-(2) in the semi-infinite interval $x \geq 0$. We apply the following area-pressure relation

$$p = p_0 + \frac{1}{nD_0} (A^n - A_0^n), \tag{5}$$

where D_0 is some constant.

For a forward traveling wave one has the relation between the wave velocity u and the luminal area A [15]

$$u = \int_{A_0}^{A} \frac{dz}{\sqrt{\rho D_0 / z^n z}}, \tag{6}$$

then the equations (1)-(2) reduce to a differential equation for A

$$\frac{\partial A}{\partial t} + \frac{1}{\sqrt{\rho D_0}} \left(\left(1 + \frac{2}{n} \right) A^{n/2} - \frac{2}{n} A_0^{n/2} \right) \frac{\partial A}{\partial x} = 0. \tag{7}$$

We supplement the equation (7) with the symmetric triangle-shaped initial-boundary condition at $x = 0$ for $t \in [0, T_0]$, where T_0 is one heartbeat

$$A(t,x)|_{x=0} = A_0 + at, \quad t \in [0, t_0), \tag{8}$$

$$A(t,x)|_{x=0} = A_0 + a(2t_0 - t), \quad t \in [t_0, 2t_0], \tag{9}$$

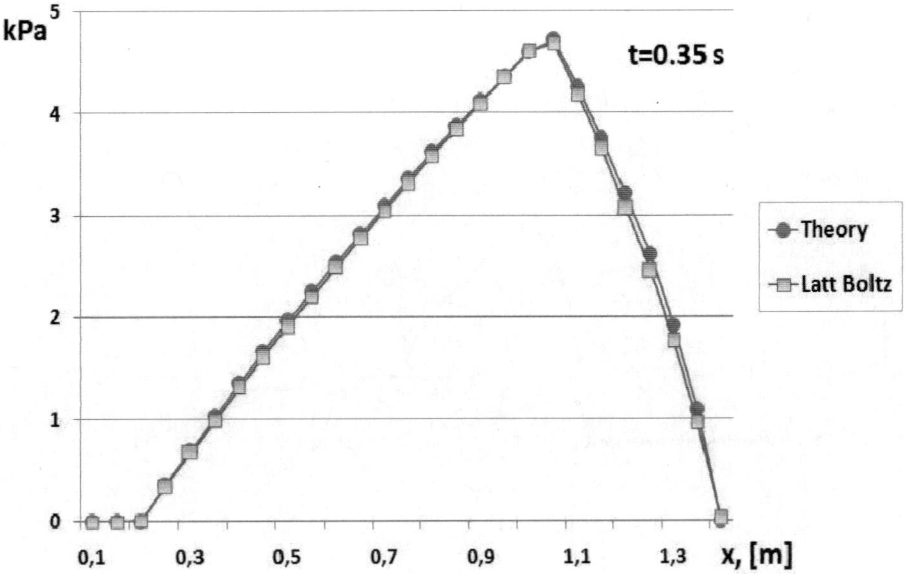

Figure 1. Two pressure profiles at $t = 0.35$. Initial pulse wave was symmetrical at $t = 0$. Both methods – Lattice Boltzmann (boxes) and analytical solution (circles) show very closed results.

$$A(t,x)|_{x=0} = A_0, \quad t \in (2t_0, T_0] \tag{10}$$

and

$$A(t,x)|_{t=0} = A_0, \quad x > 0. \tag{11}$$

We consider the case $n = 1/2$ in (5) which corresponds to the Laplace law. The analytical solutions to this problem are obtained in [15]. For the sake of brevity, we do not give them here but they will be used for benchmarking. We solve $D1Q3$ model with external force, applying the initial-boundary conditions (8)-(11) and also using (6) at the left boundary; then compare the results with the analytical solutions at different moments of time. The results show very good agreement.

The second problem is the wave propagation in a vessel with varying material properties. We consider a vessel of a length $1m$. We assume that the values of the linearized pressure pulse wave velocity which is given by the formula $\sqrt{A_0^n/\rho D_0}$, where A_0 is undisturbed vessel area are different in the considered region. We assume that in the interval $x \in [0.2m\, 0.3m]$ the linearized pulse velocity equals $10m/s$, while in the rest part of the tube the linearized pulse velocity has a value of $4m/s$ (typical physiological value). The increase of the pulse wave velocity is observed due to the smaller values of the distensiblity D_0 in the interior region in comparison with the outer part of the vessel. This can happen if the internal region is stented and has increased stiffness.

We consider the dynamics of a half-sinusoidal pressure pulse wave starting at the point $x = 0$. Assume that at the points $x = 0.15m$ ($0.05m$ before the stiff region) and at $x = 0.35m$ ($0.05m$ after) pressure recording devices are placed. The amplitude of the pressure pulse wave at $x = 0.15$ is greater than the amplitude at $x = 0.35m$ (Fig. 2). This feature is an effect of the wave reflection at the boundary where the vessel changes its elastic properties. This fact is well known in the fluid dynamics [16]. Moreover, pressure oscillations are observed which can

IOP Conf. Series: Journal of Physics: Conf. Series **1163** (2019) 012024 doi:10.1088/1742-6596/1163/1/012024

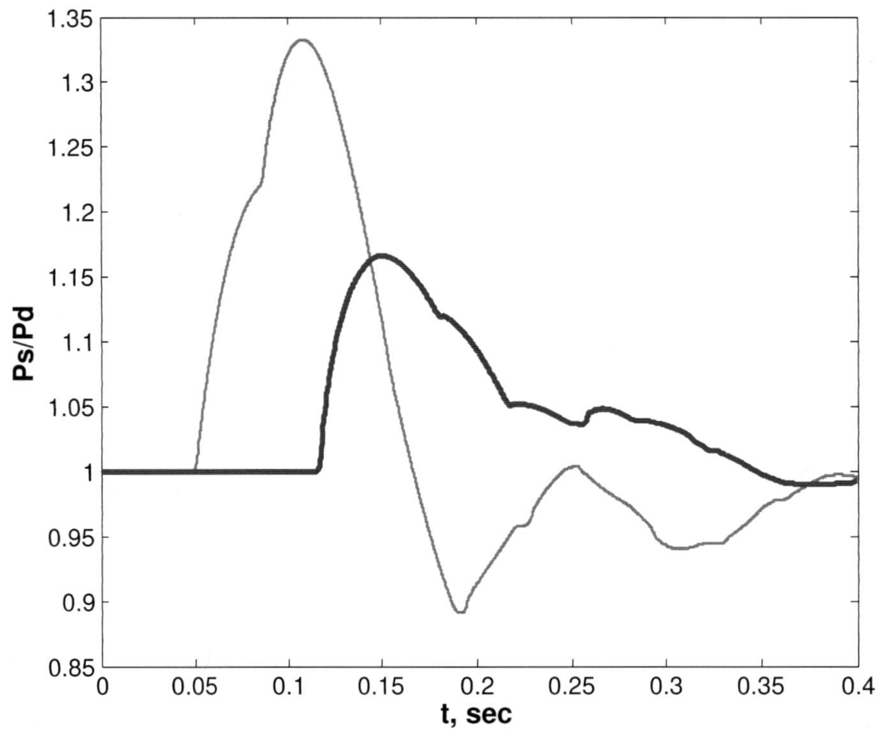

Figure 2. Two relative pressure profiles (P_s/P_d where P_s systolic pressure, P_d is the diastolic pressure) at the two spatial points: before the stented region ($x = 0.15m$, large pressure amplitudes) and after stented region ($x = 0.35m$). The wave with larger amplitude is related to the point $x = 0.15m$. The increase in amplitude is a result of the reflected wave. The further oscillations appear due to the multiple wave reflections in the stiff region.

be explained by multiple wave reflections in the interior stiff region. This result is in a good agreement with the similar simulations presented in [2].

References

[1] Taylor C and Draney M 2004 *Rev. Fluid Mech.* **36** 197
[2] Sherwin S, Franke V, Peiro J and Parker K 2003 *J. Engng. Maths* **47** 217
[3] Succi S 2018 *The Lattice Boltzmann Equation: For Complex States of Flowing Matter* (Oxford: OUP)
[4] Fang H, Wang Z, Lin Z and Liu M 2002 *Phys. Rev. E* **65** 051925
[5] Chen C, Chen H, Freed D, Shock R, Staroselsky I, Zhang R, Coşkun A, Stone P and Feldman C 2006 *Physica A* **362** 174
[6] Bisson M, Bernaschi M, Melchionna S, Succi S and Kaxiras E 2011 *Commun. Comput. Phys.* **11** 48
[7] Pontrelli G, Halliday I, Melchionna S, Spencer T and Succi S 2013 *Math. Comp. Model. of Dyn. Syst.* **20** 470
[8] Gounley J, Vardhan M and Randles A 2018 *Comput. Phys. Comm.* **In press**
[9] Hou Y and Kassab G 2007 *Am. J. Physio.l Heart. Circ. Physiol.* **292** H2623
[10] Langewouters G, Wesseling K and Goedhard W 1984 *J. Biomech.* **17** 425
[11] Cercignani C 2000 *Rarefied gas dynamics: from basic concepts to actual calculations* (Cambridge University Press)
[12] Kupershtokh A, Medvedev D and Karpov D 2009 *Comput. Math. Appl.* **58** 965
[13] Kupershtokh A 2010 *Comput. Math. Appl.* **59** 2236
[14] Kupershtokh A, Medvedev D and Gribanov I 2018 *Phys. Rev. E* **98** 023308
[15] Ilyin O 2018 *Wave Motion, accepted, to appear*
[16] Lighthill J 1978 *Waves in Fluids* (Cambridge University Press)

IOP Conf. Series: Journal of Physics: Conf. Series **1163** (2019) 012025 doi:10.1088/1742-6596/1163/1/012025

Fractal approach for determining the optimal number of topics in the field of topic modeling.

V Ignatenko[1], S Koltcov[1], S Staab[2] and Z Boukhers[2]

[1] National Research University Higher School of Economics, ul. Sedova 55/2, 192148, Saint-Petersburg, Russia

[2] Institute for Web Science and Technologies, University of Koblenz-Landau, Universitaetsstrasse 1, 56070, Koblenz, Germany

E-mail: vignatenko@hse.ru

Abstract. In this paper we apply multifractal formalism to the analysis of statistical behaviour of topic models under condition of varying number of topics. Our analysis reveals the existence of two self-similar regions and one transition region in the function of density-of-states depending on the number of topics. As earlier a function that can be expressed through density-of-states was successfully used to determine the optimal number of topics, we test the applicability of the density-of-states function for the same purpose. We provide numerical results for three topic models (PLSA, ARTM, and LDA Gibbs sampling) on two marked-up collections containing texts in two different languages. Our experiments show that the "true" number of topics, as determined by the human mark-up, occurs in the transition region.

1. Introduction

Modern information systems generate a huge number of texts such as news, blogs and comments. Analysis of big data is impossible without the construction of formalized mathematical models, and the latter gain a lot from using statistical physics. One of such a model is topic modelling. Briefly speaking, topic modeling (TM) is a family of mathematical algorithms based on the following assumptions [1]:

1. Let D be a collection of textual documents, $\tilde{W}$ be a set (vocabulary) of all unique words, and the number of elements of vocabulary be denoted by W. Each document $d \in D$ is a sequence of terms $w_1, ..., w_{n_d}$ from the vocabulary $\tilde{W}$.

2. It is assumed that there exists a finite set of topics $\tilde{T}$ with a finite number of topics T, and each entry of a word w in document d is associated with a certain topic $t \in \tilde{T}$. Topic is understood as a combination of words which often (in statistical sense) occur together in a large number of documents.

3. Collection of documents is considered a random and independent sample of triples (w_i, d_i, t_i), $i = 1, ..., n$, from a discrete distribution $p(w, d, t)$ on a finite probability space $\tilde{W} \times D \times \tilde{T}$. Words w and documents d are observable variables, topic $t \in \tilde{T}$ is a latent (hidden) variable.

4. It is assumed that word order in documents is not important for topic detection ('bag of words' model). Order of documents in a collection does not matter as well.

In TM it is assumed that probability $p(w|d)$ of term w to occur in document d can be expressed by multiplication of conditional probabilities $p(w|t)$ and $p(t|d)$. According to the formula of total probability and the hypothesis of conditional independence, we obtain the following expression [1]:

$$p(w|d) = \sum_{t\in\tilde{T}} p(w|t)p(t|d) = \sum_{t\in\tilde{T}} \phi_{wt}\theta_{td}, \quad (1)$$

where $p(w|t) := \phi_{wt}$ is the probability of word w to belong to topic t, $p(t,d) := \theta_{td}$ is the probability of topic t in document d.

Thus, to construct a topic model of data means to find the set of latent topics $\tilde{T}$ based on observable variables d and w, i.e. to find (1) a set of one-dimensional conditional probabilities $p(w|t) \equiv \phi_{wt}$ for each topic t (they form a word-topic matrix $\Phi \equiv \{\phi_{wt}\}_{w\in\tilde{W},t\in\tilde{T}}$) and (2) a set of one-dimensional conditional probabilities $p(t|d) \equiv \theta_{td}$ for each document d (they form a topic-document matrix $\Theta \equiv \{\theta_{td}\}_{t\in\tilde{T},d\in D}$).

To date, a large number of TM models have been proposed, but we focus on the following two types: 1. Models based on Gibbs sampling procedure; 2. Models based on E-M algorithm. However, although a great variety of algorithms has been proposed within these two approaches, they all share the problem of selecting the number of topics.

The rest of the paper proceeds as follows. Section 2 explains the logic of the fractal approach used for the analysis of topic models. Section 3 describes the data used and the results of numerical experiments performed to verify our approach.

2. Fractal approach

Our fractal approach is based on the following assumptions: 1. The set of documents and words is considered a mesoscopic system, where the number of elements can reach several millions [2]. 2. Collection of documents contains the finite number of topics which is unknown in advance. Let us note that variation of the number of topics in an algorithm of TM allows to regulate algorithm resolution.

Recall that under the condition of fixed number of topics, a topic solution is a matrix Φ, where $T \cdot W$ is the number of elements, T is the number of topics (the number of columns of the matrix), W is the number of unique words. Each cell of the matrix contains probability ϕ_{ij} of belonging of a word w_i to a topic t_j. The multidimensional space of words is covered by a grid of fixed size defined by matrix Φ. The size of each cell of this grid is $\epsilon = 1/(WT)$. Under the condition of fixed size of vocabulary W, the size of each cell is defined by the number of topics and if $T \to \infty$ then the size of the cell tends to zero. Let us introduce the density-of-states function which is defined according to the following formula [3]: $\rho = \frac{n}{WT}$, where n is the number of cells of Φ satisfying $\phi_{ij} > 1/W$ for $i = 1, ..., W$, $j = 1, ..., T$. It was shown in [3] that the behaviour of such function is extremely nonlinear, and the optimal number of topics for a topic model corresponds to the minimum of normalized free energy or the minimum of Massieu function. It is clear that the density-of-states function depends on the number of topics and changes in the process of topic modeling. Thus, the density-of-states function depends on the cell size and on some degree $D(\epsilon)$ [4], [5]: $\rho(\epsilon) \approx \epsilon^{D(\epsilon)}$. The distribution of fractal dimensions $D(\epsilon)$ can be found using 'box counting' algorithm. Application of this algorithm to the calculation of fractal dimensions in topic models consists of the following steps: 1. A certain number of topics is chosen. 2. Multidimensional space of words and topics is covered by a grid of fixed size (matrix Φ). 3. We calculate the number of cells satisfying $\phi_{ij} > 1/W$. 4. We calculate the value of ρ for chosen number of topics T. 5. We repeat steps 1 through 4 changing the cell size (i.e. changing the number of topics). 6. We plot a graph showing dependence of ρ in bi-logarithmic coordinates. 7. Using the method of least squares, we estimate the slope of the function. The value of the slope is the value of fractal dimension calculated according to the following formula: $D(\epsilon) = \frac{\ln(\rho(\epsilon))}{\ln(\epsilon)}$.

We assume that the optimal number of topics corresponds to the regions where the fractal

IOP Conf. Series: Journal of Physics: Conf. Series **1163** (2019) 012025 doi:10.1088/1742-6596/1163/1/012025

dimensions change since for the regions where the fractal dimension is constant, the solution of TM preserves its structure, while, the changes in structure correspond to changes in fractal dimensions.

3. Numerical experiments

In this research, the following computer experiments were executed. We used three algorithms of TM, namely, 1. PLSA [1] (E-M algorithm) as implemented in BigARTM package (http://bigartm.org/); 2. ARTM [6] (E-M algorithm) as implemented in BigARTM package; 3. LDA [7] Gibbs sampling as implemented in GibbsLDA++ package (http://gibbslda.sourceforge.net/). Two collections of textual documents were used: 1. English-language dataset '20 Newsgroups' (http://qwone.com/ jason/20Newsgroups/) containing 15404 news texts with 50948 unique words. The documents of this dataset are manually labeled with a topic class among 20 topic classes. Since some of these topics are similar and can be united, the dataset can be represented with 15 topics. 2. Russian-language dataset 'Lenta_ru' consists of 8630 documents (containing 23297 unique words) in Russian language, each of which is manually labeled with a class among 10 topic classes. However, some of these topics can be viewed as parts of other topics, therefore, the documents in this dataset can be represented with 7 distinct topics. When conducting topic modeling on these datasets, the number of topics was varied in the range T=[2;50] in the increments of 1 topic. For each topic solution the value of the density-of-states function was calculated. The obtained curves for two collections were analysed in bi-logarithmic coordinates. These collections were chosen to represent different languages and different dataset sizes.

Fig. 1 shows the distribution of fractal dimensions for the English-language dataset implemented using PLSA algorithm. The figure demonstrates that one can distinguish three regions. The first and the third regions are approximated by linear functions using the method of least squares. Linear regions correspond to the situation when density-of-states function is self-similar, i.e. $\rho(\lambda\frac{1}{WT}) = \lambda^d\rho(\frac{1}{WT})$. Thus, this function contains self-similar regions that possess the properties of fractals. The second region is a transition region, where the first fractal set transforms into another fractal set. Notice that the transition region corresponds to $T = [9, 15]$, the fractal dimension for the first region is equal approximately to 0.805 and for the third region to 0.375.

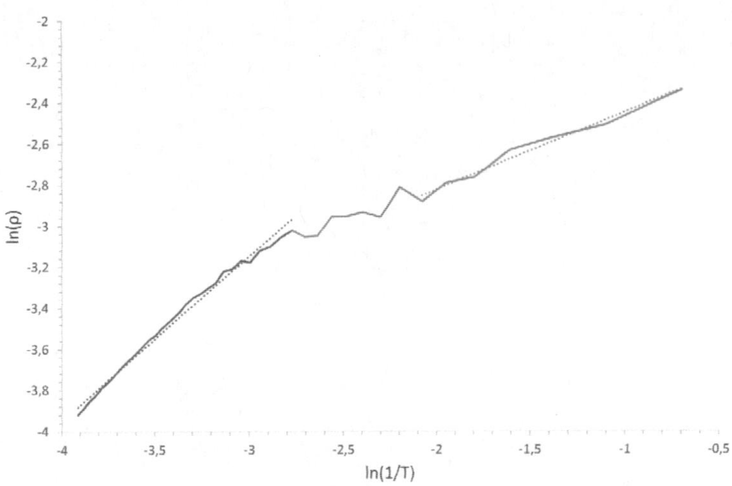

Figure 1. Distribution of fractal dimensions for English-language dataset (PLSA model).

In paper [8] a set of cluster algorithms with different types of regularization was investigated. The authors showed that optimal cluster solutions lie in the range of $[15, 20]$ clusters for '20

IOP Conf. Series: Journal of Physics: Conf. Series **1163** (2019) 012025 doi:10.1088/1742-6596/1163/1/012025

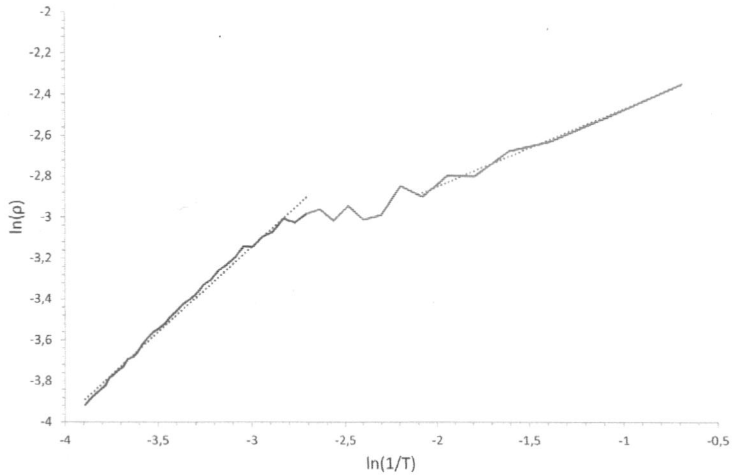

Figure 2. Distribution of fractal dimensions for English-language dataset (LDA model, $\alpha = 0.4$, $\beta = 0.5$).

Newsgroups dataset'. In our experiment, restructuring of a topic model occurs in the region of $[9, 15]$ topics which is slightly lower than the "true" number. Fig. 2 shows the distribution of fractal dimensions for the English-language dataset implemented using LDA algorithm with the following values of hyperparameters $\alpha = 0.4$ and $\beta = 0.5$. Let us note that transition region in Fig. 2 corresponds to $T = [8, 15]$. The fractal dimension for the first region of Fig. 2 equals approximately to 0.834 and for the third region to 0.381. Fig. 3 and 4 show the distribution of fractal dimensions for '20 Newsgroups dataset' implemented using ARTM algorithm with different values of regularization coefficients (sparse_phi). One can see that the location of the transition region changes significantly under variation of the values of regularization coefficient. Therefore transition region corresponds to approximately $T = 18$ on Fig. 3 and on Fig. 4 it corresponds to $T = [3, 7]$. Hence, one can conclude that values of regularizing coefficients for the ARTM model influence the structure of the results of TM significantly, and it leads to changing in the optimal number of topics. The fractal dimension for the first region of Fig. 3 equals approximately to 0.853, and for the second region to 0.347. The fractal dimension for the first region on Fig. 4 equals to 0.87, the third region consists only of two points, hence, it makes no sense to speak about fractal dimension for that region.

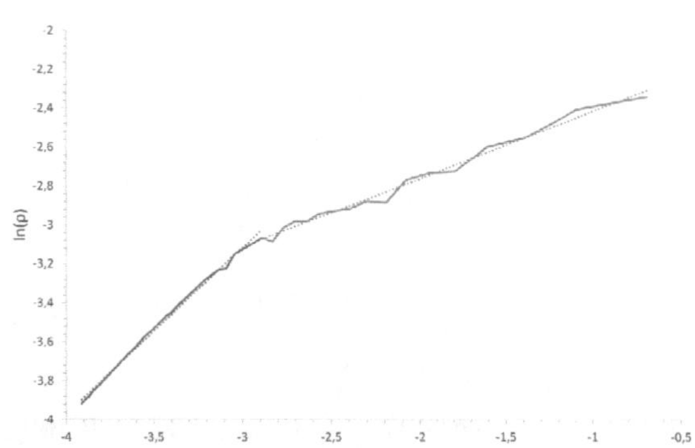

Figure 3. Distribution of fractal dimensions for English-language dataset (ARTM, sparse_phi=0.01).

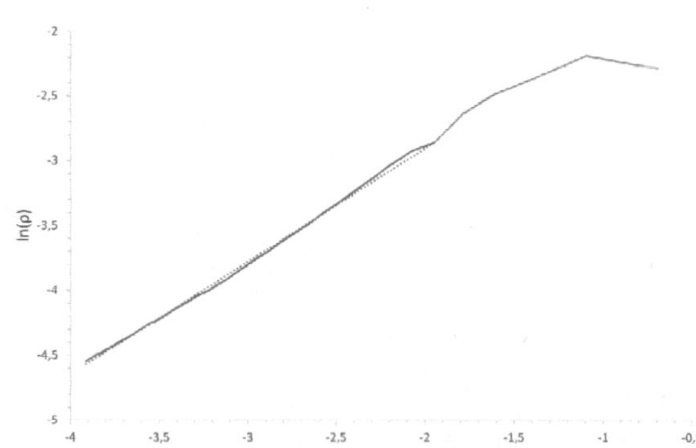

Figure 4. Distribution of fractal dimensions for English-language dataset (ARTM, sparse_phi=-10).

Let us consider numerical results for the Russian-language dataset. Fig. 5 shows the

IOP Conf. Series: Journal of Physics: Conf. Series **1163** (2019) 012025 doi:10.1088/1742-6596/1163/1/012025

distribution of fractal dimensions for topic model based on the PLSA algorithm. The transition region in this case corresponds to $T \approx 7$. The fractal dimension for the first region corresponds to $D \approx 0.688$ and for the second $D \approx 0.448$. Fig. 6 shows the distribution of fractal dimensions for LDA Gibbs sampling model with the following values of hyper-parameters $\alpha = 0.4$ and $\beta = 0.5$. Note that the transition region on Fig. 6 corresponds to $T = [6, 11]$. The fractal dimension for the first region is $D \approx 0.708$ and for the second $D \approx 0.49$. Fig. 7 and 8 demonstrate the distribution of fractal dimensions for the ARTM model with different values of "sparse_phi" coefficient. Transition region on Fig. 7 corresponds to $T = [6, 9]$, and on Fig. 8 it corresponds to $T = [3, 4]$. For the first region of Fig. 7 the fractal dimension is $D \approx 0.73$ and for the third it is $D \approx 0.447$. Finally, on Fig. 8 the fractal dimensions for the first and for the third regions are $D \approx 1.006$ and $D \approx 0.744$, correspondingly.

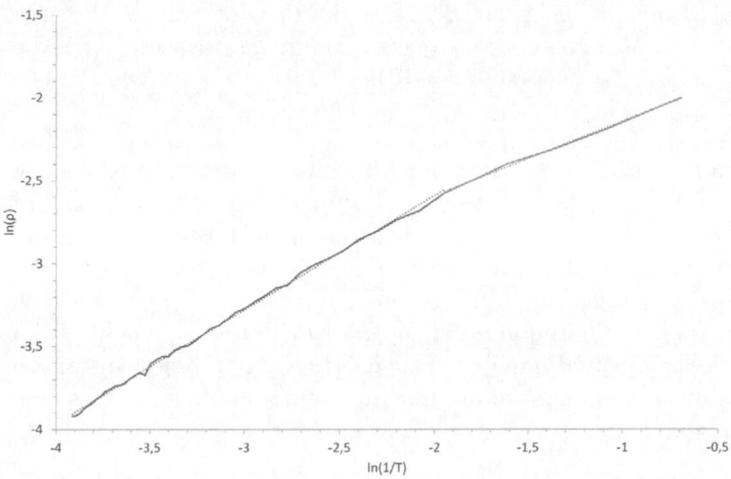

Figure 5. Distribution of fractal dimensions for Russian-language dataset (PLSA model).

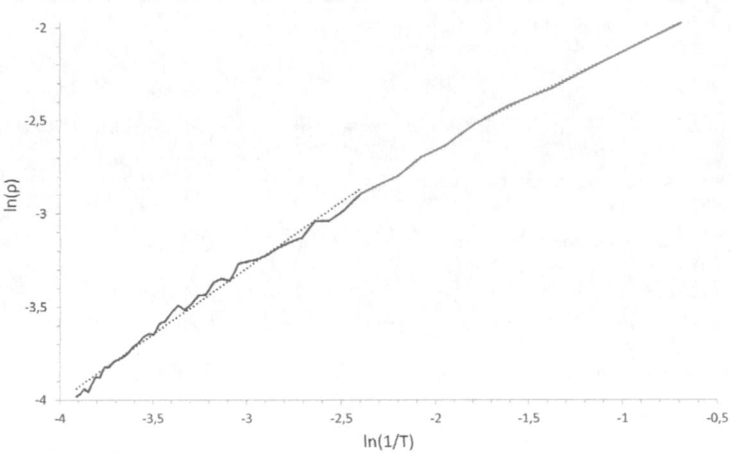

Figure 6. Distribution of fractal dimensions for Russian-language dataset (LDA, $\alpha = 0.4$, $\beta = 0.5$).

4. Conclusion

Fractal analysis of topic model behavior under the condition of changing number of topics has shown that self-similar fractal regions exist in the density-of-states function. We have also found out that this function is not a perfect fractal; instead, it always contains at least one transition region that separates self-similar parts of the curve. Next, we have tested the assumption

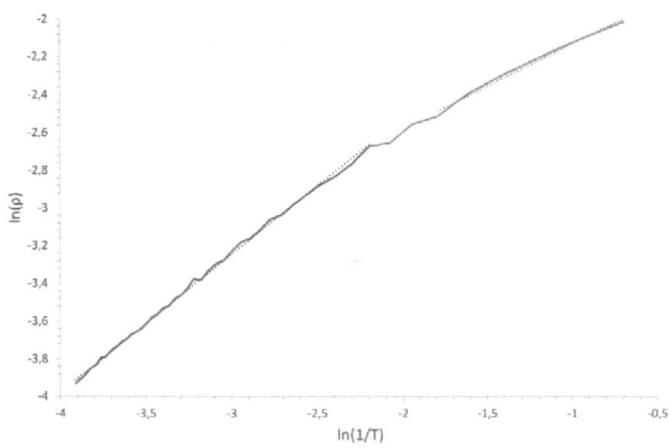

Figure 7. Distribution of fractal dimensions for Russian-language dataset (ARTM, sparse_phi=0.01).

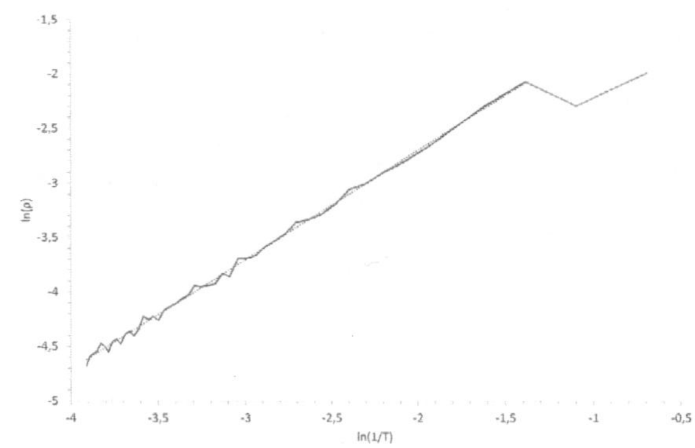

Figure 8. Distribution of fractal dimensions for Russian-language dataset (ARTM, sparse_phi=-10).

that this transition region corresponds to the "true" number of topics, as determined by the human mark-up. In our limited number of tests, the best performing algorithm was ARTM with "sparse_phi" = 0.01, although other algorithms have shown promising results, too. More experiments are needed to determine an algorithm (or algorithms) that is best suited for the usage of density-of-states function as an indicator of the optimal number of topics. If such algorithm is found that can perform well on a large number of different datasets, it will give a more solid ground for the determining the optimal number of topics through finding a transition region. This will confirm the applicability of the fractal approach to topic number optimization that has been preliminarily tested in this paper.

4.1. Acknowledgments

The study was implemented in the framework of the Basic Research Program at the National Research University Higher School of Economics (HSE) in 2018.

References

[1] Hofmann T 1999 *Proc. of the 22nd Annual Int. ACM SIGIR Conference on Research and Development in Information Retrieval (August 15 - 19, 1999, Berkeley)* (New York, NY, USA: ACM) p 50
[2] Koltcov S 2018 *Physica* A: Statistical Mechanics and its Applications **512** 1192
[3] Koltcov S 2017 *Tech. Phys. Letters* **43** 584
[4] Feder J 1988 *Fractals* (Boston: Springer)
[5] Sornette D 2006 *Critical Phenomena in Natural Sciences* (Heidelberg: Springer-Verlag)
[6] Kochedykov D, Apishev M, Golitsyn L and Vorontsov K 2017 *Proc. of the 21st Conf. of Open Innovations Association FRUCT (November 6-10, 2017, Helsinki)* (Helsinki, Finland: FRUCT Oy) pp 182–193
[7] Griffiths T and Steyvers M 2004 *PNAS* **101** 5228
[8] Basu S *et al.* (eds) 2008 *Constrained Clustering: Advances in Algorithms, Theory, and Applications* (Taylor & Francis Group)

IOP Conf. Series: Journal of Physics: Conf. Series **1163** (2019) 012026 doi:10.1088/1742-6596/1163/1/012026

Development of multiprocessor system-on-chip based on soft processor cores schoolMIPS

A E Ryazanova[1], **A A Amerikanov**[1], **E V Lezhnev**[1,2,3]

[1] National Research University Higher School of Economics, 20 Myasnitskaya Ulitsa, 101000, Moscow, Russia
[2] e-mail: elezhnev@hse.ru
[3] Author to whom any correspondence should be addressed

Abstract. This work includes a review of MIPS architecture processor cores and a review of network topology consisting of routers. It was demonstrated by realization of 2 multiprocessor systems developed on the basis of mesh topology using modified schoolMIPS soft-processor cores, in which architecture additional blocks and instructions were added, and routers with XY routing. As a result, the obtained NoC performance is up to 1.87 Gbit/s (4 processor cores), and up to 1.54 Gbit/s (10 processor cores). The extended processor core schoolMIPS consumes 452 ALMs and 1692 bits of memory; NoC of 4 processor cores takes 2223 ALMs and 9136 bits of memory; NoC of 10 processor cores – 5696 ALMs and 22840 bits of memory. The obtained results suggest that there is a possibility of NoC development with the number of nodes up to 200 nodes on Stratix IV GX EP4SGX230 (DE4).

1. Introduction

An important task is to investigate the possibilities of using different architecture processor cores. Currently, the MIPS architecture is becoming more widespread through freely distributable cores like MIPSfpga [1] or schoolMIPS [2], which are used in many colleges and universities for teaching students computer architecture, digital design and hardware-software codesign. MIPSfpga soft-processor [1] combines properties of commercial and academic cores, since it is built on the basis of the microAptiv UP core (used in the Microchip PIC32MZ microcontroller), and is also provided with extensive educational materials.

At the same time, in the development of large-scale multiprocessor systems, it is necessary to observe a balance between data transfer rate, amount of resources expended to implement the system, and ability to scale the system without cardinal changes. The network-on-chip architecture influences on all of the above. The architecture provides simultaneous interaction of multiple cores, with data packets transferred from one processor core to another processor core without interfering with each other. Applying this approach to the network of MIPS cores is a promising solution, but it is required to develop a methodology and tools to design new solutions in the field of networks-on-chip.

2. SchoolMIPS soft-processor core

The schoolMIPS processor core [2] is a simplified version of the MIPS processor developed by Sarah Harris [3] and is provided as a set of Verilog files. The basic processor implementation is a single-cycle microarchitecture with a minimal set of instructions, instruction memory, general-purpose registers, but without data memory (Figure 1). With the core source, scripts for project build, for

IOP Conf. Series: Journal of Physics: Conf. Series **1163** (2019) 012026 doi:10.1088/1742-6596/1163/1/012026

simulation in ModelSim, and for project loading on development board are provided. Despite the fact that initially schoolMIPS is a simplified processor core, its architecture can be improved to achieve the functionality that is required in a multiprocessor system. Such opportunity makes it a convenient tool for testing various architectural solutions of the network-on-chip communication subsystem.

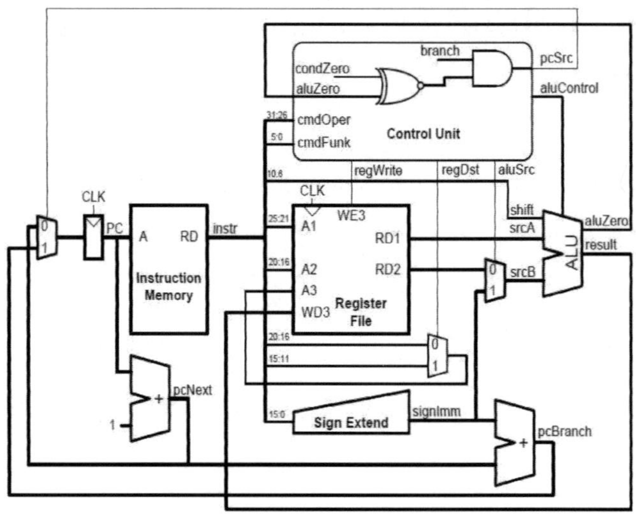

Figure 1. SchoolMIPS soft-processor core microarchitecture [3].

3. Network-on-chip architecture based on routers

Currently, there is a tendency to use systems-on-chip (SoC), which include a large number of processors, architectural blocks and peripheral nodes. In such systems, crossbar switch or bus topology as a way of interaction between processors is not effective, resulting in a new architecture called network-on-chip. In this architecture, each core or processor unit is connected to a router, through which it interacts with other units or processors. Routers are united into a network and data packets are sent from one unit or processor to another (Figure 2).

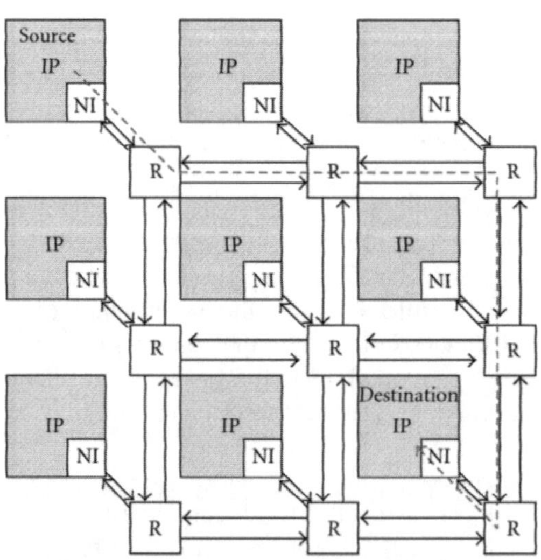

Figure 2. The interaction of processor cores in network-on-chip with XY routing [4].

In this type of communication subsystem, multiple units are able to simultaneously transmit data without interfering with each other, which greatly simplifies the topology of the chip and removes scaling limitations. In addition, power consumption is reduced and the speed of interaction increases, and this network-on-chip architecture allows to combine heterogeneous cores and blocks on a single chip. The network-on-chip is the most rational method of interaction, combining fast data exchange, resource consumption and the ability to add new processor units, without the need to drastically change the topology of the system.

4. The choice of topology for multiprocessor system

Compared to other topologies, the mesh topology [5] provides sufficient network performance and evenly distributes the load on the nodes, thus avoiding blocking, and also relatively simple scaling. Mesh topology provides sufficient network performance even with a large amount of transmitted data and an average network load. In a network with a mesh topology, a router can redirect a transmitted packet over several alternative free connections.

In this research, two networks-on-chip of 4 and of 10 processors were developed. The architecture of these networks based on the mesh topology is shown below (Figure 3-4).

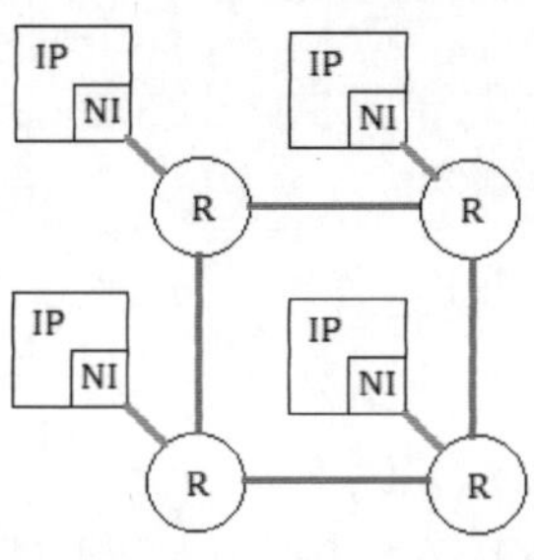

Figure 3. Network-on-chip with a mesh topology of 4 processor cores.

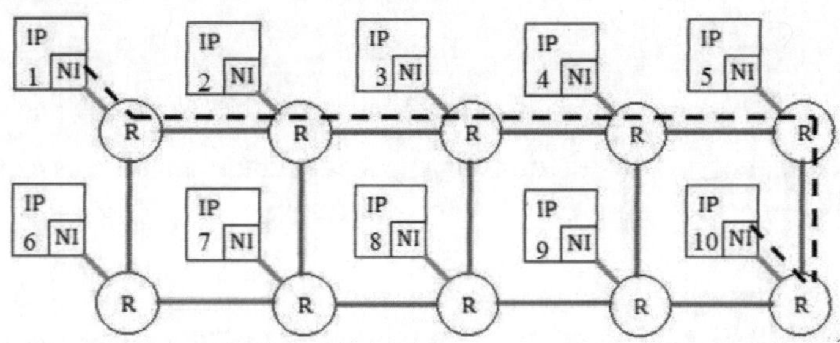

Figure 4. Network-on-chip with a mesh topology of 10 processor cores.

5. SchoolMIPS

According to the proposed method the rising complexity of the processor core for the needs of a multiprocessor system, the simplest version of the core has been selected (00_simple) [2], after which it was modified as follows [6]:

1. A block of local RAM memory was added to the microarchitecture for data storage (Figure 5).

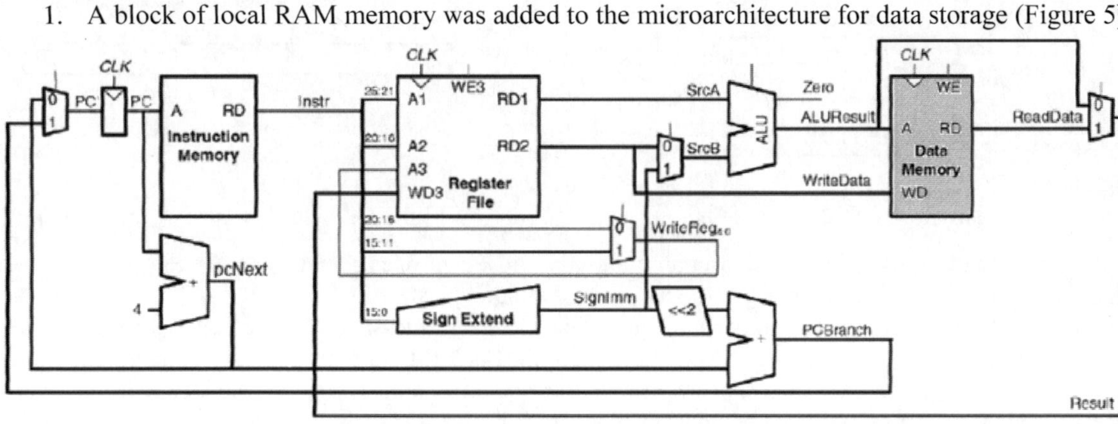

Figure 5. SchoolMIPS microarchitecture with RAM memory block.

2. For interaction of the core with the router required the presence of a network interface – an architectural unit by which the router's data is sent/received. The network interface is implemented as two separate modules for input to the processor and to output from it. The inputCPU module, by which the GET command receives data from the router, and the

IOP Conf. Series: Journal of Physics: Conf. Series **1163** (2019) 012026 doi:10.1088/1742-6596/1163/1/012026

outputCPU module, which implements transmission from the processor core, by which the SEND command transmits data to the router.

3. Support of GET and SEND instructions for receiving and sending data has been added to the microarchitecture of the processor core. The presence of these instructions is not described in the technical documentation for the schoolMIPS processor [7], and is not provided by the standard MIPS instruction set. Therefore, a 32-bit format has been independently determined for them (Figure 6), unused operation/functional codes have been allocated, registers and signals have been set, on which these instructions will be executed.

31 – 26	25 – 21	20 – 16	15 – 0
SEND 110111	rd1 00010	rd2 00011	offset
6	5	5	16

31 – 26	25 – 21	20 – 16	15 – 0
GET 111011	0 00000	rd2 00010	offset
6	5	5	16

Figure 6. 32-bit format of the SEND and GET instructions.

Format of new instructions in MIPS assembler language is as follows:

- `send $reg_data, $reg_addr`
- `get  $reg_data`

Here `reg_data` is the register with the data to be sent ([25:21] command bits; rd1 – input to the schoolMIPS), and `reg_addr` – register with the destination processor address ([20:16] command bits; rd2 – input to the schoolMIPS).

6. Structure of the router for NoC

In this research, a simple NoCSimp router [8] is used, which implements the separation of the switching part into input and output blocks connected by a FIFO buffer memory. This architecture involves the elimination of unnecessary resource costs, and also ensures the continuous use of buffer memory (Figure 7).

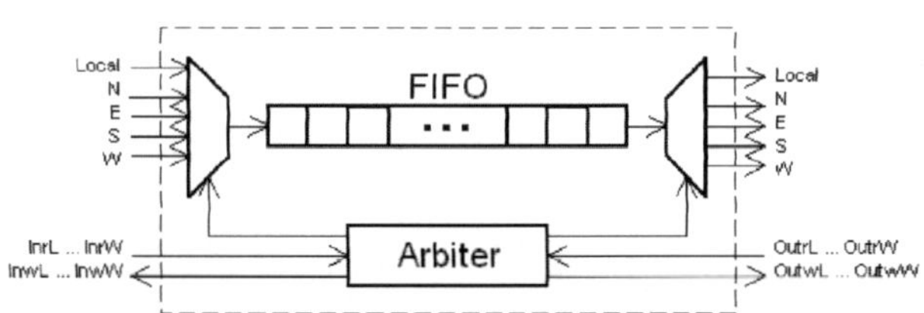

Figure 7. The structure of the router [9].

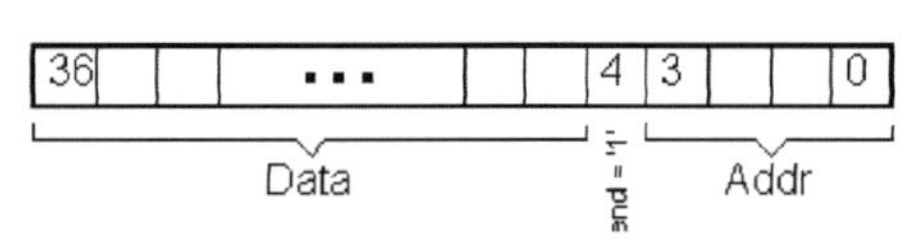

Figure 8. The structure of the flit [9].

Arbiter module through respective enable signals and acknowledgment/transmission controls the interaction with this router connected to the processor core as well as with the neighboring routers. The destination address is assigned by the source that forms the message. All messages arriving at the router are divided into packets from 1 to 8 flits in length. Each flit contains 37 bits: 32 data bits, 1 end bit, one value of which indicates the last flit in the package, and 4 bits for the destination router address, which corresponds to the destination processor (Figure 8). Thus, the network can consist of a maximum of 16 routers. The router performs packet switching based on the classic deterministic XY routing algorithm, that is, first the packets transmitted over the network in a horizontal direction, and then in the vertical direction.

7. The development of networks-on-chip

As mentioned before, in the network-on-chip architecture each IP-core or processor unit is connected to a router, and the routers themselves are connected with each other, thereby forming a network. In

IOP Conf. Series: Journal of Physics: Conf. Series **1163** (2019) 012026 doi:10.1088/1742-6596/1163/1/012026

this research, mesh topology is used, in which only interaction with neighboring routers is assumed. Based on this, the network has been designed for 4 and 10 routers.

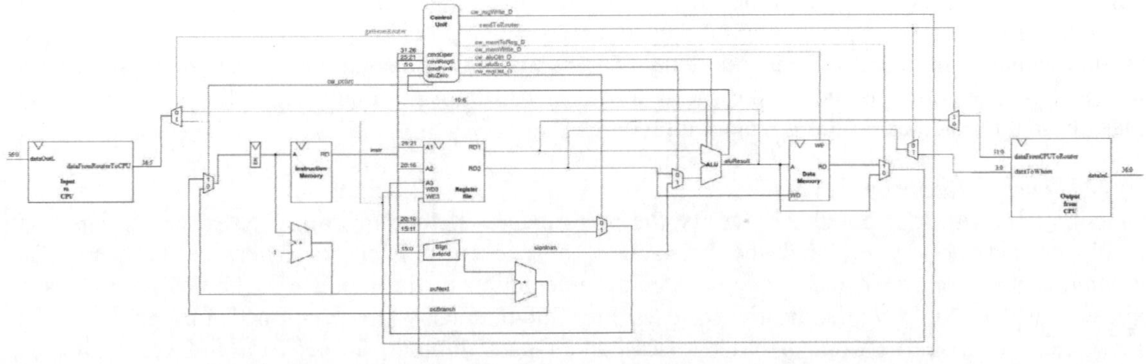

Figure 9. The core schoolMIPS with added architectural solutions.

7.1. Network-on-chip of 4 processors

In this network the routers interact in accordance with the incidence matrix (Table 1). Each router has been attached processor core so that the core network interface outputs connects to inputs of the router, and data or signals from the router output (Figure 10) come to the interface inputs.

Table 1. Incidence matrix of a network of 4 routers

Router Number	1	2	3	4
1		E	S	
2	W			S
3	N			E
4		N	W	

For example, data output from the `dataOutL[36:0]` router is connected to the input of the `dataFromRouterToCPU[36:0]` network interface, and the input of the `dataInL[36:0]` router is connected to the data output of the `dataOutCPU[36:0]` processor. The same implemented for read/write request/acknowledge signals.

7.2. Network-on-chip of 10 routers

In a network of 10 routers, the interaction is carried out in accordance with the incidence matrix (Table 2). In order to facilitate the process of designing a network of 10 processor cores, the router and its local core were combined into a single unit, thereby forming a network cell. This module combines the inputs and outputs of the processor core and router.

Table 2. Incidence matrix of a network of 10 routers

Router Number	1	2	3	4	5	6	7	8	9	10
1		E				S				
2	W		E				S			
3		W		E				S		
4			W		E				S	
5				W						S
6	N						E			
7		N				W		E		
8			N				W		E	
9				N				W		E
10					N				W	

In a network of 10 nodes with a selected topology, the transfer of packets between nodes that are far away from each other will be quite slow. a packet will have to go through 5-6 routers, each of which will first check for place in the FIFO, then exchange request/permission signals to send data to neighboring routers. In addition, central routers can be busy processing a large number of incoming packets, which significantly affects the delay of transmission. The solution of these problems can be achieved by applying instead of the mesh other, more optimal topologies, such as, for example: quasioptimal topologies [10] or circulant topologies [11].

7.3. Network-on-chip simulation

The simulation made it possible to verify the performance of the NoCSimp router, as well as each developed network-on-chip and the correctness of packet transmission over the network. The simulation also confirmed the correctness of the interaction of the modified schoolMIPS processor cores with the NoC communication subsystem. Simulations were performed for single packet transactions over the network in order to track their path through the network and verify that the routing algorithm works correctly, as well as for multiple simultaneous transactions, to ensure that the network is resilient to loads with multiple packets.

7.4. Analysis of performance and hardware costs

NoCSimp router was synthesized in Quartus Prime. Its implementation requires 142 ALMs and 592 bits of memory, which is less than 1% of the resources of the De1-SoC board. The maximum frequency of the router is 200 MHz, and the bandwidth reaches 2.13 Gbit/s. Ideally, such bandwidth should be stored in each network-on-chip, however, taking into account the network load and delays, the network of 4 routers provides a speed of 1.87 Gbit/s, and a network of 10 routers – 1.54 Gbit/s.

The original schoolMIPS processor core (Figure 1) occupies 209 ALMs and 768 bits of memory. The extended core consumes 415 ALMs (98% more) and 1,692 bits of memory (120% more), which is nevertheless still comparable to the resources consumed even by the simple NoCSimp router [8] (Table 3). Naturally, the schoolMIPS functionality is incomparably less, but for the tasks of NoC architectures prototyping and networks with a large number of cores modeling on chips of the lower and middle price range, and at the same time at high frequencies, this core fits perfectly. As the project under development expands, the amount of resources expended increases linearly, suggesting that the Terasic De1-SoC [12] board can accommodate NoC sizes up to 50 nodes and up to 200 on Stratix DE4 [13].

Table 3. Project resources consuming at the each stage of development

Stage	ALMs	Memory bits
schoolMIPS core	209	768
NoCSimp router	142	592
NoC of 4 routers	415 (1%)	2 368 (<1%)
NoC of 10 routers	1176 (4%)	5920 (<1%)
Modified schoolMIPS	452 (1,5%)	1692 (<1%)
NoC of 4 routers with connected cores	2223 (7%)	9136 (<1%)
NoC of 10 routers with connected cores	5696 (18%)	22840 (<1%)

8. Conclusion

As a result of this research, networks-on-chip of 4 and 10 NoCSimp routers were developed, after which the schoolMIPS soft-processor cores were attached to each router. Mesh topology was chosen for modeling purposes.

The upcoming strategy for developing the processor core was proposed, where only the necessary components were added to the initially simple implementation of the schoolMIPS soft-processor, such as core network interface allowing to interact with the network and hardware support needed to interact with the network communication instructions.

The analysis of the consumed resources demonstrates that such an approach allows to significantly reduce the consumption of resources in comparison with the opposite method, when the initially functional processor core is optimized for resources by abandoning part of its functionality. The results suggest the possibility of networks-on-chip development with the number of nodes of schoolMIPS to 50 on Terasic De1-SoC board and up to 200 on DE4 board at operating frequencies up to 200 MHz. This allows modeling and performing hardware prototyping of various architectural solutions for NoCs communication subsystems with a large number of nodes development.

Acknowledgments
This work is an output of a research project implemented as part of the Basic Research Program at the National Research University Higher School of Economics (HSE).

References

[1] Harris S L, Harris D M and Chaver D 2017 MIPSfpga: using a commercial MIPS soft-core in computer architecture education *IET Circuits, Devices & Systems Vol. 11(4)* pp 1–9.
[2] Zhelnio S, ‘Processor core schoolMIPS. User Guide’, 2018 [Online] Available: https://github.com/MIPSfpga/schoolMIPS (Accessed 29-Sep-2018).
[3] Harris D M, Harris S L 2012 Digital design and computer architecture *Morgan Kaufmann Vol. 2* p 712.
[4] Tsai W C, Lan Y C and Hu Y H 2012 Networks on Chips: Structure and Design Methodologies // *Journal of Electrical and Computer Engineering Vol. 4.* pp 1–15.
[5] Kumar A, Tyagi S and Jha C K 2017 Performance analysis of network-on-chip topologies *Journal of Information and Optimization Sciences Vol. 38. – No. 6.* pp 989–977.
[6] Ryazanova A ‘Multiprocessor systems based on the schoolMIPS processor core’ [Online] Available: https://github.com/ylianova0704/07_schoolMIPS_NoC (Accessed 30- Sep-2018).
[7] MIPS Technologies, Inc 2013 MIPS® Architecture For Programmers Volume II-A: The MIPS32® Instruction Set p 372.
[8] Romanov A Y, Ivannikov A D and Romanova I I 2016 Simulation and synthesis of networks-on-chip by using NoCSimp HDL library *IEEE 36th International Conference on Electronics and Nanotechnology (ELNANO)* (Kiev: IEEE) pp 300–303.
[9] Lysenko A N and Romanov A Y 2011 Resource efficient router for multiprocessor network-on-chip *Vestnik NTU "KPI": Collection of scientific papers. Thematic issue: Informatics and modeling №17* (NTU “KPI”, Kharkov) pp 86 – 92.
[10] Romanov A Y and Romanova I I 2015 Use of irregular topologies for the synthesis of networks-on-chip *2015 IEEE 35th International Conference on Electronics and Nanotechnology (ELNANO)* (Kiev) pp 445–449.
[11] Romanov A Y, Romanova I I and Glukhikh A Y 2018 Development of a Universal Adaptive Fast Algorithm for the Synthesis of Circulant Topologies for Networks-on-Chip Implementations *2018 IEEE 38th International Scientific Conference on Electronics and Nanotechnology (ELNANO)* (Kiev) pp 110–115.
[12] Terasic Technologies, Inc. 2015 De1-SoC User Manual p 115.
[13] Terasic Technologies, Inc. 2015 DE4 User Manual p 185.

IOP Conf. Series: Journal of Physics: Conf. Series **1163** (2019) 012027 doi:10.1088/1742-6596/1163/1/012027

Routing in Networks on Chip with Multiplicative Circulant Topology

M A Shchegoleva[1], A Yu Romanov[1,2,3], E V Lezhnev[1], A A Amerikanov[1]

[1] National Research University Higher School of Economics, 20 Myasnitskaya Ulitsa, 101000, Moscow, Russia
[2] e-mail: a.romanov@hse.ru
[3] Author to whom any correspondence should be addressed

Abstract. The development of multi-core processor systems is a demanded branch of science and technology. The appearance of processors with dozens and hundreds of cores poses to the developers the question of choosing the optimal topology capable to provide efficient routing in a network with a large number of nodes. In this paper, we consider the possibility of using multiplicative circulants as a topology for networks-on-chip. A specialized routing algorithm for networks with multiplicative circulant topology, taking into account topology features and having a high scalability, has been developed.

1. Introduction

Currently, one of the most important areas of research in the field of computer science and computing systems is the construction of multi-core processors. The transition to multi-core processors allows overcoming the performance degradation in the design of increasingly complex single-core systems [1]. When growing interest in the Systems on Chip (SoCs) and Multi-Processor Systems on Chip (MPSoCs) [2], Networks-on-Chip (NoCs) [3] are becoming widespread. In a multi-core processor with a small number of cores (2–8 cores), communication between IP-cores occurs via a common bus not capable to ensure communication between a large number of cores [1]. The problem of scalability can be solved by technology for NoC construction used to replace bus architectures.

One of the most urgent problems in the study of NoCs is the search for optimal topologies, since standard regular topologies (mesh [4], torus [5], hypercube [6], spidergon [7]) do not meet modern requirements for NoCs, especially with an increase in the number of nodes [8].

Circulant topologies have better characteristics than standard ones: they have better indicators of structural survivability, reliability, and connectivity, and also require fewer inter-processor exchanges in solving computing tasks, and system management tasks [9]. This allows them to be used in networks with a large number of nodes [10, 11]. To use circulant topologies as a topology for NoCs, it is necessary to develop routing algorithms in them taking into account the features of these families of graphs and the organization of NoCs, and all these determine the relevance of this work.

2. Multiplicative circulants

Let us give the definition of a circulant graph. Let $s_1, s_2, \ldots, s_k, n$ – whole numbers, such as $1 \le s_1 < s_2 < \ldots < s_k < n/2$. Graph C with set of vertices $V = \{0, 1, \ldots, n-1\}$ and set of edges$E = \{(i,j)\colon |i-j| \equiv s_l \bmod n,\ l = \overline{1,k}\}$ is called circulant [12, 13]; numbers $S = (s_1, s_2, \ldots, s_k)$ – generatrices; numbers k and n – dimension and order of the graph [14].

Circulant graphs of $C(s^k;\ 1,\ s,\ s^2,\ \dots,\ s^{k-1}) = MC(s,\ k)$ with $c = 1$ and $n = s^k$ (Figure 1) are considered to be a separate class of multiplicative circulants [15].

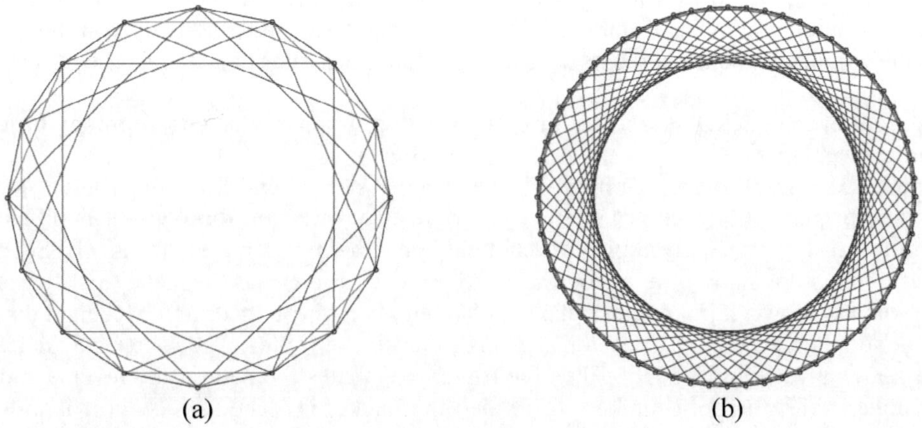

Figure 1. Multiplicative circulant $MC(2,\ 4)$ (a) and $MC(4,3)$ (b).

The most important characteristic of a graph is its diameter and average distance. The diameter of a graph G is the maximum possible distance (the length of the shortest path) between two vertices.

$$d(G) = \max_{i,j\in V} d(i,\ j),$$

where $d(i,j)$ – the length of the shortest path from vertex i to j.

In a number of works [15, 16], attempts have been made to find the most accurate estimates of the diameter and the average distance for multiplicative circulants.

More accurate estimates of the diameter (1) and mean distance (2) for the class of multiplicative circulants $MC(2,k)$ are given in [17]:

$$d(MC(2,\ k)) = \left\lceil \frac{k}{2} \right\rceil, \tag{1}$$

$$L_{av}(MC(2,\ k)) \approx \frac{k}{3}. \tag{2}$$

A comparison of the main characteristics of topologies on the basis of multiplicative circulants (1-2) and mesh-type topology with the same $n = s^k$ number of nodes is presented in Table 1.

Table 1. Comparison of characteristics of the circulant and mesh topologies.

Circulant $MC(s,k)$	Number of nodes n	Diameter $d(MC)$	Average distance L_{av} (MC)	Diameter $d(mesh)$	Average distance $L_{av}(mesh)$
MC(2,4)	16	2	1,33	6	2,50
MC(2,6)	64	3	2,00	14	5,25
MC(3,4)	81	4	2,67	16	5,93
MC(5,4)	625	8	4,80	48	16,64
MC(3,6)	729	6	4,00	52	17,98
MC(6,4)	1296	10	(5,00, 6,00)	70	23,98
MC(7,4)	2401	12	6,86	96	32,65

To calculate diameter and average distance of mesh topology, the formulas (3-4) [8, 18] are used.

$$d(mesh_n) = 2(\sqrt{n} - 1), \tag{3}$$

$$L_{av}(mesh_n) = \frac{2(n-1)}{3\sqrt{n}}, \tag{4}$$

From Table 1 it follows that even with a small network size, the circulant topology has better performance in all characteristics compared to mesh topology. Thus, multiplicative circulants are good enough for NoC designing with a large number of cores. However, it should be noted that the this topology, have a limited number of options – the number of nodes must be strictly an integer degree. Otherwise, the circulant becomes a recursive one with other properties and characteristics [19].

3. Development of the packet structure for static routing in networks with topology based on multiplicative circulants

Circulant networks use pairwise routing [14], when a packet is sent from the source node to the receiver node. For pair routing in NoCs with a multiplicative circulant topology, it is possible to use standard shortest path search algorithms, for example, breadth-first search algorithm (BFS) [20, 21]. It is suggested to choose a static type of routing in which each router has adjacency list for each node. Each router knows its own serial number, i.e. the node number whose incoming packets it distributes.

The destination node number (receiver node) is input to the router. The router calculates shortest path by the search algorithm in width. Then the router overwrites the path in the reverse and replaces the node numbers with the port numbers for which the packet is to be sent. A port is a connection between the current node and others. Each node has $2k$ ports: 2 ports of each length from S (except for circulants $MC(2,\ k)$, where one port is less). Thus, for circulant $MC(4,\ 3)$ (Figure 1b) the path $5 \rightarrow 21 \rightarrow 17$ is converted to $17 \leftarrow 21 \leftarrow 5$, then to the sequence of actions –(4), +(16), and then to 2|1 or 010|001 in binary code, if the rule for determining the port numbers is set as following: –(16) «left at 16»; –(4) «left at 4»; –(1) «left at 1»; +(1) «right at 1»; +(4) «right at 4»; +(16) « right at 16».

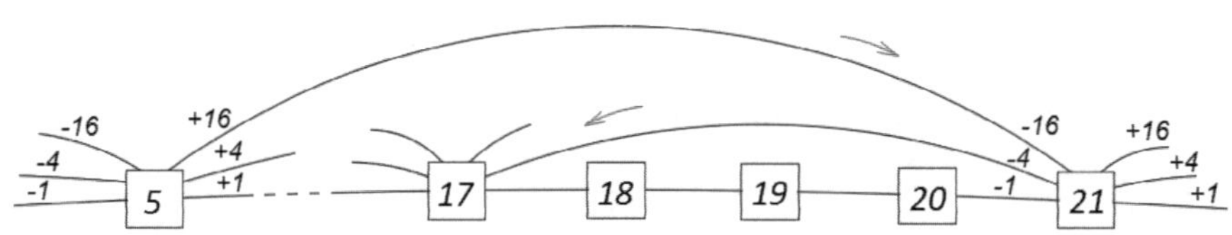

Figure 2. Path from node 5 to node 17 for circulant $MC(4,3)$.

Replacing node numbers to port numbers allows reducing size of part of the packet that is allocated for storing the path. Each node reads last k bits, shifts the path to same number of bits to the right, and passes the packet further. The package reaches the destination when the entire path is filled with zeros.

This algorithm is universal and it is suitable for networks built on the basis of any multiplicative circulants with difference s and k. The problem with this algorithm is that with the increase in the number of nodes and connections, the running time of the algorithm is significantly increased too. Also, the size of the address part of the packet increases, so for large NoCs, it is required to develop a specialized algorithm optimized for this type of topology.

4. Development of a specialized routing algorithm for NoCs based on the topology multiplicative circulant

The review of algorithms of routing used in NoCs [22], shows that the most common in networks with a mesh-like topology is the XY algorithm [23]. Multiplicative circulants also have a strict, previously known geometric form. Therefore, if we take into account the peculiarity of this topology, consisting in the fact that the lengths of generatrices are powers of one base, it is possible to propose a specialized routing algorithm that simplifies the structure of the address part of the packet and reduces its size.

The address field stores the number of the destination node; the field size can be calculated by the formula $P = \lceil \log_2 N \rceil$, where N – number of nodes in the network:

Receiving the node number where the packet is to be delivered, the router of the current node does not calculate all the way, but only the next step. In order to calculate the next step, it is enough for the router to know its own number, destination node number and circulant characteristics – s and k.

Thus, the total size of the data, stored by the network routers in bits, can be calculated by the following formula:

$$M = N * (\lfloor \log_2 N \rfloor + \lceil \log_2 N \rceil + k * (\lfloor \log_2 S^{k-1} \rfloor + 1) + 3 * \lceil \log_2 k \rceil + 2), \quad (6)$$

where N – number of nodes in the network;

$\lfloor \log_2 N \rfloor + 1$ – required amount of memory (bit) to store the number of routers in the network;

$\lceil \log_2 N \rceil$ – required amount of memory (bit) to store the network router number;

$k * (\lfloor \log_2 S^{k-1} \rfloor + 1)$ – required amount of memory (bit) to store array of the generatrices;

$\lceil \log_2 k \rceil + 1$ – required amount of memory (bit) to store the indices of the generatrices and the primary port of the router and flag of primary port;

Taking into account the fact that the circulant topology is cyclic, the algorithm is developed from the position of the null node. To do this, before starting work, the destination node number is recalculated, based on the current node number: if the number of the current node is greater than the destination node number, the recalculated destination node number is equal to the difference of the two numbers. If the destination node number is less than the current node number, the resulting difference is subtracted from the number of nodes in the network to allow for the transition through the null node. Then the algorithm determines in which direction it is better to start moving – to the left, or to the right. Since the circulant is symmetrical, it is sufficient to consider half of the circulant and by fixing the chosen direction of motion. The step is determined by the length of the generatrix closest to the destination node number. This makes it possible to take into account the cases when it is more advantageous to overstep the destination node, and then go back. To do this, a generatrix with a maximum length, not exceeding the destination node number, and next longer generatrix are chosen. Of the two generators, it is selected the one whose absolute value of difference between its length and destination node number is the minimum. The packet reaches the destination node when the current node number is equal to the node number specified in the address part of the packet. Thus, for routing in the address part of the packet, it is necessary to provide $\lceil k \cdot \log_2 s \rceil$ bits for storing the destination node number.

The algorithm, specially developed for multiplicative circulants, requires significantly less time for calculations (Table 2) and is easily scaled to larger networks. This algorithm has a linear complexity in contrast to the width-search algorithm with an exponential dependence of the search path time on the graph dimensions.

Table 2. The time spent on searching for the shortest path in multiplicative circulants.

Circulant	Number of nodes	*BFS*	Specially developed algorithm
$MC(2,4)$	16	0,000	0,000
$MC(2,5)$	32	0,002	0,000
$MC(2,6)$	64	0,013	0,001
$MC(3,4)$	81	0,029	0,000
$MC(5,3)$	125	0,340	0,001
$MC(3,5)$	243	2,096	0,004
$MC(6,3)$	216	10,578	0,005

The developed algorithm is characterized by an approach similar to the routing algorithm proposed independently for recursive circulants in [19]. Due to the fact that the algorithm, obtained in this work, was developed on the basis of the features of the structure of multiplicative circulants, which are a articular case of recursive circulants, it is characterized by a lower computational complexity, but it can only be used for this class of circulants.

5. Approbation of algorithm operation

Testing of developed algorithms was performed on FPGA Cyclone V 5CGXFC9A6U19I7 form Intel FPGA (Altera). The description of the routers was done in Verilog [24]. Testing [25] was conducted in two stages. At the first stage, the data on the occupied chip resources was obtained for one router and the network as a whole for those considered earlier (Figure 3a, Figure 3b).

Based on the data obtained, it can be concluded that the major cause for increase in the router-occupied chip resources is the number of multiplicative network generatrices.. With an increase in number of generatrices in the structure of the circulant, the number of mathematical operations for their verification, as well as memory for storing the values of the generators, increases.

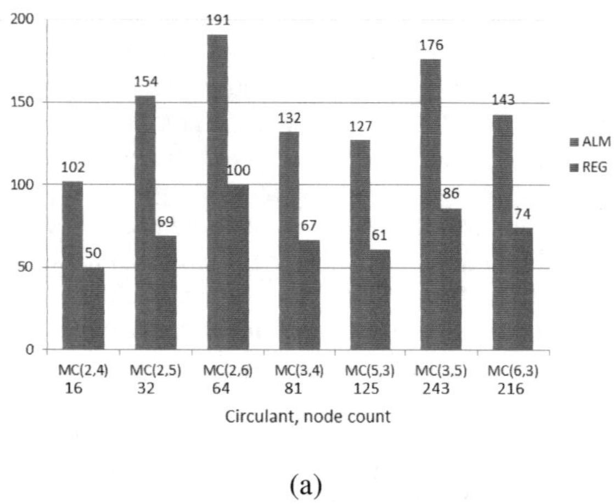

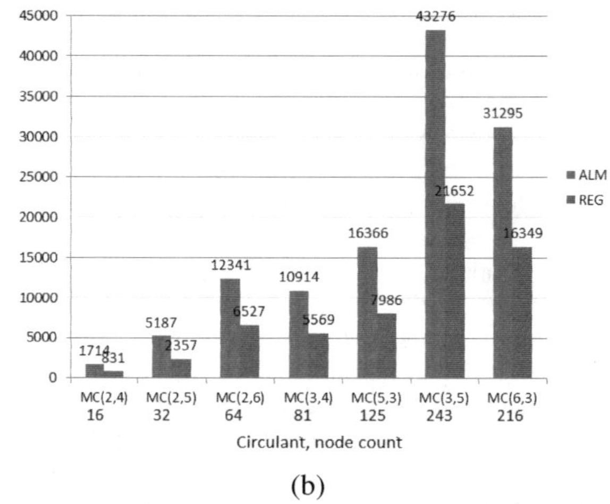

Figure 3. Number of resources occupied by the router (a) and by the network (b) on a chip.

At the second stage, for comparison with other families of circulants, the circulants, which can be described in several representations, were chosen. The comparison was made between multiplicative circulants with the number of generatrices equal to 2, and ring circulants. The number of network nodes was formed as a natural number in degree 2. As a result of RTL synthesis, the data on the number of ALM blocks and registers (Table 3, Figure 4a, b) was obtained.

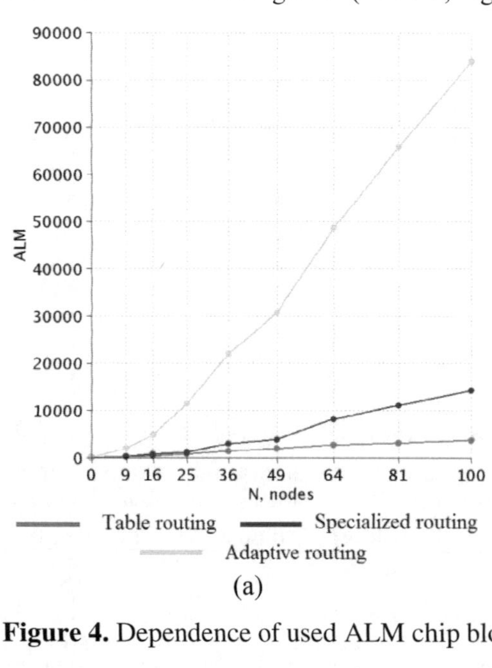

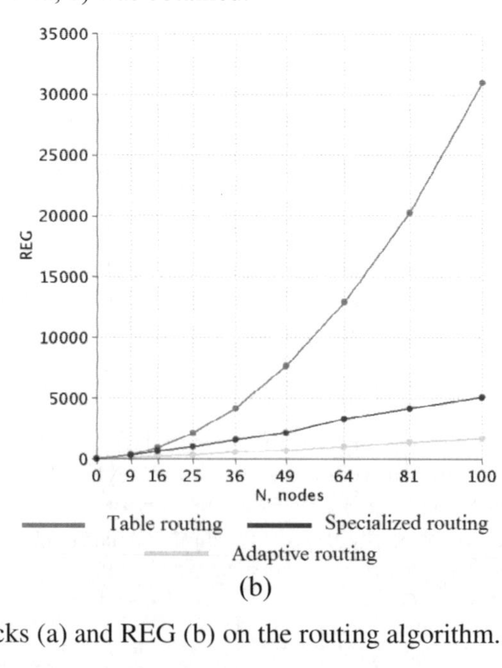

Figure 4. Dependence of used ALM chip blocks (a) and REG (b) on the routing algorithm.

Table 3. Resources occupied chip blocks for the whole network.

Circulant	Number of nodes	ALM (Table alg)	REG (Table alg)	ALM (Ring)	REG (Ring)	ALM (MC)	REG (MC)
$MC(3,2)$	9	205	292	1933	81	275	293
$MC(4,2)$	16	390	912	4737	210	732	605
$MC(5,2)$	25	753	2089	11491	301	1121	1000
$MC(6,2)$	36	1350	4158	22025	546	2906	1558
$MC(7,2)$	49	1909	7637	30709	664	3774	2118
$MC(8,2)$	64	2686	12879	48671	988	8157	3275
$MC(9,2)$	81	3055	20249	65825	1356	11121	4151
$MC(10,2)$	100	3644	30956	83938	1662	14278	5076

The columns "ALM (Table alg.)" and "REG (Table alg." represent the results of tabular routing, the columns "ALM (Ring)" and "REG (Ring)" show the results of the algorithm for describing the graph as a circular circulant [26], and in the columns "ALM (MC)" and "REG (MC)" – for multiplicative circulant. The graphs show that the rate of increase in the use of ALM blocks in the algorithm for describing the network as a multiplicative circulant is much lower than that of the algorithm for describing the network as a ring circulant. At the same time, the rate of increase in the use of registers in the algorithm, proposed for the multiplicative circulant, is slightly higher than that of the algorithm for the ring circulant. Given the fact that the logical resources of the chip are much smaller than the registers, it can be concluded that the use of multiplicative circulants is more effective than the ring ones with an equal number of generatrices.

6. Conclusion

In the conditions of inconsistency of the characteristics of common topologies to the requirements of modern networks and in view of the need to search for alternative options for building networks, it is suggested to consider as a topology a special kind of graphs – multiplicative circulants. Strict rules for the formation of the circulant structure impose restrictions on the number of nodes in the network, and make the topology data easily scalable to larger networks and allow improving such important characteristics as diameter and average network distance in comparison with classical topologies.

For multiplicative circulants, standard shortest path search algorithms are applicable, and in this paper we propose the address part of a packet to reduce its size when static routing. At the same time, most of the standard shortest path search algorithms are built around the graph; that is why with increasing complexity of the circulant, the running time of the algorithm dramatically increases. Taking into account the peculiarities of the structure of multiplicative circulants, it was developed a specialized algorithm for this class of circulants, which makes it possible to considerably simplify the structure of address part of the packet and to reduce the time for finding the optimal path in the graph.

Acknowledgment

The publication was prepared within the framework of the Academic Fund Program at the National Research University Higher School of Economics (HSE) in 2018-2019 (grant № 18-01-0074) and by the Russian Academic Excellence Project "5-100".

References

[1] Nychis G, Fallin C and Moscibroda T 2010 Next Generation On-Chip Networks: What Kind of Congestion Control Do We Need? *Hotnets-IX: Proceedings of the 9th ACM SIGCOMM Workshop on Hot Topics in Networks* (Monterey, California: ACM).

[2] Paul J, Stechele W and Oechslein B 2015 Resource-awareness on heterogeneous MPSoCs for image processing *Journal of Systems Architecture V. 61, I. 10.*(Elsevier) pp 668-680.

[3] Abdelfattah M S, Bitar A and Betz V 2017 Design and applications for embedded networks-on-chip on FPGAs *IEEE Transactions on Computers V. 66, I. 6.* pp 1008-1021.

[4] Deb D, Jose J and Das S 2019 Cost effective routing techniques in 2D mesh NoC using on-chip transmission lines *Journal of Parallel and Distributed Computing V. 123* (Elsevier) pp 118-129.

[5] Ansari A Q, Ansari M R and Khan M A 2016 Modified quadrant-based routing algorithm for 3D Torus Network-on-Chip architecture *Perspectives in Science V. 8,* pp 718-721.

[6] Marvasti M B and Szymanski T H 2012 The performance of hypermesh NoCs in FPGAs *IEEE 30th International Conference on Computer Design (ICCD)* (Montreal, Canada) pp 492-493.

[7] Bishnoi R, Kumar P and Laxmi V 2014 Distributed adaptive routing for spidergon NoC *18th International Symposium on VLSI Design and Test* (IEEE) pp 1-6.

[8] Dally W J and Towles B 2004 Principles and practices of interconnection networks. (Elsevier). p 550.

[9] Monahova Je A 2011 Strukturnye i kommunikativnye svojstva cirkuljantnyh setej *Prikladnaja diskretnaja matematika No. 3(13)* pp 92–115.

[10] Intel Corporation 'Single-chip cloud computer', 2009 [Online]. Available: https://www.intel.ru/content/dam/www/public/us/en/documents/technology-briefs/intel-labs-single-chip-cloud-overview-paper.pdf (Accessed: 20-Oct-2018).

[11] Intel Corporation 'Intel's Teraflops Research Chip' [Online]. Available: http://download.intel.com/pressroom/kits/Teraflops/Teraflops_Research_Chip_Overview.pdf (Accessed: 20-Oct-2018).

[12] Elspas B and Turner J 1970 Graphs with circulant adjacency matrices *Journal of Combinatorial Theory V. 9. Iss. 3.* (California: Stanford Research Institute / Elsevier) pp 297–307.

[13] Boesch F and Tindell R 1984 Circulants and their connectivities *Journal of Graph Theory. V. 8. I. 4.* (New Jersey: Stevens Institute of Technology) pp 487–499.

[14] Monahova Je A 2010 Mul'tiplikativnye cirkuljantnye seti // *Diskretnyj analiz i issledovanie operacij V. 17. No. 5.* pp 56–66.

[15] Stojmenovic I 1997 Multiplicative circulant networks. Topological properties and communication algorithms *Discrete Applied Mathimatics N. 77* (Ottava: University of Ottawa / Elsevier) pp 281–305.

[16] Wong C K and Coppersmith D 1974 A combinatorical problem related to multimodule memory organizations *Journal of the ACM V. 21, I. 3* (New York / ACM) pp 392–402.

[17] Arno S and Wheeler F S 1993 Signed digit representations of minimal Hamming weight *IEEE Transactions on Computers. V. 42. I. 8.* pp 1007–1010.

[18] Suboh S, Bakhouya M and Gaber J 2008 An interconnection architecture for network-on-chip systems *Telecommunication Systems V. 37. I. 1–3* pp 137–144.

[19] Park J H and Chwa K Y 1994 Recursive Circulant: a New Topology for Multicomputer Networks *Proceedings of the International Symposium on Parallel Architectures, Algorithms and Networks (ISPAN)*, (Kanazawa, Japan: IEEE Computer Society Press) p 73–80.

[20] Bundy A and Wallen L 1984 Breadth-First Search // *Catalogue of Artificial Intelligence Tools*. (Heidelberg / Springer) pp 13.

[21] Crespelle C and Gambette P 2010 Unrestricted and complete Breadth-First Search of trapezoid graphs in O(n) time *Information Processing Letters V 110 I 12-13 pp 497-502.*

[22] Gabis A.B and Koudil M 2016 NoC routing protocols – objective-based classification *Journal of Systems Architecture. V. 66–67* pp 14–32.

[23] Kuo-Shun D, Ho C T and Jyh-Jong T 1998 Matrix transpose on meshes with wormhole and XY routing *Discrete Applied Mathematics V. 83, I.1-3,* pp 41-59.

[24] IEEE Std, 2001 'IEEE Standard Verilog® Hardware Description Language'[Online] Available: http://www-inst.eecs.berkeley.edu/~cs150/fa06/Labs/verilog-ieee (Accessed 10- Oct-2018).

[25] Lezhnev E V 'NoCs with circulant toplogy comparison' [Online] Available: https://github.com/EvgenijLezhnev/NoC-Comparison (Accessed 13-Oct-2018).

[26] Romanov A Yu Development of routing algorithms in networks-on-chip based on ring circulant topologies. In Press.

IOP Conf. Series: Journal of Physics: Conf. Series **1163** (2019) 012028 doi:10.1088/1742-6596/1163/1/012028

Direct numerical simulation of bag-breakup - mechanism of sea spray generation in strong winds

A N Zotova, Yu I Troitskaya, D A Sergeev and A A Kandaurov
Institute of Applied Physics, Nizhny Novgorod, Russia
E-mail: aniazotova@yandex.ru

Abstract. For the direct numerical simulation of the bag-breakup phenomenon, the Gerris flow solver was used. As an initial configuration for numerical simulation a drop of liquid was placed in the air flow. A drop of water with a diameter of 1 cm (the characteristic size of the perturbation from which the bag arises) placed in the air stream at a speed of 20 m/s corresponds best to the experimental situation, such a system is characterized by the Weber number We = 54. Modeling the dynamics of two media with very different densities, such as water and air, require a lot of computational time, so we used liquids that differ in density by a factor of 10, but to keep the same Weber number, we changed the other parameters of the problem. The calculation was carried out with keeping the same Reynolds number and also with the Reynolds number reduced and increased by a factor of 10. In our simulation just as in an experiment under the action of an air stream a drop is blown into a micro-sail that bursts to form a micro-spray. As the Reynolds number decreases (with increasing of viscosity of both media), the process of destruction of the drop after the formation of a micro-sail from it changes.

1. Introduction

The flow of apparent and latent heat between the atmosphere and the ocean play an important role in the development of storms and tropical hurricanes [1]. A major contribution to heat transfer at the boundary between the atmosphere and the ocean is caused by sea splashes - water droplets ranging in size from 10 nm to several millimeters, injected from the sea surface as a result of waves breaking in the coastal zone or during strong winds. At the moment, there are several basic mechanisms of spray generation over the sea surface: fragmentation of liquid "fingers", rupture of underwater bubbles, and, as was shown in recent works [2], bag-breakup mechanism. In [2] the mechanisms of spray generation under hurricane winds were experimentally studied and it was found that for friction velocity $u^* > 1\ m/s$ (for wind speeds at 10 m height exceeding 20 m/s), bag-breakup determines the contribution to the formation of large droplets.

A typical event of this type begins with a small-scale increase in the water surface, which then turns into a "micro-sail", swells into a water film bounded by a thicker rim, and finally explodes, producing hundreds of droplets. The mean values of the radii R and the lifetimes τ of bags estimated as $\langle R \rangle \sim 10^{-2}\ m$ and $\langle \tau \rangle \sim 10^{-2}\ s$, as a rule, decrease with increasing wind speed [2]. In the dynamics of engineering fluids [2], this phenomenon is known as the "bag-breakup" mode of liquid fragmentation in gas streams.

Experimental study of the phenomenon of bag-breakup in full-scale conditions is extremely difficult, but under laboratory conditions such studies are conducted. A separate bag-breakup event was investigated in a high-speed wind-wave channel with a small reservoir set at a distance of 7.5 m from its beginning. The initial perturbation, from which the bag develops, was artificially created in a given place using a jet from a submerged nozzle. The surface of the water around the experimental area was covered with a damping material. Simultaneous multi-stream high-speed shooting was used. Several frames of one of the received video recordings are shown in Figure 1.

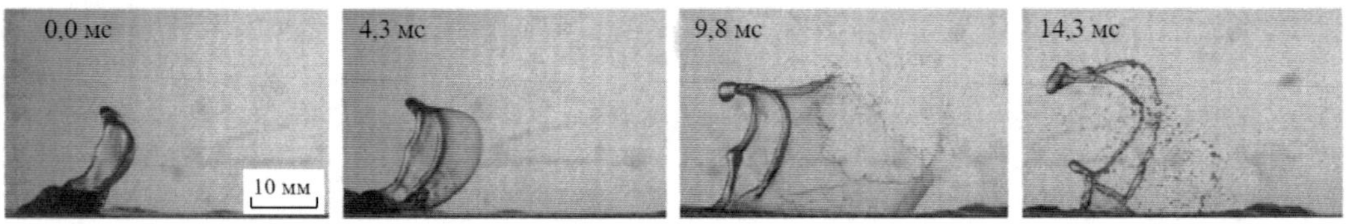

Figure 1. Forming and breaking of the bag (side view).

The processing of the data obtained is laborious and time-consuming. In this situation it is also useful to carry out a numerical experiment that will allow us to study in more detail the processes leading to droplets in the investigated regime, and also to study the influence of various external factors on these processes.

2. The Gerris flow solver

For the direct numerical simulation of the bag-breakup phenomenon, the Gerris flow solver was used [3, 4]. Gerris is a software package for solving incompressible Euler and Navier-Stokes equations, combining the quad/octree adaptive mesh refinement method, projection method and multi-level Poisson solver. Complex solid boundaries are processed by the cartesian VOF method.

In Gerris, a numerical algorithm is implemented that solves the Navier-Stokes equations for incompressible media with variable density and surface tension:

$$\rho\left(\partial_t \vec{u} + \vec{u}\nabla\vec{u}\right) = -\nabla p + \nabla\left(2\mu\vec{D}\right) + \sigma\kappa\delta_s\vec{n} \tag{1}$$

$$\partial_t\rho + \nabla\left(\rho\vec{u}\right) = 0 \tag{2}$$

$$\nabla\vec{u} = 0 \tag{3}$$

where $\vec{u} = (u, v, w)$ - the fluid velocity, ρ - the density of the liquid, μ - the dynamic viscosity and $\vec{D}$ - the deformation tensor defined as $D_{ij} \equiv \left(\partial_i u_j + \partial_j u_i\right)/2$. The Dirac distribution function δ_s expresses the fact that the surface tension term is concentrated at the media interface; σ - the surface tension, κ and $\vec{n}$ - the curvature and the normal to the interface.

For two-phase flows, the volume fraction c of the first liquid is introduced and the density and viscosity are determined as

$$\rho \equiv c\rho_1 + (1-c)\,\rho_2 \tag{4}$$

$$\mu \equiv c\mu_1 + (1-c)\,\mu_2 \tag{5}$$

where ρ_1, ρ_2 and μ_1, μ_2 are the densities and viscosities of the first and second media, respectively.

Then the advection equation for density can be replaced by the equivalent advection equation for the volume fraction of the liquid:

$$\partial_t c + \nabla\left(c\vec{u}\right) = 0 \tag{6}$$

IOP Conf. Series: Journal of Physics: Conf. Series **1163** (2019) 012028 doi:10.1088/1742-6596/1163/1/012028

To solve the advection equation for the volume fraction, the piecewise linear geometric VOF method [5] is used, which is generalized for quad/octree spatial discretization.

3. Simulation

We decided to start the simulation not with a geometrically complex configuration of the initial perturbation, from which the bag was developed in the experiment [2], but with the problem of a simpler geometry: a drop of liquid was placed in the air flow (see Figure 2). The droplet dynamics under different conditions was studied experimentally, theoretically and numerically, and it was shown [6] that the droplet destruction mechanism depends on the Weber number characterizing the system $We = \frac{\rho_a v^2 D}{\sigma}$, where ρ_a – the density of the external medium, v – the velocity of the external medium, D – the initial diameter of the drop, σ – the surface tension.

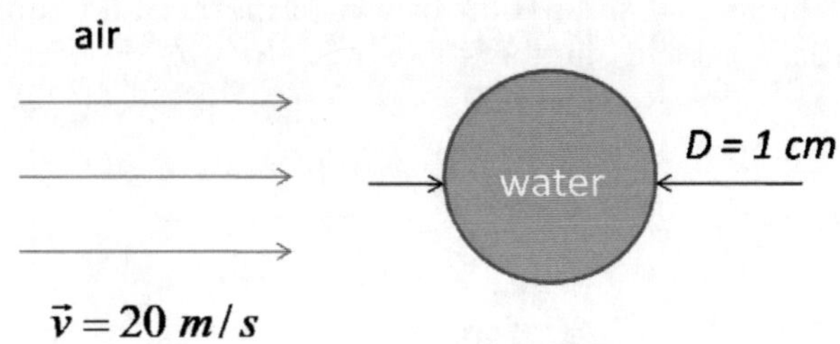

Figure 2. Configuration of the problem.

A drop of water with a diameter of 1 cm (the characteristic size of the perturbation from which the bag arises) placed in the air stream at a speed of 20 m/s corresponds best to the experimental situation, such a system is characterized by the Weber number $We = 54$ and the Reynolds number $Re = 1.1 \cdot 10^4$.

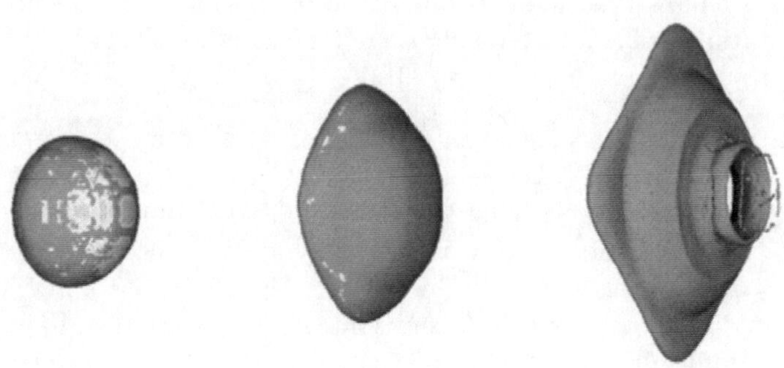

Figure 3. The result of numerical simulation of the formation and rupture of the bag with corresponding to the initial problem numbers We and Oh_d, Re increased by 10 times ($\rho'_a = \rho_a \cdot 100; v' = v/10; \mu'_a = \mu_a; \mu'_d = \mu_d; \sigma' = \sigma$).

Modeling the dynamics of two media with very different densities, such as water and air, require a lot of computational time, so we used liquids that differ in density by a factor of 10 (air density ρ_a was increased 100 times), but to keep the Weber number fixed in one case we changed the value of velocity of the external medium ($v' = v/10$).

Figure 4. The result of numerical simulation of the formation and rupture of the bag with corresponding to the initial problem numbers We and Re ($\rho'_a = \rho_a \cdot 100; v' = v/10; \mu'_a = \mu_a \cdot 10; \mu'_d = \mu_d \cdot 10; \sigma' = \sigma$).

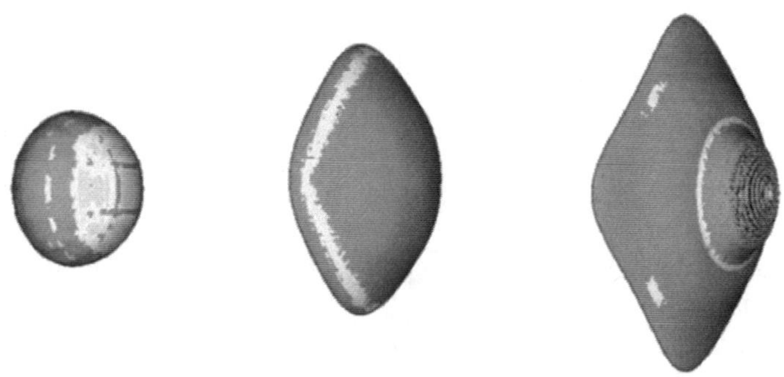

Figure 5. The result of numerical simulation of the formation and rupture of the bag with corresponding to the initial problem number We and the number Re reduced by a factor of 10 ($\rho'_a = \rho_a \cdot 100; v' = v/10; \mu'_a = \mu_a \cdot 100; \mu'_d = \mu_d \cdot 100; \sigma' = \sigma$).

The calculation was carried out with keeping the Reynolds number $Re = \frac{\rho_a D v}{\mu_a}$ fixed, see Fig. 4 ($\mu'_a = \mu_a \cdot 10; \mu'_d = \mu_d \cdot 10$), and with the Reynolds number 10 times increased, see Fig. 3 ($\mu'_a = \mu_a; \mu'_d = \mu_d$), and 10 times reduced, see Fig. 5 ($\mu'_a = \mu_a \cdot 100; \mu'_d = \mu_d \cdot 100$), compared to the Reynolds number for the experimental configuration. The Ohnesorge number $Oh_d = \frac{\mu_d}{(\rho_d D \sigma)^{1/2}}$, which determines the Weber numbers, for which there are transitions from one droplet destruction regime to another, for all this cases is ≤ 0.1, so the effect of drop viscosity is not significant [7].

Also in our calculations, keeping the Weber and Reynolds numbers fixed was provided by another set of parameters: $\rho'_a = \rho_a \cdot 100; v' = v; \mu'_a = \mu_a \cdot 100; \mu'_d = \mu_d \cdot 100; \sigma' = \sigma \cdot 100$ (see Fig. 6).

The simulation results are shown in Figures 3 - 6. It can be seen that, just as in the experiment, under the action of the air flow the drop inflates into a micro-sail, which bursts with the formation of micro-spray. The process of destruction of the drop after the formation of a micro-sail from it changes: for Reynolds numbers $Re = 1.1 \cdot 10^5$ and $Re = 1.1 \cdot 10^4$ after the formation of a break and small splashes in the center of the sail, a rim is formed along the

Figure 6. The result of numerical simulation of the formation and rupture of the bag with corresponding to the initial problem numbers We and Re ($\rho_a' = \rho_a \cdot 100; v' = v; \mu_a' = \mu_a \cdot 100; \mu_d' = \mu_d \cdot 100; \sigma' = \sigma \cdot 100$).

edge of the break, which is subsequently separated from the rest of the sail; for the Reynolds number $Re = 1.1 \cdot 10^3$ after the formation of a break in the center of the sail, the edge of the break continues to move to the edge of the micro-sail.

Also, calculations have shown that the mechanism of droplet destruction does not change when the parameters of the problem change in such a way that the Weber and Reynolds numbers remain unchanged.

Acknowledgments

Carrying out experiments themselves was supported by Russian Science Foundation (Agreement No. 14-17-00667); designing of methods of measurements including: optical and visualization scheme, methods of measurements of the air flow and wave field parameters were supported by Russian Foundation of Basic Research (No 18-55-50005, 18-05-60299, 18-35-00658, 18-35-20068); the development of the software for video processing was supported by the Grant of the President no. MK-2041.2017.5.

References

[1] Andreas E L and Emanuel K A 2001 *J. Atmos. Sci.* **58** 3741–3751
[2] Troitskaya Yu, Kandaurov A, Ermakova O, Kozlov D, Sergeev D and Zilitinkevich S 2017 *Sci. Rep.* **7** 1614
[3] Popinet S 2003 *J. Comput. Phys.* **190** 572–600
[4] Popinet S 2009 *J. Comput. Phys.* **228** 5838–5866
[5] Scardovelli R and Zaleski S 1999 *Annual Review of Fluid Mechanics* **31 (1)** 567–603
[6] Pilch M and Erdman C A 1987 *Int. J. Multiphase Flow* **13 (6)** 741–757
[7] Krzeczkowski S A 1980 *Int. J. Multiphase Flow* **6** 227–239

IOP Conf. Series: Journal of Physics: Conf. Series **1163** (2019) 012029 doi:10.1088/1742-6596/1163/1/012029

Numerical investigation of the critical behavior of the three-dimensional Ising model near the percolation threshold

M A Shlyakhtich[1,†], P V Prudnikov[2,‡], A S Papushin[2]

[1] Department of Theoretical Physics and Wave Phenomena, Siberian Federal University, 79 Svobodny Av., Krasnoyarsk, 660041, Russia

[2] Department of Theoretical Physics, Omsk State University, Mira prospekt 55-A, Omsk, 644077, Russia

E-mail: [†]`mmed@mail.ru`, [‡]`prudnikp@univer.omsk.su`

Abstract. In this work we investigate the critical behavior of the disordered three-dimensional Ising model with an impurity concentration close to the threshold of impurity percolation on high-performance computing systems. Starting from initial configurations, the system was updated with Metropolis algorithm and invaded cluster algorithm. The values of dynamic critical exponent z and static critical exponent β/ν were obtained in this paper for both algorithm.

1. Introduction

The question of how structure defects modifies a phase transition is not fully solved. Many issues is still open, even in the case of the ferromagnetic transition in the the system with quenched disorder, from the qualitative nature of the ordered phase near criticality [1, 2].

The description of phase transitions is considered to be one of the most urgent and complex problems of statistical physics. The anomalously long-lived and large fluctuations of some thermodynamic quantities observed as the phase transition point approaches are characterized by strong interactions, which creates considerable difficulties in the experimental and analytical investigation of critical behavior. Computer simulation is an independent tool for investigating the abnormal behavior of a second-order phase transition [3, 4], as evidenced by the progress in the development of various methods of computer simulation. Therefore, one of the important tasks is the development of computer simulation methods that depend weakly on the effects of critical slowdown.

In recent years, the study of various disordered models near the threshold of impurity percolation has become an interesting and actual problem [5, 6, 7]. In this paper we have investigated the critical properties of the Ising model near the percolation threshold. At a concentration close to critical, a cluster appears that connects the opposite sides of the lattice. The largest cluster has a fractal dimension at the critical concentration $p = p_c$. In the theoretical description of the behavior of such systems, the concentration of defects by a small quantity can no longer be considered. That makes their theoretical description very difficult or even impossible. A Invaded cluster algorithm was proposed in [8], which is much more effective than all previous methods near the impurity percolation threshold.

IOP Conf. Series: Journal of Physics: Conf. Series **1163** (2019) 012029 doi:10.1088/1742-6596/1163/1/012029

2. Algorithms and Methods

Anomalous properties of system in second order phase transitions is a critical slowing down of the relaxation process. Which manifests itself in a slow change over time, the characteristic thermodynamic quantity - of the order parameter (magnetization for ferromagnetic systems) tends to equilibrium:

$$\tau_{\text{rel}} \sim |T_c - T|^{-z\nu} \tag{1}$$

where τ_{rel} – the relaxation time diverges on approaching the critical temperature.

In the theoretical description of the behavior of systems with critical spin concentration $p = p_c$, the concentration of defects by a small quantity can no longer be considered. That makes their theoretical description very difficult. The algorithm proposed in [8] is much more effective than all previous methods near the impurity percolation threshold. The new algorithm called Invaded cluster algorithm. The primary goal is to reduce the effective autocorrelation time and, therefore, to improve the statistical sampling of generated configurations.

The Invaded cluster algorithm works as follows. The bonds of the lattice are given a random order starting with an Ising spin configuration S. Correlated invasion percolation clusters are grown as until one of clusters spans the system. After the growth process is terminated, cluster is flipped with probability P_n yielding a new spin configuration S'.

$$P_n = \exp\left(-\Delta E/kT\right), \tag{2}$$

where ΔE is the change in the energy of the system caused by the test configuration change.

According to Janssen's arguments [9] obtained using the ε-expansion and renormalization group method, one may anticipate a generalized scaling relation for the k-th moment of the magnetization

$$m^{(k)}(t,\tau,L,m_0) = b^{-k\beta/\nu} m^{(k)}(t/b^z, b^{1/\nu}\tau, L/b, b^{x_0} m_0). \tag{3}$$

This relation realized after a very small time t_{mic} in macroscopic sense but this time is large enough. In (3) b is a spatial rescaling factor, β and ν are static critical exponents, z is the dynamic exponent, the new independent exponent x_0 is the scaling dimension of the initial magnetization m_0 and $\tau = (T - T_c)/T_c$ is the reduced temperature.

Problem of finite size is almost absent while the correlation length is small. This happens in the early stage of system evolution. Therefore, we consider $L = 128$ large enough and solve this problem.

We can skip the argument $m_0 = 1$ in (3) like a large enough linear size L,

$$m^{(k)}(t,\tau) = b^{-k\beta/\nu} m^{(k)}(b^{-z}t, b^{1/\nu}\tau). \tag{4}$$

In this work computer simulation was carried out. It began from a completely ordered initial state. The system is simulated at critical temperature. We measured the magnetization $m(t)$ and second order moment of magnetization $m^{(2)}(t)$. One can steer clear of the main dependence from time in (4) when $b = t^{1/z}$ and $k = 1$. And we have

$$\begin{aligned} m(t,\tau) &= t^{-\beta/\nu z} m(1, t^{1/\nu z}\tau) = \\ &= t^{-\beta/\nu z}(1 + a t^{1/\nu z} + O(\tau^2)). \end{aligned} \tag{5}$$

For critical point $\tau = 0$

$$m(t) \sim t^{-\beta/\nu z}. \tag{6}$$

In order to calculate the critical exponent z we have used the second order Binder cumulant

$$U_2(t) = \frac{[\langle m^{(2)} \rangle]}{[\langle m \rangle]^2} - 1 \sim t^{d/z}. \tag{7}$$

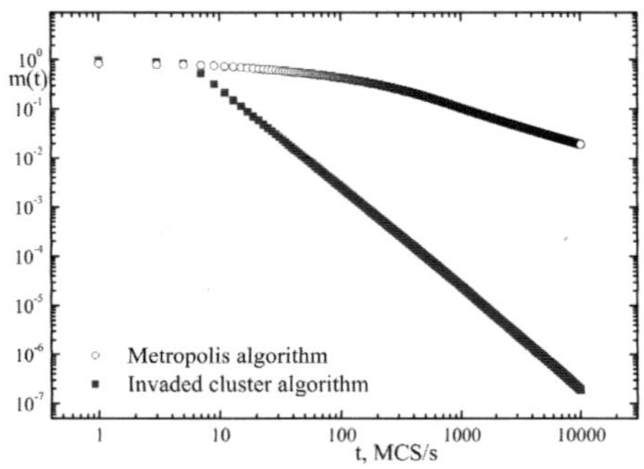

Figure 1. Time dependencies of magnetization.

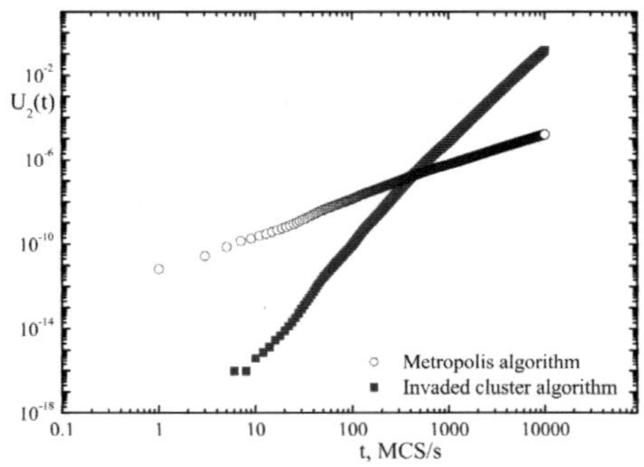

Figure 2. Time dependencies of second order Binder cumulant $U_2(t)$.

3. Results of Monte Carlo modeling

For calculating the critical exponents of disordered Ising model we considered simple cubic lattices with linear size $L = 128$.

In this work computer simulation was carried out. It began from a completely ordered initial state. The system is simulated at critical temperature $T_c = 3.091266(20)$. We measured the magnetization $m(t)$ and second order moment of magnetization $m^{(2)}(t)$. Simulations have been performed up to $t = 10000$ MCS/s. The final lines were obtained after averaging more than 600 samples. Starting from initial configurations, the system was updated with Metropolis algorithm and Invaded cluster algorithm.

For ferromagnetic films the order parameter can be defined as magnetization and calculate by equation

$$m^{(k)}(t) = \left[\left\langle \left(\frac{1}{N_s} \left| \sum_{i=1}^{N_s} p_i S_i(t) \right| \right)^k \right\rangle \right]. \tag{8}$$

In (8), the angular and square brackets designate statistical averaging and averaging over the different impurity configurations, respectively, $N_s = pL^3$ is a number of spins in the simple cubic lattice, and $p = p_c = 0.688313500$ is a spin concentration.

Time dependencies of magnetization are presented on figure 1 both for Metropolis algorithm and Invaded cluster algorithm. We can estimate the value of critical non-equilibrium exponent $\beta/\nu z$ from slope of magnetization at low-temperature initial state on $t \in [1000; t_{\text{right}}]$. The minimum error of determining the critical exponent $\beta/\nu z$ is reached on time interval $[1000; t_{\text{right}}]$ with $t_{\text{right}} = 5400$ for Metropolis algorithm and $t_{\text{right}} = 3100$ for invaded cluster algorithm. We got the values of critical exponent $\beta/\nu z = 0.7225$ for Metropolis algorithm and $\beta/\nu z = 2.1792$ for invaded cluster algorithm.

From the slope of time dependence of the second order Binder cumulant $U_2(t)$ (figure 2) we can estimate the value of critical exponent d/z. The time interval $[t_{\text{left}}; 9800]$ for $t_{\text{left}} = 3700$ gives the minimum of errors for exponent d/z for Metropolis algorithm. The time interval $[t_{\text{left}}; 2900]$ for $t_{\text{left}} = 1300$ gives the minimum of errors for exponent d/z for invaded cluster algorithm. We got the values of critical exponent $d/z = 1.3835$ for Metropolis algorithm and $d/z = 4.1528$ for invaded cluster algorithm.

We have regarded the influence of a principal corrections to the scaling on asymptotic values of exponents [4] in order to get exact values of the critical exponents.We have used the next

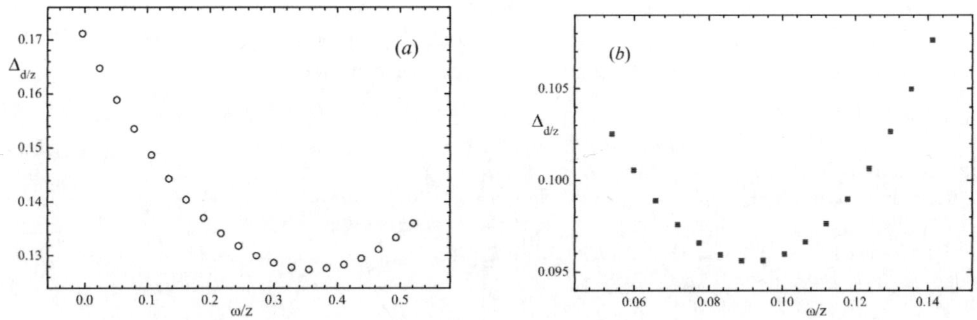

Figure 3. Dependencies of global mean-square error $\Delta_{d/z}$ for all time intervals as a function of the exponent ω/z: (a) Metropolis algorithm, (b) Invaded cluster algorithm.

relation for the observable $X(t)$:

$$X(t) = A_x t^{\delta}(1 + B_x t^{-\omega/z}), \tag{9}$$

where ω is an exponent of the principal corrections to scaling, A_x and B_x are fitting parameters. In (9) $\delta = \beta/\nu z$ when $X \equiv m(t)$, $\delta = d/z$ when $X \equiv U_2(t)$. We can determine the exponents δ and ω/z from minimum of the mean square errors σ_δ of this fitting procedure. after that all minimal values of δ are averaged by all intervals of time. The global mean-square error Δ_δ (Figure 3) is determined from averaged values of δ [4].

4. Conclusion

The value of exponent $\beta/\nu z = 0.6518(609)$ was calculated using $\omega/z = 0.0860$, value of $d/z = 1.1838(109)$ was calculated using $\omega/z = 0.3275$ for Metropolis algorithm. The value of $\beta/\nu z = 2.2547(327)$ was calculated using $\omega/z = 0.2700$, value of $d/z = 4.0994(881)$ was calculated using $\omega/z = 0.0975$ for invaded cluster algorithm. We defined the final values of z and β/ν using the values given above. The calculated values of z and β/ν are presented in table 1.

Table 1. The values of critical exponents z and β/ν for Ising disordered system

Algorithm	z	β/ν
Metropolis	2.534(234)	1.652(37)
Invaded cluster	0.732(16)	1.650(60)

The value of dynamic critical exponent z obtained by the invaded cluster algorithm three times smaller than the value for the Metropolis algorithm. That's why the invaded cluster algorithm can be successfully applied to introduce a disordered system into an equilibrium state.

Acknowledgments

This research was supported by the grants 17-02-00279, 18-42-55003 of Russian Foundation of Basic Research and by the grant MD-6868.2018.2 of Russian Federation President.

The simulations were supported in through computational resources provided by the Shared Facility Center "Data Center of FEB RAS" (Khabarovsk), by the Supercomputing Center of Lomonosov Moscow State University, by Moscow Joint Supercomputer Center and by St. Petersburg Supercomputer Center of the Russian Academy of Sciences.

References

[1] Tarjus G and Dotsenko V 2002 *J. Phys.* A **35** 1627
[2] Krzakala F, Ricci-Tersenghi F, Sherrington D, and Zdeborova L 2011 *J. Phys.* A **44** 042003
[3] Prudnikov V V, Prudnikov P V, Krinitsyn A S, et. al. 2010 *Phys. Rev.* E **81** 011130
[4] Prudnikov P V, Medvedeva M A 2012 *Progr. Theor. Phys.* **127** 369
[5] Balog I, Uzelac K 2007 *Phys. Rev.* E **76** 011103
[6] Faraggi E 2008 *Phys. Rev.* B **78** 134416
[7] Dixon K, Venus D 2017 *Phys. Rev.* B **95** 245438
[8] Machta J, Choi Y S, Lucke A, Schweizer T 1995 *Phys. Rev. Lett.* **15** 2792
[9] Janssen H K, Schaub B, Schmittmann B 1989 *Z. Phys.* B **73** 539

IOP Conf. Series: Journal of Physics: Conf. Series **1163** (2019) 012030 doi:10.1088/1742-6596/1163/1/012030

Intermediate Lattice Boltzmann-BGK method and its application to micro-flows

Oleg Ilyin

Dorodnicyn Computing Center of Federal Research Center Computer Science and Control of Russian Academy of Sciences, Vavilova - 40, Moscow, 119333, Russia

E-mail: oilyin@gmail.com

Abstract. A novel discretization approach for the Bhatnager-Gross-Krook (BGK) kinetic equation is proposed. The method inherits properties of the Lattice Boltzmann (LB) method like linear streaming step, conservation of moments. Similar to the finite-difference methods for the BGK model the presented approach describes high-order moments of the distribution function with controllable error. Two test problems are considered: the Poiseuille flow between two parallel plates and plane Couette flow. Significant increase in precision over the conventional LB models is observed.

1. Introduction

The Lattice Boltzmann (LB) method is supposed to be an adequate approach for the modeling of non-equilibrium flows and kinetic effects beyond the Navier-Stokes level [1]. Nevertheless, when the flow is rarefied and the role of high-order moments of the velocity distribution in a rarefied media increases the precision of conventional LB models go down. In the real world, the large values of the rarefaction parameter - the Knudsen number can be observed in the micro, nano-flows, in shale gas flow inside porous media, water transport inside nano-graphene membranes [2].

The Lattice Boltzmann method is based on the idea of the full discretization (in velocity and physical space) of the BGK kinetic equation [3], which reads as

$$\frac{\partial f}{\partial t} + \boldsymbol{\xi}\frac{\partial f}{\partial \boldsymbol{x}} = \frac{1}{\tau}(\Phi - f), \tag{1}$$

where $f(t, x, \boldsymbol{\xi})$ is the gas distribution function, $t, \boldsymbol{x}, \boldsymbol{\xi}$ are the time, spatial coordinate, particle velocity respectively; $\Phi(t, x, \boldsymbol{\xi}) = \frac{\rho}{(2\pi\theta)^{D/2}} \exp\left(-\frac{|\boldsymbol{\xi}-\boldsymbol{u}|^2}{2\theta}\right)$ is the local Maxwell state and $\rho(t, \boldsymbol{x}), \mathrm{u(t,x)}, \theta(\mathrm{t,x})$ are the macro-parameters (the density, bulk velocity, temperature), D is the number of spatial dimensions, τ is the relaxation time.

In the conventional LB method the discretization is performed in a such way that the LB distribution function (DF) is equivalent to the solution for the BGK equation projected on a finite basis in velocity space spanned by the Hermite polynomials (the Grad expansion). This equivalence is achieved via the Gauss-Hermite quadratures [4, 5, 3]. Since the first moments of the LB local equilibrium are reproduced in the same form as for the local Maxwell state then LB method is conservative by the construction. Moreover, the streaming step is modeled in a

IOP Conf. Series: Journal of Physics: Conf. Series **1163** (2019) 012030 doi:10.1088/1742-6596/1163/1/012030

simple linear form, the particles in this step travel from one lattice site to another. The latter guarantees second order accuracy in physical space.

Instead of the exact reproduction of the first moments for the distribution function, the finite-difference methods for the BGK model are aimed to recover the distribution function in overall [6], including high-order moments. The cost to pay is the usage of relatively large discrete velocity sets, moreover conservation of moments requires some additional efforts.

The present research is aimed to develop a new LB based discrete velocity (DV) method which is able to cope with the kinetic high-order moments in the rarefied flow but without drastic increase in number of discrete velocities.

2. Hierarchy LB models

2.1. The construction of LB hierarchy

The simplest LB one-dimensional discretization of the BGK equation (1) ($D = 1$) which is able to reproduce the Navier-Stokes equations at the limit of low Mach numbers is D1Q3 model (D is the number of dimensions, Q is the number of the discrete velocities) [1]

$$f_{-1}(t+\Delta t, x - c\Delta t) - f_{-1}(t,x) = \frac{1}{\tau}\left(f_{-1}^{eq}(t,x) - f_{-1}(t,x)\right),$$

$$f_{0}(t+\Delta t, x) - f_{0}(t,x) = \frac{1}{\tau}\left(f_{0}^{eq}(t,x) - f_{0}(t,x)\right),$$

$$f_{+1}(t+\Delta t, x + c\Delta t) - f_{+1}(t,x) = \frac{1}{\tau}\left(f_{+1}^{eq}(t,x) - f_{+1}(t,x)\right),$$

where τ is the relaxation time, $f_{\pm 1}, f_0$ are the lattice distribution functions corresponding to the lattice velocities $\pm c, 0$. The equilibrium states $f_{\pm 1}^{eq}, f_0^{eq}$ are defined as

$$f_{\pm 1}^{eq}(t,x) = \frac{\rho(t,x)}{6}\left(1 \pm 3\frac{u(t,x)}{c} + 3\frac{u(t,x)^2}{c^2}\right),$$

$$f_{0}^{eq}(t,x) = \frac{4\rho(t,x)}{6}\left(1 - 3\frac{u(t,x)^2}{2c^2}\right),$$

where the macroscopic parameters ρ, u are defined from the expressions

$$\rho(t,x) = f_{-1}(t,x) + f_0(t,x) + f_{+1}(t,x), \quad \rho(t,x)u(t,x) = -f_{-1}(t,x)c + f_{+1}(t,x)c,$$

moreover the full energy is calculated as follows

$$\rho(t,x)(u(t,x)^2 + c_s^2) = (f_{-1}(t,x) + f_{+1}(t,x))c^2,$$

where the sound velocity $c_s = \sqrt{\frac{1}{3}}c$, i.e. the model describes only isothermal flows.

Now lets associate with the local equilibrium distribution function for D1Q3 model a random variable $X(N)$, here $N = 1$ (the first step in the hierarchy). Assume that this random variable has three outcomes $-c, 0, +c$. We also assume that this random variable has the distribution function same as the local equilibrium distribution for D1Q3 model. At the present moment this procedure is formal and does not give any additional information. The underlying reason for the introduction of the random variable $X(1)$ will be clear at the next step.

Next, consider two independent identically distributed (i.i.d.) random variables $X_1(2), X_2(2)$ (here $N = 2$, the second step) in a such way that their sum has the same expected value (the bulk velocity u) and the same variance (the temperature c_s^2) as for $N = 1$ case.

IOP Conf. Series: Journal of Physics: Conf. Series **1163** (2019) 012030 doi:10.1088/1742-6596/1163/1/012030

Now, generalizing the previous step I consider a sum of N i.i.d. random variables such that their sum has again an expected value u and variance c_s^2. The variable $\sum_j^N X_j(N)$ has $2N+1$ possible outcomes $-Nc, (-N+1)c \dots Nc$. Since this sum is composed from the independent identical random variables the corresponding distribution function can be calculated in an exact closed form (see below the expression for $Prob\left(\sum_j^N X_j = nc\right)$). The Central Limit Theorem (CLT) for sequences in the form of Lyapunov or Lindeberg-Feller guarantees that the sequence of the distribution functions in the hierarchy converges to the Gaussian distribution (Maxwell state with bulk velocity u and temperature c_s^2).

Finally, at N-s step the following $2N+1$ discrete velocity model can be introduced for $-Nc, (-N+1)c \dots Nc$ lattice

$$f_n(t+1, x+c_n t) - f_n(t,x) = \frac{1}{\tau}\left(f_{n;N}^{eq} - f_n\right)(t,x), \quad n = -N \dots N, \quad c_n = nc \tag{2}$$

$$f_n^{eq} = \rho Prob\left(\sum_j^N X_j = nc\right) \tag{3}$$

$$Prob\left(\sum_j^N X_j = nc\right) = \sum_{m=0}^{\lfloor \frac{N-n}{2} \rfloor} \frac{N!}{(n+m)!m!(N-n-2m)!} P_N(c)^{n+m} P_N(-c)^m P_N(0)^{N-n-2m}, \tag{4}$$

for $n \geq 0$, and where

$$P_N(\pm c) = \frac{1}{2c^2(N)N}\left(c_s^2 \pm cu + \frac{u^2}{N}\right), \quad P_N(0) = 1 - P(c) - P(-c) \tag{5}$$

and

$$c_s^2 = \frac{Nc^2}{3}. \tag{6}$$

To keep the temperature constant $c_s^2 = \theta_0$ at the all levels of the hierarchy it should be required that

$$c \equiv c(N) = \sqrt{3\theta_0/N},$$

thus the lattice step is decreasing when N grows.

Similar expressions can be obtained for $n < 0$ by taking $|n|$ instead of n and changing $P_N(\pm c)$ by $P_N(\mp c)$.

The model (2)-(6) is the main result of the paper. It can considered as a new LB type discretization for the BGK kinetic equation (for isothermal flows) since the CLT guarantees the convergence of the local equilibrium to the local Maxwell distribution.

2.2. Some analytical properties

The equilibrium states in the hierarchy contain $P(\pm c), P(0)$ functions. This states are non-negative if $P(\pm c), P(0)$ are non-negative. This requirement leads to the following inequality

$$u \leq \sqrt{\frac{2}{3}} Nc.$$

Therefore, the domain of non-negativity is growing when N increases. Potentially this property can result in better stability for the models with $N > 1$ in comparison with the conventional $D1Q3$.

The straightforward computation of the moment generating function (MGF) defined by $M(s)$ is complicated. One has

$$M(s) \equiv < e^{ns} >= \sum_{n=0}^{N} e^{ns} \sum_{m=0}^{\lfloor \frac{N-n}{2} \rfloor} \frac{N!}{(n+m)!m!(N-n-2m)!} P_N(c)^{n+m} P_N(-c)^m P_N(0)^{N-n-2m}$$

and the further convolution of the sums seems to be problematic. Nevertheless, the result can be obtained if one takes in account the fact that the local equilibrium is related to a sum $Y = \sum_j^N X_j(N)$ of independent random variables $X_j(N)$. Then the moment generating function for Y is a product of the moment generating functions for $X_j(N)$, they are defined as $M_X(s)$. One has

$$M(s) = M_X(s)^N = (P_N(-c)e^{-cs} + P_N(0) + P_N(c)e^{cs})^N, \tag{7}$$

then any moment m_k of order k can be calculated from Eq. (7) using the formula $m_k = \frac{d^k M(s)}{ds^k}|_{s=0}$. Now taking logarithm from the MGF function (7) and using the expressions (5) for $P_N(\pm c), P_N(0)$ one obtains after some lengthy algebra

$$\log(M(s)) = \frac{\theta s^2}{2} + us + \left\{(9/6! - 1/16)s^6 + \ldots + (-u^3/2)s^3\right\} \frac{1}{N^2} + O\left(\frac{1}{N^3}\right)$$

where $\frac{\theta s^2}{2} + us$ is the logarithm of MGF for the Gaussian distribution and $\theta = \frac{Nc^2}{3}$. Therefore, the difference between the logarithms of MGF for the intermediate LB local equilibrium state and the local Maxwell state is of $O(1/N^2)$ order. Then one concludes that the moments for the presented intermediate LB models converge to the local Maxwell ones with the error decreasing as $O(1/N^2)$.

2.3. Models in several dimensions

The models in several dimensions can be constructed as a tensor product of 1D models. Since all the models in the discussed above hierarchy have the same order of isotropy ($D1Q3$, $D1Q5$, $D1Q7$ and so on) then one can construct 3D models in the form $D1Qn \times D1Qm \times D1Qk$ with $n * m * k$ velocities preserving the order of isotropy of 1D models. The local equilibrium takes the product form of equilibrium states in 1D models. This product form approach is based on the ideas from [7]. For instance, one can construct 15 velocity model $D2Q15$ composed by the $D1Q5$ (the formal sum of $D1Q3$ and $D1Q3$) and $D1Q3$ model.

2.4. Ballistic streamers removal

As was discussed above, 2D and 3D models can be obtained by taking a tensor product of 1D LB models. This models have velocities parallel to the axis - a streaming directions which do not touch a wall if the wall is placed parallel to the axis (ballistic streamers effect) [8]. The removal of zero lattice velocity mitigates the problem and significantly increases the calculation precision for several problems [8, 9].

Having $(2N+1)$ velocity model in the hierarchy for $[-Nc \ldots 0 \ldots Nc]$ lattice I construct $(2N+2)$ velocity model (zero velocity is removed) in $[-(N+0.5)c, -(N-0.5)c, \ldots (N-0.5)c, (N+0.5)c]$ lattice, the local equilibrium is given by the formulas

$$\widetilde{f^{eq}}(nc + \frac{1}{2}c) = \frac{1}{2}\{f^{eq}(nc+c) + f^{eq}(nc)\}, \quad 0 \le n < N; \quad \widetilde{f^{eq}}(Nc + \frac{1}{2}c) = \frac{1}{2} f^{eq}(Nc),$$

$$\widetilde{f^{eq}}(mc - \frac{1}{2}c) = \frac{1}{2}\{f^{eq}(mc-c) + f^{eq}(mc)\}, \quad -N < m \le 0; \quad \widetilde{f^{eq}}(-Nc - \frac{1}{2}c) = \frac{1}{2} f^{eq}(-Nc),$$

the temperature θ is now equals to

$$\theta = c_s^2 + \frac{c^2}{4}$$

and remembering that $c_s^2 = \frac{Nc^2}{3}$ (in the case of $2N+1$ velocity models the temperature is equal to c_s^2) one has

$$\theta = \frac{4N+3}{12}c^2,$$

then the requirement of constant temperature $\theta = \theta_0$ at the all levels of the hierarchy leads to the following formula for the lattice velocities

$$c = \sqrt{\frac{12}{4N+3}\theta_0}.$$

Finally, it is worth to mention that the error terms in the third moment for $2N+2$ hierarchy in comparison to the Maxwell distribution are the same as for $2N+1$ hierarchy.

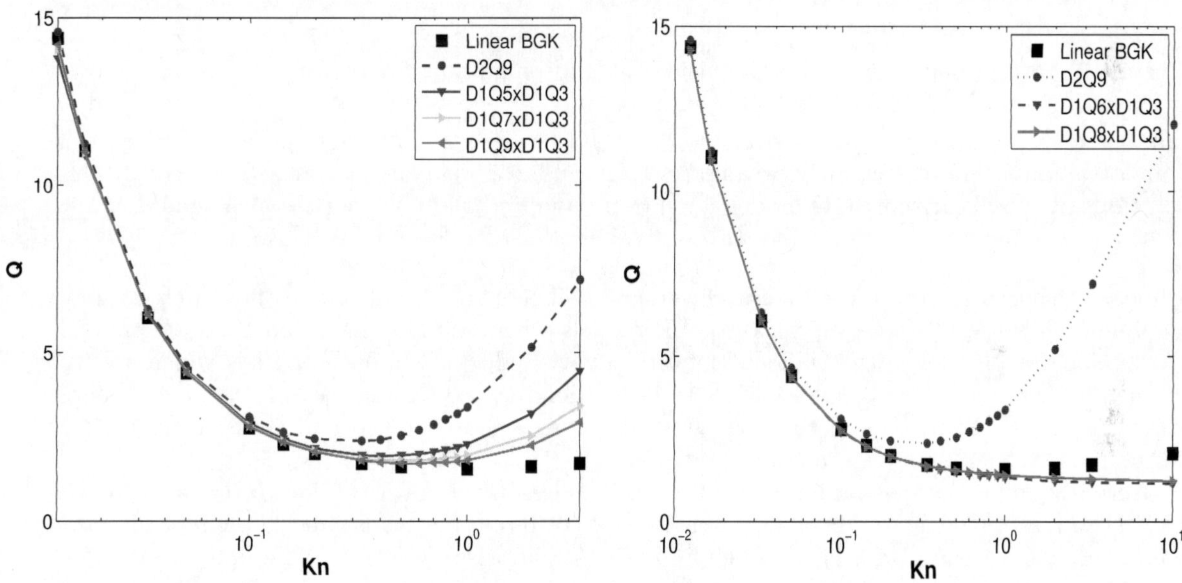

Figure 1. Poiseuille flow across flat walls and Knudsen paradox. Volumetric flow vs Knudsen number is presented. The left slide corresponds to the models from $2N+1$ hierarchy, right slide for the models from $2N+2$ hierarchy. The increase of N obviously leads to the convergence of the LB results to the solution for the linearized BGK model. The Knudsen number (Kn) is defined as $Kn = \sqrt{\frac{\pi T_0}{2}}\frac{\tau}{H}$, where T_0 is unit temperature, τ is the relaxation time and H the distance between the walls. The volumetric flow rate results for the linearized BGK solution are taken from [10]

3. Numerical validation

The first test problem is the Poiseuille flow between two parallel plates. The convergence to the theoretical results can be observed (Fig. 1). For instance, $D1Q7 \times D1Q3 = D2Q21$ model shows approximately 10 percent departure from the linearized BGK solution up to Kn=0.5; In ballistic regime runaway effects prevent the correct computation of the flux for $2N+1$ hierarchy

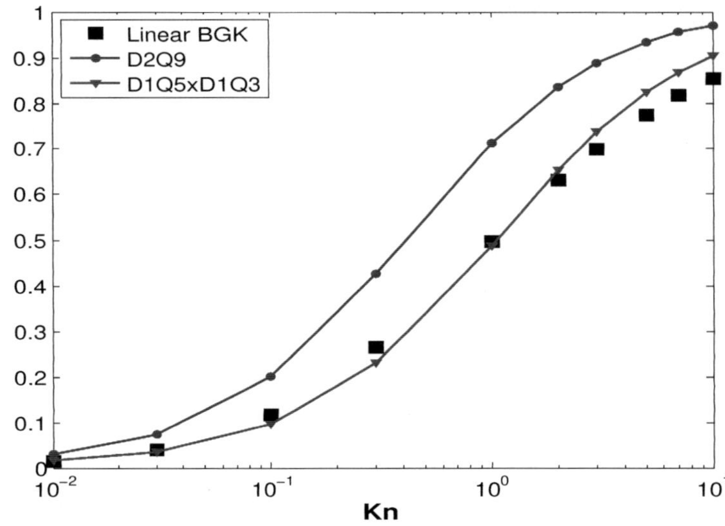

Figure 2. Plane Couette flow. Slip velocities vs Knudsen number are presented. The application of the model $D1Q5 \times D1Q3$ significantly improves the precision over $D2Q9$ model. The Knudsen number for Couette flow is defined as $Kn = \frac{\sqrt{2T_0}}{H}\tau$, where T_0, H, τ are the wall temperature equals to unity, the distance between the walls and the relaxation time respectively. The solutions of the linearized BGK model [11, 12] are adopted as benchmark (black squares).

while for the models from $2N+2$ hierarchy runaway effects are absent but the flow does not have minimum. This result is very typical for LB models with a single relaxation time: the models with velocities parallel to the wall suffer from runaway effects, while the models which are free of such lattice velocities do not reproduce the minimum at $Kn \sim 1$, see [13, 9]. Nevertheless, the accuracy of the models in the present study is very good for slip and transitional regimes which are most important in microfluidics (Fig. 1).

Here it is important to emphasize that for 2D models in the form $D1Q5 \times D1Q3$ and etc, the part $D1Q5, D1Q6, D1Q7$ and etc is responsible for the dynamics transverse to the flow direction while the component $D1Q3$ is responsible to the streamwise direction. The kinetic boundary conditions for high-order lattices are stated at the walls [14, 15]. For the Poiseuille flow the periodic boundary conditions with pressure variations [16] and the addition of pressure as a force term were tested. The both methods give very closed results. For the numerical study 200 spatial cross-stream nodes are taken, the boundary conditions are taken in a such form that the flow is very slow, $Ma \sim 10^{-4}$.

The last test case is the plane Couette flow between parallel walls. Again, the kinetic boundary boundary are adopted [14, 15]. The wall velocity is Ma^{-4}. The application of the model $D1Q5 \times D1Q3$ significantly improves the precision over $D2Q9$ model (Fig. 2).

The results regarding the Poiseuille and Couette flows can be compared with the results obtained using high-order and regularized LB models [13, 17, 18, 19], on-lattice and off-lattice models with correct half-fluxes [9, 20].

4. Conclusion

The new discretization approach for the kinetic BGK model is proposed. The method is based on the ideas of the LB method (linear streaming step, conservation of moments) and the discrete velocity method for the BGK model (the description of high moments with controllable error).

Two test case problems are considered (the plane Poiseuille flow, the plane Couette flow), the method shows better results in the description of the Knudsen paradox and the slip velocities than the conventional LB approach.

The presented method is restricted to slow isothermal flows, the further developments towards transonic and supersonic flows are interesting.

References

[1] Succi S 2018 *The Lattice Boltzmann Equation: For Complex States of Flowing Matter* (Oxford: OUP)
[2] Montessori A and Falcucci G 2018 *Lattice Boltzmann Modeling of Complex Flows for Engineering Applications* (Morgan Claypool Publishers)
[3] Shan X, Yuan X and Chen H 2006 *J. Fluid Mech.* **550** 413–441
[4] He X and Luo L 1997 *Phys. Rev. E* **55** R6333–R6336
[5] Shan X and He X 1998 *Phys. Rev. Lett.* **80** 65–67
[6] Mieussens L 2000 *Math. Models Meth. Appl. Sci.* **10** 1121–1149
[7] Karlin I and Asinari P 2010 *Physica A* **389** 1530–1548
[8] FToschi and Succi S 2005 *Europhys Lett.* **69** 549
[9] Feuchter C and Schleifenbaum W 2016 *Phys. Rev. E* **94** 013304
[10] Cercignani C, Lampis M and Lorenzani S 2004 *Phys. Fluids* **16** 3426
[11] Li W, Luo L and Shen J 2015 *Comput. Fluids* **111** 18
[12] Jiang S and Luo L 2016 *J. Comput. Phys.* **316** 416
[13] Montessori A, Prestininzi P, Rocca M and Succi S 2015 *Phys. Rev. E* **92** 043308
[14] Meng J and Zhang Y 2014 *J. Comput. Phys.* **0021-9991** **258** 601–612
[15] Ansumali S and Karlin I *Phys. Rev. E* **66** 026311
[16] Kim S and Pitsch H 2007 *Phys. Fluids* **19** 108101
[17] Zhang R, Shan X and Chen H *Phys. Rev. E* **74** 046703
[18] Kim S, Pitsch H and Boyd I 2008 *J. Comput. Phys.* **227** 8655
[19] Ansumali S, Karlin I, Arcidiacono S, Abbas A and Prasianakis N 2007 *Phys. Rev. Lett.* **98** 124502
[20] Ambrus V and Sofonea V 2014 *Phys. Rev. E* **89** 041301 (R)

IOP Conf. Series: Journal of Physics: Conf. Series **1163** (2019) 012031 doi:10.1088/1742-6596/1163/1/012031

HIGH RESOLUTION WAVES AND WEATHER FORECASTS USING ADAPTED WAVEWATCH III AND WRF MODELS

A Kuznetsova[1, 2], G Baydakov[1], D Sergeev[1, 2], Yu Troitskaya[1, 2]

[1] Institute of Applied Physics of the Russian Academy of Sciences, 46 Ulyanov st, 603950, Nizhny Novgorod, Russian Federation
[2]Nizhny Novgorod State University, 23 Gagarina av, 603950, Nizhny Novgorod, Russian Federation

alexandra@ipfran.ru

Abstract. High-resolution waves forecast for inland water body was proposed with the use of WAVEWATCH III wave model forced with wind data from WRF atmospheric model. Different surface layer parameterizations and planetary boundary layer parameterizations were tested within WRF. The results were compared with the results of the in-situ measurements.

1. Introduction

High resolution waves forecast is in demand for nearshore areas and for inland water bodies [1]. Nowadays wave models (e.g. WAVEWATCH III) are usually forced with wind field from reanalysis. The disadvantage of the reanalysis is its low resolution which is too big to be used for the considered conditions. To increase the wind input field resolution, the WRF atmospheric model can be used. In this paper, we propose a method of a wave forecast that uses the adapted to the conditions of middle-sized water body global wave model WAVEWATCH III (WW3) [2]. Initially the model was tuned to the conditions of the seas and oceans. The adaptation of WW3 to the conditions of the middle-sized water body and coastal area was discussed in our previous papers [3-5]. Then the atmospheric model WRF [6] is used as input data to force WW3. It takes into account changes in wind distribution and its heterogeneity over the water area so this setting of the wind provides a more precise prediction of the wave parameters in the reservoir.

Comparing to our previous work [7], where WRF v3.6 was used, new version of WRF model is implemented. WRF v.3.9.1.1 has a number of updates that influence the results of the modeling in the conditions of the area containing a water body. Among them there the most important are the use of new topographical MODIS data with an algorithmically added lake category, better quality of the input reanalysis data and its interpolation method. These solutions helped to make more accurate forecasts presented in this paper.

The PBL and surface layer parameterizations are tested within WRF v.3.9.1. The results are compared with in-situ measurements held by our group in the test days in 2016 and 2017.

2. Simulation

The study is applied to the area containing a middle-sized water body. Gorky Reservoir is considered as a real-case test area. Gorky Reservoir is an artificial lake in the central part of the Volga River formed by a hydroelectric dam of Gorky Hydroelectric Station. Its lake part extends for ~80 km, ~10 km wide. Our group has a huge amount of data from the field experiments carried out in the south part of the Gorky reservoir from the boat. In the course of the experiment, the profiles of wind speed and surface wave spectra were simultaneously measured. The technique of the field experiment is presented in [3]. Test days undertaken in 2016 and 2017 gave the data for the verification of the simulations.

IOP Conf. Series: Journal of Physics: Conf. Series **1163** (2019) 012031 doi:10.1088/1742-6596/1163/1/012031

Special attention is paid to the spatial resolution of the modeling because of the small size of the object of the study. The grid resolution used in WW3 wave model is 0.00833^0, thus it is expected that the wind forcing obtained from the reanalysis data is too coarse to create accurate forecasts. Today, the modern reanalysis datasets include the ECMWF reanalysis (European Center for Medium-Range Weather Forecasts, UK), NCEP/NCAR reanalysis (National Center for Environmental Prediction, National Center for Atmospheric Research, USA) and JRA-25 reanalysis (Japan Meteorological Agency, Japan). The minimal possible resolution of the reanalysis data is presented by NCEP Climate Forecast System Version 2 (CFSv2) and it is 0.205^0. Better spatial and temporal variability may be obtained by the use of the Weather Research and Forecasting (WRF) atmospheric model. It uses reanalysis data to compute fluxes and presses at different levels and calculates many parameters including 10 m wind speed and its direction. Thus the high resolution wave forecast can be obtained with WRF wind forcing. In this paper, we compare the wave forecasts forced by CFSv2 reanalysis data with $0{,}205^0$ resolution and by wind field from WRF output.

Each step of the WRF model run is performed by a special module. First, the preprocessing of the data to prepare input to the real program for real-data simulations is realized. The geogrid module interpolates the geographical data for the lakes description «modis_lakes» on 4 nested domains in study area. The minimum cell size of the fourth nested domain is 0.00833^0, and the same cell size of the topographical data is used in WW3. Then ungrib module unpacks the loaded meteorological data. CFSv2 with 1^0 resolution is used as the initial meteorological data. Metgrid module produces a horizontal interpolation of the extracted meteorological data on the domains grid. Then the simulations of real.exe module and wrf.exe module are held in MPI option on the IAP RAS cluster. Different Planetary Boundary Layer (PBL) and surface layer parameterizations are tested in WRF by changing namelist.input for the real.exe and wrf.exe modules, and the chosen option is used for the WRF simulation. The output wind parameters (u10 and v10) are used to force WW3 simulations.

WW3 run is performed by module ww3_grid with GLOBE bathymetry data (0.00833^0 resolution), ww3_strt with idealized initial conditions, ww3_prnc for the preprocessing of wind forcing (WRF output wind field parameters or reanalysis parameters), ww3_shel for the simulations held in MPI option on the IAP RAS cluster, and ww3_ounf to get the output wave parameters.

3. Results of the simulation and discussions

Some packages for the calculation of the atmosphere parameters within WRF are intercompared. Among them there are option 1 Yonsei University Scheme (YSU) [8], option 2 Mellor–Yamada–Janjic Scheme (MYJ) [9], option 5 Mellor–Yamada Nakanishi Niino (MYNN) Level 2.5 Scheme [10] and Large Eddy Simulation (LES) option [11, 12].

The LES option let the first three domains to be calculated within the Yonsei University scheme (YSU) for the PBL and within the surface layer scheme based on the Monin-Obukhov similarity theory, taking into account the viscous Carlson-Boland sublayer form. Simulation of the wind speed in the fourth domain were performed with the PBL parameterization turned off, but with taking into account turbulent flows (LES).

The comparison of the behavior of the 10 m wind speed and direction calculated with different parameterizations is shown in the Fig.1. The results are obtained in the simulation for the test day 08.08.17. The comparison shows that the use of the considered WRF parameterizations shows better agreement with the experiment than the direct use of the reanalysis. Among the tested parameterizations, LES package has good both temporal and spatial variability and is chosen for the further simulations.

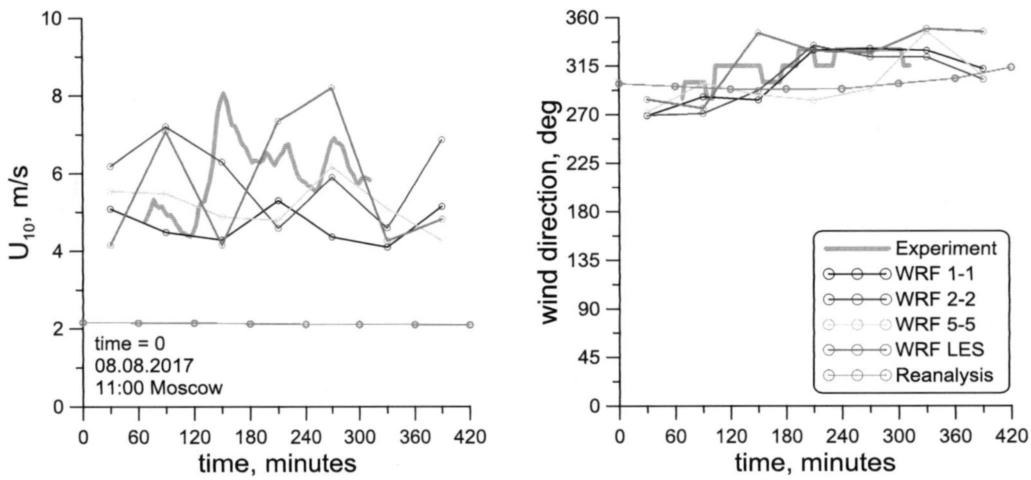

Figure 1. WRF simulation of wind speed (left) and direction (right) with option 1_1 – MM5 similarity, 2_2 – Eta similarity, 5_5 – MYNN for Nakanishi & Niino, option of WRF LES in comparison with the in-situ measurements in the test day 08.08.2017.

Figure 2 shows trial weather forecasts of the 10 m wind speed distribution obtained within the new version of the WRF v.3.9.1 model. The new version of the model allowed to achieve better spatial and temporal variability comparing with the work [7] both in wind speed and direction. An increase in wind speed over the water surface by a factor of 2-3 is shown comparing with the wind speed over the land. This result favorably emphasizes the use of wind forcing from the atmospheric model instead of the reanalysis data, which do not have spatial variability that is sufficient for the conditions of inland waters and the coastal zone.

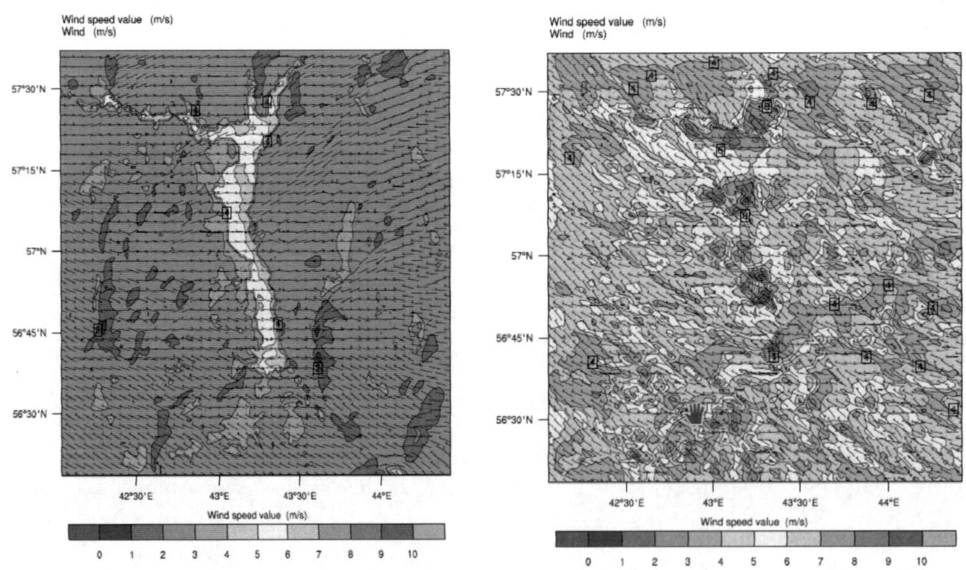

Figure 2. The distribution of the wind over the water area of the Gorky reservoir in the test day 08.08.17: night time (left) and day time (right).

The key result is the implementation of wind forcing from the atmospheric model WRF to the wave model WW3. Taking into account the accurate forecast of the state of the atmosphere in the framework of the modern atmospheric model WRF allowed us to obtain qualitative data for calculating the waves parameters, taking into account the heterogeneity of the wind field and the topography of the terrain.

The verification of the wind field calculated in WRF and of then mean wave parameters calculated in the adapted WW3 with WRF forcing is made on the basis of the experimental data. Simulated significant wave heights and mean wave periods are compared with the measured ones. Both in the model and in the processing of the experimental results, the calculation of H_S is performed by the formula:

$$H_S = 4\sqrt{E}\,. \tag{1}$$

The calculation of mean wave period T_m is performed by the formula:

$$T_m = T_{m0,-1} = \left(\int_{f_{\min}}^{f_r} E(f)df \right)^{-1} \int_{f_{\min}}^{f_r} E(f)f^{-1}df\,. \tag{2}$$

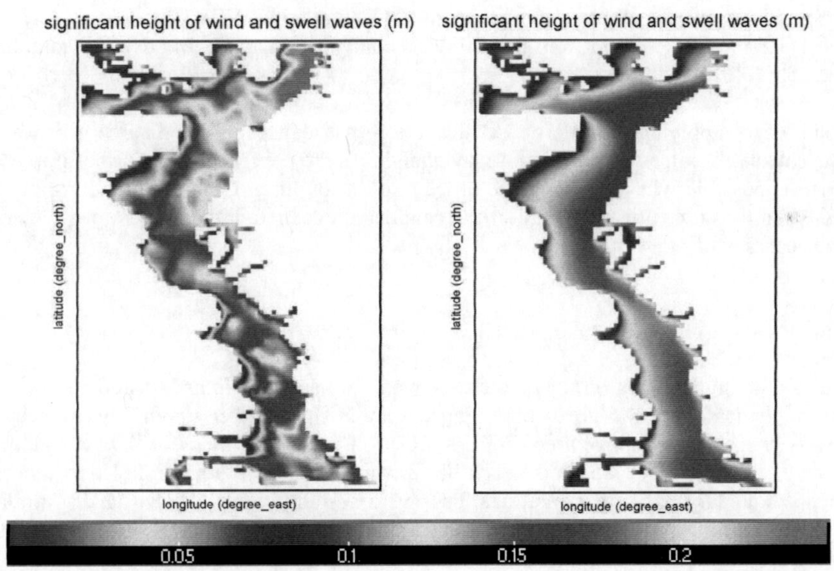

Figure 3. The distribution of the significant wave height on the Gorky reservoir in the test day 08.08.17 a) WRF LES wind forcing, b) CFSv2 reanalysis wind forcing.

Then all simulated data is obtained at the point corresponding to the point of observations.

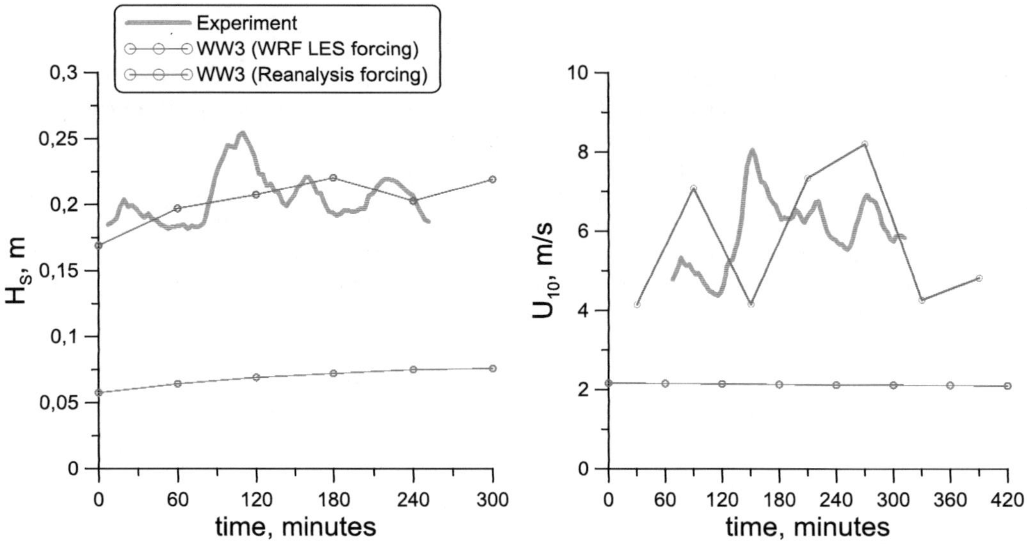

Figure 4. Significant wave heights (left) calculated by the adapted WW3 model using wind forcing (right) from WRF model (red) and from the CFSv2 reanalysis (magenta) and experimental data (grey) for a test day.

As a result of the applying of the proposed wave forecast technique, good agreement was obtained between the calculated values of significant wave heights and the experiment. Further improvement of the technique is possible when applying the procedure of coupling. Coupled modeling of WW3 and WRF is based on the automatic exchange of the calculated ocean and atmosphere parameters and can improve the accuracy of the simulations.

4. Conclusion

Technique of a high resolution wave forecast was proposed for inland water bodies. The WRF model was implemented in the area containing a middle-sized water body. The area with Gorky Reservoir was chosen as a control area. WRF simulation was performed for 4 nested domains with min cell size of 1 km. Initial reanalysis was CFSv2 winds. Different surface layer parameterizations and planetary boundary layer parameterizations were tested within WRF: MM5 similarity, Eta similarity, MYNN for Nakanishi & Niino PBL parameterization, and Large Eddy Simulation case. The results were compared with the results of the in-situ measurements held by our group. LES package was chosen for the further simulations.

The WRF model wind calculation results were used as wind forcing of the WW3 wave model. A comparison of the calculated mean wave parameters with the ones measured in the experiment at the Gorky Reservoir was made.

This investigation will be continued with the attempt of coupled modeling based on the automatic exchange of the calculated ocean and atmosphere parameters.

References

[1] Janssen P A E M, Bidlot J - R, // Procedia IUTAM Symposium Wind Waves, 4-8 September 2017, London, UK, pp 14 - 29.

[2] Tolman H and WAVEWATCH III Development Group. User manual and system documentation of WAVEWATCH III version 4.18. //Environmental Modeling Center,

Marine Modeling and Analysis Branch. 282 pp. + Appendices. 2014

[3] Kuznetsova A M, Baydakov G A, Papko V V, Kandaurov A A, Vdovin M I, Sergeev D A, Troitskaya Yu I, // Geography, environment, sustainability, Vol. 9, № 2, 2016, DOI: 10.15356/2071-9388_02v09_2016_02

[4] Kuznetsova A M, Baydakov G A, Papko V V, Kandaurov A A, Vdovin M I, Sergeev D A, Troitskaya Yu I, // Russian Meteorology and Hydrology, 2016, №2, c.85-97

[5] Kuznetsova A M, Baydakov G A, Papko V V, Kandaurov A A, Vdovin M I, Sergeev D A, Troitskaya Yu I, // Hindawi Publishing Corporation, Advances in Meteorology, 2016, vol. 1, article ID 574602, pp. 1-13

[6] Skamarock W C, Klemp J B, Dudhia J, Gill D O, Barker D M, Duda M G, Huang X-Y, Wang W, and Powers J G, 2008: A Description of the Advanced Research WRF Version 3. *NCAR Tech. Note NCAR/TN-475+STR,* 113 pp

[7] Kuznetsova A, Baydakov G, Sergeev D, Troitskaya Yu, // Journal of Physics: Conference Series, Volume 955 (2018), 012014, 2018, doi :10.1088/1742-6596/955/1/012014.

[8] Hong Song–You, Yign Noh, Dudhia J, // Mon. Wea. Rev., 2006, **134**, 2318–2341.

[9] Janic, Zavisa I., // *Mon. Wea. Rev.*, 1994, **122**, 927 - 945.

[10] Nakanishi M. and Niino H., // *Bound. Layer Meteor.* 2006, **119**, 397–407

[11] Moeng C-H, Dudhia J, Klemp J B, and Sullivan P, // Mon. Wea. Rev., 2007, **135**, 2295–2311.

[12] Mirocha J D, Lundquist J K and Kosović B, // Mon. Wea. Rev., 2010, 138, 4212–4228

Acknowledgments

An optimal configuration of the WRF model was created with the support of the RFBR grant 18-35-00602 (mol_a). Field measurements at the Gorky reservoir were supported by a grant from the Russian Science Foundation (Agreement No. 15-17-20009). The simulations of surface waves within the adapted WAVEWATCH III model was partially supported by the RFBR grant 17-05-41117.

Construction of computing balancing model in the Internet of Things devices system

Aleksandr Sidorov, Sergey Slastnikov

National Research Institute High School of Economics, Moscow, Russia

E-mails: asidorov@hse.ru, sslastnikov@hse.ru

Abstract: This article is dedicated to our work in field of research and development of a math model for load balancing in the Internet of things (IoT). Here, we perform analysis and classification of tasks in the IoT devices system. We split subprograms in following systems on balancing and non-balancing. Next, we classify balancing subprograms by different parameters. Then, based on those classifications, we construct model for load balancing in IoT. With that model, in future we can perform modeling IoT devices systems and found pros and contras of each balancing algorithm.

1. Introduction

At this work, we start to research development and construction of a math model for load balancing in the Internet of Things devices system. Suggested model is useful to perform experiments of modeling the parameters and behavior of the IoT devices system with different scripts of data loading. The main goals of current work are to reduce delays in the performance of tasks, to increase battery life and energy efficiency. The objectives are to describe tasks performed in IoT systems, to summarize definitions of tasks balancing methods, to develop a math model in the IoT devices system and to design various balancing methods for this model.

At first, we decide to perform analysis and classification of tasks for the IoT devices system. Based on this classification, we assume that not each kind of tasks (and associated computations) could be reassigned to another node in system.

At second part, we compare existing methods of computing balancing for IoT systems. We classify methods by its centralization, scaling and flexibility. Here we introduce our balancing method based on the computation offloading at heterogeneous systems with different levels of performance. This method allows moving computation to any performance level node. There are some works of other authors [1] [2], describing similar methods, but mostly concentrated on transfer tasks to higher performance level (e.g. cloud center).

At the end, we develop the mathematical model of the Internet of Things devices system. It is particularly based on the model proposed in work by Elizarov M. [3]. Model extended for balancing processes and node health check.

In advance, results of this work could be used for modeling and comparative analysis of balancing methods. Further, we will compare obtained results with other results, based on experiments on the virtual and physical system of devices to determine the quality of the constructed model.

2. Classification of calculations

At the beginning, we classify calculations, performed in considered systems and its balancing methods. Under term "calculation", we mean processing methods (as Object-Oriented Programing term) and subprograms. In real usage every subprogram in Internet of Things devices system unites in computation chain with others. This chain is called – task.

Our main goal in this section is assign calculations in two groups – balancing and non-balancing, and find their distinguish parameters.

At first, we will describe some rules that are necessary for computing balancing.

1. Runtime environment equality. Subprogram may be balanced between nodes *A* and *B* if subprogram is executable on these nodes (hardware and software).
2. Execution equality. Subprogram may be balanced between nodes *A* and *B* if result of execution (output parameters) will be equal at both nodes.
3. Data availability. Subprogram may be balanced between nodes *A* and *B*, working with data on node *C* if from both nodes *A* and *B* subprogram could access (read or write) to the node *C* data.

Consider subprogram assignment by following parameters:

- function;
- connectivity.

Each program in task performs specific use, for example:

- collecting data from sensors;
- transmitting control signals to actuators (for example, electric drives);
- data's zip and unzip;
- nodes sync and etc.

We can develop uncountable set of subprograms with different usage. In addition, one subprogram can unite usages of two different subprograms, for example, message-sending subprogram can zip transmitted data.

However, we can select type of local subprogram. These local subprograms use local physical node's parameters as input or output parameters (reading sensors, transmitter, self-diagnose subprogram, etc.). Obvious that these types of subprograms couldn't be balance, because results on different nodes differ or their execution fail.

Subprogram connectivity describes by intensiveness of data exchange between neighbor subprograms in task. The more intensiveness of data exchange goes to the more connectivity between tasks. Obvious, that subprogram with high-level intensiveness should be performed on one node. It could reduce network load. At the following works, we are going to find connectivity level threshold, after that we should execute them on one node.

Therefore, we can select subprograms with high-level connectivity and local subprograms as non-balanced subprograms. Non-local subprograms with low-level connectivity may be rebalanced when higher described rules are respected. Balancing is allowed if it leads to improvement of system's quality parameters – decrease net load, summary task execution time, nodes load, etc.

Although calculations perform in method or subprogram, we can divide subprograms on two types:

- movable;
- fixed.

Movable subprograms can move their source code between nodes in net and execute on other system's nodes. These subprograms does not require their copy on the receiver node and can copy their source code to receiver node while performing balancing. However, this type may have difficulties with capability between source code and hardware on receiving node, which cause problems in heterogeneous systems for compiled subprograms. Solution for this problem is in using interpreted languages or virtual machines but these cause higher load on memory and CPU and may increase task's execution time.

Fixed subprograms can't move their source code. For this type of subprograms calls redirection between nodes performs while calculations balancing. It reduces balancing time because we don't need to copy subprogram source. These subprograms are preinstalled on nodes, so they can be optimized for node's hardware, improving CPU load and performance. Lack of fixed subprograms is that calculations can be moved only to the node where this or similar subprogram is installed. Also, node store every subprogram even if its unused, which cause higher memory use.

3. Classification of balancing methods

Different algorithms can balance subprograms. For cloud computing great taxonomy of resource allocation is defined in [4]. We accept this taxonomy for Internet of Things systems:

- By balancing order - we can split methods on preventive, on detect and mixed. Preventive methods perform tasks distribution once at their start moment. "On detect" algorithm performs tasks redistribution on detection of disbalance or overload. Mixed method uses both sides, and as result, it requires more resources than both.
- By search area - we can split methods on global and local. Global balancing methods are trying to find optimal solution using all nodes in the system. Local methods try to improve current system's configuration using limited count of surrounded nodes to the origin. Obvious that local methods take less time for perform balancing. On the other side, in general local search does not get adorable results.
- By relocation count - we split methods on single and group. Single relocation methods are aimed to reduce specific node's load. They start after critical overload on one of the optimized parameters. Group methods try to achieve optimal load between nodes. They trigger when occurs critical difference between minimum and maximum nodes' load in system.

Also, we should highlight static tasks allocation (STA). STA appears when developer or administrator distributes tasks to nodes manually while adding. It's effectivity depends on developer skills, but it is much more complicated with huge task amount. Obvious, that STA doesn't use background tasks for load control and rebalance, so higher effectivity of load using is achieved.

This classification doesn't define how algorithms decide which subprograms are going to balance. For example, in [5] and [6] authors used fuzzy logic. Various decision algorithms achieve different effectiveness. In future work, we will compare different decision algorithms such a fuzzy logic, least connected, most CPU usage etc.

In Table 1 we define algorithms for each combination of parameters. STA is defined as a_0.

Table 1. Balancing methods classification

		Preventive		**On detect**		**Mixed**	
		Single	**Group**	**Single**	**Group**	**Single**	**Group**
Global	**Movable**	a_1	a_2	a_3	a_4	a_5	a_6
	Fixed	a_7	a_8	a_9	a_{10}	a_{11}	a_{12}
Local	**Movable**	a_{13}	a_{14}	a_{15}	a_{16}	a_{17}	a_{18}
	Fixed	a_{19}	a_{20}	a_{21}	a_{22}	a_{23}	a_{24}

4. Model of IoT system's for calculations' balancing

Here we describe model for IoT devices system:

1. Each subprogram is described as {memory usage, operations count, input parameters, output parameters, trigger event, execution node}
$$SubP = \{M, Op, In,\ Out,\ trigger, node\}$$
2. Each task is ordered set of subprograms
$$Task = \{SubP_1 \dots SubP_n\}$$
3. Subprograms are connected through output parameters and input parameters or trigger event

IOP Conf. Series: Journal of Physics: Conf. Series **1163** (2019) 012032 doi:10.1088/1742-6596/1163/1/012032

$$SubP_n.Out \rightarrow \{SubP_{n+1}.In; SubP_{n+1}.trigger\}$$

4. Nodes performance is count of operations per time and memory value
$$Node.p = \{OPS, M\}$$
5. Balancer – special task that can move subprograms among the nodes
$$TaskBalancer = SubP.node\{v_1 \rightarrow v_2\}$$
6. Network interconnection is based on model by M. Elizarov [3].
7. Balancer execution time consist of:
   a. $t_{discover}$ – time, that balancer needs to find disbalance from time it appears;
   b. t_{search}– time, that balancer needs to find receiver node;
   c. t_{move}– time needed for copy source codes of subprogram;
   d. $t_{reconnect}$– time needed for restore connections with relocated subprogram;
8. System's state is scored by following parameters:
   a. By nodes' load:
      i. On CPU:
         1. CPU_load_{Me} – median CPU load of node;
         2. CPU_load_{max} – maximum CPU load of node;
      ii. On memory:
         1. MEM_load_{Me} – median memory load of node;
         2. MEM_load_{max} – maximum memory load of node;
      iii. On network:
         1. Net_load_{Me} – median network load;
         2. Net_load_{max}– maximum network load.
   b. By tasks' execution time:
      i. t_diff_{Me} – median difference between real time of tasks' execution time and reference execution time;
      ii. t_diff_{max} – maximum difference between real time of tasks' execution time and reference execution time.

Reference execution time – task's execution time in the system without other tasks (as it was only one executed in the system).

At first stage of modelling we suggest that nodes have stable connections with each other and available bandwidth is much higher than actual data rates between nodes, i.e. network errors' influence is minimal.

5. Summary

In this article we perform classification of subprograms in Internet of Things to select computations that available for balancing. This classification doesn't performed yet.

We select parameters of calculation balancing methods. These parameters are close to taxonomy for cloud computing balancing. By these parameters we can find balancer execution time easier because of their classification.

We describe model for IoT devices systems computing balancing. We decide to use median scoring instead of average as more objective score.

In following work, we continue to research balancing methods. We will find formulas for execution time of computing balancing, create configurations for modeling systems, analyse results and create experiment with virtual machines to estimate model.

IOP Conf. Series: Journal of Physics: Conf. Series **1163** (2019) 012032 doi:10.1088/1742-6596/1163/1/012032

References

1. Flores H., Xiang S., Kostakos V., Yi Ding A., Nurmi P., Tarkoma S., Hui P., Li Y. 2017 IEEE International Conference on Pervasive Computing and Communications Workshops (PerCom Workshops) // Large-scale offloading in the Internet of Things. Kona, HI, USA. 2017.

2. Samie F., Tsoutsouras V., Bauer L., Xydis S., Soudris D., Henkel J. 2016 IEEE 3rd World Forum on Internet of Things (WF-IoT) Reston, VA, USA. 2016. pp. Computation offloading and resource allocation for low-power IoT edge devices.

3. Elizarov M. MODELS AND ALGORITHMS OF INFORMATION INTERACTION IN THE INTERNET OF THINGS NETWORKS. Saint-Petersburg: 2017.

4. Mustafa S., Nazir B., Hayat A., Khan A.U.R., Madani S.A. A survey and taxonomy on energy efficient resource allocation techniques for cloud computing systems // Comput. Electr. Eng. Oct 2015. pp. 186–203.

5. Hatti D.I., Sutagundar A.V. International Conference on Recent Advances in Electronics and Communication Technology // Agent Based Job Classification and Resource Allocation in IoT. 2017.

6. Sutagundar A.V., Manvi S.S. The Third Advanced International Conference on Telecommunications // Agent Based Approach to Information Fusion in Wireless Sensor Networks. 2017.

A simple description of turbulent transport in a stratified shear flow devoted to the simulation of thermohydrodinamics of inland waters

D S Gladskikh [1, 2], I A Soustova [1], Yu I Troitskaya [1], S S Zilitinkevich [3, 4], D S Sergeev [1]

[1] Institute of Applied Physics of the Russian Academy of Sciences, Nizhny Novgorod, Russia
[2] Nizhny Novgorod State Technical University n.a. R.E. Alekseev, Russia
[3] Finnish Meteorological Institute, Helsinki, Finland
[4] Nizhny Novgorod State University n.a. N.I. Lobachevsky, Russia

daria.gladskikh@gmail.com

Abstract. The paper presents three major models used to describe the thermohydrodynamics of inland water objects: the standard one-dimensional E-ε (k-ε) model, the EFB-model by S. S. Zilitinkevich, and RANS-type model of turbulent transfer by L. O. Ostrovsky and Yu. I. Troitskaya. For models that take into account the two-sided transformation of the kinetic and potential energies of turbulent pulsations, the dependences of the turbulent Prandtl number on the gradient Richardson number are provided. The obtained dependences were used in the E-ε model to parameterize the coefficient of heat transfer eddy diffusivity in order to take into account stratification when calculating the thermohydrodynamic regimes of inland waters. The results of verification of the modified E-ε model based on experimental data are presented.

1. Introduction

By now, the role of inland water bodies in the processes of general circulation of the atmosphere, the ocean and land, their thermohydrodynamic and biological properties in weather forecasts, and also in climate change has been established [1-6]. The basis of physico-mathematical simulation of thermodynamics and hydrodynamics of inland waters is RANS (Reynolds-averaged Navier-Stokes equations) [7]. Since this system contains unknown quantities (turbulent flows), it is necessary to involve additional hypotheses for its closure. Practically significant is the so-called E-ε (k-ε) model based on the equations for the kinetic energy of turbulence E and its dissipation rate ε. The ratio of the coefficient of momentum eddy diffusivity to the coefficient of heat transfer eddy diffusivity (the turbulent Prandtl number, *PrT*) is assumed to be constant. This, in particular, limits the description of the known effect [8-9] associated with the existence of turbulence for large gradient Richardson numbers *Ri*. At the same time, both laboratory and field measurements demonstrate the dependence of *PrT* on *Ri*. In this connection, special attention in the simulation of the thermohydrodynamic regime of inland water bodies is paid to stratification and, in particular, to its role in the processes of turbulent mixing, thermocline dynamics, etc. Recently, modernized approaches to the description of geophysical turbulence are actively developing, taking into account stable and unstable stratification, as well as internal waves interacting with turbulence and even amplifying it.

Authors propose geophysical turbulence models that take into account the two-sided transformation of the kinetic and potential energies of turbulent pulsations. This is the so-called EFB model [10] and the closed Reynolds system of equations (RANS) [11] obtained on the basis of the sequential kinetic approach. These models allowed, in particular, to explain the case of maintaining turbulence observed in laboratory and field conditions by weak velocity shears, as well as to calculate the dependence of *PrT* on *Ri*. In the present work, a parameterization of the coefficient of heat transfer eddy diffusivity for the E-ε model is proposed in order to take into account stratification when calculating the thermohydrodynamic regimes of inland waters. Verification of the proposed parameterization is presented.

2. E-ε (k- ε) model for the description of thermohydrodynamics of inland waters

In this paper, the model is used in the form proposed in [12]:

$$\frac{\partial E}{\partial t} = \frac{\alpha_E}{h^2}\frac{\partial}{\partial \xi}k\frac{\partial E}{\partial \xi} + \frac{\xi}{h}\frac{dh}{dt}\frac{\partial E}{\partial \xi} + P - \varepsilon, \tag{2.1}$$

$$\frac{\partial \varepsilon}{\partial t} = \frac{\alpha_\varepsilon}{h^2}\frac{\partial}{\partial \xi}k\frac{\partial \varepsilon}{\partial \xi} + \frac{\xi}{h}\frac{dh}{dt}\frac{\partial \varepsilon}{\partial \xi} + C_1\frac{\varepsilon}{E}(P - \varepsilon), \tag{2.2}$$

where a_E and a_ε are dimensionless constants, P is total generation of turbulence energy through the shear of velocity and effect of density stratification, C_l is the function of Reynolds number. The beginning of the vertical downward z-coordinate is supposed to be coupled with water surface $z=h(t)$, so we can turn from z to the new independent variable $\xi = \frac{z}{h}$, where $h=h(t)$ is a depth of water body, and t is a time.

The coefficient of momentum eddy diffusivity is as follows:

$$k = C_e\frac{E^2}{\varepsilon},$$

where C_e is dimensionless constant.

Vertical heat exchange, expressed as the coefficient of heat transfer eddy diffusivity λ, is defined as:

$$\lambda = c\rho k. \tag{2.3}$$

It is used in the calculations of the temperature distribution (see below) and the parameter P in equations (2.1-2.2)

The numerical model for calculating the temperature distribution is based on solving a one-dimensional heat equation:

$$c\rho\frac{\partial T}{\partial t} = \frac{1}{h^2}\frac{\partial}{\partial \xi}\left(\lambda\frac{\partial T}{\partial \xi}\right) + c\rho\frac{dh}{dt}\frac{\xi}{h}\frac{\partial T}{\partial \xi} - c\rho\frac{1}{h}B_w\frac{\partial T}{\partial \xi} - \frac{1}{h}\frac{\partial S}{\partial \xi},$$

where c is heat capacity of the water, ρ – its density, λ – the coefficient of heat transfer eddy diffusivity, T – temperature, $B_w = dh_0/dt$ – water balance at free surface, S – solar radiation flux penetrating the water body.

As for the water object dynamics, in [12] the zonal u and the meridional v components of the flow velocity used to compute the parameter P and the gradient Richardson number Ri in our study, are found using the following equations:

$$\frac{\partial u}{\partial t} = \frac{1}{h^2}\frac{\partial}{\partial \xi}k\frac{\partial u}{\partial \xi} + \frac{\xi}{h}\frac{dh}{dt}\frac{\partial u}{\partial \xi} - fv,$$

$$\frac{\partial v}{\partial t} = \frac{1}{h^2}\frac{\partial}{\partial \xi}k\frac{\partial v}{\partial \xi} + \frac{\xi}{h}\frac{dh}{dt}\frac{\partial v}{\partial \xi} + fu,$$

where f is the Coriolis frequency. Components of friction stress are used to calculate the boundary conditions for u and v according to the Monin-Obukhov similarity theory at the upper boundary (water – atmosphere) and according to the Chezi formulas [13] taking into account the surface roughness at the lower boundary (water – soil).

3. EFB-model and parameterization of turbulent Prandtl number

This model, presented in [9-10], is based on the balance equations for the kinetic energy of turbulence E_K, potential energy E_P, turbulent fluxes of momentum τ_{i3} and potential temperature F_z, as well as on the relaxation equation for the turbulent time scale t_T:

$$\frac{DE_K}{Dt} - \frac{\partial}{\partial z} K_E \frac{\partial E_K}{\partial z} = -\tau_{i3} \frac{\partial U_i}{\partial z} + \beta F_z - \frac{E_K}{t_T}, \tag{3.1}$$

$$\frac{DE_P}{Dt} - \frac{\partial}{\partial z} K_E \frac{\partial E_P}{\partial z} = -\beta F_z - \frac{E_P}{C_P t_T}, \tag{3.2}$$

$$\frac{D\tau_{i3}}{Dt} - \frac{\partial}{\partial z} K_{FM} \frac{\partial \tau_{i3}}{\partial z} = -2E_z \frac{\partial U_i}{\partial z} - \frac{\tau_{i3}}{C_\tau t_T} \quad (i = 1, 2), \tag{3.3}$$

$$\frac{DF_z}{Dt} - \frac{\partial}{\partial z} K_{FH} \frac{\partial F_z}{\partial z} = -2(E_z - 2C_\theta E_P) \frac{\partial \Theta}{\partial z} - \frac{F_z}{C_F t_T}, \tag{3.4}$$

$$\frac{Dt_T}{Dt} - \frac{\partial}{\partial z} K_T \frac{\partial t_T}{\partial z} = -C_R \left(\frac{t_T}{t_{TE}} - 1 \right). \tag{3.5}$$

Here D/Dt is the substantial derivative, K_E, K_{FM}, K_{FH}, K_T are the coefficients of turbulent transport, β is the buoyancy parameter, E_z is the vertical component of the kinetic energy, C_P, $C\tau$, C_θ, C_F, C_R are constants, Θ is the average potential temperature, t_{TE} is the equilibrium relaxation time.

It should be noted that this model is rather difficult for numerical simulation, but it allows to calculate various characteristics of turbulence in the conditions from neutral stratification to extremely stable one. The model of gradient transfer under the condition of equilibrium versions of balance equations for turbulent fluxes of momentum and potential temperature follows from the system (3.1-3.5). In this case:

$$F_z = -K_H \frac{\partial \theta}{\partial z}, \qquad \tau_{i3} = -K_M \frac{\partial U_i}{\partial z}, \tag{3.6}$$

$$K_H = 2C_F E_z t_T \left(1 - C_\theta \frac{E_P}{E_z} \right), \qquad K_M = 2C_\tau E_z t_T, \tag{3.7}$$

K_H is the coefficient of heat transfer eddy diffusivity (is equivalent to λ), K_M is the coefficient of momentum eddy diffusivity (is equivalent to k), $E_z = A_z E_K$, $A_z = A_z(\Pi)$, $\Pi = {}^{E_P}/_{E_K}$.

Let us consider the following simple task of the evolution of homogeneous turbulence in a stationary shear flow with a given velocity profile $\vec{u} = U_0(z)\vec{x}$, where $\partial U_0/\partial z = const$. Shear flow is stably stratified and with a constant Brunt–Väisälä frequency N: $N^2 = \beta\, \partial\Theta/\partial z$. The turbulent regime of such a fluid will be described by a system of equations of the first order in time:

$$\frac{DE_K}{Dt} = -\tau_{13} \frac{\partial U_1}{\partial z} + \beta F_z - \frac{E_K}{t_T}, \qquad \frac{DE_P}{Dt} = -\beta F_z - \frac{E_P}{C_P t_T}, \tag{3.8}$$

The system has two equilibrium states: an unstable one $E_P = E_K = 0$ and stable one with a non-zero value of the energy parameter of the stratification $\Pi = E_P/E_K$, defined by the relation:

$$\Pi(Ri) = \frac{C_P \frac{C_F}{C_\tau} \left(1 - C_\theta \frac{\Pi(Ri)}{A_z(\Pi)} \right) Ri}{1 - \frac{C_F}{C_\tau} \left(1 - C_\theta \frac{\Pi(Ri)}{A_z(\Pi)} \right) Ri}, A_z(\Pi) = \frac{\Pi_\infty \left(C_r - 3\frac{\Pi}{C_P} \right)(C_P + \Pi) - 2C_r C_0 (C_P + \Pi_\infty)\Pi}{3\Pi_\infty (1 + C_r)(C_P + \Pi) - 2C_r (1 + C_0)(C_P + \Pi_\infty)\Pi}, \tag{3.9}$$

where Π_∞ is the limiting value of the parameter equal to 0.287.

So, from the definition of the turbulent Prandtl number and (3.7–3.9) we get:

$$PrT(Ri) = \frac{C_P\, Ri \left(1 + \frac{\Pi(Ri)}{C_p}\right)}{\Pi(Ri)}. \tag{3.10}$$

IOP Conf. Series: Journal of Physics: Conf. Series **1163** (2019) 012033 doi:10.1088/1742-6596/1163/1/012033

In Fig. 1 (see below) the dependence of the turbulent Prandtl number on the gradient Richardson number is presented. You can see that *PrT(Ri)* is an unboundedly increasing function, and for large $Ri \gg 1$, the Prandtl number grows linearly.

4. Turbulent transfer model by L. Ostrovsky and Yu. Troitskaya

On the basis of the kinetic approach, the expressions for turbulent fluxes and average values of the corresponding hydrodynamic quantities were obtained in [11]. This allowed us to close the RANS equations:

$$\frac{\partial\langle u_i\rangle}{\partial t}+\langle u_j\rangle\frac{\partial\langle u_i\rangle}{\partial x_j}+\frac{1}{\rho_0}\frac{\partial\langle\rho\rangle}{\partial x_i}+g_i\frac{\langle\rho\rangle-\rho_0}{\rho_0}=\frac{\partial}{\partial x_j}\left(L\sqrt{b}\left(\frac{\partial\langle u_i\rangle}{\partial x_j}+\frac{\partial\langle u_j\rangle}{\partial x_i}\right)\right), \tag{4.1}$$

$$\frac{\partial\langle\rho\rangle}{\partial t}+\langle u_i\rangle\frac{\partial\langle\rho\rangle}{\partial x_i}=2\frac{\partial}{\partial x_i}L\sqrt{b}\left(\frac{\partial\langle\rho\rangle}{\partial x_i}+\frac{3}{2b\rho_0}(g_i\langle\rho'^2\rangle+g\beta_i)\right), \tag{4.2}$$

$$\frac{\partial b}{\partial t}+\langle u_i\rangle\frac{\partial b}{\partial x_i}-L\sqrt{b}\left(\frac{\partial\langle u_i\rangle}{\partial x_j}+\frac{\partial\langle u_j\rangle}{\partial x_i}\right)^2-\frac{g}{\rho_0}L\sqrt{b}\times\left(\frac{\partial\langle\rho\rangle}{\partial z}+\frac{3}{2b\rho_0}(g\langle\rho'^2\rangle+g\beta_z)\right)-\frac{Cb^{3/2}}{L}=\frac{5}{3}\frac{\partial}{\partial x_i}\left(L\sqrt{b}\frac{\partial b}{\partial x_i}\right), \tag{4.3}$$

$$\frac{\partial\langle\rho'^2\rangle}{\partial t}+\langle u_i\rangle\frac{\partial\langle\rho'^2\rangle}{\partial x_i}-2\frac{\partial\langle\rho\rangle}{\partial x_i}L\sqrt{b}\times\left(\frac{\partial\langle\rho\rangle}{\partial x_i}+(g_i\langle\rho'^2\rangle-g\beta_i)\frac{3}{2b\rho_0}\right)=\frac{\partial}{\partial x_i}L\sqrt{b}\frac{\partial\langle\rho'^2\rangle}{\partial x_i}, \tag{4.4}$$

The equation (4.3) is equivalent to (3.1), (4.4) – to (3.2), the kinetic energy of turbulence is denoted by the symbol *b*, and the potential energy is considered through density fluctuations.

In the framework of this approach, it is possible to calculate, in particular, the dependence of the turbulent Prandtl number on the gradient Richardson number:

$$PrT(Ri)^{-1}=\frac{(4-3R)+\frac{1}{Ri}-\left(\left((4-3R)+\frac{1}{Ri}\right)^2-\frac{4}{Ri}\right)^{1/2}}{2}, \tag{4.5}$$

where *R* is a parameter characterizing the anisotropy of small-scale turbulence and depending on the ratio of the vertical and horizontal scales of correlation field of density fluctuations. Values of *R* equal to 0.3 and 0.5 were considered, *PrT (Ri)* is presented in Fig. 1.

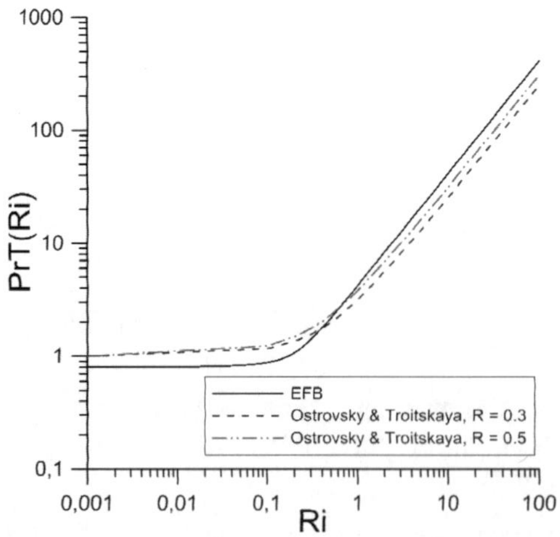

Figure 1. The dependence of the turbulent Prandtl number *PrT* on the gradient Richardson number *Ri* in different models.

5. Results of calculation

In the previous work of the group of authors [14] fairly accurate results for calculating the thermodynamics of inland water bodies were obtained with the standard E-ε model. They were used to verify the results of the current study. The verification was based on experimental data obtained at the Gorky reservoir experimental site in the period 2014-2017.

In the present work, the parameterization of the coefficient of heat transfer eddy diffusivity for the E-ε model is proposed in order to take into account stratification when calculating the thermo-hydrodynamic regimes of inland water bodies. Formula (2.3) turns into the following form:

$$\lambda = c\rho k \cdot PrT^{-1}, \tag{5.1}$$

where (3.10) and (4.5) are used as the turbulent Prandtl number.

Within the framework of verification, the temperature distribution profiles obtained as a result of simulation using the standard model and the ones with parameterizations were compared. Several cases were considered, two examples are presented in the Fig. 2.

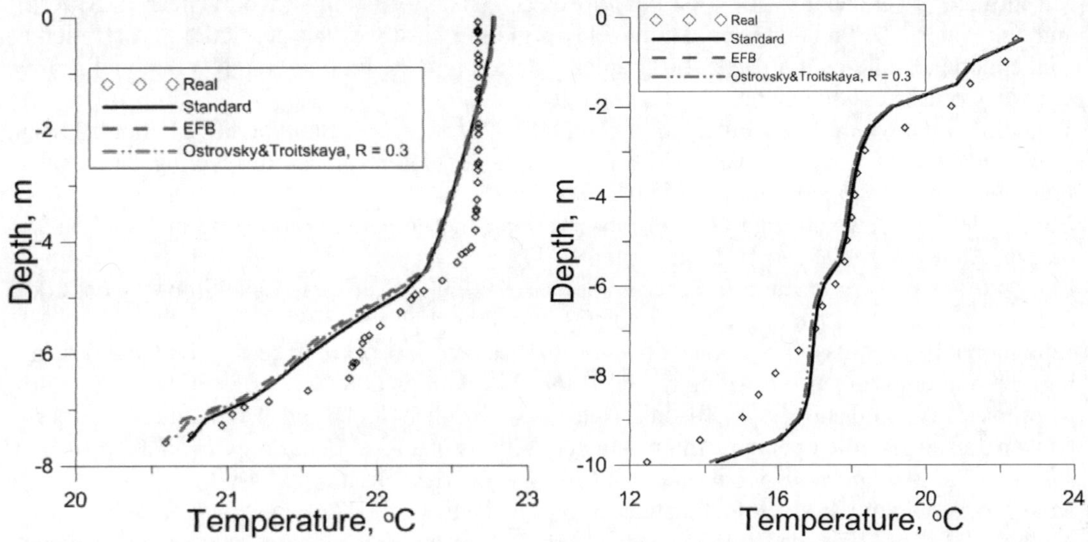

Figure 2. Comparison of the temperature distribution profiles obtained as an experimental data and as a result of simulation with the standard E-ε model and the models with parameterizations.

It is demonstrated that the use of parameterizations made it possible to obtain no less accurate results, but, as the analysis of the standard deviation (see Table 1) showed, even more accurate ones.

Table 3. Standard deviation (in °C) obtained with the standard E-ε model and with the parameterizations

	Standard	EFB	Ostrovsky & Troitskaya, *R=0.3*
Case 1 (Fig. 1a)	0.23	0.19	0.20
Case 2 (Fig. 1b)	0.63	0.60	0.60

6. Conclusion

In this paper, we consider three main models of turbulence, which are used to describe the thermohydrodynamics of both inland waters, and the ocean and atmosphere. For models that take into account the two-sided transformation of the kinetic and potential energies of turbulent pulsations, the

dependences of the turbulent Prandtl number on the gradient Richardson number are provided. The obtained dependences were used in the E-ε model to parameterize the coefficient of heat transfer eddy diffusivity in order to take into account stratification when calculating the thermohydrodynamic regimes of inland waters. Based on the previously obtained results [14], it was found that the use of parameterizations allows reducing the standard deviation from the real values
Due to the fact that such parameterization allows to correctly describe the processes of turbulence in inland waters, in the future they are planned to be used for the conditions of the ocean and atmosphere, including for large values of the gradient Richardson number.

The work was partly supported by Russian Science Foundation project 15-17-20009 and RFBR project 18-05-00292.

.

References

[1] Mironov D V, Terzhevik A, Beyrich F, Golosov S, Haise E, Kirillin G, Kourzeneva E, Ritter B, and Schneider N 2006 Parameterization of lakes in numerical weather prediction: description of a lake model, single-column tests, and implementation into the limited-area NWP model *Bound. Lay. Meteor.* Spec. issue 56 pp
[2] Tranvik L J, Downing J A, Cotner J B, Loiselle S A, Striegl R G, Ballatore T J, Dillon P, Knoll L B, Kutser T et al 2009 Lakes and reservoirs as regulators of carbon cycling and climate *Limnology and Oceanography* vol 54 pp 2298-2314
[3] Comer Neil T, McKendry Ian G 1993 Observations and numerical modelling of Lake Ontario breezes *Atmosphere-Ocean* vol 31 no 4 pp 481-499
[4] Long Z, Perrie W, Gyakum J, Caya D, Laprise R 2007 Northern Lake Impacts on Local Seasonal Climate *Journal of Hydrometeorology* vol 8 no 4 pp 881-896
[5] Ljungemyr P, Gustafsson N, Omstedt A 1996 Parameterization of lake thermodynamics in a high-resolution weather forecasting model *Tellus A.* vol 48 pp 608-621
[6] Mironov D V, Golosov S D, Zilitinkevich S S, Kreiman K D and Terzhevik A Yu 1991 Seasonal changes of temperature and mixing conditions in a lake *Modelling air-lake interaction. Physical Background* ed. S S Zilitinkevich (Springer-Verlag, Berlin) pp 74-90
[7] Monin A S, Yaglom A M 1965 Statistical Fluid Mechanics: v. 1 *Moscow: Nauka* p 640
[8] Mellor C L and Yamada T 1975 A hierarchy of turbulence closure models for planetary boundary layers *J. Atmos. Sci.* vol 31 pp 1791-1806
[9] Zilitinkevich S S, Elperin T, Kleeorin N, Rogachevskii I 2007 Energy- and Flux-Budget (EFB) turbulence closure models for stably-stratified flows. Part I: Steady-state, homogeneous regimes 2007 *Bound.-Layer Meteorol.* v 125 pp 167-191
[10] Zilitinkevich S S, Elperin T, Kleeorin N, Rogachevskii I, and Esau I 2013 A Hierarchy of Energy- and Flux-Budget (EFB) Turbulence Closure Models for Stably-Stratified Geophysical Flow *Bound.-Layer Meteorol.* v 146 pp 341–373
[11] Ostrovsky L A, Troitskaya Yu I 1987 Model of turbulent transition and turbulence dynamics in stratified shear flow *Izvestiya RAN Atmospheric and Ocean Physics* v 23 №10 pp 1031-1040
[12] Stepanenko V M 2005 Numerical modeling of thermal regime in shallow lakes *Computational technologies* vol 10 (Special issue. Proceedings of the International Conference and the School of Young Scientists "Computational and informational technologies for environmental sciences" (CITES 2005) Tomsk, 13-23 March 2005) part 1 pp 100-106
[13] Chebotarev A E 1975 Obshchaya gidrologiya (Leningrad)
[14] Gladskikh D, Sergeev D, Baydakov G, Soustova I, Troitskaya Yu 2018 Numerical modeling of thermal regime in inland water bodies with field measurement data *Journal of Physics: Conference Series* v 955

IOP Conf. Series: Journal of Physics: Conf. Series **1163** (2019) 012034 doi:10.1088/1742-6596/1163/1/012034

Computer simulation of thermal processes in water bodies under different hydrometeorological conditions

Ya N Parshakova[1], **T P Lyubimova**[1, 2], **Yu S Lyakhin**[3], **A P Lepikhin**[3]
[1] Institute of Continuous Media Mechanics UB RAS, Acade. Koroleva 1, Perm 614013, Russia
[2] Perm State University, Bukireva 15, Perm 614990, Russia
[3] Mining Institute UB RAS, Sibirskaya 78a, Perm 614007, Russia

E-mail: parshakova@icmm.ru

Abstract. Key to the solution of a wide range of technological and ecological problems is getting comprehensive and reliable estimates of the parameters of temperature fields generated by waste water discharges taking into account a set of technological and hydrometereological parameters. In this work, the problem under consideration is analysed, using as an example the Magnitogorsk Iron and Steel Works (MMK), one of the world's largest steel producers and a leading Russian metal company. The discharge channels of MMK drain warm water in the Magnitogorsk reservoir of the Ural River.

1. Introduction

At present, the relationship between power and water resources is the burning issue of the renovation projects for water supply systems of many industrial plants. A rapid development of the industrial sector led to an increase in electric power demands. A successful implementation of this task implies the reconstruction of existing and the creation of new thermal power plants and nuclear power stations with a marked increase of their power generation capacity, which is naturally associated with a significant increase in the volume of cooling water required for their operation. For economy and safety reasons most power plants are located in the coastal zones and operate on permanent watercourse systems, in which water is taken directly from the reservoirs that act as coolers. At such plants, a thermal plume heated to high temperatures is emitted into the atmosphere and is carried directly into the reservoir [1, 2]. A huge amount of heated water is discharged into the receiving reservoir, which causes the natural temperature of the receiving water to increase. This has a direct or indirect effect on the ecological status of the water environments known as thermal effects [3]. This is the reason why the relationship between power and water resources is considered to be the problem at the focus of researchers; attention [4–8]. Different research methods are aimed at solving the problems of thermal pollution. The spread of the heat spot released by shore-based power plants is usually controlled by measuring the water temperature at the observation points. Although the measurement data obtained by this method are highly accurate, this method also requires large resources, since due to the limited sample size the method yields only discrete data, which cannot reflect the spatial change in the effect of the thermal plume [9]. An important research method is numerical simulation, which is necessary for understanding the effects produced on the environment by thermoelectric power plants [2, 10].

A three-dimensional numerical simulation of turbulent mixing of water masses at different temperatures was carried out in [13–15] in the framework of the LES turbulence model described in [11,12] by solving the Navier -Stokes equations. Modelling of great water masses was accomplished with the use of the parallel computation technique. It was shown that the computation time depends on the number of grid nodes and processors.

In this paper, the hydrodynamic models of the surface layers of water bodies were developed based on the three-dimensional numerical modelling [2, 16-18]. The numerical calculations were performed for the environmental conditions of the Magnitogorsk Iron and Steel Works (*Magnitogorskiy Metallurgicheskiy Kombinat*, abbreviated as **MMK)**. Hot water from this plant is discharged directly into the Magnitogorsk reservoir of the Ural River. The study was carried out for different technological and hydrometeorological conditions.

2. Computational technique.

The 3D hydrodynamical model was built for the part of Magnitogorsk reservoir with linear dimensions of 2000 m adjacent to the MMK and including the locations of water intake and water discharge channels to/into the cooling pool (Fig.1). The Magnitogorsk Iron and Steel Works discharges warm waste water through the special channels directly into the Magnitogorsk reservoir of the Ural River.

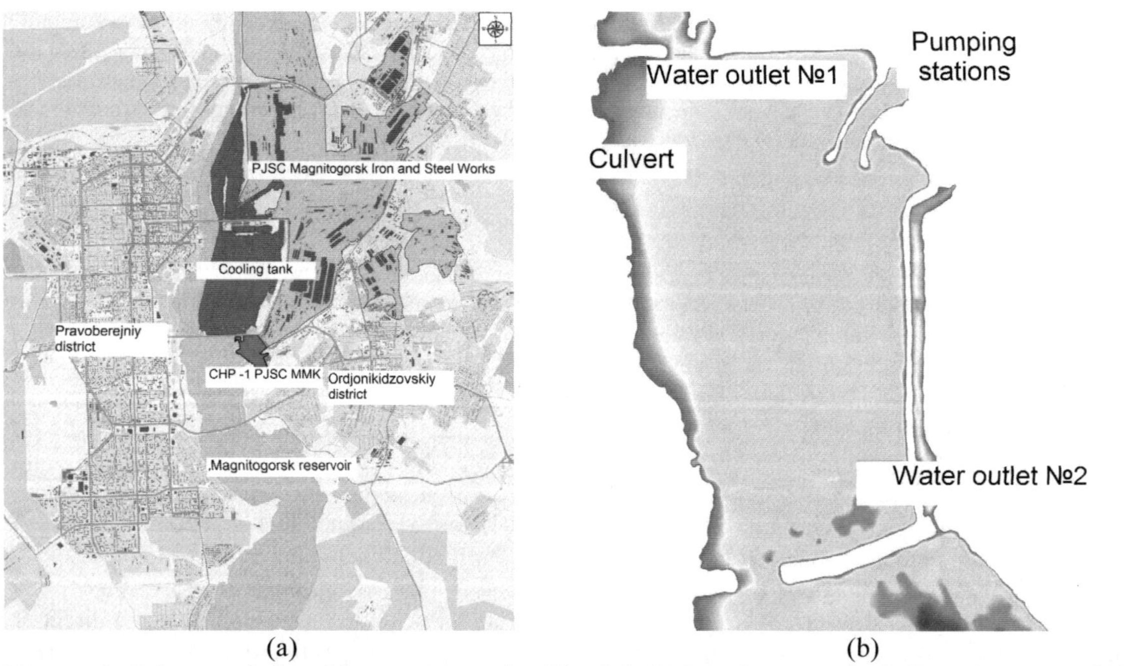

Figure 1. Scheme of the object under study. The MMH location map including the area of the Magnitogorsk reservoir (a). Computational domain diagram (b).

The ANSYS Fluent software package was used to make 3D simulation on the computer cluster URAN at the IMM UB RAS. The problem was solved in the framework of the $k-\varepsilon$ model describing turbulent pulsations based on the non-stationary non-isothermal approach. We use the Reynolds-averaged Navier-Stokes equations:

$$\frac{\partial \rho}{\partial t}+\frac{\partial}{\partial x_i}(\rho u_i)=0 \quad (1)$$

$$\frac{\partial}{\partial t}(\rho u_i)+\frac{\partial}{\partial x_j}(\rho u_i u_j)=-\frac{\partial p}{\partial x_i}+\frac{\partial}{\partial x_j}\left[\mu\left(\frac{\partial u_i}{\partial x_j}+\frac{\partial u_j}{\partial x_i}-\frac{2}{3}\delta_{ij}\frac{\partial u_l}{\partial x_l}\right)\right]+$$
$$+\frac{\partial}{\partial x_j}\left[\mu_t\left(\frac{\partial u_i}{\partial x_j}+\frac{\partial u_j}{\partial x_i}\right)-\frac{2}{3}\left(\rho k+\mu_t\frac{\partial u_l}{\partial x_l}\right)\delta_{ij}\right]+\rho g_i \quad (2)$$

Here, ρ is the density, x_i are the coordinates (we use the Cartesian coordinate system), u_i are the velocity components, μ is the kinematic viscosity, μ_t is the turbulent viscosity.

The turbulence kinetic energy k and the rate of its dissipation ε are obtained from the following transport equations:

$$\frac{\partial}{\partial t}(\rho k)+\frac{\partial}{\partial x_i}(\rho k u_i)=\frac{\partial}{\partial x_j}\left[\left(\mu+\frac{\mu_t}{\sigma_k}\right)\frac{\partial k}{\partial x_j}\right]+G_k+G_b-\rho\varepsilon \quad (3)$$

$$\frac{\partial}{\partial t}(\rho\varepsilon)+\frac{\partial}{\partial x_i}(\rho\varepsilon u_i)=\frac{\partial}{\partial x_j}\left[\left(\mu+\frac{\mu_t}{\sigma_\varepsilon}\right)\frac{\partial\varepsilon}{\partial x_j}\right]+C_{1\varepsilon}\frac{\varepsilon}{k}(G_k+C_{3\varepsilon}G_b)-C_{2\varepsilon}\rho\frac{\varepsilon^2}{k} \quad (4)$$

In equations (3)-(4), G_k represents the generation of turbulence kinetic energy due to the mean velocity gradients $G_k=\mu_t S^2$ where S is the modulus of the mean strain rate tensor, defined as $S=\sqrt{2S_{ij}S_{ij}}$, $S_{ij}=0.5\left(\partial u_j/\partial x_i+\partial u_i/\partial u_i\right)$, G_b is the turbulence kinetic energy due to buoyancy, which is calculated as

$$G_b=g_i\left(\beta\frac{\mu_t}{\Pr_t}\frac{\partial T}{\partial x_i}\right) \quad (5)$$

where μ_t is the turbulent viscosity determined as: $\mu_t=\rho C_\mu k^2/\varepsilon$, where C_μ is a constant.

Simulation of the turbulent heat transfer similarly to that of turbulent momentum transfer is erformed using the Reynolds model. Hence, the equation of energy is written as

$$\frac{\partial}{\partial t}(\rho E)+\frac{\partial}{\partial x_i}\left[u_i(\rho E+p)\right]=\frac{\partial}{\partial x_j}\left(k_{eff}\frac{\partial T}{\partial x_j}+u_i(\tau_{ij})_{eff}\right) \quad (6)$$

where $E=ch+\frac{p}{\rho}$ denotes the total energy, $h=C_pT$ is the system enthalpy, k_{eff} is the effective thermal conductivity, and $(\tau_{ij})_{eff}$ is the stress tensor deviator defined as

$$(\tau_{ij})_{eff}=\mu_{eff}\left(\frac{\partial u_j}{\partial x_i}+\frac{\partial u_i}{\partial x_j}\right)-\frac{2}{3}\mu_{eff}\frac{\partial u_k}{\partial x_k}\delta_{ij} \quad (7)$$

The model constants $\Pr_t$, $G_{1\varepsilon}$, $C_{2\varepsilon}$, C_μ, σ_k and σ_ε were taken to have the following values [19]: $\Pr_t=0.85$, $C_{1\varepsilon}=1.44$, $C_{2\varepsilon}=1.92$, $C_\mu=0.09$, $\sigma_k=1.0$, $\sigma_\varepsilon=1.3$.

The spatial discretization scheme of second-order accuracy was applied. Simulation of temporal evolution was carried out using an explicit second-order scheme.

The boundary conditions at the edges of the computation domain were taken as follows. At the bottom and at the banks of the reservoir the no-slip conditions and fixed temperature were imposed: $u_1=u_2=u_3=0,\quad T=T_0$,

At the inlet of the computational domain the main flow velocity was assumed to have one nonzero component, which was invariable over the inlet cross-section, and the temperature was assumed to be equal to the background temperature of the reservoir water: $u_i = V_i$, $T = T_0$,

At water intake and water discharge points, the water velocity and temperature were assumed to be constant: $u_i = V_{\text{intake}_i}$, $T = T_0$ at the entrance of the discharge channel and $u_i = V_{\text{entrance}_i}$, $T = T_{\text{entrance}}$ at the exit of the channel. The upper boundary of the fluid was free and non-deformable, and the effect of wind was taken into account through the evaluation of the tangential stresses by the Ekman formula [20]. Heating of the water surface by the surrounding air was estimated by applying the linear heat transfer law and the heat transfer coefficient was selected based on the analysis of the in-situ measurement data.

The computational grid was generated using the Gambit 2.4 package of ANSYS Fluent. The number of nodes through the depth of the computational domain was taken to be 21. The non-uniform grid was constructed using the bottom morphometric data obtained from the in-situ measurements in 2014. In the horizontal direction, the computational grid consisted of tetragonal elements distributed uniformly along the entire length, with the characteristic linear size of 20m. The computational grid included 400 hundred thousands of nodes.

The complex geometry of the computational domain was simulated in the following way. To adapt the morphological data available in the coordinate-depth format to the capabilities of the mesh generator, the bottom of the reservoir was represented as a set of simple geometrical objects of some specified resolution, which were then introduced into the file. Based on the array of data describing the reservoir bottom morphology, a code was written to produce a batch file for the grid generator of the ANSYS Fluent package. The proposed code is of general character and can be applied to the construction of similar geometries and to other tasks.

3. Formulation of the problem

The Magnitogorsk Metallurgical Works is currently implementing a set of environmental initiatives aimed at reducing the man-made impact on water bodies. In the near future, the company plans a complete elimination of water draining into the Magnitogorsk reservoir. At present, according to this plan, the MMK is completing the implementation of the main phase of the project for the reconstruction of the circulating water supply system including the expansion of the cooling reservoir. The project envisages the construction of a dam separating the reservoir from the influence of the MMK circulating water supply system. This will stop the discharge of waste waters from the cooling reservoir to the Magnitogorsk reservoir. Completion of the construction and installation works is scheduled for the end of October. In parallel, works on the formation of the dam crest are currently under way. Considerable efforts are being made to arrange its landscaping and stocking the reservoir with fish. The purpose of this work is to assess the parameters and modes for monitoring the process of water cooling during the operation of the MMK in the presence of the dam. The problem was solved within the framework of the model described in [2]. This model can be applied to the above problem because its thermal and hydrodynamic characteristics are identical to those used in [2] for estimating the zones of thermal pollution formed in the Kama reservoir due to discharges of waste waters of the Permskaya GRES.

Numerical simulation was carried out for the model scenario. The objective of the numerical simulation in the context of the proposed scenario was to determine the operating capacity of the water intake stations during the hottest summer months at temperatures not exceeding 28 °C when using the circulating water supply system.

According to the field surveys conducted in 2018, it was found that the bottom water temperature of the Magnitogorsk reservoir in the region of the culvert unit does not exceed 23 °C. The water temperature at the waste water outlet №1 was taken equal to 300C at the discharge flow rate of

IOP Conf. Series: Journal of Physics: Conf. Series **1163** (2019) 012034 doi:10.1088/1742-6596/1163/1/012034

11.6 m^3/s, and at the outlet №2 – 40 °C at the discharge flow rate of 3.7 m^3/s. The flow rate through the culvert was taken to be 3.9 m^3/s.

The computation was made taking into account the interaction between the water mass and atmosphere temperatures. The data used were based on the decadal maximum air temperature, which was observed in the 2nd decade of July 2012 and was equal to 25.3 °C under calm conditions.

In addition to water temperature, the results of model calculations in 3D could be used to evaluate the dynamics of heat fluxes in the water area and the depth of the cooling tank.

4. Computational Results

The results of simulation presented below are the outcome of a rather cumbersome process, associated not only with the specifics of setting the initial information and building an adapted computational grid, but also with the computation process itself. In particular, to obtain a relative stationary behavior of flows under the existing multi-factor conditions, the calculations were performed for three days since the beginning of the waste water discharge.

The results obtained clearly demonstrate the presence of significant stratification in the formation of temperature fields (Fig. 2). Figures 2a and 2b show the temperature distribution fields. It can be seen that the greatest temperature impact on the water area is caused by the discharge of water from the outlet № 2, while water discharge from the outlet № 1 is cooled due to the inflow of cooler water through the culvert from the bottom horizon of the Magnitogorsk reservoir.

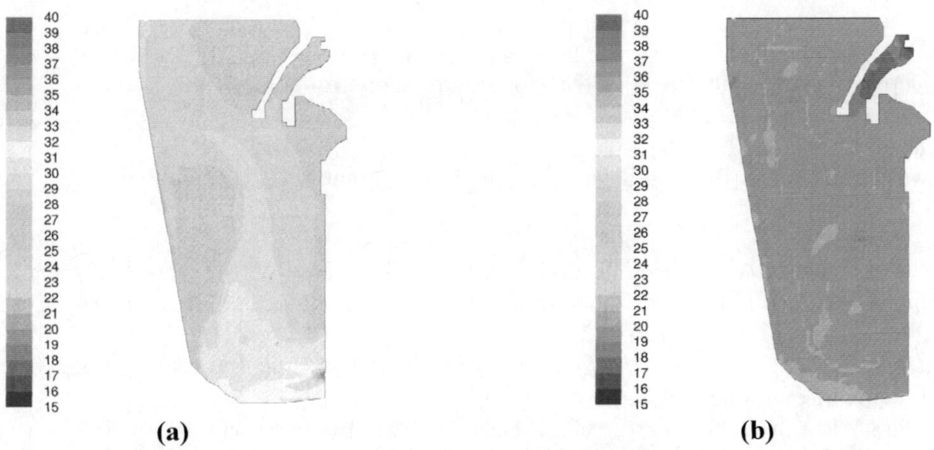

Figure 2. Temperature field (°C) in the surface layer of the Magnitogorsk reservoir (a), at a distance of 3 meters from the surface (b).

For elaboration of specific technical solutions, as well as recommendations of how to improve the operational efficiency of the structure under development, it was necessary to determine the horizon of the transition layer. To this end, 6 vertical reference lines were drawn on the water area and diagrams of the depth-wise water temperature distribution were constructed (Figure 3).

It should be noted that, depending on the location of the water outlets, the surge layer is located, on average, at a depth of 2-4 m. This is an additional substantiation of the fact that deeper (colder) water masses are poorly involved in the cooling process. This finding paves the way to studying possible measures of improving the efficiency of reservoir-coolers, over a relatively short period of maximum air temperature.

5. Conclusion

In this work, a numerical simulation of the distribution of thermal pollution in water bodies was carried out using as an example the water cooling system of the Magnitogorsk Metallurgical Plant. The discharge channel of this plant drains warm water in the Magnitogorsk reservoir of the Ural River.

The study was conducted for extreme technological and hydrometeorological conditions. The results of numerical computation revealed significant temperature stratification, the presence of which should be taken into account when designing new technological water intake structures. Water intakes should have such structures as to prevent water from flowing at a depth of 4 meters from the surface, because water at a given depth cannot be used in a direct-flow cooling process.

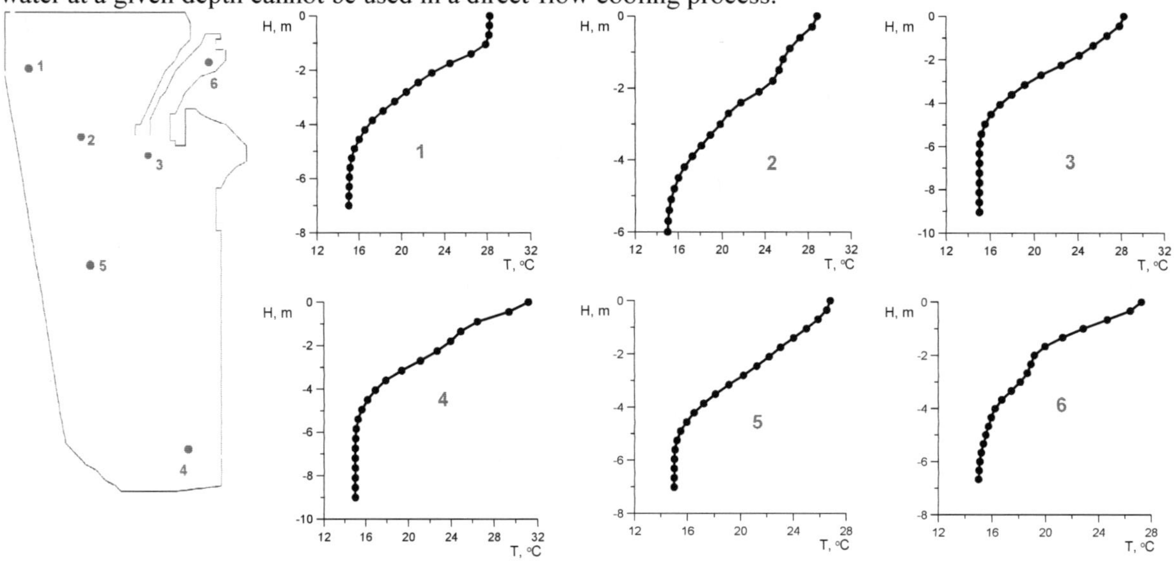

Figure 3. Depth-wise distributions of water temperature at control points 1-6.

Acknowledgments
The work was supported by Russian Scientific Foundation (grant No. 17-77-20093).

References

[1] Laws E A 2000 *Aquatic Pollution: An Introductory* Text. New York: John Wiley and Sons 672
[2] Lyubimova T, Lepikhin A, Parshakova Ya, Lyakhin Yu, Tiunov A 2018 *International Journal of Heat and Mass Transfer* **126** 342
[3] Walter K, Dodds M R 2010 *Freshwater Ecology Concepts and Environmental Applications of Limnology*. Elsevier Inc. All 829
[4] Perrone D, Murphy J, Hornberger G 2011 *Environ. Sci. Technol.* **45** (10) 4228
[5] Scott C A, Pierce S A, Pasqualetti M J, Jones A L, Montz B E, Hoover J H 2011 *Energy Policy* **39** (10) 6622
[6] Chandel M. K, Pratson L F, Jackson R B 2011 *Energy Policy* **39** (10) 6234
[7] Hussey K, Pittock J 2012 *Ecol. Soc.* **17** (1) 31
[8] Sikdar S K, Agrawal R 2014 *Curr. Opin. Chem. Eng.* 5
[9] Thomas A C, Byrne D, Weatherbee R 2002 *Remote Sens Environ* **81** 262
[10] Duran-Colmenares A, Barrios-Pina H, Ramirez-Leon H. *Water Resources* **8** (11) 482
[11] Issakhov A 2011 *J. of Physics* **318** (4) 42051
[12] Issakhov A 2013 *Power, Control and Optimization* **239** 165
[13] Issakhov A 2014 *Scientific World J.* doi 10.1155/2014/678095
[14] Tennekes H, Lumley J L 1972 *The MIT Press*
[15] Lesieur M, Metais O, Comte P 2005 *Large Eddy Simulation of Turbulence*
[16] Lepikhin A P, Lyubimova T P, Parshakova Y N, Tiunov A A 2012 *J. of Min. Science* **48** (2) 390
[17] Lyubimova T, Lepikhin A, Konovalov V, Parshakova Ya, Tiunov A 2014 *J. Hydrology* **508** 328
[18] Lyubimova T, Lepikhin A, Parshakova Ya, Tiunov A 2016 *J. of Hydrology*. **534** 579
[19] Launder B E, Spalding D B 1972 Lectures in Mathematical Models of Turbulence
[20] Welander P 1968 *Tellus XX* **1** 1

Wave number selection under the action of accelerated rotation

O Krivonosova[1], D Zhilenko[1] and M Gritsevich[2,3]

[1] Moscow State University, GSP-1, Leninskie Gory, 119991,Moscow, Russia
[2] Department of Physics, University of Helsinki, Gustaf Hällströmin katu 2a, FI-00014 Helsinki, Finland
[3] Institute of Physics and Technology, Ural Federal University, Mira str. 19, 620002, Ekaterinburg, Russia

E-mail: olga@imec.msu.ru

Abstract. Unsteady viscous incompressible flows in a spherical layer due to an increase in the rotation velocity of the inner sphere with constant acceleration are investigated. The acceleration starts at the Reynolds numbers Re corresponding to a stationary flow and ends at Re higher than the stability limit of the stationary flow, whereupon the rotation velocity of the inner sphere remains constant. The outer sphere is fixed and the spherical layer thickness is equal to the inner sphere radius. The inner sphere acceleration effect is studied on both the formation of one of two possible secondary-flow structures after the acceleration has been stopped, namely, traveling azimuthal waves with wavenumbers of 3 or 4, and on the change in the flow structure during the action of the acceleration. It is shown that not only an increase in the acceleration but also a decrease in Re corresponding to the acceleration onset can lead to an increase in the deviation of the instantaneous velocity profiles from their stationary values and can be accompanied by a variation in the secondary flow wavenumber.

1. Introduction

After the loss of stability three-dimensional viscous incompressible flows can be nonunique, namely, several different flow patterns may exist at the same Reynolds numbers Re but different flow prehistory. In spherical layers it is the influence of a constant acceleration of the inner sphere on transition to different secondary flows that is most completely experimentally studied. Thus, at the same Reynolds numbers but different accelerations of the inner sphere secondary flows with different numbers of the Taylor rings can be formed in thin layers [1], while in wide layers azimuthal waves with different wave numbers can be generated [2]. In the case of thin layers the experimentally obtained results on the formation of different stationary patterns, symmetric about the axis of rotation, under the action of acceleration are confirmed by direct calculations [3]. In the case of wide layers the numerical results are restricted to the calculations of the structures of unsteady axisymmetric flows formed in the process of internal sphere acceleration [4] and to an analysis of their stability in the quasi-stationary approximation [5]. In the wide layer with relative gap size $\delta = (r_2 - r_1)/r_1 = 1$ under stationary boundary conditions the growth rate of mode 4 is higher than that of mode 3 over a wide Re range [5,6]; however, in stationary flows developing after a stepwise change in the inner sphere velocity the relation between the growth rates can be opposite [5].

IOP Conf. Series: Journal of Physics: Conf. Series **1163** (2019) 012035 doi:10.1088/1742-6596/1163/1/012035

At present, the available experimental results on the inner sphere acceleration effect on the secondary flow formation in a wide $\delta = 1$ layer have made it possible to establish the following patterns. Under the influence of the acceleration the mode realized in the secondary flow exceeds the noise level and starts to grow earlier than the other modes [7]. The same effect was previously numerically found in the case of axisymmetric secondary flows in the cylindrical Couette flow [8]. The existence of the critical acceleration is established, such that a four-vortex flow pattern is realized below it and a three-vortex pattern above it. For small values of Re_i, that is, the Reynolds number associated with the acceleration onset, its small variations lead to large variations in the critical acceleration, whereas at high Re_i the critical acceleration depends on it only slightly. It was found that there exist indefiniteness domains, that is, the Re_i ranges on which the four-vortex pattern is formed at small accelerations, while with increase in the acceleration the azimuthal wavenumber is chosen in a random fashion. It was shown that an increase in the acceleration is accompanied by an increase in the nonequilibrium of the azimuthal velocity profiles, that is, the deviation from the profiles corresponding to stationary Re values [9].

2. Formulation of the problem

Isothermal flow of viscous incompressible fluid is described by the Navier–Stokes and continuity equations:

$$\frac{\partial \boldsymbol{U}}{\partial t} = \boldsymbol{U} \times \mathrm{rot}\boldsymbol{U} - \mathrm{grad}\left(\frac{p}{\rho} + \frac{\boldsymbol{U}^2}{2}\right) - \nu \mathrm{rotrot}\boldsymbol{U}, \quad \mathrm{div}\boldsymbol{U} = 0.$$

In the case of a flow in a spherical layer it is natural to use a spherical coordinate system with the radial (r), polar (θ), and azimuthal (φ) directions, corresponding velocity $\boldsymbol{U}$ components are u_r, u_θ and u_φ. The no-slip and impermeability conditions are imposed on the boundaries $u_\varphi(r = r_{1,2}) = \Omega_{1,2}(t) r_{1,2}\sin\theta$, $u_r(r = r_{1,2}) = 0$, $u_\theta(r = r_{1,2}) = 0$, where subscripts 1 and 2 correspond to the inner and outer spheres, respectively. Because of $\Omega_2 = 0$ in this work the subscript "1" will be omitted both for inner sphere rotation velocity Ω_1 and Reynolds number Re_1. Before the acceleration onset and after it has stopped the inner sphere rotation velocity is constant and equal to Ω_i and Ω_f, respectively; during the acceleration it varies in accordance with the linear law $\Omega = \Omega_i + a(t - t_0)$, $t - t_0 < (\Omega_f - \Omega_i)/a$, here, a is the inner sphere acceleration. As initial conditions the stationary flow at the Re_i was taken without introducing any disturbances.

The computational algorithm is presented in detail in [10]; it is based on a conservative difference scheme for discretizing the Navier–Stokes equations in space and a semi-implicit third-order Runge–Kutta scheme for integrating in time. The discretization in space was performed using grids nonuniform in the radial and meridional directions (maximum-to-minimum cell dimension ratio in each direction varied from 1 to 4) The calculations were carried out for the dimensional parameters corresponding to the experimental conditions in [7,9], namely, for sphere radii r_1 =0.075 m and r_2 =0.150 m and the fluid kinematic viscosity in the layer $\nu = 5\times10^{-5}$ m^2/s.

In our investigations the complete system of Navier-Stokes equations and reduced system with conditions of symmetry about the axis of rotation and the equatorial plane were solved. The latter system describes a two-dimensional flow which under nonstationary boundary conditions is time-dependent. The linear stability analysis of these unsteady flow was carried out, as in [5], in the quasi-stationary approximation, in which the flowfield stability is determined for a current moments of time.

3. Axisymmetric flow structure under the action of acceleration

The accelerated rotation of the inner sphere leads to a situation in which the flow is no longer symmetric about the axis of rotation even before the stability limit has been reached, though it still remains symmetric about the equatorial plane. This was experimentally established both in the case of acceleration with alternating signs (oscillations relative the axis of rotation with zero mean velocity) [11] and in the case of an acceleration constant in the value and the sign [7]. At the same time, the experiment shows that in both cases the "asymmetry" about the axis of rotation is chiefly concentrated in a very narrow region near the equatorial plane, while outside of this zone the flow is axisymmetric. Precisely this makes it possible to investigate the effect of the acceleration and the initial Reynolds number Re_i on the variation in the flow structure in the process of acceleration and after its termination in the axisymmetric formulation. In presenting the calculated results we use the following dimensionless parameters: the dimensionless time $\tau = t/\left(r_1^2/\nu\right)$ and the dimensionless acceleration of the inner sphere $d\mathrm{Re}/d\tau = \left(\mathrm{Re}_f - \mathrm{Re}_i\right)\left(t_f - t_i\right)^{-1}\left(r_1^2/\nu\right)$, here Re_f is the Reynolds number corresponding to the termination of acceleration).

In Figure 1 the angular velocity of the inner sphere Ω_1 and the ratio E_ψ/E_φ are presented. Before the acceleration onset all the quantities have the values corresponding to the stationary flow with Rei and after the acceleration termination they tend to the stationary values corresponding to Re_f. Before the acceleration termination it monotonically increases. The acceleration being stopped, after a lapse of time the ratio E_ψ/E_φ reaches a maximum, which means a more rapid attainment of the stationary value by the meridional component of the kinetic energy, as compared with the azimuthal component. The ratio E_ψ/E_φ changes in a narrow time interval, not greater than 0.1 of the acceleration duration.

It is well known that under stationary boundary conditions the intensity of the meridional circulation increases with increase in the relative layer thickness δ, whereas the wavenumber of the secondary flow at the stability limit reduces: thus, in the δ = 1layer a flow with m = 4 is attained, while in the layer with δ = 1.27 it is the flow with m = 3, which is confirmed by both the experiment [12] and by direct calculations [6,10]. The same is true for the flow at the end of acceleration action: the intensity of the meridional circulation increases, the greater the acceleration the greater this increase. Because of this, in the experiment an increase in the acceleration higher than certain values leads to a reduction in the secondary flow wavenumber. Thus, precisely the variation in the flow structure in the form of spatial-temporal redistribution of the circulation intensity in the meridional plane is the main cause of the reduction in the secondary flow wavenumber under the action of the acceleration.

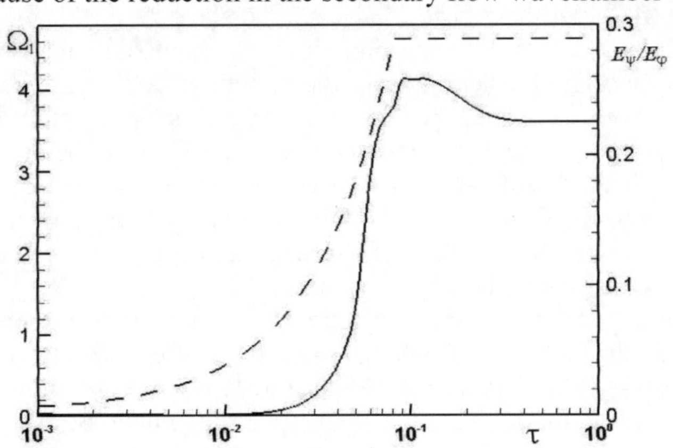

Figure 1. Flow in the process of acceleration for $Re_i = 10$, $Re_f = 520$, and dRe/dτ = 6328; dashed line – angular velocity of the inner sphere; solid line – the ratio E_ψ/E_φ;

4. Secondary nonaxisymmetric flows

The nonlinear calculations were performed for $Re_i = 10$, 200, 450, and 480 and $Re_f = 520$. Both finite accelerations dRe/dτ ranging from 6 to 6×10^7 and the infinite acceleration in the form of a stepwise velocity change were considered. Both in the calculations and in earlier experiments in the process of acceleration the flow remains symmetric about the equatorial plane but loses the symmetry about the axis of rotation. Depending on the acceleration, transition from the axisymmetric to the nonaxisymmetric flow occurs either during the acceleration or after its termination. The absence of the axial symmetry of the flow is particularly clearly expressed near the equatorial plane Figure 2. Precisely in this region the meridional flow velocity is maximum and the nonuniform rotation of the inner sphere leads to the formation of wave structures in the radial direction.

IOP Conf. Series: Journal of Physics: Conf. Series **1163** (2019) 012035 doi:10.1088/1742-6596/1163/1/012035

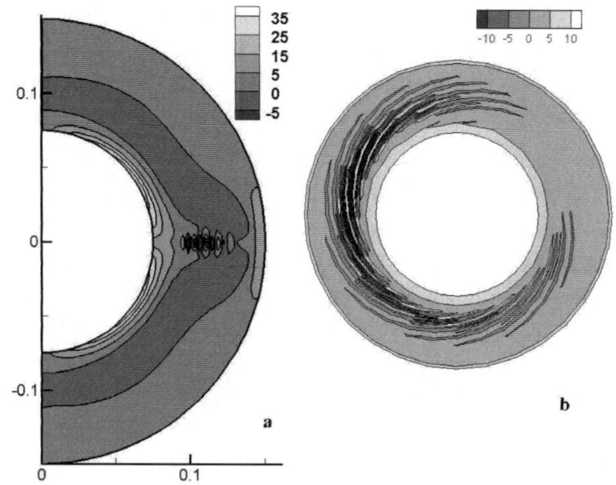

Figure 2. Meridional vorticity component in the meridional (a) and equatorial (b) planes of the flow during the acceleration.

With further development of the flow the nonuniformity in the radial direction vanishes but an asymmetry about the axis of rotation with an amplitude of about 10^{-8} is retained in the equatorial plane. Depending on the acceleration and Re_i the wavenumber in this structure can vary from 1 to 4. If an m < 3 mode is established, then with time it successively increases up to 3 or 4. If a structure with the wavenumber equal to 3 or 4 is formed at once, then the corresponding mode is developed in the secondary flow. In the velocity spectrum there appear peaks on the frequencies corresponding to the m = 3 and 4 modes. The amplitudes of these peaks increase with time, one of them attaining a constant value, whereas the other starts to decrease. The similar process of simultaneous growth of two competing modes was previously observed in the experiment [7].

Figure 3 represent the summary of calculations results of the secondary flows. For fixed values of Re_i equal to 200 and 450 an increase in the acceleration leads, as in the experiments [7,9], to the change of the four-vortex flow pattern to the three-vortex regime. In certain cases, (for example, for $Re_i = 450$) this flow pattern is conserved at arbitrarily large accelerations. For small $Re_i = 10$ the point of the flow pattern change could not be determined: despite very small acceleration values, the three-vortex secondary flow is formed. With increase in Re_i from 200 to 450 the acceleration needed for the change of the flow pattern increases by an order. The same effect of a strong Re_i-dependence of the critical acceleration was previously observed in the experiment [7].

As was shown by present calculations the dependence of the secondary flow wavenumber on the acceleration may be more complicated. This can be clearly seen in Figure 3 for $Re_i = 200$: with increase in the acceleration on the $6.3 \leq dRe/d\tau \leq 63.0$ interval the first transition from m = 4 to m = 3 is observable. Then the m = 4 pattern is restored on the $6.3\times10^3 \leq dRe/d\tau \leq 2.5\times10^6$ range. So on an acceleration range wider than in the experiments the secondary flow pattern can vary two times and, correspondingly, two critical values of the acceleration exist. Thus, the calculated results exhibit the same tendency in the dependence of the wavenumber on Re_i and on $dRe/d\tau$. Both with increase in the acceleration and with decrease in Re_i the wavenumber first reduces from 4 to 3 and then an inverse change of the wavenumber can occur on certain Re_i and $dRe/d\tau$ ranges.

Obtained results make it possible to explain certain experimentally observed features. In the case of infinite acceleration, the change of the flow pattern with variation in Re_i physically means that the selection of the secondary flow wavenumber is very sensitive to this quantity. The noise inevitable in experiment (deviation of the rotation velocity of the sphere from given values) and its amplification with increase in the acceleration lead to a situation in which in the vicinity of Re_i corresponding to the

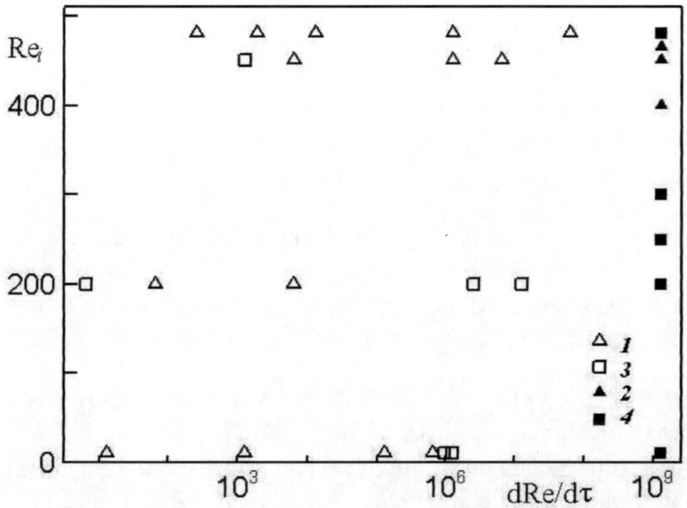

Figure 3. Dependence of the secondary flow wavenumber on Rei and dRe/dτ; (*1*) three-vortex flow; (*2*) four-vortex flow; (*3*, *4*) relate to stepwise variation in the velocity (shown conditionally for the acceleration value of 10^9).

change of the flow pattern the wavenumber selection will occur in a random fashion, since the flow will "forget" the exact value of Re_i under the noise influence. Thus, it may be inferred that the experimentally observed indefiniteness domain is formed as a result of the joint influence of two factors, namely, the noise and the presence of a boundary of the wavenumber change with variation in Re_i. In the experiment [7] only one indefiniteness domain was observed, while in the calculations two domains were found to exist between the domains of existence of different wavenumbers. Thus, it can be inferred that one more indefiniteness domain can also exist in the experiment, on the range or higher Re_i. If this domain will be experimentally found in the future, this will serve as an additional verification of the calculated results.

5. Summary

Unsteady nonequilibrium flows of a viscous incompressible fluid in a spherical layer formed both during inner sphere rotation at a constant acceleration and after its termination are calculated. It is established that the meridional component of the kinetic energy reaches a stationary value more rapidly than the azimuthal component, which can be considered as a short-duration increase in the relative thickness of the layer. It is inferred that under the action of acceleration, periodic in value and direction, the flow in the spherical layer can be destabilized.

A direct calculation of nonaxisymmetric secondary flows in the spherical layer formed after the termination of acceleration is carried out for the first time. The calculated dependence of the secondary flow wavenumber on both the acceleration and the initial Reynolds number turns out to be more complicated than that previously obtained in the experiments. Thus, for certain values of the acceleration and the initial Reynolds number one more domain of the $m = 4$ mode existence was found. In this case, the domain of the $m = 3$ mode existence separates two domains of the $m = 4$ mode existence. The possibility of the experimental verification of this result is shown and an explanation of the experimentally observed indefiniteness domains with random selection of the secondary flow wavenumber is presented.

Acknowledgments
This work was supported by the Russian Foundation for Basic Research Project No. 16-05-00004 and 18-08-00074. MG acknowledges,in part, support from the ERC Advanced Grant No. 320773. Research at the Ural Federal University is supported by the Act 211 of the Government of the Russian Federation, agreement No 02.A03.21.0006.

References

[1] Wimmer M 1976 Experiments on a Viscous Fluid Flow between Concentric Rotating Spheres *J. Fluid Mech.* **78** 317

[2] Yavorskaya I M, Belyaev Yu N and Monakhov A A 1975 Experimental Investigation of Spherical Couette Flow *Dokl. Akad. Nauk SSSR* **221** 1059

[3] Buhler K 1990 Symmetric and Asymmetric Taylor Vortex Flow in Spherical Gaps *Acta* Mech. **81** 3

[4] Peralta C, Melatos A, Giacobello M, and Ooi A 2008 Superfluid Spherical Couette Flow *J. Fluid Mech.* **609** 221

[5] Zhilenko D Yu and Krivonosova O E 2009 Linear Mode Competition in the Spherical Couette Flow after an Instantaneous Increase in the Rotation Velocity of the Inner Sphere *Dokl. Ross. Akad. Nauk* **429** 332

[6] Dumus G 1991 *Study of Spherical Couette Flow via 3-D Spectral Simulation: Large and Narrow-Gap Flows and their Transitions* Ph.D. Thesis Calif. Inst. Technol.

[7] Zhilenko D Yu and Krivonosova O E 2007 Selection of the Flow Pattern in a Wide Spherical Layer under the Action of the Inner Sphere Acceleration *Dokl. Ross. Akad. Nauk* **415** 754

[8] Rigopoulos J, Sheridan J and Thompson M 2003 State Selection in Taylor-Vortex Flow with an Accelerated Inner Cylinder *J. Fluid Mech.* **489** 79

[9] Zhilenko D Yu and Krivonosova O E 2009 Effect of Fluid Temperature and Uniformly Accelerated Rotation of the Inner Sphere on the Selection of the Flow Pattern in a Spherical Layer *Fluid Dynamics* **44** 842

[10] Nikitin N 2006 Finite-Difference Method for Incompressible Navier–Stokes Equations in Arbitrary Orthogonal Curvilinear Coordinates *J. Comput. Phys.* **217**, 759

[11] Hollerbach R, Wiener R J, Sullivan I S, Donnelly R J and Barenghi C F 2002 The Flow around a Torsionally Oscillating Sphere *Phys. Fluids* **14** 4192

[12] Belyaev Yu N and Yavorskaya I M 1980 Viscous Flows in Rotating Spherical Layers and Their Stability *Advances in Science and Engineering* (*Fluid Mech.Series.* vol. 15) (Moscow All-Union Institute of Science and Technical Information) pp 3-80

IOP Conf. Series: Journal of Physics: Conf. Series **1163** (2019) 012036 doi:10.1088/1742-6596/1163/1/012036

Numerical simulation of the motion of a free rising air bubble

K O Vlasov[1]

[1]Department of fundamental research and perspective science-consuming developments, RFNC "VNIIEF", 607188, Sarov, Russia

E-mail: konstmephi@gmail.com

Abstract. In this work the results of numerical simulation of the motion of a free rising air bubble in the water are presented. The calculations were performed for bubbles with different diameters in the range from 1 to 10 mm. Calculated parameters of the trajectory and the average rising velocity were compared with the experimental data. The vortex flow behind the bubble was analyzed. Multidirectional circulations were found. The imbalance between these circulations causes the vertical deviation in the trajectory.

1. Introduction

Today, the process of bubble's rising is not well studied and described, it is caused by a variety of physical phenomena in this problem and significant size differences. Experimental study of this problem requires the using of modern, high-precision detection tools [1, 2] and the numerical simulation is impossible without large computational resources [4, 5].

Small-size bubbles have great importance at the industry. Bubble flows can be observed in the various industrial processes: bubble column, intermixing reactor, flotation, sedimentation. Cavitation is the main rupture source for pipelines and screw propellers. All these processes depend on the concentration, sizes and rising velocities of the bubbles.

2. Physical model and numerical method

2.1. Governing equations

To simulate the rising bubble in water, we assumed that the rising occurs in an incompressible viscous liquid at a constant temperature. The movement of the liquid around the bubble is described by the solution of the equation of continuity, Navier-Stokes equations and the continuity equation for the volume concentration of water:

$$\nabla \vec{U} = 0 \tag{1}$$

$$\frac{\partial}{\partial t}(\rho \vec{U}) + \nabla(\rho \vec{U} \times \vec{U}) = -\nabla P + \nabla(\overset{=}{\tau}) + \rho \vec{g} + \vec{F}_{vol} \tag{2}$$

$$\frac{\partial}{\partial t}(c_i \rho_i) + \nabla \cdot (c_i \rho_i \vec{U}) = 0 \tag{3}$$

where ρ – density, U – velocity, P – pressure, $\overset{=}{\tau} = \mu(\nabla\vec{U} + \nabla\overrightarrow{U^T})$ – viscous stress tensor, μ – dynamic viscosity, g – acceleration of gravity, $\vec{F}_{vol}$ – exterior volume force, ρ_i – density of ith phase, c_i –

IOP Conf. Series: Journal of Physics: Conf. Series **1163** (2019) 012036 doi:10.1088/1742-6596/1163/1/012036

concentration of ith phase. The air concentration in each control volume was calculated based on the normalization condition per unit: $\rho = \rho_i \cdot c_i + \rho_j(1-c_i)$. The action of surface tension forces is described by the value of continuous volume force:

$$\vec{F}_{vol} = \sigma\rho\vec{\kappa}\nabla c_{AIR}(0.5(\rho_{H_2O} + \rho_{AIR}))^{-1}, \tag{4}$$

where σ – surface tension coefficient, $\kappa = \nabla \cdot \hat{n}$ – surface curvature, $\hat{n} = \vec{n} / |n|$ – unit surface normal. This method is called Continuum Surface Force Model (CSF) [6].

The SIMPLE method was used to solve the Navier-Stokes equations. To calculate the motion of a multiphase liquid and determine the phase interface, the VOF method was used [7]. When calculating the flow of two-phase liquids there is a problem of blurring the phase boundary due to the diffusion of one liquid to another. To reduce this effect, different techniques are used. For example, in [8] and [9], special terms are added to the continuity equation for the volume concentration of the phase (3). These terms act on the interface and reduce the diffusion of one phase to another. Therefore equation (3) takes the following form:

$$\frac{\partial c}{\partial t} + \nabla(c\vec{U}) + \nabla[c(1-c)\vec{U}_c] = 0 \tag{5}$$

where c – concentration of first phase, $\vec{U}_c = f(\nabla\alpha)$ – compressive velocity. Compaction of the liquid-gas interface is determined by the multiplier $c(1-c)$, which has a maximum value at the phase boundary (0.25) and tends to zero when c tends to 0 or 1 [8].

Figure 1 illustrates the schematic diagram of the numerical simulation, where d_b – bubble diameter, h – height of bubble initialization, H – mesh height, L – base width, ρ_1, μ_1 – air parameters, ρ_2, μ_2 – water parameters.

The sizes were chosen with the view of minimizing the influence of boundaries on the dynamics of bubble's rising and minimizing the size of the grid. The boundary conditions were set as follows: on the side boundaries – the symmetry condition, on the upper boundary – the velocity inlet condition, on the lower boundary – the free outflow condition.

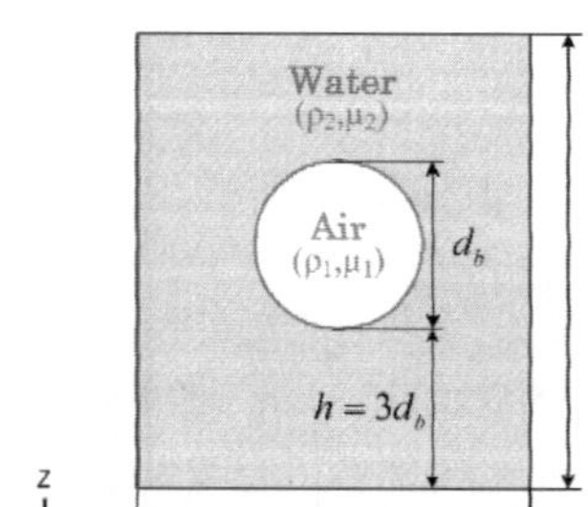

Figure 1. Schematic diagram of the numerical simulation.

In all simulations were used the following material properties: ρ_1=1.225 kg/m^3, μ_1=1.789·10^{-2} mPa·s, ρ_2=998.2 kg/m^3, μ_2=1.003 mPa·s, σ=0.072 N/m.

2.2. Moving frame

When modeling the bubble's rising to a height of more than 10 cm, it is necessary to use grids with the number of cells about 10 million. To reduce the mesh, a moving frame approach was used.

Let XYZ is a fixed coordinate system, X'Y'Z' – a moving coordinate system, U – the velocity in a fixed coordinate system, U' – the velocity in a moving coordinate system, and U_m – the velocity of a moving coordinate system, then:

IOP Conf. Series: Journal of Physics: Conf. Series **1163** (2019) 012036 doi:10.1088/1742-6596/1163/1/012036

$$\vec{U}(x,y,z,t)=\vec{U}'(x',y',z',t)+\vec{U}_m(t) \tag{6}$$

Since the X'Y'Z' coordinate system is noninertial, the equation (2) takes the form:

$$\frac{\partial}{\partial t}(\rho\vec{U}')+\nabla(\rho\vec{U}'\vec{U}')=-\nabla P+\nabla(\bar{\bar{\tau}})+\rho\vec{g}+\vec{F}_{vol}-\rho\frac{d}{dt}\vec{U}_m(t) \tag{7}$$

The value $U_m(t)$ was calculated as the velocity of the bubble center of mass in the X'Y'Z' coordinate system, and was used as a flow boundary condition with a reverse sign. As a result, the velocity in the fixed coordinate system, calculated by the formula (7), was close to zero, and the bubble actually rested. The use of a moving coordinate system significantly increased the accuracy and reduced the calculation time. This approach allowed carrying out calculations for bubbles with diameters of 1 and 2 mm in a reasonable time.

3. Numerical results

3.1. Parameters of the trajectory

Figure 2 shows the bubble center of mass trajectories for different diameters in the XYT coordinate system. It can be seen that for bubbles with diameters of 3; 4 and 5 mm there is a stable spiral motion. The bubbles with diameters of 2 and 6 mm have a combined trajectory: a zigzag mode at the initial part of the rising that switches to a stable spiral mode. For a 10 mm bubble, the trajectory looks like a spiral. For a 1 mm diameter bubble, the trajectory is close to the vertical, so its trajectory is not shown in figures 2 and 3.

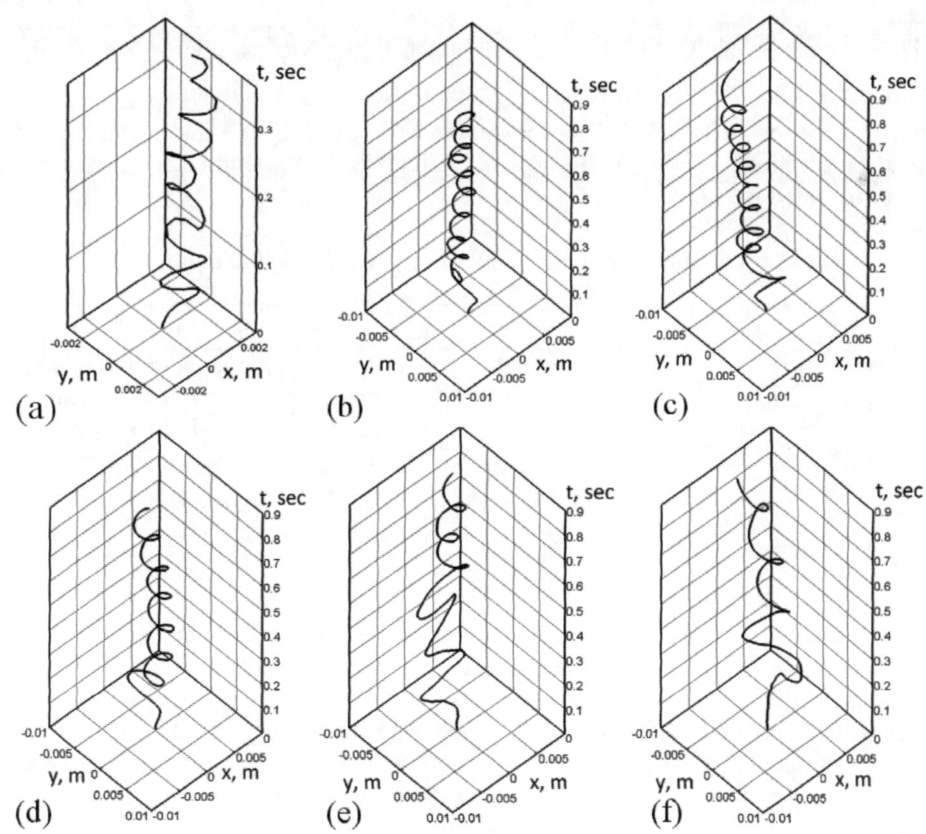

Figure 2. The paths of the bubbles in XYT: (a) $d_b = 2$ mm, (b) $d_b = 3$ mm, (c) $d_b = 4$ *mm*, (d) $d_b = 5$ mm, (e) $d_b = 6$ mm, (f) $d_b = 10$ mm.

The projections of the bubble's paths on the planes XZ, YZ was obtained. Comparison of the dependences X(t) and Y(t) shows the oscillation dynamics (figure 3).

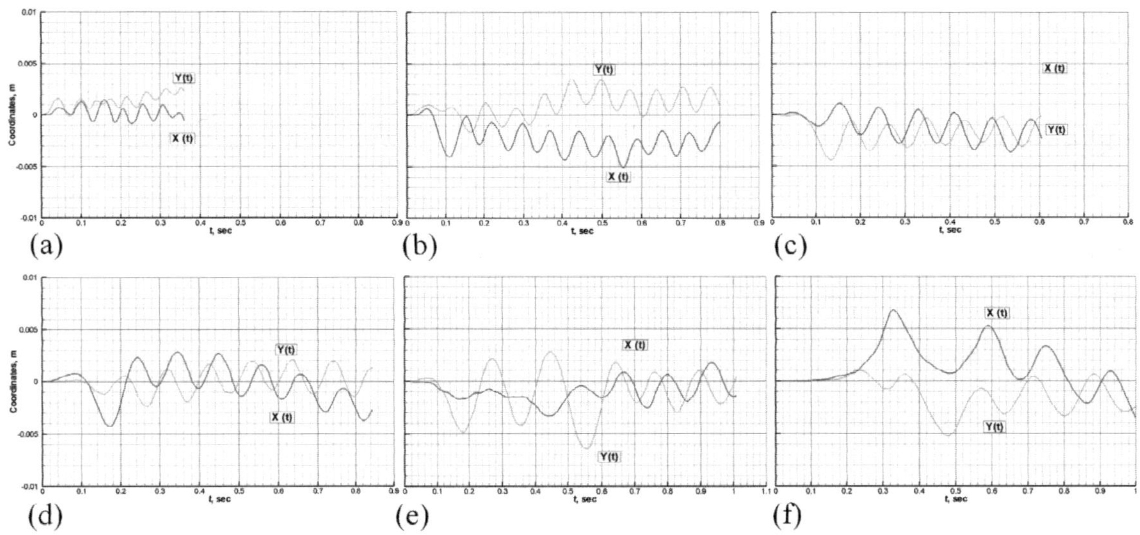

Figure 3. The dependences X(t) and Y(t): (a) d_b = 2 mm, (b) d_b = 3 mm, (c) d_b = 4 mm, (d) d_b = 5 mm, (e) d_b = 6 mm, (f) d_b = 10 mm.

Table 1 presents the calculated values of the average amplitude ($\bar{X}_{max}$, $\bar{Y}_{max}$) and the period of the bubble's oscillations. The value A, which is presented in the table is defined as: $A = \sqrt{\bar{X}_{max}^{2} + \bar{Y}_{max}^{2}}$. Analyzing the dependences X(t) and Y(t) for the bubble with diameter of 3 mm presented in [2], it was concluded that the amplitude and the period of oscillations in the experiments are greater due to the imperfect purity of water and air.

Table 1. Parameters of the trajectory for various bubble diameters.

d_b (mm)	$\bar{X}_{max}$ (mm)	$\bar{Y}_{max}$ (mm)	A (mm)	T (sec)	Path form
2.0	1.0	0.5	1.1	0.05	Zigzag and helical motion
3.0	2.0	1.5	2.0	0.08	Helical motion
4.0	2.5	1.8	3.1	0.09	Helical motion
5.0	1.75	1.75	2.5	0.10	Helical motion
6.0	1.65	1.65	2.3	0.12	Zigzag and helical motion
10.0	1.43	1.41	2.0	0.16	Helical motion

3.2. The average rising speed

Figure 4 shows a comparison of the calculated values of the average bubble rising speed (U_b) with experimental data [1,2]. For both areas of experimental data (blue and red curves), the upper curve corresponds to pure water experiments, the lower curve corresponds to experiments with contaminated water. The calculated values of U_b, lie within the range of the experimental data. In the calculations, there are no bifurcations of the speed obtained experimentally. [1-3]. Traditionally this bifurcation has been explained by the presence of surfactants (contaminated water) or by the variations of forms of the initial bubbles (sphere or ellipsoid). In the calculations carried out, the bubble during initialization is an ideal sphere, and impurities in the water are not modeled. Reproduction of bifurcation in the calculations will be very important to explain the experiments. The table in the right of the figure 4 shows the average values of U_b for different bubble's diameters. The good agreement of numerical calculation with experiment for all range of calculated values is seen. The value of the Z- component of the velocity of the bubble center of mass for a bubble of arbitrary diameter increases rapidly at first, and then fluctuates near a constant value. In the article [2] the average bubble rising speed (d_b = 3 mm) of the order of ~ 20 cm/s, which is in good agreement with the calculated value (see figure 4). The period of oscillation of U_z in the experiments: 0.08–0.10 s; in the results of calculations: ~0.06 s.

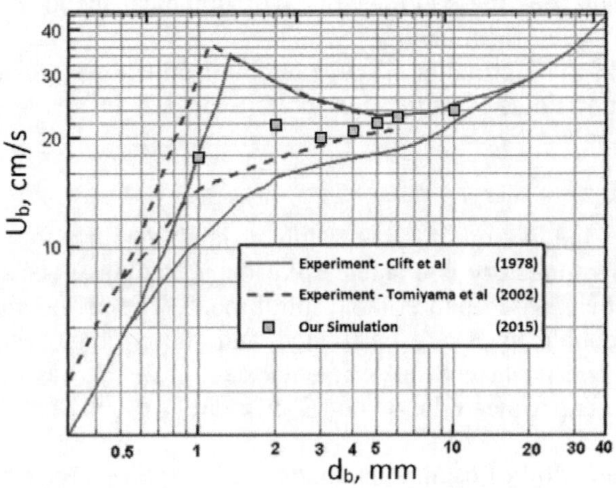

d_b, mm	U_b, cm/s
1.0	17.8
2.0	21.8
3.0	20.2
4.0	21.1
5.0	22.1
6.0	23.0
10.0	24.0

Figure 4. Comparison calculated values of U_b with experimental data.

4. Analysis of vortex flow behind the bubble

For the analysis of vortex flow behind the bubble was considered a three-dimensional field of Z-components of vorticity in the wake of the rising bubble. Figure 5(a) shows the formation of "vortex tails" rotating around the bubble's trajectory. The scale of "vortex tails" grows after the beginning of the rising.

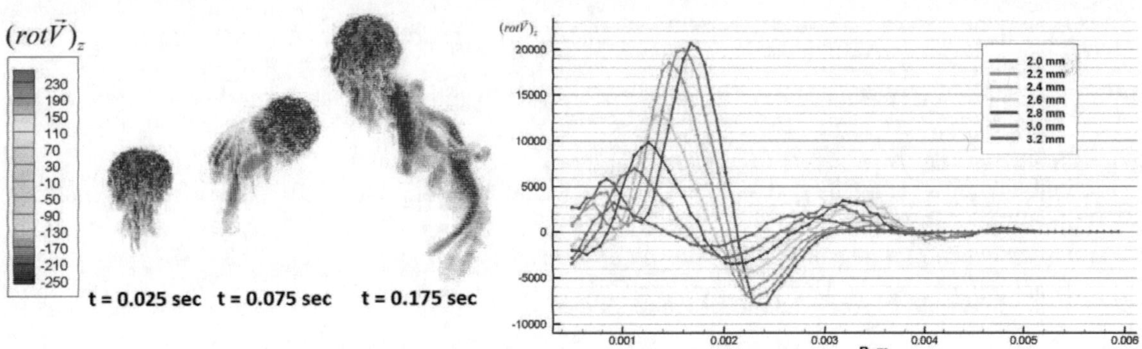

Figure 5. Analysis of vortexes behind the bubble with d_b = 3 mm: (a) 3D field of $(rot\vec{V})_z(R)$ near the bubble at different times, (b) Dependence $(rot\vec{V})_z$ for shifts from the bubble equal 2.0–3.2 mm.

Figure 5(b) shows a graph of change of the circulation velocity behind the bubble for different horizontal cross sections (shift from the bubble bottom is 2.0–3.2 mm) depending on the radius of the contour R. The contours with R of 0.5 to 6.0 mm were considered. The graph shows that one direction of the rotation is dominant and that the tail with the opposite direction with the growth of the distance to the bubble becoming inside the more powerful tail. These multidirectional circulations of the velocity in the horizontal plane generate the imbalance in flow around and behind the bubble and cause the vertical deviation in the trajectory.

5. Numerical data postprocessor

In the course of the calculations there was a problem of processing huge amounts of data (data size of the one calculation is about 100 Gb). Initially, the data storage format was replaced by the binary format CGNS (CFD General Notation System). But the time of manual processing for one calculation was several hours as before. Therefore own program for post-processing of results was created. Using the postprocessor has reduced the time of data processing to 10 min. All data processing algorithms are written in C++ programming language with Qt Framework. The graphical shell (GUI) is created in Qt Designer. The program provides a wide functionality for post-processing of calculation results and can be quickly adapted to other tasks. The program allows obtaining trajectories, shapes and vortex traces for rising bubbles. All graphs and fields of values presented in this article are obtained using this postprocessor.

6. Numerical results

The work on numerical simulation of the motion of a free rising air bubble with different diameters is carried out. It is shown that bubble's trajectory is a stable spiral or a combined trajectory: a zigzag mode at the initial part of the rising that switches to a stable spiral mode. The period and amplitude of the trajectories for all simulated bubble sizes are calculated and the average rising velocity is calculated. The obtained values are consistent with experimental data [1,2]. However, the amplitude and the period of oscillations in the experiments is greater than in CFD calculations due to foreign substances in real water and air.

Calculations have shown that the period of oscillations increases with the diameter of the bubble, while the amplitude of oscillations with the diameter increases first and after it reaches a maximum then falls. This behavior of the amplitude is due to the fact that with the increase of the diameter bubbles becoming more flattened and the increased drag force stabilizes the rising and not allows the bubble to deviate from the vertical stronger. The initiation of multidirectional circulations of the velocity in the horizontal plane behind the bubble was shown. It is assumed that the imbalance between these circulations causes the vertical deviation in the trajectory.

References

[1] Clift R, Grace J R and Weber M E 1978 *Bubbles, Drops, and Particles* (New York: Academic Press)
[2] Tomiyama A, Celata G P, Hosokawa S and Yoshida S 2002 *International Journal of Multiphase Flow* **28** 1497–1519
[3] Wu M and Gharib M 2002 *Physics of Fluids.* **14** 49–52
[4] Dijkhuizen W, Roghair I, Van Sint Annaland M and Kuipers J A M 2010 *Chemical Engineering Science* **65** 1427–1437
[5] Hua J, Stene J F and Lin P 2008 *Journal of Computational Physics* **227** 3358–3382
[6] Brackbill J U, Kothe D B and Zemach C A 1992 *Journal of Computational Physics* **100** 335–354
[7] Hirt C W and Nichols B D 1981 *Journal of Computational Physics* **39** 201–225
[8] Albadawi A, Donoghue D B, Robinson A J, Murray D B and Delaure Y M C 2014 *International Journal of Multiphase Flow* **65** 82–97
[9] Marschall H, Hinterberger K, Schuler C, Habla F and Hinrichsen O 2012 *Chemical Engineering Science* **78** 111–127

IOP Conf. Series: Journal of Physics: Conf. Series **1163** (2019) 012037 doi:10.1088/1742-6596/1163/1/012037

A comparative example of cyclostationary description of a non-stationary random process

Timofey Shevgunov[1,2]

[1] Moscow Aviation Institute (National Research University), 4 Volokolamskoe shosse, Moscow, 125993, Russia
[2] National Research University Higher School of Economics, 20 Myasnitskaya ulitsa, Moscow, 101000, Russia
E-mail: shevgunov@gmail.com

Abstract. The paper deals with cyclostationarity as a natural extension of stationarity as the key property in designing the widely-used models of random processes. The comparative example of two processes, one is wide-sense stationary and the other is wide-sense cyclostationary, is given in the paper and reveals the lack of the conventional stationary description based on one-dimensional autocorrelation functions. It is shown that two significantly different random processes appear to be characterized by exactly the same autocorrelation function while their two-dimensional autocorrelation functions provide outlook where the difference between processes of two above-mentioned classes becomes much clearer. More concise representation by expanding the two-dimensional autocorrelation function to its Fourier series where the cyclic frequency appears as the transform parameter is illustrated. The closed-form expression for the components of the cyclic autocorrelation function is also given for the random process which is an infinite pulse train made of rectangular pulses with randomly varying amplitudes.

1. Introduction

The stationary model for description of random processes could not be the best choice in the case of process containing a structural periodicity which is impossible to be reliably identified by the conventional Fourier analysis or extracted by a direct subtraction. The vital alternative for the description of these processes could be the global probabilistic model with the hidden periodicity taking into account the regular change of their statistical properties. The cyclostationarity (CS), a term originally introduced by W. Bennet [1] in the 1950s primarily to describe digitally modulated pulses, provides the formal description for the concept of the hidden periodicity including but not limited to the prominent case of periodically correlated time series and continuous signals. The cyclostationarity, which is a natural yet not evident expansion of the stationarity, has been developing by W. Gardner [2], A. Napolitano [3], H. Hurd [4], I. Javorsky [5] and others [6], although the main results obtained over few decades are still not familiar to the significant majority of modern researchers in various fields.

In general, the process $x(t)$ will be named cyclostationary (of order K) if there exists its transformation of the k-th degree such that its expectation (via ensemble averaging) is a periodic in time function. Considering a simpler case of a wide-sense cyclostationary process, one could obtain its characterization through the varying yet periodic mean $m_x(t)$, which is the first-order transformation:

$$m_x(t) = E[x(t)] = E[x(t+T)] = m_x(t+T), \qquad (1)$$

and two-dimensional autocorrelation function $R_x(t, \tau)$, which is a possible second-order transformation:

$$R_x(t,\tau) = E\left[x(t+0.5\tau)x^*(t-0.5\tau)\right], \quad R_x(t,\tau) = R_x(t+T,\tau). \tag{2}$$

The CS approach extends the statistical description based on the habitual characteristics of the well-known stationary model by taking into account periodic change in time of the chosen characteristics, e.g., the mean (1) and the auto-correlation function (2). One of the advantages of the CS approach is that the characteristics of the stationary model become the essential and additive parts of the corresponding characteristics of the more general CS model.

The paper is organized as follows. A comparative example of two processes sharing the same autocorrelation function is introduced in section 2. The importance of two-dimensional correlation function is shown in section 3 together with its link to conventional one. In section 4, the cyclic autocorrelation is described for the strictly periodic two-dimensional correlation function and gives a concise explanation for the feature of cyclostationarity. The paper ends with the conclusion.

2. Insufficiency of stationary description

One possible and relatively easy understandable example showing the clear difference between wide-sense stationary (WSS) and wide-sense cyclostationary processes (WSCS) are presented below. At first, let us define a random process $x(t)$ which is generated as

$$x(t) = \sum_{n=-\infty}^{+\infty} AX_n g(t-nT), \tag{3}$$

where T is a period, A is an amplitude multiplier, $\{X_n\}$ is a set of independent identically distributed (*i.i.d.*) random numbers defined on the same probability space and $g(t)$ is a deterministic signal of finite energy. For further simplification, the distribution for $\{X_n\}$ is chosen to be Bernoulli taking values from a set $\{-1, 1\}$ with equal probabilities. Similarly, the waveform of $h(t)$ is chosen to be a symmetric rectangule of the width Δ, such as $\Delta \leq T$:

$$g(t) = rect\left(t\Delta^{-1}\right). \tag{4}$$

A typical realization of $x(t)$ is shown in figure 1, where one can see the time sequence of the same width rectangular pulses with altering amplitudes.

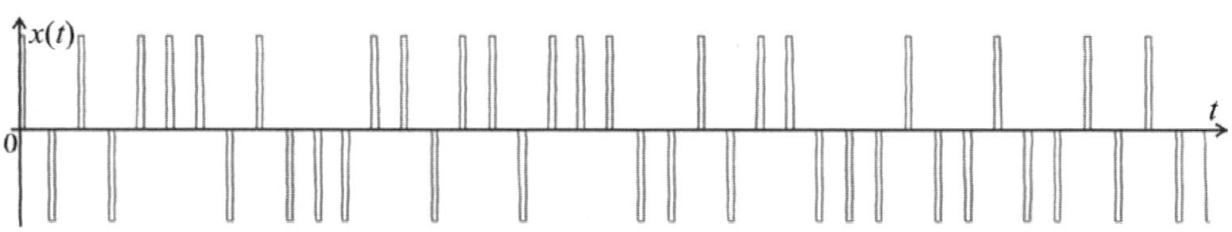

Figure 1. A typical realization of the process belonging to the cyclostationary class.

The second process being introduced as $y(t)$ has a different origin. Its typical realization is shown in figure 2. It is generated as the output of the linear time-invariant system with impulse response $h(t)=g(t)/\Delta$, where $g(t)$ is defined by equation (4), when it is fed with a white Gaussian noise $n(t)$ with uniform two-sided power spectral density $N_0=P\Delta$, where $P=A^2\Delta/T$ is the average power of the process $x(t)$. The assumption about N_0 has been made to make the average powers of both processes equal.

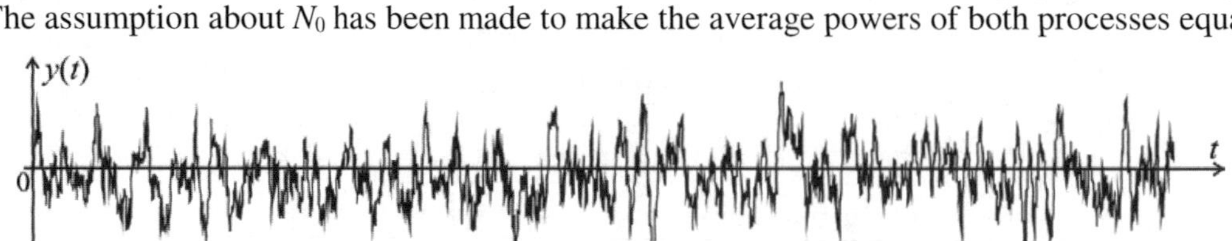

Figure 2. A typical realization of the stationary process.

The conventional description of a random process could be obtained via its autocorrelation function or power spectral density as the Fourier-counterpart in frequency domain. The autocorrelation function $R_y(\tau)$ for the process $y(t)$ could be obtained using well-known convolution relation:

$$R_y(\tau) = R_n(\tau) * R_h(\tau), \tag{5}$$

where $R_n(\tau)=N_0\delta(\tau)$ is the autocorrelation function of white noise and $R_h(\tau)$ is the autocorrelation function of the impulse response $h(t)$. The autocorrelation function of $y(t)$ is the triangle waveform shown in figure 3 together with is power spectral density $S_y(f) = P\Delta\mathrm{sinc}^2(\pi f\Delta)$ plotted in figure 4.

The correlation function $R_x(\tau)$ of the process $x(t)$ can be derived via its probability density as the expectation of the product of two observation $x^*(t)$ and $x(t+\tau)$, here superscript '*' denotes a complex conjugation. After $R_x(\tau)$ is obtained one will come to the fact, that if the process $x(t)$ is assumed to be stationary – it is not indeed but is thought in the vast majority cases – both processes will share exactly the same autocorrelation function as well as the same power spectral density.

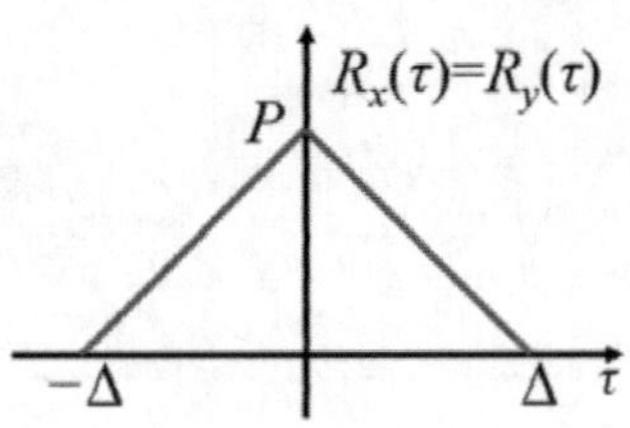

Figure 3. Autocorrelation function of both random processes.

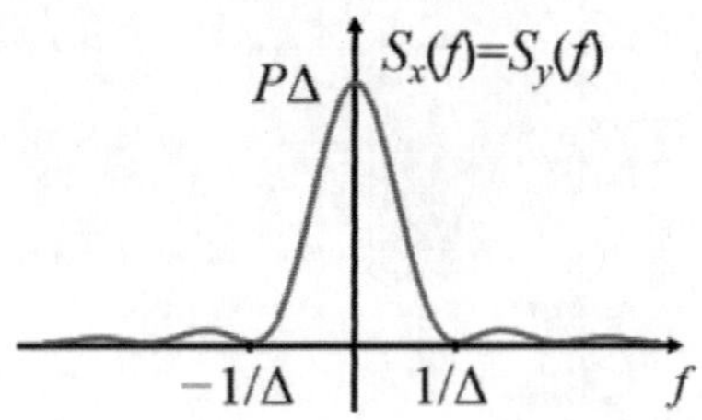

Figure 4. Power spectral density of both random processes.

A simple look at two realization in figures 1 and 2 can reveal the dissimilarity of those two processes despite the fact they have the same correlation functions. However, if those processes had been superimposed with some noise, the distinction would not remain so noticeable any more. Moreover, it is no doubt impossible to rely on such an argumentation as visible structure during an automated or machine processing of signals. The cyclostationary theory is expected to be a perfect tool for solving this and other related problems.

3. Two-dimensional correlation function

3.1. Function of two instant time arguments

The first characteristic one should consider when they move to the cyclostationary description of some process $s(t)$ is its two-dimensional autocorrelation function define via probabilistic expectation as follows:

$$R_s(t_1,t_2) = E[s(t_1)s^*(t_2)], \tag{6}$$

where t_1 and t_2 are two instants taken at two moments along the observation time.

The two-dimensional autocorrelation functions of processes $x(t)$ and $y(t)$ are shown in figures 5 and 6 correspondingly by means of the planar color diagrams which third dimension (values of the correlation) is drawn by the colour intensity. So the higher intensity is, the larger value of correlation function it expresses. The process $x(t)$ exhibits a periodical correlation pattern that is a sequence of identical prisms along the diagonal $t_2=t_1$ depicted as squares with sharp edges due to the rectangular waveform of $g(t)$ given by equation (4).

The two-dimensional ACF of $y(t)$ is different. It could basically be describe with the lines $t_1-t_2=\tau$ where the correlation remains the same and equal to $R_y(\tau)$ along. The diagram in figure 6 displays the linear gradient decay from the main diagonal if one moves in the perpendicular direction.

3.2. Symmetric form of ACF

Another form of two-dimensional ACF could be obtained by the coordinate transformation, where each point is defined by (t, τ) rather than by pair (t_1, t_2). In the former pair, τ stands for a difference, or

IOP Conf. Series: Journal of Physics: Conf. Series **1163** (2019) 012037 doi:10.1088/1742-6596/1163/1/012037

time shift, between moments of the observations while t is the current time that could be chosen in different ways. The choice enforcing useful symmetry means that t relates to the centre of the time interval (t_2, t_1):

$$\tau = t_1 - t_2, \quad t = 0.5\left(t_1 + t_2\right). \tag{7}$$

In fact, this transformation leads to the function which is generally periodic only in one dimension denoted by t. Thus, the ACF $R_x(t, \tau)$ of the process $x(t)$, which is shown in figure 7, exhibits periodicity in t with the same period as in its model given by equation (1):

$$R_x(t,\tau) = R_x(t+T,\tau). \tag{8}$$

The squares from figure 5 have changed their form to diamonds according to the transformation given by equation (7). In contrast, for the process $y(t)$, its ACF $R_y(t, \tau)$ remains the same while t is varying.

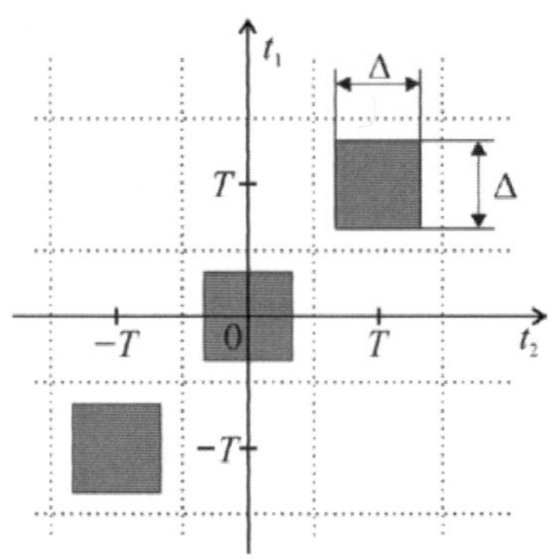

Figure 5. Two-dimensional autocorrelation function of process $x(t)$.

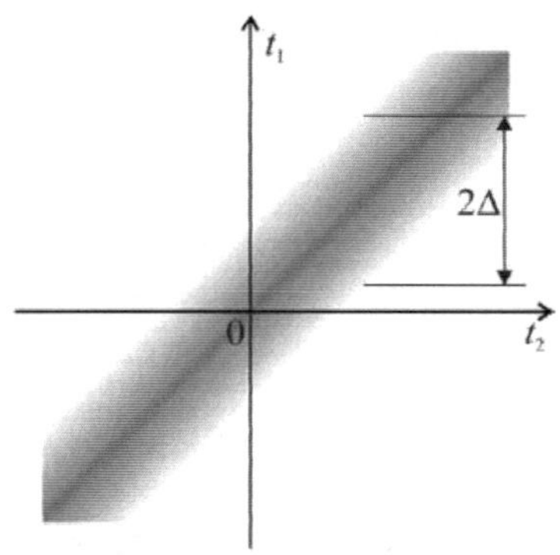

Figure 6. Two-dimensional autocorrelation function of process $y(t)$.

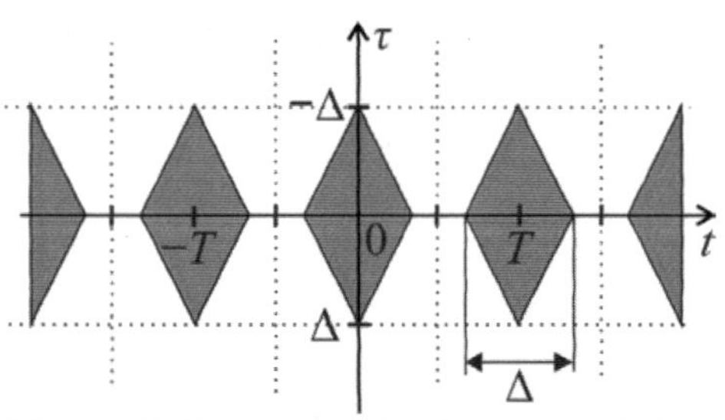

Figure 7. Symmetric (t, τ) autocorrelation function of process $x(t)$.

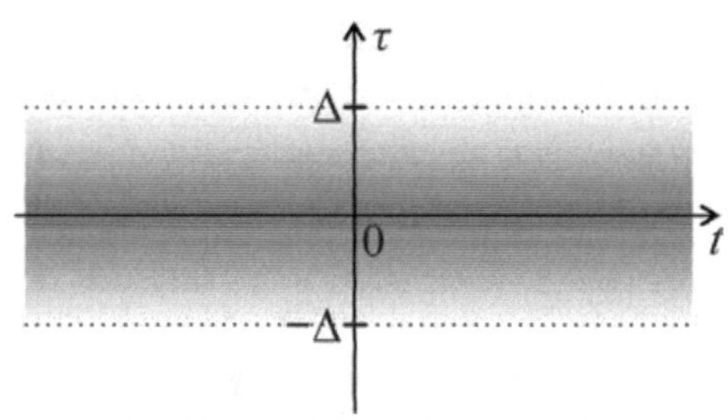

Figure 8. Symmetric (t, τ) autocorrelation function of process $y(t)$.

3.3. Averaging and stationarization

The link from two-dimensional ACF described above to the conventional one-dimensional ACF can be established using the averaging over t:

$$R_s(\tau) = \left\langle R_s(t,\tau)\right\rangle = \lim_{B\to\infty}\frac{1}{B}\int_{-B/2}^{+B/2} R_s(t,\tau)dt. \tag{9}$$

The averaging performed on $R_y(t, \tau)$, which is a constant over t for each τ, just keeps its triangle profile (see figures 8 and 3) having integrated the variable t out: $R_y(t,\tau) = R_y(\tau)$

The infinite-time averaging (7) applied to the strictly periodic function $R_x(t, \tau)$ can be simplified to the averaging over one its period:

$$R_x(\tau) = \langle R_x(t,\tau) \rangle_T = \frac{1}{T} \int_{-T/2}^{+T/2} R_x(t,\tau) dt . \tag{10}$$

The figure 7 can be easily used to conduct this averaging. Thus, for any τ such that $|\tau|<\Delta$ the cross-section of ACF is a rectangular function of t with width $\Delta-|\tau|$. If, for instance, the value $\tau=0$ is taken, than $R_x(0)=A^2\Delta/T=P$. Again, if $\tau=\pm\Delta/2$, $R_x=A^2\Delta/(2/T)=P/2$ since the width of the underlying rectangle turns to be twice shorter. Thus, the conventional ACF $R_x(\tau)$ comes as a function of time shift τ only

There is a concise interpretation of time averaging named a stationarization. The latter means that a random delay θ, which may be modelled as a random number distributed uniformly over the interval [0; T), affects each single observation of the process $x(t)$. Making ensemble averaging over θ one will come to the stationarized process made of the original $x(t)$. In turns, the stationarized process is well-described by $R_x(\tau)$, but this function is not enough to characterization of $x(t)$ itself.

4. Cyclostationary description

Since the two-dimensional ACF $R_x(t, \tau)$ is a periodic function of t as well as a constant over t, it could be represented as its Fourier series

$$R_x(t,\tau) = \sum_{\alpha \in \mathrm{A}} R_x^{\alpha}(\tau) \exp(\mathrm{j}2\pi\alpha t), \tag{11}$$

where α is called cyclic frequency and the set of cyclic frequencies $\mathrm{A}=\{\alpha : \alpha=m/T, m\in\mathbb{Z}\}$ consisting of multiples of the fundamental frequency $1/T$; j is the imaginary unit here, i.e. $\mathrm{j}^2=-1$.

The terms standing before the complex exponentials are functions of the time shift τ. Basically, they can be obtained as the coefficient of Fourier series:

$$R_x^{\alpha}(\tau) = \langle R_x(t,\tau)\exp(-\mathrm{j}2\pi\alpha t) \rangle_T = \frac{1}{T} \int_{-T/2}^{+T/2} R_x(t,\tau)\exp(-\mathrm{j}2\pi\alpha t) dt . \tag{12}$$

The set of all non-zero functions $\{R_x^{\alpha}(\tau)\}$ represents the cyclic autocorrelation function (CACF) of random process $s(t)$. This process will be called wide-sense stationary (WSS) if and only if its cyclostationary representation consists of $R_x^0(\tau)$ only, otherwise it can be called wide-sense cyclostationary (WSCS). From that point of view, it is easy to conclude that WSS is a subclass of WSCS. As a matter of fact, all components belonging to cyclic frequencies other than zero, $\alpha\neq 0$, describe the varying part of $R_x(t, \tau)$ and relate to the non-stationarity in the process.

Returning to the pulse amplitude-modulated [7] process $x(t)$, its CACF could be expressed [8], in a closed form for each integer number m:

$$R_x^{m/T}(\tau) = \begin{cases} P\left(1-|\tau|\Delta^{-1}\right)\operatorname{sinc}\left[\pi m(\Delta-|\tau|)T^{-1}\right], & |\tau|<\Delta; \\ 0, & otherwise. \end{cases} \tag{13}$$

where sinc(x) defines sin(x)/x with value 1 at x=0. The components of cyclic correlation associated with zero cyclic frequency α=0 (m=0) and m=±1, ±2, ±3 are shown by means of three-dimensional plot in figure 9. The rich presence of cyclic components reveals the fact that $x(t)$ belongs to WSCS class rather than WSS.

On the contrary, the WSS process $y(t)$ is described by a single function associated with m=0 only:

$$R_y^0(\tau) = \begin{cases} P\left(1-|\tau|\Delta^{-1}\right), & |\tau|<\Delta; \\ 0, & otherwise. \end{cases} \tag{14}$$

The cyclic decomposition of $y(t)$ shown in figure 10 is a single triangle at α=0. The dashed lines are shown only for the comparison with $R_x^{\alpha}(\tau)$ since no guess about period could be possible for $y(t)$.

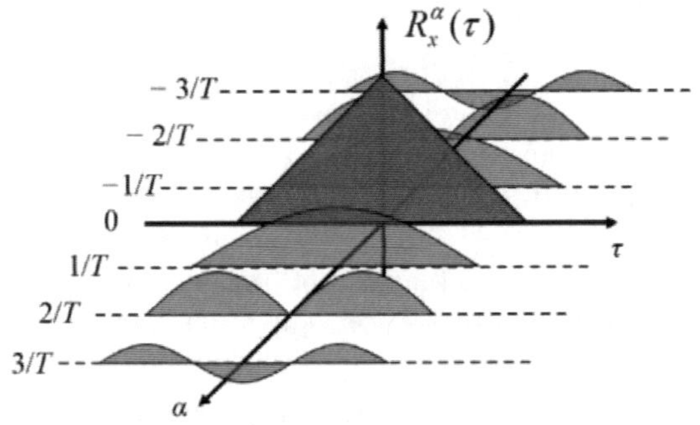

Figure 9. Cyclic autocorrelation function of the WSCS process: zero and first three components.

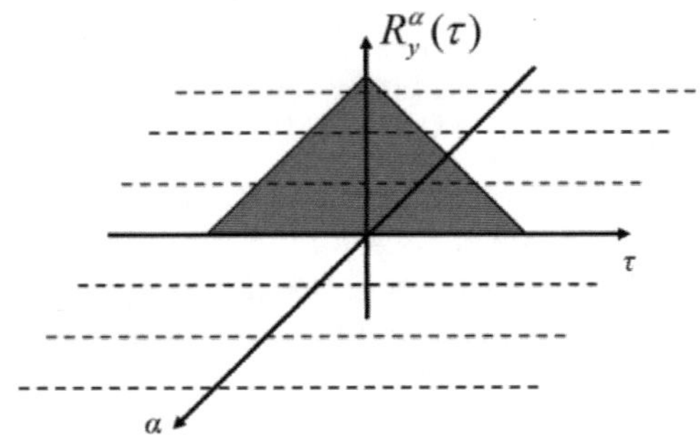

Figure 10. Cyclic autocorrelation function of process y(t). Only zero component presents.

5. Conclusion

The comparative example of wide-sense cyclostationary process considered in the paper can give one a simple introduction to cyclostationarity with revealing the reasonable distinction between two types of processes – wide-sense stationary and wide-sense cyclostationary. The essential feature standing the WSCS processes out against the stationary is the behaviour of its two-dimensional autocorrelation function. For WSCS, it basically exhibits a periodic pattern depending on both time variables rather than a one-dimensional profile extended in the cross-dimension holding for WSS.

Actually, the two-dimensional autocorrelation function could be expanded into a Fourier series with respect to the current-time variable in the case of its strict periodicity. A countable set of functional coefficients forms the representation of cyclic autocorrelation function. Its necessary part associated with zero cyclic frequency is just a conventional autocorrelation function while the other parts, if they present, are in charge of the non-stationary behaviour.

6. Acknowledgement

This work was supported by state assignment of the Ministry of Education and Science of the Russian Federation (project 8.8502.2017/BP).

7. References

[1] Bennett W R 1958 Statistics of regenerative digital transmission *Bell Syst. Tech. J.* **37** pp 1501–42

[2] Gardner W A 1990 *Introduction to random processes: with application to signals and systems* (McGraw-Hill Inc.) p 560

[3] Napolitano A 2012 *Generalizations of Cyclostationary Signal Processing: Spectral Analysis and Applications* (Wiley) p 492

[4] Hurd H L and Miamee A 2007 *Periodically Correlated Random Sequences: Spectral Theory and Practice* (Wiley-Interscience) p 384

[5] Javorskyj I, Yuzefovych R, Matsko I, Zakrzewski Z and Majewski J 2017 Coherent covariance analysis of periodically correlated random processes for unknown non-stationary period *Digital signal Processing* **65** (Elsevier) pp 27-51

[6] Gardner W A, Napolitano A. and Paura L 2006 Cyclostationarity: Half a century of research *Signal Processing* **86** (Elsevier) pp 639–697

[7] Shevgunov T, Efimov E. and Zhukov D 2018 Averaged absolute spectral correlation density estimator *Proc. Moscow Workshop on Electronic and Networking Tech.* pp 1-4

[8] Efimov E, Shevgunov T and Kuznetsov Y 2018 Time delay estimation of cyclostationary signals on PCB using spectral correlation function *Proc. Baltic URSI Symp. (Poznan)* pp 184-187

IOP Conf. Series: Journal of Physics: Conf. Series **1163** (2019) 012038 doi:10.1088/1742-6596/1163/1/012038

Calculation of the third-order elastic constants of charge-stabilized colloidal crystals with monatomic hexagonal crystal lattice

P E Dyshlovenko

Department of Applied Mathematics and Informatics, Ulyanovsk State Technical University, 32 Severny Venets Street, Ulyanovsk 432027, Russia

E-mail: p.dyshlovenko@mail.ru

Abstract. Elastic constants up to the third order of a charge-stabilized colloidal crystal with monatomic hexagonal crystal lattice are obtained numerically. The properties of the crystal are described by the Poisson-Boltzmann nonlinear differential equation for the electric potential. The elastic constants are obtained from the stress-strain relations for the crystal treated as a solid medium with an initial stress.

1. Introduction

Colloidal crystals are dispersions of solid particles in a liquid medium in which the particles are spatially ordered and form a crystal structure of a certain type. Interest in colloidal crystals has recently increased significantly due to the possibility of their use in creation of photonic crystals and self-assembling structures [1]. They can also serve as a model in study of conventional molecular crystals as well as disordered colloidal systems.

In charge stabilized colloidal crystals, the particles are electrically charged, while the liquid medium is usually an electrolyte. We describe a model crystal which is formed by electrically charged rigid disks arranged into a hexagonal crystal lattice and immersed into an electrolyte with two species of valences +1 and −1 (1:1 electrolyte). Electric potential in the electrolyte is described by the nonlinear Poisson-Boltzmann (PB) differential equation in its standard form [2]. It is appropriate for the systems with a 1:1 electrolyte, in which the correlation effects are rather small. Correlation effects that have been intensively studied over the past twenty years [3-9] may possibly be included in the model in the form of a modified Poisson-Boltzmann equation. The van der Waals forces are not included in the model to keep it simple and to keep track of the electrostatic contributions solely into the elastic properties of the crystal.

To calculate elastic constants, we treat the crystal as a solid medium with an initial stress [10]. External pressure in equilibrium is necessary here for stability of the crystal and is provided by the external boundaries. One can also consider closed areas like a drop [11]. An alternative mechanism of stabilization is described in [12].

We propose the numerical method and calculate elastic constants of the model crystal up to the third order. Elastic constants are derived from the stress-strain relations obtained numerically. We employ the relations between the second Piola-Kirchhoff stress tensor and the Lagrange strain tensor [13]. Elastic constants up to the second order for this crystal were obtained earlier in [14] by an alternative method. We compare both sets of the results.

IOP Conf. Series: Journal of Physics: Conf. Series **1163** (2019) 012038 doi:10.1088/1742-6596/1163/1/012038

2. Description of the Model

The unit cell of the crystal is shown in the figure 1(a). The model has three parameters: the lattice parameter a and the radius of the particles R and the surface charge density on the particles σ_0.

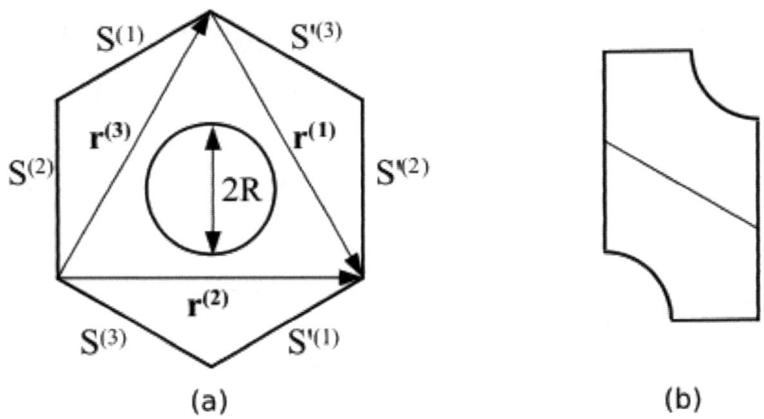

Figure 1. The unit cell of the hexagonal crystal with a circular particle at the centre (a) and domain for the numerical calculations (b). Under strain, both the cell and domain are deformed accordingly. Symbols are described in the text.

Electric potential φ in the electrolyte at the absolute temperature T with the bulk concentration of the species n_0 and the relative dielectric permittivity ε obeys the PB equation

$$\nabla^2\varphi = \frac{2q_e n_0}{\varepsilon_0\varepsilon}\sinh\left(\frac{q_e\varphi}{kT}\right). \tag{1}$$

Using the Debye length $\kappa^{-1} = \left(2n_0 q_e^2/\varepsilon_0\varepsilon kT\right)^{-1/2}$ as a distance unit and expression kT/q_e for the electric potential, one can reduce the equation (1) to its dimensionless form

$$\nabla^2\varphi = \sinh\varphi. \tag{2}$$

The dimensionless quantities which are convenient for computer calculations are only used hereafter in the present paper.

We assume that the dielectric permittivity of the electrolyte is big and the approximation of the infinite dielectric permittivity can be used. This is a good approximation for aqua solutions. Then, the domain of the problem includes only the electrolyte region and the following nonhomogeneous Neumann boundary conditions (dimensionless) hold on the surface of the particle:

$$-\nabla\varphi\cdot\mathbf{n} = \sigma_0, \tag{3}$$

where $\mathbf{n}$ is an external unit normal vector of the particle. Electric potential inside the particle can be calculated separately if needed.

The domain for numerical calculations is shown in the figure 1(b). It is only a part of the whole unit cell. The straight external sides of the domain are all the mirror axes of the crystal. Thus the simple homogeneous Neumann boundary conditions hold on them:

$$-\nabla\varphi\cdot\mathbf{n} = 0, \tag{4}$$

where $\mathbf{n}$ is an external unit normal vector of the domain. This is valid for the deformed configurations as well since only the stretching in x and y directions are used in the numerical experiments (see below).

Equations (2)–(4) together with the domain constitute the boundary value problem to be solved in any given spatial configuration.

3. Numerical Experiment

We define elastic constants from the following stress-strain relations [13]:

$$\Sigma_{ij} = C_{ij}\delta_{ij} + C_{ijkl}\eta_{kl} + \tfrac{1}{2}C_{ijklmn}\eta_{kl}\eta_{mn} + \ldots, \tag{5}$$

where dots designate cubic and higher order terms omitted, Σ_{ij} is the second Piola-Kirchhoff stress tensor, η_{kl} is the Lagrange strain tensor, C_{ij}, C_{ijkl} and C_{ijklmn} are tensors of the elastic constants, δ_{ij} is a Kronecker's delta-symbol . The subscripts go over the values 1, 2 that correspond to the x, y directions respectively and summation over repeating subscripts is implied. Elastic constants are symmetric with respect to the interchanges of subscripts in each pair as well as of different pairs of subscripts.

Some elastic constants are identically equal to zero while the non-zero constants are not fully independent due to the high symmetry of the hexagonal crystal lattice. The choice of independent non-zero constants is not unique. Our choice in the present paper is the following: one elastic constant of the first order, C_{11}, two constants of the second order, C_{1111} and C_{1122}, and three constants of the third order, C_{111111}, C_{111122} and C_{222222}. In the Voigt's notation they are C_1, C_{11}, C_{12}, C_{111}, C_{112}, C_{222} respectively. The rest non-zero constants can be expressed through the independent ones.

The independent elastic constants are obtained from the two series of deformation: stretching in *x*-direction (zig) and unibiaxial (bi) stretching in *xy*-plane. The first series gives two stress-strain relations

$$\Sigma_1^{zig} = C_1 + C_{11}\eta + \tfrac{1}{2}C_{111}\eta^2 + \ldots, \tag{6a}$$

$$\Sigma_2^{zig} = C_1 + C_{12}\eta + \tfrac{1}{2}C_{112}\eta^2 + \ldots, \tag{6b}$$

where $\eta \equiv \eta_1$ is just a strain parameter. The second series gives

$$\Sigma_1^{bi} = \Sigma_2^{bi} = C_1 + (C_{11} + C_{12})\eta + \tfrac{1}{2}(2C_{111} - C_{222} + 3C_{112})\eta^2 + \ldots, \tag{6c}$$

where $\eta \equiv \eta_1 = \eta_2$.

The stress tensor Σ_{ij} (tensor notation) is related to the Cauchy stress tensor T_{kl} as follows:

$$\Sigma_{ij} = J\gamma_{ik}\gamma_{jl}T_{kl}\ , \tag{7}$$

where γ_{ij} are components of the tensor inverse to the deformation gradient tensor α_{ij}, $J = \det(\alpha_{ij})$, and $\alpha_{ij} = \delta_{ij} + u_{ij}$, where u_{ij} is a displacement gradient. Since $\eta_{ij} = 1/2(u_{ij} + u_{ji} + u_{ki}u_{kj})$, the only non-zero components of tensor u_{ij} are the following: $u_{11} = -1 + \sqrt{1+2\eta}$ in the first experiment and $u_{11} = u_{22} = -1 + \sqrt{1+2\eta}$ in the second one.

To calculate the Cauchy stress tensor T_{kl} we need some designations shown in the figure 1(a). The sides of the hexagon, both in the initial and deformed configurations, can be split into three pairs, $(S^{(m)}, S'^{(m)})$, $m = 1,2,3$. The opposite sides in each pair are separated by the corresponding vectors of elementary translations, $\mathbf{r}^{(m)}$, $m = 1,2,3$. These vectors for the initial configuration written in the Cartesian coordinate system are the following: $\mathbf{r}^{(1)} = \left(\tfrac{1}{2}, -\tfrac{\sqrt{3}}{2}\right)a$, $\mathbf{r}^{(2)} = (1,0)a$, $\mathbf{r}^{(3)} = \left(\tfrac{1}{2}, \tfrac{\sqrt{3}}{2}\right)a$. In the deformed configurations, the vectors $\mathbf{r}^{(m)}$ are transformed according to the strain.

The Cauchy stress tensor T_{ik} in an arbitrary configuration is calculated then in the following way [15]:

$$T_{ik} = \frac{1}{V_c}\sum_{m=1}^{3} r_k^{(m)} \int\limits_{S'^{(m)}} \Pi_{ij} da_j\ , \tag{8}$$

where integration is carried out over the sides $S'^{(m)}$ pointed by the vectors $\mathbf{r}^{(m)}$, $r_k^{(m)}$ are components of the vectors $\mathbf{r}^{(m)}$, V_c is an area of the unit cell and Π_{ij} is the stress tensor of the PB equation for the 1:1 electrolyte:

$$\Pi = \nabla\varphi \otimes \nabla\varphi - (\tfrac{1}{2}|\nabla\varphi|^2 + \cosh\varphi - 1)I, \tag{9}$$

where I is a unit tensor.

We obtain stress-strain relations (6) for the crystals with $R=1$, $\sigma_0 = 2$ and the lattice parameter a in the range 2.3 to 6.0. Strain parameter η in (6) varies from –0.1 to +0.1 with step 0.001. Polynomial approximation of the experimental stress-strain curves gives the elastic constants.

The boundary value problem for any given set of parameters R, σ_0, a and η is solved by the finite element method with triangular free meshes and quadratic Lagrangian finite elements. Calculations were supported by the Supercomputing Center of Lomonosov Moscow State University [16].

4. Results and Discussion

Table 1 contains numerical results for the elastic constants of the crystal with parameters $R=1$, $\sigma_0 = 2$ and $a=3$.

Table 1. Elastic constants of the third order for $R=1$, $\sigma_0=2$, $a=3$.

Constant	C	ΔC_{LMS} (0.95)	C from [14]
C_1	-1.532724906	$2.0\ 10^{-8}$	1.53272540
C_{11}	6.257451	$2.0\ 10^{-6}$	6.257492
C_{12}	1.723446	$2.0\ 10^{-6}$	1.723484
C_{111}	-43.4522	0.0012	–
C_{112}	-2.9900	0.0009	–
C_{222}	-36.5130	0.0012	–

Elastic constants of the first and second orders are very close to the same constants obtained earlier in [14] by a different method. The least mean square errors are small although increase with the order of the constants. The third-order elastic constants are all negative making the stress-strain curves convex up. The C_{222} is not equal to the C_{111} in contrast to the second-order case where always $C_{22} = C_{11}$ in the hexagonal lattice. That means that elastic properties of the crystal are not isotropic provided the third-order elastic constants are taken into account.

The third-order elastic constants for the whole range of the lattice parameter a are shown in the figure 2. The absolute value of the constants falls monotonically with the growth of a. The curves do not cross each other so that the elastic behavior of the crystals is qualitatively the same for any density.

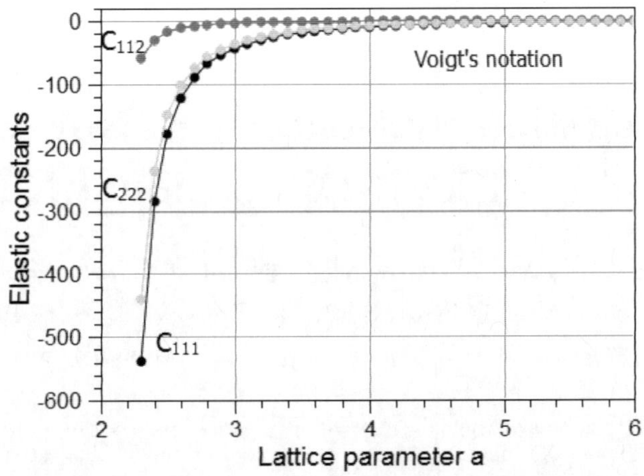

Figure 2. The third-order elastic constants of the crystal.

5. Conclusion

The third-order elastic constants of the model crystals were calculated for the first time. The proposed numerical method proves to be good for calculation of elastic constants of the two-dimensional colloidal crystals up to the third order and possibly higher.

Acknowledgements

The research is carried out using the equipment of the shared research facilities of HPC computing resources at Lomonosov Moscow State University.

References

[1] Joannopoulos J D, Villeneuve P R and Fan S H 1997 *Nature* **386** 143.
[2] Belloni L 2000 *J. Phys.: Condens. Matter* **12** R549.
[3] Abrashkin A, Andelman D and Orland H 2007 *Phys. Rev. Lett.* **99** 077801.
[4] Grochowski P and Trylska J 2008 *Biopolymers* **89**, 93.
[5] Naji A, Kanduč M, Forsman J and Podgornik R 2013 *J. Chem. Phys.* **139** 150901.
[6] Budkov Y A, Kolesnikov A L and Kiselev M G 2015 *Europhys. Lett.* **111** 28002.
[7] Budkov Y A, Kolesnikov A L and Kiselev M G 2016 *J. Chem. Phys.* **144** 184703.
[8] Budkov Y A 2018 *J. Phys. Condens. Matter* **30** 344001.
[9] Budkov Y A, Kolesnikov A L, Goodwin Z A H, Kiselev M G and Kornyshev A A 2018 *Electrochimica Acta* **284** 346.
[10] Wallace D C 1965 *Rev. Mod. Phys.* **37** 57.
[11] Keim P, Maret G, Herz U and von Grünberg H H 2004 *Phys. Rev. Lett.* **92** 215504.
[12] Budkov Y A, Frolov A I, Kiselev M G and Brilliantov N V 2013 *J. Chem. Phys.* **139** 194901.
[13] Holzapfel G A 2000 *Nonlinear Solid Mechanics: A Continuum Approach for Engineering* (Chichester: Wiley).
[14] Gladkova E V and Dyshlovenko P E 2015 *Bulletin of Chelyabinsk State University* **22 (377)** 110.
[15] Dyshlovenko P E 2010 *Colloid J.* **72** 627.
[16] Sadovnichy V, Tikhonravov A, Voevodin Vl and Opanasenko V 2013 *"Lomonosov": Supercomputing at Moscow State University*. In Contemporary High Performance Computing: From Petascale toward Exascale (Chapman & Hall/CRC Computational Science), (Boca Raton, USA: CRC Press) 283-307.

IOP Conf. Series: Journal of Physics: Conf. Series **1163** (2019) 012039 doi:10.1088/1742-6596/1163/1/012039

Disorder Aggregation by Vortices in Non-Equilibrium Critical Annealing of Two-Dimensional XY-model

Ivan S Popov[†], **Anna P Popova**[‡], **Pavel V Prudnikov**[§]
Department of Theoretical Physics, Omsk State University, Omsk 644077, Russia
E-mail: [†]diphosgen@mail.ru, [‡]popovaap@omsu.ru, [§]prudnikp@univer.omsk.su

Abstract. It is shown that the non-equilibrium vortex annealing of structural disorder in a two-dimensional XY-model leads to a non-equilibrium critical coarsening. Large clusters are separated into an individual subsystem, the relaxation dynamics of which is much slower than the relaxation dynamics of the cluster subsystem as a whole. Despite the fact that the main equilibrium critical properties, even the Berezinskii-Kosterlitz-Thouless phase transition temperature $T_{\rm BKT}(p)$, of the two-dimensional XY-model with disorder thermalization do not change as compared with the system with quench disorder, the non-equilibrium critical relaxation shows significant differences.

The study of critical behavior of disordered systems causes considerable fundamental and applied scientific interest [1, 2]. The critical phenomena are characterized by a strong fluctuation behavior. In the system at the critical state there are anomalously large and long-lived fluctuations of the main thermodynamic quantities, characterized by strong interaction. Lowering the dimension of the system leads to increased fluctuation effects. The systems with continuous symmetry of the ground state are usually characterized by much stronger fluctuation effects than systems with discrete symmetry. Therefore, low-dimensional systems with continuous symmetry are characterized by a significant contribution of fluctuations to critical properties [3]. In two-dimensional systems with continuous symmetry, the long-range order (LRO) is destroyed by anomalously strong transverse fluctuations of the spin density at all non-zero temperatures $T \neq 0$. However, in such systems there are more exotic forms of ordering. The main features of such systems are associated with the existence of topological excitations in them, such as vortices and other topological defects [4–6].

A special place among two-dimensional systems with continuous symmetry is occupied by a two-dimensional XY-model [3–5]. Despite the absence of a long-range order at all non-zero temperatures $T \neq 0$, there exists a Berezinskii-Kosterlitz-Thouless (BKT) topological phase transition at temperature $T = T_{\rm BKT}$ in it. In this system there is a low-temperature Berezinskii phase $T < T_{\rm BKT}$, where each temperature T is a critical point, i.e. there is a continuous set of fixed points of the renormalization group transformation. This makes it possible to study the critical behavior of the system not only in one isolated point $T_{\rm C}$ (as, for example, in the Ising model), but in the whole temperature range $T \leq T_{\rm BKT}$. The low-temperature phase has a quasi-long-range order (QLRO). In this case, the spatial correlation function decreases by power-law to zero with increasing distance. The critical exponents of the two-dimensional XY-model depend on temperature, for example $\nu = \nu(T)$. The phase transition is physically related to dissociation of the coupled vortex pairs at the transition point. In the two-dimensional XY-model in the

IOP Conf. Series: Journal of Physics: Conf. Series **1163** (2019) 012039 doi:10.1088/1742-6596/1163/1/012039

state of thermodynamic equilibrium in the low-temperature Berezinskii phase $T < T_{\mathrm{BKT}}$, all the vortices are connected in coupled pairs, and in the high-temperature phase $T \geq T_{\mathrm{BKT}}$ the vortices are uncoupled.

The two-dimensional XY-model is used to describe the critical properties of a wide range of real physical systems [3], such as critical properties of ultra-thin magnetic films [7] and extensive class of easy plane planar magnets [8,9], including quite specific ones Rb_2CrCl_4 [10], $(C_6H_5CH_2NH_3)_2CrBr_4$ [11], $(CH_3NH_3)_2MnCl_4$ [12], $(\mathrm{tetrenH}_5)_{0.8}Cu_4[W(CN)_8]_4 \cdot 7.2H_2O$ [13]; singularities in the critical properties of superconducting thin films [14,15] and superfluid thin films of liquid helium [16–19]; arrays of Josephson junctions [3,20,21] and SFS-junctions [22–24]; two-dimensional crystals [5] and smectic liquid crystals [25–29]; some correlation properties of two-dimensional turbulence [30]; melting of several layers of sorbed xenon in single-crystal graphite [31]; the process of sorbing hydrogen on tungsten W(011) $p(2 \times 2)$ [32]; and some properties of many other physical systems [3,33]. Also, to the phenomena described by the XY-model, it is possible to relate dynamic properties of such exotic systems as flocks of birds [34] and some other forms of collective behavior of living organisms [35,36].

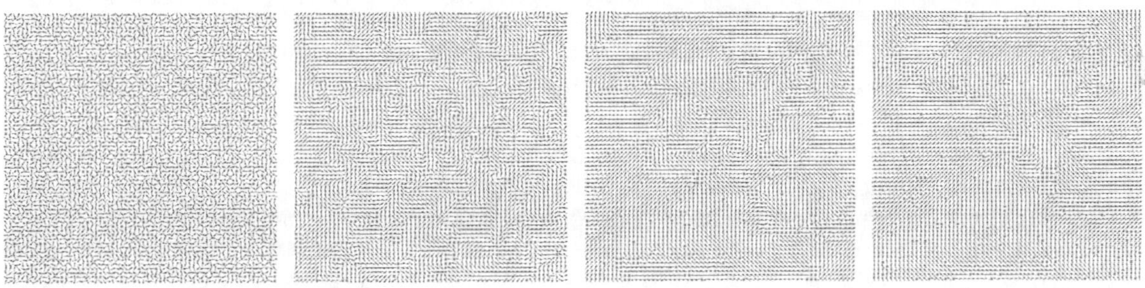

Figure 1. Snapshot configurations of the system with linear size $L = 64$ and spin concentration $p = 0.9$ at times 0, 500, 2000 and 10000 MCS/s. Arrows indicate the spins, red squares indicate the defects.

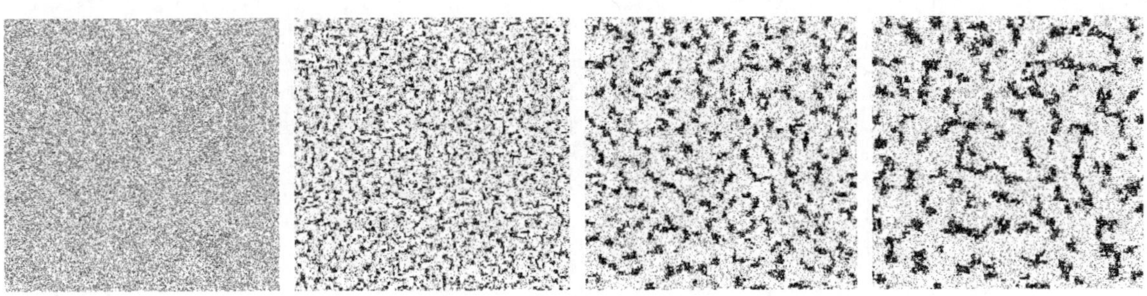

Figure 2. Snapshot configurations of the defects in the system with linear size $L = 512$ and spin concentration $p = 0.9$ at times 0, 5000, 20000 and 100000 MCS/s.

The non-equilibrium critical relaxation of a two-dimensional XY-model can have vortex and spin-wave characters [37–41]. It depends on the initial non-equilibrium state. If the initial state is the low-temperature one, prepared in contact with a thermostat at zero temperature $T_0 = 0$, then the relaxation dynamics will be spin-wave [37,38,40]. In the case of the initial high-temperature state the system is prepared at a high temperature $T \gg T_{\mathrm{BKT}}$ thermostat, and the relaxation is accompanied by vortex dynamics [37–39]. Inclusion the structural disorder in the model leads to a significant qualitative change in critical behavior [39–41]. The effect of structure

IOP Conf. Series: Journal of Physics: Conf. Series **1163** (2019) 012039 doi:10.1088/1742-6596/1163/1/012039

defects on spin-wave relaxation is not considered in this work, details are given in another our work [40]. The effect of structure defects on the non-equilibrium critical vortex dynamics is accompanied by the appearance of non-equilibrium pinning of vortices on defects [39, 41].

Existing research (see [2, 39–42] and refs. in them) focuses on the study of the quenched disorder influence on the vortex dynamics of a two-dimensional XY-model. The quenched disorder model assumes that the relaxation time of the defect subsystem is significantly longer than the relaxation time of the spin subsystem. In the papers [43, 44] it was shown that the vortex relaxation of the diluted model occurs extremely slowly. This particularly applies to the vortex concentration and geometric characteristics of QLRO-areas. Therefore, there is an interest in the study of the structural disorder thermalization in the diluted two-dimensional XY-model.

In a diluted XY-model with a quenched structural disorder in the process of non-equilibrium pinning, vortices are attracted by defects. If defects are defrosted, a reverse process will occur, when the defects are attracted by vortices. Defects will be collected in vortex cores, thereby creating non-equilibrium defect aggregations. The first produced simple Monte-Carlo simulation shows that this effect does take place (see Fig. 1 and Fig. 2). When the system goes into the thermodynamic equilibrium state, the vortices will gather into the vortex pairs. In this case defects can not be pinned with the vortices, and the defect clusters will be destroyed. Thus, the non-equilibrium annealing of defects by vortices in the diluted two-dimensional XY-model will be accompanied by a substantially non-equilibrium effect associated with non-equilibrium aggregation. This paper is devoted to research of this disorder aggregation by vortices in non-equilibrium critical annealing of diluted two-dimensional XY-model.

The Hamiltonian of the two-dimensional XY-model in this work was chosen as

$$H[p, \mathbf{S}] = -\frac{1}{2}\sum_{\langle i,j\rangle} p_i p_j \mathbf{S}_i \mathbf{S}_j = -\frac{1}{2}\sum_{\langle i,j\rangle} p_i p_j \cos\left(\varphi_i - \varphi_j\right), \tag{1}$$

where $\mathbf{S} = \{\mathbf{S}_i(t)\}$ and $p = \{p_i(t)\}$ – spin and defect lattice fields; $\mathbf{S}_i = (S_{i,x}, S_{i,y}) \equiv (\cos\varphi_i, \sin\varphi_i)$ is a classical planar spin which is associated with i-node of a square lattice with the linear size L; φ_i is a phase of spin $\mathbf{S}_i$; p_i is an occupation number of i-node: if $p_i = 1$ then i-node is occupied by a spin, else, if $p_i = 0$, by a defect; $\sum_{\langle i,j\rangle}$ is a summation over all pairs of the nearest neighbors. Defects on the lattice are distributed uniformly at time $t = 0$ with probability $P(p_i) = (1-p)\delta(p_i) + p\delta(1-p_i)$, where p – spin concentration, i.e. $c_{\text{imp}} = 1 - p$ is the concentration of impurity. Simulation of critical dynamics of the system was carried out by Metropolis algorithm, as it adequately sets the dynamic properties of this system [45]. To include the mobility of defects, the elementary step of the Metropolis algorithm was modified. If a randomly selected node is occupied by a spin, then the classical elementary step is produced. But if a node is occupied by a defect, then an attempt is made to swap this defect with a random neighboring spin. Similarly, the calculation of ΔE is carried out and the elementary step of Metropolis algorithm is used. For analysis of defect cluster structure the Hoshen-Kopelman algorithm [46–48] is used. The study was carried out for spin concentrations $p = 0.9$, 0.8, 0.7 and 0.6. Spin concentration $p = 0.6$ is close to the spin percolation threshold $p_c \simeq 0.592745(2)$ [49] by the nearest neighbors on the two-dimensional square lattice.

Initially in this work the temperatures of Berezinskii-Kosterlitz-Thouless phase transition $T_{\text{BKT}}(p)$ were determined for considered spin concentrations p. To do this, we used Binder cumulants $U_4 = \frac{1}{2}(3 - [\langle M^4\rangle]/[\langle M^2\rangle]^2)$ and spatial correlation function $C(r)$ ratio at two different scales $R = [\langle C(L/2)\rangle]/[\langle C(L/4)\rangle]$. The temperature dependences of these quantities for three different values of L have triangles in the region $T_{\text{BKT}}(p)$. As a result of simulation, $T_{\text{BKT}}(p = 0.9) = 0.679(6)$, $T_{\text{BKT}}(p = 0.8) = 0.487(9)$, $T_{\text{BKT}}(p = 0.7) = 0.339(9)$ and $T_{\text{BKT}}(p = 0.6) = 0.161(5)$. Also it is important to note that the temperatures $T_{\text{BKT}}(p)$ for spin concentrations $p = 0.9$, 0.8, 0.7 and 0.6 are equal within the statistical error to the corresponding

IOP Conf. Series: Journal of Physics: Conf. Series **1163** (2019) 012039 doi:10.1088/1742-6596/1163/1/012039

values for the two-dimensional XY-model with quenched disorder [50]. However, this result is not unexpected. It is well known that annealed structural disorder does not significantly affect the equilibrium critical behavior [51]. In this case, the influence of annealed structural disorder did not even lead to a change in the temperature of the Berezinskii-Kosterlitz-Thouless phase transition $T = T_{\mathrm{BKT}}(p)$.

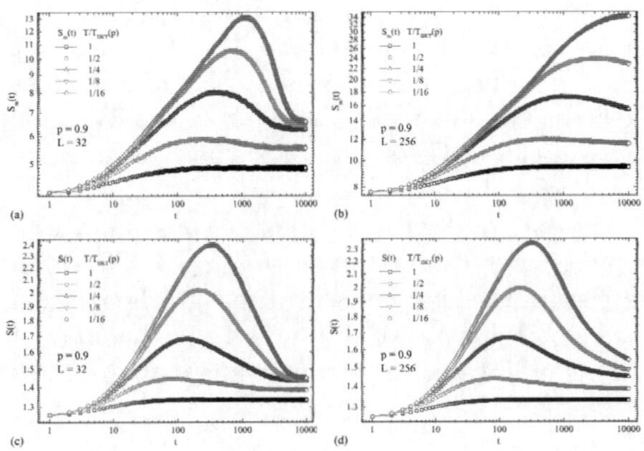

Figure 3. The dynamic dependences of the size of the largest cluster of defects $S_{\mathrm{m}}(t)$ (a, b) and of the average size of clusters of defects $S(t)$ (c, d) for the system with spin concentrations $p = 0.9$ and linear sizes $L = 32$ (a, c) and $L = 256$ (b, d).

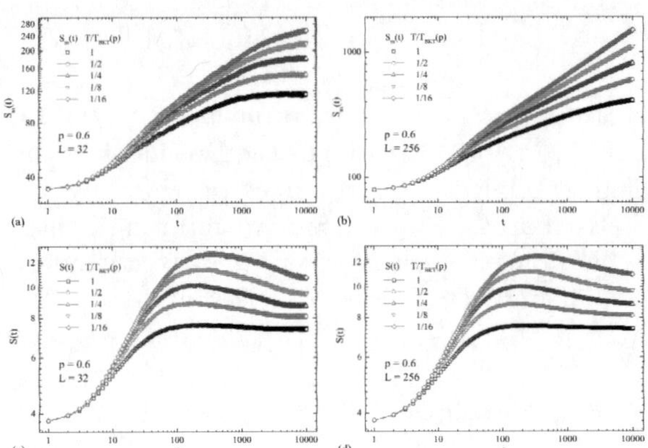

Figure 4. The dynamic dependences of the size of the largest cluster of defects $S_{\mathrm{m}}(t)$ (a, b) and of the average size of clusters of defects $S(t)$ (c, d) for the system with spin concentrations $p = 0.6$ and linear sizes $L = 32$ (a, c) and $L = 256$ (b, d).

Subsequent research in this work was focused on the study of the non-equilibrium critical annealing of defects by vortices. The expected behavior of the subsystem of defects showed that there is a non-equilibrium aggregation of defects in vortex cores (see Fig. 1 and Fig. 2). In the vortex cores, defects are collected in clusters. Therefore, the geometric characteristics of defect clusters were chosen to characterize the non-equilibrium vortex annealing of defects. In works [44, 52–54] it was shown that non-equilibrium vortex dynamics in a two-dimensional model is accompanied by coarsening of the vortex subsystem. During coarsening, the concentration of the vortices decreases with time, and the effective "size" of the vortices increases. For a detailed description of the clustering of defects in a non-equilibrium vortex annealing, we calculated the average values of the sizes $S(t)$ of defect clusters and the sizes $S_{\mathrm{m}}(t)$ of the largest defect clusters. Fig. 3 and Fig. 4 show the results for $p = 0.8$ and 0.6. Immediately, attention should be paid to an important feature of the defect clustering dynamics in the system: dynamic dependences for the average cluster sizes do not depend on the linear size L of the system. This is manifested in the absence of changes for systems with linear sizes $L = 32$ and $L = 256$. Dynamic dependences of the sizes of defect clusters clearly demonstrate the process of coarsening of the subsystem of defects. One can immediately notice the inertial feature of cluster growth. At first, the sizes of defect clusters reach the maximum value, the peak, and then decrease to their equilibrium values. The initial value ($t = 0$) of cluster size can be considered equal to the size of similar defect clusters in the simulation of quenched structural disorder, since the preparation of the defect distribution in both cases is identical. The equilibrium values ($t \to \infty$) of cluster sizes do not equal the initial values. This suggests that in the equilibrium state, the defects are distributed ununiformly over the lattice, despite the absence of uncoupled vortices in the system. However, since in this work the dynamic dependences of the vortex concentration are not investigated, one can say that few unpaired vortices remain and exist in the system longer than was simulated in this work. With increasing system temperature T, the peak amplitude decreases, and at the phase transition point $T = T_{\mathrm{BKT}}(p)$ the peak disappears from the dependences. Dynamic dependences of the average values of the sizes $S(t)$ of defect clusters and the sizes $S_{\mathrm{m}}(t)$ of the largest defect clusters in Fig. 3 for system with linear size $L = 32$ show a clear time retard in the dynamics of cluster growth on average $S(t)$ and large clusters $S_{\mathrm{m}}(t)$. In the low-temperature region, for example at a temperature $T = T_{\mathrm{BKT}}(p)/16 \ll T_{\mathrm{BKT}}(p)$, the dynamic dependence for the average cluster size peaks at times of 200-300 MCS/s, while for large clusters the peak occurs at time 1000 MCS/s. For large linear sizes L, this effect is enhanced. In particular, for the linear size of $L = 256$ the time to reach the peak of the dynamic dependence for the average cluster size $S(t)$ is also 200-300 MCS/s, while for the sizes of large clusters $S_{\mathrm{m}}(t)$ the peak occurs at times of 10000 MCS/s. With increasing system temperature T, the scale of subsystem of large clusters retard, as compared with the cluster subsystem as a whole, gradually decreases and disappears at the phase transition point $T = T_{\mathrm{BKT}}(p)$. With an increase in the defects concentration c_{imp} (i.e. decrease in the spin concentration p), the process of relaxation of the subsystem of structural defects slows down. At the same time the sizes of clusters $S(t)$ and $S_{\mathrm{m}}(t)$ become larger. Thus, the non-equilibrium vortex annealing demonstrates many interesting features, in particular, requiring further study.

Conclusion

In conclusion, in this work a study of the non-equilibrium vortex annealing of defects in a two-dimensional model was carried out. It is shown that the "defrosting" of defects does not change the values of the Berezinskii-Kosterlitz-Thouless phase transition temperature $T_{\mathrm{BKT}}(p)$. The study revealed that the non-equilibrium vortex annealing of structural disorder in a two-dimensional XY-model leads to a non-equilibrium critical coarsening. In this case, large clusters are separated into an individual subsystem, the relaxation dynamics of which is much slower than the relaxation dynamics of the cluster subsystem as a whole.

Acknowledgments

The reported study was supported by RFBR according to the research projects 17-02-00279, 18-42-550003, 18-32-00814 and grants MD-6868.2018.2, MK-4349.2018.2 of the President of the Russia. The simulations were supported in through computational resources provided by the Shared Facility Center "Data Center of FEB RAS" (Khabarovsk), by the Supercomputing Center of Lomonosov Moscow State University, by Moscow Joint Supercomputer Center and by St. Petersburg Supercomputer Center of the Russian Academy of Sciences.

References

[1] U.C. Tauber, Critical Dynamics: A Field Theory Approach to Equilibrium and Non-Equilibrium Scaling Behavior, (Cambridge University Press, Cambridge, 2014).
[2] V.V. Prudnikov, P.V. Prudnikov, M.V. Mamonova, Phys. Usp. **60**, 762 (2017).
[3] S.E. Korshunov, Phys. Usp. **49**, 225 (2006).
[4] V.L. Berezinskii, Sov. Phys. JETP **32**, 493 (1971).
[5] V.L. Berezinskii, Low-Temperature Properties of Two-Dimensional Systems (Fizmatlit, Moscow, 2007) [in Russian].
[6] I.S. Popov, P.V. Prudnikov, A.N. Ignatenko, A.A. Katanin, Phys. Rev. B, **95**, 134437 (2017).
[7] C.A.F. Vaz, J.A.C. Bland, G. Lauhoff, Rep. Progr. Phys. **71**, 056501 (2008).
[8] C. Kawabat, A.R. Bishop, Solid State Commun. **60**, 167 (1986).
[9] H.-J. Elmers, J. Hauschild, G.H. Liu, U. Gradmann, J. Appl. Phys. **79**, 4984 (1996).
[10] J. Als-Nielsen, S.T. Bramwell, M.T. Hutchings, G.J. McIntyre, D. Visser, J. Phys.: Condens. Matter. **5**, 7871 (1993).
[11] C. Bellitto, P. Filaci, S. Patrizio, Inorg. Chem. **26**, 191 (1987).
[12] A. Paduan-Filho, C.C. Becerra, J. Appl. Phys. **91**, 8294 (2002).

[13] F.L. Pratt, P.M. Zieliński, M. Balanda, R. Podgajny, T. Wasiutyński, B. Sieklucka, J. Phys.: Condens. Matter. **19**, 456208 (2007).
[14] R. Ganguly, D. Chaudhuri, P. Raychaudhuri, L. Benfatto, Phys. Rev. B. **91**, 054514 (2015).
[15] H.N. Wolfgang, Y.K. Na, G. Roumpos, C. Schneider, M. Kamp, S. Höfling, A. Forchel, Y. Yamamoto, Phys. Rev. B. **90**, 205430 (2014).
[16] Y.S. Karimov, Y.N. Novikov, Sov. Phys. – JETP Lett. **19**, 159 (1974).
[17] D.J. Bishop, J.D. Reppy, Phys. Rev. Lett. **40**, 1727 (1978).
[18] D.J. Bishop, J.D. Reppy, Phys. Rev. B. **22**, 5171 (1980).
[19] S. Misra, L. Urban, M. Kim, G. Sambandamurthy, A. Yazdani, Phys. Rev. Lett. **110**, 037002 (2013).
[20] M.R. Beasley, J.E. Mooij, T.P. Orlando, Phys. Rev. Lett. **41**, 1165 (1979).
[21] A.F. Hebard, A.T. Fiory, Phys. Rev. Lett. **44**, 291 (1980).
[22] L.N. Bulaevskii, V.V. Kuzii, A.A. Sobyanin, JETP Lett. **25**, 314 (1977).
[23] A.I. Buzdin, L.N. Bulaevskii, S.V. Panyukov, JETP Lett. **35**, 147 (1982).
[24] A.I. Buzdin, B. Bujicic, M.Yu. Kupriyanov, Zh. Eksp. Teor. Fiz. **101**, 231 (1992).
[25] A.N. Pargellis, S. Green, B. Yurke, Phys. Rev. E. **49**, 4250 (1994).
[26] W. Bietenholz, U. Gerber, F.G. Rejón-Barrera, J. Stat. Mech. P12009 (2013).
[27] A. Singh, S. Ahmad, S. Puri, S. Singh, Euro. Phys. Lett. **100**, 36004 (2012).
[28] A.J. Bray, Adv. Phys. **43**, 357 (1994).
[29] A.J. Bray, Adv. Phys. **51**, 481 (2002).
[30] P. Tabeling, Phys. Rep. **362**, 1 (2002).
[31] W.J. Nuttall, D.Y. Noh, B.O. Wells, R.J. Birgeneau, J. Phys.: Condens. Matter. **7**, 4337 (1995).
[32] I.F. Lyuksyutov, A.G. Fedorus, Sov. Phys. JETP. **53**, 1317 (1981).
[33] A. Taroni, S.T. Bramwell, P.C.W. Holdsworth, J. Phys.: Condens. Matter. **20**, 275233 (2008).
[34] Y. Tu, J. Toner, Phys. Rev. Lett. **75**, 4326 (1995).
[35] A. Cavagna, I. Giardina, T.S. Grigera, A. Jelic, D. Levine, S. Ramaswamy, M. Viale, Phys. Rev. Lett. **114**, 218101 (2015).
[36] A.P. Solon, H. Chaté, J. Tailleur, Phys. Rev. Lett. **114**, 068101 (2015).
[37] L. Berthier, P.C.W. Holdsworth, M. Sellitto, J. Phys. A. **34**, 1805 (2001).
[38] S. Abriet, D. Karevski, Euro. Phys. J. B. **37**, 47 (2004).
[39] P.V. Prudnikov, V.V. Prudnikov, I.S. Popov, JETP Lett. **101**, 596 (2015).
[40] V.V. Prudnikov, P.V. Prudnikov, I.S. Popov, JETP **126**, 369 (2018).
[41] P.V. Prudnikov, I.S. Popov, J. Phys.: Conf. Series **510**, 012014 (2014).
[42] G. Odor, Rev. Mod. Phys. **76**, 663 (2004).
[43] I.S. Popov, P.V. Prudnikov, Solid state phenomena **233-234**, 8 (2015).
[44] I.S. Popov, P.V. Prudnikov, V.V. Prudnikov, J. Phys.: Conf. Series **681**, 012015 (2016).
[45] V.V. Prudnikov, P.V. Prudnikov, S.V. Alekseev, I.S. Popov, Phys. Met. Metalogr. **115**, 1186 (2014).
[46] J. Hoshen, R. Kopelman, Phys. Rev. B, **14**, 3438 (1976);
[47] J. Hoshen, R. Kopelman, J. Stat. Phys., **19**, 219 (1978);
[48] J. Hoshen, R. Kopelman, J. Stat. Phys., **21**, 583 (1979).
[49] R.M. Ziff, B. Sapoval, J. Phys. A: Math. Gen. **19** L1169 (1986).
[50] B. Berche, A.I. Fariñas-Sánchez, Yu. Holovatch, and R. Paredes V., Eur. Phys. J. B **36**, 91 (2003).
[51] S.-K. Ma, Modern theory of critical phenomena, (Westview Press, New York, 2000).
[52] B. Yurke, A.N. Pargellis, T. Kovacs, D.A. Huse, Phys. Rev. E **47**, 1525 (1993).
[53] H. Qian. G.F. Mazenko, Phys. Rev. E **68**, 021109 (2003).
[54] K. Nam, W.B. Baek, B. Kim, S.J. Lee, J. Stat. Mech., P11023 (2012).

IOP Conf. Series: Journal of Physics: Conf. Series **1163** (2019) 012040 doi:10.1088/1742-6596/1163/1/012040

Quasi-3D TCAD modeling of STI radiation-induced leakage currents in SOI MOSFET structure

K. O. Petrosyants, D. A. Popov, D. V. Bykov

National Research University Higher School of Economics (Moscow Institute of Electronics and Mathematics) Russia, Moscow, 123458 Tallinskaya str. 34

kpetrosyants@hse.ru

Abstract. Quasi-3D model for calculation of radiation leakage currents in modern submicron Silicon-on-Insulator (SOI) metal–oxide–semiconductor field-effect transistor (MOSFET) structures is proposed. The solution of a complex 3-dimensional problem is reduced to solving two tasks: 2D modeling of the traditional MOSFET cross-section and 3D modeling of the side parasitic transistor. The radiation-induced leakage current simulation in the 0.35 μm SOI MOSFET structure with taking account ionizing radiation with a dose of up to 500 krad was simulated. The results of the simulation show that in comparison with the traditional fully 3D modeling, which requires 11 hours of computer time, the computer time for the I_dV_g characteristic was reduced to 71 minutes (i.e. the computer time decreased by 9 times).

1. Introduction

Modern complementary metal-oxide-semiconductor (CMOS) SOI technology is a basic technology for manufacturing high-performance radiation-hardened Very Large Scale Integration (VLSI) due to the advantages of dielectric isolation on a semiconductor chip.

2D and 3D Technology Computer-Aided Design for semiconductor manufacturing technology (TCAD) modeling is an effective tool for investigating the radiation hardness of various designs and technological solutions of SOI MOSFETs. Radiation degradation assessments of such important MOSFET parameters as threshold voltage (V_{th}), cut-off amplification frequency (f_{max}/f_T), transconductance (S), leakage current (I_{leak}) can be obtained using TCAD modeling. The traditional 2D MOSFET modeling is sufficient to estimate the radiation dependencies of V_{th}, and f_{max}/f_T. However, the 3D description of the MOSFET structure should be used to determine the leakage current I_{leak}. This is due to the fact that under the ionizing irradiation the positive charge Q_{ot} accumulates in a thick SiO_2 layer of shallow trench isolation (STI) and a buried oxide layer (BOX) and also surface states with densities N_{it} are induced at the interface oxide-silicon (SiO_2-Si). Moreover, the value of Q_{ot} and N_{it} depends on the intensity of irradiation. This leads to inversion of the near-surface layer along STI and BOX, and as the result to the formation of parasitic channels for the leakage current (see Fig. 1).

2. General approach to the leakage current simulation in SOI MOSFET structures

The STI areas surrounding the active MOSFET structure on the sides are embedded in the layers of the buried oxide in its lower part (see Fig. 2). This way, the side and bottom parasitic MOSFET are connected in parallel and their leakage currents are summed $I_{leak} = I_{leakSTI} + I_{leakBOX}$. It was previously shown that the main component of the leakage current is the STI component [1, 2].

IOP Conf. Series: Journal of Physics: Conf. Series **1163** (2019) 012040 doi:10.1088/1742-6596/1163/1/012040

For modern submicron MOSFET structures, the leakage current of the top active transistor can be neglected. First, in a thin gate oxide layer (from 1 to 10 nanometers), the radiation-induced positive charge in the SiO_2 layer or/and the negative charge in the HfO_2 layer is quite small. Secondly, for a well-developed technology, the density of radiation-induced defects N_{it} at the oxide-silicon interface is

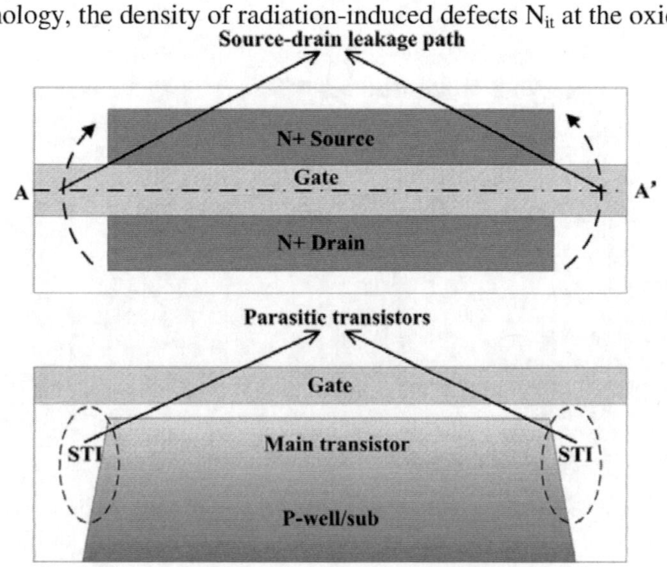

Figure 1. Active transistor (top) and sidewall parasitic transistor (bottom).

also quite small. As a result, the drain leakage current and the threshold voltage shift on input I_dV_g characteristic of the active transistor are negligibly small compared to the parasitic side and bottom transistor structures. As an example, the input I_dV_g characteristics before and after irradiation for the active MOSFET (gate) and the parasitic MOSFET (field or edge) are shown in Fig. 3 [2].

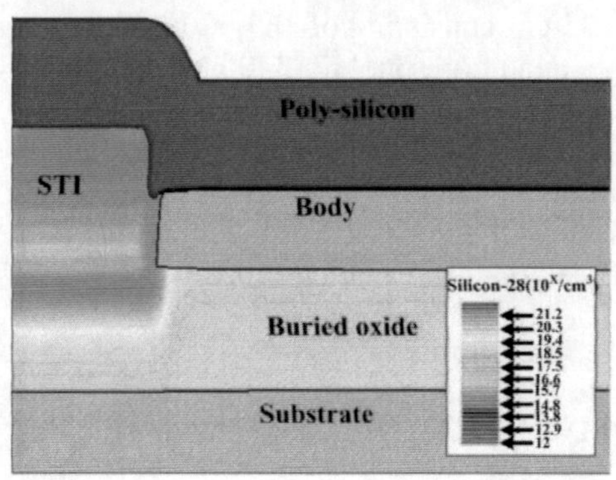

Figure 2. The location of the side and buried oxide relative to the active area [1].

The threshold voltage of the side transistor is quite large due to the large thickness of the side oxide before irradiation, but after irradiation a positive charge accumulates in the thick side oxide. This charge is enough to cause a noticeable shift in the threshold voltage. The threshold voltage shift can be so large that the current flowing along the sidewall will make a significant contribution to the total leakage current at V_{gs}=0 V (OFF regime). It should be noted that the current flowing through such a parasitic transistor depends on the gate voltage, since the leakage area is located in the corner between the STI and the gate oxide (see Fig. 2). As for the main transistor, the threshold voltage shift ΔV_{th} and

IOP Conf. Series: Journal of Physics: Conf. Series **1163** (2019) 012040 doi:10.1088/1742-6596/1163/1/012040

the magnitude of the leakage current in the OFF regime are significantly less than for a parasitic transistor.

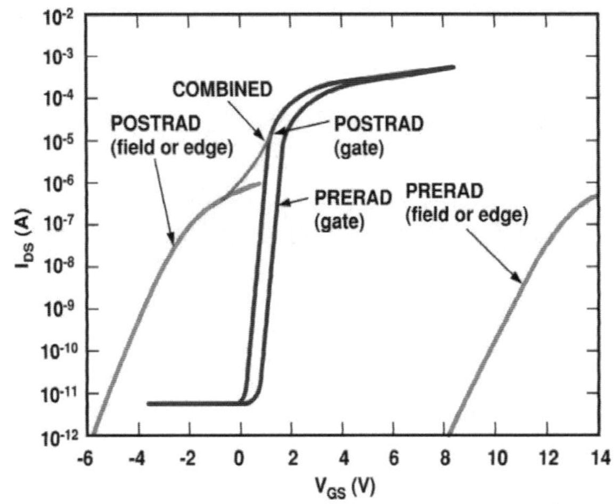

Figure 3. I_dV_g characteristics before and after irradiation for the active MOSFET (gate) and the parasitic MOSFET (field or edge).

As noted above, it is impossible to take into account the influence of the side parasitic transistor in the case for 2D simulation of a MOSFET. Thereby, 3D modeling of the complete MOSFET structure together with STI region (fully 3D) must be carried out. Synopsys Sentaurus TCAD allows carrying out for fully 3D simulation. However, the number of the difference grid elements for calculating the structure is tens of thousands, which, together with the complexity of modeling the charge capture process in the oxide and inversion of the layer in the border region, leads to very large computer time expenditures. For example, the average simulation time of the I-V characteristic for the 3D MOSFET structure with a size W/L=1.5/0.35 μm consisting of 165,000 difference grid elements (see Fig. 4) was 10 hours [3] on a computer with a quad-core Intel i7 (3400 MHz) processor and 16 GB RAM.

This example, as well as the results of other works ([1, 4]), show that fully 3D, taking into account the real transistor structure configuration and the physical effects, is a useful tool for studying various constructive-technological solutions of MOSFET structures. However, this approach requires large labor cost a preparatory description of the device structure, a lot of computer time for simulation, processing and interpretation of the obtained calculation results.

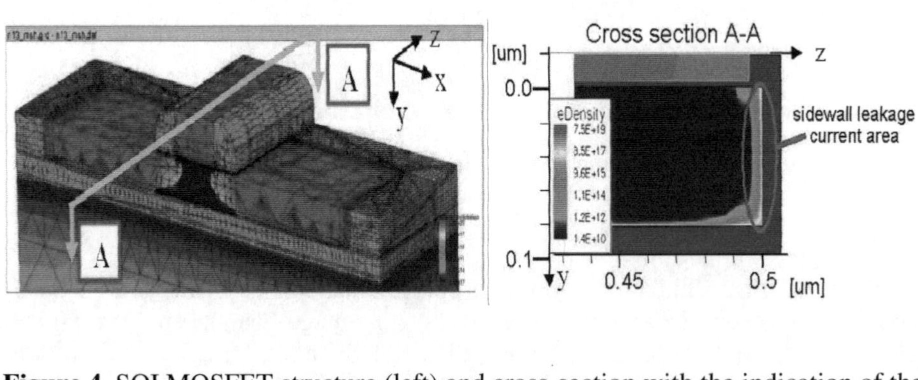

Figure 4. SOI MOSFET structure (left) and cross-section with the indication of the leakage area along the sidewall at the STI/Si interface (right).

IOP Conf. Series: Journal of Physics: Conf. Series **1163** (2019) 012040 doi:10.1088/1742-6596/1163/1/012040

3. Quasi 3D approach for leakage current modeling in SOI MOSFETs

As an instrument of a development engineer, fully 3D modeling is unsuitable for the rapid evaluation of the device characteristics in real operating conditions and for calculating a large number of variants, primarily because of the very high expenditure of computer time. First version of Quasi 3D model for calculation of radiation leakage currents in submicron SOI MOSFET structures is proposed in [5]. The solution of a complex 3-dimensional problem is reduced to solving a combination of two interconnected, much simpler and less time-consuming 2-dimensional tasks. The first problem is solved in the XY plane of the traditional device cross-section. The second problem is solved in the XZ plane of the longitudinal device cross-section (section A-A in Fig. 4). The simulation time has decreased by 10 times compared to the fully 3D approach for the 0.35 μm structure of the SOI MOSFET shown in Fig. 4. The simulation error in the range of ionization doses up to 300 krad did not exceed 15%.

The disadvantage of the previously proposed Quasi 3D model is the presence an intermediate stage of the merger the 2D structure simulation results in two different planes XZ and XY. The radiation-induced charge distributed over the sidewall in the XZ plane (see Fig. 4) needs to be integrated and transferred into the 2D model of the active transistor in the XY plane to calculate the component of the side leakage current.

A further development of the proposed method is 3D modeling of the leakage current for the side parasitic MOSFET structure. This current is determined directly from the simulation results of an "ultra-narrow" parasitic MOSFET, which is formed at the Si/STI vertical interface (see. Fig. 5). This transistor has a polysilicon gate, a thick "gate" oxide t_{STI}, a thin ΔZ_{SIDE} silicon layer in which radiation-induced charge is localized and also fictitious drain, source, substrate, and well contacts. In such the design, the currents of the side parasitic transistor are simulation on the contacts in the Sentaurus TCAD for any applied voltages and ionization doses.

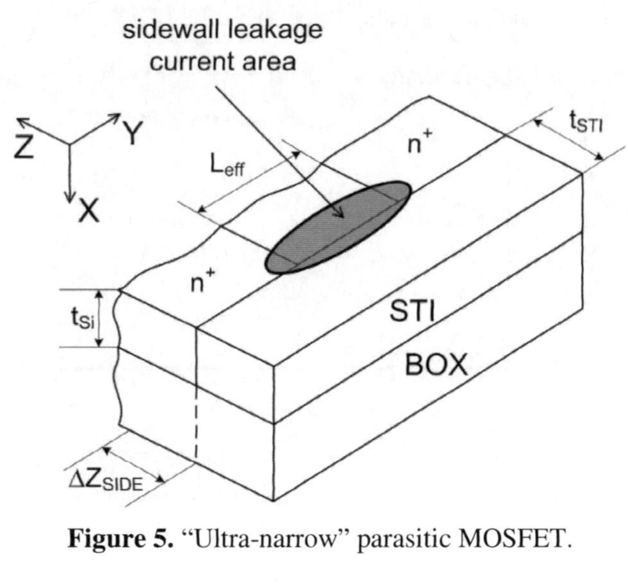

Figure 5. "Ultra-narrow" parasitic MOSFET.

4. Simulation results by Quasi-3D TCAD model

0.35 μm SOI MOSFET manufactured using 0.13 μm technology with the following structure parameters L/W=0.35/10 μm, t_{ox}=7 nm, t_{Si}=100 nm, t_{BOX}=145 nm, t_{STI}=130 nm, $N_A=8\cdot 10^{17}$ cm^{-3}under the influence of ionizing radiation with a dose of up to 500 krad was simulated. Radiation models based on experimental data obtained for 0.13 μm CMOS technologies were used in modeling to

IOP Conf. Series: Journal of Physics: Conf. Series **1163** (2019) 012040 doi:10.1088/1742-6596/1163/1/012040

determine the radiation-induced density of N_{ot} defects in SiO_2 oxide layers for BOX and STI regions and on Si/SiO_2 interfaces – N_{it} [4, 6].

For all structures, the following physical models were used:

- Recombination: SRH(DopingDep), Auger;
- Mobility: PhuMob, CarrierCarrierScattering, DopingDep, Enormal;
- Band gap narrowing model.

The I_dV_g simulation results are shown in the Fig. 6 and Fig. 7. The computer time for the quasi-3D simulation of the SOI MOSFET with the above parameters is 71 minutes, which is 9 times less than the time costs of a fully-3D calculation, which are about 11 hours. Simulation on a Quasi 3D model are in good agreement with experimental data [7], the calculations error does not exceed 10%.

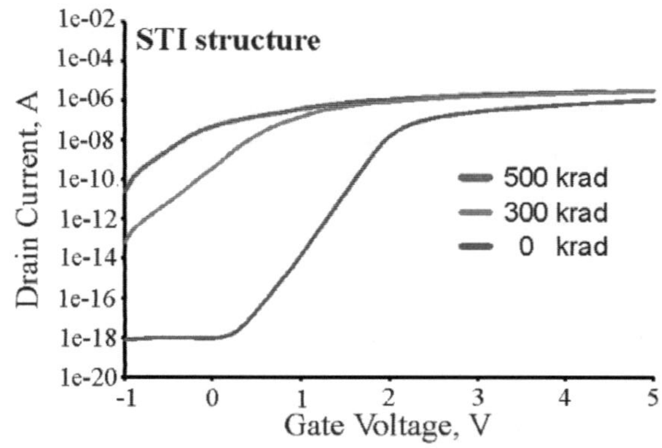

Figure 6. Simulated I_dV_g characteristics sidewall parasitic MOSFET before and after irradiation with a dose of 500 krad.

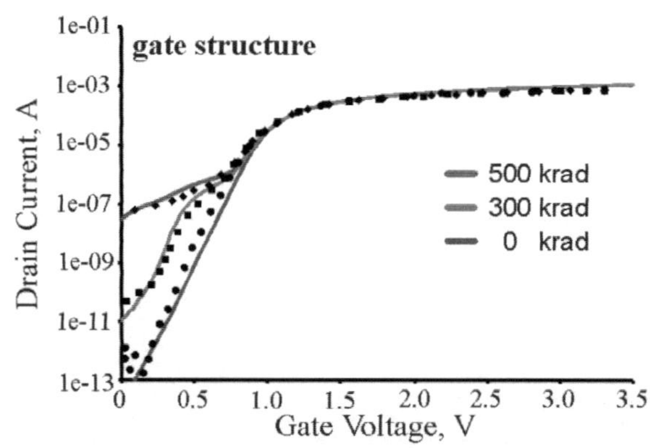

Figure 7. Comparison of experimental (points) [7] and simulating (lines) I_dV_g characteristics before and after irradiation with a dose of 500 krad.

5. Conclusions

The Quasi 3D simulation of the parasitic MOSFET leakage current is proposed. This will make it possible to reduce the computing time by a dozen times and substantially simplify the preprocessor description of the device structure in the TCAD tool. The traditional 3D modeling of MOSFET structure in the Synopsys TCAD has replaced by the two standard subtasks by modeling the primary MOSFET and parasitic lateral STI structure. Both subtasks are solved independently, the components of the active transistor ($I_{leakTOP}$) and side parasitic transistor ($I_{leakSTI}$) leakage currents are determined, and then these components are summed to determine the total leakage current $I_{leak}=I_{leakTOP}+I_{leakBOX}+I_{leakSTI}$.

The results of the simulation show that in comparison with the traditional fully 3D modeling, which requires 11 hours of computer time, the computer time for the I_dV_g characteristic was reduced to 71 minutes (i.e. the computer time decreased by 9 times). Simulation for the quasi-3D model are in good agreement with the experimental data, the error in the calculations does not exceed 10%.

The research was supported by Basic Research Program at the National Research University Higher School of Economics in 2018, grant No. TZ-94 and by Russian Foundation for Basic Research, grant No. 18-07-00898 A.

References

[1] Peng C., Hu Z. En Y., Chen Y., Lei Z., Zhang Z., Zhang Z., Li B. Radiation Hardening by the Modification of Shallow Trench Isolation Process in Partially-Depleted SOI MOSFETs // IEEE Trans. Nucl. Sci., Vol. 65, No. 3, Jan. 2018, pp. 877–883.

[2] Schwank J.R. et al. Radiation Effects in MOS Oxides // IEEE Trans. Nucl. Sci., Vol. 45, No. 4, August 2008, pp. 1833-1853.

[3] Petrosyants K.O., Orekhov E.V., Sambursky L.M., Kharitonov I.A., Yatmanov A.P. Three-dimensional simulation of radiation leakage currents in submicron MOSFET with a silicon-on-insulator structure // Izvestiya Vysshikh Uchebnykh Zavedenii. Elektronika, 2010. №2 (82). pp. 81-83 (in Russian).

[4] Shaneyfelt M. R., Dodd P. E., Draper B. L., R. S. Flores. Challenges in hardening technologies using shallow-trench isolation // IEEE Trans. Nucl. Sci., Vol. 45, No. 6, Dec. 1998, pp. 2584–2592.

[5] Petrosyants K.O., Kharitonov I.A., Orekhov E.V., Yatmanov A.P. Quasi 3D TCAD Modeling of SOI MOSFET with taking into account stationary ionization radiation // Electronics, micro and nanoelectronics: Collection of scientific papers / Edited by Stenin – M.: MEPHI, 2009, pp. 241-249 (in Russian).

[6] Barnaby H.J., McLain M.L., Esqueda I.S., Chen X.J. Modeling Ionizing Radiation Effects in Solid State Materials and CMOS Devices // IEEE Transactions on Circuits and Systems I: Regular Papers, Vol. 56, No. 8, August 2009, pp. 1870-1883.

[7] Peng C., Hu Z., Zhang Z., Huang H., Ning B., Bi D., Zou Z. A New Method for Extracting the Radiation Induced Trapped Charge Density Along the STI Sidewall in the PDSOI NMOSFETs // IEEE Trans. Nucl. Sci., Vol. 60, No. 6, Dec. 2013, pp. 4697–4704.

IOP Conf. Series: Journal of Physics: Conf. Series **1163** (2019) 012041 doi:10.1088/1742-6596/1163/1/012041

Dimensional Crossover in Critical Behavior of Thin XY-films: Equilibrium and Non-equilibrium Properties

Ivan S Popov†, **Anna P Popova**‡, **Pavel V Prudnikov**§

Department of Theoretical Physics, Omsk State University, Omsk 644077, Russia

E-mail: †diphosgen@mail.ru, ‡popovaap@omsu.ru, §prudnikp@univer.omsk.su

Abstract. In the present work it is studied the dimensional crossover in the critical behavior of the thin XY-films. Increasing the thickness N of XY-film leads to a gradual transition from the Berezinskii-Kosterlitz-Thouless phase transition in the two-dimensional XY-model to the "ferromagnetic-paramagnetic" phase transition in the three-dimensional XY-model. It is investigated how phenomena associated with the topological phase disappears when film thickness N is increased. The main consideration is focused on the dependence of the spin stiffness ρ_s on the thickness N. In this work a generalization of Pokrovsky-Uimin equation for spin-wave approximation is introduced. The dynamical dependencies of thermodynamic values in non-equilibrium relaxation from low-temperature and high-temperature initial states are investigated.

1. Introduction

The study of critical behavior causes considerable fundamental and applied scientific interest. In recent years, of particular interest is the study of quasi-two-dimensional systems [1], in particular, the question of the dimensional crossover in the transition from two-dimensional to three-dimensional systems [2]. In this aspect, the XY-model has an important and interesting property. In the two-dimensional system the long-range order is destroyed by transverse fluctuations of the spin density at all nonzero temperatures, but there exists a topological Berezinskii-Kosterlitz-Thouless (BKT) phase transition [3–5] at temperature $T = T_{\mathrm{BKT}}$, and there is a Berezinskii low-temperature phase at $T < T_{\mathrm{BKT}}$, where all temperatures T are critical points. There is a continuous set of fixed points of the renormalization-group transformation for the two-dimensional case, and a quasi-long-range order (QLRO) is present in the system. On the other hand, the critical behavior of the three-dimensional XY-model in vicinity of $T = T_{\mathrm{C}}$ is described by a fixed point of the "ferromagnetic-paramagnetic" phase transition and a classical long-range order (LRO) is present at $T < T_{\mathrm{C}}$ [7–9]. It is interesting to investigate the crossover in the XY-model when changing dimension of the system. The study of the critical behavior of a thin XY-film, or a layered XY-model, is well suited for this. With increasing system thickness, a transition to the behavior of the three-dimensional system should be observed. However, with a small thickness, the system should behave as a two-dimensional or quasi-two-dimensional one.

The XY-model is used to describe the critical properties of a wide range of real physical systems [3], such as critical properties of ultra-thin magnetic films, in particular [10] 1-2.5

ML (atomic layers) Fe/Au(100) at temperatures T 300-500 K; 2 ML Fe/W(100) at T 180-220 K; Ag/2.2 ML Fe/W(100) at T 270-330 K; 2 ML Co/Cu(100) at T 230-410 K and 3-6.2 ML Ni/Cu(100) at T 210-388 K. Critical properties of an extensive class of "easy plane" planar magnets [11,12], including quite specific ones Rb_2CrCl_4 [13], $(C_6H_5CH_2NH_3)_2CrBr_4$ [14], $(CH_3NH_3)_2MnCl_4$ [15], $(tetrenH_5)_{0.8}Cu_4[W(CN)_8]_4 \cdot 7.2H_2O$ [16], are described using XY-model. Singularities in the critical properties of superconducting thin films [17,18]; arrays of Josephson junctions [3,19,20] and SFS-junctions [21–23]; two-dimensional crystals [5] and smectic liquid crystals [24–28]; some correlation properties of two-dimensional turbulence [29]; singularities in the critical properties of superfluid thin films of liquid helium [30–33]; melting of several layers of sorbed xenon in single-crystal graphite [34]; the process of sorbing hydrogen on tungsten W(011) $p(2 \times 2)$ [35]; and some properties of many other physical systems [3,36] are described using XY-model. Also, to the phenomena described by the XY-model, it is possible to relate dynamic properties of such exotic systems as flocks of birds [37] and some other forms of collective behavior of living organisms [38,39].

2. Model and methods

The Hamiltonian of the system in this work was chosen in the form

$$H[\mathbf{S}] = -\frac{1}{2}\sum_{\langle i,j\rangle} \mathbf{S}_i\mathbf{S}_j = -\frac{1}{2}\sum_{\langle i,j\rangle} \cos\left(\varphi_i - \varphi_j\right), \tag{1}$$

where $\mathbf{S} = \{\mathbf{S}_i\}$ is a lattice spin field; $\mathbf{S}_i = (S_i^{(x)}, S_i^{(y)}) \equiv (\cos\varphi_i, \sin\varphi_i)$ is a classical planar spin which is associated with i-node of a cubic lattice with the plane linear size L and the thickness of the film N; φ_i is a phase of spin $\mathbf{S}_i$; $\sum_{\langle i,j\rangle}$ is a summation over all pairs of the nearest neighbors. Periodic boundary conditions were set in the film plane and free boundary conditions in the perpendicular direction.

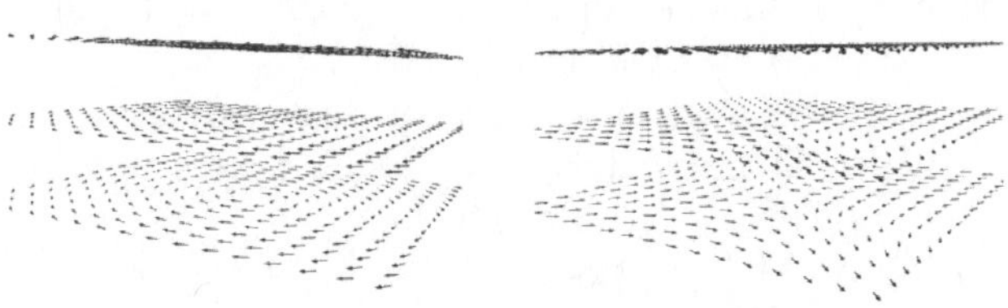

Figure 1. Vortex-loop configurations in a thin XY-film with thickness $N = 3$.

The simulation of critical dynamics and relaxation properties of the system was carried out by Metropolis algorithm.

The dynamics of the relaxation model is characterized by the existence of topological defects. In the two-dimensional case with $N = 1$, they are vortices and antivortices. In quasi-two-dimensional $N > 1$ and three-dimensional $N \to \infty$ cases, they are vortex-loop configurations (see Fig. 1).

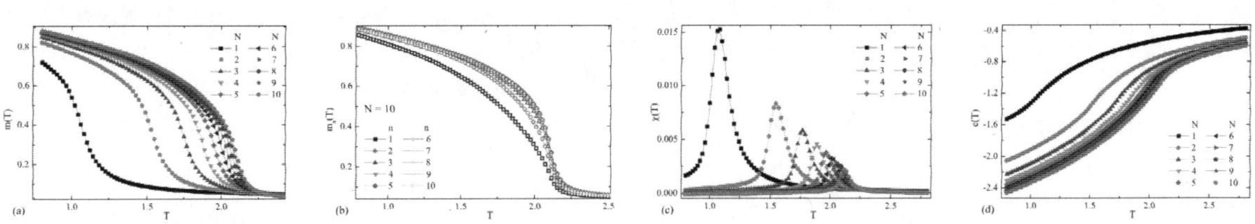

Figure 2. The temperature dependences of equilibrium values for thin XY-films with linear size $L = 32$: the magnetization m (a); layered magnetization m_n for the film with thickness $N = 10$ (b); magnetic susceptibility χ (c); energy e (d).

The simulation of equilibrium properties of the system was carried out by Wolf algorithm. A random unit vector $\mathbf{r}$, determining the direction, was chosen for each cluster. The spin $\mathbf{S}_j$,

IOP Conf. Series: Journal of Physics: Conf. Series **1163** (2019) 012041 doi:10.1088/1742-6596/1163/1/012041

a neighbour of spin $\mathbf{S}_i$ in the cluster, may join the growing cluster provided both spins lie at the same side of the line perpendicular to the unit vector $\mathbf{r}$, that is, $(\mathbf{S}_i \cdot \mathbf{r})(\mathbf{S}_j \cdot \mathbf{r}) > 0$. The probability of a spin joining a growing cluster was determined by the expression $P(\mathbf{r}; \mathbf{S}_i, \mathbf{S}_j) = 1 - \exp\big(-2(\mathbf{S}_i \cdot \mathbf{r})(\mathbf{S}_j \cdot \mathbf{r})/T\big)$. The difference between the Wolf algorithm used in this work and the standard implementation is the choice of neighboring spins for surface nodes where free boundary conditions are established.

The temperature dependences of equilibrium values of the total magnetization m, layered magnetization m_{n}, magnetic susceptibility χ and energy e for thin XY-films are shown in Fig. 2.

The main results for equilibrium values for the layered XY-model are obtained using Metropolis algorithm [6]. Therefore, dependence of phase transition temperature $T_{\mathrm{BKT}}(N)$ on thickness N was calculated for testing and verification of the Wolff algorithm developed in this work.

3. Dependence of phase transition temperature $T_{\mathrm{BKT}}(N)$ on thickness

The temperature of BKT phase transition in the two-dimensional XY-model, with $N = 1$, $T_{\mathrm{BKT}} = 0.893(4)$ [40, 41]. In the three-dimensional XY-model there is a "ferromagnetic-paramagnetic" phase transition with a critical temperature $T_{\mathrm{C}} = 2.2018(5)$ [6–9]. For definiteness we will designate the temperature of phase transition at all thicknesses N, except the value $N \to \infty$, as $T_{\mathrm{BKT}}(N)$.

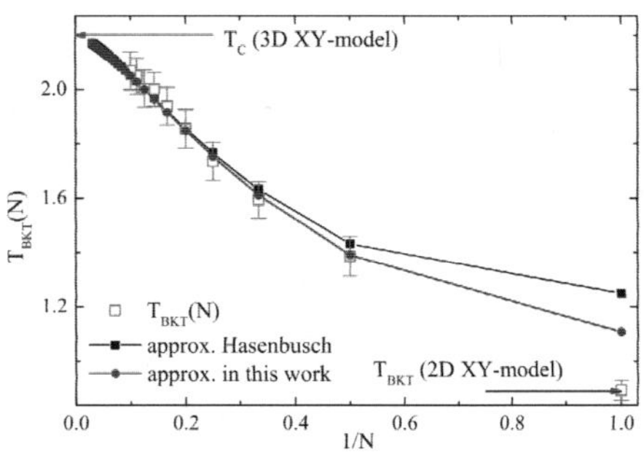

Figure 3. The dependence of the temperature $T_{\mathrm{BKT}}(N)$ on the thickness N of a thin XY-film. Approximation Hasenbusch is taken from work [6]. The arrows indicate the positions of T_{BKT} for the two-dimensional XY-model and T_{C} for the three-dimensional XY-model.

The methods of finite-size scaling analysis for magnetic susceptibility χ and specific heat C, method of fourth-order Binder cumulants U_4 and spatial correlation function ratio R on two different scales were used to determine the dimensional dependence of phase transition temperature $T_{\mathrm{BKT}}(N)$.

The simulations for dimension dependent phase transition temperature $T_{\mathrm{BKT}}(N)$ determination were performed for thicknesses of the system $N = 1 \ldots 10$. Linear sizes $L = 12 \ldots 36$ were used for the methods of finite-size scaling analysis. The Wolf algorithm was used to obtain equilibrium values. One MCS/s corresponded to 20 cluster flips. The relaxation time was chosen equal to 1000 MCS/s for the initial temperature step and 100 MCS/s for subsequent temperature steps. The averaging time was chosen equal to 10000 MCS/s. Statistical averaging was performed over 100 initial configurations.

The dependence of the phase transition temperature $T_{\mathrm{BKT}}(N)$ on the thickness N is shown in Fig. 3. Also a comparison with the dimensional dependence from the work [6] is given. Approximation of $T_{\mathrm{BKT}}(N)$ has the form $T_{\mathrm{BKT}}^{-1}(N) = T_{\mathrm{C}}^{-1} + a(1.0 - bN^{-\omega})N^{-1.0/\nu}$, where $T_{\mathrm{C}}^{-1} = 0.4541655$, $a = 1.15(2)$ and $b = 0.61(1)$. $\nu = 0.6717(1)$ [7–9] is a critical exponent of the correlation length of the three-dimensional XY-model and $\omega = 0.785(1)$ is an exponent of finite-size scaling corrections. These parameters in [6] are: $a = 1.1545(46)$ and $b = 0.70(2)$. The dependence $T_{\mathrm{BKT}}(N)$ obtained in this work agrees well with the results of the work [6].

4. Transverse spin stiffness and Berezinskii-Kosterlitz-Thouless phase transition

The BKT topological phase transition in the XY-model is physically associated with the dissociation of coupled vortex-antivortex pairs at the transition point [3, 5].

IOP Conf. Series: Journal of Physics: Conf. Series **1163** (2019) 012041 doi:10.1088/1742-6596/1163/1/012041

A formal condition [4, 42, 43] is representable, as a condition of zeroing the free energy of a vortex, in the form $\rho_s(T_{BKT}(N)) = 2T_{BKT}(N)/\pi$ for the two-dimensional XY-model. However, for the three-dimensional limit there is a condition $\rho_s(T_{BKT}(N \to \infty)) = 0$ which is the case for the "ferromagnetic-paramagnetic" phase transition, and $T_{BKT}(N \to \infty) \equiv T_C$ is a critical temperature for the three-dimensional XY-model.

In a general case of an arbitrary thickness N of a thin XY-film the transverse spin stiffness at the transition temperature $\rho_s(T_{BKT}(N))$ should change from $2T_{BKT}(N)/\pi$ for $N = 1$ to 0 for $N \to \infty$. To study the dimensional crossover in critical behavior of thin XY-films, in this paper we introduced the relation

$$\Im(N) = \frac{\rho_s(T_{BKT}(N))}{T_{BKT}(N)}. \tag{2}$$

The relation $\Im(N)$ is equal to $2/\pi$ for the two-dimensional case with $N = 1$ and is equal to 0 for the three-dimensional case with $N \to \infty$.

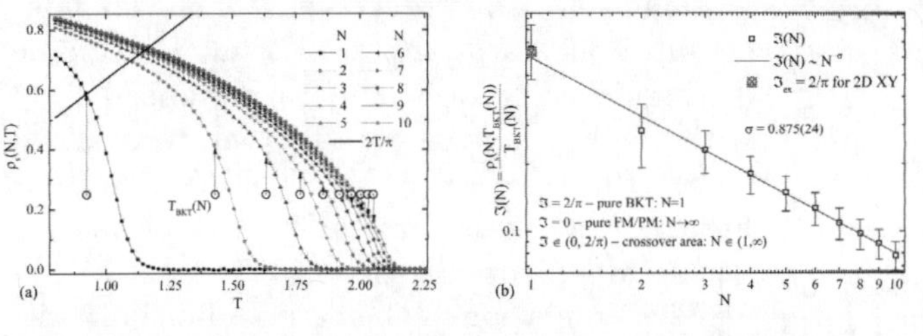

Figure 4. Fig. (a): The temperature dependence of the transverse spin stiffness ρ_s of the thin XY-film with $L = 128$ for various thicknesses N. The arrows indicate the positions of $T_{BKT}(N)$. Fig. (b): The dependence of the ratio $\Im(N)$ on the thickness N.

The simulations for the temperature dependence of the transverse spin stiffness ρ_s determination were performed for thicknesses of the system $N = 1 \ldots 10$ and linear size $L = 128$. As before in this work, the Wolf algorithm was used to obtain equilibrium values. The relaxation time was chosen equal to 5000 MCS/s for the initial temperature step and 500 MCS/s for subsequent temperature steps. The averaging time was chosen equal to 50000 MCS/s. Statistical averaging was performed over 1000 initial configurations. The transverse spin stiffness ρ_s was calculated by the method [44].

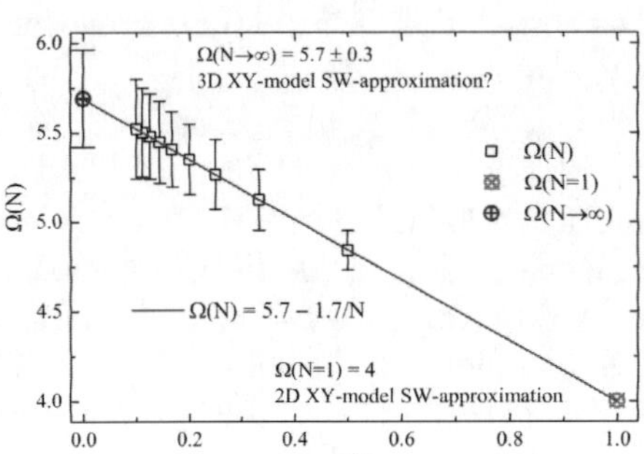

Figure 5. The dimensional dependence of the value $\Omega(N)$.

The temperature dependence of the transverse spin stiffness ρ_s for various thicknesses N of the film with $L = 128$ is shown in Fig. 4 (a). The results show that the spin stiffness ρ_s at the temperature $T_{BKT}(N)$ of BKT topological phase transition decreases with increasing thickness N of the system. The dimension dependence of the relation $\Im(N)$ for these results is shown in Fig. 4 (b). As can be seen from the results, there is a power law of decay with increasing thickness N for relation $\Im(N) = \frac{2}{\pi}N^{-\sigma}$ with exponent $\sigma = 0.875(24)$. Thus, we can conclude that the transverse spin stiffness $\rho_s(T_{BKT}(N))$ associated with the decay of coupled vortex-antivortex pairs at the transition point $T_{BKT}(N)$, decreases as a power law with increasing system thickness N and asymptotically tends to zero in the case of a three-dimensional system with $N \to \infty$.

5. Pokrovsky-Uimin equation and spin-wave approximation

The Pokrovsky-Uimin equation was introduced in works [42, 43] and allows to obtain a spin-wave approximation of the temperature dependence of the transverse spin stiffness $\rho_s(T)$ for the two-dimensional XY-model. This transcendental algebraic equation has a simple form

$\rho_s = \exp(-T/(4\rho_s))$. To study the dimensional crossover in critical behavior of thin XY-films, in this paper we introduced a generalization of Pokrovsky-Uimin equation in the form $\rho_s = \exp(-T/(\Omega(N)\rho_s))$ with dimensional parameter $\Omega(N)$. Numerical calculations of this equation's roots were carried out for different values $\Omega(N)$ for the purpose of sewing the spin-wave solution with numerical results in Fig. 4 (a), to determine the dimensional dependence of the parameter $\Omega(N)$ on the thickness N. The dimensional dependence of the parameter $\Omega(N)$ on $1/N$ obtained in this work is shown in Fig. 5. The dependence obtained is well approximated by a linear relation $\Omega(N) = 5.7 - 1.7/N$. If $N = 1$, as in the two-dimensional XY-model, $\Omega(N = 1) = 4$, as it was in the original Pokrovsky-Uimin equation. For a three-dimensional system, the value $\Omega(N \to \infty) = 5.7 \pm 0.3$.

6. Relaxation dynamical dependences of thin XY-films

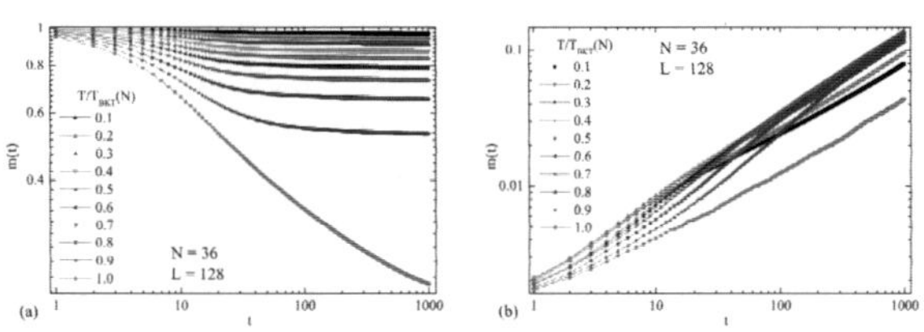

Figure 6. The dynamical dependences of the magnetization $m(t)$ for thin XY-films with thickness $N = 36$ and linear size $L = 128$ for temperatures $T \leq T_{\mathrm{BKT}}(N)$ in relaxation from different non-equilibrium initial states: low-temperature (a) and high-temperature (b).

The non-equilibrium critical relaxation depends on the selected initial state of the system. Two different types of initial non-equilibrium states in the XY-model can be distinguished: a low-temperature initial state with $T_0 = 0$ and a high-temperature initial state with $T_0 \gg T_{\mathrm{BKT}}(N)$. The initial state is prepared by thermalization with a thermostat with temperature T_0.

The simulation of the non-equilibrium critical relaxation of the system was performed using the Metropolis algorithm. Systems with thicknesses $N = 1 \ldots 36$ were studied. The dynamical dependencies of the magnetization $m(t)$ and its dispersion $D_m(t)$, the second and fourth order cumulants $U_2(t)$ and $U_4(t)$ of magnetization for non-equilibrium relaxation of the system from the high-temperature and low-temperature initial states were investigated. The dynamical dependences of the magnetization $m(t)$ of the system with thickness $N = 36$ and linear size $L = 128$ are presented in Fig. 6. A detailed discussion of the dimensional crossover in the relaxation dynamics of the XY-films is planned in future works.

7. Conclusion

In this paper, we study the dimensional crossover in the critical behavior of the thin XY-films. Increasing the XY-film thickness N leads to a gradual transition from the Berezinskii-Kosterlitz-Thouless phase transition in the two-dimensional XY-model to the "ferromagnetic-paramagnetic" phase transition in the three-dimensional XY-model. It is shown that this manifests itself in the dependence of the spin stiffness $\rho_s(T_{\mathrm{BKT}}(N))$ on the thickness N of the film in the asymptotics $N \to \infty$ with the form $\Im(N) \equiv \rho_s(T_{\mathrm{BKT}}(N))/T_{\mathrm{BKT}}(N) = \frac{2}{\pi}N^{-\sigma}$. The dependence of the temperature $T_{\mathrm{BKT}}(N)$ of the phase transition on the film thickness N is defined, and it was shown that $T_{\mathrm{BKT}}(N \to \infty) \equiv T_C$ for the three-dimensional XY-model. In this work a generalization of Pokrovsky-Uimin equation for spin-wave approximation $\rho_s = \exp(-T/(\Omega(N)\rho_s))$ was introduced, with $\Omega = \Omega(N)$, and it was shown that the case of film thickness N is well approximated by expression $\Omega(N) = 5.7 - 1.7/N$. The dynamical dependencies of thermodynamic values in non-equilibrium relaxation from the low-temperature, with $T_0 = 0$, and high-temperature, with $T_0 \gg T_{\mathrm{BKT}}(N)$, non-equilibrium initial states were investigated.

Acknowledgments

The reported study was supported by RFBR according to the research projects 17-02-00279, 18-42-550003, 18-32-00814 and grants MD-6868.2018.2, MK-4349.2018.2 of the President of the Russia. The simulations were supported in through computational resources provided by the Shared Facility Center "Data Center of FEB RAS" (Khabarovsk), by the Supercomputing Center of Lomonosov Moscow State University, by Moscow Joint Supercomputer Center and by St. Petersburg Supercomputer Center of the Russian Academy of Sciences.

References

[1] V.V. Prudnikov, P.V. Prudnikov, M.V. Mamonova, Phys. Usp. **60**, 762 (2017).
[2] P.V. Prudnikov, V.V. Prudnikov, M.A. Menshikova, N.I. Piskunova, J. Magn. Magn. Mater. **387**, 77 (2015).
[3] S.E. Korshunov, Phys. Usp. **49**, 225 (2006).
[4] V.L. Berezinskii, Sov. Phys. JETP **32**, 493 (1971).
[5] V.L. Berezinskii, Low-Temperature Properties of Two-Dimensional Systems (Fizmatlit, Moscow, 2007) [in Russian].
[6] M. Hasenbusch, J. Stat. Mech., P02005 (2009).
[7] M. Campostrini, A. Pelissetto, P. Rossi, E. Vicari, Phys. Rev. B. **61**, 5905 (2000).
[8] M. Campostrini, A. Pelissetto, P. Rossi, E. Vicari, Phys. Rev. B. **62**, 5843 (2000).
[9] M. Campostrini, M. Hasenbusch, A. Pelissetto, P. Rossi, E. Vicari, Phys. Rev. B. **63**, 214503 (2001).
[10] C.A.F. Vaz, J.A.C. Bland, G. Lauhoff, Rep. Progr. Phys. **71**, 056501 (2008).
[11] C. Kawabat, A.R. Bishop, Solid State Commun. **60**, 167 (1986).
[12] H.-J. Elmers, J. Hauschild, G.H. Liu, U. Gradmann, J. Appl. Phys. **79**, 4984 (1996).
[13] J. Als-Nielsen, S.T. Bramwell, M.T. Hutchings, G.J. McIntyre, D. Visser, J. Phys.: Condens. Matter. **5**, 7871 (1993).
[14] C. Bellitto, P. Filaci, S. Patrizio, Inorg. Chem. **26**, 191 (1987).
[15] A. Paduan-Filho, C.C. Becerra, J. Appl. Phys. **91**, 8294 (2002).
[16] F.L. Pratt, P.M. Zieliński, M. Balanda, R. Podgajny, T. Wasiutyński, B. Sieklucka, J. Phys.: Condens. Matter. **19**, 456208 (2007).
[17] R. Ganguly, D. Chaudhuri, P. Raychaudhuri, L. Benfatto, Phys. Rev. B. **91**, 054514 (2015).
[18] H.N. Wolfgang, Y.K. Na, G. Roumpos, C. Schneider, M. Kamp, S. Höfling, A. Forchel, Y. Yamamoto, Phys. Rev. B. **90**, 205430 (2014).
[19] M.R. Beasley, J.E. Mooij, T.P. Orlando, Phys. Rev. Lett. **41**, 1165 (1979).
[20] A.F. Hebard, A.T. Fiory, Phys. Rev. Lett. **44**, 291 (1980).
[21] L.N. Bulaevskii, V.V. Kuzii, A.A. Sobyanin, JETP Lett. **25**, 314 (1977).
[22] A.I. Buzdin, L.N. Bulaevskii, S.V. Panyukov, JETP Lett. **35**, 147 (1982).
[23] A.I. Buzdin, B. Bujicic, M.Yu. Kupriyanov, Zh. Eksp. Teor. Fiz. **101**, 231 (1992).
[24] A.N. Pargellis, S. Green, B. Yurke, Phys. Rev. E. **49**, 4250 (1994).
[25] W. Bietenholz, U. Gerber, F.G. Rejón-Barrera, J. Stat. Mech. P12009 (2013).
[26] A. Singh, S. Ahmad, S. Puri, S. Singh, Euro. Phys. Lett. **100**, 36004 (2012).
[27] A.J. Bray, Adv. Phys. **43**, 357 (1994).
[28] A.J. Bray, Adv. Phys. **51**, 481 (2002).
[29] P. Tabeling, Phys. Rep. **362**, 1 (2002).
[30] Y.S. Karimov, Y.N. Novikov, Sov. Phys. – JETP Lett. **19**, 159 (1974).
[31] D.J. Bishop, J.D. Reppy, Phys. Rev. Lett. **40**, 1727 (1978).
[32] D.J. Bishop, J.D. Reppy, Phys. Rev. B. **22**, 5171 (1980).
[33] S. Misra, L. Urban, M. Kim, G. Sambandamurthy, A. Yazdani, Phys. Rev. Lett. **110**, 037002 (2013).
[34] W.J. Nuttall, D.Y. Noh, B.O. Wells, R.J. Birgeneau, J. Phys.: Condens. Matter. **7**, 4337 (1995).
[35] I.F. Lyuksyutov, A.G. Fedorus, Sov. Phys. JETP. **53**, 1317 (1981).
[36] A. Taroni, S.T. Bramwell, P.C.W. Holdsworth, J. Phys.: Condens. Matter. **20**, 275233 (2008).
[37] Y. Tu, J. Toner, Phys. Rev. Lett. **75**, 4326 (1995).
[38] A. Cavagna, I. Giardina, T.S. Grigera, A. Jelic, D. Levine, S. Ramaswamy, M. Viale, Phys. Rev. Lett. **114**, 218101 (2015).
[39] A.P. Solon, H. Chaté, J. Tailleur, Phys. Rev. Lett. **114**, 068101 (2015).
[40] R. Gupta, C.F. Baillie, Phys. Rev. B **45**, 2883 (1992).
[41] P.V. Prudnikov, I.S. Popov, J. Phys.: Conf. Ser. **510**, 012014 (2014).
[42] V.L. Pokrovsky, G.V. Uimin, Phys. Lett. A **45**, 467 (1973).
[43] V.L. Pokrovsky, G.V. Uimin, Sov. Phys. JETP **38**, 847 (1974).
[44] I.S. Popov, P.V. Prudnikov, V.V. Prudnikov, J. Phys.: Conf. Ser. **510**, 012015 (2016).

IOP Conf. Series: Journal of Physics: Conf. Series **1163** (2019) 012042 doi:10.1088/1742-6596/1163/1/012042

Dynamical Scaling in Time dependence of Correlation Length in Non-equilibrium Critical Relaxation of Pure and Site-diluted 2D XY-model

Ivan S Popov†, **Anna P Popova**‡,
Pavel V Prudnikov§, **Vladimir V Prudnikov**⋆

Department of Theoretical Physics, Omsk State University, Omsk 644077, Russia

E-mail: †diphosgen@mail.ru, ‡popovaap@omsu.ru,
§prudnikp@univer.omsk.su, ⋆prudnikv@univer.omsk.su

Abstract. Structural disorder in two-dimensional XY-model creates an anomalous slowing down in the process of critical relaxation. Dynamic dependences of the correlation length $\xi(t)$ show slow growth with an "effective" dynamical critical exponent $z > 2$. For a pure system, it has been shown that a detailed consideration of the vortex dynamics leads to deviations from the known dependence $\xi(t) \sim (t/\ln t)^{1/2}$.

The study of critical behavior of disordered systems causes considerable fundamental and applied scientific interest [1,2]. The critical phenomena are characterized by a strong fluctuation behavior. The low-dimensional systems and the systems with continuous symmetry of the ground state are usually characterized by much stronger fluctuation effects than bulk systems and systems with discrete symmetry. Therefore, low-dimensional systems with continuous symmetry are characterized by a significant contribution of fluctuations to critical properties [3]. In two-dimensional systems with continuous symmetry, the long-range order (LRO) is destroyed by anomalously strong transverse fluctuations of the spin density at all non-zero temperatures. However, in such systems there are more exotic forms of ordering, which are associated with the existence of topological excitations in them, such as vortices [4–6].

A special place among two-dimensional systems with continuous symmetry is occupied by a two-dimensional XY-model [3–5]. In the two-dimensional XY-model take place a Berezinskii-Kosterlitz-Thouless (BKT) topological phase transition at temperature $T = T_{\mathrm{BKT}}$ and a low-temperature Berezinskii phase $T < T_{\mathrm{BKT}}$, where each temperature T is a critical point (there is a continuous set of fixed points of the renormalization group transformation) and a quasi-long-range order (QLRO) occurs. This makes it possible to study the critical behavior of the system not only in one isolated point T_{C} (as, for example, in the Ising model), but in the whole temperature range $T \leq T_{\mathrm{BKT}}$. The phase transition is physically related to dissociation at the transition point of the coupled vortex pairs. A two-dimensional XY-model is used to describe the critical properties of a wide range of real physical systems [3], such as critical properties of ultra-thin magnetic films [7] and extensive class of easy plane planar magnets [8–13]. The singularities in the critical properties of superconducting [14, 15] and superfluid helium [16–19] thin films are described by a two-dimensional XY-model. The arrays of Josephson [3, 20, 21] and SFS-junctions [22–24]; two-dimensional crystals [5] and smectic liquid crystals [25–29]; some

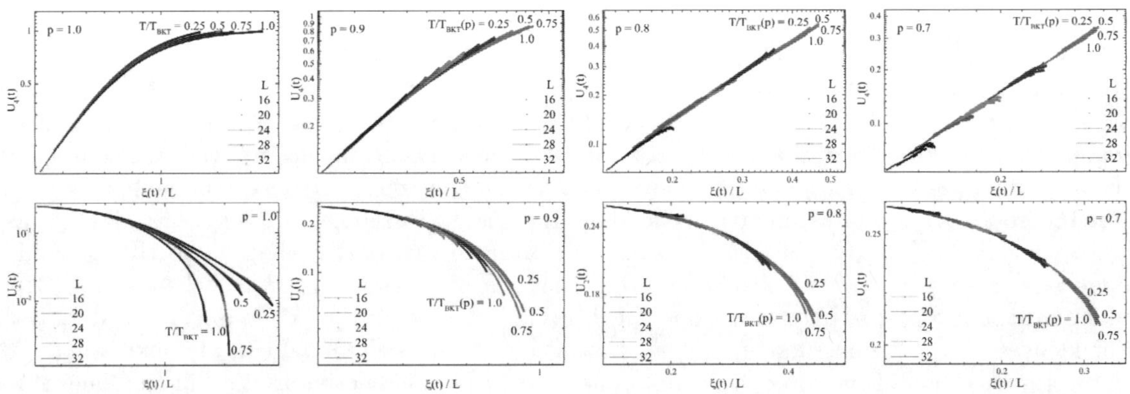

Figure 1. Parametric dependences of the second-order $U_2(t)$ and fourth-order $U_4(t)$ magnetization cumulants on the reduced correlation length $\xi(t)/L$ for linear sizes $L = 12$, 16, 20, 24 and 32 in the process of non-equilibrium critical relaxation of a two-dimensional pure and diluted XY-model. Spin concentrations $p = 1.0$ (pure system), 0.9, 0.8 and 0.7, and temperatures $T = 0.25$, 0.5, 0.75 and 1.0 $T_{\text{BKT}}(p)$.

correlation properties of two-dimensional turbulence [30]; melting of several layers of sorbed xenon in single-crystal graphite [31]; the process of sorbing hydrogen on tungsten [32]; and properties of many other physical systems [3, 33] are described using a two-dimensional XY-model. Also, to the phenomena described by the XY-model, it is possible to relate dynamic properties of such exotic systems as collective behavior of living organisms [34–36].

The non-equilibrium critical relaxation of a two-dimensional XY-model can be a vortex and spin-wave depending on the initial non-equilibrium state [37–41]. The low-temperature initial state is prepared in contact with a thermostat at zero temperature $T_0 = 0$ and the relaxation dynamics will be spin-wave [37,38,40]. The high-temperature initial state is prepared at a high temperature $T \gg T_{\text{BKT}}$ thermostat, and the relaxation is accompanied by vortex dynamics [37–39]. Inclusion the structural disorder in the model leads to a significant change in critical behavior of a two-dimensional XY-model [39–41]. The effect of structure defects on spin-wave relaxation is not considered in this work, details are given in our other work [40]. The effect of structure defects on the non-equilibrium critical vortex dynamics is accompanied by the appearance of non-equilibrium pinning of vortices on defects [39, 41].

The relaxation dynamics of the two-dimensional XY-model is accompanied by slow dynamics and aging effects [2, 37–40]. It was shown in the papers [42, 43] that the vortex relaxation of the diluted model occurs extremely slowly. This particularly applies to the vortex concentration and geometric characteristics of QLRO-areas. In the works [39, 40] it was shown that the inclusion of structural disorder into the system leads to changes in the aging effects. For spin-wave relaxation, structural disorder leads to the violation of the canonical dynamic scaling of the autocorrelation function $C(t, t_w)$ and to the occurrence of superaging and subaging effects [40]. Previously, such properties were observed only in complex disordered systems, such as spin glasses [44]. However, in recent years, similar behavior has been found in the critical dynamics of the diluted three-dimensional Ising model [45–49] in the process of relaxation from the initial low-temperature state. It also recently emerged that superaging occurs in the critical relaxation of the two-dimensional diluted Ising model [50]. It is clear that the two-dimensional diluted XY-model is no exception [40], especially considering the much slower dynamics of this system compared with the diluted Ising model.

The dynamic scaling was considered based on the dynamic dependence of the correlation length $\xi(t)$ of the system, that was used in all the above works, for both two-dimensional XY-model [2, 37–40] and the Ising model [45–50]. Thus, the dynamic dependence of the correlation length $\xi(t)$ becomes the "stumbling block" (or "hot spot") of all studies of slow dynamics and aging in non-equilibrium critical relaxation of systems in the critical state. The dynamic dependence of two-dimensional XY-model of the correlation length $\xi(t)$ in the spin-wave relaxation has the form $\xi(t) \sim t^{1/z}$, where $z = 2$ is the dynamical critical exponent, following from the considerations of dimension and dynamic scale invariance. In the work [51] it was shown that the vortex dynamics is characterized by a logarithmic correction to the dynamic dependence $\xi(t) \sim (t/\ln t)^{1/z}$. However, this result is not based on direct calculation of the dynamic dependence of the correlation length $\xi(t)$, but only on the collapse of the dynamic dependences of the cumulants of the magnetization of the system and is also explained by solving the equation of motion of a vortex under the action of effective friction in the long-time mode. However, this dependence was obtained and, mostly, well-proven for studying processes in a pure system. In this paper, a direct calculation of the dynamic dependence of the correlation length $\xi(t)$ of a pure and diluted two-dimensional XY-model was performed.

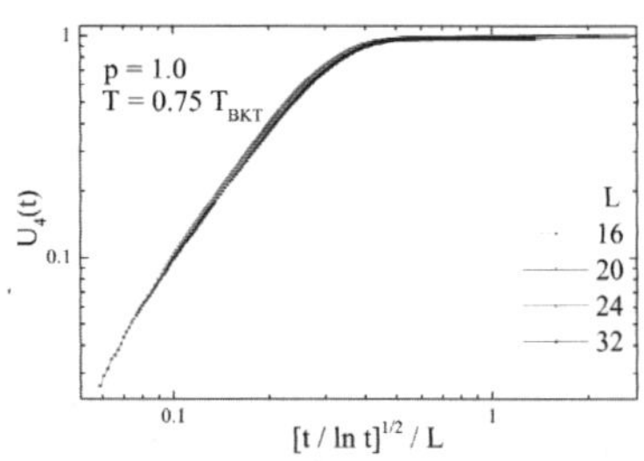

Figure 2. Parametric scaling dependences of the fourth-order magnetization cumulant $U_4(t)$ on $\sqrt{t/\ln t}/L$. There is good agreement with the results of the work [51].

The Hamiltonian of the two-dimensional XY-model in this work was chosen as

$$H[p,\mathbf{S}] = -\frac{1}{2}\sum_{\langle i,j\rangle} p_i p_j \mathbf{S}_i \mathbf{S}_j, = -\frac{1}{2}\sum_{\langle i,j\rangle} p_i p_j \cos(\varphi_i - \varphi_j), \quad (1)$$

where $\mathbf{S} = \{\mathbf{S}_i\}$ and $p = \{p_i\}$ are the spin and defect lattice fields; $\mathbf{S}_i = (S_{i,x}, S_{i,y}) \equiv (\cos\varphi_i, \sin\varphi_i)$ is a classical planar spin which is associated with i-node of square lattice with the linear size L; φ_i is a phase of spin $\mathbf{S}_i$; p_i is an occupation number of i-node: if $p_i = 1$ then i-node is occupied by spin, else if $p_i = 0$ by defect; $\sum_{\langle i,j\rangle}$ is the summation over all pairs of the nearest neighbors. Defects on the lattice are distributed uniformly at time $t = 0$ with spin concentration p, i.e. $c_{\text{imp}} = 1 - p$ is the concentration of impurity.

Simulation was carried out in the low-temperature Berezinskii phase at "freezing" temperatures $T \leq T_{\text{BKT}}(p)$ by using Metropolis algorithm, as it adequately sets the dynamic properties of this system [52]. The study was carried out for the spin concentrations $p = 1.0$ (pure system), 0.9, 0.8 and 0.7 (diluted systems). The values of the phase transition $T_{\text{BKT}}(p)$ temperature were chosen from the works [41, 53].

The correlation length was calculated from the Ornstein-Zernike relation (see, for example, [54])

$$\xi(t) = \frac{1}{2\sin(\pi/L)}\sqrt{\frac{\chi(t)}{\Phi(t)} - 1}, \quad (2)$$

where $\chi(t) = \sum_i [\langle p_i \mathbf{S}_i^2 \rangle]$ and $\Phi(t)$ is the structure factor:

$$\Phi(t) = \frac{1}{2}\Big[\Big\langle \sum_{n=1,2} \Big|\sum_i p_i \mathbf{S}_i \exp\Big(i\frac{2\pi x_n}{L}\Big)\Big|^2\Big\rangle\Big]; \quad (3)$$

where the square brackets $[\ldots]$ denote averaging over the initial configurations of the distribution of impurities in the system and the angle brackets $\langle\ldots\rangle$ denote averaging over various initial spin configurations within the created configuration of structural disorder. The dynamic scaling were determined from the considerations of the collapse of the dynamic dependences of the Binder magnetization cumulants $U_2(t) = [\langle \mathbf{M}^2 \rangle]/[\langle \mathbf{M} \rangle]^2 - 1$ and $U_4(t) = 2 - [\langle \mathbf{M}^4 \rangle]/[\langle \mathbf{M}^2 \rangle]^2$, where $\mathbf{M} = \sum_i p_i \mathbf{S}_i$ is the magnetization of system, in reduced variables $\xi(t)/L$ for various linear sizes L.

The averaging of the calculated values for diluted system was carried out over 50000 impurity configurations and 25 statistical passes for each impurity configuration. A large amount of statistical sampling is associated with a bad convergence of dynamic dependences with averaging for diluted system and with the slowness of the relaxation process, which required the study of large time scales. This allowed the use of linear sizes $L = 16$, 20, 24 and 32 to study the non-equilibrium critical relaxation of the diluted system, and the observation time was chosen equal to 20000 MCS/s.

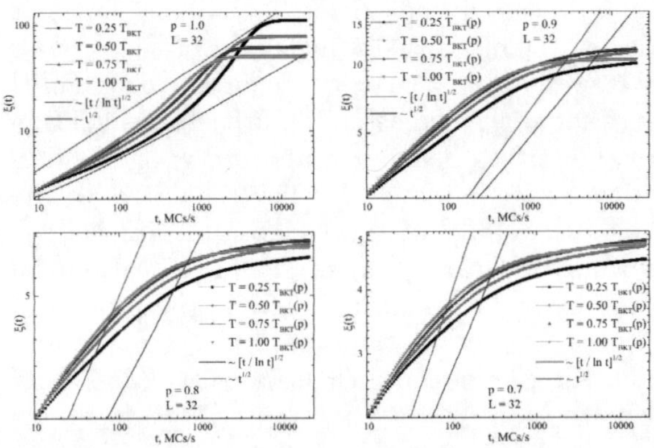

Figure 3. Dynamic dependences of the correlation length $\xi(t)$ in the process of non-equilibrium critical relaxation of a two-dimensional pure and diluted XY-model. Spin concentrations $p = 1.0$ (pure system), 0.9, 0.8 and 0.7, and temperatures $T = 0.25$, 0.5, 0.75 and 1.0 $T_{\mathrm{BKT}}(p)$.

The obtained parametric dependences of the fourth-order magnetization cumulant U_4 on $\xi(t)/L$ (see Fig. 1) for a diluted system demonstrate the collapse of curves for different linear sizes L. The results for a pure system show excellent performance of finite-size dynamic scaling dependence $U_4(t) = U_4(\xi(t)/L)$ for all considered "freezing" temperatures up to the phase transition point T_{BKT}. The insertion of disorder leads to a slower relaxation of the system, while during the observation time of 20000 MCS/s, the quantities of $U_4(t)$ do not reach the asymptotic value $U_4(t \to \infty) \approx 1$. Parametric dependences U_4 on $\xi(t)/L$ do not have a power growth range for pure system ($p = 1.0$); and the curves are clearly separated for different temperatures T. With the insertion of disorder, a power stage of growth occurs $U_4 \sim (\xi(t)/L)^a$. The duration of this power stage increases with increasing impurity concentration c_{imp}, and the separation of the curves for different temperatures T disappears. It should be noted that the parametric dependence for a pure system is somewhat different from that obtained in the work [51]. To verify the results obtained, curves were constructed for an explicitly defined dynamic dependence $\xi(t) \sim (t/\ln t)^{1/2}$ (Fig. 2). This demonstrates good agreement with the results [51]. However, the results in Fig. 1 for a pure system ($p = 1.0$) demonstrate a significantly better collapse of the curves $U_4(\xi(t)/L)$ than those shown in Fig. 2 and in work [51]. The obtained parametric dependences of the second-order cumulant $U_2(t)$ on $\xi(t)/L$ (see Fig. 1) for a diluted system also demonstrate the collapse of curves for different linear sizes L.

The obtained dynamic dependences of the correlation length $\xi(t)$ are shown in Fig. 3. For a pure system, one can immediately note that the dependence $\xi(t)$ explicitly calculated by the formula (2) cannot be correctly approximated by dependence $(t/\ln t)^{1/2}$ at all time scales. If we conditionally introduce the dependence $\xi(t) \sim t^{b(t)}$ or $\xi(t) \sim (t/\ln t)^{b(t)}$ (qualitative aspect of the case does not change the selection) and the "effective exponent" $b(t)$ of the slope of the curve, then we get $db(t)/dt > 0$. In this case, starting from a certain time t (separately selected for each temperature T), we can get $b(t) > 1/2$, i.e. "effective" dynamical critical exponent $z < 2$. This immediately raises many questions about the singularity of the kinetic coefficient. The obtained dynamic dependence $\xi(t)$ conveys in detail the features of ordering in the system, while for the qualitative performance of the collapse (Fig. 2), a rather simple relation $\xi(t) \sim (t/\ln t)^{1/2}$. In this case, the actual values $z < 2$ may not occur. This is due to the detail of the description embedded in the dependence according to the formula (2). In the work [51], the dynamic dependence $\xi(t) \sim (t/\ln t)^{1/2}$ is obtained as a solution to the equation $d\xi/dt \sim \rho_s F/\xi \ln(\xi/a)$ on time $t \gg a^2/(4\rho_s\Gamma)$. This equation is obtained from considering the motion of a free vortex in the presence of friction, which only effectively includes the processes of interaction of vortices, such as annihilation. However, it is clear that this approach does not take into account the whole picture of the dynamics of vortex relaxation, and the dependences obtained in this work take into account (of course, within the framework of the designated model representations). We associate the features of the deviation of dynamical dependence $\xi(t)$ from the dependence introduced in the work [51] with a detailed account of the full picture of the non-equilibrium vortex dynamics at different stages of the relaxation dynamics, and not only the effective friction of an isolated free vortex in the process of movement.

The insertion of disorder into the system leads to a significant slowing down of the process of critical relaxation. For illustration in Fig. 3 curves are shown corresponding to dynamic dependencies $t^{1/2}$ and $(t/\ln t)^{1/2}$. It is clearly seen that with increasing impurity concentration c_{imp}, the "effective exponent" $b(t)$ (see above) of the curves decreases, and $b(t) < 1/2$. This may indicate that the "effective" dynamical critical exponent $z > 2$. The reliability of the obtained dynamic dependences can be ensured by the quality of the collapse of dimensional dynamic scaling curves (Fig. 1). An anomalous slowing down in the dynamics of relaxation can be associated with pinning of the vortices on the defects, which indirectly affects the process of coarsening of the quasi-long-order areas [42,43]. It was shown [43] that the characteristic times of relaxation processes in a two-dimensional diluted XY-model can increase almost unlimitedly even for small linear sizes of the system ($L \approx 256$).

Conclusion

The insertion of disorder into the system leads to an anomalous slowing down of the process of critical relaxation. The resulting dynamic dependences of correlation length $\xi(t)$ show slow growth with an "effective" dynamical critical exponent $z > 2$, but this does not necessarily apply to the true value of the dynamic critical exponent z. For a pure system, it has been shown that a detailed consideration of the vortex dynamics leads to deviations from the known dependence $\xi(t) \sim (t/\ln t)^{1/2}$.

Acknowledgments

The reported study was supported by RFBR according to the research projects 17-02-00279, 18-42-550003, 18-32-00814 and grants MD-6868.2018.2, MK-4349.2018.2 of the President of the Russia. The simulations were supported in through computational resources provided by the Shared Facility Center "Data Center of FEB RAS" (Khabarovsk), by the Supercomputing Center of Lomonosov Moscow State University, by Moscow Joint Supercomputer Center and by St. Petersburg Supercomputer Center of the Russian Academy of Sciences.

References

[1] U.C. Tauber, Critical Dynamics: A Field Theory Approach to Equilibrium and Non-Equilibrium Scaling Behavior, (Cambridge University Press, Cambridge, 2014).
[2] V.V. Prudnikov, P.V. Prudnikov, M.V. Mamonova, Phys. Usp. **60**, 762 (2017).
[3] S.E. Korshunov, Phys. Usp. **49**, 225 (2006).
[4] V.L. Berezinskii, Sov. Phys. JETP **32**, 493 (1971).
[5] V.L. Berezinskii, Low-Temperature Properties of Two-Dimensional Systems (Fizmatlit, Moscow, 2007) [in Russian].
[6] I.S. Popov, P.V. Prudnikov, A.N. Ignatenko, A.A. Katanin, Phys. Rev. B, **95**, 134437 (2017).
[7] C.A.F. Vaz, J.A.C. Bland, G. Lauhoff, Rep. Progr. Phys. **71**, 056501 (2008).
[8] C. Kawabat, A.R. Bishop, Solid State Commun. **60**, 167 (1986).

[9] H.-J. Elmers, J. Hauschild, G.H. Liu, U. Gradmann, J. Appl. Phys. **79**, 4984 (1996).
[10] J. Als-Nielsen, S.T. Bramwell, M.T. Hutchings, G.J. McIntyre, D. Visser, J. Phys.: Condens. Matter. **5**, 7871 (1993).
[11] C. Bellitto, P. Filaci, S. Patrizio, Inorg. Chem. **26**, 191 (1987).
[12] A. Paduan-Filho, C.C. Becerra, J. Appl. Phys. **91**, 8294 (2002).
[13] F.L. Pratt, P.M. Zieliński, M. Balanda, R. Podgajny, T. Wasiutyński, B. Sieklucka, J. Phys.: Condens. Matter. **19**, 456208 (2007).
[14] R. Ganguly, D. Chaudhuri, P. Raychaudhuri, L. Benfatto, Phys. Rev. B. **91**, 054514 (2015).
[15] H.N. Wolfgang, Y.K. Na, G. Roumpos, C. Schneider, M. Kamp, S. Höfling, A. Forchel, Y. Yamamoto, Phys. Rev. B. **90**, 205430 (2014).
[16] Y.S. Karimov, Y.N. Novikov, Sov. Phys. – JETP Lett. **19**, 159 (1974).
[17] D.J. Bishop, J.D. Reppy, Phys. Rev. Lett. **40**, 1727 (1978).
[18] D.J. Bishop, J.D. Reppy, Phys. Rev. B. **22**, 5171 (1980).
[19] S. Misra, L. Urban, M. Kim, G. Sambandamurthy, A. Yazdani, Phys. Rev. Lett. **110**, 037002 (2013).
[20] M.R. Beasley, J.E. Mooij, T.P. Orlando, Phys. Rev. Lett. **41**, 1165 (1979).
[21] A.F. Hebard, A.T. Fiory, Phys. Rev. Lett. **44**, 291 (1980).
[22] L.N. Bulaevskii, V.V. Kuzii, A.A. Sobyanin, JETP Lett. **25**, 314 (1977).
[23] A.I. Buzdin, L.N. Bulaevskii, S.V. Panyukov, JETP Lett. **35**, 147 (1982).
[24] A.I. Buzdin, B. Bujicic, M.Yu. Kupriyanov, Zh. Eksp. Teor. Fiz. **101**, 231 (1992).
[25] A.N. Pargellis, S. Green, B. Yurke, Phys. Rev. E. **49**, 4250 (1994).
[26] W. Bietenholz, U. Gerber, F.G. Rejón-Barrera, J. Stat. Mech. P12009 (2013).
[27] A. Singh, S. Ahmad, S. Puri, S. Singh, Euro. Phys. Lett. **100**, 36004 (2012).
[28] A.J. Bray, Adv. Phys. **43**, 357 (1994).
[29] A.J. Bray, Adv. Phys. **51**, 481 (2002).
[30] P. Tabeling, Phys. Rep. **362**, 1 (2002).
[31] W.J. Nuttall, D.Y. Noh, B.O. Wells, R.J. Birgeneau, J. Phys.: Condens. Matter. **7**, 4337 (1995).
[32] I.F. Lyuksyutov, A.G. Fedorus, Sov. Phys. JETP. **53**, 1317 (1981).
[33] A. Taroni, S.T. Bramwell, P.C.W. Holdsworth, J. Phys.: Condens. Matter. **20**, 275233 (2008).
[34] Y. Tu, J. Toner, Phys. Rev. Lett. **75**, 4326 (1995).
[35] A. Cavagna, I. Giardina, T.S. Grigera, A. Jelic, D. Levine, S. Ramaswamy, M. Viale, Phys. Rev. Lett. **114**, 218101 (2015).
[36] A.P. Solon, H. Chaté, J. Tailleur, Phys. Rev. Lett. **114**, 068101 (2015).
[37] L. Berthier, P.C.W. Holdsworth, M. Sellitto, J. Phys. A. **34**, 1805 (2001).
[38] S. Abriet, D. Karevski, Euro. Phys. J. B. **37**, 47 (2004).
[39] P.V. Prudnikov, V.V. Prudnikov, I.S. Popov, JETP Lett. **101**, 596 (2015).
[40] V.V. Prudnikov, P.V. Prudnikov, I.S. Popov, JETP **126**, 369 (2018).
[41] P.V. Prudnikov, I.S. Popov, J. Phys.: Conf. Series **510**, 012014 (2014).
[42] I.S. Popov, P.V. Prudnikov, Solid state phenomena **233-234**, 8 (2015).
[43] I.S. Popov, P.V. Prudnikov, V.V. Prudnikov, J. Phys.: Conf. Series **681**, 012015 (2016).
[44] M. Henkel, M. Pleimling, Non-Equilibrium Phase Transitions, vol. 2, (Springer Netherlands, 2010).
[45] V.V. Prudnikov, P.V. Prudnikov, E.A. Pospelov, P.N. Malyarenko, JETP Lett. **102**, 167 (2015).
[46] P.V. Prudnikov, V.V. Prudnikov, E.A. Pospelov, P.N. Malyarenko, A.N. Vakilov, Progr. Theor. Exp. Phys. **2015** 053A0120 (2015).
[47] P.V. Prudnikov, V.V. Prudnikov, E.A. Pospelov, P.N. Malyarenko, Mat. Science Forum. **845** 101 (2016).
[48] P.V. Prudnikov, V.V. Prudnikov, E.A. Pospelov, J. Stat. Mech. 043303 (2016).
[49] P.V. Prudnikov, V.V. Prudnikov, P.N. Malyarenko, JETP **125** 1102 (2017).
[50] V.V. Prudnikov, P.V. Prudnikov, E.A. Pospelov, P.N. Malyarenko, JETP Lett. **107**, 569 (2018).
[51] A.J. Bray, A.J. Briant, and D.K. Jervis, Phys. Rev. Lett. **84**, 1503 (2000).
[52] V.V. Prudnikov, P.V. Prudnikov, S.V. Alekseev, I.S. Popov, Phys. Met. Metalogr. **115**, 1186 (2014).
[53] B. Berche, A.I. Fariñas-Sánchez, Yu. Holovatch, and R. Paredes V., Eur. Phys. J. B **36**, 91 (2003).
[54] M. Wintel, H.U. Everts, W. Apel, Phys. Rev. B **52**, 13480 (1995).

IOP Conf. Series: Journal of Physics: Conf. Series **1163** (2019) 012043 doi:10.1088/1742-6596/1163/1/012043

Monte Carlo simulation of colloidal particles dynamics in a drying drop

K. S. Kolegov

Volga State University of Water Transport, Caspian Institute of Maritime and River Transport, 2 Kostina St., Astrakhan, 414014, Russia

Astrakhan State University, 20A Tatishchev St., Astrakhan, 414056, Russia

Landau Institute for Theoretical Physics Russian Academy of Sciences, 1-A Academician Semenov Ave., Chernogolovka, 142432, Russia

E-mail: konstantin.kolegov@asu.edu.ru

Abstract. The work is devoted to particles dynamics simulation in a colloidal drop, when it dries on a substrate and the triple-phase boundary is fixed. Experimental observations [Deegan R. D. et. al., 2000] show a ring deposition on a solid substrate after full droplet desiccation. This phenomenon of macrolevel is known as coffee ring effect. There are other experiments that show a co-effect at the microlevel. We are talking about the formation of a quasicrystalline structure on the outer part of the ring and occurrence of an amorphous inner zone of ring [Marín Á. G. et. al., 2011]. The goal of this work is to check the hypothesis of other authors that this phenomenon is explained by the competition between the characteristic times of diffusional displacement of particles and their transfer by a compensation flow [Marín Á. G. et. al., 2011]. Numerical calculations using a two-dimensional model built on the basis of such assumptions and effects did not show formation of the amorphous zone. Probably, it is necessary to take into account the possibility of forming a multilayer structure of particles (3D-model).

1. Introduction

The processes of heat and mass transfer in drops drying on substrates are interesting because of their different applications. We list only some of them, for example, evaporation of picoliter droplets under reduced pressure for cooling the surface [1], technology for assessing the quality of drinks [2], removing nanoparticles from a solid surface [3], and so on.

The coffee ring effect is studied theoretically and experimentally in [4]. Capillary flows transfer colloidal particles to the three-phase (solid–liquid–gas) contact line during the process of the droplet drying on the substrate. The formation of a ring deposition is observed, if the three-phase boundary has been fixed throughout the process (pinning). There are many continual models describing this phenomenon at the macrolevel, for example, [4–7] etc. However, this approach does not allow to predict the structure of the particles relative to each other (microlevel). The first few rows of polystyrene particles near the contact line according to the experiment [8] form a kind of quasicrystal (particles are close packed). There is a transition to the amorphous zone at a short distance from the contact line, where the particles are located more or less chaotic. The authors [8] explain this phenomenon by the ratio of the characteristic times of diffusion ordering of particles and their transfer by the compensation flow of fluid. This flow is directed to the edge of the droplet where intense evaporation occurs. The velocity of capillary flow in the last about

5% of the process time increases by more than an order of magnitude [9]. The authors [8] believe that this rush-hour leads to the fact that the particles do not have time to pack themselves in the process of Brownian motion near the triple line. Thus, there is an amorphous zone of deposition. The authors of another experiment [10] also study the reasons for the transition from an ordered structure of particles to an unordered one. Their opinion is different, because they believe that the particles at high velocity begin to pile on the formed layers, so there is amorphous deposition. In addition, the authors [10] suggest that the liquid surface covering the particles can influence the final form of the structure due to the surface tension when the film rupture. Also, some additional influence may be exerted by the Marangoni flow in the case of a heated substrate [10]. The effect of particle shape on the resulting structure is studied in [11, 12]. Understanding this kind of processes is extremely important, for example, in micro- and nanostructuring applications.

There are several discrete models that explicitly describe the dynamics of each individual particle. These models are usually based on Monte Carlo or molecular dynamics methods [13,14]. Some models based on the Monte Carlo method are lattice ones [15–17], others are not [18]. Often the fluid flow is described continually. Typically, the flow velocity is calculated analytically for simple special cases [13, 16–19]. The calculated velocity field is superimposed on the particle dynamics. Particle adhesion, wetting and substrate roughness are taken into account in [13]. The simulation results show that the coffee ring effect is the result of compensation flow, low roughness and adhesion. The method used in [13] requires a large computing power and is more suitable for a small number of particles in the system. The formation of branched aggregates of nanoparticles in the droplet drying on the substrate is simulated in [15]. The models [16, 19] describe the formation of ring depositions during the evaporation of a droplet using a cubic lattice. Cubic particles attached to each other by side faces [16, 19] do not allow to describe the phenomenon of transition from a quasicrystal to an amorphous structure [8]. The paper [18] compares Monte Carlo method and lattice Boltzmann equations. The semi-cylindrical geometry of the droplet is considered. The authors [18] numerically describe the process of deposition of particles on the substrate as a result of their transfer by the compensation fluid flow and their Brownian motion. A computational experiment is carried out for different contact angles (hydrophilic and hydrophobic substrates) and modes of three-phase boundary (pinning and depinning). Exclusion of particle collisions (their intersection or overlapping) is not described in [18]. It is a disadvantage of the model in case if this condition is not implemented in the algorithm, because such model does not allow to obtain the form of the structure at the microlevel. At least such results are not presented in [18]. The lattice model is used to mathematically describe the coffee ring effect in [17]. The circular lattice used in [17] by definition does not allow to explain the reason for the formation of a quasicrystal at the drop edge, since this structure is already predetermined by the symmetrical arrangement of cells.

The aim of this paper is to determine whether the formation of a structure in which the ordered arrangement of particles (quasicrystal) transforms into an amorphous form is a consequence of advection and diffusion only.

2. Problem and method description

2.1. Physical formulation

Consider an axisymmetric thin drop with an initial contact angle $\theta \approx \pi/6$ and the radius of the base $R \approx 0.5$ mm [8]. We mean base is the plane of contact of the liquid and the substrate. Three-phase boundary is fixed, $R = \text{const}$ and $\theta = \theta(t)$, where t is process time. The duration of the evaporation process up to the complete drying of the droplet $t_{max} \approx 1750$ s [8]. Capillary forces prevail over gravitational forces in a drop of this size, so the geometry of the free surface shape of the liquid corresponds to a spherical segment.

Let us assume that the particles are in suspension. Brownian motion of particles dominates their sedimentation due to the small size ones (particle radius $r_p \approx 1$ μm [8]), so gravity is not taken into account. The particles are neutrally charged, so the question of interparticle interaction in this work is not affected. We consider the case when the temperature difference of the liquid, substrate and air is insignificant, so thermocapillary flows are excluded from attention [20].This paper does not address the description of processes when the substrate is heated [21] or there is some external heat source [22]. Evaporation is relatively slow in our case, so we assume that the process is isothermal. Let us consider that the dynamic viscosity of the liquid η is a constant in the zero approximation. The diffusion coefficient D is also assumed to be a constant. The diffusion coefficient of particles is determined by Einstein's formula $D = kT/(6\pi\eta r_p)$, where the Boltzmann constant is denoted by k, T is the liquid temperature. Substituting the water viscosity at room temperature, we obtain the value of the diffusion coefficient $D \approx 2 \cdot 10^{-13}$ m^2/s.

2.2. Mathematical model

It is convenient to use polar (r, ϕ) and Cartesian coordinates (x, y) at the same time in this problem. We are describing the particle dynamics only in the horizontal plane xy, since the thickness of the liquid layer in a thin drop is much less than the diameter of its base. The problem will be solved not in the whole area, but only in the selected sector of the circle due to the axial symmetry. Let us use periodic boundary conditions on the side borders of the sector. There is also a movable boundary R_f in the plane xy despite that the case with a pinned contact line is considered. Let the radius of fixation R_f is such a boundary, where the particle size and the thickness of the liquid layer are comparable ($2r_p \approx h$), at that $R_f < R$. Particles cannot reach the R boundary in a thin drop, since near this boundary the thickness of the liquid layer is small, $h \to 0$. Therefore, the particles will not move further than R_f because behind this boundary $h < 2r_p$. The force of the surface tension of the fluid will restrain them. Write an expression for the shape of the droplet surface $h(r,t) = \theta(t)\left(R^2 - r^2\right)/(2R)$, where the radial coordinate $r = \sqrt{x^2 + y^2}$. Hence, we obtain the dependence of the fixation radius on time

$$R_f(t) = \sqrt{R^2 - \frac{4r_pR}{\theta(t)}}. \tag{1}$$

The height of the droplet during evaporation decreases, in consequence of which the contact angle decreases too. This evolution of contact angle is explain the movement of the boundary R_f. Note that most of the time the value of R_f changes slightly. This radius begins to decrease rapidly only at the end of the process. The contact angle θ decreases linearly over time, $\theta \to 0$ when $t \to t_{max}$ [9].

The radial fluid flow velocity, averaged over the height of the drop, is calculated based on the results from the model [4],

$$\bar{v}_r(r,t) = \frac{R}{4(r/R)(t_{max} - t)}\left(\frac{1}{\sqrt{1-(r/R)^2}} - \left[1 - (r/R)^2\right]\right). \tag{2}$$

Let us simulate diffusion as a random walk of particles by the Monte Carlo method.

2.3. Problem solving algorithm

We take into account that there must be several modes of starting the calculation module («diffusion + advection», «diffusion» and «advection») when developing the program. Thus, it will allow to consider two types of mass transfer separately and both together. Let each particle be in one of three states. The state «motion» is characterized by both diffusion and advective transfer. The status «diffusion only» implies only one Brownian motion. Particle is stationary if its status is «stop».

First, we uniformly generate random coordinates of particles (x, y) in our domain. Second, we set initial values of their status («motion» or «diffusion only» depending on the program start mode) for all particles. Then we go in a cycle on temporary layers with a step $\Delta t = 0.1$ ms to achieve time t_{max}. The time step size is choiced on the basis of a series of computational experiments in such a way that the Einstein relation for the mean square of the displacement $\overline{(\Delta l)^2} = 2Dt_{max}$ is carried out, where

$$\overline{(\Delta l)^2} = \sum_{j=1}^{t_{max}/\Delta t} \frac{1}{N_p} \sum_{i=1}^{N_p} \left((x_i^j - x_i^{j-1})^2 + (y_i^j - y_i^{j-1})^2 \right),$$

N_p is the number of particles.

The program recalculate the value of R_f according to (1) at each time step and iterate through all the particles. It is depended on status of particle to shift it or not. The flow velocity for the current coordinate r and time t is calculated based on the expression (2). We calculate the distance of the particle displacement by the fluid flow in one time step Δt by the formula $\Delta t \bar{v}_r(r, t)$. The diffusion displacement distance is defined as $\sqrt{2D\Delta t}$. The value of the polar angle ϕ for the diffusion motion of a particle is generated randomly. This step is not executed if the movement is impossible due to the coming overlapping with any other particle.

Let us use the following rules. First, a particle goes into the state «stop» if this one was in contact with the radius of fixation R_f and had the status «motion» or «diffusion only». A particle cannot move beyond R_f because it is prevented by the surface tension of the liquid. Second, if the moving particle could potentially intersect with the stationary one, it changes the status «motion» to the value «diffusion only». Third, if a diffusion particle has moved away from the fixation radius by a distance of more than $10r_p$, then it becomes mobile again (status «motion»). The last condition is valid for the case when the program is started in the «diffusion + advection» mode. The program algorithm is written with C++ language.

2.4. Numerical results

The fig. 1 shows the simulation results of the case when the particles only diffuse. This is possible with a uniform evaporation of the liquid along the free surface of the drop [4]. Therefore, compensation flows are absent. The arrangement of the particles in several consecutive moments of time is shown in the fig. 1a. It can be seen that the final structure ($t = 1750$ s) is characterized by a more or less uniform distribution. In addition, in this structure we observe a thin layer of particles near the contact line (1c). These particles hit the fixation radius as a result of chaotic Brownian motion, when the R_f boundary moved very slowly. The number of particles per unit area (density distribution) is shown in the fig. 1b. The density of the particle distribution is approximately 1050 ± 150 at all considered time intervals.

The calculation result of particle transfer by a single fluid flow is shown in the fig. 2. We observe the formation of dendritic-like structures near the drop edge during the whole process in this case (fig. 2a). The particles do not mix randomly when diffusion is absence. This leads to the fact that the particle, bumping with another sticked particle in front, stops moving further to the fixation boundary. This sticked particle interferes with another particles behind the current one to move relative to the direction of flow. Their tree-like shape distribution (fig. 2c) is formed in this way. The density distribution of particles in the central part of the domain decreases, and their number near the three-phase boundary increases (fig. 2b).

The numerical data for the third case are shown in fig. 3. Advection and diffusion leads to the formation of an ordered ring deposition (fig. 3a). The number of particles near the contact boundary increases significantly over time (fig. 3b). These numerical results correspond

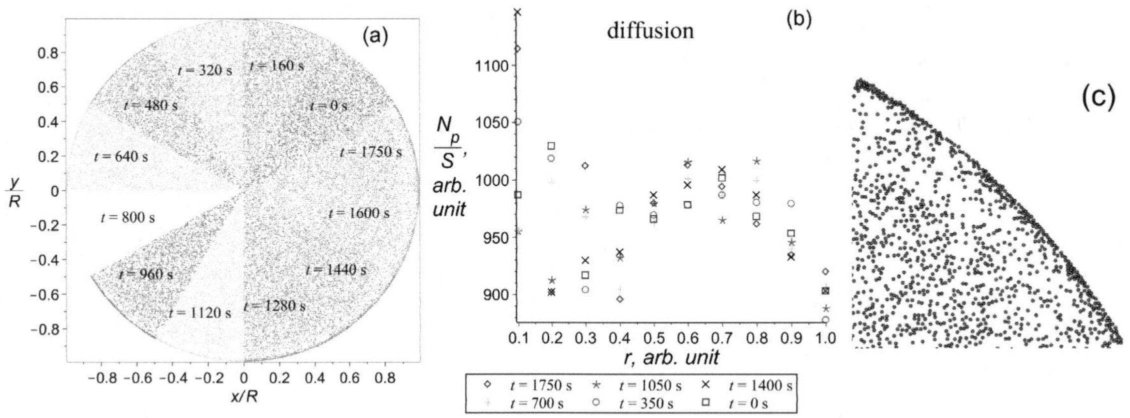

Figure 1. Dynamics of colloidal particles in the case of diffusion motion: (a) arrangement of particles at different times, (b) the density distribution of particles and (c) final structure near the contact line.

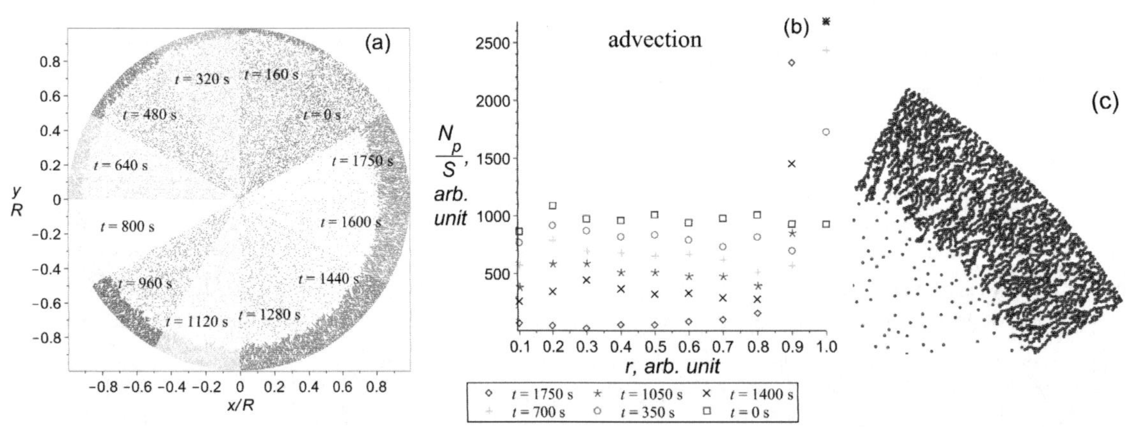

Figure 2. Dynamics of colloidal particles in the case of advective transfer: (a) arrangement of particles at different times, (b) the density distribution of particles and (c) final structure near the contact line.

qualitatively to the experimental observations of the coffee ring effect [4]. But the change of close packing of particles (quasicrystal) to the inner amorphous part of the ring when moving from the edge of the drop towards its center is not observed here (fig. 3c). These simulation results differ from experimental data [8].

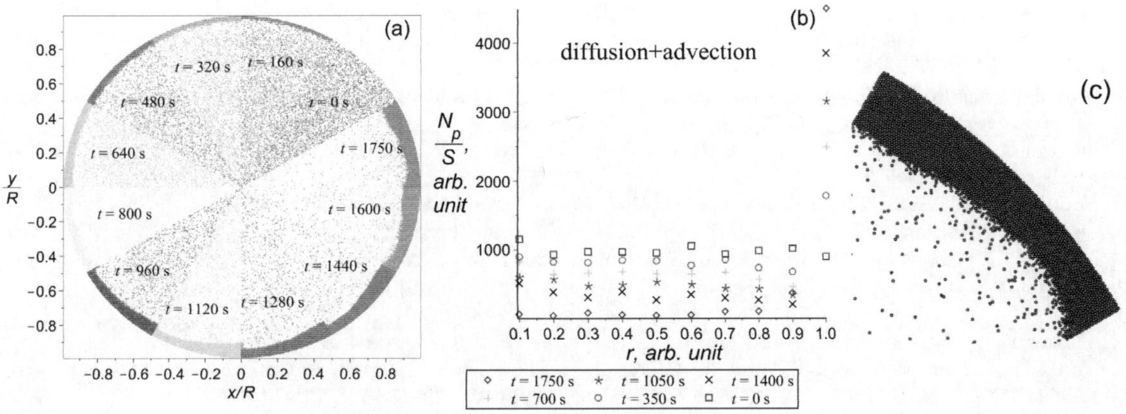

Figure 3. Dynamics of colloidal particles in the case of advective and diffusion transfer: (a) arrangement of particles at different times, (b) the density distribution of particles and (c) final structure near the contact line.

3. Conclusion

Discrete (non-lattice) mathematical model describing the mass transfer of colloidal particles during the evaporation of a droplet on a substrate in the pinning mode is constructed. The concept of a fixation radius R_f, which represents a moving boundary, in the region of which the thickness of the liquid layer is comparable with the particle size, is introduced. The Brownian motion of particles is modeled by the Monte Carlo method. The radial flow velocity of the liquid is calculated analytically. The cases are considered separately taking into account only diffusion, advection and both effects at the same time. The final particle distribution after the droplet drying for these cases is obtained. The resulting structure resembles a spot if there is no fluid flow and the particles move randomly. The same was observed in the experiment [4]. The model predicts the formation of dendritic structures in the periphery region in the presence of only one advection in the system. Numerical calculations taking into account both factors (diffusion and advection) show the formation of a ring structure of close-packed particles at the place of a dried drop, which corresponds to experimental observations [4]. Unfortunately, it was not possible to mathematically describe the effect of the formation of the amorphous structure in the inner zone of ring [8] on the basis of the proposed model. The video (in the supplement) shows that a wide quasicrystalline zone is formed much earlier than the radius of fixation begins to decrease noticeably. Obviously, the model should take into account the possibility of forming a multilayer structure of particles (three-dimensional model). Thus, the question remains open. In addition, it is still not clear what effect the surface tension of the fluid has on the formation of the structure [23].

Acknowledgments

The problem statement belongs to Yu. Yu. Tarasevich. The author is grateful to L. Yu. Barash, I. V. Vodolazskaya, A. V. Eserkepov and Yu. Yu. Tarasevich for fruitful discussions and useful advice. This work was supported by the Russian Science Foundation (project 18-71-10061).

IOP Conf. Series: Journal of Physics: Conf. Series **1163** (2019) 012043 doi:10.1088/1742-6596/1163/1/012043

References

[1] Saverchenko V I, Fisenko S P and Khodyko Y A 2015 *Colloid J.* **77** 71–76
[2] Yakhno T, Sanin A, Yakhno V, Kazakov V, Pakhomov A, Guguchkina T and Markovsky M 2018 Chapter 13 - drying drop technology in wine and hard drinks quality control *Food Control and Biosecurity* Handbook of Food Bioengineering ed Holban A M and Grumezescu A M (Academic Press) pp 451 – 480 ISBN 978-0-12-811445-2
[3] Mahdi F, Record T, Amadi C, Emmanuel F, Ivanova N, Trybala A and Starov V 2015 *Colloids Interface Sci. Commun.* **6** 13–16 ISSN 2215-0382
[4] Deegan R D, Bakajin O, Dupont T F, Huber G, Nagel S R and Witten T A 2000 *Phys. Rev. E* **62** 756–765
[5] Fischer B J 2002 *Langmuir* **18** 60–67
[6] Vodolazskaya I V and Tarasevich Y Y 2011 *Mod. Phys. Lett. B* **25** 1303–1310
[7] Ozawa K, Nishitani E and Doi M 2005 *Jap. J. Appl. Phys.* **44**(6A) 4229–4234
[8] Marín A G, Gelderblom H, Lohse D and Snoeijer J H 2011 *Phys. Rev. Lett.* **107**(8) 085502
[9] Hamamoto Y, Christy J R E and Sefiane K 2011 *Phys. Rev. E* **83**(5) 051602
[10] Zhang W, Yu T, Liao L and Cao Z 2013 *AIP Advances* **3** 102109
[11] Yunker P J, Still T, Lohr M A and Yodh A G 2011 *Nature* **476** 308–311 ISSN 0028-0836
[12] Yunker P J, Lohr M A, Still T, Borodin A, Durian D J and Yodh A G 2013 *Phys. Rev. Lett.* **110**(3) 035501
[13] Lebedev-Stepanov P and Vlasov K 2013 *Colloids Surf., A* **432** 132–138 ISSN 0927-7757
[14] Shi B and Webb E B 2016 *Phys. Rev. E* **94**(1) 012614
[15] Zhang H, Shan Y, Li L, Lu M and Li R 2016 *Appl. Therm. Eng.* **94** 650–656 ISSN 1359-4311
[16] Kim H S, Park S S and Hagelberg F 2011 *J. Nanopart. Res.* **13** 59–68 ISSN 1572-896X
[17] Jung N, Yoo C S and Leo P H 2014 *J. Phys. Chem. B* **118** 2535–2543
[18] Petsi A J, Kalarakis A N and Burganos V N 2010 *Chem. Eng. Sci.* **65** 2978–2989 ISSN 0009-2509
[19] Crivoi A and Duan F 2014 *Sci. Rep.* **4** 4310
[20] Barash L Y, Bigioni T P, Vinokur V M and Shchur L N 2009 *Phys. Rev. E* **79** 046301
[21] Gatapova E Y, Semenov A A, Zaitsev D V and Kabov O A 2014 *Colloids Surf., A* **441** 776–785
[22] Kolegov K S 2018 *Microgravity Sci. Technol.* **30** 113–120 ISSN 1875-0494
[23] Kralchevsky P A and Nagayama K 1994 *Langmuir* **10** 23–36

Algorithmic method for modeling the optimal treatment of patients with HIV

Anna P. Presnova

National Research University Higher School of Economics, Tallinskaya street 34, Moscow, Russia

E-mail: apresnova@hse.ru

Abstract. The mathematical model describing the dynamics of HIV in the human body is a nonlinear system of differential equations. This model takes into account the effect of drugs on the body. Thus, it is possible to obtain "optimal" treatment regimens for patients, which cause minimal harm to the body. In the work for constructing suboptimal control of the supply of drugs, the method of "extended linearization" is used, which makes it possible to switch from a nonlinear model to a linear model, but with parameters that depend on the state. To solve the resulting equation Riccati and search for control actions, a method is proposed for the formation of optimization algorithms for nonlinear control systems based on the application of functions of admissible values of control actions.

1. Introduction

The problem of optimal control synthesis for a nonlinear system of differential equations describing some real process is relevant today. Since the theory of control of linear systems is sufficiently developed, various linearization methods are often used to synthesize nonlinear control systems. In this paper, we use the extended linearization method, which allows us to go from the initial nonlinear model to a linear model, but with parameters depending on the state. In the literature, such a representation of the original model is called the SDC representation [1]. Such a transition makes it possible to find the optimal control not by using the partial derivative solution of the Hamilton-Jacobi-Bellman equation, but to proceed to solving a Riccati-type equation with state-dependent parameters. Solving this equation at the pace of system operation is also not an easy task. To solve the resulting equation, an algorithmic method of parametric optimization of the object is proposed. This method is based on the necessary conditions for optimality of the system, which consist in a certain behavior of the Hamiltonian on the optimal trajectory.

The proposed algorithm is applied to a mathematical model of the human immune system in the presence of HIV in the body. The synthesis of suboptimal control, controlling the flow of drugs. The simulation was carried out in the package MATLAB Simulink. The simulation results confirm the correct operation of the algorithm.

2. Nonlinear system and optimal control

Consider a controllable non-linear system described by the differential equation

$$\begin{aligned} \frac{d}{dt}x(t) &= f(x(t)) + B(x(t))u(t), \\ x(t_0) &= x_0, \end{aligned} \tag{1}$$

where $x(t) \in R^n$ - the state of the system; $x_0 \in X_0$ - the set of possible initial conditions of the system; $y(t) \in R^m, m \leq n$ - system output; $u(t) \in R^r$ - control to be found; $f(x), B(x)$ - real continuous matrix functions of appropriate dimensions.

Introduce the cost functional

$$J(x(\cdot), u(\cdot)) = \frac{1}{2}\int\limits_{t_0}^{t_f} \{ x^{\mathrm{T}}(t) Q\, x(t) + u^{\mathrm{T}}(t) R\, u(t) \} dt\,. \tag{2}$$

In functional (2) a symmetrical matrices Q and R are positive definite. Limitations on control actions are taken into account when assigning a positive definite matrix R. The problem consists in construction of an optimal control $u(t)$ minimizing a functional of (2) on the object (1).

The system (1) is assumed controllable and observable. It is also assumed that the condition $x = 0_n$ is the equilibrium point of the system where $u = 0$ so that $f(0) = 0$ and $B(x) \neq 0$.

Suppose that function $f(x(t))$ - continuous differentiable with respect to $x \in \Omega_x$ i.e. $f(\cdot) \in C^1(\Omega_x)$. In addition, we shall assume that the functions $f(x(t))$, $B(x(t))$ such that for any initial conditions $(t_0, x_0) \in R_+ \times \Omega_x$ only one solution $x(t, t_0, x_0)$ of (1) is possible.

Optimal control for the initial system, as is known from the theory of analytical construction, is determined by the expression

$$u(t) = -R^{-1}B^{\mathrm{T}}(x(t))\left\{\frac{\partial V(t, x(t))}{\partial x}\right\}^{\mathrm{T}}, \tag{3}$$

where vector $\left\{\frac{\partial V(t,x)}{\partial x}\right\}^{\mathrm{T}}$ is a solution of the Hamilton-Jacobi-Bellman equation [2]:

$$\begin{gathered} \frac{\partial V(t,x(t))}{\partial t} + \frac{\partial V(t,x(t))}{\partial x} f(x(t)) - \frac{1}{2}\frac{\partial V(t,x(t))}{\partial x} B(x(t))R^{-1}B^{\mathrm{T}}(x(t))\left\{\frac{\partial V(t,x(t))}{\partial x}\right\}^{\mathrm{T}} + \\ + \frac{1}{2}x^{\mathrm{T}}(t)Qx(t) = 0, \\ V(t_f, x(t_f)) = 0, \end{gathered} \tag{4}$$

where $V(t, x(t))$ - function, defined as

$$V(s, x(\cdot)) \triangleq \inf_{u(\cdot)\in U} \left[\frac{1}{2}\int\limits_{s}^{t_f} \{ x^{\mathrm{T}}(t)Qx(t) + u^{\mathrm{T}}(t)R\,u(t)\}dt \right]. \tag{5}$$

The realization of controls in the form (3) encounters difficulties connected with solving a scalar partial differential equation (4).

IOP Conf. Series: Journal of Physics: Conf. Series **1163** (2019) 012044 doi:10.1088/1742-6596/1163/1/012044

3. The method of extended linearization and SDRE

Using the extended linearization method, we will present the original model in the form

$$\begin{aligned}\frac{d}{dt}x(t) &= A(x(t))x(t) + B(x(t))u(t),\\ x(t_0) &= x_0.\end{aligned} \tag{6}$$

It should be noted that such a presentation is not the only one. The problem of choosing the optimal representation has not been solved yet. To find solutions of the Hamilton-Jacobi-Bellman equation (4), we introduce the relation

$$\{\partial V(t, x(t))/\partial x\}^{\mathrm{T}} = S(x(t))x(t), \tag{7}$$

where $S(x(t))$ - a positive definite symmetric matrix, allows when searching for optimal controls, to pass from the solution of the Hamilton-Jacobi-Bellman equation in partial derivatives (4) to an equation of the form:

$$\begin{aligned}&\Bigg[\frac{d}{dt}S(x(t)) + A^{\mathrm{T}}(x(t))S(x(t)) + S(x(t))A(x(t)) - S(x(t))B(x(t))R^{-1}B^{\mathrm{T}}(x(t))S(x(t)) +\\ &+ Q + \left\{\left[\frac{\partial A(x(t))}{\partial x}\right]^{\mathrm{T}} \otimes x(t) + \left[\frac{\partial B(x(t))}{\partial x}\right]^{\mathrm{T}} \otimes u(t)\right\} S(x(t))\Bigg] x(t) = 0\,,\end{aligned} \tag{8}$$

where $\otimes$ - Kronecker multiplication symbol. The solution of the resulting equation (8) also presents a certain difficulty, therefore, we consider the case with an unlimited transition time. In this case (8) can be written as algebraic equation of Riccati type with the state dependent coefficient [3]:

$$A^{\mathrm{T}}(x(t))\hat{S}(x(t)) + \hat{S}(x(t))A(x(t)) - \hat{S}(x(t))B(x(t))R^{-1}B^{\mathrm{T}}(x(t))\hat{S}(x(t)) + Q = 0. \tag{9}$$

And the control will be written as

$$u(t) = -R^{-1}B^{\mathrm{T}}(x(t))\hat{S}(x(t))x(t). \tag{10}$$

Since the control action $u(t)$ is determined by the expression (10), into which the matrix $\hat{S}(x)$ enters, which in the general case is not equal to $S(x)$, the solution of the equation (8), then the control (10) is more correctly called suboptimal. In [4], a theorem is proved that the resulting suboptimal control provides the system (1) local asymptotic stability.

4. The parametric optimization algorithms

In this paper, to solve the resulting equation (9) and synthesize suboptimal controls of a nonlinear object, an algorithmic method for constructing control actions based on the behavior of the Hamiltonian along the optimal trajectory will be proposed.

For this, we consider a controllable object described by a nonlinear equation of a general form

$$\begin{aligned}\frac{d}{dt}x(t) &= f(x(t), u(t), \eta(t), a(t)),\\ x(t_0) &= x_0,\end{aligned} \tag{11}$$

where $x \in R^n, u \in R^r, \eta \in R^k$ - vector of object parameters exposed to uncontrolled disturbances, $a \in R^p$ - vector of system optimization object parameters. We note that in the general case $k \geq p$. The parameters of the system selected for parametric optimization can be found both in the object itself and in the regulator.

IOP Conf. Series: Journal of Physics: Conf. Series **1163** (2019) 012044 doi:10.1088/1742-6596/1163/1/012044

Let us consider the behavior of the Hamiltonian on the optimal trajectory. Differentiating $H(t,x,u,\lambda)$ in respect of time, taking into account the possibility of transition to an open field of control actions, we obtained:

$$\frac{d}{dt}H(t,x(t),u(t),\lambda(t)) = \frac{\partial H(t,x(t),u(t),\lambda(t))}{\partial t}. \tag{12}$$

The behavior of the Hamiltonian under optimal control changes during a transient process along a completely defined trajectory determined by solving the differential equation (12). This property of the Hamiltonian was used as the basis for the design of algorithms for optimizing the control system.

The necessary conditions for the minimum of the quality functional, expressed in the behavior of the Hamiltonian on the optimal trajectory, are written with the help of a scalar function $\Re(t)$ such as

$$\Re^0(t) = H^0(t) + \varphi(t) = 0, \tag{13}$$

where the scalar function $\varphi(t)$ takes specific values for each problem under consideration and $H^0(t) = H(t,x^0,u^0,\lambda^0) = \varphi(t)$ - is the value of the Hamiltonian at each instant of control time in the absence of parametric disturbances (or when they are completely parried) under optimal control and the corresponding trajectory of the system. It is obvious that, if in $a(t) \neq \eta(t)$, then (13) will not be satisfied, i.e. $\Re(t) = H(t) + \varphi(t) \neq 0$. Let us take the Lyapunov function into consideration

$$V_L(\eta,a) = \frac{1}{2}\{\Re(t) - \Re^0(t)\}^2 = \frac{1}{2}\{\Re(t)\}^2. \tag{14}$$

Then for local asymptotic stability, its derivative should be nonpositive:

$$\frac{d}{dt}V_L(\eta,a) = \Re(t)\left\{\frac{\partial H(t,x(t),u(t),\lambda(t))}{\partial \eta(t)}\frac{d}{dt}\eta(t) + \frac{\partial H(t,x(t),u(t),\lambda(t))}{\partial a(t)}\frac{d}{dt}a(t)\right\} \leq 0, \tag{15}$$

since

$$\frac{\partial H(t,x(t),u(t),\lambda(t))}{\partial t} = \frac{\partial H^0(t,x(t),u(t),\lambda(t))}{\partial t}, \frac{\partial \varphi(t)}{\partial \eta(t)} = 0, \frac{\partial \varphi(t)}{\partial a(t)} = 0. \tag{16}$$

Let the algorithm of parametric optimization have the form

$$\begin{gathered}\frac{d}{dt}a(t) = -\left\{\frac{\partial H(t,x(t),u(t),\lambda(t))}{\partial a(t)}\right\}^{\mathrm{T}}\Re(t),\\ a(t_0) = a_0.\end{gathered} \tag{17}$$

Consider another parametric optimization algorithm, in the form

$$\frac{d}{dt}a(t) = \Im(t)\Re(t), \tag{18}$$

where the function $\Im(t)$ determines the direction and rate of change of the optimization parameters. Assigning the Lyapunov function in the same form as in the previous case, condition (15) is obtained as

$$\Re(t)\frac{\partial H(t,x(t),u(t),\lambda(t))}{\partial \eta(t)}\frac{d}{dt}\eta(t) - \Re^2(t)\frac{\partial H(t,x(t),u(t),\lambda(t))}{\partial a(t)}\Im(t) \leq 0. \tag{19}$$

Then the algorithm (17) takes the form

$$\begin{gathered}\frac{d}{dt}a(t) = -\left\{\frac{\partial H(t,x(t),u(t),\lambda(t))}{\partial a(t)}\right\}^{\mathrm{T}}\Re^2(t),\\ a(t_0) = a_0.\end{gathered} \tag{20}$$

These algorithms are assigned in this form is not accidental, with the same values, they have unequal efficiency. In the case of a function $|\Re(t)| < 1$, the restructuring of the parameters occurs faster with the algorithm (17). If the values of the function $|\Re(t)| > 1$, then the speed of parameter adjustment is higher when using the algorithm(20).

5. Mathematical model of the human immune system with HIV

The model consists of five differential equations [5]:

$$\begin{aligned}
&\frac{d}{dt}i(t) = \lambda - di(t) - \beta(1-\eta u)i(t)y(t),\\
&\frac{d}{dt}y(t) = \beta(1-\eta u)i(t)y(t) - ay(t) - p_1 z_1(t)y(t) - p_2 z_2(t)y(t),\\
&\frac{d}{dt}z_1(t) = c_1 z_1(t)y(t) - b_1 z_1(t),\\
&\frac{d}{dt}w(t) = c_2 i(t)y(t)w(t) - c_2 qy(t)w(t) - b_2 w(t),\\
&\frac{d}{dt}z_2(t) = c_2 qy(t)w(t) - hz_2(t),
\end{aligned} \tag{21}$$

where i - concentration of uninfected cells of the immune system, T-helpers; λ - the production rate of T-helpers in the body; d - rate of natural death of T-helpers. When a virus enters the bloodstream, T-helper cells become infected at a rateβ and become infected cells (y), that is the HIV. The infected cells naturally die at the rate a, in addition T-killers (z_1) kill them at the rate p_1, and immunoglobulins (z_2) kill the infected cells at the rate p_2. B-lymphocytes (w) are activated in the body at the speed c_2, and at the speed q they turn into immunoglobulins. The maximum effectiveness of drugs is expressed by the coefficient η, u - the dose of the drug administered, that is, our intervention. The values of the parameters are taken from [6].

It is for this model we shall synthesize the suboptimal administration of drug delivery, found by means of the algorithmic method. To do this, the initial nonlinear model (21) using the method of "extended linearization" will be represented as (6). Here the matrix has an ambiguous representation, as mentioned earlier.

To find a suboptimal control in the form of (10), it is necessary to find the matrix $S(x(t))$, which is a solution of the Riccati equation with state-dependent parameters (9), which is not an easy task. To find the matrix $S(x(t))$ we use the algorithmic method. Let us present the matrix $S(x(t))$ in the form

$$S(x) = S_0 + s(t), \tag{22}$$

here the matrix S_0 is found from the solution of the Riccati equation with constant parameters (at $x(t_0) = x_0$). In this problem $H\left\{x^0, u^0, S(x^0)x^0\right\} = 0$ - is the behavior of the Hamiltonian on the optimal trajectory. In accordance with the above method of forming the parametric optimization algorithms (17) and (20), where $\Re(t) = H\left\{x^0, u^0, S(x^0)x^0\right\} - H\left\{x, u, [S_0 + s(t)]\, x\right\}$, the algorithms takes the form

$$\begin{aligned}
&\frac{d}{dt}s(t) = -\left\{\frac{\partial H\left\{x, u, [S_0 + s]\, x\right\}}{\partial s(t)}\right\}^{\mathrm{T}} H\left\{x, u, [S_0 + s]\, x\right\}, \quad \text{if} \, |H| \le 1;\\
&\frac{d}{dt}s(t) = -\left\{\frac{\partial H\left\{x, u, [S_0 + s]\, x\right\}}{\partial s(t)}\right\}^{\mathrm{T}} H^2\left\{x, u, [S_0 + s]\, x\right\}(t), \quad \text{if} \, |H| > 1;\\
&s(t_0) = 0.
\end{aligned} \tag{23}$$

6. Computer simulation and results

Computer simulation of the synthesized control actions was carried out in the MATLAB Simulink package. For example, a very weak patient with an HI virus is considered. The simulation was carried out for a given initial state in two modes: in the treatment synthesized by the algorithmic method, and in the absence of any treatment, that is, when $u = 0$. The obtained

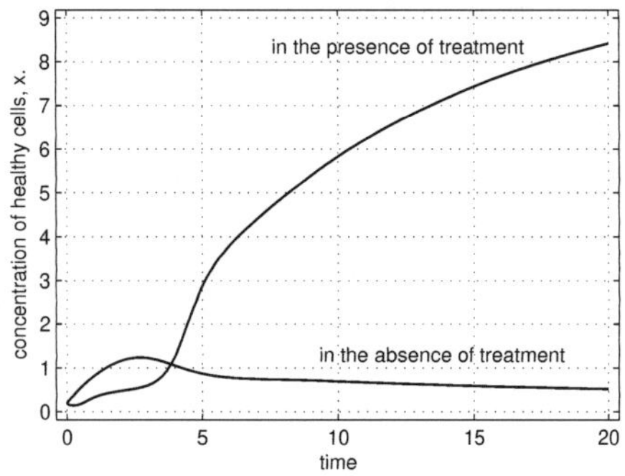

Figure 1. The concentration of healthy cells of the immune system in the presence of HIV.

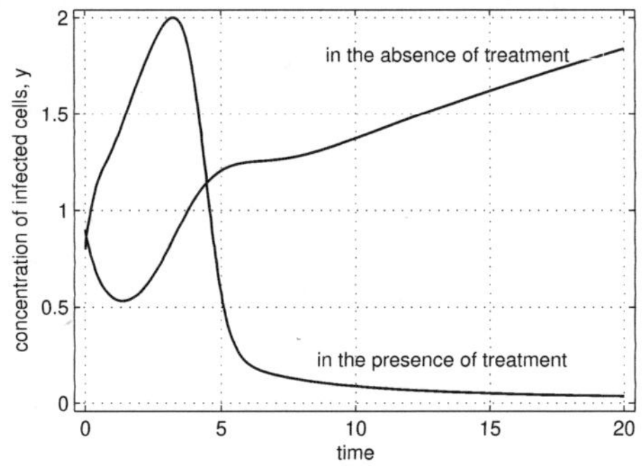

Figure 2. The concentration of cells of the immune system infected with HIV in the human body.

simulation results for the chosen mathematical model of the immune system of a person with HIV demonstrate the success of the constructed control using the algorithmic method.

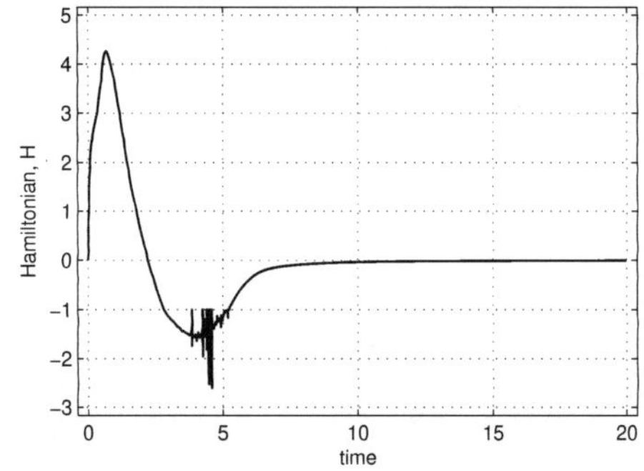

Figure 3. Hamiltonian.

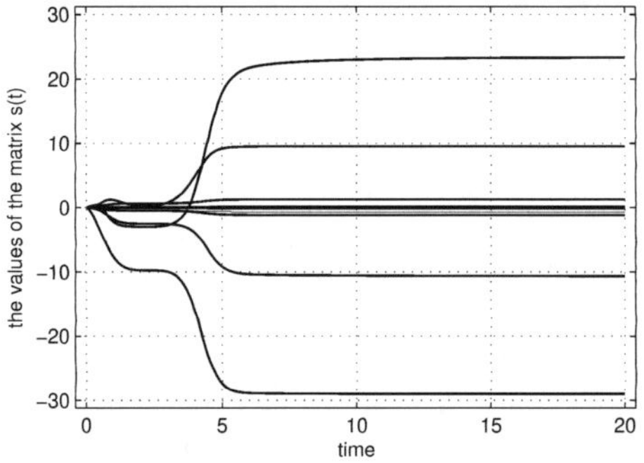

Figure 4. Matrix $s(t)$.

References

[1] Pearson J D 1962 Approximation methods in optimal control *Journal of Electronics and Control* B **12** 453-69
[2] Afanasyev V N, Kolmanovskyi V B, Nosov V R 2003 *Mathematical theory of control system design* (Moscow: High school)
[3] Cimen T D 2008 State-Dependent Riccati Equation (SDRE) Control: A Survey *Proc. 17th World Conf. IFAC* 3771-75
[4] Afanasyev V N 2015 *Control of nonlinear uncertain dynamic objects*(Moscow: URSS)
[5] Wodarz D and Nowak M A 1999 Specific therapy regimes could lead to long-term immunological control of HIV *Proceedings of the National Academy of Sciences* vol. 96 **6** 14464-69
[6] Zurakowski R and Teel A 2006 A model predictive control based scheduling method for HIV therapy *Journal of Theoretical Biology* **238** 368-82

IOP Conf. Series: Journal of Physics: Conf. Series **1163** (2019) 012045 doi:10.1088/1742-6596/1163/1/012045

Estimation of Lyapunov exponents for quasi-stable attractors of dynamical systems with time delay

S V Aleshin, D S Glyzin, S D Glyzin, V E Goryunov

P.G. Demidov Yaroslavl State University, Sovetskaya str., 14, Yaroslavl, 150000, Russia
Scientific Center in Chernogolovka RAS, Lesnaya str., 9, Chernogolovka, Moscow region, 142432, Russia

E-mail: fktiby@yandex.ru, glyzin@uniyar.ac.ru, glyzin@gmail.com, salkar@ya.ru

Abstract. Lyapunov exponents play an important role among the invariant characteristics of dynamical systems. Analysis of the spectrum of Lyapunov exponents is widely used to study the complex dynamics in systems of ordinary differential equations and models that can be reduced to maps. In this paper we consider the question of numerical evaluation of Lyapunov exponents for delay differential equations. We used a new algorithm with FFT, tested it on Hutchinson equation with known Lyapunov exponents spectrum and compared with results of the old algorithm. The so-called quasi-stable behavior is typical for a number of gene networks models and neuronal associations that have been studied recently. The phenomenon of cycle (k-dimensional torus) quasistability in dynamical system is characterized by the fact that some of its multipliers are asymptotically close to the unit circle, and the remaining multipliers are modulo less than one (with the exception of a simple root 1 (k such roots)). In some cases, it is possible to prove the existence and give an asymptotic estimate of the multipliers of considered system by using the large parameter methods. However, if these methods are not applicable, it is necessary to obtain a tool for the numerical estimation of multipliers. Such a tool is provided by algorithms for Lyapunov exponents estimation. Given that the equations with delay are often applied in the models of neural and gene networks, the algorithm for Lyapunov exponents estimation for such systems will be in demand.

1. Introduction

Lyapunov exponents and Lyapunov dimension play an important role among the invariant characteristics of dynamical systems attractors. The definition of Lyapunov exponents can be found in the book [1]. For a linear system of n equations written in vector form

$$\dot{x} = A(t)x, \tag{1}$$

where $x \in \mathbb{R}^n$ and $A(t)$ is $n \times n$ matrix, Lyapunov exponents are determined by formula $\lambda(x) = \overline{\lim}_{t\to\infty} t^{-1} \ln |x(t)|$. Analysis of Lyapunov exponents spectrum is widely used to study complex dynamics in systems of ordinary differential equations and models that can be reduced to maps. Cases where they can be found analytically are quite rare. The Benettin method is commonly used to calculate the largest exponent [2]. Further development of this method was obtained in [3]. Authors added renormalization of the initial conditions by the Gram–Schmidt algorithm [4], which allowed to calculate the spectrum of Lyapunov exponents. As follows from the Oseledets theorem [5] in the finite-dimensional case, the linearized on attractor system of the form (1) is always Lyapunov proper and thus the upper limit may be replaced by a normal limit, that allows us to effectively compute Lyapunov exponents [6]. This theorem cannot be proved in the cases of equations with delay and boundary value problems. Therefore, during the development of Lyapunov exponents estimation algorithms it is important to have a model equation with delay, which spectrum of Lyapunov exponents can be computed in some other way. It allows us to test the algorithm and make sure that it is correct. Spectrum of Lyapunov exponents has been estimated in [7, 8]. However, authors do not justify their algorithms and testing examples.

In this article we describe a standard algorithm extension for computing the Lyapunov exponents spectrum for systems of differential equations with delay. We show the testing results of the algorithm applied to Hutchinson equation as an example [9], and illustrate the algorithm application to Lyapunov exponents spectrum computing for some systems described in [10, 11, 12, 13].

In the strict sense, the estimated invariant characteristics of the maps are not Lyapunov exponents. But for the case of Hutchinson equation equilibrium state, as will be shown below, we obtained results that can illustrate the “proximity” of the estimated characteristics and Lyapunov exponents. It makes it possible to suppose that our assumptions hold in other cases.

2. Description of Lyapunov exponents evaluation algorithm for delay differential equations

Let us describe an algorithm for obtaining the first K Lyapunov exponents for systems of differential equations with time delay of the following form:

$$\dot{x} = F\big(t, x, x(t-h_1), x(t-h_2), \ldots, x(t-h_s)\big), \tag{2}$$

where $\forall t\ x(t) \in \mathbb{R}^n$, n is a system dimension, $h_i \in \mathbb{R}^n$ $(i = 1, \ldots, s)$, $h_1 > h_2 > \ldots > h_s > 0$. We take the space of continuous functions on the interval $[-h_1, 0]$ in $\mathbb{R}^n$, namely $C([-h_1, 0]; \mathbb{R}^n)$, as the phase space. As a numerical method for solving the system (2) with initial conditions

$$x(\varphi) = f(\varphi), \quad \varphi \in [-h_1, 0], \quad f(\varphi) \in \mathbb{R}^n, \tag{3}$$

we choose the fifth-order Dormand–Prince method (DOPRI54) [14] with variable time step.

We solve the system (2) with the corresponding initial condition (3) by the chosen method until the time Θ when the solution is sufficiently close to attractor. Next we obtain the function $x^0(t) \in \mathbb{R}^n$ on the interval $t \in [\Theta - h_1, \Theta]$, which will subsequently become the new initial condition. Using it we get the solution $x_*(t) \in \mathbb{R}^n$, on which we will estimate Lyapunov exponents.

We add to the system (2) with the initial condition $x^0(t)$ the following K identical systems

$$\begin{aligned} \dot{u}_j = \left.\frac{\partial F}{\partial x}\right|_{x=x_*(t)} \cdot u_j + \left.\frac{\partial F}{\partial x(t-h_1)}\right|_{x(t-h_1)=x_*(t-h_1)} \cdot u_j(t-h_1) + \ldots + \\ \left.\frac{\partial F}{\partial x(t-h_s)}\right|_{x(t-h_s)=x_*(t-h_s)} \cdot u_j(t-h_s), \end{aligned} \tag{4}$$

where $j = 1, \ldots, K$. They represent the systems (2) linearized on solution $x_*(t)$.

For each system (4) we use the initial conditions in the form of orthonormal impulse functions, for example:

$$u_j(\varphi) = \begin{cases} \sqrt{\dfrac{K}{nM}} & \text{at } \varphi \in \Big[(\Theta - h_1) + (j-1)h_1/K, (\Theta - h_1) + jh_1/K\Big], \quad j = 1, \ldots, K, \\ 0 & \text{else.} \end{cases} \tag{5}$$

Here M is a number of solution partition points on the interval with length equals to h_1, $K \le nM$. Also we can use a system of orthonormal trigonometric functions:

$$u_j = \begin{cases} 1, \quad j = 1, \\ \sin\left(j\pi t/h_1\right)/\sqrt{2}, \quad j = 2, 4, 6, \dots \\ \cos\left((j-1)\pi t/h_1\right)/\sqrt{2}, \quad j = 3, 5, 7, \dots \end{cases} \tag{6}$$

Solving together the system (2) with the initial condition $x(\varphi) = x^0(\varphi)$ and the system (4) with the initial conditions (5) or (6) on the interval $t \in [\Theta, \Theta + T]$, $T \ge h_1$, we obtain for linearized systems $u_j^1(t) \in \mathbb{R}^n (j = 1, \dots, K)$. Considering that u_j^1 usually behave exponentially, it is necessary to renormalize them from time to time. Note that the problem is the solution's unlimited growth or vanishing. Time of renormalization T can be chosen constant or variable [15].

Further, on the interval $t \in [\Theta + T - h_1, \Theta + T]$ we discretize the obtained solutions into M equal parts and use one of the algorithms described below.

Impulse functions method. We do orthonormalization of $u_j^1(t)$, $j = 1, \dots, K$ by Gram–Schmidt method [4]. At the same time, after the orthogonalization procedure and before the normalization, we calculate and keep the values

$$\xi_j^1 = \|u_{\text{ort } j}^1(t)\|. \tag{7}$$

Then we re-solve the systems (2), (4) with obtained orthonormal solutions used as initial conditions.

Trigonometric functions method. Before the orthogonalization procedure we apply the Fast Fourier Transform to u_j^1, obtaining the complex coefficients $c_j = \langle c_{j,i} \rangle$, $i = 1, \dots, M/2+1$. Then we split coefficients into real and imaginary parts (at cosine and sine) $d_j = \langle d_{j,i} \rangle$, $i = 1, \dots, M+2$. We apply the Gram–Schmidt method to d_j. At the same time, after the orthogonalization procedure and before the normalization, we compute and keep the values

$$\xi_j^1 = \|d_{\text{ort } j}(t)\|. \tag{8}$$

Finally, we use the inverse Fast Fourier Transform to obtain the initial conditions on the time delay interval.

Thus, we obtain an approximation of Lyapunov exponents:

$$\lambda_j = \lim_{L \to \infty} \frac{\sum_{k=1}^{L} \xi_j^k}{TL}, \quad j = 1, \dots, K. \tag{9}$$

Let us proceed to the results of the testing algorithms.

3. Algorithms test results

Computational experiments were carried out for the Hutchinson equation [9]:

$$\dot{x} = rx(t)(1 - x(t-1)), \quad r > 0. \tag{10}$$

Non-zero solutions of the equation (10) are asymptotically stable for $r \in [0, \pi/2]$ (monotonically stable for $r \in [0, e^{-1}]$) and tend to 1 in an oscillatory manner for $r \in [e^{-1}, \pi/2]$. In this case, the Lyapunov exponents of (10) are equal to real parts of the roots of characteristic quasi-polynomial $P(\lambda) \equiv \lambda + r\exp(-\lambda)$, $\lambda = \tau + i\omega$. To calculate them, a system of algebraic equations is used:

$$\tau + re^{-\tau}\cos\omega = 0, \quad \omega - re^{-\tau}\sin\omega = 0. \tag{11}$$

IOP Conf. Series: Journal of Physics: Conf. Series **1163** (2019) 012045 doi:10.1088/1742-6596/1163/1/012045

Table 1. The first 10 computed Lyapunov exponents ($\check{\lambda}_i$ – by impulse functions method, and $\tilde{\lambda}_i$ – by trigonometric functions method) for the equation (10) at $r = 1.0$ and relative difference σ_i between them and reference values.

i	τ_i	$M = 100$				$M = 2000$			
		$\check{\lambda}_i$	$\check{\sigma}_i$	$\tilde{\lambda}_i$	$\tilde{\sigma}_i$	$\check{\lambda}_i$	$\check{\sigma}_i$	$\tilde{\lambda}_i$	$\tilde{\sigma}_i$
1	−0.3181	−0.3227	1.4461	−0.3104	2.4206	−0.3184	0.0943	−0.3177	0.1257
2	−0.3181	−0.3228	1.4775	−0.3106	2.3577	−0.3185	0.1257	−0.3176	0.1572
3	−2.0623	−2.0760	0.6643	−2.0555	0.3297	−2.0630	0.0339	−2.0618	0.0242
4	−2.0623	−2.0761	0.6692	−2.0555	0.3297	−2.0630	0.0339	−2.0618	0.0242
5	−2.6532	−2.6693	0.6068	−2.6496	0.1357	−2.6540	0.0302	−2.6528	0.0151
6	−2.6532	−2.6693	0.6068	−2.6496	0.1357	−2.6540	0.0302	−2.6528	0.0151
7	−3.0202	−3.0375	0.5728	−3.0195	0.0232	−3.0210	0.0265	−3.0198	0.0132
8	−3.0202	−3.0375	0.5728	−3.0198	0.0132	−3.0210	0.0265	−3.0198	0.0132
9	−3.2878	−3.3057	0.5444	−3.2902	0.0730	−3.2885	0.0213	−3.2874	0.0122
10	−3.2878	−3.3057	0.5444	−3.2903	0.0760	−3.2886	0.0213	−3.2874	0.0122

The rounded components τ_i of this system's solution at $r = 1.0$ are presented in the second column of Table 1. We will call them reference values.

Linearized equations for (10) have the form: $\dot{u}_j = r(1 - x(t-1))u_j(t) - rx(t)u_j(t-1)$.

For all experiments the following parameters were used: number of computed Lyapunov exponents $K = 10$, time to approach the attractor $\Theta = 150$, renorming time $T = 4$, number of Lyapunov exponents recomputations $L = 5000$, as initial conditions we used the system (5) in the case of impulse functions method and the system (6) in the case of trigonometric functions method.

As shown in Table 1, the accuracy of exponents computing depends on the chosen partition points count. With an increase of M by 20 times from 100 to 2000, the relative difference decreased by an order. When the partition points number is equal to the number of computed Lyapunov exponents, the achieved accuracy is low.

4. Lyapunov exponents computing in case of dynamical systems with a quasi-stable attractor

The study of dynamic systems from neurobiological and physical applications leads to singularly perturbed systems, which are characterized by so-called relaxation oscillations. Recently, among the problems with relaxation dynamics, we can distinguish a subclass of equations, which stable modes have multipliers that are modulo less than 1, but still close (sometimes asymptotically) to the unit circle on the complex plane. Attractors having this property can be called quasi-stable [10]. This fact make numerical analysis of such problems difficult, since the approach to the stable mode is very slow and, based on numerical analysis, it is not possible to find out the form of stable modes. At the same time, even a long computation can lead to incorrect conclusions about the stability of some part of the transition process.

In such situation, computing the Lyapunov exponents along the studying problem solutions can help. The presence of any number of values close to zero in the Lyapunov exponents spectrum indicates the problem's quasi-stability. In many cases the model's quasi-stability allows to make conclusions about adequacy or inadequacy of this problem to the model.

As an example of this situation, we consider a mathematical model of synaptically connected

IOP Conf. Series: Journal of Physics: Conf. Series **1163** (2019) 012045 doi:10.1088/1742-6596/1163/1/012045

Table 2. The first 4 calculated Lyapunov exponents of different modes of the system (13) at $a = 1.5$, $b = 7$, $c = -4.5$, $h = 7$, $\lambda = 100$ after 4000 steps of renorming rounded to the 4th digit.

	λ_1	λ_2	λ_3	λ_4		λ_1	λ_2	λ_3	λ_4
a)	0.0000	0.0000	0.0000	−1.9968	d)	0.0000	0.0000	−1.7336	−1.8855
b)	0.0000	0.0000	0.0000	−1.9799	e)	0.0000	−0.0001	−1.7149	−1.8853
c)	0.0000	0.0000	0.0000	−1.9769	f)	−0.0004	−1.5190	−1.5236	−2.0897

neural oscillators, based on the idea of fast threshold modulation:

$$\begin{aligned} \dot u_1 &= [\lambda f(u_1(t-1)) + b\,g(u_2(t-h))\ln(u_*/u_1)]u_1, \\ \dot u_2 &= [\lambda f(u_2(t-1)) + b\,g(u_1(t-h))\ln(u_*/u_2)]u_2. \end{aligned} \tag{12}$$

This model was proposed in [12] (see also [13]). There $u_1(t), u_2(t) > 0$ are normalized membrane potentials of neurons. Each neuron is modeled by singularly perturbed differential-difference equation with delay $\dot u = \lambda f(u(t-1))u$, the parameter $\lambda \gg 1$ characterizes the rate of electrical processes, the term $bg\big(u_{j-1}(t-h)\big)\ln(u_*/u_j)u_j$ models the synaptic interaction. Communication is assumed to be a threshold with a time lag $h > 1$. Parameter $b = \text{const} > 0$, $u_* = \exp(c\lambda)$ is an interaction control threshold, $c = \text{const} \in R$. Functions $f(u), g(u) \in C^2(R_+)$ with $R_+ = \{u \in R : u \geq 0\}$ are such that $f(0) = 1$, $g(0) = 0$, $f(u) + a$, $uf'(u)$, $g(u) - 1$, $ug'(u) = O(u^{-1})$ at $u \to +\infty$, where $a = \text{const} > 0$. Examples of functions that satisfy these properties are $f(u) = (a - au)(a + u)^{-1}$, $g(u) = u(1 + u)^{-1}$. Solutions of the system (12) at large λ are intense peaked oscillations, which are alternated by area where the solution is close to zero (see [10, 11, 12, 13]). For simplicity of the problem's numerical analysis, we perform an exponential substitution $u_j = \exp(\lambda x_j)$, $j = 1, 2$, and the system (12) takes the form

$$\begin{aligned} \dot x_1 &= f(\exp(\lambda x_1(t-1)) + b\,(c - x_1)g(\exp(\lambda x_2(t-h)), \\ \dot x_2 &= f(\exp(\lambda x_2(t-1)) + b\,(c - x_2)g(\exp(\lambda x_1(t-h)). \end{aligned} \tag{13}$$

For the system (13) at $a = 1.5$, $b = 7$, $c = -4.5$, $h = 7$, $\lambda = 100$ we have found at least 6 (without symmetry) different coexisting stable cycles. However, the transition process leading to the first five of them takes a lot of time. The use of the methods described in the previous paragraph allows us to understand the reasons for this phenomenon. As it turned out, in addition to one zero exponent, which is determined by the autonomy of the system (13), the first three self-symmetric cycles have two additional exponents close to zero (Fig. 1 a), b), c)) and non-symmetric cycles have one additional exponent close to zero (Fig. 1 d), e)). The case presented in Fig. 1 f) stands alone, as it corresponds to only one zero exponent and the transition to this cycle from its attraction area is incomparably faster than in previous cases.

5. Conclusions

Quasi-stable structures are detected in mathematical models with relaxation behavior of various nature. Often enough, the models in such problems are equations or systems of equations with time delay. In this regard, the tools of numerical analysis of such systems are very important. Calculating the spectrum of Lyapunov exponents makes it possible to understand that the studied attractor is quasi-stable and we have to do sufficiently long computing and use special initial conditions to get reliable results. Note that in the case when dynamical system attractors are cycles or tori, there is no need in such statements as Oseledets theorem. The significant problems with the described algorithms applicability can arise only in the case of chaotic behaviour of corresponding system solutions.

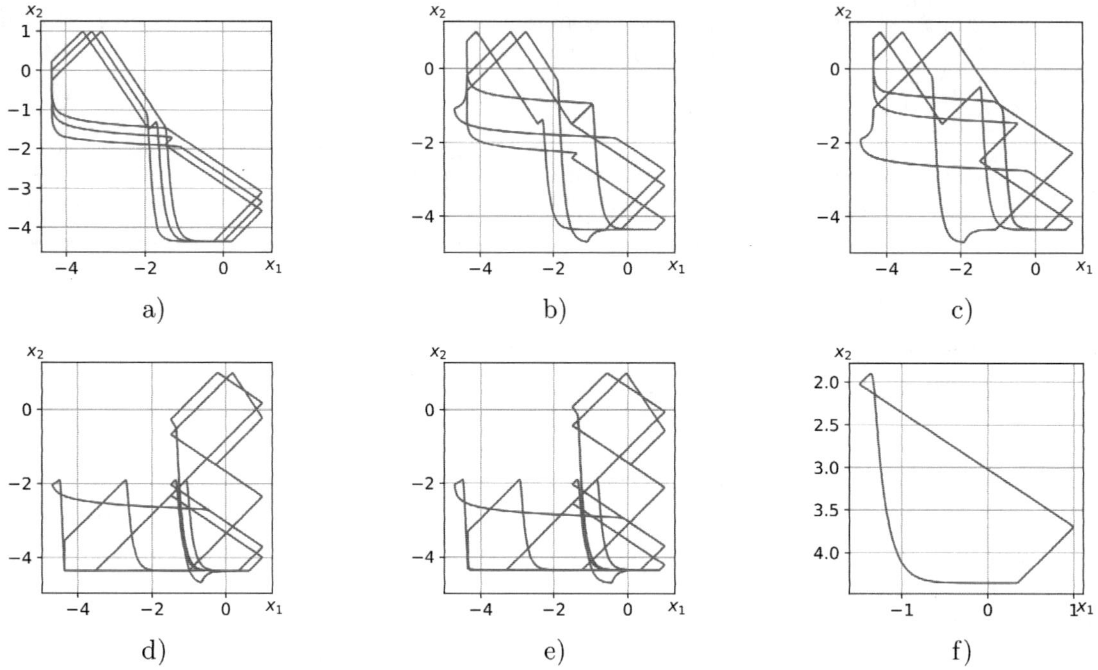

Figure 1. Different modes of the system (13) at $a = 1.5$, $b = 7$, $c = -4.5$, $h = 7$, $\lambda = 100$.

Acknowledgments
This work is supported by the Russian Science Foundation (project no. 14-21-00158).

References

[1] Bylov B F, Vinograd R E, Grobman D M and Nemytskii V V 1966 *Theory of Lyapunov Exponents and Its Applications to Stability Problems* Moscow: Nauka
[2] Benettin G, Galgani L, Giorgilli A and Strelkyn J M 1980 *Meccanica* **15** 21–30
[3] Wolf A, Swift J B, Swinney H L and Vastano J A 1985 *Physica D: Nonlinear Phenomena* **16** 285–317
[4] Cheney W and Kincaid D 2009 *Linear Algebra: Theory and Applications* Sudbury, Ma: Jones and Bartlett
[5] Oseledets V I 1968 *Trans. Moscow Math. Soc.* **19** 197–231
[6] Kuptsov P V 2010 *Appl. Nonlinear Dyn.* **18** 80–92
[7] Balyakin A A and Ryskin N M 2007 *Izv. VUZ. Applied Nonlinear Dynamics* **15**
[8] Koloskova A D, Moskalenko O I and Koronovskii A A 2018 *Pisma v ZHTF* **44** 19–25
[9] Hutchinson G E 1948 *Ann. N.Y. Acad. Sci.* **50** 221–46
[10] Glyzin S D, Kolesov A Yu and Rozov N Kh 2018 *Comp. Math. and Math. Phys.* **58** 659–79
[11] Glyzin S D, Kolesov A Yu and Rozov N Kh 2013 *Mathematical Notes* **93** 676–90
[12] Glyzin S D, Kolesov A Yu and Rozov N Kh 2013 *Diff. Eq.* **49** 1227–44
[13] Glyzin S D, Kolesov A Yu and Rozov N Kh 2015 *Rus. Math. Surveys* **70** 383–452
[14] Hairer E, Norsett S P and Wanner G 2008 *Solving Ordinary Differential Equations I: Nonstiff Problems* Springer-Verlag Berlin Heidelberg
[15] Glyzin D S, Glyzin S D, Kolesov A Yu and Rozov N Kh 2005 *Diff. Eq.* **41** 284–89
[16] Kaplan J L and Yorke J A 1979 *Func. Diff. Eq. and Approx. of Fixed Points* **730** 204–27
[17] Kuptsov P V and Kuznetsov S P 2016 *Phys. Rev. E* **94**

Phase diagrams of polarized ultra-cold gases on attractive-U Hubbard ladders

E.A. Burovski[1], R.Sh. Ikhsanov[1], A.A. Kuznetsov[1] and M.Yu. Kagan[1,2]

[1] National Research University Higher School of Economics, 101000 Moscow, Russia
[2] P.L. Kapitza Institute for Physical Problems, Russian Academy of Sciences, 119334, Moscow, Russia

rihsanov@hse.ru

Abstract. We consider a quasi-one-dimensional model of a two-component Fermi gas at zero temperature on one, two and three-leg attractive-U Hubbard ladders. We construct the grand canonical phase diagram of a two-component spin-polarized gas. We find that the structure of the phase diagram of the attractive-U Hubbard model for two and three leg ladders significantly differs q from the structure of the phase diagram of a single chain. We argue that the single chain model is a special case, and that multichain ladders display qualitative features of the 1D-to-3D crossover, observed in experiments with trapped ultracold gases.

1. Introduction

Recent advances in experimental techniques of creation and manipulation of strongly correlated systems using ultra-cold gases in traps with varying forms and effective dimensionality of the trapping potential [1, 2, 3, 4] stimulates an interest in studying quasi-one-dimensional Fermi gases at ultralow temperatures. In such systems a motion of the particles in two spatial dimensions out of three is frozen out by dimensional quantization. The possibility of experimental engineering of one-dimensional and quasi-one-dimensional geometries opens up a unique opportunity for studying polarized superfluids, or, more specifically, the one-dimensional analog [5] of the long-elusive Fulde-Ferrell-Larkin-Ovchinnikov polarized superfluid phase [6, 7], and the dimensional crossover between the strictly one-dimensional (1D) and three-dimensional (3D) phenomena.

An important feature of the experimental studies of ultracold gases is that the (strongly interacting) atomic clouds are subject to an external confining potential. This leads to a non-trivial spatial structure of the atomic cloud, which is directly observable in either time-of-flight experiments or, recently, with an *in situ* imaging. For polarized Fermi gases, the spatial structure of the atomic cloud was found to be significantly different for quasi 1D cigar-shaped magnetic traps [4] and 3D spherically symmetric traps [3]. A recent experimental effort, Ref. [1], investigates a crossover between the 1D and 3D in an array of cigar-shaped tubes with varying the intertube tunneling amplitudes.

In this Paper, we use a quasi-one-dimensional model of a two-component Fermi gas at zero temperature to map out the grand canonical phase diagram of the uniform system. With the local density approximation, the grand canonical phase diagram provides the shell structure of the density profile of a trapped gas, which can be compared to experimental measurements. Specifically, we consider an attractive Hubbard model on two- and three-leg ladders and compare the behavior of the model to the known case of a strictly 1D single chain lattice [8–11]. We concentrate on studying the shell structure of the density profile, and leave aside the details of the algebraic ordering in 1D versus the true long-range ordering in 3D. We use the numerically exact density matrix renormalization group (DMRG) simulations. This way, our work complements a recent study [12], which used a mean field approximation.

2. Model and methods

We consider the attractive-U Hubbard model ($U<0$) defined by the Hamiltonian,

$$H_0 = -t\sum_{i,\sigma}\left(c_{i,\sigma}^{+}c_{i+1,\sigma} + h.c.\right) + U\sum_{i} n_{i\uparrow}n_{i\downarrow}, \tag{1}$$

where $c_{i,\sigma}^{+}$ is the creation operator for a fermion with spin σ at site i, $n_{i,\sigma}=c_{i,\sigma}^{+}c_{i,\sigma}$ is the number of particles operator; summations run over $L\times w$ sites of a w-leg ladder lattice of length L. U is the onsite Hubbard interaction between fermions with opposite spins, and t is the hopping amplitude, which we set $t=1$ without loss of the generality.

We only consider zero temperature ($T=0$) and compute the ground state energies using the DMRG method [13, 14], as implemented in the ALPS package [15].

In the canonic ensemble, the ground state energy $E\left(N_\uparrow, N_\downarrow\right)$ of the model (1) is the function on the particle numbers for spin-up and spin-down fermions, $N_\uparrow$ and $N_\downarrow$, respectively. For convenience, we also define the total number of particles, $N_\Sigma = N_\uparrow + N_\downarrow$, and total polarization, $N_\Delta = N_\uparrow - N_\downarrow$. It is enough to consider only the case $N_\Sigma < L\times w$ because of the particle-hole symmetry of the model (1) and $N_\Delta > 0$. The case $N_\Delta < 0$ corresponds to a trivial relabelling of up and down spins: $\uparrow\leftrightarrow\downarrow$.

Given the energies, we can transform to the grand canonic ensemble, which corresponds to the change of variables: the convenient variables being total chemical potential μ, and the effective magnetic field, h, are given by the derivatives of the energy with respect to N_Σ and N_Δ, respectively: $\mu = \left(\frac{\partial E}{\partial N_\Sigma}\right)_{N_\Delta}$, $h = \left(\frac{\partial E}{\partial N_\Delta}\right)_{N_\Sigma}$.

Since our main interest is the grand canonical phase diagram, we compute the ground state energies in the canonical ensemble, and approximate the derivatives with finite differences. The possible phases of the model (1) are well documented in the literature (see, e.g. [9] and references therein). The following phases are of interest:

(i) vacuum (V), which corresponds to $N_\uparrow = N_\downarrow = 0$;

(ii) equal densities (ED), $N_\uparrow = N_\downarrow$;

(iii) the partially polarized phase (PP) with $N_\uparrow > N_\downarrow$;

(iv) the fully polarized phase (FP$_1$), $N_\downarrow = 0$.

In the presence of the lattice, there are additional phases, which correspond to a full band fillings (e.g., FP$_2$: $N_\uparrow = L\times w$ and $N_\downarrow = 0$). The ED phase is expected to have the Bardeen-Cooper-Schrieffer (BCS) type of algebraic ordering, while the PP phase is expected to display the the FFLO type order, where the superconding correlations decay algebraically in real space and are modulated with the modulation length given by the mismatch of the Fermi momenta [5].

To locate the phase boundaries, we compare the energies in different phases:

a) The boundary between ED and PP. At the boundary we have $N_\uparrow = N_\downarrow$, therefore the chemical potential and the effective magnetic field are given by:

$$\mu = \left(\frac{\partial E}{\partial N_\Sigma}\right)_{N_\Delta} = \frac{E(N_\uparrow+1, N_\downarrow+1) - E(N_\uparrow, N_\downarrow)}{2}, \quad h = \left(\frac{\partial E}{\partial N_\Delta}\right)_{N_\Sigma} = \frac{E(N_\uparrow+1, N_\downarrow-1) - E(N_\uparrow, N_\downarrow)}{2}.$$

On the $\mu - h$ plane, the boundary is parameterized by $N_\uparrow$.

b) The boundary between ED and the vacuum is the related to the binding energy of a single pair, and therefore on the $\mu - h$ plane the boundary is nothing but the horizontal line $\mu = E(1,1)/2$.

c) The boundary between the vacuum and FP_1 phase is defined by the bottom of the single-particle band, so that the boundary is given by the straight line $\mu = -h + E(1,0)$.

d) The boundary between the partially polarized phase PP and fully polarized phase FP_1. The boundary is defined by:

$$\mu = \left(\frac{\partial E}{\partial N_\Sigma}\right)_{N_\Delta} = \frac{E(N_\uparrow + 1, 1) - E(N_\uparrow, 0)}{2}, \; h = \left(\frac{\partial E}{\partial N_\Delta}\right)_{N_\Sigma} = \frac{E(N_\uparrow - 1, 1) - E(N_\uparrow, 0)}{-2}.$$

On the $\mu - h$ plane, the boundary is parameterized by $N_\uparrow$.

e) The boundary between the fully polarized phases FP_1 and FP_2 is given by the straight line $\mu = -h - E(1,0)$.

3. Results and Discussion

We map the grand canonical phase diagram for the attractive-U Hubbard model on the chain and two- and three-leg ladders with up to 120 sites and varying the Hubbard interaction parameter up to $U/t = -7$. In our DMRG simulations we keep up to 400 states.

Our numerical results are summarized in Fig. 1(a)-(c). For $w = 1$ our results agree with the Bethe Ansatz based calculations [9, 16, 17]. For the two-leg ladders ($w = 2$) our results agree with the DMRG simulations of Ref [8].

The grand canonical phase diagrams, Fig. 1, define the shell structure of the atomic clouds, which are directly observable in experiments. In the local density approximation, a weak spin-independent confining potential $V(x)$ is equivalent to replacing $h \to h - V(x)$, which graphically corresponds to a vertical slice in Fig. 1 [9].

Several features of the phase diagrams are interesting and important. First, the non-trivial boundaries (ED-PP, line 1, and PP-FP1, line 4) are non-monotonic for $w > 1$ (for two and three-leg ladders). This is a purely lattice effect due to filling of the higher bands at finite values of the filling fraction [8]. Experimentally, the most relevant part of the phase diagram is the vicinity of the multicritical point *O*, since it corresponds to the low filling fraction, i.e. to the continuum limit. A qualitative feature is the slope of the line 1, which separates the equal-density phase and the partially polarized phase PP: for $d\mu/dh < 0$ in the vicinity of the multicritical point *O*, the PP phase occupies the center of the trap, and the ED phase is either absent, or occupies the wings of the atomic cloud [18]. This behavior has been observed for strictly one-dimensional problem ($w = 1$) in Bethe Ansatz calculations in the continuum space [18] and on the lattice [9], also numerically in the DMRG calculations [9] and experimentally in tight cigar-shaped traps [4].

For both $w = 2$ and $w = 3$, we find that the slope of the phase boundary is positive, $d\mu/dh > 0$, see Fig 1(b) and 1(c). In the local density approximation, this translates into a different scenario for the shell structure of the atomic cloud, where the structure of the cloud is inverted: the center of the trap has equal densities, while the wings can be partially polarized. This is closer to the symmetric 3D trap results of Ref. [3], which observed the phase separation between the ED core and FP wings.

While our calculations are only done for two- and three-leg ladders, it is thus natural to conjecture that this scenario is generic, and that the strictly 1D, single-chain is special in this respect.

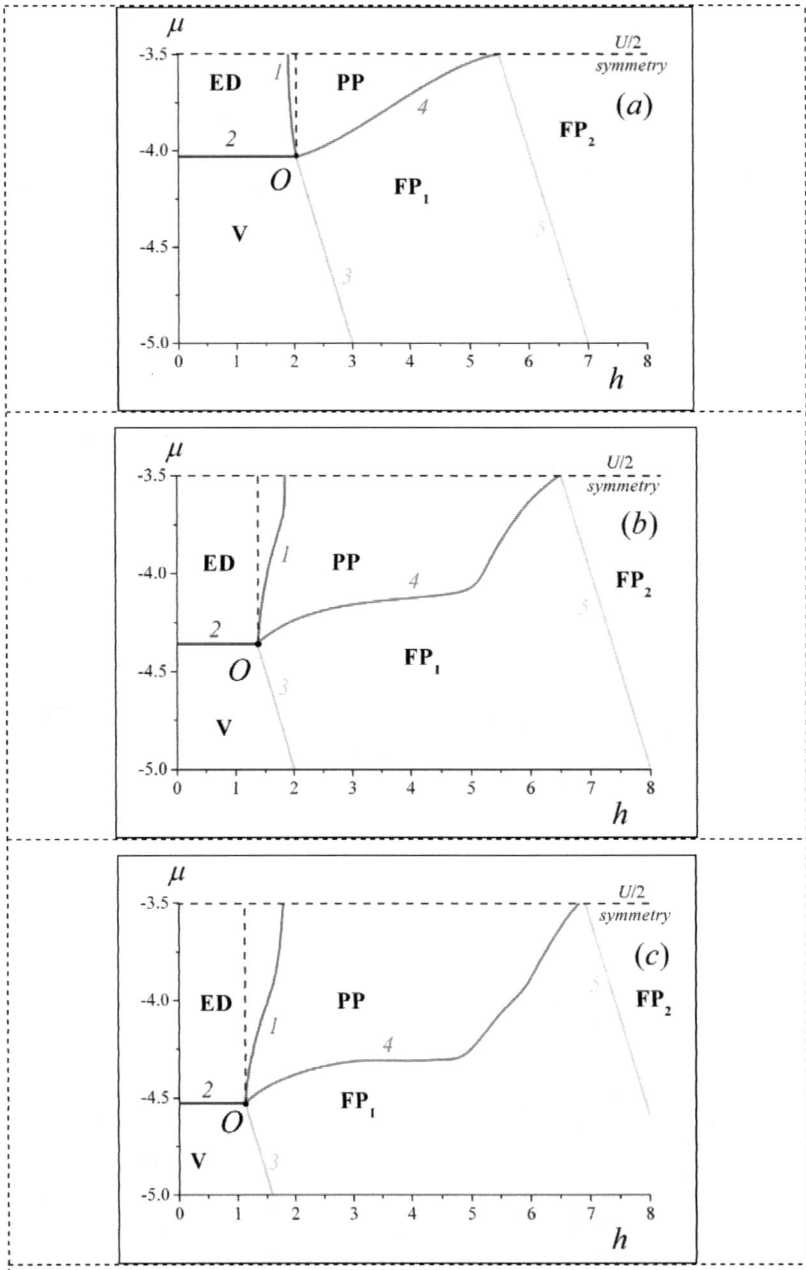

Figure 1. (color online) Grand-canonical phase diagrams of the attractive-U Hubbard model for $U=-7t$ and varying number of legs w. We only show the $\mu<U/2$, $h>0$ part of the phase diagram since the $h<0$ part is symmetric with respect to the $h\rightarrow -h$ symmetry, and $\mu>U/2$ is symmetric with respect to the particle-hole symmetry of the Hubbard model (1). Solid lines are simulation results, the vertical dashed line is guide to an eye only. See the text for the discussion. (a) $w=1$, $L=40$; (b) $w=2$, $L=40$; (c) $w=3$, $L=40$.

4. Conclusions and Outlook

We perform numerically exact DMRG simulations of the attractive-U Hubbard model on a single chain and on two- and three-leg ladders, which complement the mean-field studies of Ref. [12]. We construct the grand canonical phase diagrams, which allow "reading off" the shell structure of the atomic cloud in the experiments with quasi-one-dimensional polarized ultracold gases. We find that the structure of the phase diagram for multi-leg ladders has qualitative differences from the single chain case, which is consistent with recent experimental results for arrays of weakly coupled 1D tubes [1]. It would be interesting to study the effect of the varying interchain hopping amplitude to mimic the experimental setup, which we reserve for a future study.

5. Acknowledgments

E.A.B. and M.Yu.K. acknowledge the support of the Academic Fund Program at the National Research University Higher School of Economics (HSE) in 2018–2019 (grant № 18-05-0024) and by the Russian Academic Excellence Project "5-100". The work of R.S.I. was financially supported by joint Russian-Greek projects RFMEFI61717X0001 and T4ΔΡΩ-00031 "Experimental and theoretical studies of physical properties of low-dimensional quantum nanoelectronic systems". We are grateful also to RFBR grant № 17-02-00135.

References

[1] Revelle M C, Fry J A, Olsen B A and Hulet R G 2016 *Phys. Rev. Lett.* **117** 235301
[2] Onofrio R 2016 *Phys. Usp.* **59** 1129
[3] Shin Y, Zwierlein M W, Schunck C Y et al 2006 *Phys. Rev. Lett.* **97** 030401
[4] Liao Y, Rittner A S C, Paprotta T et al 2010 *Nature* **467** 567
[5] Yang K 2001 *Phys. Rev. B* **63** 140511
[6] Fulde P and Ferrell R A 1964 *Phys. Rev.* **135** A550
[7] Larkin A I and Ovchinikov Yu N 1965 *Sov. Phys. JETP* **20** 762
[8] Feiguin A E and Heidrich-Meisner F 2009 *Phys. Rev. Lett.* **102** 076403
[9] Heidrich-Meisner F, Orso G and Feiguin A E 2010 *Phys. Rev. A* **81** 053602
[10] Kagan M Yu, Haas S and Rice T M 1999 *Physica C: Superconductivity* **317-318** 185
[11] Burovski E, Orso G and Jolicoeur Th 2009 *Phys. Rev. Lett.* **103** 215301
[12] Dutta S and Mueller E J 2016 *Phys. Rev. A* **94** 063627
[13] White S R 1992 *Phys. Rev. Lett.* **69** 2863;
[14] Schollwöck U 2005 *Rev. Mod. Phys.* **77** 259
[15] Bauer B et al. (ALPS collaboration) 2011 *J. Stat. Mech.* P05001
[16] Woynarovich F and Penc K 1991 Zeitschrift f. Physik B-Cond. Mat. **85** 269
[17] Essler F, Frahm H, Goehmann F, Kluemper A and Korepin V E 2005 *The one-dimensional Hubbard model* (Cambridge University Press)
[18] Orso G 2007 *Phys. Rev. Lett* **98** 070402

IOP Conf. Series: Journal of Physics: Conf. Series **1163** (2019) 012047 doi:10.1088/1742-6596/1163/1/012047

Multiuser wireless channel simulation for communication systems with nonparametric reception

D. Osipov[1, 2]

[1]Institute for Information Transmission Problems of the Russian Academy of Sciences (IITP RAS), 127051, 19 bld.1 Bolshoy Karetny Lane, Moscow, Russian Federation
[2]National Research University Higher School of Economics, 20 Myasnitskaya Str., Moscow, 101000, Russian Federation

E-mail: d_osipov@iitp.ru

Abstract. When combined with error-correction codes reception techniques based on nonparametric hypothesis testing and order statistics provide strong immunity to different types of interference including multiuser interference. That makes communication systems using such reception techniques most appealing candidates for various applications such as Machine-to-Machine (M2M) communications and Internet Of Things (IOT). Unfortunately analytical treatment of communication systems with nonparametric reception remains a cumbersome task. Therefore simulation remains the main tool for the development of such systems. For multiuser systems supporting hundreds of active users and operating in fading channels (e.g. IOT) time spending grows drastically hampering the design process. Thus the development of simplified multiuser channel models is of great interest. In this paper two simplified mathematical models of multiuser interference for the case of a single user nonparametric reception are proposed. The effectiveness of the proposed models is compared by means of modulation. Special attention is paid to the problem of software implementation of the models proposed.

1. Introduction

Reception techniques that are used in modern communication systems rely on the assumption that channel state information (CSI) and thus decision statistics distributions at the receiver side are known to the receiver or at least can be estimated with desired accuracy. In many real life applications that is not true. This is particularly true for M2M and IOT systems in which hundreds of devices send short packets simultaneously and thus it is difficult to estimate the CSI with desired accuracy. Thus reception techniques based on nonparametric statistical tests and the use of order statistics not requiring CSI and able to withstand severe interference are of great interest.

Unfortunately very little is known of the analytical description of systems making use of such reception techniques. Thus simulation remains the primary method for effective design and investigation of various communication systems with nonparametric reception. However in case of multiuser systems (which is probably the most interesting case) straightforward simulation of hundreds of users each having a different channel due to the difference in their location is both a very computational- and time-consuming operation. Consequently simplified channel models need to be proposed and investigated in order to reduce the computation burden and speed up the development process. In what follows two simplified channel models for the simulation of the coded DHA FH OFDMA systems [1] with nonparametric reception are considered.

2. A coded DHA FH OFDMA system description

Let us consider the coded DHA FH OFDMA in more detail. First and foremost let us consider the transmission procedure. The information to be transmitted is first encoded with a $C_q(n,k)$ code. Each symbol of the resulting (say i th) codeword $\overline{v}^i = [v_0^i, v_1^i, \ldots, v_{n-1}^i]$ is mapped into a weight 1 column vector of length q. Thus the codeword is represented by a binary n by q matrix B^i with exactly one nonzero entry in each column. An all zero n by $Q-q$ ($Q >> q$) matrix Z is appended to the matrix B^i forming the matrix $X^i = \begin{bmatrix} B^i \\ Z \end{bmatrix} = \left[\overline{x}_0^i, \overline{x}_1^i, \ldots, \overline{x}_{n-1}^i\right]$.

Each column of the matrix X^i is then permuted independently (permutations $\pi_0, \pi_1, \ldots, \pi_{n-1}$ are chosen equiprobably from the set of all possible permutations.) The columns of the resulting matrix $\Omega^i = \left[\pi_0(\overline{x}_0^i), \pi_1(\overline{x}_1^i), \ldots, \pi_{n-1}(\overline{x}_{n-1}^i)\right]$ are then fed into the OFDM modulator. In what follows we shall assume that apart from the signal from the user under consideration K interfering signals are transmitted via the same channel.

Let us now consider the reception process. Within the scope of this process n OFDM symbols are to be received. Thus a matrix $F^i = \left[\overline{f}_0^i, \overline{f}_1^i, \ldots, \overline{f}_{n-1}^i\right]$ corresponds to each transmitted codeword $\overline{v}^i$ (here $\overline{f}_j^i$ is the column vector corresponding to the j th OFDM symbol (and thus the j th symbol of this codeword)). The receiver applies inverse permutations to each column of the matrix F^i the obtained matrix being given by $\Xi^i = \left[\overline{\chi}_0^i, \overline{\chi}_1^i, \ldots, \overline{\chi}_{n-1}^i\right]$ where $\overline{\chi}_j^i = \pi_j^{-1}(\overline{f}_j^i)$ and $\pi_j^{-1}()$ stands for the inverse permutation for the permutation $\pi_j()$. Each element of the first q rows of the matrix Ξ^i is then squared. The elements of the resulting matrix Y^i are given by the following equation

$$\begin{gathered} \forall\, j \in \{1,\ldots,q\} \;\; \kappa \in \{0,\ldots,n-1\} \\ Y^i(j,\kappa) = (\chi_j^i(\kappa))^2 \end{gathered} \qquad (1)$$

where $\chi_j^i(k)$ is the k th element of the column vector $\overline{\chi}_j^i$. Please note that the column-wise permutation at the transmitter side corresponds to the hopping procedure whereas the inverse permutations correspond to dehopping. Since inverse permutations are applied at the receiver side the first q rows of the matrix Ξ^i (and thus corresponding rows of the matrix Y^i) correspond to the first q rows of the transmitted matrix X^i (and thus corresponding rows of the matrix B^i.) The matrix Y^i is then used for detection i.e. decsion making on the codeword transmitted.

In what follows two nonparametric detectors will be considered: the RANKSUM detector [2] and the α detector [3]

3. Wireless channel simulation

A common approach to simulation of a wireless radio channel is to distinguish between the three different types of factors that influence the signal while its propagation:

- path loss (which is treated as a deterministic process depending on the distance between the receiver and the transmitter)
- slow fading (a relatively slow varying random process due to shadowing)
- fast fading (a faster random process due to scattering, reflections etc.)

Probably one of the most general models that can describe the first two factors is the log-normal model. Within the scope of this model the power loss (in dB) is given by

$$PL(d) = 10\log_{10} P_t - 10\log_{10} P_r = PL_{fs}(d_0) - 10\gamma\log_{10}\left(\frac{d}{d_0}\right) + X_N(0,\sigma_F) \quad (2)$$

where P_t is the transmitted power, P_r is the power at the receiver, $PL_{fs}(\ )$ is the path loss (in dB) in free space, d is the distance between the receiver , d_0 is the reference distance, γ is the path loss exponent, that depends on the scenario under consideration, $X_N(0,\sigma_F)$ is the Gaussian random variable with mean 0 and standard deviation σ_F (in dB) which depends on the scenario under consideration.

The log-normal model possesses several advantages. It is more realistic than deterministic path loss models such as free space or two-ray model. On the other hand, unlike path loss models based on experimental data such as Longley-Rice, Hata (or Okumura-Hata), Walfisch-Bertoni et al. [4] it holds for a wide range of scenarios and parameters and thus needs not to be restricted to any specific kind of a scenario.

Several methods have been proposed in order to incorporate the power degradation due to fast fading into the mathematical model of the path loss presented above. The simplest way to do so is to introduce additional term with exponential distribution into the equation (2). However it is difficult to derive the parameters corresponding to real life channels for that approach. Thus in what follows we shall be using another approach that is based on the assumption made by Suzuki [5]: the signal undergoes lognormal shadowing when traveling to the zone where it experiences fading due to multiple scattering. Assume a FH OFDMA (or a DHA FH OFDMA) system where K users transmit in the uplink . The amplitude of the received signal from the j th user transmitting a signal on the k th subcarrier is then given by

$$\tilde{A}_k^j = \Lambda^j \left|H_k^j\right| A^j \quad (3)$$

where $\Lambda^j = 10^{\frac{PL(d_j)}{20}}$ is the signal amplitude decrease due to shadow fading (here d_j is the distance between the transmitter of the j th user and the receiver and PL is given by (2)),

$H_k^j = \sum_{m=0}^{Q-1} h^j(m) e^{-i\frac{2\pi km}{Q}}$ is the gain factor corresponding to k th subcarrier of the channel.

A^j is the amplitude of the signal transmitted by the j th on the k th subcarrier

4. Models development

Let us consider the main assumptions to be used hereinafter. First and foremost we are interested in simulation of the single user reception of the signal transmitted in a multi-user coded DHA FH OFDMA using one of the detectors described above. Uplink transmission is considered and although we assume that all the K active users have the same transmitted power we also assume that no power control is used in the system under consideration. Thus the received power of each user (including the user under consideration) depends solely on the power degradation due to propagation and the transmitted power.

Since the goal of the communication system design is to guarantee reliable communication to be available within the certain zone (commonly referred to as transmission zone i.e. the zone within which communication with Word Error Probability (WER) less or equal to certain predefined threshold values is guaranteed) we are interested in the worst case scenario. Thus our main assumption is “pessimistic” i.e. we assume that the user under consideration is at the edge of the zone i.e. the

IOP Conf. Series: Journal of Physics: Conf. Series **1163** (2019) 012047 doi:10.1088/1742-6596/1163/1/012047

distance between the receiver and the transmitter of the other users (hereinafter they will be referred to as "interfering users") is either the same or smaller than that between the receiver and the transmitter of the user under consideration and the same holds for the average power of the respective signals.

4.1. The Pessimistic Path Loss Model (PPLM) and the Pessimistic Shadowed Fading Model (PSFM)

The first model to be considered is the Pessimistic Path Loss Model. Within the scope of this model it is assumed that path loss plays predominant part in signal power degradation and thus if we are interested in the worst case scenario it is sufficient to model the path loss by assuming that the power of the signal received form each of the interfering users is κ ($\kappa \gg 1$) times greater than that of the user under consideration while the power degradation due to fading is negligible.

The Pessimistic Shadowed Fading Model is more realistic. In this model it is assumed that the signal from each user undergoes power degradation due to both path loss and shadowing and fading as explained in Section 3. Thus both lognormal path loss and gain factors need to be calculated for each user for each OFDM symbol.

4.2. Implementation issues

One of the most challenging problems commonly associated with wireless channel simulation for multiuser communication systems is the need for multiple convolutions calculation. In fact the propagation of every single OFDM symbol sent by every active user can be described as linear convolution in time domain. Thus for straightforward simulation of the communication system with hundreds of active users in time domain both the computation burden and time spending are very high. That is why in what follows we choose to carry up channel simulation in frequency domain only. Although this approach has some drawbacks since some of the effect at the receiver end cannot be addressed directly for OFDM-based systems it reduces the time spending drastically since each codeword can now be represented as a sparse matrix and numerous convolutions can be omitted as well as multiple calculations of Fast Fourier Transform and Inverse Fast Fourier Transform. However for the PSFM channel impulse responses or frequency transfer functions are to be generated which is also a time consuming process. Thus the following approach was used: multiple channel impulse responses for both the user under consideration and the interfering users were generated in advance and pre-sorted in files. Please note that since we were interested in a DHA FH OFDMA simulation the only interfering signals that can affect the system performance and thus are of interest to us are those transmitting in the hopset of the user under consideration at the moment under consideration. So even for hundreds of users there is no need to generate signals and channel gains for each user, but rather for a small fraction of them transmitting via the subcarriers that are in the hopset. Thus fading channel simulation for the interfering users is done in the following way: for each user a random subcarrier number is chosen, then for those u active users that transmit via one of the subcarriers currently in the hopset of the user under consideration a random file with pre-stored channel impulse responses is chosen. For each of the u active users a random channel response is then chosen and the gains corresponding to the subcarrier chosen by the respective user is calculated. Thus although used fixed set of pre-stored impulse channel responses is used the choice is randomized by the three random steps discussed above. Please note that we choose to store impulse responses rather than transfer functions in order to minimize both disc space consumption and computational burden due to multiple reads from the disc. This approach turns out to be very beneficial for the software that is optimized for the FFT and sparse matrix storage implementation.

5. Simulation

To validate the potential of the PPLM model simulation has been carried out. The total number of subcarriers has been fixed to $Q = 4096$, while the instantaneous hopset cardinality was set to $q = 16$. As an example in what follows we present results for a rate 1/6 MDS code obtained from RS code (15,2,14) by puncturing. Channel impulse responses were generated using 3GPP Typical Urban model for various rates (5 km/h, 20 km/h, 40 km/h, 60 km/h, 100 km/h, 200 km/h), carrier frequency $f_c = 2.4$ GHz and bandwidth 20 MHz. For interfering users 1000 files each containing 150 channel impulse responses were written, while for the user under consideration 200 files each containing 1000 channel impulse responses were written. For lognormal path loss simulation the following parameters were used: it was assumed that the distance between the user under consideration and the receiver is equal to d, whereas the distance between each of the interfering users and the receiver is equiprobably chosen from the set $S_d = \{d, 0.99d, 0.98d, \ldots, 0.01d\}$, the path loss exponent was fixed to $\gamma = 3.5$ (this value correspond to urban radio) and the standard deviation is given by $\sigma_F = 10\,dB$. Within the scope of this model an assumption was made that apart form the multi-user interference the received signal is affected by the background noise which can be described by the SNR ratio (in dB) given by

$$SNR = 10\log_{10}\left(\frac{E_s}{\log_2(q)E_N}\right) \tag{4}$$

were E_s is the transmitted energy (per symbol) and E_N is the noise energy (in the entire band available to the users). In the next section simulation results for this scenario are presented.

5.1. Simulation results

In Figure 1 dependencies of the Word Error Rate (WER) provided by the α detector (for α=4) on the number of interfering users for different values of the SNR are presented

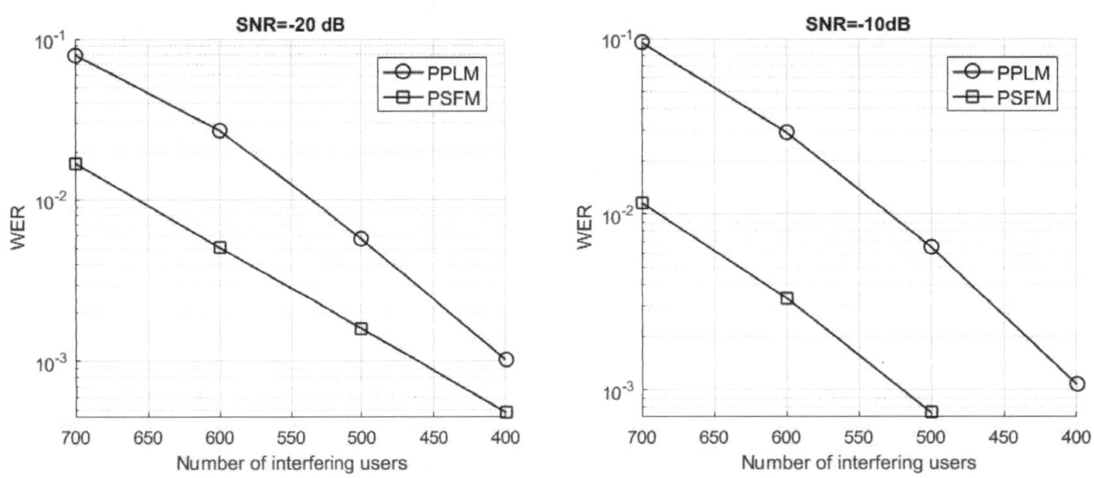

Figure 1 WER vs. number of interfering users for the α detector ($\alpha = 4$).

As can bee seen the PPLM model indeed gives a very pessimistic estimation of the WER (although for higher SNR values the gap is less substantial.) Moreover the curves presented show that for the PPLM the detector demonstrates almost no sensitivity to SNR value, whereas the more realistic PSFM shows greater sensitivity (although the difference in WER for different SNR values is moderate due to the robustness of the detector itself.)

In Figure 2 dependencies of the Word Error Rate (WER) provided by the RANKSUM detector on the number of interfering users for different values of the SNR are presented. Similarly to the previous case one might note that for SNR=-10 dB the PPLM model provides a pessimistic (i.e. upper) estimate on the value of WER. However the case where SNR=-20 dB is different since for relatively low values of the number of active users the value of the WER for PSFM is greater than that for the PPLM. This effect shows that for a relatively low number of interfering users and low SNR the influence of the background noise is dominant. Again similarly to the previous case the RANKSUM detector shows greater sensitivity if the PSFM is used. Thus not only PSFM is more realistic it is also better suited to demonstrate the background noise effect. Although PPLM is less time consuming it provides less accurate estimates.

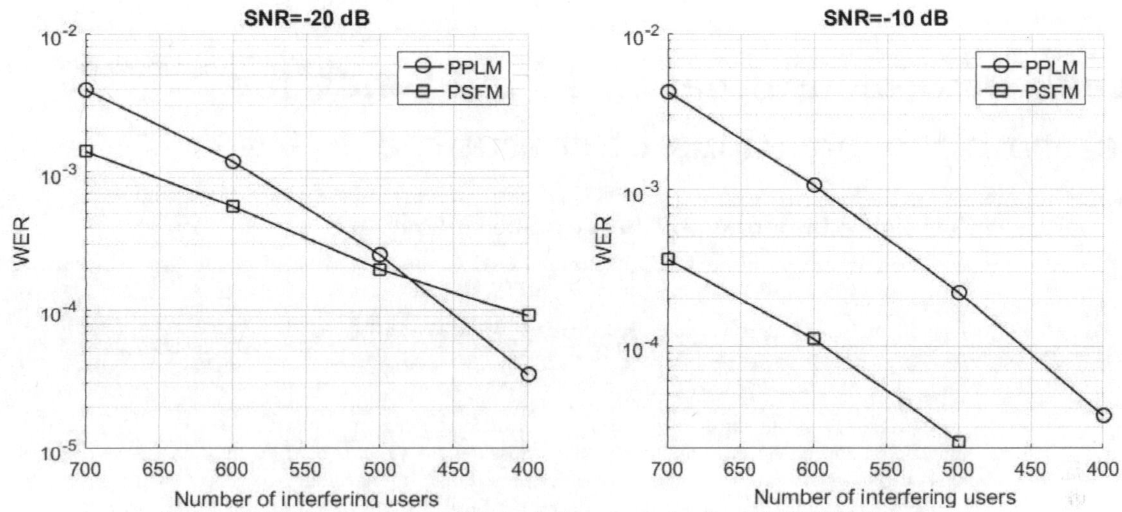

Figure 2 WER vs. number of interfering users for the RANKSUM detector.

6. Conclusion

Hereinabove two simplified models of the multi-user system employing DHA FH OFDMA were considered. It has been shown that although the PPLM is less computationally expensive it is less accurate and thus in most scenarios it is preferable to use PSFM. It has been shown that for the DHA FH OFDMA systems the computational burden of the PSFM implementation can be reduced drastically thus making the use of the PSFM practical.

Acknowledgment

This work has been supported in part by the Russian Foundation for Basic Research, project no. 18-07-01409 a

References

[1] Osipov D 2012 *Proc. Int. Work. on Multiple Access Communications (Maynooth)* (Berlin, Heidelberg: Springer) pp. 37-48
[2] Viswanathan R and Gupta S C 1985 *IEEE Trans. Comm.* **33** pp. 178-84
[3] Osipov D 2014 *Proc. Int. Work. on Multiple Access Communications (Halmstad)* (Berlin, Heidelberg: Springer) pp 29-34
[4] Barsocchi P 2006 *Channel models for terrestrial wireless communications: a survey* National Research Council – ISTI Institute Technical Report (Pisa: CNR-ISTI)
[5] Suzuki H 1977 *IEEE Trans. Comm.* **25** pp. 673-80

IOP Conf. Series: Journal of Physics: Conf. Series **1163** (2019) 012048 doi:10.1088/1742-6596/1163/1/012048

Analysis of neural networks efficiency for determining positions of corrupted bytes

Vladislav Lupanov and Sergey Slastnikov

National Research University Higher School of Economics, 20 Myasnitskaya Street, Moscow 101000 Russia

E-mail: velupanov@gmail.com, sslastnikov@gmail.com

Abstract.

A lot of files and data, in general, are transferred throughout the networks. But the data may be corrupted by intrusions or package loss so, the executable files may be marked as non-executable and violate the local network policy. Thus, it's necessary to detect such files. In this paper, we present a novel method for detecting broken bytes of a file, so the corrupted files may be detected. Also, the positions of wrong bytes might be helpful in restoring the original file content.

This work is devoted to study of modern neural network models applied to detect corrupted bytes of a file problem. Since recurrent neural networks (RNNs) seem to be well suited for such tasks, the main tasks of this work are to analyze the efficiency of popular state-of-the-art RNNs solving the problem mentioned above and to compare the results of different models. We use data consisting of the most popular file types collected from the Internet and manually randomly added noise to that data to test our models. An experiment on this data demonstrates the advantages and disadvantages of the considered models.

1. Introduction

Finding positions of broken bytes in the file and determining if a file has broken bytes at all are very import problems nowadays. Bytes of a file may be changed by intruders or virus, so additional checks may be required if unusual bytes are detected during a file check. Also, this method may be useful to detect if significant bytes of a file are changed, which can lead to errors while opening the file. Moreover, the firewall restrictions and policies may be bypassed by intruders by changing the file header. In addition, detecting positions of broken bytes is helpful while restoring the original file content. It can be used to help restoring the file and to validate the restoration process not damaging the uncorrupted bytes.

The goal of this paper is to study the efficiency of applying existing neural network models to solve the determining wrong bytes positions problem. The main tasks of this paper are study, search and analysis of neural network models, that seem to be suitable for solving the stated problem. Neural networks were chosen because this field becomes more and more popular every day and has a rapid growth. Besides, lots of papers on this theme have already been published and are still been published. Furthermore, there are a lot of successful solutions of such a problem class - when the input and output sizes are dynamic and output is generated or chosen from a huge sample.[1]

2. Bytes embeddings

In this section, we describe and study modern methods of obtaining vector representations of bytes. Most of them used to obtain embeddings for words. So, we decided to try these models too due to the fact that relative positions of a word are taken into account in them. We trained all models to obtain vector representations of bytes and N-gram of bytes. N took the following values: 2, 3, 4, 8, 12 and 16. Immediately check the quality of vectors from trained models is a hard task. Thus, the results of using each model will be calculated later, while using each model to determine broken bytes positions.

2.1. Word2Vec

One of the first with a complicated structure and, perhaps, the most popular among embedding models. Was introduced by T. Mikolov et al. in their work [2]. In our work we use the *Continuous Bag Of Words (CBOW)* model from the stated work. Because the another model - *Skip-gram* is better in semantic problems. Fig. 1 shows the model architecture. The model is trying to predict the word using it's *context* - the words, that stay near, in a fixed-size window. The windows size is a parameter. In this work, we use a window size equal to five, what means that every byte is predicted using four surrounding its bytes. Instead of words, we used not only bytes but the N-gram concatenations of bytes. Bytes were converted to a hexadecimal system and concatenated to use them in a model. For example, bytes 255 and 15 are converted to "FF" and "0F" respectively, then, they are joined to "FF0F".

2.2. FastText

Very popular model made by researchers from Facebook Inc. Its use cases are both acquiring vector representations of words and classification. Was introduced by P. Bojanowski et al. in their work [3]. Besides, this model provides functionality to get not only word vectors but vectors for sentences and parts of words. Unlike *Word2Vec*, which was described earlier, *FastText* has no dictionary and may calculate embeddings for words, that was not used in the training dataset. In the training process, the model is trying to predict the context of a word by the word themself. Moreover, *FastText* is using not only words but word parts too: N-grams for N from 3 to 6 are taken. Special symbols are added to words to distinguish them from character N-grams and to understand when the word is over. For example, "<her>" is a word and "her" is a 3-gram in word "<where>".

So-called *negative sampling* is used in *FastText* to achieve better results. Negative examples are words that never appear in current word context, they are used in loss function with opposite sign. That means that the likelihood of dissimilar words are minimized, while the distance between them is maximized. The window size is also a parameter like it was in *Word2Vec*. In our work, we use a window size equal to 5. Which means that every byte (or concatenated N-gram of bytes) is used to predict four surrounding it bytes (or N-grams). Bytes were processed like in *Word2Vec* described above.

2.3. Autoencoder

Autoencoder is an artificial neural network that uses unsupervised learning to get effective vector representations of input data, most of all, with dimensions reduction. Was firstly introduced by the authors in their work [4]. The simple architecture is used in the most cases - three-layer dense feed-forward network (three-layer *perceptron*). Typical *Autoencoder* is shown in Fig. 2. Many modern authors in their works about text processing, clustering, and classification report about the efficiency of *Autoencoders* applied to reduce the input dimension and to get vector representations of words. In this work, we use the complicated architecture of a model. Similar models are described in different works about text processing and good results in comparison with statistical methods of vector dimensions reduction (principal component analysis and other) are shown.[5]

A vector of size 256 is passed through the input layer of the model. Input vector consists of zeros except the one 1 that stays in a position corresponding to a decimal byte value. For example, '0F' is 16 in a decimal system, so the 1 would be in 16th position. This method of data encoding is called *one-hot encoding*. The *Autoencoder* model is split into two parts: encoder and decoder. The first part is the encoder: several layers of the model reduce the input shape to 32, which is the shape of vector representation. Next is a decoder part. Decoder recovers the encoded input to the source vector. Thus, the input and output of the model are source bytes encoded in *one-hot* encoding. And the *Autoencoder* model is learning to map the input vector to the desired dimension and to translate it back to source input.

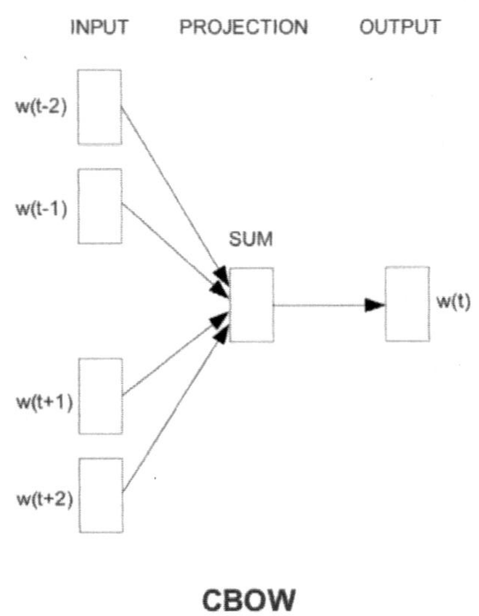

Figure 1. The *CBOW* model architecture. The image was taken from [2].

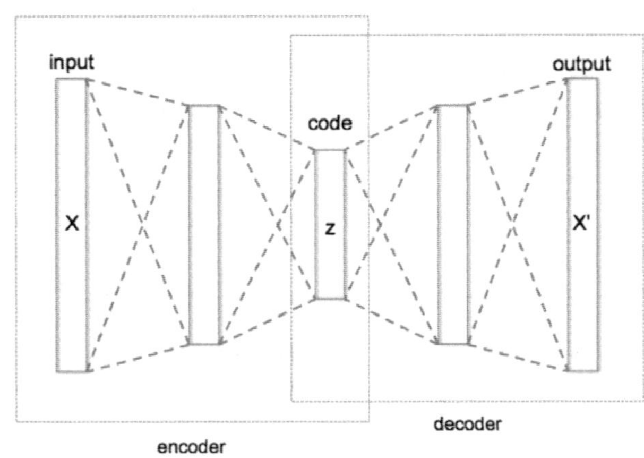

Figure 2. The *Autoencoder* model architecture: three-layer *perceptron*.

3. Model architectures

Many different types of neural networks were proposed by many authors. We tested a lot of models in this work, but we have chosen recurrent neural networks (RNNs) class due to their usability in processing sequences. RNNs seem to be well suited for such tasks because the input and the output of the model are sequences and we need to convert input sequence of bytes to output sequence with probabilities of corrupted bytes in sequence.

Recurrent neural networks - class of neural networks where connections between vertices (neurons) form a directed graph (sequence). That's why it may process time series and sequences. In this work, we study the possibility of predicting the positions of broken bytes with neural networks, so the use of the RNN class seems to be an efficient way, because of their ability in processing sequences.

LSTM - Long Short-Term Memory is a kind of RNN that has an ability to remember input values for both short and long terms. Each LSTM cell consists of three gates: the input gate, which controls the weight of new value in memory. Next, the forget gate, which controls the extent of remains the values in the cell. And finally, the output gate which controls the extent of the value in the cell used to compute the output function. Fig. 3 shows the architecture of one LSTM cell.

3.1. Sequence LSTM

The most simple among considered neural network models. The main idea is to pass the windows of size N bytes to the input layer. Since files have large sizes in bytes, it would be an inefficient use of resources to pass the full file to the neural network input layer. In this work, we use the window size from 4 to 256 bytes. The input vectors are bytes of the corrupted file, while the output vectors contain the probabilities for every byte to be broken. In this case, when we have positions of corrupted bytes, it could be additional information for the restoration of source file content using the additional models or utilities. Furthermore, the efficiency of the model will be tested both with different vector representation models and without them. The architecture of naive sequence LSTM model is changed depending on window size and embeddings.

3.2. Stateful LSTM

The main difference between stateful LSTM and common LSTM networks is that neurons states are not reset each time the network get a new input. Which means that the memory is reset only when it is needed. In our work we use different approaches:

- Feed N-grams of bytes to input and reset memory after feeding a file completely
- Feed N-grams of bytes to input and reset memory after feeding M bytes

For example, the input window size is 16 and M is equal to 256, which means that memory is reset every 256 bytes. Fig. 5 presents one of our tested model architecture, where the window size of 4 is fed to the network and states of neurons are remembered. States are reset every 256 bytes which are equal to 16 input windows.

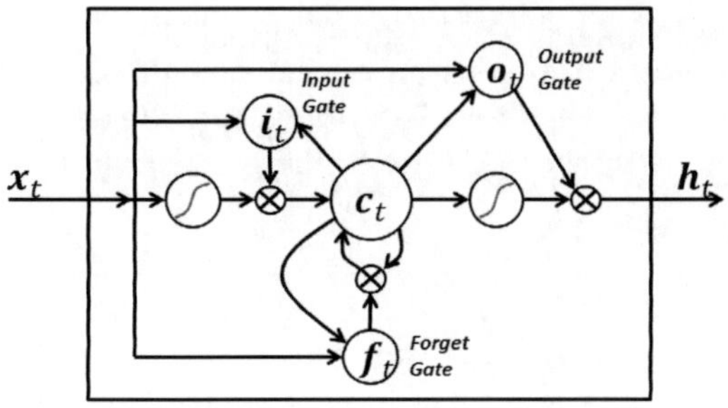

Figure 3. The *LSTM* cell scheme.

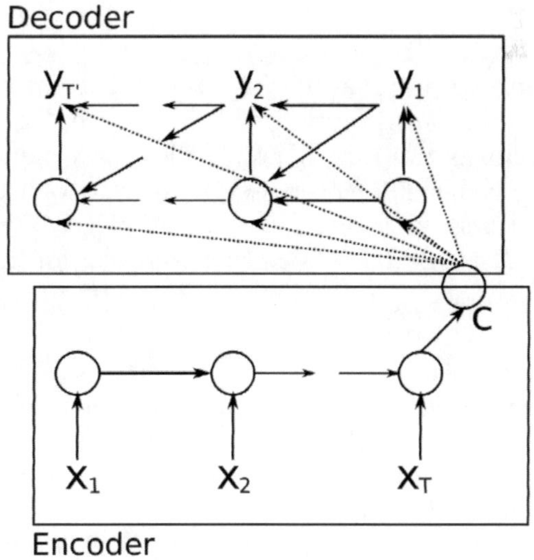

Figure 4. General seq2seq model architecture.

IOP Conf. Series: Journal of Physics: Conf. Series **1163** (2019) 012048 doi:10.1088/1742-6596/1163/1/012048

3.3. Seq2seq

Sequence To Sequence (seq2seq) is a complex model that consists of many LSTM cells and has a complicated structure. It was proposed by I. Sutskever et al. in their work [6]. This model became very popular because of machine translation problem. Generally, there are two parts: encoder and decoder. Both encoder and decoder are plenty of LSTM cells. The input vector is fed to the encoder after that to the decoder and finally, the output is generated. A special symbol is used in the training process to mark the end of a sequence. It is also fed to a decoder to show the end of input from an encoder and the beginning of the generation process. Fig. 4 shows the architecture of seq2seq model.

In our work, we use a window size of 4 to 64 bytes as input and the output is a vector with probabilities of each byte to be corrupted. In addition, the ability of feeding the full file and get all wrong positions was also studied. What became possible due to seq2seq architecture. The model takes a lot of computing resources if we pass the whole file through it, so we discarded this approach. Fig. 6 presents the simplified example of the used in this work seq2seq model input and output.

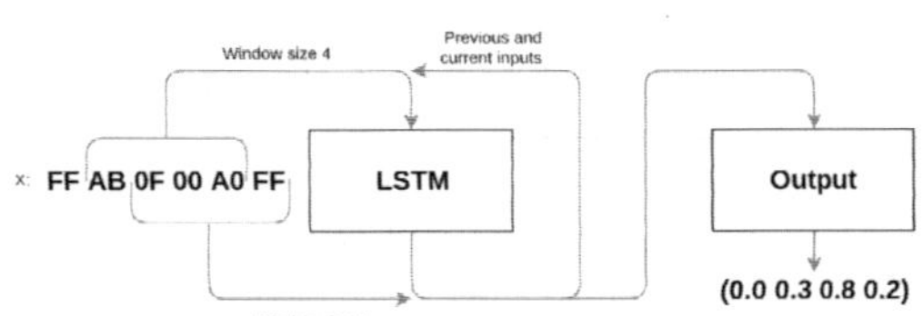

Figure 5. Stateful LSTM with window size 4 input and output example. State is reset every 256 bytes.

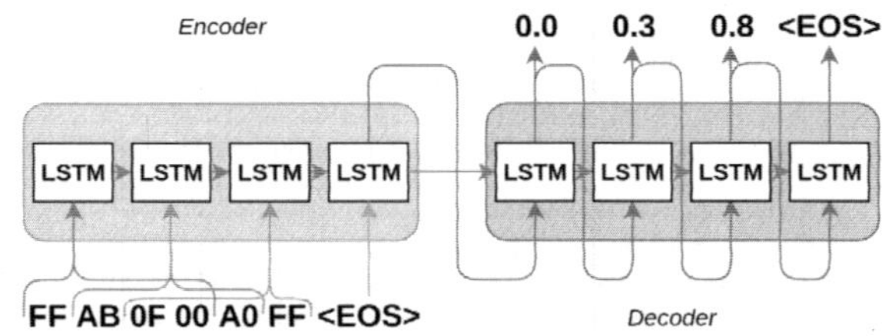

Figure 6. Seq2seq model input and output example. Window size is equal to 4, end of stream 'symbol' is 'EOS'.

4. Results

To compare the quality of different recurrent neural network models and approaches for getting a vector representation of bytes we use the same data on each combination. Also, we use embeddings trained on another file types to understand the impact of knowing the file extension on the result. Moreover, we try to train vector representation models on all file types to avoid the necessity of knowing the type of file. We try different models described above to understand if the N plays role while using N-grams for getting embeddings. Also, we change the window size to understand the impact of it on the quality of a model.

4.1. Used data

We use data collected from the Internet using Google search by map and extensions. Also, we processed a large dataset of imaged called "ImageNet" which was introduced as a challenge in [7]. All files were processed and a noise was added: we randomly change every byte of a file with probability p, where p are: $0.01, 0.05, 0.10, 0.15$ and 0.30. To understand the training size it is better to provide not a count of files but the total size of datasets of each file type. The numbers of bytes are shown in Table 1. All the data is shuffled split into a training and test sets. The test size is 20% of dataset size.

4.2. Comparison of results for different file types

Table 2 illustrates results of best models applied for the stated problem for different file types. For "exe" file extension models scores are slightly better than a random pick. So, some upgrades

Table 1. Distribution of file types

File type	Bytes count, million
EXE	3633
JPG (JPEG)	3200
MP3	2715
PDF	4075

Table 2. Precision of models for different file types

Error rate	File type			
	jpg	pdf	exe	mp3
1%	0.9951	0.9928	0.9900	0.9932
5%	0.9732	0.9711	0.9501	0.9801
10%	0.9291	0.9246	0.9004	0.9202
15%	0.8948	0.8788	0.8505	0.8890
20%	0.8744	0.8592	0.8009	0.8530
30%	0.7782	0.7523	0.7013	0.7637

are needed to achieve better results. We got an averaged 0.8923 precision and 0.9991 recall. Such precision means that if you take any byte that is marked by a model as corrupted, this is a probability for a byte to be corrupted really. Similarly for recall: it means the probability that a marked as a correct byte is actually a corrupted one.

5. Conclusion

In this paper we studied the quality of different recurrent neural network applied to determine broken bytes positions. Also, we described what changes are needed to apply these models to determine the positions of corrupted bytes. In addition, we considered the most popular models for getting vector representation of the word and used them to get an *embeddings* for bytes and *N-grams* of bytes. All needed transformations to feed the bytes to the vector representation models are provided. Next, comparison of getting embeddings using different approaches is shown in plots. As well as tables showing the quality of all described neural networks applied to the stated problem is demonstrated. Finally, all sort of combinations of vector representation models and *RNNs* were compared and results are shown in the table.

To sum it up, recurrent neural networks proposed in this paper are well suited for the file analysis and processing due to their mechanisms which allow variable size inputs. The solution of the determining the corrupted bytes positions problem may be used in such applications, as network security, file validation, file recovery and so on.

References

[1] Raff E, Barker J, Sylvester J, Brandon R, Catanzaro B and Nicholas C 2017 *ArXiv e-prints* (*Preprint* 1710.09435)

[2] Mikolov T, Chen K, Corrado G S and Dean J 2013 *CoRR* **abs/1301.3781**

[3] Bojanowski P, Grave E, Joulin A and Mikolov T 2017 *Transactions of the Association for Computational Linguistics* **5** 135–146 ISSN 2307-387X

[4] Hinton G E and Salakhutdinov R R 2006 *Science* **313** 504–507 ISSN 0036-8075 (*Preprint* http://science.sciencemag.org/content/313/5786/504.full.pdf)

[5] Xie J, Girshick R and Farhadi A 2016 *Proceedings of the 33rd International Conference on International Conference on Machine Learning - Volume 48* ICML'16 (JMLR.org) pp 478–487

[6] Sutskever I, Vinyals O and Le Q V 2014 *Proceedings of the 27th International Conference on Neural Information Processing Systems - Volume 2* NIPS'14 (Cambridge, MA, USA: MIT Press) pp 3104–3112

[7] Russakovsky O, Deng J, Su H, Krause J, Satheesh S, Ma S, Huang Z, Karpathy A, Khosla A, Bernstein M, Berg A C and Fei-Fei L 2014 *ArXiv e-prints* (*Preprint* 1409.0575)

Monte Carlo simulation of amorphous magnets with random exchange interactions

A V Bondarev, I L Bataronov and I M Pashueva

Voronezh State Technical University
14 Moskovskii prospekt, 394026 Voronezh, Russian Federation

E-mail: bondarev_a_v@mail.ru

Abstract. Using the Monte Carlo method, the computer simulation of magnetic properties of pure amorphous Gd and Re-Gd amorphous alloys was performed. For the model of amorphous Gd the temperature dependencies of magnetization and magnetic susceptibility were calculated at different values of the ratio of exchange interaction integrals wihin the first and the second coordination spheres $J_1/|J_2|$. The magnetic phase diagram in coordinates $T - J_1/|J_2|$ was constructed. For the models of Re-Gd amorphous alloys the dependence of spin-glass transition temperature on concentration of Gd atoms was calculated. For all our models the hysteresis loops were calculated at different temperatures.

1. Introduction

Binary amorphous alloys of heavy rare-earth metals with nonmagnetic transition metals are of great interest due to their unique magnetic properties [1, 2]. In these systems the transition into the spin-glass state takes place. In amorphous alloys of the rhenium-gadolinium (Re-Gd) system the typical for spin glasses maximum on the temperature dependence of magnetic susceptibility $\chi(T)$ and irreversibility of magnetization $M(T)$ were experimentally revealed [3]. The spin-glass state occurs due to random exchange interactions in this system.

The nature of the spin-glass state on the microscopic level is studied insufficiently. In this connection, the computer simulation of atomic structure and magnetic properties of amorphous magnetic materials is perspective direction.

In this work we report on the Monte Carlo simulation of the magnetic properties of pure amorphous Gd and amorphous alloys of the Re-Gd system which are not studied so far.

2. Simulation technique

Using the molecular dynamics method, we constructed the models of atomic structure of pure amorphous Gd and Re-Gd amorphous alloys. Each model contained 100 000 atoms in a cubic cell with periodic boundary conditions. The interatomic interaction was described by an empirical polynomial potential [4].

The magnetic moments of Gd atoms in the process of simulation were situated in the corresponding atomic locations in the amorphous structure. The coordinates of the atoms were previously obtained by the molecular dynamics method. Magnetic properties of these models were studied using the Monte Carlo method within the standard Metropolis algorithm [5] in the framework of the Heisenberg model. After formation of a random initial spin configuration, an attempt of changing the configuration is made, i.e. the randomly chosen spin randomly changes its spherical coordinates θ and φ. If the energy of the system is decreases after this change, then the new configutation is accepted, otherwise it is accepted with the probability $W = \exp(-\Delta E/k_B T)$ where ΔE is the energy variation. This procedure repeats for calculation of average values of physical quantities.

The Hamiltonian describing the interaction of the magnetic moments of gadolinium atoms was written in the following form [6, 7]:

$$E = -\frac{1}{2}\sum_{i,j} J_1(\mathbf{S}_i \cdot \mathbf{S}_j) - \frac{1}{2}\sum_{i,k} J_2(\mathbf{S}_i \cdot \mathbf{S}_k) - \mu H \sum_i S_i^z \qquad (1)$$

where $J_1 > 0$ is the exchange interaction integral between the spins S_i and S_j within the distances not exceeding $r_1^{\min} = 0,455$ nm – the position of the first minimum of the pair RDF $g(r)$; $J_2 < 0$ is the exchange interaction integral between the spins within the distances in the interval from $r_1^{\min}$ to $r_2^{\min} = 0,77$ nm – the first and the second minima of the pair RDF $g(r)$; $\mu = 7,98\mu_B$ is the effective magnetic moment of the Gd atom; H is the strength of the external magnetic field.

The value of the exchange integral within the first coordination sphere was chosen $J_1 = 45,7$ K. The value of the exchange integral within the second coordination sphere J_2 varied in the following range: $J_1/|J_2| = 2$, 6, 8, 9, 10, 11, 12, 13, 14 and 16.

During the cooling process the temperature dependencies of potential energy of the system (1), magnetization $M(T)$, Edwards–Anderson order parameter $q(T)$ and magnetic susceptibility $\chi(T)$ were calculated.

3. Results and discussion

For the model of pure amorphous Gd, we calculated the temperature dependencies of spontaneous magnetization *M*, where M_S is saturation magnetization, at various values of the $J_1/|J_2|$ ratio ($J_1/|J_2| = 10$, 11, 12, 13 and 14). In the Fig. 1 we present the dependence of maximum spontaneous magnetization (at T = 1 K) on the $J_1/|J_2|$ ratio for the model of amorphous Gd.

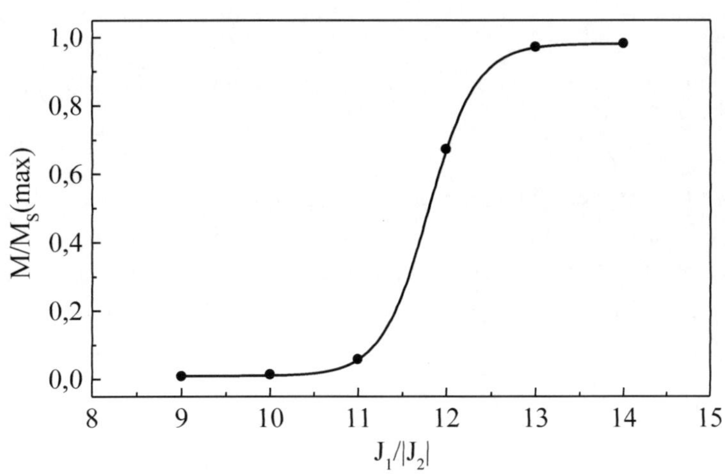

Fig. 1. Dependence of maximum spontaneous magnetization (at T = 1 K) on the $J_1/|J_2|$ ratio for the model of amorphous Gd

As it is seen from Fig. 1, at the low temperatures at $J_1/|J_2| \geq 13$ the ferromagnetic state takes place, at $11 < J_1/|J_2| < 13$ – the asperomagnetic state, and at $J_1/|J_2| \leq 11$ the system transits into the spin-glass state. In the asperomagnetic state the spontaneous magnetization differs from zero but the magnetic moments are non-collinear, arranged randomly as in spin glasses. The type of a phase was determined by the value of M/M_S at $T = 1$ K.

In Fig. 2 we present the temperature dependencies of magnetic susceptibility for the models of pure amorphous gadolinium at different values of the $J_1/|J_2|$ ratio equal to 8, 10, 12, 14 and 16. The values of susceptibility were calculated during cooling the model from the paramagnetic state in the temperature interval $T = 100 - 1$ K with the step $\Delta T = 5$ K in the absence of magnetic field. At each temperature the values of susceptibility were averaged over 10 cycles of 10^3 MC-steps/spin. In all these $\chi(T)$ curves a distinct maximum is observed that proves the presence of a magnetic phase transition. Position of the maximum corresponds to the phase transition temperature T_f.

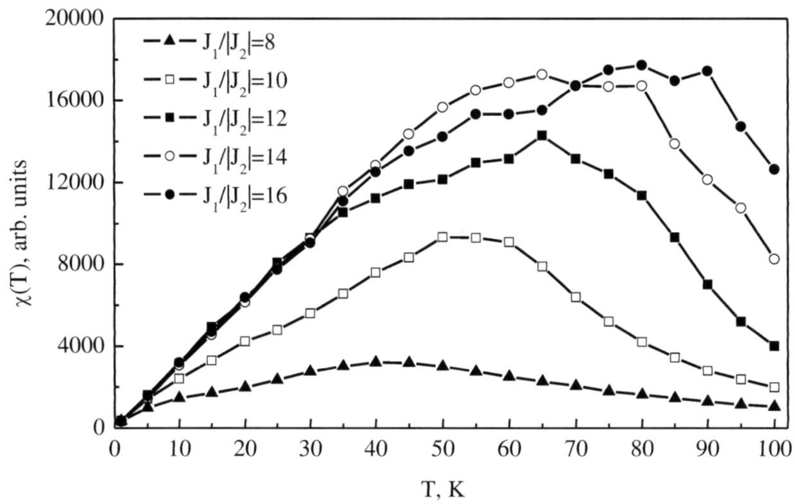

Fig. 2. Temperature dependence of magnetic susceptibility for the model of amorphous Gd at different values of the $J_1/|J_2|$ ratio

Thus, we obtained the magnetic phase diagram for amorphous Gd in the $J_1/|J_2| - T$ coordinates (Fig. 3). It allows one to determine the phase state of the system depending on the temperature and the exchange integral within the second coordination sphere J_2. The temperature of the magnetic phase transition increases monotonically with increasing the $J_1/|J_2|$ value and reaches the constant value $T_f \approx 75$ K at $J_2 \to 0$ (in this case $J_1/|J_2| \to \infty$).

We studied the temperature dependencies of magnetic susceptibility $\chi(T)$ for amorphous alloys $Re_{100-x}Gd_x$ (x = 4, 5, 7, 8, 10, 12, 37, 61 and 93 at. %) with $J_1/|J_2| = 10$. At $x > 7$ at. % Gd in all the $\chi(T)$ curves the distinct maxima are observed. The temperatures of the spin glass transition T_f were determined by the positions of the maxima. At $x \leq 7$ at. % Gd a maximum in the $\chi(T)$ curves disappears and magnetic susceptibility monotonically increases with decreasing the temperature, what is typical for paramagnetic state.

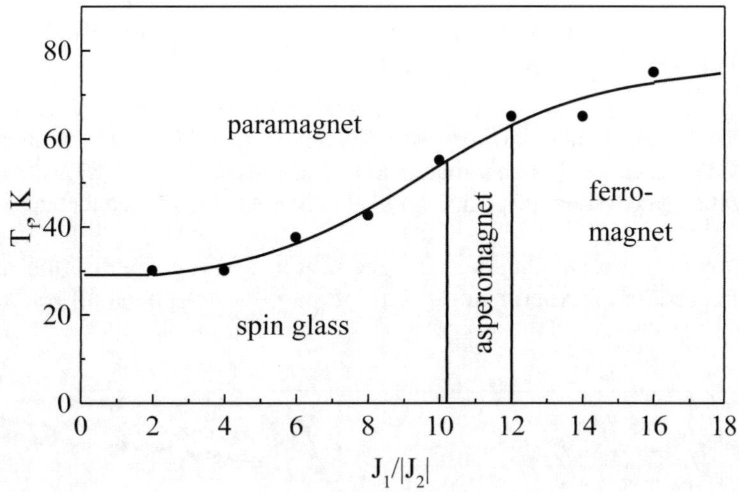

Fig. 3. Magnetic phase diagram $T_f\left(J_1/|J_2|\right)$ for the model of amorphous Gd

The minimal concentration of gadolinium atoms at which the spin glass transition takes place is above the percolation threshold in this system that is 4 at. % Gd. Thus, the spin glass transition in the Re-Gd system takes place only above the percolation threshold in this sysytem, i.e. at $x > 7$ at. % Gd.

The temperature dependence of the transition temperature (Fig. 4) is linear and agrees well with the experimental results [3].

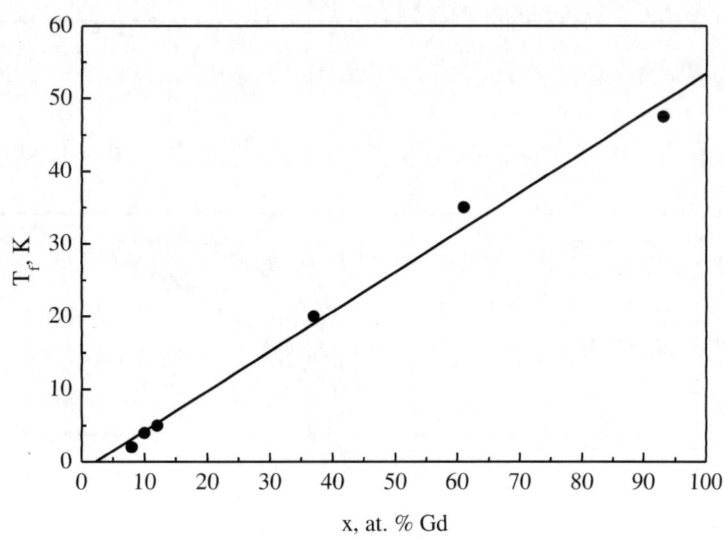

Fig. 4. Dependence of the spin-glass transition temperature for the models of $Re_{100-x}Gd_x$ amorphous alloys on the concentration of Gd atoms

We studied the behaviour of the models of amorphous Gd and Re-Gd amorphous alloys with application of an external magnetic field. The hysteresis loops for the model of amorphous Gd were calculated at $J_1/|J_2| = 10$ (this corresponds to the spin-glass state) and temperatures $T = 1$, 30, 50 K (Fig. 5). The external magnetic field was varied from –100 to 100 kOe with a step 5 kOe. The coercive field at $T = 1$ K is ~10 kOe and the remanent magnetization is ~0,55M_S. With increasing the temperature, the coercive field and the remanent magnetization monotonically decrease and reduce to zero at the spin-glass transition temperature $T_f = 50$ K.

The hysteresis loops for the models of $Re_{100-x}Gd_x$ amorphous alloys at $T = 1$ K were also calculated (Fig. 6). The external magnetic field was also varied from –100 to 100 kOe with a step 5 kOe. At x = 12–61 at. % Gd the coercive field is less than 5 kOe and only at x = 93 % it is about 10 kOe. The remanent magnetization monotonilcally increases with increasing the concentration of gadolinium atoms.

The results of Monte Carlo simulation of magnetization and remagnetization processes coincide qualitatively with the results of experimental study of various amorphous alloys based on rare-earth metals [8, 9].

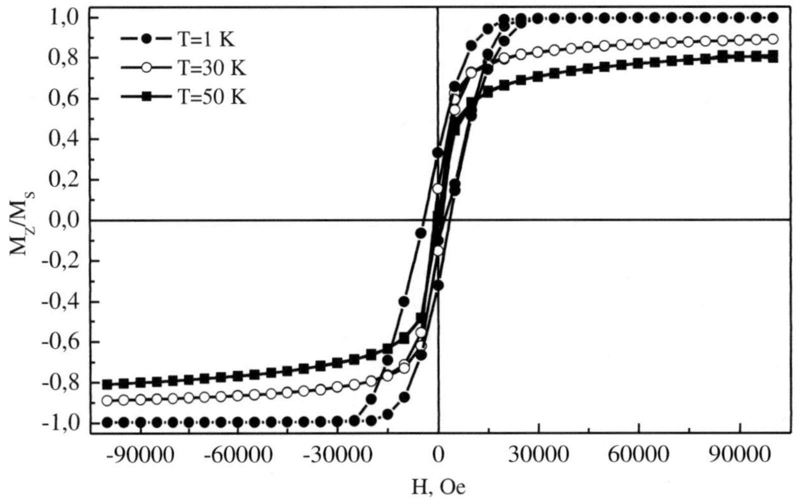

Fig. 5. Hysteresis loops for the model of amorphous Gd at $J_1/|J_2| = 10$ and different temperatures

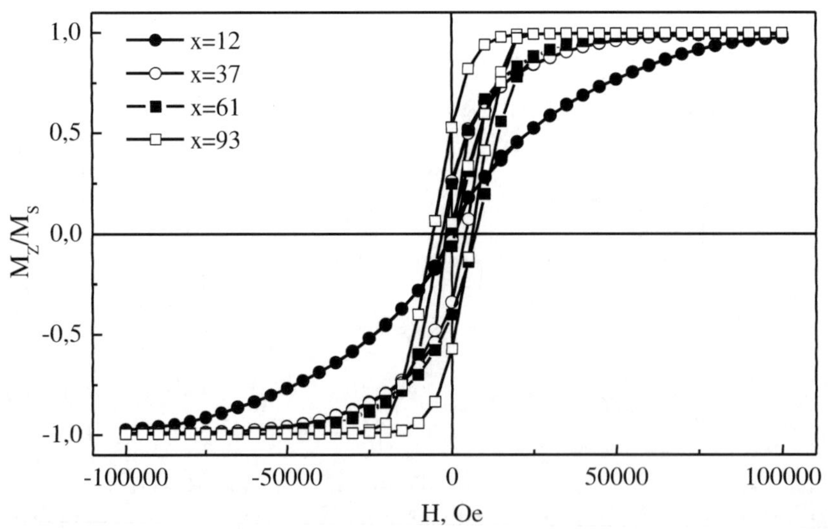

Fig. 6. Hysteresis loops for the models of $Re_{100-x}Gd_x$ amorphous alloys at $T = 1$ K

IOP Conf. Series: Journal of Physics: Conf. Series **1163** (2019) 012049 doi:10.1088/1742-6596/1163/1/012049

4. References

[1] Fernandez-Baca J A and Ching W-Y (ed.) 1995 *The Magnetism of Amorphous Metals and Alloys*, (Singapore: World Scientific)

[2] Chudnovsky E M 1988 *J. Appl. Phys.* **64** 5770

[3] Barmin Yu V, Balalaev S Yu, Bondarev A V et al. 2006 *Bull. Russ. Acad. Sci. Physics.* **70** 1308

[4] Bataronov I L, Bondarev A V and Barmin Yu V 2000 *Bull. Russ. Acad. Sci. Phys.* **64** 1329

[5] Binder K 1997 *Rep. Prog. Phys.* **60** 487

[6] Bondarev A, Bataronov I 2018 *EPJ Web of Conferences* **955** 012010

[7] Bondarev A V, Ozherelyev V V, Bataronov I L, Barmin Yu V and Chetkin D A 2010 *Bull. Russ. Acad. Sci. Phys.* **74** 1474

[8] Wang L, Ding J, Kong K Z, Li Y and Feng Y P 2001. *Phys. Rev. B* **64** 214410

[9] Hattori Y, Fukamichi K, Suzuki K et al. 1995 *J. Phys.: Cond. Mat.* **7** 4193

Automated Robot's Workspace Approximation [1]

Mikhail Posypkin

Dorodnicyn Computing Centre, Federal Research Center Computer Science and Control of Russian Academy of Sciences

E-mail: mposypkin@gmail.com

Abstract. The workspace of a robot is defined as a set of positions that robot or its part can take. The workspace size maximization is often one of the design goals: the larger workspace the bigger area a robot can serve. A manual workspace determination can be an error-prone and time-consuming process. In this work, we propose a method for an automated constructing of the robot's workspace by a paving composed from boxes. Two different ways of obtaining a paving that covers the solution set of kinematic equations are considered. The first method overestimates the solution set of kinematic equations by comparing interval bounds on the left parts of equations against zero. Though this method constructs a valid paving the approximation can be quite inaccurate. We propose a way for improving the quality of the approximation based on a local solution search method. We compare both ways of constructing the workspace approximation on a real robot.

1. Introduction

A robot's workspace [1] is one of its key characteristics. It is used for optimal path planning. The size and the shape of the workspace determine the area that the robot can serve. For simple cases, a workspace can be easily constructed manually. However, for more complex cases, this process should be automated.

The topic has been investigated thoroughly for decades, thus an interested reader should be referred to [2, 3]. For instance, in [2] the authors suggest the following classification of the workspace assessment techniques, which can be applied to parallel robots: geometrical, algebraic, and discretization, though the majority of these techniques are intended to be used only for robots within the class they were developed for. The geometric techniques are quite efficient, though they can be applied to relatively simple robots. The techniques that based on discretization can be applied to a broader class of parallel robots. However, these techniques are time-consuming and intended to be used in cases when the forward or the inverse kinematic problems have a simple solution [3, 4].

In [5, 6] authors propose to reduce kinematic equations to a set of inequalities and then use the branch-and-bound technique to construct a set of boxes enclosing the solution set (so-called *paving*). The drawback of this approach is that it requires additional efforts for obtaining a system of inequalities from kinematic equations. For complex robots, this reduction is rather complicated or even impossible. In this paper, we show how to avoid this step and obtain a paving directly from kinematic equations.

[1] This work was supported by the Russian Science Fund, project 16-19-00148

2. The outline of the approach

The problem under consideration consists in finding a projection of a set of points defined by a system of equations. Let $I \subseteq \overline{1,n}$ be a subset of indices. Without loss of generality we assume that $I = \{\overline{1,k}\}$, $k < n$. Consider a system of equations:

$$\begin{cases} f_1(x) = 0, \\ \dots \\ f_m(x) = 0, \end{cases} \quad (1)$$

where $f_i(x)$, $i = 1, \dots, m$ are continuous functions defined on $\mathbb{R}^n$, $m \leq n$. We assume that parameters satisfy interval constraints:

$$x_i \in [\underline{a_i}, \overline{a_i}], i = \overline{1,n}. \quad (2)$$

A set defined by constraints (2) is a box denoted by $\mathbf{a}$ in a sequel. Equations (1) together with interval constraints (2) define a (possibly empty) set $X \subseteq \mathbb{R}^m$. This set consists of all points x satisfying (1) and (2). We are interested in finding the projection X_I of the set X on a subspace $\mathbb{R}^k$, spanning the first k parameters $x_1, \dots, x_k$,

Example 1. Let $m = 1, n = 2$, $f_1(x) = x_1^2 + x_2^2 - 1$, $k = 1$, $x_i \in [-2, 2], i = 1, 2$. Then X is a circle centered at $(0, 0)$ with the radius 1 and X_I is an interval $\{x_1 : -1 \leq x_1 \leq 1\}$.

In this paper, we propose an algorithm to construct an outer approximation of the set X_I. An approximation is searched as a set of boxes. Following [5] we call this approximation a *paving*. More formally a set of boxes P is called a paving for a set S if the following two properties hold:

(i) $S \subseteq \cup_{\mathbf{x} \in B}$;
(ii) boxes in P can't share common internal points (may have common boundary points).

Pavings have some nice properties. For instance, the volume of a set $\cup_{\mathbf{x} \in P}$ is a sum of volumes of boxes from P. When approximating a working set of a robot pavings can be used for path planning [1].

The *accuracy* acc(P) of a paving is defined as a Hausdorff distance between the set S and the paving P:

$$\text{acc}(P) = d_H(S, P).$$

The maximal width of a box from a paving P is called *granularity* and is denoted as gran(P).

For a box $\mathbf{x}$ define vectors $m(\mathbf{x})$ and $M(\mathbf{x})$ as follows:

$$m_i(\mathbf{x}) = \min_{x \in \mathbf{x}} f_i(x), \;\; M_i(P) = \max_{x \in \mathbf{x}} f_i(x), i = 1, \dots, m. \quad (3)$$

The following fact directly follows from the continuity of functions $f_i(x)$, $i = 1, ..., m$.

Proposition 1. *If* $\max_{i \in \overline{1,n}} m_i(\mathbf{x}) > 0$ *or* $\min_{i \in \overline{1,n}} M_i(\mathbf{x}) < 0$ *then* $\mathbf{x} \cap X = \emptyset$.

In practice finding an exact range $[m_i(\mathbf{x}), M_i(\mathbf{x})]$ for a function $f_i(x)$ over a box $\mathbf{x}$ can be problematic. However it is almost always possible to obtain an outer approximation of this range $[\hat{m}_i(\mathbf{x}), \hat{M}_i(\mathbf{x})]$ with the help of interval analysis [7] or some other bounding techniques. Values $\hat{m}_i(\mathbf{x}), \hat{M}_i(\mathbf{x})$ obey the following property:

$$\hat{m}_i(\mathbf{x}) \leq \min_{x \in \mathbf{x}} f_i(x), \;\; \hat{M}_i(\mathbf{x}) \geq \max_{x \in \mathbf{x}} f_i(x), i = 1, \dots, m. \quad (4)$$

In a same way as the Proposition 1 we can prove the following proposition.

Proposition 2. *If* $\max_{i \in \overline{1,n}} \hat{m}_i(\mathbf{x}) > 0$ *or* $\min_{i \in \overline{1,n}} \hat{M}_i(\mathbf{x}) < 0$ *then* $\mathbf{x} \cap X = \emptyset$.

IOP Conf. Series: Journal of Physics: Conf. Series **1163** (2019) 012050 doi:10.1088/1742-6596/1163/1/012050

The Proposition 2 leads to the following simple algorithm BASECOV to construct a paving (Algorithm 1). For the sake of clarity we first consider the case $I = \overline{1,n}$, i.e. $X = X_I$.

```
Input: a, δ
Output: A
L := {a}
A := ∅
while L ≠ ∅ do
    remove a box x from L
    if max_{i∈1,n} m̂_i(x) ≤ 0 and min_{i∈1,n} M̂_i(x) ≥ 0 then
        if d(x) ≤ δ then
            A := A ∪ x
        else
            split x into x′, x″
            L := L ∪ {x′, x″}
        end
    end
end
```

Algorithm 1: The basic coverage algorithm BASECOV

The algorithm maintains two lists: L for storing processed boxes and A for storing the resulting approximation. The list A is empty at the beginning and the list L initially contains only the box $\mathbf{a}$. At each iteration, the main while loop takes a box $\mathbf{x}$ from the list L. If this box satisfies the proposition 2 it is discarded. Otherwise its diameter $\mathrm{d}(\mathbf{x})$ is checked: if it is less than the prescribed accuracy δ the box is stored in the list A otherwise it is split into two equal boxes across the longest edge. The loop terminates when the list L becomes empty.

The list of boxes A constructed by the BASECOV algorithm satisfies both paving's properties. Unfortunately, it is difficult to estimate the accuracy of this paving. One of the ways to provide a guaranteed accuracy is to ensure that every box in the approximation contains at least one point from X. We'll call pavings satisfying this property *tight*. The following proposition holds for tight pavings.

Proposition 3. *Let P be a tight paving with a granularity δ. Then $acc(P) \le \delta$.*

The algorithm 2 (ADVCOV) constructs a tight paving. The difference between BASECOV and ADVCOV algorithms is that ADVCOV runs a local search for each box with a width less or equal to δ. A local algorithm LS searches for a solution z of a system (1) starting from the center $\mathrm{mid}(\mathbf{x})$ of a box $\mathbf{x}$. If z lies inside $\mathbf{x}$ then the box is added to the constructed paving A. Otherwise the box $\mathbf{x}$ is split into two equal boxes $\mathbf{x}', \mathbf{x}''$ that replace it in the current list of boxes L.

Input: $\mathbf{a}$, δ
Output: A
$L := \{\mathbf{a}\}$
$A := \emptyset$
while $L \neq \emptyset$ **do**
  remove a box $\mathbf{x}$ from L
  if $\max_{i \in \overline{1,n}} \hat{m}_i(\mathbf{x}) \leq 0$ *and* $\min_{i \in \overline{1,n}} \hat{M}_i(\mathbf{x}) \geq 0$ **then**
    if $d(\mathbf{x}) \leq \delta$ **then**
      $z := LS(\text{mid}(\mathbf{x}))$
      if $z \in \mathbf{x}$ **then**
        $A := A \cup \mathbf{x}$
      else
        split $\mathbf{x}$ into $\mathbf{x}', \mathbf{x}''$
        $L := L \cup \{\mathbf{x}', \mathbf{x}''\}$
      end
    else
      split $\mathbf{x}$ into $\mathbf{x}', \mathbf{x}''$
      $L := L \cup \{\mathbf{x}', \mathbf{x}''\}$
    end
  end
end

Algorithm 2: The advanced coverage algorithm ADVCOV

Any algorithm able to locate a local solution of the system (1) can be taken as a local search method LS. We used a box-constrained coordinate descent method applied to a convolution $\phi(x) = \sum_{i=1}^{m} f_i(x)^2$.

Usually, we are interested in a projection of a workspace to a set of coordinates. Since the paving is composed by boxes the projection can be obtained easily as a union of projections of all boxes in it. However, this can be a very resource-demanding process because the number of boxes in a paving can be huge.

3. Experimental evaluation

We applied the proposed algorithm to the problem of finding a working space of a parallel robot called DexTAR [8]. This robot is formed by 4 links AB, BC, CD, DE, connected with passive joints B, C, D and driven by two rotating actuators A, E (figure 1). The working tool of a robot is located in C. The workspace is defined as a set of all possible positions of the working tool.

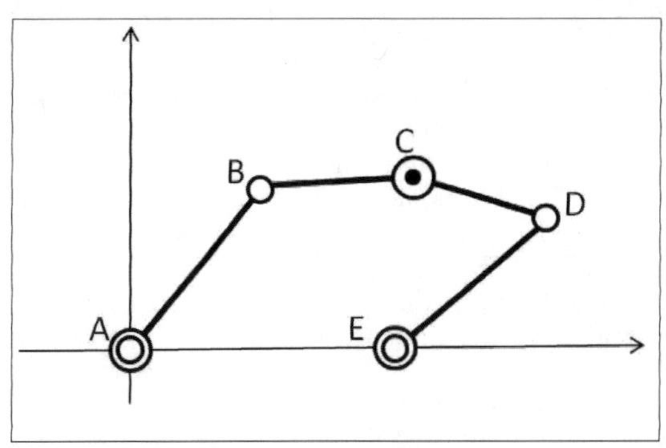

Figure 1. The DexTAR robot's picture and the scheme

IOP Conf. Series: Journal of Physics: Conf. Series **1163** (2019) 012050 doi:10.1088/1742-6596/1163/1/012050

We considered the following configuration: $|AB| = |ED| = 8$, $|BC| = |CD| = 6$, $|AE| = 6$. The workspace and its approximations obtained by algorithms BASECOV and ADVCOV are depicted at the figure 2. The value of δ was set to $5 \cdot 10^{-2}$ for BASECOV and $1.25 \cdot 10^{-1}$ for ADVCOV. The precisions were chosen to obtain a comparable number of boxes in pavings. Notice that the actual size of boxes in a paving constructed by ADVCOV can be less due to splitting of boxes where the local solution wasn't found. With this settings, the numbers of boxes obtained by the BASECOV and ADVCOV methods were 5563 and 3417 respectively. It is visually evident that the paving obtained by ADVCOV is more accurate w.r.t. the paving obtained by BASECOV, though the number of boxes in the latter is bigger.

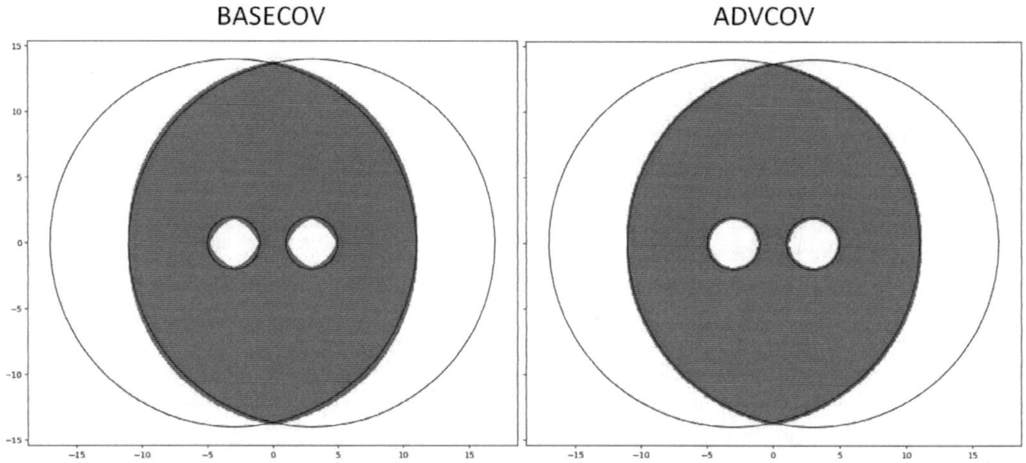

Figure 2. The comparison of pavings

4. Conclusions

In this work, we addressed a problem of workspace approximation. An algorithm to construct an approximation of the workspace defined by a set of kinematic equations is proposed. We show that employing the local search significantly improves the quality of the approximation. However, the computational complexity of the proposed method is still significant. In the future we plan to try more efficient local search methods, e.g. Newton method [9] and apply parallel computing.

References

[1] Merlet J P 2006 *Parallel robots* vol 128 (Springer Science & Business Media)
[2] Jha R, Chablat D, Rouillier F and Moroz G 2016 *Robotics and Mechatronics* (Springer) pp 121–130
[3] Snyman J, Du Plessis L and Duffy J 2000 *Journal of Mechanical Design* **122** 447–456
[4] Bohigas O, Manubens M and Ros L 2012 *IEEE Transactions on Robotics* **28** 993–1006
[5] Jaulin L 2001 *Applied interval analysis: with examples in parameter and state estimation, robust control and robotics* vol 1 (Springer Science & Business Media)
[6] Evtushenko Y, Posypkin M, Rybak L and Turkin A 2018 *Journal of Global Optimization* **71** 129–145
[7] Hansen E and Walster G W 2003 *Global optimization using interval analysis: revised and expanded* vol 264 (CRC Press)
[8] Sergiu-Dan Stan V M and Balan R 2008 *Frontiers in Evolutionary Robotics* 465–490
[9] Rheinboldt W C 1998 *Methods for solving systems of nonlinear equations* vol 70 (Siam)

IOP Conf. Series: Journal of Physics: Conf. Series **1163** (2019) 012051 doi:10.1088/1742-6596/1163/1/012051

Calculation of magnetic moments and lattice parameters Co-based Heusler alloys with determination of their energy favorable structure

[1]**Marina V. Mamonova**, [2]**Irina A. Prudnikova**

[1]Department of Theoretical Physics, Omsk State University, Omsk , Russia
[2]Omsk State Agrarian University, Omsk , Russia

E-mail: [1]mamonovamv@omsu.ru [2]ia.prudnikova@omgau.org

Abstract. In this work we present results of ab-initio investigation atomic structures and magnetic properties of Cobalt-based full-Heusler compounds with composition Co_2YZ by using VASP software package. The lattice parameters of Co_2FeAl(CFA), Co_2FeSi(CFS), $Co_2FeAl_{0.5}Si_{0.5}$(CFAS) Heusler alloys depending on the parameters of convergence are calculated. Investigation of CFA and CFS alloys in $B2$-type and $L2_1$-type structures with determination of their energy efficiency of the formation are carried out. The values of the magnetic moment of CFAS atoms in the bulk and in CFAS/Ag(100)/CFAS multilayer structure taking into account the relaxation effects are determined.

1. Introduction

Cobalt-based full-Heusler compounds with composition Co_2YZ (where Y is a transition metal and Z is a main group element) are attracting attention due to their predicted half-metallic behaviour, a much desired property for spintronic devices [1-2]. Knowledge of the basic magnetic properties of these materials, especially in the form of thin films, is required both to exploit these promising materials and to understand the properties of magnetic multilayer structures based on them. Prominent are the Co2-based Heusler compounds, which gained a particular interest and became most widely studied in this field due to their high Curie temperatures, a requirement for spin-dependent electron transport devices. Tezuka et al. [3] reported tunnel junctions built using the iso-electronic compound $Co_2FeAl_{0.5}Si_{0.5}$. The junctions exhibited tunnel magnetoresistance (TMR) ratios for the $L2_1$ structure of 51% and 78% at 300 K and 5 K, respectively. In this work, we use the spin-density functional method for theoretical description and calculation of energy and magnetic characteristics of Co_2FeAl(CFA), Co_2FeSi(CFS), $Co_2FeAl_{0.5}Si_{0.5}$(CFAS) Heusler alloys.

2. Model and methods

We studied Cobalt-based full-Heusler compounds with the use of VASP (Vienna Ab-Initio Simulation Package) software package [4] by means of the Projector Augmented Wave (PAW) method with PBE [5] version of generalized gradient approximation(GGA) to describe the exchange-correlation interactions.

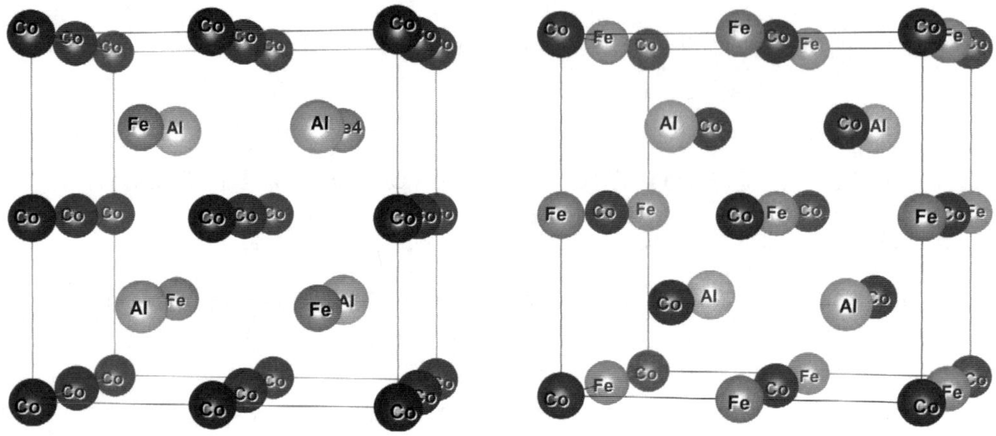

Figure 1. a) $L2_1$-type and b) $B2$-type ordered structure of Co_2FeAl-alloy

Table 1. Result of calculation cell total energy and lattice parameters for $B2$ and $L2_1$-ordered structures of Co_2FeAl-alloy.

Type ordering	Energy, eV	a,Å	a_{lit},Å	a_{exp},Å [6]
B2	-107.1048(64)	5.6800(400)	—	5.7300
L21	**-109.8830(20)**	**5.6965(9)**	**5.6900**	**5.7030**

Heusler compounds with the general formula Co_2YZ crystallize in the $L2_1$ and the $B2$ structures. At $L2_1$ structure (figure 1a) the cubic unit cell consists of four interpenetrating fcc sublattices, two of which are occupied by Co atoms and the other two by the Y and Z atoms, respectively. At $B2$ (figure 1b) structure in (0,0,0) and (1/2,1/2,1/2) sites are randomly occupied by Y and Z with an equal probability.

Since the half-metallicity of full-Heusler alloys depends on their atomic ordered crystal structures, determination of the energy efficiency of the formation of various ordered crystal structures are necessary. We obtained the total energy and lattice parameters as the functions of convergence parameters. We confined ourselves to values of the plane waves cut-off energy $E_{cut} = 400$eV and Monkhorst-Pack grid size 16x16x16.

The results of our calculations have shown that the bulk energy of CFS alloy $L2_1$-type structure is less than in the case of in B2-type structure , this suggests that the $L2_1$-type structure is energetically more favorable. The values of the lattice constant were found from the condition of the minimum energy of the unit cell. The calculated bulk lattice parameters of $B2$ and $L2_1$-ordered structures of Co_2FeAl Heusler compound are listed in table 1 and are in good agreement with the experimental values.

Similar calculations total energy and lattice parameters were conducted for another Heusler alloy Co_2FeSi and are presented in table 2. These results determines that the formation of a $L2_1$-ordered structure for Co_2FeSi-alloy is also energetically more favorable as for Co_2FeAl. The agreement of our calculated values with experimental ones demonstrate the reliability of our calculations.

Comparison of the magnetic moments of the atoms in both $L2_1$ and B2-type ordered structures Co_2FeSi alloy, which values are presented in table 3, allows to conclude that as

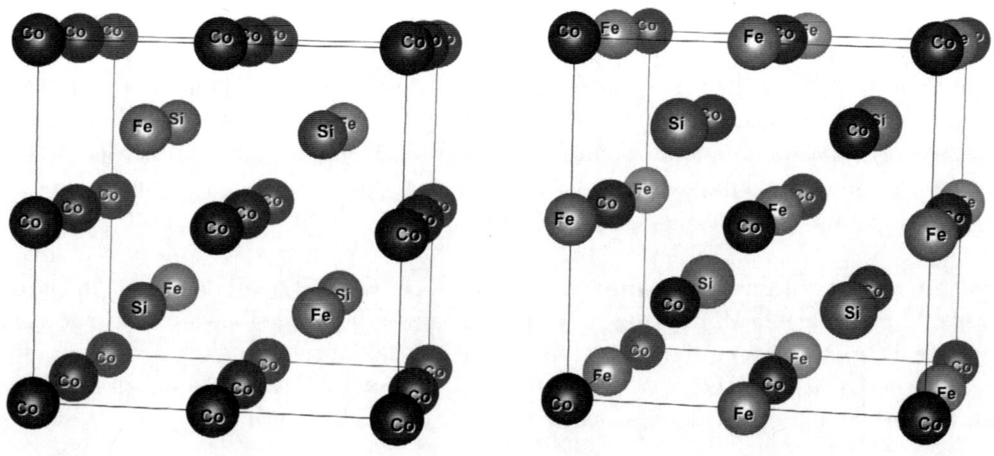

a) b)

Figure 2. a) $L2_1$-type and b) $B2$-type ordered structure of Co_2FeSi-alloy

Table 2. Result of calculation cell total energy and lattice parameters for $B2$ and $L2_1$-type ordered structures of Co_2FeSi-alloy.

Type ordering	Energy, eV	a,Å	a_{lit},Å [7]	a_{exp} ,Å [8]
B2	-115.0021(88)	5.5732(139)	5.5780	—
L21	**-116.5981(76)**	**5.6224(6)**	**5.6310**	**5.6400**

Table 3. Result of calculation magnetic moment of atoms μ, μ_B and cell total magnetic moment μ_{tot},μ_B for Co_2FeSi-alloy.

Type ordering	μ_{Co}	μ_{Fe}	μ_{tot}
B2	0.9200 1.5990	1.6040	16.1056
L21	1.3130	2.7840	21.6396

Table 4. Comparison results of calculation cell total energy, lattice parameters and magnetic moment of atoms for $L2_1$-type ordered structure Co-based Heusler alloys.

Alloy	Energy, eV	a, Å	μ_{Co}, μ_B	μ_{Fe}, μ_B	μ_{tot}, μ_B
CFAS	-113.41573(965)	5.579(12)	1.305	2.812	21.700
CFS	-116.5981(76)	5.6224(6)	1.3130	2.7840	21.6396
CFA	-109.8830(20)	5.6965(9)	1.187	2.751	19.940

magnetic moments of individual Co and Fe atoms and cell total magnetic moment in $L2_1$-type structure is greater than corresponding values in the case of B2-type structure. This is due to the fact that the atoms of iron and cobalt in the case of B2-type structure are in the same layer.

$Co_2FeAl_{0.5}Si_{0.5}$(CFAS) Heusler alloy is more interesting as material for thin film in multilayer structure, since, for example, values of the tunnel magnetoresistance (TMR) ratio CFAS are larger than those found for pure Co_2FeAl or Co_2FeSi alloys [3]. Since the previous calculations showed that the $L2_1$-type ordered structure is more energetically favorable for the alloys of the CFA and CFS, it was decided to consider it immediately for the CFAS. Results of calculation cell total energy, lattice parameters and magnetic moment of atoms for $L2_1$-type ordered structure Co-based Heusler alloys (CFA, CFS and CFAS) are combined in table 4. Among these alloys, CFAS has the lowest lattice constant, but the largest magnetic moment.

3. Investigation of magnetic properties CFAS/Ag/CFAS multilayer structure

In the case of CFAS/Ag/CFAS multilayer structure, the Ag lattice cell parameter is too small to accommodate a CFAS compound in the cube-on-cube epitaxial relationship. However, as shown in figure 3, if the CFAS cube edge grows at an angle of 45 with respect to an in-plane Ag(001) direction, a good lattice match is obtained, enabling epitaxial growth.

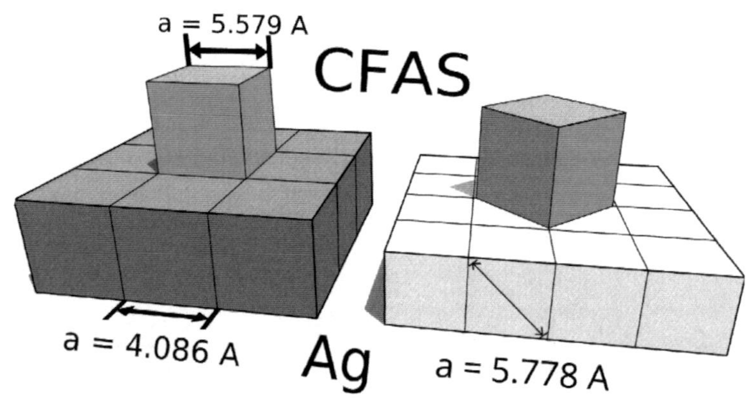

Figure 3. Epitaxial relationships between CFAS and Ag(100)

The supercell of CFAS/Ag/CFAS structure is consist of the 13 monatomic layers and vacuum layer with thickness 5Å. The layers of CFAS are placed on each side of the Ag five-layer slab, as shown in figure 4. The magnetic moment of atoms is directed collinearly along the z axis. The dimension of the k-point grid was selected 21x21x1. Calculations of this system consisting of 52 atoms and thickness 28Åwere carried out on the basis of the Center for Data of the Far Eastern Branch of the Russian Academy of Sciences in the city of Khabarovsk, since these calculations require a large amount of RAM. Total CPU time at 8 cores OpenPower8 was 26623sec $\simeq$ 7hours and maximum memory used was 10137.22 KBytes.

Figures 4 present the arrangement of atoms projected along the z axis and result of the calculation magnetic moment of atoms. Comparison of the values of the magnetic moment of CFAS atoms in the bulk (table 4.) and in the film adsorbed on Ag shows, that the Co atoms most distant from the substrate have greatest magnetic moment. For Fe atoms we obtained that magnetic moment for atoms in the bulk cell are slightly larger then magnetic moment for atoms located on substrate.

This figure 5 represents the comparative arrangement of the atomic layers and the change in the interlayer distance of the system before and after relaxation. An analysis of the figures 5 shows that the interlayer distance of outermost layers are compressed and the remaining especially silver interlayer distance are expanded.

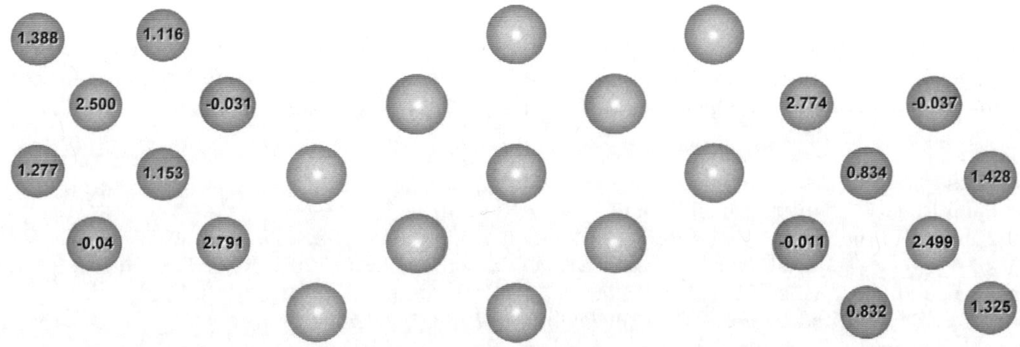

Figure 4. System CFAS/Ag/CFAS (Co - pink, Fe - yellow, Al - blue, Ag - gray)

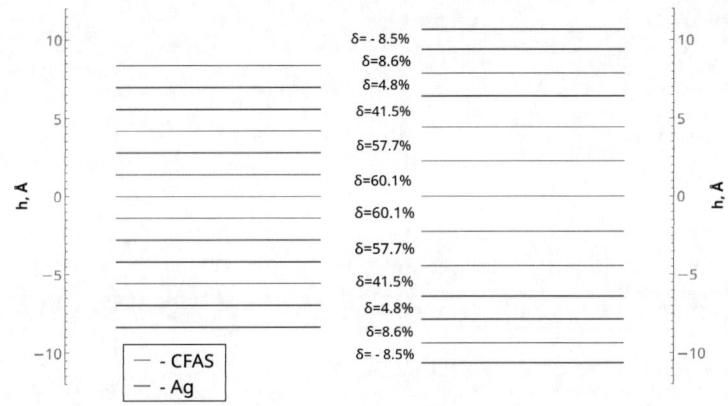

Figure 5. Atomic layers position for CFAS/Ag/CFAS system without and with according relaxation.

4. Conclusions

In this work we present results of ab-initio investigation atomic structures and magnetic properties of Cobalt-based full-Heusler compounds Co_2FeAl(CFA), Co_2FeSi(CFS), $Co_2FeAl_{0.5}Si_{0.5}$(CFAS) by using VASP software package.

The results of our calculations have shown that the bulk energy of CFA and CFS alloys in B2-type structure is greater than in the case of $L2_1$-type structure, this suggests that the $L2_1$-type structure is energetically more favorable. Also total magnetic moment in $L2_1$-type structure is greater than in the case of B2-type structure.

Investigation of the alloys lattice parameter dependence on the parameters of convergence made it possible to choose the optimal values of the plane waves cut-off energy $E_{cut} = 400$ eV and Monkhorst-Pack grid size 16x16x16. The optimized lattice constants for Co-based Heusler alloys $a^{CFA} = 5.6965(9)$ Å, $a^{CFS} = 5.6224(6)$Å, $a^{CFAS} = 5.5790(4)$ Åobtained by us are in good agreement with the experimental values.

Comparison of the values of the magnetic moment of CFAS atoms in the bulk and in the film adsorbed on Ag shows, that the Co atoms most distant from the substrate have greatest magnetic moment. For Fe atoms we obtained that magnetic moment for atoms in the bulk cell are slightly larger then magnetic moment for atoms located on substrate.

These results can be applied in numerical simulation by the Monte Carlo methods of the nonequilibrium behavior of multilayer magnetic superstructures

Acknowledgments

This research was supported by the grants 17-02-00279, 18-32-00814 of Russian Foundation of Basic Research. The simulations were supported in through computational resources provided by the Shared Facility Center "Data Center of FEB RAS" (Khabarovsk).

References

[1] Bass J and Pratt W P 1999 *J. Magn. Magn. Mater.* **200** 274–89
[2] Prudnikov V V, Prudnikov P V and Romanovskiy D E 2016 *J. Phys. D: Appl. Phys.* **49** 235002
[3] Tezuka N, Ikeda N, Miyazaki A, Sugimoto S, Kikuchi M and Inomata K 2006 *Appl. Phys. Lett.* **89** 112514
[4] Kresse P G and Furthmuller J 1996 *Phys. Rev.* B **54** 11169
[5] Perdew J P, Burke K and Ernzerhof M 1996 *Phys. Rev. Lett.* **77** 3865
[6] Elmers H J, Wurmehl S, Fecher G H et al. 2004 *Appl. Phys.* A **79** 557
[7] Amari S, Dahmane F, Bin Omran S et al. 2016 *Journal of the Korean Physical Society* **69** 1462–68
[8] Wurmehl S, Fecher G H, Kandpal H C, Ksenofontov V 2005 *Phys. Rev.* B **72** 184434

IOP Conf. Series: Journal of Physics: Conf. Series **1163** (2019) 012052 doi:10.1088/1742-6596/1163/1/012052

Simplification of tensor expressions in computer algebra

A.Kryukov and G.Shpiz
Skobeltsyn Institute of Nuclear Physics Lomonosov Moscow State University, 119992 Moscow, Russia
E-mail: kryukov@theory.sinp.msu.ru

Abstract.
Computer algebra is widely used in various fields of mathematics, physics and other sciences. The simplification of tensor expressions is an important special case of computer algebra. In this paper, we consider the reduction of tensor polynomials to canonical form, taking into account the properties of symmetry under permutations of indices, the symmetries associated with the renaming of summation indices, and also linear relations between tensors of a general form. We give a definition of the canonical representation for polynomial (multiplicative) expressions of variables with abstract indices, which is the result of averaging of the original expression by the action of some finite group (the signature stabilizer). In practice, the proposed algorithms demonstrate high efficiency for expressions made of Riemann curvature tensors.

1. Introduction

Tensor calculations are widely used in theoretical physics, solid state physics, mechanics and many other fields.

There are three main approaches to the simplification of tensor expressions which are used in computer algebra: (i) component calculations, (ii) manipulations with tensors with abstract indices and (iii) abstract tensor calculations. All of them have some advantages and disadvantages. The easiest approach to the tensor simplification is the component calculations. With this approach we have to choose some basis. All further calculations in this basis are the calculations of components of available tensors in a specific representation. Thus this approach reduces the task to calculations of scalar values the number of which is equal to the number of tensor components. The disadvantage of this approach is the huge number of tensor components that we must calculate. Also we cannot use special tensor properties, such as symmetry, automatically.

As examples of such calculations, we can specify the programs xAct [1], DifferentialGeometry [2], GRTensor [3], Atlas 2 [4],CTENSOR [5].

The last type of tensor simplification is purely abstract calculus. A good example of this approach is exterior algebra. We will not discuss it here. Examples of such programs are xTerior [6] in the system xAct [1], or ITENSOR [5] in the CAS MAXIMA [7].

In this paper we consider the second approach, in which a tensor is considered as an abstract object with indices, possessing various symmetry properties. We briefly discuss some computer algebra algorithms for simplification of tensor polynomials. Also we present a sketch of an algorithm based on stabilizers of permutation subgroups and on isomorphism of graphs corresponding to tensor monomials.

In the paper we will use the Einstein notations implying that two identical indices in a monomial means summation over all their values:

$$T(i,i) \equiv \sum_{i=1}^{i=n} T(i,i)$$

For simplicity, we will not distinguish upper and lower indices. Also we will not be interested in the transformation properties of tensors under coordinate transformations.

The structure of the paper is as follows. Section 2 briefly describes the use of the coset approach for tensor simplification. Section 3 is devoted to the group algebra approach. In Section 4, we give necessary definitions and statements. In Section 5, we describe the approach based on stabilizers of groups and isomorphism of graphs. We also give a sketch of the algorithm based on the ideas above. In conclusion, we summarize the results obtained and present plans for further development.

2. Double coset

We will consider a tensor as an abstract object with indices $T(i,j,k,\ldots)$, which has certain symmetry properties with respect to the permutation of the indices.

For example, let us consider the Riemann tensor $R(i,j,k,l)$. The Riemann tensor has the following symmetry properties:

$$R(i,j,k,l) = -R(j,i,k,l), \tag{1}$$
$$R(i,j,k,l) = R(k,l,i,j). \tag{2}$$

To give a general example, we can say that all symmetries of this kind form a subgroup of the permutation group of indices:

$$S_N \subset P_N,$$

where N is the order of the permutation group.

Another type of symmetry arising in tensor expressions is the symmetry associated with summation of indices. For example, consider the scalar curvature R:

$$R = R(i,i), \ \ \textit{where Ricci tensor} \ \ R(i,j) = R(k,i,k,j).$$

The scalar curvature R has an additional symmetry associated with the renaming of summation indices

$$R = R(k,i,k,i) = R(i,k,i,k) = R(i_1,i_2) = R(i_2,i_1), \tag{3}$$

where i_1 and i_2 are summation indices which were temporally marked by subindices 1 and 2 for illustration.

So summation indices generate another symmetry group which is a subgroup of the permutation group

$$D_N \subset P_N$$

and it acts from the right.

The problem of simplification of a tensor monomial which has both types of symmetry is reduced to the finding the double coset:

$$S_N \backslash T / D_N.$$

This approach was developed in [8] and is widely used in most tensor simplification algorithms (see, for example, [9, 10]).

The problem is how to deal with so-called multi-term identities. For example, the Riemann tensor satisfies the following Bianchi identity:

$$R(i,j,k,l) + R(i,l,j,k) + R(i,k,l,j) = 0.$$

Such identities destroy the group structure and must be treated by some other method.

3. Group algebra

Another approach to the simplification of tensor expressions was proposed by V. Ilyin and A. Kryukov [11]. The package ATENSOR implements this idea in CAS REDUCE[12]. In this approach all relations between tensor monomials both following from symmetries and multi-term identities were considered as vectors in some linear vector space V with dimension $N!$.

All the identities generate a hyperplane $H \subset V$. The hyperplane H defines a set of tensor expressions equivalent to zero. Thus, the problem of simplifying a tensor expression is reduced to finding the (orthogonal) hyperplane K that is supplementary to H.

Let us illustrate this point on the simplest example of symmetric tensor of the 2-nd rank $\delta(i,j) = \delta(j,i)$. We have two basis vectors: $e_1 = \delta(i,j)$ and $e_2 = \delta(j,i)$. Let t denote the vector $t = \alpha e_1 + \beta e_2$. The canonical representation of t_k is a projection of t on K

$$t = \alpha e_1 + \beta e_2 = \frac{1}{2}(\alpha - \beta)(e_1 - e_2) + \frac{1}{2}(\alpha + \beta)(e_1 + e_2)$$

The first monomial is equal to zero. Thus the canonical form of t is the following: $t = \frac{1}{2}(\alpha + \beta)(e_1 + e_2) \in K$.

The disadvantage of the group algebra approach is the huge dimension of the vector space equal to $N!$ where N is the dimension of the full permutation group of tensor expressions. For example, a product of two Riemann tensors $R(i,j,k,l) \cdot R(k,l,m,n)$ has dimension of space equal to $8! = 40320$.

4. Definitions and statements

Let us fix two linearly ordered sets: the set of types $\mathbf{\Theta}$ and the set of indices $\mathbf{I}$. In the set $\mathbf{I}$, we fix a subset $\mathbf{D} \subset \mathbf{I}$ of the summation indices.

We will call the **signature** the pair (t, i), where $t \in \mathbf{\Theta}$, and $i = (i_1, \ldots, i_n)$ is non-decreasing sequence of indices.

Elementary symbol of order n is an expression e of the form $e = t(i)$, where $t \in \mathbf{\Theta}$,ı $= (i_1, \ldots, i_n)$,$i_k \in \mathbf{I}$.

Signature of symbol $e = t(i_1, \ldots, i_n)$ will be the pair (t, i'), where $t \in \mathbf{\Theta}$, and $i' = (i'_1, \ldots, i'_n)$ ordered by ascending sequence of the indices $i = (i_1, \ldots, i_n)$.

Let $\mathbf{S}$ denote the set of all symbol signatures.

The set of sequences of indices is linearly ordered with respect to the lexicographic order. The set of signatures and the set of elementary symbols are also linearly ordered with respect to the lexicographic order.

The vector space of formal linear combinations of elementary symbols will be denoted by $\mathbf{Sym}$, and its elements will be called indexed values.

Let $s = (t, \mathbf{i}) \in \mathbf{S}$ be a signature.

We denote by $\mathbf{Sym}(s) \subset \mathbf{Sym}$ the vector space generated by all elementary symbols with the signature s, that is, from characters of type t and a sequence of indices obtained from the sequence $i = (i_1,\ dots, i_n)$ by some permutation of elements.

IOP Conf. Series: Journal of Physics: Conf. Series **1163** (2019) 012052 doi:10.1088/1742-6596/1163/1/012052

The group G of permutations of the set of indices $\mathbf{I}$ naturally acts on the set of all elementary symbols and, accordingly, in the space $\mathbf{Sym}$. In the following, we will call it the index renaming group. If $g \in G$, then $\mathbf{Sym}(ge) = g(\mathbf{Sym}(e))$.

Obviously, the index renaming group G translates symbols with the same signature into symbols with the same signature, and its natural action on the set of signatures $\mathbf{S}$ is defined.

Let $D \subset \mathbf{I}$ be some subset of indices. We denote by $G(D)$ a subgroup of the index renaming group that converts the set D into itself leaving indices from the complement of D in place.

The group of renaming of summation indices will be denoted G_d.

We assume that all relations between elementary symbols are reduced to some linear relations between symbols with the same signature, and the set of relations is invariant with respect to the index renaming group.

Let $\mathbf{R}(s)$ denote the space of all (linear) relations in the space $\mathbf{Sym}(s)$. We will require that the correspondence $s \to \mathbf{R}(s)$ be invariant with respect to the renaming of the indices.

Let $\mathbf{R}$ denote the subspace in § generated by all $\mathbf{R}(s)$ in the space $\mathbf{Sym}$. By construction, $\mathbf{R}$ is invariant with respect to renaming of indices and $\mathbf{R}(s) = \mathbf{Sym}(s) \cap \mathbf{R}$.

In the space $\mathbf{Sym}(s)$ there is a single sequence $\mathbf{b}(s) = e_1, \ldots, e_k$ of elementary symbols forming the basis in the factor space $\mathbf{V}(s) = \mathbf{Sym}(s)/\mathbf{R}(s)$, which is minimal among such sequences in the sense of lexicographical order.

The space generated by the basis $\mathbf{b}(s)$ is naturally identified by the quotient $\mathbf{Sym}(s)/\mathbf{R}(s)$ and will be denoted $\mathbf{V}(s)$.

We will call $\mathbf{b}(s)$ the **canonical basis**, and elements of the canonical basis will be called **canonical symbols**.

Example. Let $s = (Ri, (1, 2, 3, 4))$ be the signature of the Riemann tensor. Then the canonical basis in $\mathbf{V}(s)$ consists of two symbols: $Ri(1, 2, 3, 4)$ and $Ri(1, 3, 2, 4)$.

Since the space of relations $\mathbf{R}(s)$ is invariant with respect to the group $\mathbf{G}(s)$ of renaming the indices entering s, this group acts in $\mathbf{V}(s) = \mathbf{Sym}(s)/\mathbf{R}(s)$. Note that with this action, the elementary symbol turns, generally speaking, into a linear combination of canonical symbols.

Monomial is the formal commutative product of elementary symbols. We will require that the summation indices and only they be in the set $\mathbf{D}$.

Let $\mathbf{M}$ denote the set of all monomials, and $\mathbf{P}$ be a vector space with basis $\mathbf{M}$. Elements of the space $\mathbf{P}$ are naturally called **polynomials**.

The Signature $\mathbf{s}(m)$ of a monomial $m = e_1 \cdot \ldots \cdot e_k$ is the sequence of signatures of its factors, ordered in ascending order.

The set of all possible signatures of monomials will be denoted by $\mathbf{S_m}$.

Let S be some set of monomial signatures. Let $\mathbf{M}(S)$ denote the set of all monomials whose signatures lie in S and $\mathbf{P}(S)$ be a vector space with basis $\mathbf{M}(S)$.

Let $m = e_1 \cdot \ldots \cdot e_k$ be a monomial. Denote by $\mathbf{S}(m)$ the set of signatures of all monomials obtained from $\mathbf{s}(m)$ by renaming the summation indices, that is, the orbit of $\mathbf{s}(m)$ with respect to the action of the group $\mathbf{G}_d$.

There are three types of linear relations between monomials.

(i) Relations generated by relationships within one of the factors, with the other factors fixed.

(ii) Equality of monomials obtained by permuting the factors

(iii) Equality of monomials obtained by renaming summation indices (that is, by the action of the group$\mathbf{G}_d$).

Denote by $\mathbf{P}(m)$ the vector space generated by all monomials whose signatures lie in $\mathbf{S}(m)$, and $\mathbf{R}(m) \subset \mathbf{P}(m)$ is the subspace generated by relations of the types listed. The subspace generated by relations of type (i) and (ii) will be denoted by $\mathbf{R}_0(m)$.

The spaces $\mathbf{P}(m)$,$\mathbf{R}(m)$ and $\mathbf{R}_0(m)$ are invariant with respect to the action of the group $\mathbf{G}_d$. In particular, this group acts on the factor spaces $\mathbf{P}(m)/\mathbf{R}_0(m)$ and $\mathbf{P}(m)/\mathbf{R}(m)$.

IOP Conf. Series: Journal of Physics: Conf. Series **1163** (2019) 012052 doi:10.1088/1742-6596/1163/1/012052

We call a monomial **reduced** if all its factors are canonical symbols and the sequence of factors is ordered in ascending order.

The space generated by all reduced monomials in $\mathbf{P}(m)$ will be denoted by $\mathbf{P}_r(m)$.

Any monomial in the space $\mathbf{P}(m)$ can be represented up to a relation in $\mathbf{R}_0(m)$ as a linear combination of reduced monomials. The space $\mathbf{P}_r(m)$ is naturally identified with the factor-spaces $\mathbf{P}(m)/\mathbf{R}_0(m)$.

We denote by $\rho : \mathbf{P}(m) \to \mathbf{P}_r(m)$ the natural projector onto the factor space $\mathbf{P}(m)/\mathbf{R}_0(m)$.

Note that the image $\rho(m)$ of the monomial m is generally a polynomial.

5. Reducing monomials to the canonical form

The canonical representation is a linear operator $\kappa : \mathbf{P} \to \mathbf{P}$ such that $x = \kappa(x)\ mod(\mathbf{R})$ and $\kappa(x) = \kappa(y)$ if and only if $x = y\ mod(\mathbf{R})$ for $x, y \in \mathbf{P}$.

We define the operator $\alpha : \mathbf{P} \to \mathbf{P}$, assuming for arbitrary $p \in \mathbf{P}$ that

$$\alpha(p) = \frac{1}{N} \sum_{g \in \mathbf{G}_d} g(p) \tag{4}$$

where N is the number of elements in the group G_d.

The operator $\alpha\rho = \rho\alpha$ is the canonical representation.

Unfortunately, this canonical representation is too cumbersome for real computations. Therefore, we modify its structure.

The set $\mathbf{S_m}$ of the signatures of all monomials is a subset of the set of finite sequences of signatures of elementary symbols linearly ordered with respect to the lexicographic order. Accordingly, the set $\mathbf{S_m}$ is linearly ordered with respect to the inherited order.

The group $\mathbf{G}_d$ of renaming of summation indices acts naturally on the set of monomials and on the set of their signatures.

The minimal element in the orbit of the signature $\mathbf{s}(m)$ of the monomial m with respect to the action of the group $\mathbf{G}_d$ will be denoted by $\mathbf{s}_{min}(m)$.

The signature of the monomial m will be called **extremal** one if it cannot be reduced by any renaming of the summation indices, that is, it coincides with $\mathbf{s}_{min}(m)$.

A monomial will be called **extremal** if its signature is the extremal one.

The space generated by all extremal reduced monomials in $\mathbf{P}$ will be denoted by $\mathbf{P}_{min}$. It coincides with the subspace in $\mathbf{P}_{min}(m)$ generated by all monomials with signature $\mathbf{s}_{min}(m)$.

Any monomial $m \in \mathbf{P}_{min}$ can be represented up to a relation in $\mathbf{R}$ as a linear combination of extremal reduced monomials with the signature $\mathbf{s}_{min}(m)$.

Let the set of monomials $M \in \mathbf{P}_{min}$ contain one monomial from $\mathbf{P}_{min}$ for each signature. The space $\mathbf{P}_{min}$ is a direct sum

$$\bigoplus_{m \in M} \mathbf{P}_{min}(m)$$

and

$$\mathbf{R} \cap \mathbf{P}_{min} = \bigoplus_{m \in M} \mathbf{R} \cap \mathbf{P}_{min}(m).$$

For the monomial $m \in \mathbf{P}$ we set $\mathbf{r}(m) = \{x \in \mathbf{P}_{min}(m) : x = m\ mod(\mathbf{R})\}$. Thus, $\mathbf{r}(m)$ is the set of all polynomials from $\mathbf{P}_{min}(m)$ that coincide with m modulo relations from $\mathbf{R}$.

Denote by $\mathbf{k}(m)$ the arithmetic mean of the elements from the set $\mathbf{r}(m)$.

Obviously, the function $\mathbf{k}(m)$ extends to a linear operator on the space $\mathbf{P}$, which we will denote by the same symbol $\mathbf{k} : \mathbf{P} \to \mathbf{P}_{min}$.

The operator $\mathbf{k}$ is a canonical representation.

Note that the dimension of the space $\mathbf{P}_{min}$ is significantly less than the dimension of the space $\mathbf{P}$, and the signature stabilizer $\mathbf{s}_{min}(m)$ in the group $\mathbf{G}_d$ is much smaller than the group itself.

5.1. The algorithm

We assume that at each step of the simplifications: (1) all elementary tensors are reduced to the canonical form; (2) there are no contraction of indices into an elementary tensor. We also use lexicographically order in the tensors.

We will map each monomial to the colored graph. The vertices of such graphs correspond to elementary symbols that are factors of monomials, the colors of the vertices being the signatures of the corresponding factors from which the summation indices are excluded, and the edges correspond to the summation indices.

If the monomials are mapped into two non-isomorphic graphs, then the monomials lie in non-intersecting spaces modulo the relations (independent).

5.2. The algorithm(sketch)

(i) Transform each tensor term to the canonical form with respect to its symmetry properties. Every step should be finished with the canonization of tensor itself and ordering the tensors in the monomials (so called pre-canonical form).

(ii) Apply the allowed transformation of summation indices to the monomial.

(iii) Calculate the average over all the monomials which have isomorphic graphs with the initial one.

(iv) The obtained polynomial is invariant with respect to the transformations from the equivalence class of the initial monomial.

Using the obtained invariants we can get the canonical form of the monomials.

Conclusion

The proposed algorithm significantly reduce the amount of computations in the case of large groups.

We have implemented a prototype of the program on Python. The program is working rather good for tensor expressions in practical cases where monomials contain about 10–15 factors and each factor contains up to 8–10 indices. The next step is a further program code refinement and optimization.

Our nearest plan is to finish the development of the program and to carry out a detailed investigation of the algorithm as well as to compare it with other algorithms for tensor simplifications available on the market.

Acknowledgments

The authors are grateful to V. Ilyin and A.Demichev for his interest in the work and useful discussions. We are also grateful to J. Dubenskaya for help of preparing of the paper.

References

[1] Martin-Garcia J M, Yllanesa D and Portugal R 2008 *Comp. Phys. Comm.* **179** 597
[2] Anderson I M and Torre C G 2012 *J. Math. Phys.* **53** 013511
[3] GRTensorII URL:http://grtensor.phy.queensu.ca/
[4] Atlas 2 URL:http://digi-area.com/Maple/atlas/
[5] Toth V 2008 URL:http://www.vttoth.com/
[6] xTerior URL:http://www.xact.es/xTerior/index.html
[7] MAXIMA URL:http://maxima.sourceforge.net/
[8] Rodionov A Y and Taranov A Y 1987 *Lecture Notes in Computer Science* **378** 192–201
[9] Butler G 1991 *Fundamental Algorithms for Permutation Groups* (Springer-Verlag, Berlin)
[10] Manssur R U and Portugal R 2004 *Comp. Phys. Comm.* **157** 173
[11] Ilyin V A and Kryukov A R 1996 *Comp. Phys. Comm.* **96** 36–52
[12] Hearn A C and Schpf R *REDUCE Users Manual* URL:https://reduce-algebra.sourceforge.io/manual/manual.html

IOP Conf. Series: Journal of Physics: Conf. Series **1163** (2019) 012053 doi:10.1088/1742-6596/1163/1/012053

Computer modeling and simulation of electrically charged nano/micro-particles interaction into aqueous solution by the Brownian dissipative dynamics of their atomic counterions ensemble

P. V. Lebedev-Stepanov[1,2]

[1]Photochemistry Center of FSRC Crystallography & Photonics RAS Novatorov St., 7a/1, Moscow, 119421, Russia
[2]National Research Nuclear University MEPhI (Moscow Engineering Physics Institute), 31 Kashirskoe shosse, 115409, Moscow, Russia

E-mail: petrls@yandex.ru

Abstract. On the basis of free energy functional of charged ions and equation of diffusion of ions in self-consistent electric potential, a method of Brownian dissipative dynamics of ions near electrically charged colloidal micro/nano-particles in aqueous solution or dusty plasma particles in the air is proposed for investigation of ionic atmosphere structure near charged surfaces, and electric interparticle force is calculated.

1. Introduction

Formation of ionic crystals (such as NaCl), colloidal crystals self-assembly, and dusty plasma particles' (macroions) ordered structures indicates the existence of an energy minimum for ion-counterion system, and hence the attraction of similarly charged particles mediated by their counterions. Some experiments show that interacting electrically charged colloidal particles with diameters of 10-1000 nm in aqueous solution with small compensative counter-ions demonstrate the long range attraction and form the colloidal crystals if the concentration of additional electrolyte (salt) is low enough. The dusty plasma particles (macroions) demonstrate the analogous behavior and form the ordered structures in ionized air.

The first part of researchers argues that the attraction describes by only dispersion forces. Another part has suggested the non-DLVO (Derjaguin-Landau-Verwey-Overbeek) approaches with electrical long range attraction obtained from Poisson-Boltzmann equation (PBE) solution [1] or from novel free energy of ions consideration [2-5].

A simple model of salt-free electrolyte with a strong asymmetry of charge and size of the constituent particles, macroions and counterions, is in the area of fundamental interests of many modern theoretical and experimental researches. For example, work [6] reports a novel phenomenon of a surface-induced phase transition in salt-free solutions of charged colloids. Experimental investigation the properties and applications of aqueous charge-stabilized colloidal solution without side electrolyte was reported in [7].

PBE solution shows that the attraction is possible if the counterions are concentrated mostly between the interacting colloidal particles, and electric interparticle force can be calculated analytically. Two flat parallel plates, having the same sign of surface charge density, electrically attract each other if the space lying between plates is filled with the neutralizing counterions. The attractive force acting to the unit of surface area is $f = -\sigma^2(2\varepsilon\varepsilon_0)^{-1}$, where σ is surface charge density distributed on the plate; ε and ε_0 are the dielectric constants of continuous media around the plates and of vacuum respectively. On the contrary, an electric repulsion takes place always if these counterions have filled both the space between the plates and all space outside the plates. In analogous case, analytical solution of PBE does not exist for charged cylinders or spheres.

Currently there are many interesting works dedicated to derivation the ordinary PBE and its modifications from statistical field theory. The PBE is a mean-field approximation of statistical ensemble of point-like ions. The modifying PBE in the context of dense ionic liquids where steric effects become important was studied in [8]. Authors of that work elucidate how equivalent convex free energy functions can be constructed to describe steric effects in a manner, which is more convenient for numerical minimisation. In work [5], a non-linear integro-differential equation was obtained with respect to the mean-field electrostatic potential, generalizing the Poisson-Boltzmann-Langevin equation for the point-like dipoles obtained first in [9].

We guess, it is clear that self-consistent insertion the PBE into framework of the classic electromagnetic theory of field could be useful due to opportunity to develop well-justified further modifications of the PBE with taking into account the features, which did not included in ordinary PBE. In this work, it is shown that ordinary PBE can be obtained in form of Lagrange equation by standard Lagrange's procedure of the field theory ([10], p.77). The Lagrangian of equilibrium system of charged particles and counterions with Maxwell-Boltzmann energy distribution of counterions is considered. This Lagrangian was derived from the chemical potential of the counterion including the electrostatic term (Coulomb interaction of the given counterion charge with all other charges of the system) and entropy term which corresponds to an ideal gas (or ideal solute). In such a model, steric effects between neighboring counterions, which are treated as point-like particles, do not take into account. The medium is modeled by a homogeneous and isotropic dielectric constant.

Here is shown, that ordinary PBE corresponds to mean-field of the equilibrium state of Brownian ensemble of small counterions in the self-consistent electric field of these counterion and macroions (micro/nano-particles).

In our model, the steric effects can be taken into account by introducing of the finite radius of counterion. On the basis of free energy functional of charged ions and equation of diffusion of ions in self-consistent electric potential, a method of Brownian dissipative dynamics of ions near electrically charged colloidal micro/nano-particles in aqueous solution or dusty plasma particles in the air is proposed for investigation of ionic atmosphere structure near charged surfaces, and electric interparticle force is calculated.

2. Model

Let us consider dilute aqueous solution of colloidal particles (R denotes the radius of particle) with fixed uniformly distributed surface electric charge $Q= Ze$, where e is elementary charge, $Z >>1$. The total charge of solution is zero due to the presence of small counterions with the charge of $q=-ze$, where z is small natural number; then $Q= -Nq$, where $N = Zz^{-1}$. Let a denotes the radius of counterion; $a <<R$, so that one can suppose that the Brownian motion of the colloidal particles is negligibly small, but each counterion moves in self-consistent electric field of particles and other counterons under action of Brownian force. In frameworks of such adiabatic approximation, one can assume that counterions are founding the quasi-stationary atmosphere around motionless colloidal particle (figure 1).

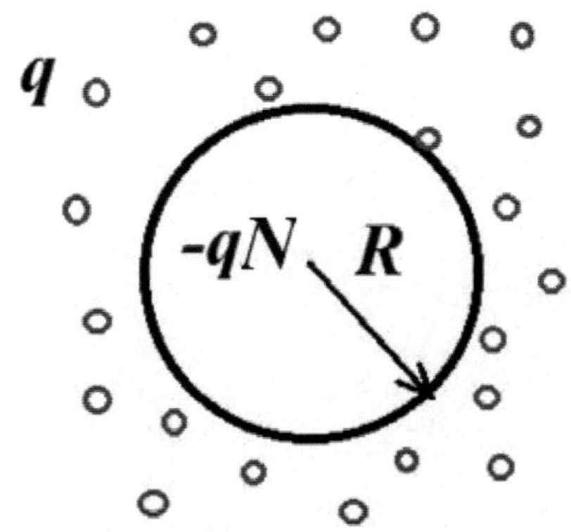

Figure 1. The big ion (colloidal particle) and its counterions atmosphere.

The methods determining the average concentration of counterions near particle can be divided into two categories: kinetic, based on equation of counterions diffusion in self-consistent electric potential of all counterions and the big motionless colloidal particle; dynamical, based on Lagrange equations of motion obtained from the Lagrange's variation principle. Counterionic current density can be written as

$$\mathbf{j} = -D\nabla n - nb\nabla U\ , \tag{1}$$

where D is diffusion coefficient, n is counterions concentration, U is potential energy of self-consistent electric field, $b = \frac{D}{kT}$ is counterion mobility describing by Einstein equation (k is Boltzmann constant, T is temperature). Equation of continuity of current (1) is

$$\frac{\partial n}{\partial t} = \nabla\left[nD(\nabla \ln n + (kT)^{-1}\nabla U)\right]. \tag{2}$$

Chemical potential of counterion is given by

$$\mu = kT\ln\left(\frac{n}{n_0}\right) + q\varphi_0 + \frac{1}{2}q\varphi_i\,. \tag{3}$$

where φ_0 is Coulomb potential of spherical particle (figure 1), φ_i is self-consistent electric potential of counterions atmosphere. Poisson equation gives

$$\nabla^2\varphi_i = -\frac{qn}{\varepsilon\varepsilon_0}, \tag{4}$$

where ε is dielectric permittivity of solution.

To determine the potential energy U which corresponds to Eq. (3), we have to obtain the Lagrange's equations for next Lagrangian

$$L = n\mu = n\left(kT\ln\left(\frac{n}{n_0}\right) + q\varphi_0 + \frac{1}{2}q\varphi_i\right) \tag{5}$$

The variational procedure, describing by equation

$$\delta\int_V L dV = \delta\int_V \left(nkT\ln\left(\frac{n}{n_0}\right) + \rho\varphi_0 + \frac{1}{2}\rho\varphi_i\right)dV = 0\,, \tag{6}$$

minimizes the chemical potential of counterions system in the external electric field of colloidal particle. With account of electric field strength determination according by equations

$$\mathbf{E}_i = -\nabla\varphi_i, \quad \mathbf{E}_0 = -\nabla\varphi_0, \tag{7}$$

after integration by parts, integral in Eq. (6) can be rewritten as

$$\delta\int_V \left(\frac{\varepsilon\varepsilon_0 kT}{q} \nabla\mathbf{E}_i \ln\left(\frac{\varepsilon\varepsilon_0}{qn_0} \nabla\mathbf{E}_i \right) + \varepsilon_0\varepsilon\mathbf{E}_i\mathbf{E}_0 + \frac{1}{2}\varepsilon_0\varepsilon E_i^{\,2} \right) dV = 0, \tag{8}$$

where we substituted concentration of counterions from Poisson equation, $\nabla\mathbf{E}_i = \dfrac{qn}{\varepsilon\varepsilon_0}$, by formula

$$n = \frac{\varepsilon\varepsilon_0}{q}\nabla\mathbf{E}_i \tag{9}$$

Lagrange equations for variational principle given by Eq. (8) has a form

$$\frac{\partial\hat{L}}{\partial E_{i\alpha}} = \frac{\partial}{\partial x_\beta}\left(\frac{\partial\hat{L}}{\partial\left(\dfrac{\partial E_{i\alpha}}{\partial x_\beta} \right)} \right), \tag{10}$$

where modified Lagrangian, $\hat{L}$, is integrand of Eq. (8); α and β can independently take the natural values 1,2,3, which denote the Cartesian space projections of vector. One can easily show that solution of Eq. (10) determines by expression

$$n = n_0 \exp\left(-\frac{q(\varphi_0 + \varphi_i)}{kT} \right). \tag{11}$$

On the other hand, the equilibrium state requires the condition $\mathbf{j} = 0$ in Eq. (1), which gives the analogous expression

$$n = n_0 \exp\left(-\frac{U}{kT} \right), \tag{12}$$

whence

$$U = q\varphi_0 + q\varphi_i. \tag{13}$$

Substituting (10) in Poisson equation (3), one can obtain the Poisson-Boltzmann equation

$$\nabla^2\varphi_i = -\frac{\rho_0}{\varepsilon\varepsilon_0}\exp\left(-\frac{q(\varphi_0 + \varphi_i)}{kT} \right). \tag{14}$$

Thus, it was shown that PBE can be considered as equilibrium form of equation of charged particles diffusion in their self-consistent electric field. The general case of non-equilibrium ensemble of counterions evolution is describing by Eq. (2). The Langevin equation which corresponds to diffusion equation (2) can be written as

$$m\frac{d\mathbf{v}}{dt} = -\frac{mkT}{D}\mathbf{v} - q\nabla(\varphi_0 + \varphi_i) + \mathbf{F}_{ne} + \mathbf{F}_B, \tag{15}$$

where m is an effective mass of given counterion; $\mathbf{F}_{ne}$ is the sum of all non-electrostatic forces including the rigid wall interacting potential between particle and counterion; $\mathbf{F}_B$ is the

Brownian force which describes by Markov's random process with zero mean, and it satisfies the equation

$$< F_{B\alpha}(t), F_{B\beta}(t') >= 2kT\delta(t-t')\delta_{\alpha\beta}, \tag{16}$$

where α and β designate the Cartesian projections symbols, and the brackets denote the average over an equilibrium ensemble.

3. Calculation results

Geometry of model system is presented in figure 2 a. In the numerical experiments we used the next values of parameters: H = 1000 nm, R=500 nm, 0.5 R +L= 1000 nm, counterion radius is 0.5 nm ; counterion charge is -2 e , total number of counterions is 4000, dielectric permittivity of solvent is 81 (aqueous solution) with temperature of 295 K. Visualization (by the program VMD 1.8) of counterions ensemble near interacting particles obtained in numerical experiment is given in figure 2 b).

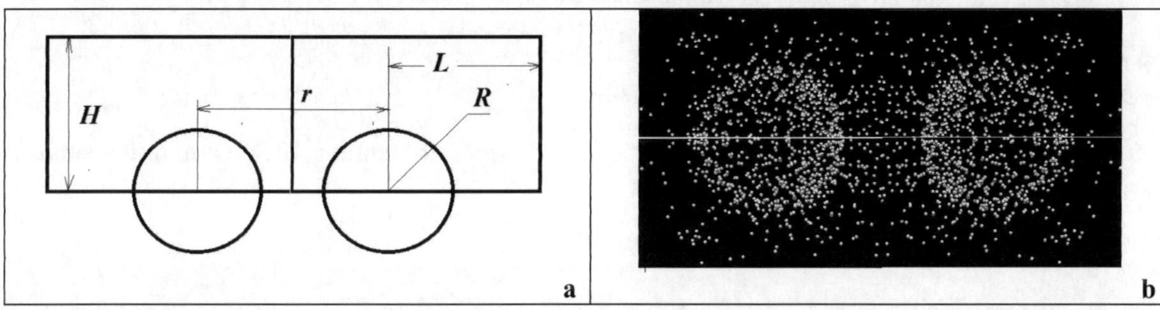

Figure 2. a) Geometry of space region occupied by both interacting particles and their counterions. b) Visualization of counterions ensemble near interacting particles.

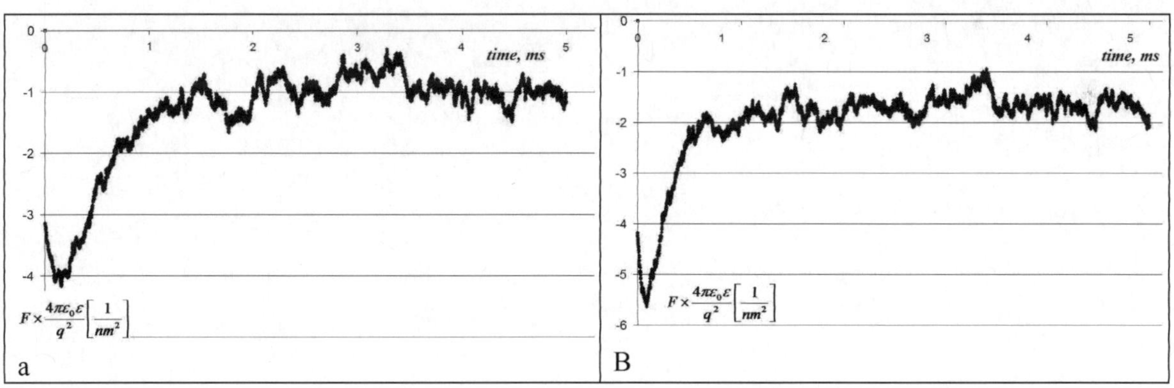

Figure 3. Electrostatic force acting to the particle if interparticle distance is 2400 μm (a) and 1400 μm (b).

The value of interaction force between two particles with account of counterions 'atmosphere' is shown in figure 3 for two distances between the centers of the particles. The negative sign of the force means the effective attraction between the particles due to counterions screening of their initial Coulomb repulsion. Our result confirms the conclusions of the work [1]. The attraction of equally charged spheres in a cloud of their counterions is possible.

In this paper, we have tested a method for calculating the interaction of colloids using the Brownian dynamics of counterions. Obviously, it was shown that this approach can be applied to the solution of the problem. However, obtained result has a preliminary character, and it has to be

IOP Conf. Series: Journal of Physics: Conf. Series **1163** (2019) 012053 doi:10.1088/1742-6596/1163/1/012053

rechecked and corrected during more detail calculations including total free energy surface drawing, entropy and adsorption terms estimations. We are supposing to do that in our next works.

Acknowledgements

This work was partially supported by the Ministry of Science and Higher Education within the State assignment FSRC «Crystallography and Photonics» and by the "Improving of the competitiveness" program of the National Research Nuclear University MEPhI (Moscow Engineering Physics Institute).

References

[1] Gundienkov V.A., Yakovlenko S.I. *Technical Physics Letters* 2002 **28**(11), pp 919-922
[2] Sogami I., Ise N. *J. Chem. Phys*. 1984 **81**(12), pp. 6320-6332
[3] Tokuyama M. *Phys. Rev. E*. 1999 **59**(3), pp 2550-2553
[4] Terada Y., Tokuyama M. *Physica A* 2004 **334**, pp. 327-334.
[5] Budkov Yu.A. *J. Phys.: Condens. Matter* 2018 **30,** p. 344001
[6] Budkov Yu.A., Frolov A.I., Kiselev M.G., Brilliantov N.V. *J. Chem. Phys*. 2013 **139**, p.194901.
[7] Yakimansky A.V., Menshikova A.Yu., Shevchenko N.N. et al. *Polym. Adv. Technol*. 2009 **20** (6), pp. 581-588.
[8] Maggs A.C., Podgornik R. *Soft Matter*. 2015 **12**, pp 1219-1229.
[9] Abrashkin A., Andelman D., Orland H. *Phys. Rev. Lett*. 2007 **99,** p. 077801.
[10] Landau L.D., Lifshitz E.M. *The classical theory of field*. Third edition. 1971. Oxford. Pergamon Press Ltd. 386 p.

IOP Conf. Series: Journal of Physics: Conf. Series **1163** (2019) 012054 doi:10.1088/1742-6596/1163/1/012054

WIND WAVES MODELING UNDER HURRICANE WIND CONDITIONS

A Kuznetsova, G Baydakov, A Dosaev, D Sergeev, Yu Troitskaya

[1] Institute of Applied Physics of the Russian Academy of Sciences, 46 Ulyanov st, 603950, Nizhny Novgorod, Russian Federation

E-mail: alexandra@ipfran.ru

Abstract. The CFSv2 reanalysis wind field for the real-case condition of the Irma hurricane that took place in the 31/08/2017 - 13/09/2017 was used as the wind forcing for WAVEWATCH III model. Wave parameters under Irma hurricane were calculated and compared with NDBC buoys data. The spectra of surface waves created by the hurricane allowed the evaluating the air-sea exchange processes parameters at hurricane wind.

1. Introduction

Tropical cyclones, or hurricane phenomena directly affect the lives of people in coastal regions and can cause destruction [1]. Nowadays the average storm intensity is increasing due to climate change [2]. Hurricanes are followed by strong winds that create large waves that can flood coastal areas and cause shoreline erosion. At the same time, high quality weather and sea state forecasts will make it possible to secure the population and minimize losses from such natural phenomena. Performing numerical simulations, validating it and improving physical parameterizations in forecast models are essential for correct hurricane prediction. Forecasting hurricane requires accurate prediction of hurricane-induced ocean surface waves.

The wave modelling may be implemented in the framework of modern third generation wave model WAVEWATCH III [3]. It has various packages of physical processes parameterizations, among them there are parameterizations that are able to reproduce correctly waves in storm conditions [4, 5]. The resulting values of mean wave and wind parameters allowed the evaluating of the air-sea exchange processes parameters at hurricane wind.

2. Simulation

In this paper, we focus on numerical simulation of a hurricane Irma that took place in 30/08/2017 - 12/09/2017 in the North Atlantic Ocean. Hurricane Irma was an extremely powerful and catastrophic Cape Verde hurricane. Irma developed from a tropical wave near Cape Verde on August 30, then it rapidly intensified into a Category 3 hurricane on the Saffir–Simpson wind scale on August 31. The intensity fluctuated on the next days, but on September 4, Irma resumed intensifying, becoming a Category 5 hurricane on the next day. The best track of the hurricane is made by USA National Hurricane Center and is shown on the Fig.1 taken from the tropical cyclone report [6].

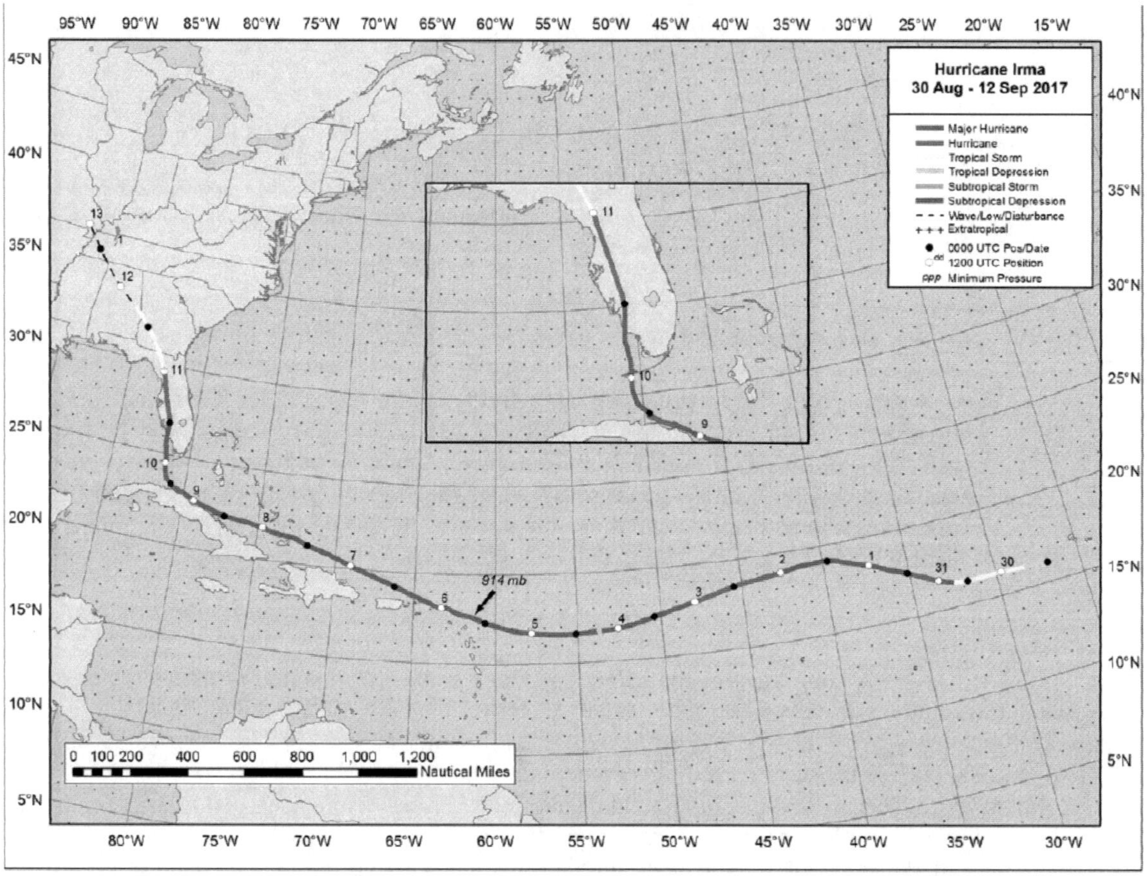

Figure 1. Best track positions for Hurricane Irma, 30 August–12 September 2017 taken from [6].

The numerical simulations of waves under hurricane Irma conditions are performed within WAVEWATCH III v.5.16 model (WW3). The modelling is performed for the period of 01/09/2017 – 04/09/2017. The area with longitudes varying from 14 up to 24 degrees and latitudes varying from 280 up to 330 degrees is considered. The ETOPO1 data is used to create bathymetry and mask files. The JONSWAP initial conditions are used to start the model. The subsequent development of the waves is forced by the wind field from the CFSv2 reanalysis with $0{,}205^0$ resolution. The simulations are carried out at the IAP RAS cluster using pure MPI option of the model code.

Two parameterizations of wind input (ST4 - Ardhuin et al [7] and ST6 - Zieger et al [8]) are tested. The resulting significant wave height distribution is obtained for two types of wind input parameterizations: ST4 and ST6. The model reproduces the rotating and motion of the hurricane. On the Fig.2 the typical phases of a hurricane development and the increase of its intensification are shown. The asymmetry of distribution of the surface wave field is obtained. ST6 parametrization gives values higher than ST4 on approximately 25%.

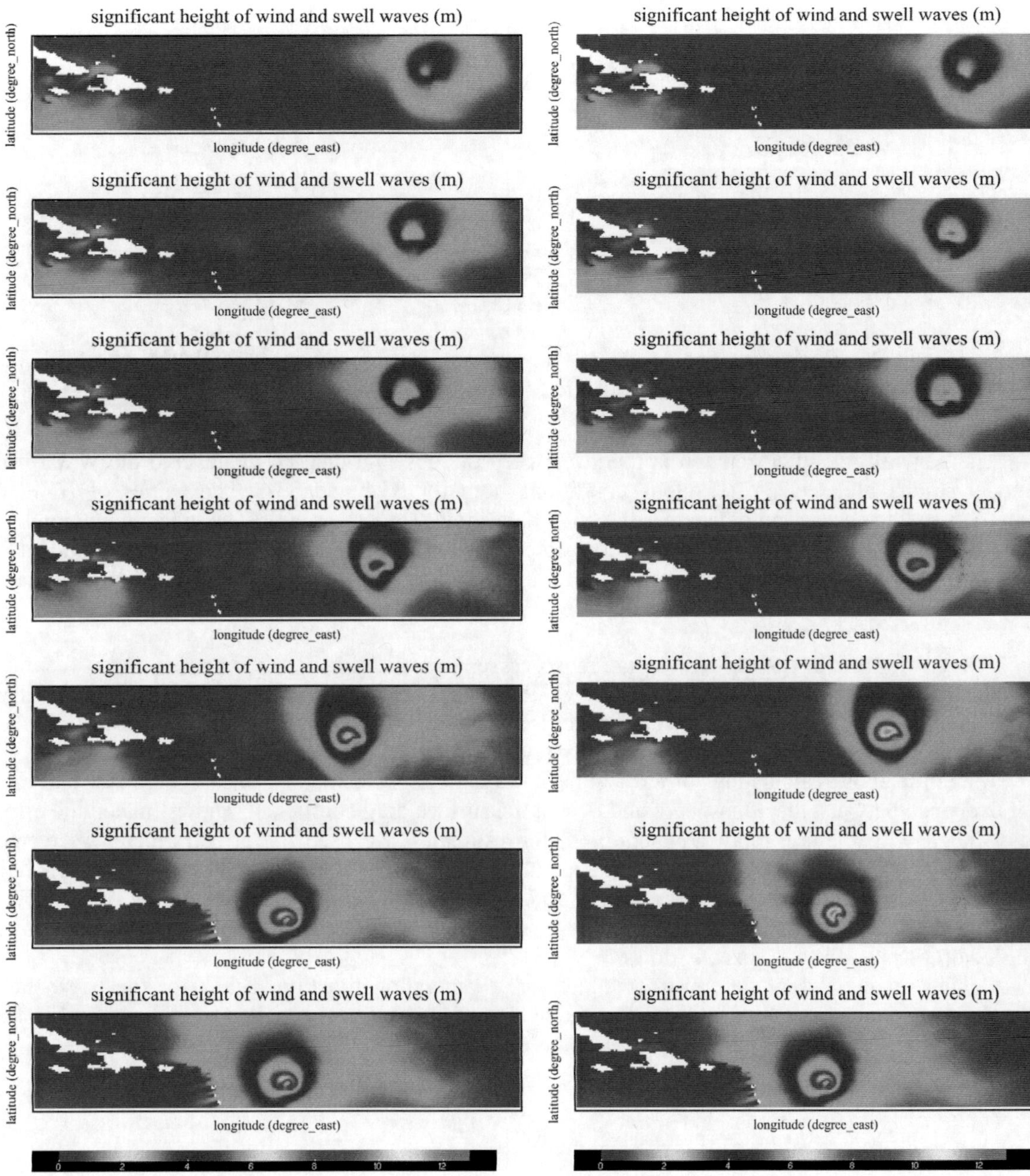

Figure 2. The evolution of significant wave height distribution under hurricane Irma conditions simulated in WW3 with ST4 wind input (left) and ST6 wind input (right) from 02/09/2017 to 04/09/2017 at the fixed time.

The calculated significant wave height of surface waves induced by the hurricane is compared with the NDBC buoys data. We examine significant wave height (H_s) at three NDBC buoys. The buoy 41044 is located at (301.26, 21.71); the buoy 41040 is located at (307.28, 14.45); the buoy 41060 is located at (308.983, 14.83). The NDBC buoys have a frequency range from 0.02 to 0.485 Hz. The wave instrumentation and data acquisition of the NDBC buoys are described in [9].

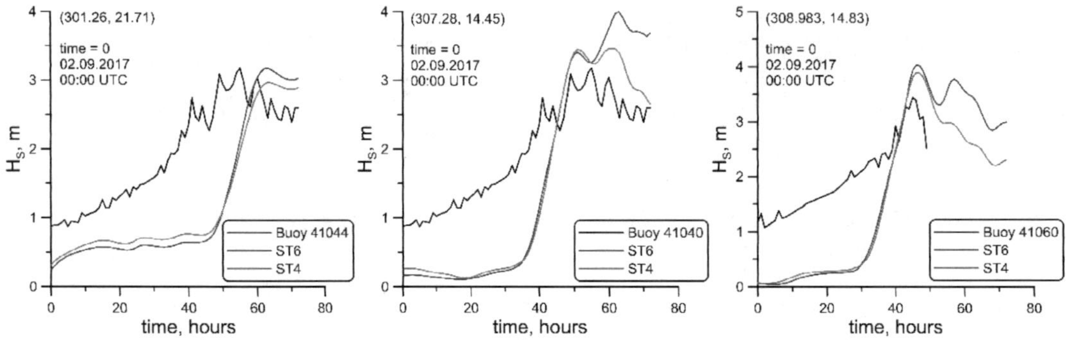

(a) (b) (c)

Figure 3. Significant wave height at three buoys location simulated in WW3 with ST4 (red), ST6 (blue), and buoy data (black) for the period from 02/09/2017 to 04/09/2017. (a) – buoy 41044, (b) – buoy 41040, (c) – buoy 41060.

The generally good correlation is found between the observational H_s values and the WW3 wave model results under tropical cyclone conditions both for ST4 and ST6. The values of H_s in the simulations increased rapidly whereas the buoy data shows smooth growth. The underestimation of H_s for the moderate wind speed in the first hours of the calculation may be due to the idealized JONSWAP initial conditions. To include the waves coming from the outside of the considered area, the run for the nested domains may be applied. It is the topic for the further investigations.

3. Estimation of the air-sea exchange processes parameters

In this last concluding part of the paper we demonstrate the skill of the coupled model to estimate the parameters of the air-sea exchange processes for hurricane wind. These novel results are very preliminary and will be extended and possibly revised in future.

Using the above calculations of the wind and wave field, we estimated one of the major parameter of exchange between the atmosphere and ocean, the surface drag coefficient, and estimated the effects associated with the sea spray production. In the estimates, we used the effect which was recently identified as the dominant spray production mechanism at high winds, the "bag-breakup" fragmentation (see [10-12]). This is inflating and consequent blowing of short-lived, sail-like pieces of the water-surface film, "bags". The process is similar to "bag-breakup" mode of fragmentation of liquid droplets and jets in gaseous flows.

Basing on these data we constructed the spray generation function (SGF) for the bag-breakup mechanism. And then we suggested the fetch dependent SGF valid both for laboratory and field conditions using the scaling by the dimensionless wind-sea Reynolds number (Re) (this term was suggested later in [13]):

$$\mathrm{Re}_B = {u_*^2} \big/ {\omega_p \nu} . \tag{1}$$

Where u_* is the friction velocity, ω_p is the peak frequency in the spectrum of surface wind waves, and ν is the kinematic viscosity of the air. The best fit of the experimental data has shown that the specific number of "bags" can be fitted by the following equation (see Fig.4(a))

$$\langle N \rangle = M_0\, Re_B^{2/3}\, exp\left(-\frac{M_1}{Re_B^{2/3}} \right), \tag{2}$$

where M_0=2.08, M_1=972.

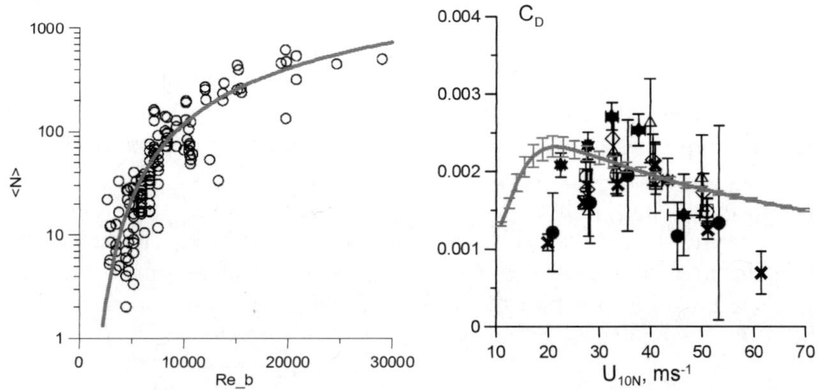

(a) (b)

Figure 4. (a) The specific number of "bags" versus the wind-sea Reynolds number. (b) The sea-surface drag coefficient versus the surface wind speed. Red line - estimations within the model from [14, 15] for the wave age parameter Ω=3, symbols are available experimental data [16, 17, 18, 19].

With the use of the statistics of the spray-generation events, the contribution of the spray to the air-sea exchange processes can be estimated (see details in [14, 15]). In particular, it enables one to explain the unusual peaking dependence of the sea-surface drag coefficient on the surface wind speed (see Fig.4b)

The curve in Fig. 4 (b) was obtained for the assumption of a certain wave-age parameter, defined as the ratio of the surface wind speed U_{10}, to the phase velocity of the waves corresponding to the peak in the frequency spectrum of the wind surface waves c_p, $\Omega = U_{10}/c_p$. Here we estimated the surface drag coefficient for the parameters of the waves calculated within the WW3 model for the major hurricane Irma. Fig. 5 shows the examples of the maps of the parameters controlling the spray production and associated exchange parameters. Fig.5a shows the surface wind velocity field and the wind direction in the hurricane Irma according to the reanalysis CFSv2. Fig 5b demonstrates the spectral peak frequency map calculated with the use of the WW3 model. Fig. 5c shows the map of the surface drag coefficient estimated for the parameters of Figs 5 a, b taking into account the contribution of sea spray. One can see the reduction of the surface drag coefficient in the area of the strong winds. The pronounced asymmetry of the surface drag distribution with respect to the direction of motion of the storm center reflects the asymmetry of distribution of the surface wind and wave field.

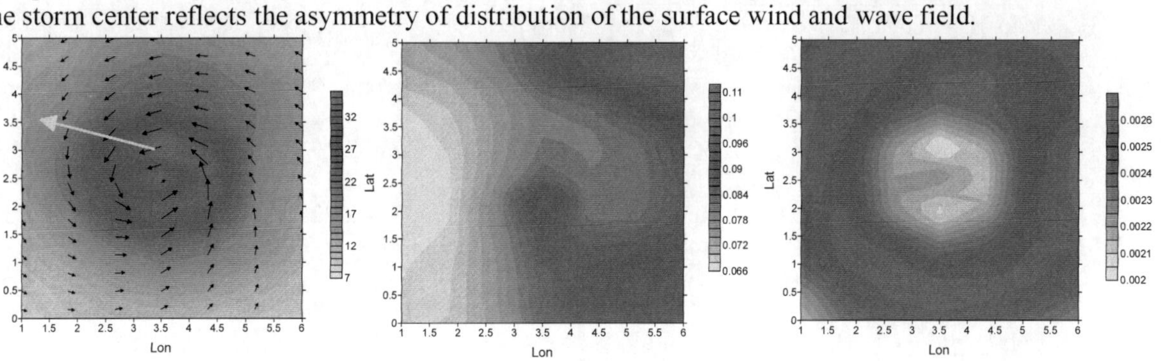

(a) (b) (c)

Figure 5. The maps of wind field (a), the spectral peak frequency (b) and the surface drag coefficient (c). In the panel (a) the black arrows show the wind direction, the grin large arrow sows the direction of the storm motion, the color scales are in m/s. In the panel (b) the color scale is in Hz. In the panel (c) the color scale is dimensionless. the horizontal axis is the relative longitude, the vertical axes is the relative latitude, both are in degrees.

4. Conclusion

In this paper, we use WW3 to simulate wave parameters under hurricane Irma wind field. The model reproduces the rotating and motion of the hurricane. The asymmetry of distribution of the surface wave field is obtained. The calculated significant wave height of surface waves induced by the hurricane is compared with the NDBC buoys data. The generally good correlation is found between the observational H_s values and WW3 wave model results under tropical cyclone conditions both for ST4 and ST6 parametrizations. With the use of the statistics of the spray-generation events, the contribution of the spray to the air-sea exchange processes is estimated. The reduction of the surface drag coefficient in the area of the strong winds is obtained. The asymmetry of the surface drag distribution with respect to the direction of motion of the storm center reflects the asymmetry of distribution of the surface wind and wave field.

References

[1] Sivakumar M.V. (2005) Impacts of Natural Disasters in Agriculture, Rangeland and Forestry: an Overview. // *Springer, Berlin, Heidelberg*
[2] Emmanuel K // Nature, 436 (7051) (2005), pp. 686-688
[3] Tolman H and WAVEWATCH III Development Group. User manual and system documentation of WAVEWATCH III version 4.18. //*Environmental Modeling Center, Marine Modeling and Analysis Branch*. 282 pp. + Appendices. 2014
[4] Port J, Hara T, Reichl B G, Ginis I, // American Geophysical Union, Ocean Sciences Meeting 2016, abstract #A54C-2735
[5] Liu Q, Babanin A, Fan Y, Zieger S, Guan C, Moon I-J, // Ocean Modeling, Vol 118, pp 73-93.
[6] https://www.nhc.noaa.gov/data/tcr/AL112017_Irma.pdf
[7] Ardhuin F, Chapron B and Collard F, // Geophys. Res. Lett., 2009a, 36.
[8] Babanin A V // Cambridge University Press, 2011, 480 pp
[9] Steele K E, Chi-Kin Lau, Hsu Y-H L // IEEE J of Oc. Eng., Vol OE-10, NO4, 1985.
[10] Troitskaya Y, Kandaurov A, Ermakova O, Kozlov D, Sergeev D, and Zilitinkevich S // Sci. Rep., 2017, 7, 1614.
[11] Troitskaya Y I, Ermakova O S, Kandaurov A A, Kozlov D S, Sergeev D A, and Zilitinkevich S S // Dokl. Earth Sc., 2017, 477, 1330
[12] Troitskaya Y, Druzhinin O, Kozlov D, and Zilitinkevich S //. J. Phys. Oceanogr., 2018, https://doi.org/10.1175/JPO-D-17-0104.1.
[13] Zhao D, Toba Y, Sugioka K, and Komori S // J. Geophys. Res., 2006, C02007, doi:10.1029/2005JC002960.
[14] Troitskaya Y I, Ermakova O S, Kandaurov A A, Kozlov D S, Sergeev D A, and Zilitinkevich S S // Dokl. Earth Sc., 2017, 477, 1373
[15] Troitskaya Y, Druzhinin O, Kozlov D, and Zilitinkevich S // J. Phys. Oceanogr., 2018, https://doi.org/10.1175/JPO-D-17-0105.1
[16] Powell M D, Vickery P J, and Reinhold T A // Nature, 2003, 422, 279–283.
[17] Holthuijsen L H, Powell M D, and Pietrzak J D // J. Geophys. Res. Ocean., 2012, 117, C09003, https://doi.org/10.1029/2012JC007983.
[18] Jarosz E, Mitchell D A, Wang D W, and Teague W J // Science, 2007, 315, 1707–1709, https://doi.org/10.1126/science.1136466.
[19] Richter D H, Bohac R, and Stern D P // J. Atmos. Sci., 2016, 73, 2665–2682, https://doi.org/10.1175/JAS-D-15-0331.1.

Acknowledgments

This work is supported by Russian Science Foundation (Agreement No. 14-17-00667). The work of G.Baydakov on the estimation of air-sea exchange processes parameters is partially supported by RFBR grant 18-05-60299.

IOP Conf. Series: Journal of Physics: Conf. Series **1163** (2019) 012055 doi:10.1088/1742-6596/1163/1/012055

Numerical simulations of complex nonequilibrium flows in finite regions on the basis of the Boltzmann kinetic equation

V.V. Aristov, S.A. Zabelok, A.A. Frolova

Dorodnicyn Computing Center, Federal Research Center of Computer Science and Control of Russian Academy of Sciences. Ul. Vavilova 40, 119333, Moscow, Russia.

aristovvl@yandex.ru

Abstract. A new formulation of the inner boundary problem for description complex nonequilibrium flows between membranes is presented. Solutions by means of the Boltzmann equation and the model kinetic equation for mixtures with chemical reactions demonstrate interesting physical properties in this open nonequilibrium system. Possibility of the anomalous thermal transport where the heat flux and the temperature gradient are of the same sign, is studied. Influence of boundary conditions on the parameters of the structure is considered as well.

Introduction

We study strong nonequilibrium gas flows in which effects of the anomalous heat transfer do not correspond to a continuum law. Note that some researches have investigated nonequilibrium flow regimes and considered questions concerning the correct transition from the microscopic description to the macroscopic description, see for example [1-3]. We use the nonlinear Boltzmann equation and the model kinetic equations. Direct methods for numerical solving these equations are applied [4] for studying new nonequilibrium flows in the boundary steady problems. For different problems with nonequilibrium boundary conditions (see [5, 6]) nonclassical transport in flows appears (this effect for the first time has been described in [7]). In particular, heat can be transferred from the region with the lesser temperature to the region with the greater temperature. Now we consider a problem with "membrane-like" boundary conditions. In the simplest case particles leaving the region under consideration do not collide with particles entering this region. For such a situation even equilibrium distributions for the boundary conditions can lead to nonclassical anomalous transport mentioned above due to complex interaction of the oppositely directed flows. In a more complex situation a part of gas can be reflected from the membrane molecules with the diffuse condition. This problem for a mixture of chemically reacting gases is also solved using the kinetic model equations [8,9] which we have previously applied in study of nonequilibrium structures in the nonuniform relaxation problem [10].

1. Formulation of the problems and governing equations

1.1. Flows of a simple gas in a closed region with membrane type boundary conditions

The stationary flow of a rarefied gas is investigated in 1D geometry that represents an interval with length L. At the ends of the interval, membrane-type conditions are set. In this case it is assumed that the counter flows of particles do not interact at the boundaries. In other words, the unchanged velocity distribution functions are specified on the left and right boundaries for particles with velocities directed inside the region (the distribution functions can be as equilibrium as nonequilibrium). When solving the problem, the stationary Boltzmann kinetic equation for the velocity distribution function (VDF) $f(x,\boldsymbol{\xi})$ is used in the following form:

$$\xi_x \frac{\partial f}{\partial x} = I(f,f),$$

where x is the coordinate of the physical space, $\boldsymbol{\xi} = (\xi_x, \xi_y, \xi_z)$ is the velocity vector, and $I(f,f)$ is the integral of elastic collisions. The boundary conditions of membrane-type are as follows

$$f(0,\boldsymbol{\xi}) = F_0(\boldsymbol{\xi}), \forall \xi_x > 0,\ f(L,\boldsymbol{\xi}) = F_1(\boldsymbol{\xi}), \forall \xi_x < 0.$$

Here F_0 and F_1 are the distribution functions (generally nonequilibrium). Note that the problem on evaporation and condensation is a particular case of this problem. The gas is assumed to be monatomic with the interaction law being the model of hard spheres. Macroscopic variables, such as particle number density n, velocity vector $\mathbf{u}$, temperature T, heat flux vector $\mathbf{q}$ and pressure tensor $\mathbf{P}$ are determined by integrating the VDF with the corresponding weights over the three-dimensional velocity space.

1.2. Flow of a chemically reacting gas in a closed region with membrane type boundary conditions

A more difficult case can be investigated for a chemically interacting gas. We study the flow of a chemically reacting gas of four components A^i with masses of molecules m^i (i = 1, ..., 4) and the energy of a chemical bond E^i. The regime of slow chemical reactions is considered, that is, it is assumed that the relaxation time due to elastic collisions is much lesser than the relaxation time during chemical interaction. The gas components are involved into a reversible bimolecular reaction, and the change in the internal chemical energy ΔE of the mixture is assumed to be positive.

$$A^1 + A^2 \Leftrightarrow A^3 + A^4,\ \Delta E = -\sum_{i=1}^{4} \lambda^i E^i = E^4 + E^3 - E^2 - E^1 > 0,$$

where λ^i is the stoichiometric coefficients, respectively (1. 1, -1, -1).

The formulation of a problem with membrane-type boundary conditions for a one-dimensional flow of a gas mixture of four components is similar to that for a single-component gas. Also, as above, it is assumed that the outgoing gas does not interact with the incoming gas.

The solution of the problem is determined by the system of Boltzmann equations, but the calculations of the system of Boltzmann equations are extremely complicated therefore, the use of model equations in the case of a chemically reacting gas is reasonable. The approximating system of BGK type equations is proposed in [8, 9], in which the integral terms are replaced by one relaxation operator for each component of the mixture

$$\frac{\partial f^i}{\partial t} + \left(\boldsymbol{\xi}, \frac{\partial f^i}{\partial \mathbf{x}}\right) = Q^i(f) \equiv \nu_i (f_{Mi} - f^i),\quad i = 1, \ldots, 4, \tag{1}$$

where, $f_{M,i}(\xi) = n_i \left(\frac{m^i}{2\pi k T_i}\right)^{3/2} \exp\left(-\frac{m^i(\boldsymbol{\xi} - \mathbf{u}_i)^2}{2kT_i}\right),\quad i = 1, \ldots, 4.$

The free five parameters of the equilibrium function for each reacting component, namely the particle number density n_i, the three components of velocity $\mathbf{u}_i$ and temperature T_i are determined from the equality of momentum, energy and density exchanges for each component in (1) and the

IOP Conf. Series: Journal of Physics: Conf. Series **1163** (2019) 012055 doi:10.1088/1742-6596/1163/1/012055

system of complete Boltzmann equations. The boundary conditions are analogous to the mentioned above but now should be accepted for each component.

2. The problem with membrane-like boundary conditions for a simple monatomic gas

The results of computations for the Boltzmann equation for the different variants boundary conditions are presented in figures 1-2. We will mark by a color (blue in the case of monatomic gas and yellow in the case of mixture) the zones with the anomalous thermal transport in the region under consideration.

The results shown in figure 1 are obtained for Knudsen number Kn=1. The boundary condition for the left plot are Maxwellians $F_L(\xi) = f_M(1.0, 3.0, 1.0)$ and $F_R(\xi) = f_M(1.5, -1.5, 2.5)$. It is interesting that for equilibrium boundary conditions and moderate Knudsen number there is a significant anomalous zone due to the nonequilibrium flow in region under consideration. In the right plot the results are obtained with $F_L(\xi) = f_M(1.0, 3.0, 1.0)$ and $F_R(\xi) = f_M(1., 0.5, 3.0) + f_M(1, -1.5, 3.0)$, the anomalous heat transport is in the entire region (here is not marked by the color). In figure 2 (the left plot) the dependence of the value of the anomalous zone on Knudsen numbers is presented for the boundary conditions $F_L(\xi) = f_M(1.0, 2.0, 1.0)$ and $F_R(\xi) = f_M(1.0, -0.5, 1.0)$. The assumption of a full permeability of the membrane can be reduced, namely the coefficient of the permeability P can be introduced similar to [9]. In this case a parameter a=1-P is related to the part of molecules which are reflected from the membrane with its temperature and with the accommodation. In figure 2 (the right) the dependence of the size of anomalous zone on the coefficient of the reflection is shown for the boundary conditions $F_L(\xi) = f_M(1.0, 3.0, 1.0)$ and $F_R(\xi) = f_M(1.0, -1.5, 1.0)$.

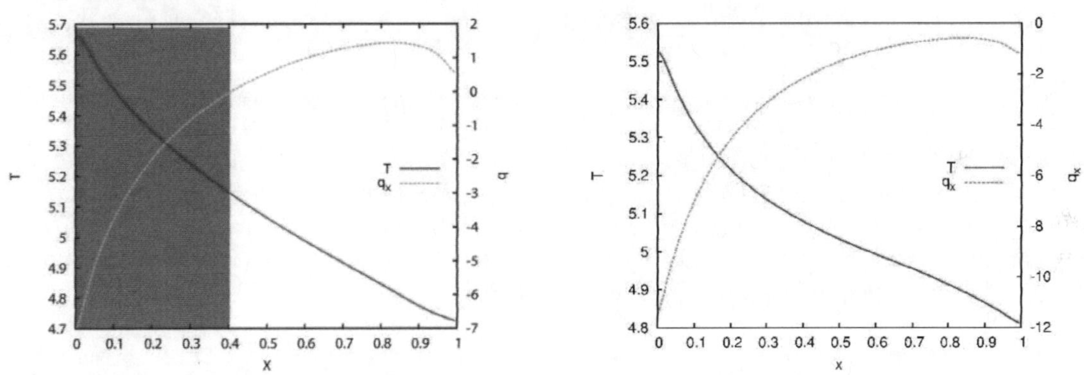

Figure 1. The demonstration of the anomalous heat transport for a problem with two variants of the "membrane-like" boundary conditions.

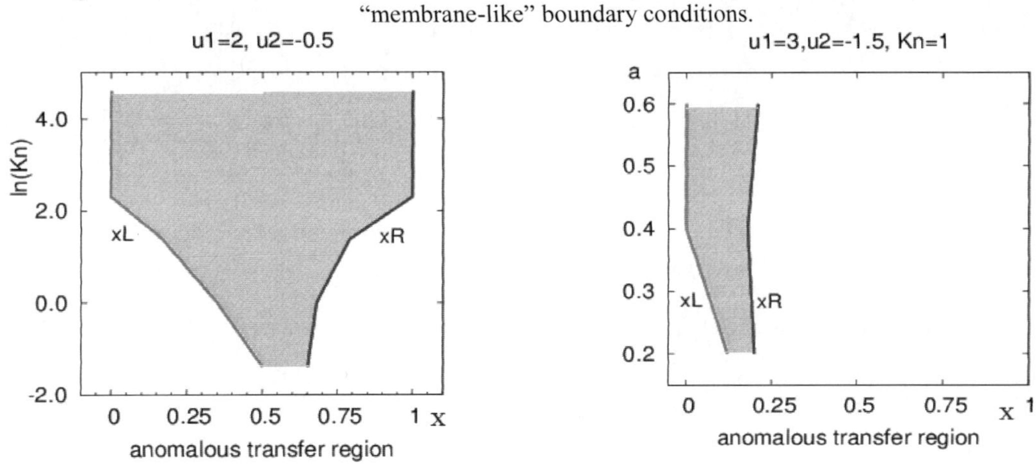

Figure 2. The dependence of the spatial positions of the anomalous zones on the parameter ln(Kn) for Knudsen numbers Kn=0.25, 0.1, 1.0, 10, 100 (left) and on the parameter of reflection a (right).

3. The problem with membrane-like boundary conditions for a mixture with chemical reactions

Two main purposes of study of the problem with the "membrane-like" boundary conditions:

1) to find an anomalous transport in the region;

2) to form complex nonequilibrium flows especially for mixtures with chemical reactions, in this case the boundary conditions can influence the structure in this "chemical reactor".

In figures. 3-5 we show the results of calculations with different boundary conditions and different regimes of rarefaction for a mixture of gases undergo slow chemical reactions (with a value $\Delta E = 1$).

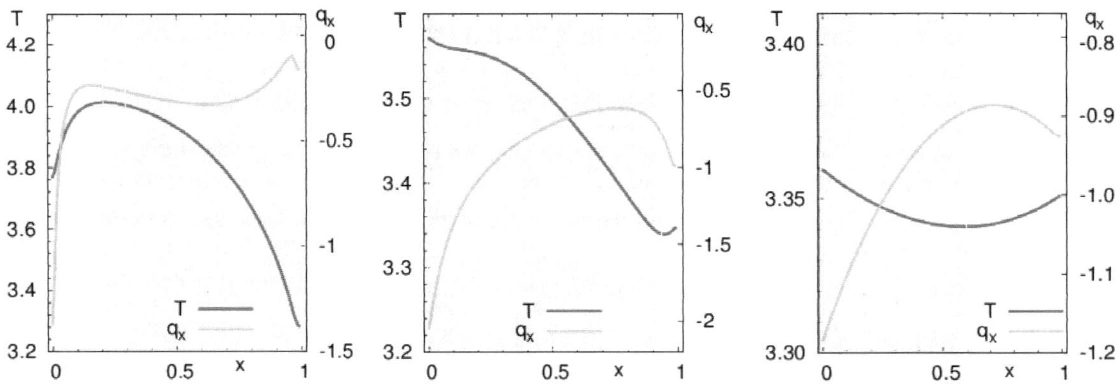

Figure 3. The profiles of heat flux q_x and temperature T in for gas mixtures with chemical reactions, Kn=0.1 (left), Kn=1.0 (middle), Kn=10 (right).

In figure 3 the following model variant of molecular component masses was used: $m_1 = 2.2$, $m_2 = 0.9$, $m_3 = 1.5$, $m_4 = 1.6$. For the left and right boundaries we use equilibrium distribution functions for all components. For the left boundary the velocities of components were as follows: $u_L = \{1.5; 1.0; 2.5; 0.5\}$ and for the right boundary $u_R = \{-0.5; -2.5; -1.5; -1.0\}$

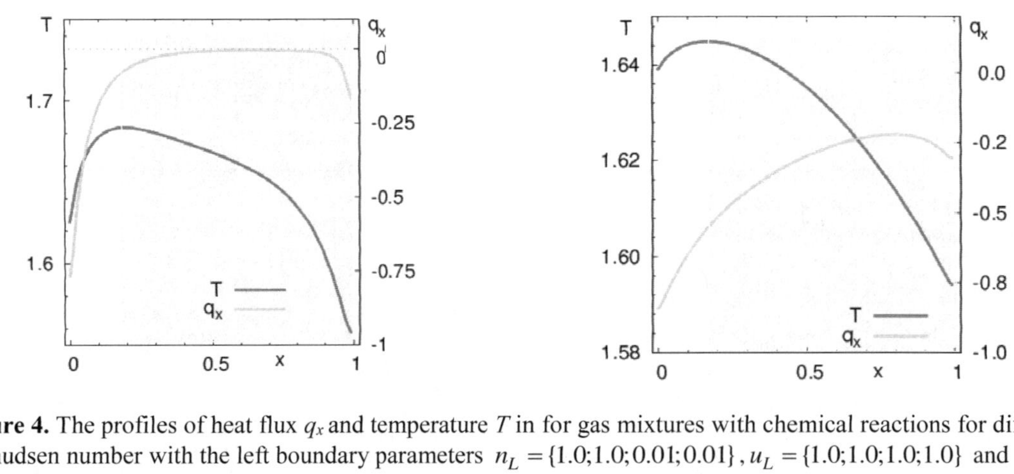

Figure 4. The profiles of heat flux q_x and temperature T in for gas mixtures with chemical reactions for different Knudsen number with the left boundary parameters $n_L = \{1.0; 1.0; 0.01; 0.01\}$, $u_L = \{1.0; 1.0; 1.0; 1.0\}$ and right values $n_R = \{0.5; 1.0; 0.01; 0.01\}$ $u_R = \{-0.5; -0.5; -0.5; -0.5\}$ and T=1 for Kn = 0.1 (the left plot), and Kn = 1 (the right plot).

In figure 4 results for the boundary conditions with parameters $n_L = \{1.0;1.0;0.01;0.01\}$, $u_L = \{1.0;1.0;1.0;1.0\}$, $n_R = \{0.5;1.0;0.01;0.01\}$, $u_R = \{-0.5;-0.5;-0.5;-0.5\}$ and temperature on both sides T=1 are presented , one can see that in this case the change of temperature is smaller, nevertheless the anomalous zone is significant.

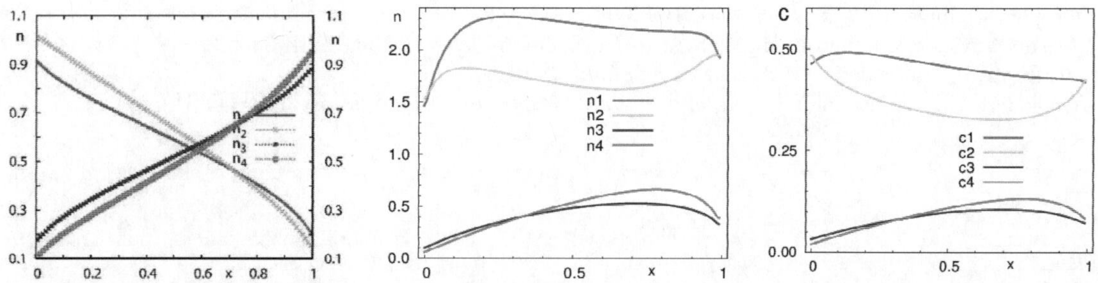

Figure 5. The profiles of densities and concentrations. Left plot: densities of mixture components for the variant with the following left boundary conditions $n_L = \{1.0;1.0;0.01;0.01\}$, $u_L = \{0.05;0.1;0.05;0.05\}$ and right values $n_R = \{0.01;0.01;1.0;1.0\}$ $u_R = \{-0.1;-0.05;-0.05;-0.05\}$, temperature T=1.0. Densities (middle) and concentrations (right) of mixture components for boundary conditions as in Fig. 4 for Kn = 0.1.

In figure 5 profiles of densities and concentrations for four components of the mixture for two variants of the boundary conditions, it is seen their influence on the behavior of the quantities.

4. Concluding remarks

A new type of nonequilibrium transport effect is confirmed by the direct solutions for the Boltzmann and the other kinetic equation for the problem with the "membrane-like" boundary conditions. The anomalous thermal transfer does not contradict to the second law of thermodynamics in the form of the H-theorem. This fact has been proved numerically.

There is an interesting question devoted to the experimental tests of this effect. Nonequilibrium flows can be prepared in the region between two thin permeable membranes with the equilibrium boundary distributions. Particles entering the region form the nonequilibrium state. The issue concerns the reliability of these "membrane like" boundary conditions. Possible experimental tests are discussed. For these purposes mixed cellulose ester (MCE) membranes could be used as a material for the boundaries. As reported in [12], a pore-size of these membranes is 25 nm, so the appropriate Knudsen number is more than unity at the atmospheric pressure. As an method of diagnostics of the nonequilibrium distribution function in the region under consideration could be proposed the Electron Beam Fluorescence (EBF). This problem for the chemical mixture allows us to create different (qualitatively and quantitatively) nonequilibrium structures.

Acknowledgments

This work by the first (V.A.) and second (S.Z.) authors was supported by Russian Foundation for Basic Research, grant 18-01-00899. The calculations were carried out by means of Joint Supercomputer Center of RAS MVS10p multiprocessor computer.

References

[1] Gorban A.N, Karlin I.V. *Physica A.* **206** (1994) 401-20.
[2] Gorban A.N., Karlin I.V. *Physica A.* **360** (2006) 325-64.
[3] SoneY, *et al. Phys.Fluids.* **8** (1996) 628-38.

IOP Conf. Series: Journal of Physics: Conf. Series **1163** (2019) 012055 doi:10.1088/1742-6596/1163/1/012055

[4] Kolobov V.I., Arslanbekov R.R., Aristov V.V., Frolova A.A., Zabelok S.A. *J. Comp. Phys.* **223** (2007) 589-608.
[5] Aristov V.V., Zabelok S.A., Fedosov M.A., Frolova A.A. *Comp. Math. Math. Phys.* **56** (2016) 854-63.
[6] Aristov V.V., Zabelok S.A., Frolova A.A. *Doklady Physics*. **62** (2017) 149-53.
[7] Aristov V.V. *Phys. Letters A.* **250** (1998) 354-9.
[8] Rossani A., Spiga G. *Physica A.*.**272** (1999) 563-73.
[9] Groppi M., Spiga G. *Physics of Fluids.* **16** (2004) 4273-84.
[10] Aristov V.V., Frolova A.A., Zabelok S.A. *EPL (Europhys. Letters).* **106** (2014) 20002.
[11] Erofeev A.I., Fridlender O.G., *Fluid Dynamics.* **45** (2010) 965-74.
[12] Gupta N.K., Gianchandani Y.B. *Microporous and Mesoporous Materials.* **142** (2011) 535-41.

IOP Conf. Series: Journal of Physics: Conf. Series **1163** (2019) 012056 doi:10.1088/1742-6596/1163/1/012056

Entanglement studies of interacting fermionic models

F Parisen Toldin and F F Assaad

Institut für Theoretische Physik und Astrophysik, Universität Würzburg, Am Hubland, D-97074 Würzburg, Germany

E-mail: francesco.parisentoldin@physik.uni-wuerzburg.de

Abstract. Recent advances in the field of strongly correlated electron systems allow to access the entanglement properties of interacting fermionic models, by means of Monte Carlo simulations. We briefly review the techniques used in this context to determine the entanglement entropies and correlations of the entanglement Hamiltonian. We further apply these methods to compute the spin two-point function of entanglement Hamiltonian for a stripe embedded into a correlated topological insulator. Further we discuss a recent method that allows an unbiased, numerically exact, direct determination of the entanglement Hamiltonian by means of auxiliary field quantum Monte Carlo simulations.

1. Introduction

In recent years, a growing number of studies have focused on the quantum entanglement in many-body systems. Its characterization has provided a variety of new insights in old and new problems, complementing the information obtained by more traditional approaches [1]. In this context, quantum entanglement also plays an important role in Density-Matrix Renormalization Group (DMRG) studies, as it essentially gives a limitation to the various algorithms [2]. The simplest setup to investigate entanglement is the so-called bipartite entanglement, where one divides a system into two parts, conventionally indicated by A and B. By tracing out the degrees of freedom of B, one obtains the reduced density matrix $\hat{\rho}_A$ of the remaining subsystem A. In this framework, the most studied quantities are the von Neumann entropy $S_{\rm vN}$ and especially the Renyi entropies $S_{\rm R}^{(n)}$, defined as

$$S_{\rm vN} \equiv -\,{\rm Tr}_A\{\hat{\rho}_A \ln \hat{\rho}_A\}, \tag{1}$$

$$S_{\rm R}^{(n)} \equiv \frac{1}{1-n} \ln {\rm Tr}_A\{\hat{\rho}_A^n\}, \tag{2}$$

where $n > 1$ is an integer. In the ground state, entanglement entropies generically satisfy an area law, i.e., they are to the leading order proportional to the area between the two subsystems [3]. Especially of interest are corrections to the area law, which can in principle reveal topological properties [4–6] (see, however, Ref. [7]), signal the presence of Goldstone modes [8], and, most notably, allow to extract the central charge of models described by a 1+1 dimensional Conformal Field Theory (CFT) [9, 10].

Additional information is contained in the entanglement (or modular) Hamiltonian $\hat{H}_E$, defined by[1]

$$\hat{\rho}_A \equiv e^{-\hat{H}_E}. \tag{3}$$

[1] Strictly speaking, Eq. (3) defines the modular Hamiltonian, which equals the entanglement Hamiltonian only in the ground state. Here, in agreement with recent literature, we shall make no distinction and always use the term entanglement Hamiltonian.

The notion of entanglement Hamiltonian is of great interest for several reasons. The spectrum of $\hat{H}_E$, known as "entanglement spectrum", has been shown to display the edge physics of topologically ordered fractional Quantum Hall state [11], and of symmetry-protected topological states [12–15]. The entanglement Hamiltonian plays a central role in the so-called first law of entanglement [16]. Moreover, an explicit determination of $\hat{H}_E$ opens the possibility of studying the entanglement using statistical mechanics tools, so as, for instance, to possibly characterize the reduced density matrix as a thermal state. Furthermore, an explicitly known entanglement Hamiltonian could be engineered so as to provide a direct experimental measure of entanglement in many-body systems, an otherwise very difficult task [17]. We also notice that almost all entanglement measures can be obtained from the entanglement Hamiltonian. For example, the expectation value of $\hat{H}_E$ equals the von Neumann entanglement entropy S_{vN}, which is generically not accessible in numerical Monte Carlo (MC) simulations; in contrast, S_{vN} can be computed by methods which access to the ground state wave function, such as DMRG or Exact Diagonalization.

Compared to the entanglement entropies, much less results are known for the entanglement Hamiltonian. Its explicit determination has proven to be a considerably more difficult problem, where only a few solvable results are available. Besides easily solvable limiting cases, such as in the absence of interactions between the two subsystems, or the high-temperature limit, a particularly important case is that of a relativistic field theory with Hamiltonian density $\hat{H}(x)$ in flat $d-$dimensional Minkowski space. The entanglement Hamiltonian for a semi-infinite subspace $A = \{(x_1, \ldots, x_d) \in \mathbb{R}^d, x_1 > 0\}$ is [18,19]

$$\hat{H}_E = 2\pi \int_A d^d x \left(x_1 \hat{H}(x) \right) + c, \tag{4}$$

where c is a constant. The right-hand side of Eq. (4) is known as Bisognano-Wichmann (BW) form of the entanglement Hamiltonian, or Rindler Hamiltonian [20]. With additional conformal symmetry, a mapping of A to a ball allows again to express $\hat{H}_E$ as an integral of $\hat{H}(x)$ [21]. Ref. [22] reviews the cases in $1+1$ dimensional CFT, where $\hat{H}_E$ can be expressed as an integral. On the lattice, a few cases are exactly known. For a noncritical one-dimensional transverse-field Ising model, and for the XXZ model in the massive phase, the entanglement Hamiltonian for a semi-infinite line subsystem has been exactly determined [23, 24]. For a free (nonrelativistic) fermionic system the entanglement Hamiltonian is [25]

$$\hat{H}_E = \ln \left[\left(G_A^T \right)^{-1} - \mathbb{1} \right] - \ln \det(\mathbb{1} - G_A), \qquad (G_A)_{ij} \equiv \langle \hat{\boldsymbol{c}}_i^\dagger \hat{\boldsymbol{c}}_j \rangle, \tag{5}$$

where $\hat{\boldsymbol{c}}_i^\dagger$, $\hat{\boldsymbol{c}}_j$ are the fermionic creation and annihilation operators in the subsystem A, and G_A is the Green's function matrix of the model, restricted to A. Despite the exact result of Eq. (5), an explicit computation of the entanglement Hamiltonian in the deceptively simple case of a free fermionic chain, and for a subsystem consisting in a segment, has eluded an analytical treatment so far. In this case, even a numerical computation of Eq. (5) is technically hard, because G_A is plagued by eigenvalues close to 0 and 1, rendering the formula of Eq. (5) numerically unstable. The lattice models mentioned so far share a common property of being described by a CFT in the low-energy limit. Thus, one may conjecture that the corresponding entanglement Hamiltonian attains the BW form, up to a lattice discretization. Indeed, this is the case for the aforementioned exact results. Nevertheless, for a free fermionic chain, the entanglement Hamiltonian for a segment displays intriguing corrections to the BW form which, remarkably, persist even in the limit of a long segment [26]. Recent studies have provided further numerical evidence in support of a lattice-discretized BW form of the entanglement Hamiltonian for various one- and two-dimensional lattice models [17,27–30]. In Ref. [31] we have introduced a method which allows a numerically exact determination of the entanglement Hamiltonian for generic interacting fermionic models, exploiting the auxiliary field Quantum Monte Carlo (QMC) method [32–34].

In this paper we focus on the QMC studies of entanglement in interacting fermionic models. In Sec. 2 we review the techniques used to compute the Renyi entropies. In Sec. 3 we present some results for the correlations of the entanglement Hamiltonian in a correlated topological insulator. In Sec. 4 we briefly discuss the method introduced in Ref. [31] to explicitly determine the entanglement Hamiltonian. We conclude in Sec. 5 with an outlook to future research topics.

IOP Conf. Series: Journal of Physics: Conf. Series **1163** (2019) 012056 doi:10.1088/1742-6596/1163/1/012056

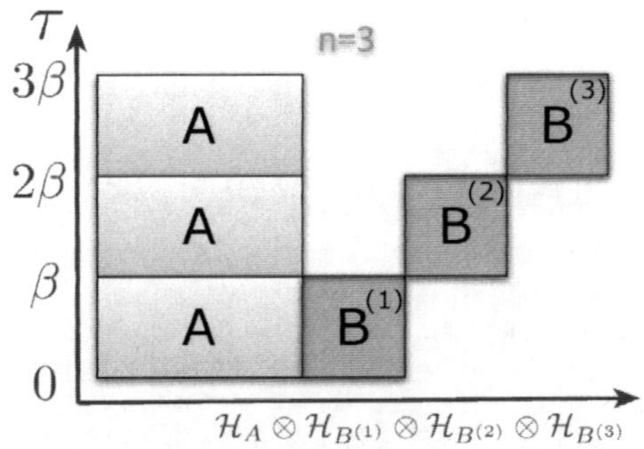

Figure 1. Illustration of the replica trick for $n = 3$. The effective time-dependent Hamiltonian vanishes in the white regions.

2. Entanglement entropies

The auxiliary field QMC method is a widely used MC method for fermionic models [32–34]. It can be conveniently programmed using the ALF package [35]. Its formulation relies in a Trotter decomposition, followed by a Hubbard-Stratonovich (HS) transformation, which decouples the interactions via HS fields $\{s\}$. The MC consists then in a stochastic sampling of the resulting probability distribution $P(\{s\})$ of the HS fields. At each $\{s\}$ configuration the system corresponds to a free fermionic model, hence by using Eq. (5) one obtains an expression for the reduced density matrix $\hat{\rho}_A$ [36]

$$\hat{\rho}_A = \int d\{s\} P(\{s\}) \det\left(\mathbb{1} - G_A(\{s\})\right) e^{-\hat{c}_i^\dagger h_{ij}(\{s\})\hat{c}_j}, \qquad h(\{s\}) = \ln\left((G_A(\{s\})^T)^{-1} - \mathbb{1}\right), \quad (6)$$

where now the restricted Green's function matrix $G_A(\{s\})$ explicitly depends on the HS fields configuration. To compute the nth Renyi entropy, one introduces n parallel simulations with HS fields $\{s_i\}$. Using Eq. (2), $S_{\text{R}}^{(n)}$ is obtained by sampling an observable depending on all $\{s_i\}$:

$$e^{-(n-1)S_{\text{R}}^{(n)}} = \int \prod_{i=1}^{n} d\{s_i\} P(\{s_i\}) \left[\prod_{i=1}^{n} \det\left(\mathbb{1} - G_A(\{s_i\})\right)\right] \det\left(1 + \prod_{i=1}^{n} \frac{G_A(\{s_i\})}{\mathbb{1} - G_A(\{s_i\})}\right). \quad (7)$$

Using Eq. (7), Renyi entropies have been computed in Refs. [14, 36–38]. While this approach is simple compared to other methods, it suffers from two main issues. The first one is that the sampled observable appears to be affected by considerably large fluctuations, increasing with the interaction strength, such that $S_{\text{R}}^{(n)}$ is easily computed in the weak-coupling regime only [14, 37]. A solution to this issue, obtained within the framework of the Hybrid QMC, where also Eq. (6) holds, consists in reformulating Eq. (7) as an integral over an auxiliary parameter of an observable sampled with a modified measure [37, 38]. The replica method discussed below provides another solution to the problem. The second issue in using Eq. (7) is due to the fact that, in sampling the ground state, $G_A(\{s_i\})$ shows eigenvalues close to 1, giving a singularity on the right-hand side of Eq. (7) for $n > 2$. (For $n = 2$ Eq. (7) can be easily simplified, avoiding the singularity [36].) The singularity can be avoided by introducing a thermal broadening, as we discuss in Ref. [14], or by recasting the right-hand side of Eq. (7) as a determinant of a bigger matrix [38].

An alternative, more general, framework to study the entanglement is the so-called replica trick. It has been used in fermionic [15, 39–41], bosonic [6], spin systems [42, 43], and it is also a starting point for a field-theory approach [9, 10]. This method consists in a representation of $\hat{\rho}_A^n = \exp(-n\hat{H}_E)$, obtained by introducing n replicas of the subsystem B, enlarging the original Hilbert space $\mathcal{H} = \mathcal{H}_A \otimes \mathcal{H}_B$ to $\mathcal{H}_R = \mathcal{H}_A \otimes \mathcal{H}_{B^{(1)}} \otimes \ldots \otimes \mathcal{H}_{B^{(n)}}$. In $\mathcal{H}_R$ one defines a time-dependent Hamiltonian, such that the subsystem A evolves in the imaginary-time interval $[0, n\beta]$, whereas each replica evolves in an interval $[i\beta, (i+1)\beta]$. An illustration of the replica trick for $n = 3$ is shown in Fig. 1. The replica trick, combined with the swap algorithm [43], allows to compute the Renyi entropies in fermionic models, avoiding the aforementioned sampling issues [39], but with an increased computational cost $\propto n^3$. As we discuss in Ref. [15], the approach based on Eq. (7) and the replica trick are formally equivalent.

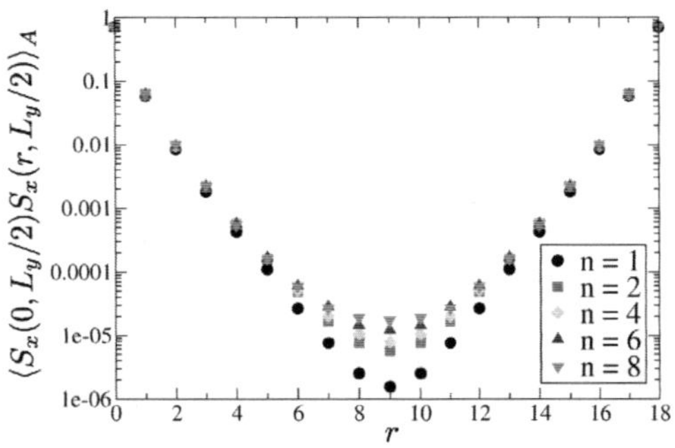

Figure 2. Semilogarithmic plot of the spin-spin correlations of the entanglement Hamiltonian in the middle of a stripe.

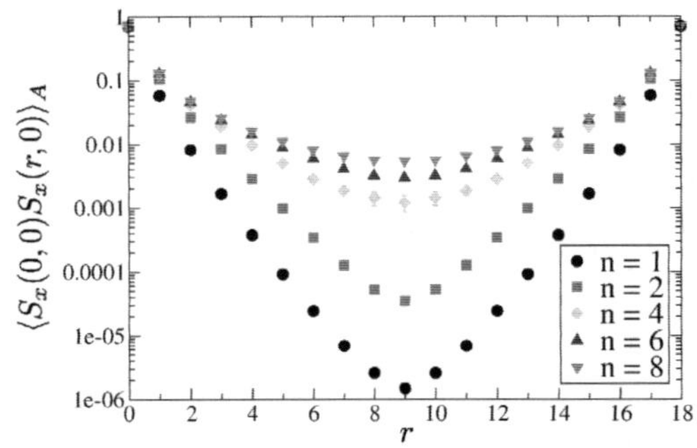

Figure 3. Same as Fig. 2 for spin-spin correlations along the edge of the stripe.

3. Correlations of the Entanglement Hamiltonian

By sampling a Gibbs measure $\sim \exp(-n\hat{H}_E)$, the replica trick allows to compute the correlations of the entanglement Hamiltonian at an effective integer inverse temperature n. It is important to note that the entanglement Hamiltonian also depends implicitly on the physical inverse temperature β. (See Eq. (3)). We apply this technique to the Kane-Mele-Hubbard model, a paradigmatic model to study the interplay of topology and correlations [44–52]. The model is defined on a honeycomb lattice, with Hamiltonian

$$\hat{H} = -t \sum_{\langle i\, j\rangle,\sigma} \hat{\boldsymbol{c}}^\dagger_{i,\sigma}\hat{\boldsymbol{c}}_{j,\sigma} + i\lambda \sum_{\langle\langle i\, j\rangle\rangle} \hat{\boldsymbol{c}}^\dagger_i(\vec{\nu}_{ij}\cdot\vec{\sigma})\hat{\boldsymbol{c}}_j + U\sum_i \left(\hat{\boldsymbol{n}}_{i,\uparrow} - \frac{1}{2}\right)\left(\hat{\boldsymbol{n}}_{i,\downarrow} - \frac{1}{2}\right), \tag{8}$$

with $\hat{\boldsymbol{c}}^\dagger_i \equiv (\hat{\boldsymbol{c}}^\dagger_{i,\uparrow}, \hat{\boldsymbol{c}}^\dagger_{i,\downarrow})$, $\hat{\boldsymbol{c}}_i \equiv (\hat{\boldsymbol{c}}_{i,\uparrow}, \hat{\boldsymbol{c}}_{i,\downarrow})$ the creation and annihilation operator vectors for nonrelativistic spinful fermions, $\hat{\boldsymbol{n}}_{i,\sigma} \equiv \hat{\boldsymbol{c}}^\dagger_{i,\sigma}\hat{\boldsymbol{c}}_{i,\sigma}$ the number operator, and $\vec{\sigma}$ the vector of Pauli matrices. In Eq. (8) the first term represents hoppings between nearest-neighbor sites, the second term couples next-nearest-neighbors sites and represents the spin-orbit interaction, and the third term is a on-site Hubbard repulsion. The unit vectors $\vec{\nu}_{ij}$ take values $\vec{\nu}_{ij} = \pm\vec{e}_z$ depending on the sublattice, spin, and direction of hoppings; see, e.g., Ref. [52] for a definition. At half-filling the Kane-Mele-Hubbard Hamiltonian can be simulated using the auxiliary field QMC without encountering the sign problem. For $\lambda = 0$ the model reduces to the Hubbard model and exhibits a quantum phase transition in the ground state at $U/t = 3.80(1)$ [52] in the Gross-Neveu-Heisenberg universality class [52–55]. For $\lambda > 0$ the model shows a quantum phase transition in the classical 3D XY universality class, between a quantum spin-Hall insulator phase at small values of U/t and an antiferromagnetic insulating phase at large values of U/t [47, 48, 52, 56].

We consider the model in the ground state for $U/t = 4$ and $\lambda/t = 0.2$, in the quantum spin-Hall phase, and compute the spin-spin correlations of the entanglement Hamiltonian for a stripe of width $W_A = 3$, translationally invariant in one direction, cut inside a finite 9×18 lattice with periodic boundary conditions. In Fig. 2 and Fig. 3 we show the results for the spin-spin correlations in the middle of the stripe, and along its edge, for various values of the replica index n. For $n = 1$ by definition the correlations of $\hat{H}_E$ are identical to those of $\hat{H}$, and show the expected exponential decay of the gapped spin-Hall insulating phase. Upon increasing n the correlations in the middle do not change qualitatively, whereas those on the edge of the strip display a slower decay. For $n = 8$ we approach the ground state properties of the entanglement Hamiltonian: while the correlations in the middle of the strip are still exponentially decaying, those on the edge support a power-law decay, analogous to the gapless modes found at the physical edge of a correlated quantum spin-Hall insulator [46]. A similar correspondence is found by studying the imaginary-time displaced Green's function of the entanglement Hamiltonian [15]. Time-displaced correlation functions of the entanglement Hamiltonian can be also used to compute the low-lying entanglement spectrum [14, 15].

4. Determination of the Entanglement Hamiltonian

Eq. (6) is the starting point for a method that we have introduced in Ref. [31] to explicitly determine the entanglement Hamiltonian for interacting fermionic models, within the framework of auxiliary field QMC simulations. We briefly describe here the main ideas of the method. The exponential appearing on the right-hand side of Eq. (6) admits an exact expansion in terms of normal-ordered many-body operators, whose coefficients can be sampled using the auxiliary field QMC, obtaining a MC determination of the reduced density matrix $\hat{\rho}_A$. We then compute the negative logarithm of the sampled $\hat{\rho}_A$, and express it in terms of fermionic creation and annihilation operators. This results in an unbiased, numerically exact determination of the entanglement Hamiltonian, as a sum of normal-ordered operators. We refer to Ref. [31] for more details on the method.

We have applied the technique to the Hubbard chain model, obtaining the one-body term of $\hat{H}_E$ for a segment of length $L = 8$ embedded in a chain of size $L = 32$, and at inverse temperatures $1 \leq \beta/t \leq 3$. The resulting nearest-neighbor hopping terms in $\hat{H}_E$ qualitatively resemble the BW form, while other hopping terms are negligible. In Ref. [31] we have also determined the entanglement Hamiltonian for a two-legs Hubbard chain. This model exhibits a single gapped phase [57, 58], ensuring a fast approach to the thermodynamic limit in the linear size [59, 60]. The entanglement Hamiltonian for a single leg shows an interesting temperature dependence. At high temperatures $\hat{H}_E$ matches a Hubbard Hamiltonian, in line with general arguments. Upon lowering the temperature $\hat{H}_E$ exhibits a crossover to a Heisenberg-like Hamiltonian, characterized by a large U/t ratio, and the emergence of additional nearest-neighbors antiferromagnetic and next-nearest-neighbors ferromagnetic spin-spin interactions. This result resembles what is found in the two-legs antiferromagnetic Heisenberg model upon tracing out one leg [61–65].

5. Outlook

Recent progresses in the entanglement studies of correlated fermionic systems form a basis for promising future research. The established techniques for entropies can be used to compute other entanglement measures, such as the mutual information. At the same time, investigations of the entanglement Hamiltonian via its correlations, together with its explicit determination, can foster a fruitful exchange with statistical mechanics and field theory approaches, enabling a deeper understanding of correlated many-body systems.

Acknowledgments

FPT thanks the German Research Foundation (DFG) through the grant No. AS120/13-1 of the FOR 1807. FFA thanks the DFG for financial support through the SFB 1170 ToCoTronics. We acknowledge the computing time granted by the John von Neumann Institute for Computing (NIC) and provided on the supercomputer JURECA [66] at the Jülich Supercomputing Center.

References

[1] Laflorencie N 2016 *Phys. Rep.* **646** 1–59
[2] Schollwöck U 2005 *Rev. Mod. Phys.* **77** 259–315
[3] Eisert J, Cramer M and Plenio M B 2010 *Rev. Mod. Phys.* **82** 277–306
[4] Kitaev A and Preskill J 2006 *Phys. Rev. Lett.* **96** 110404
[5] Levin M and Wen X G 2006 *Phys. Rev. Lett.* **96** 110405
[6] Isakov S V, Hastings M B and Melko R G 2011 *Nat. Phys.* **7** 772–775
[7] Budich J C, Eisert J and Bergholtz E J 2014 *Phys. Rev. B* **89** 195120
[8] Metlitski M A and Grover T 2011 *ArXiv:1112.5166*
[9] Calabrese P and Cardy J 2004 *J. Stat. Mech.* P06002
[10] Calabrese P and Cardy J 2009 *J. Phys. A: Math. Theor.* **42** 504005

[11] Li H and Haldane F D M 2008 *Phys. Rev. Lett.* **101** 010504
[12] Fidkowski L 2010 *Phys. Rev. Lett.* **104** 130502
[13] Turner A M, Zhang Y and Vishwanath A 2010 *Phys. Rev. B* **82** 241102
[14] Assaad F F, Lang T C and Parisen Toldin F 2014 *Phys. Rev. B* **89** 125121
[15] Assaad F F 2015 *Phys. Rev. B* **91** 125146
[16] Blanco D D, Casini H, Hung L Y and Myers R C 2013 *J. High Energy Phys.* **8** 60
[17] Dalmonte M, Vermersch B and Zoller P 2018 *Nat. Phys.* **14** 827–831
[18] Bisognano J J and Wichmann E H 1975 *J. Math. Phys.* **16** 985–1007
[19] Bisognano J J and Wichmann E H 1976 *J. Math. Phys.* **17** 303–321
[20] Susskind L and Lindesay J 2006 *The Holographic Universe: An Introduction to Black Holes, Information and the String Theory Revolution* (New Jersey: World Scientific)
[21] Casini H, Huerta M and Myers R C 2011 *J. High Energy Phys.* **5** 36
[22] Cardy J and Tonni E 2016 *J. Stat. Mech.* 123103
[23] Itoyama H and Thacker H B 1987 *Phys. Rev. Lett.* **58** 1395–1398
[24] Peschel I, Kaulke M and Legeza Ö 1999 *Ann. Phys. (Leipzig)* **8** 153–164
[25] Peschel I 2004 *J. Stat. Mech.* P06004
[26] Eisler V and Peschel I 2017 *J. Phys. A: Math. Theor.* **50** 284003
[27] Kim P, Katsura H, Trivedi N and Han J H 2016 *Phys. Rev. B* **94** 195110
[28] Kosior A, Lewenstein M and Celi A 2018 *arXiv:1804.11323*
[29] Zhu W, Huang Z and He Y c 2018 *arXiv:1806.08060*
[30] Giudici G, Mendes-Santos T, Calabrese P and Dalmonte M 2018 *Phys. Rev. B* **98** 134403
[31] Parisen Toldin F and Assaad F F 2018 *Phys. Rev. Lett.*
[32] Blankenbecler R, Scalapino D J and Sugar R L 1981 *Phys. Rev. D* **24**(8) 2278–2286
[33] White S, Scalapino D, Sugar R, Loh E, Gubernatis J and Scalettar R 1989 *Phys. Rev. B* **40**(1) 506–516
[34] Assaad F F and Evertz H 2008 World-line and Determinantal Quantum Monte Carlo Methods for Spins, Phonons and Electrons *Computational Many-Particle Physics* (*Lecture Notes in Physics* vol 739) ed Fehske H, Schneider R and Weiße A (Berlin Heidelberg: Springer) pp 277–356 ISBN 978-3-540-74685-0
[35] Bercx M, Goth F, Hofmann J S and Assaad F F 2017 *SciPost Phys.* **3** 013
[36] Grover T 2013 *Phys. Rev. Lett.* **111** 130402
[37] Drut J E and Porter W J 2015 *Phys. Rev. B* **92** 125126
[38] Drut J E and Porter W J 2016 *Phys. Rev. E* **93** 043301
[39] Broecker P and Trebst S 2014 *J. Stat. Mech.* P08015
[40] Wang L and Troyer M 2014 *Phys. Rev. Lett.* **113** 110401
[41] Broecker P and Trebst S 2016 *Phys. Rev. E* **94** 063306
[42] Hastings M B, González I, Kallin A B and Melko R G 2010 *Phys. Rev. Lett.* **104** 157201
[43] Humeniuk S and Roscilde T 2012 *Phys. Rev. B* **86** 235116
[44] Assaad F F, Bercx M and Hohenadler M 2013 *Phys. Rev. X* **3** 011015
[45] Bercx M, Hohenadler M and Assaad F F 2014 *Phys. Rev. B* **90** 075140
[46] Hohenadler M and Assaad F F 2013 *J. Phys.: Condens. Matter* **25** 143201
[47] Hohenadler M, Lang T C and Assaad F F 2011 *Phys. Rev. Lett.* **106** 100403; *Phys. Rev. Lett.* **109** 229902
[48] Hohenadler M, Meng Z Y, Lang T C, Wessel S, Muramatsu A and Assaad F F 2012 *Phys. Rev. B* **85** 115132
[49] Hohenadler M, Parisen Toldin F, Herbut I F and Assaad F F 2014 *Phys. Rev. B* **90** 085146
[50] Kane C L and Mele E J 2005 *Phys. Rev. Lett.* **95** 226801
[51] Kane C L and Mele E J 2005 *Phys. Rev. Lett.* **95** 146802
[52] Parisen Toldin F, Hohenadler M, Assaad F F and Herbut I F 2015 *Phys. Rev. B* **91** 165108
[53] Herbut I F 2006 *Phys. Rev. Lett.* **97** 146401
[54] Herbut I F, Juričić V and Roy B 2009 *Phys. Rev. B* **79** 085116
[55] Herbut I F, Juričić V and Vafek O 2009 *Phys. Rev. B* **80** 075432
[56] Hohenadler M, Lang T C and Assaad F F 2012 *Phys. Rev. Lett.* **109** 229902
[57] Noack R M, White S R and Scalapino D J 1996 *Physica C* **270** 281–296
[58] Weihong Z, Oitmaa J, Hamer C J and Bursill R J 2001 *J. Phys.: Condens. Matter* **13** 433–448
[59] Neuberger H 1989 *Phys. Lett. B* **233** 183–186
[60] Wang Z, Assaad F F and Parisen Toldin F 2017 *Phys. Rev. E* **96** 042131
[61] Poilblanc D 2010 *Phys. Rev. Lett.* **105** 077202
[62] Cirac J I, Poilblanc D, Schuch N and Verstraete F 2011 *Phys. Rev. B* **83** 245134
[63] Peschel I and Chung M C 2011 *EPL* **96** 50006
[64] Läuchli A M and Schliemann J 2012 *Phys. Rev. B* **85** 054403
[65] Schliemann J and Läuchli A M 2012 *J. Stat. Mech.* P11021
[66] Jülich Supercomputing Centre 2016 *J. Large-Scale Res. Facil.* **2** A62

IOP Conf. Series: Journal of Physics: Conf. Series **1163** (2019) 012057 doi:10.1088/1742-6596/1163/1/012057

A variational lower bound on the ground state of a many-body system and the squaring parametrization of density matrices

F. Uskov[1], **O. Lychkovskiy**[1,2]

[1] Skolkovo Institute of Science and Technology, Nobel street 3, Moscow 121205, Russia
[2] Steklov Mathematical Institute of Russian Academy of Sciences, Gubkina str., 8, Moscow 119991, Russia

E-mail: fel1992@mail.ru

Abstract. A variational *upper* bound on the ground state energy $E_{\rm gs}$ of a quantum system, $E_{\rm gs} \leqslant \langle\Psi|H|\Psi\rangle$, is well-known (here H is the Hamiltonian of the system and Ψ is an arbitrary wave function). Much less known are variational *lower* bounds on the ground state. We consider one such bound which is valid for a many-body translation-invariant lattice system. Such a lattice can be divided into clusters which are identical up to translations. The Hamiltonian of such a system can be written as $H = \sum_{i=1}^{M} H_i$, where a term H_i is supported on the i'th cluster. The bound reads $E_{\rm gs} \geqslant M \inf\limits_{\rho_{cl} \in \mathbb{S}_{cl}^G} {\rm tr}_{cl}\, \rho_{cl}\, H_{cl}$, where $\mathbb{S}_{cl}^G$ is some wisely chosen set of reduced density matrices of a single cluster. The implementation of this latter variational principle can be hampered by the difficulty of parameterizing the set $\mathbb{M}$, which is a necessary prerequisite for a variational procedure. The root cause of this difficulty is the nonlinear positivity constraint $\rho > 0$ which is to be satisfied by a density matrix. The squaring parametrization of the density matrix, $\rho = \tau^2/\,{\rm tr}\,\tau^2$, where τ is an arbitrary (not necessarily positive) Hermitian operator, accounts for positivity automatically. We discuss how the squaring parametrization can be utilized to find variational lower bounds on ground states of translation-invariant many-body systems. As an example, we consider a one-dimensional Heisenberg antiferromagnet.

1. Introduction

The ground state of a many-particle system is one of the central objects studied in condensed matter physics. The ground state energy as a rule cannot be calculated exactly. In strongly correlated systems, it is also difficult to apply the perturbation theory. A common way to asses the ground state energy is via variational methods. An upper bound on the ground state energy, $E_{\rm gs} \leqslant \langle\Psi|H|\Psi\rangle$, is well-known. It is often desirable to supplement the latter with a *lower* bound. Methods for obtaining lower bounds on ground state energies of many-body systems exist [1, 3, 2, 4, 5], but they are much less developed than standard variational methods. In this paper we suggest one such method applicable to translation-invariant lattice systems with local interactions. The method is applied to a simple system, and its merits and prospects are discussed.

IOP Conf. Series: Journal of Physics: Conf. Series **1163** (2019) 012057 doi:10.1088/1742-6596/1163/1/012057

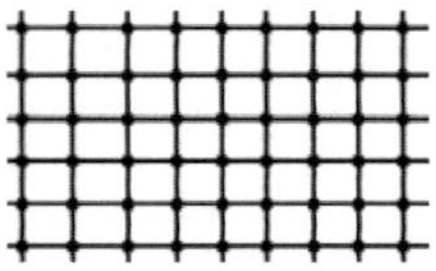

Figure 1. Square lattice.

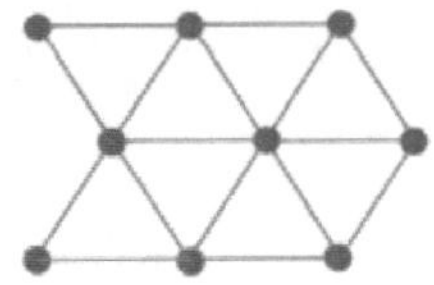

Figure 2. Triangular lattice.

2. Lower bound on the ground state energy of a translation-invariant lattice system

2.1. Our lower bound

We consider a system of spins on a lattice with N sites. The lattice is invariant with respect to the group of translation and, for two- and three-dimensional lattices, rotations. For example, this can be a linear chain in one dimension, a square or a triangular lattice in two dimensions (see figures 1, 2) *etc.* Due to the symmetry, a lattice can be divided into identical clusters. The Hamiltonian of the system is defined on the lattice and is invariant under a group G which contains the symmetries of the lattice and, in general, some other symmetries,

$$U\,H\,U^\dagger = H \qquad \forall\ U \in G. \tag{1}$$

The Hamiltonian can be written as

$$H = \sum_{i=1}^{M} H_i, \tag{2}$$

where H_i is the local Hamiltonian of the i'th cluster, and the total number of clusters is M. In what follows we will use a special notation $H_{cl} \equiv H_1$ for the first cluster. Local terms H_i can be transformed one to another by group actions, i.e.

$$H_i = U\,H_j\,U^\dagger \quad \text{for some} \quad U \in G. \tag{3}$$

Now let us derive our variational lower bound. It is well known that $E_{gs} = \inf_{\Psi}\langle\Psi|H|\Psi\rangle$, where infimum is taken over all normalized vectors Ψ of the Hilbert space. We observe that this equality can be alternatively formulated in terms of density matrices ρ of the closed quantum system under considerations. Remind that a density matrix is an operator which satisfies three conditions,

$$\rho^\dagger = \rho; \qquad \operatorname{tr}\rho = 1; \qquad \rho > 0. \tag{4}$$

One can replace optimization over vectors Ψ by optimization over density matrices ρ:

$$E_{gs} = \inf_{\rho\in\mathbb{S}^G} \operatorname{tr} H\rho, \tag{5}$$

Here $\mathbb{S}^G$ is a set of density matrices ρ invariant under the group G, i.e. in addition to conditions (4) the density matrices from the set $\mathbb{S}^G$ satisfy

$$U\,\rho\,U^\dagger = \rho \quad \text{for all} \quad U \in G. \tag{6}$$

Indeed, the density matrix which saturates the infimum in eq. (5) is simply the normalized projection onto the ground state subspace. This observation justifies eq. (5). It should be stressed that density matrices appear here and in what follows as a formal tool for calculating a bound on the ground state of a *closed* system. Whether a system can be actually prepared in mixed states described by these density matrices is of no relevance in our argument.

Further, using eqs. (2) and (3), we get

$$E_{gs} = M \inf_{\rho\in\mathbb{S}^G} \operatorname{tr}_{cl}\left(H_{cl}\operatorname{tr}_{\overline{cl}}\,\rho\right), \tag{7}$$

where tr_{cl} and $\mathrm{tr}_{\overline{cl}}$ are partial traces over the cluster and its complement, respectively. Note that while $\rho_{cl} \equiv \mathrm{tr}_{\overline{cl}}\,\rho$ is just the reduced density matrix of the cluster, variation in eq. (7) is not performed over the set of *all* ρ_{cl}. Instead, the minimization is performed over those ρ_{cl} which can be obtained from $\rho \in \mathbb{S}^G$. The set of ρ_{cl} satisfying the latter condition is unknown. However, we can lower bound E_{gs} by performing minimization over a larger set $\mathbb{S}^G_{cl}$ of the reduced density matrices of the cluster symmetric under the group G. This way we obtain our variational lower bound

$$E_{gs} \geq M \inf_{\rho_{cl}\in\mathbb{S}^G_{cl}} \mathrm{tr}_{cl}\, H_{cl}\rho_{cl}. \tag{8}$$

This bound is the main general result of the present paper. It should be stressed that H_{cl} is *not* invariant under the group G, in contrast to $\rho_{cl} \in \mathbb{S}^G_{cl}$.

We further observe that this bound can be enhanced by requiring that ρ_{cl} satisfies local sum rules which follow from the anti-Hermitian Stationary Schrödinger equation [6].

2.2. *Density matrix parametrization*

In order to be able to perform minimization in eq. (8) one needs to parameterize the set $\mathbb{S}^G_{cl}$ of density matrices. Among the conditions which determine this set, the positivity condition, $\rho_{cl} > 0$, is the most problematic due to its nonlinear nature. A squaring parametrization has been developed in [7] which automatically accounts for the positivity as well as other conditions of the type (4),(6). In addition, it is well-suited for many-body systems. Its main idea is that if we take an arbitrary hermitian matrix τ and square and normalize it, we get a valid density matrix:

$$\rho = \frac{\tau^2}{\mathrm{tr}\,\tau^2}, \qquad \text{where} \qquad \tau^\dagger = \tau. \tag{9}$$

We will use the squaring parametrization of density matrices to practically apply the bound (8).

2.3. *Comparison to the Anderson bound*

Let us remind the Anderson bound, which is arguably the first and the most widely used lower bound on E_{gs} [1]. It is based on a simple fact that an infimum of a sum is greater than the sum of infima. This leads to the bound

$$E_{gs} \geq M \inf_{\rho_{cl}} \mathrm{tr}_{cl}\, H_{cl}\rho_{cl}. \tag{10}$$

The infimum here is taken over all density matrices ρ_{cl} of a cluster. For this reason, the Anderson bound is weaker than our bound (8).

3. Application to a system of spins 1/2 with Heisenberg interactions

3.1. *Spin systems with Heisenberg interactions*

In the present section we demonstrate how the bound (8) can be applied to a system of N spins with the Heisenberg interaction. The Hamiltonian of this system reads

$$H = \sum_{<i,j>} (\boldsymbol{\sigma}_i \boldsymbol{\sigma}_j)\,, \tag{11}$$

where the sum is taken over all neighbouring sites on the lattice, $\boldsymbol{\sigma}_i$ is the vector consisting of three Pauli matrices of the i'th spin and $(\boldsymbol{\sigma}_i\boldsymbol{\sigma}_j)$ is the corresponding scalar product of sigma-matrices. This Hamiltonian is invariant with respect to a global $SU(2)$ symmetry, in other words, to the simultaneous rotations of all spins. It is also invariant under inversion of time and respects the spatial symmetries of the lattice.

IOP Conf. Series: Journal of Physics: Conf. Series **1163** (2019) 012057 doi:10.1088/1742-6596/1163/1/012057

As is discussed in details in [7, 6], a density matrix of the system (11) invariant under $SU(2)$ rotations and time inversion can be expressed in terms of scalar products of sigma matrices; the same is true for the auxiliary matrix τ from eq. (9):

$$\rho = 2^{-N} \sum_{\mathcal{A}} a_{\mathcal{A}} A_{\mathcal{A}}, \qquad \tau = \sum_{\mathcal{A}} b_{\mathcal{A}} A_{\mathcal{A}}, \tag{12}$$

$$\{A_{\mathcal{A}}\} = \{1,\ (\boldsymbol{\sigma}_j \boldsymbol{\sigma}_k),\ (\boldsymbol{\sigma}_j \boldsymbol{\sigma}_k)(\boldsymbol{\sigma}_l \boldsymbol{\sigma}_m)\ ,\ \ldots\},\quad i, k, l, m, \ldots = 1, 2, \ldots N. \tag{13}$$

Here $\mathcal{A}$ is a multi-index which enumerates the set (13), b_i are arbitrary real numbers while a_i are some functions of b_i determined by eq. (9).

3.2. Properties of the set $\{A_{\mathcal{A}}\}$

Here we discuss some properties of the set $A_{\mathcal{A}}$ and a related set $B_{\mathcal{A}}$ (see below). Some of this properties have immediate consequences for the implementation of our variational lower bound, while others may prove useful in further developments.

First, we list useful relations [7]

$$\begin{aligned}
(\boldsymbol{\sigma}_1\boldsymbol{\sigma}_2)^2 &= 3 - 2(\boldsymbol{\sigma}_1\boldsymbol{\sigma}_2)\\
(\boldsymbol{\sigma}_1\boldsymbol{\sigma}_2)(\boldsymbol{\sigma}_2\boldsymbol{\sigma}_3) &= (\boldsymbol{\sigma}_1\boldsymbol{\sigma}_3) - i(\boldsymbol{\sigma}_1\boldsymbol{\sigma}_2\boldsymbol{\sigma}_3)\\
(\boldsymbol{\sigma}_1\boldsymbol{\sigma}_2)(\boldsymbol{\sigma}_1\boldsymbol{\sigma}_2\boldsymbol{\sigma}_3) &= -(\boldsymbol{\sigma}_1\boldsymbol{\sigma}_2\boldsymbol{\sigma}_3) - 2i(\boldsymbol{\sigma}_1\boldsymbol{\sigma}_3) + 2i(\boldsymbol{\sigma}_2\boldsymbol{\sigma}_3)\\
(\boldsymbol{\sigma}_1\boldsymbol{\sigma}_2\boldsymbol{\sigma}_3)(\boldsymbol{\sigma}_1\boldsymbol{\sigma}_2) &= -(\boldsymbol{\sigma}_1\boldsymbol{\sigma}_2\boldsymbol{\sigma}_3) + 2i(\boldsymbol{\sigma}_1\boldsymbol{\sigma}_3) - 2i(\boldsymbol{\sigma}_2\boldsymbol{\sigma}_3)\\
(\boldsymbol{\sigma}_1\boldsymbol{\sigma}_2)(\boldsymbol{\sigma}_2\boldsymbol{\sigma}_3\boldsymbol{\sigma}_4) &= (\boldsymbol{\sigma}_1\boldsymbol{\sigma}_3\boldsymbol{\sigma}_4) - i(\boldsymbol{\sigma}_1\boldsymbol{\sigma}_3)(\boldsymbol{\sigma}_2\boldsymbol{\sigma}_4) + i(\boldsymbol{\sigma}_1\boldsymbol{\sigma}_4)(\boldsymbol{\sigma}_2\boldsymbol{\sigma}_3)\\
(\boldsymbol{\sigma}_2\boldsymbol{\sigma}_3\boldsymbol{\sigma}_4)(\boldsymbol{\sigma}_1\boldsymbol{\sigma}_2) &= (\boldsymbol{\sigma}_1\boldsymbol{\sigma}_3\boldsymbol{\sigma}_4) + i(\boldsymbol{\sigma}_1\boldsymbol{\sigma}_3)(\boldsymbol{\sigma}_2\boldsymbol{\sigma}_4) - i(\boldsymbol{\sigma}_1\boldsymbol{\sigma}_4)(\boldsymbol{\sigma}_2\boldsymbol{\sigma}_3)\\
(\boldsymbol{\sigma}_1\boldsymbol{\sigma}_2\boldsymbol{\sigma}_3)^2 &= 6 - 2(\boldsymbol{\sigma}_1\boldsymbol{\sigma}_2) - 2(\boldsymbol{\sigma}_1\boldsymbol{\sigma}_3) - 2(\boldsymbol{\sigma}_2\boldsymbol{\sigma}_3)\\
(\boldsymbol{\sigma}_1\boldsymbol{\sigma}_2\boldsymbol{\sigma}_3)(\boldsymbol{\sigma}_1\boldsymbol{\sigma}_2\boldsymbol{\sigma}_4) &= -(\boldsymbol{\sigma}_1\boldsymbol{\sigma}_3)(\boldsymbol{\sigma}_2\boldsymbol{\sigma}_4) - (\boldsymbol{\sigma}_1\boldsymbol{\sigma}_4)(\boldsymbol{\sigma}_2\boldsymbol{\sigma}_3) + 2(\boldsymbol{\sigma}_3\boldsymbol{\sigma}_4) + i(\boldsymbol{\sigma}_1\boldsymbol{\sigma}_3\boldsymbol{\sigma}_4) + i(\boldsymbol{\sigma}_2\boldsymbol{\sigma}_3\boldsymbol{\sigma}_4)\\
(\boldsymbol{\sigma}_1\boldsymbol{\sigma}_2\boldsymbol{\sigma}_3)(\boldsymbol{\sigma}_1\boldsymbol{\sigma}_4\boldsymbol{\sigma}_5) &= +(\boldsymbol{\sigma}_2\boldsymbol{\sigma}_4)(\boldsymbol{\sigma}_3\boldsymbol{\sigma}_5) - (\boldsymbol{\sigma}_2\boldsymbol{\sigma}_5)(\boldsymbol{\sigma}_3\boldsymbol{\sigma}_4) - i(\boldsymbol{\sigma}_1\boldsymbol{\sigma}_2)(\boldsymbol{\sigma}_3\boldsymbol{\sigma}_4\boldsymbol{\sigma}_5) + i(\boldsymbol{\sigma}_1\boldsymbol{\sigma}_3)(\boldsymbol{\sigma}_2\boldsymbol{\sigma}_4\boldsymbol{\sigma}_5),
\end{aligned} \tag{14}$$

where $(\boldsymbol{\sigma}_1\boldsymbol{\sigma}_2\boldsymbol{\sigma}_3)$ is the mixed product of sigma matrices. Further, a product of two mixed products can always be represented through scalar products:

$$(\boldsymbol{\sigma}_1\boldsymbol{\sigma}_2\boldsymbol{\sigma}_3)(\boldsymbol{\sigma}_4\boldsymbol{\sigma}_5\boldsymbol{\sigma}_6) = \det \begin{pmatrix} (\boldsymbol{\sigma}_1\boldsymbol{\sigma}_4) & (\boldsymbol{\sigma}_2\boldsymbol{\sigma}_4) & (\boldsymbol{\sigma}_3\boldsymbol{\sigma}_4) \\ (\boldsymbol{\sigma}_1\boldsymbol{\sigma}_5) & (\boldsymbol{\sigma}_2\boldsymbol{\sigma}_5) & (\boldsymbol{\sigma}_3\boldsymbol{\sigma}_5) \\ (\boldsymbol{\sigma}_1\boldsymbol{\sigma}_6) & (\boldsymbol{\sigma}_2\boldsymbol{\sigma}_6) & (\boldsymbol{\sigma}_3\boldsymbol{\sigma}_6) \end{pmatrix}. \tag{15}$$

One can introduce scalar product on the space of operators according to

$$(X, Y) \equiv \mathrm{tr}(X^\dagger Y) \tag{16}$$

(not to be confused with the scalar product of sigma matrices). We consider the case when $X, Y \in \{A_{\mathcal{A}}\}$. If supports of X and Y on a lattice do not coincide, then $(X, Y) = 0$. If X and Y

IOP Conf. Series: Journal of Physics: Conf. Series **1163** (2019) 012057 doi:10.1088/1742-6596/1163/1/012057

Table 1. The size $K(N)$ of the overcomplete set $\{A_{\mathcal{A}}\}$. One can see that for $N \lesssim 30$ this size is below the number of real parameters of the corresponding density matrix.

N	2	3	4	5	10	15	20	30	40	50	60
$K(N)$	1	3	9	25	9495	1E+7	2E+10	6E+17	7E+25	2E+34	2E+43
4^N	16	64	256	1024	1048576	1E+9	1E+12	1E+18	1E+24	1E+30	1E+36

have the same support, then $(X, Y) = 2^N 3^C$, where C is the number of cycles arising when the bonds contained in X are superimposed on the bonds contained in Y (*cf.* ref. [8]). In particular,

$$\mathrm{tr}(\ (\boldsymbol{\sigma}_i\boldsymbol{\sigma}_j)(\boldsymbol{\sigma}_j\boldsymbol{\sigma}_k)...(\boldsymbol{\sigma}_l\boldsymbol{\sigma}_m)(\boldsymbol{\sigma}_m\boldsymbol{\sigma}_i)\ ) = 3 \cdot 2^N \tag{17}$$

(one cycle) and

$$\mathrm{tr}(\ (\boldsymbol{\sigma}_i\boldsymbol{\sigma}_j\boldsymbol{\sigma}_k)(\boldsymbol{\sigma}_i\boldsymbol{\sigma}_j\boldsymbol{\sigma}_n)\ (\boldsymbol{\sigma}_k\boldsymbol{\sigma}_l)...(\boldsymbol{\sigma}_m\boldsymbol{\sigma}_n)\ ) = \mathrm{tr}(\ (\boldsymbol{\sigma}_i\boldsymbol{\sigma}_j\boldsymbol{\sigma}_k)(\boldsymbol{\sigma}_i\boldsymbol{\sigma}_j\boldsymbol{\sigma}_k)\ ) = 6 \cdot 2^N. \tag{18}$$

Let us consider a set $B_{\mathcal{B}}$ of the following form:

$$\{B_{\mathcal{B}}\} = \{(\boldsymbol{\sigma}_p\boldsymbol{\sigma}_r\boldsymbol{\sigma}_s),\ (\boldsymbol{\sigma}_p\boldsymbol{\sigma}_r\boldsymbol{\sigma}_s)(\boldsymbol{\sigma}_j\boldsymbol{\sigma}_k),\ (\boldsymbol{\sigma}_p\boldsymbol{\sigma}_r\boldsymbol{\sigma}_s)(\boldsymbol{\sigma}_j\boldsymbol{\sigma}_k)(\boldsymbol{\sigma}_l\boldsymbol{\sigma}_m)\ ,\ ...\}$$

Elements of sets $\{A_{\mathcal{A}}\}$ and $\{A_{\mathcal{A}}, B_{\mathcal{B}}\} \equiv A_{\mathcal{A}} \bigcup B_{\mathcal{A}}$ are not mutually orthogonal. For example,

$$\begin{aligned} \mathrm{X} &= (\boldsymbol{\sigma}_1\boldsymbol{\sigma}_2)(\boldsymbol{\sigma}_3\boldsymbol{\sigma}_4) \\ \mathrm{Y} &= (\boldsymbol{\sigma}_1\boldsymbol{\sigma}_3)(\boldsymbol{\sigma}_2\boldsymbol{\sigma}_4) \\ \mathrm{Z} &= (\boldsymbol{\sigma}_1\boldsymbol{\sigma}_4)(\boldsymbol{\sigma}_2\boldsymbol{\sigma}_3) \\ g &= \begin{pmatrix} (XX) & (XY) & (XZ) \\ (YX) & (YY) & (YZ) \\ (ZX) & (ZY) & (ZZ) \end{pmatrix} = \begin{pmatrix} 9 & 3 & 3 \\ 3 & 9 & 3 \\ 3 & 3 & 9 \end{pmatrix} \end{aligned} \tag{19}$$

Further, the set $\{A_{\mathcal{A}}\}$ (and, consequently, $\{A_{\mathcal{A}}, B_{\mathcal{B}}\}$) is overcomplete. However, we put forward a hypothesis that all linear dependencies within $\{A_{\mathcal{A}}, B_{\mathcal{B}}\}$ can be described by

$$+(\boldsymbol{\sigma}_1\boldsymbol{\sigma}_2)(\boldsymbol{\sigma}_3\boldsymbol{\sigma}_4\boldsymbol{\sigma}_5) - (\boldsymbol{\sigma}_1\boldsymbol{\sigma}_3)(\boldsymbol{\sigma}_2\boldsymbol{\sigma}_4\boldsymbol{\sigma}_5) + (\boldsymbol{\sigma}_1\boldsymbol{\sigma}_4)(\boldsymbol{\sigma}_2\boldsymbol{\sigma}_3\boldsymbol{\sigma}_5) - (\boldsymbol{\sigma}_1\boldsymbol{\sigma}_5)(\boldsymbol{\sigma}_2\boldsymbol{\sigma}_3\boldsymbol{\sigma}_4) = 0 \tag{20}$$

for elements with an odd number of spins, and by

$$\det \begin{pmatrix} (\boldsymbol{\sigma}_1\boldsymbol{\sigma}_5) & (\boldsymbol{\sigma}_2\boldsymbol{\sigma}_5) & (\boldsymbol{\sigma}_3\boldsymbol{\sigma}_5) & (\boldsymbol{\sigma}_4\boldsymbol{\sigma}_5) \\ (\boldsymbol{\sigma}_1\boldsymbol{\sigma}_6) & (\boldsymbol{\sigma}_2\boldsymbol{\sigma}_6) & (\boldsymbol{\sigma}_3\boldsymbol{\sigma}_6) & (\boldsymbol{\sigma}_4\boldsymbol{\sigma}_6) \\ (\boldsymbol{\sigma}_1\boldsymbol{\sigma}_7) & (\boldsymbol{\sigma}_2\boldsymbol{\sigma}_7) & (\boldsymbol{\sigma}_3\boldsymbol{\sigma}_7) & (\boldsymbol{\sigma}_4\boldsymbol{\sigma}_7) \\ (\boldsymbol{\sigma}_1\boldsymbol{\sigma}_8) & (\boldsymbol{\sigma}_2\boldsymbol{\sigma}_8) & (\boldsymbol{\sigma}_3\boldsymbol{\sigma}_8) & (\boldsymbol{\sigma}_4\boldsymbol{\sigma}_8) \end{pmatrix} = 0 \tag{21}$$

IOP Conf. Series: Journal of Physics: Conf. Series **1163** (2019) 012057 doi:10.1088/1742-6596/1163/1/012057

Table 2. Anderson bound compared to the bound (8) for a translation-invariant linear chain with the Heisenberg Hamiltonian (23). Given are the lower bounds on the ground state energy per spin, E_{gs}/N. The exact Bethe ansatz result reads $E_{gs}/N = 1 - 4\log 2 \simeq -1.77259$.

cluster size	Anderson bound	bound (8)
3	-2.0	-2.0
4	-2.1547	-2.0
5	-1.9279	-1.8685
6	-1.9947	-1.8685
7	-1.8908	-1.8255

for elements with an even number of spins. Observe that the latter formula is a consequence of eqs. (15) and (20). We tested this hypothesis for up to 10 spins.

The number of elements in the basis $\{A_{\mathcal{A}}\}$ without taking into account overcompleteness is equal to

$$K(N) = \sum_{k=0}^{[N/2]} C_N^{2k}\,(2k-1)!!\,, \tag{22}$$

where $[N/2]$ is the integer part of $N/2$. $K(N)$ grows faster then exponentially with N (see table 1), however it can be rather small for $N \sim 10$.

3.3. An example

Here we consider a one-dimensional lattice of N spins 1/2 with the nearest-neighbour Heisenberg Hamiltonian

$$H = \sum_{j=1}^{N} (\boldsymbol{\sigma}_j \boldsymbol{\sigma}_{j+1})\,, \tag{23}$$

where $\boldsymbol{\sigma}_{N+1} \equiv \boldsymbol{\sigma}_1$ which ensures translation invariance. This system is integrable by means of Bethe ansatz and the ground state energy per spin in the thermodynamic limit reads $E_{gs}/N = 1 - 4\log 2$ [9]. Our goal is to lower bound E_{gs}/N by means of our method and to compare this bound to the exact value and to the Anderson bound [1].

We start from considering a cluster with 4 spins. Its Hamiltonian reads

$$H_{cl} = (\boldsymbol{\sigma}_1\boldsymbol{\sigma}_2) + (\boldsymbol{\sigma}_2\boldsymbol{\sigma}_3) + (\boldsymbol{\sigma}_3\boldsymbol{\sigma}_4) \tag{24}$$

The basis supported by the cluster reads

$$\begin{aligned}\{A_k\} = \{&1,\ (\boldsymbol{\sigma}_1\boldsymbol{\sigma}_2),\ (\boldsymbol{\sigma}_1\boldsymbol{\sigma}_3),\ (\boldsymbol{\sigma}_1\boldsymbol{\sigma}_4),\ (\boldsymbol{\sigma}_2\boldsymbol{\sigma}_3),\ (\boldsymbol{\sigma}_2\boldsymbol{\sigma}_4),\ (\boldsymbol{\sigma}_3\boldsymbol{\sigma}_4),\\ &(\boldsymbol{\sigma}_1\boldsymbol{\sigma}_2)(\boldsymbol{\sigma}_3\boldsymbol{\sigma}_4),\ (\boldsymbol{\sigma}_1\boldsymbol{\sigma}_3)(\boldsymbol{\sigma}_2\boldsymbol{\sigma}_4),\ (\boldsymbol{\sigma}_1\boldsymbol{\sigma}_4)(\boldsymbol{\sigma}_2\boldsymbol{\sigma}_3)\},\quad k = 0, 1, , ...9.\end{aligned} \tag{25}$$

We apply the squaring parametrization [7] in the form

$$\rho_{cl} = \tau^2. \tag{26}$$

Here the reduced density matrix, ρ_{cl}, and the axillary matrix, τ, are expanded in the basis (25) as

$$\rho_{cl} = 2^{-4} a_k A_k, \qquad \tau = b_k A_k, \tag{27}$$

where summation over repeating indices is implied. The normalization of ρ_{cl} is imposed by the constraint

$$a_0 = 1. \tag{28}$$

The squaring parametrization (26) implies that each coefficient a_k is a quadratic function of coefficients b_k.

Translational invariance implies

$$a_{(1,2)} = a_{(2,3)} = a_{(3,4)}, \quad a_{(1,3)} = a_{(2,4)}. \tag{29}$$

Here we use self-explanatory notations for indexes, e.g. $a_{(1,2)}$ is the coefficient in front of $(\boldsymbol{\sigma}_1 \boldsymbol{\sigma}_2)$. The Hamiltonian (24) of the cluster is not translationally invariant, but it posses a remaining mirror symmetry. For this reason two of the above equalities can be satisfied seamlessly:

$$b_{(1,2)} = b_{(3,4)}, \quad b_{(1,3)} = b_{(2,4)} \quad \Rightarrow \quad a_{(1,2)} = a_{(3,4)}, \quad a_{(1,3)} = a_{(2,4)}. \tag{30}$$

The condition

$$a_{(1,2)} = a_{(2,3)} \tag{31}$$

remains and should be accounted for during optimization.

Finally we perform a numerical search for a minimum of $\operatorname{tr} H_{cl} \tau^2$ with the constraints (28) and (31), which are equivalent to $\operatorname{tr} \tau^2 = 1$ and $\operatorname{tr}(\boldsymbol{\sigma}_1, \boldsymbol{\sigma}_2)\tau^2 = (\boldsymbol{\sigma}_2, \boldsymbol{\sigma}_3)\tau^2$, respectively. The resulting bound is presented in table 2, along with the analogous bounds with other cluster sizes. One can see that for a given cluster size the bound (8) outperforms the Anderson bound. The caveat here is that for a given cluster size the calculations for our bound (8) require much more resources than those for the Anderson bound. Whether the bound (8) is able to compete with the Anderson bound in practical numerical computations is a question open for future research.

4. Summary

We have derived a variational *lower* bound (8) on the ground state energy of a quantum system possessing symmetries. The variation should be performed over a certain set of reduced density matrices. Technically, the variation can be performed by means of the squaring parametrization of density matrices [7] which automatically satisfies the positivity condition. We have discussed how this bound can be applied for translation-invariant spin systems with the Heisenberg interaction. Promising results has been obtained in a simple example of a linear chain, see table 2.

Acknowledgements.

We are grateful to E. Shpagina and N. Il'in for useful discussions. The work was supported by the Russian Science Foundation under the grant No. 17-71-20158.

References

[1] Anderson PW 1951 Limits on the energy of the antiferromagnetic Ground State *Phys. Rev.* **83(6)** 1260
[2] Mattis DC and Pan CY 1988 Ground-State Energy of Heisenberg antiferromagnet for Spins s=1/2 and s=1 in d=1 and 2 Dimensions *Phys.Rev.Lett.* **61(4)** 463
[3] Nishimori H and Ozeki Y 1989 Ground-State Long-Range Order in the Two-Dimensional XXZ Model *J. Phys. Soc. Jpn.* **58** 1027
[4] Schwerdtfeger CA and Mazziotti DA 2009 Convex-set description of quantum phase transitions in the transverse Ising model using reduced-density-matrix theory. *The Journal of chemical physics* **130(22)** 224102

[5] Baumgratz T and Plenio MB 2012 Lower bounds for ground states of condensed matter systems *New Journal of Physics* **14(2)** 023027
[6] Shpagina E, Uskov F, Il'in N, and Lychkovskiy O 2018 Stationary Schrdinger equation with density matrices instead of wave functions, *submitted to present proceedings*
[7] Il'in N, Shpagina E, Uskov F and Lychkovskiy O 2018 Squaring parametrization of constrained and unconstrained sets of quantum states *J. Phys.* A: Math. Theor. **51**.085301
[8] Beach KSD and Sandvik AW 2006 Some formal results for the valence bond basis *Nuclear Physics* B **750(3)** 142178
[9] Bethe H 1931 Zur theorie der metalle *Zeitschrift fr Physik* **71** 205

IOP Conf. Series: Journal of Physics: Conf. Series **1163** (2019) 012058 doi:10.1088/1742-6596/1163/1/012058

Singularity of optimal control in the problem of stabilizing a nonlinear inverted spherical pendulum

Larisa Manita[1] **and Mariya Ronzhina**[2,3]

[1] National Research University Higher School of Economics, Moscow Institute of Electronics and Mathematics, Russia

[2] Lomonosov Moscow State University, Faculty of Mechanics and Mathematics, 119991, GSP-1, Moscow, Russia

[3] Gubkin Russian State University of Oil and Gas (National Research University), Moscow, Russia

E-mail: lmanita@hse.ru, maryaronzhina@gmail.com

Abstract. We study an optimal control problem for a nonlinear spherical inverted pendulum on a movable base. As the cost functional, the mean-squared deviation of the pendulum from the upper equilibrium is considered, so optimal controls stabilize the pendulum at the unstable upper position. We show that the problem under consideration posses a singular point of the second order and there are spiral-similar solution which attains the singular point in finite time.

1. Introduction

Models of inverted pendulum systems are widely used to study the dynamics of different real complex nonlinear objects. There are many papers in which computer modeling and simulation are used for controlling and stabilizing the inverted pendulum systems. In [1, 2] for a spherical inverted pendulum a stabilizing controller is designed that brings the pendulum from any initial condition in the upper hemisphere to the preferred upright position. In [3] the controller stabilizing a spherical inverted pendulum-cart system can bring the pendulum to the unstable upright equilibrium point with the position of the cart at a desirable target position. In [4] it was proposed a controller for the tracking of elliptical paths. Many methods to control pendulum systems use an optimal control technique: a stabilizing controller is derived from minimization of a quadratic cost functional (LQR controllers) [5]– [7]. But applied to real physical systems, such controllers produced a relatively large deviations of pendulum systems from the upper equilibrium position. To improve this, there are different approaches, i.e., in [5] it was proposed to combine optimal and neural network technics. In present work we study an optimal synthesis in the minimization problem of the mean square deviation of a spherical inverted pendulum from the upper equilibrium position over an infinite time interval. For the linearized model optimal solutions were found in the form of logarithmic spirals [11, 13] that hit the origin in a finite time T. The corresponding optimal controls perform an infinite number of rotations along the circle S^1. We generalize the results obtained for the linear model to the nonlinear case.

IOP Conf. Series: Journal of Physics: Conf. Series **1163** (2019) 012058 doi:10.1088/1742-6596/1163/1/012058

2. Mathematical model of a spherical inverted pendulum

We consider a mathematical model of a spherical inverted pendulum with a moving support point (Fig.1). The pendulum consists of the point mass B on the end of the rigid massless rod of length l. The rod is attached by a hinge to a moving support point S. We suppose that there is no friction in the hinge. The support S can move in the horizontal plane under the action of an external planar control force $u = (u_1, u_2) \in \mathbb{R}^2$.

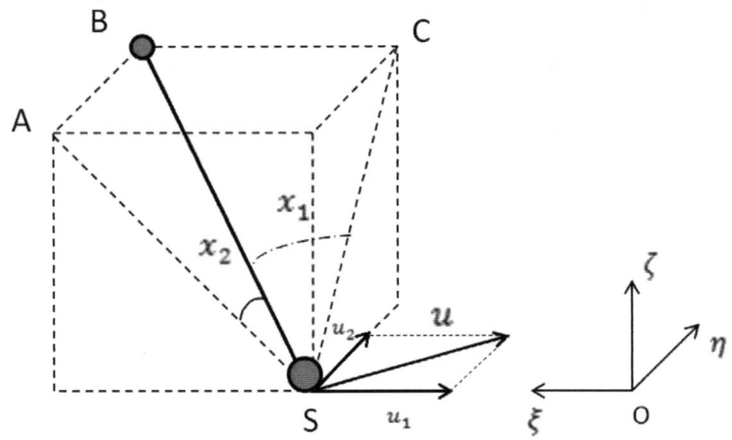

Figure 1. The spherical inverted pendulum

Let $O\xi\eta\zeta$ be a fixed coordinate system. The position of S is described by $(\xi, \eta, 0)$. The position of the pendulum B is described by (x_1, x_2). Using the Euler-Lagrange equations we obtain nonlinear equations of motion of the spherical inverted pendulum system [11]:

$$\begin{gathered}(M+m)\ddot{\xi} + ml\ddot{x}_1 \cos x_1 - ml\dot{x}_1^2 \sin x_1 = u_1 \\ (M+m)\ddot{\eta} + ml\ddot{x}_2 \cos x_2 - ml\dot{x}_2^2 \sin x_2 = u_2 \\ \ddot{\xi}b_{11} + \ddot{x}_1 b_{13} + \ddot{x}_2 b_{14} + \dot{x}_1^2 a_{11} + \dot{x}_2^2 a_{12} + \dot{x}_1\dot{x}_2 a_{13} + c_1 = 0 \\ \ddot{\eta}b_{22} + \ddot{x}_1 b_{23} + \ddot{x}_2 b_{24} + \dot{x}_1^2 a_{21} + \dot{x}_2^2 a_{22} + \dot{x}_1\dot{x}_2 a_{23} + c_2 = 0\end{gathered} \tag{1}$$

where m – mass of B, M – mass of S, b_{ij}, a_{ij}, c_i are some functions of $\cos x_k$ and $\sin x_k$. Eliminating the variables ξ and η we can rewrite (1) in the following form

$$\begin{gathered}\dot{x}_1 = y_1, \quad \dot{x}_2 = y_2 \\ \dot{y}_1 = \alpha_{11}u_1 + \alpha_{21}u_2 + \beta_1, \\ \dot{y}_2 = \alpha_{12}u_1 + \alpha_{22}u_2 + \beta_2\end{gathered} \tag{2}$$

Here α_{ij}, β_k are functions of x_i, y_i. We will study solutions of the following optimal control problem for nonlinear spherical inverted pendulum: to minimize the mean square deviation of the pendulum from the upper unstable equilibrium position over an infinite time interval.

3. Optimal Control problem

Denote $x(t) = (x_1(t), x_2(t))$, $y(t) = (y_1(t), y_2(t))$. Assume that initial states of the system $(x(0), y(0))$ are in a sufficiently small neighbourhood of the upper unstable equilibrium position $x = y = 0$. Our goal is to find the bounded control function $u(t)$

$$u(t) \in U = \left\{ u_1^2 + u_2^2 \leq 1 \right\} \tag{3}$$

and the corresponding state function $(x(t), y(t))$ which satisfies (2) and minimizes

$$\int_0^\infty (x_1^2(t) + x_2^2(t))\, dt \tag{4}$$

Following the Pontryagin maximum principle we define the Hamiltonian

$$H(x, y, \phi, \psi, u) = -\frac{1}{2}\langle x, x\rangle + \langle y, \phi\rangle + \langle \beta, \psi\rangle + u_1\langle \alpha_1, \psi\rangle + u_2\langle \alpha_2, \psi\rangle = H_0 + u_1 H_1 + u_2 H_2 \tag{5}$$

where $\langle \cdot, \cdot \rangle$ is the scalar product in $\mathbb{R}^2$, ϕ, ψ are adjoint variables that satisfy the adjoint equation

$$\dot{\phi} = -\frac{\partial H}{\partial x}, \quad \dot{\psi} = -\frac{\partial H}{\partial y}, \tag{6}$$

Assume that $(x^*(t), y^*(t), u^*(t))$ is an optimal solution for (2)–(4). The optimal control is determined from the maximum condition:

$$u^*(t) = \underset{u \in U}{\operatorname{argmax}}\, H(x^*(t), y^*(t), \phi(t), \psi(t), u) \tag{7}$$

(3), (5) and (7) imply that $u_i^* = H_i / \left(H_1^2 + H_2^2\right)^{1/2}$, $H_1^2 + H_2^2 \neq 0$. If $H_1^2 + H_2^2 = 0$ over some interval (θ_1, θ_2) than any admissible control meets the maximum condition. In this case there is a singular arc. We prove that in the problem (2)–(4) there is a second-order singular point. We show that there is a family of solutions that approach the singular point $\left(x^0, y^0\right)$ in finite time T. If $t \to T$ then $(x^*(t), y^*(t)) \to \left(x^0, y^0\right)$ and switches to the singular mode. The corresponding control $u^*(t)$ has a singularity of the second type at T and makes countably many rotations along the circle S^1 in finite time as $t \to T$.

4. Singular solution

Denote by $q = (x, y)$, $p = (\phi, \psi)$. Let $(\mathrm{ad}A)F$ be the Poisson bracket of functions A and F: $(\mathrm{ad}A)F = \{F, A\}$.

Definition. Point $(q_0, p_0) \in \mathbb{R}^8$ is the *singular point of second order*, if the following conditions are satisfied:

1) Functions
H_i, $(\mathrm{ad}H_k)H_i$, $(\mathrm{ad}H_l)(\mathrm{ad}H_k)H_i$, $(\mathrm{ad}H_j)(\mathrm{ad}H_l)(\mathrm{ad}H_k)H_i$, $(i = 1, 2,\ j, k, l = 0, 1, 2)$
are equal to zero at the point (q_0, p_0). The set of their differentials at the point (q_0, p_0) has a constant rank.

2) The bilinear form $B = (B_{ij})_{i,j=1,2}$: $B_{ij} = \mathrm{ad}H_i(\mathrm{ad}H_0)^3 H_j\big|_{(q_0,p_0)}$, $i, j = 1, 2$, has rank 2 and is symmetric and negative definite.

3) All other the Poisson brackets of the fifth order (independent of the listed) from the functions H_j $(j = 0, 1, 2)$ are equal to zero at the point (q_0, p_0).

***Theorem* 1.** Point $(q_0, p_0) = (0, 0)$ is a singular point of the second order of the control Hamiltonian system (2), (5)–(7).

IOP Conf. Series: Journal of Physics: Conf. Series **1163** (2019) 012058 doi:10.1088/1742-6596/1163/1/012058

Direct calculations show that the point satisfies the conditions 1-3 of the definition with $B = -E$, where E is the identity matrix.

To study the behaviour of the solutions of (2), (5)–(7) in the neighborhood of the singular point (q_0, p_0) we use the construction of the descending system of Poisson brackets of the functions H_i. We call the brackets H_i, $(\mathrm{ad}H_0)H_i$, $(\mathrm{ad}H_0)^2H_i$, $(\mathrm{ad}H_0)^3H_i\,(i = 1, 2)$ *primary* brackets. The other brackets of order not exceeding 4 are called *secondary.*

Denote $z_{1i} = H_i, z_{2i} = (\mathrm{ad}H_0)H_i, z_{3i} = (\mathrm{ad}H_0)^2H_i, z_{4i} = (\mathrm{ad}H_0)^3H_i(i = 1, 2)$. Introduce new local coordinates z_{ij}, $(i = 1, \ldots, 4,\ j = 1, 2)$ in the neighborhood of the point (q_0, p_0). In new coordinates the Hamiltonian system (2), (5)–(6) has the following form:

$$\begin{aligned}
\frac{d}{dt}z_{1i} &= z_{2i} + f_{1i1}(z)u_1 + f_{1i2}(z)u_2 \\
\frac{d}{dt}z_{2i} &= z_{3i} + f_{2i1}(z)u_1 + f_{2i2}(z)u_2 \qquad (8)\\
\frac{d}{dt}z_{3i} &= z_{4i} + f_{3i1}(z)u_1 + f_{3i2}(z)u_2 \\
\frac{d}{dt}z_{4i} &= (\mathrm{ad}H_0)z_{4i} + (\mathrm{ad}H_1)z_{4i}u_1 + (\mathrm{ad}H_2)z_{4i}u_2 \\
u_i &= z_{1i}/\left(z_{11}^2 + z_{12}^2\right)^{1/2}, \quad i = 1, 2 \qquad (9)
\end{aligned}$$

Here $f_{ijk}(z)$, $(i = 1, \ldots, 4,\ j = 1, 2,\ k = 1, 2)$ are the secondary brackets. It was proved [8] that the first order approximation of the behavior of the Hamiltonian system (8) in the neighborhood of the singular extremal is fully determined by the primary Poisson brackets. The first-order approximation of the behavior of the primary Poisson brackets coincides with the optimal synthesis in the model problem provided below. Under first approximation we understand an approximation of the order $o(\lambda)$ with respect to the homothety g_λ which is used for blowing up the singularity.

5. Model Problem

$$\int_0^\infty \|x(t)\|^2\, dt \to \inf$$

$$\dot{x} = y, \quad \dot{y} = u, \quad \|u(t)\| \le 1, \quad x(0) = x^0, \quad y(0) = y^0 \qquad (10)$$

Here $x,\ y,\ u \in \mathbb{R}^2$, $\|\cdot\|$ means the standard Euclidean norm on $\mathbb{R}^2$. The Hamiltonian system of the maximum principle for the problem is given by

$$\dot{\phi} = x, \quad \dot{\psi} = -\phi, \quad \dot{x} = y, \quad \dot{y} = \hat{u} \qquad (11)$$

$$\hat{u}(t) = argmax_{\|u\|\le 1} H(x(t), y(t), \phi(t), \psi(t), u) \qquad (12)$$

where

$$H(x, y, \phi, \psi) = -\frac{1}{2}\langle x, x\rangle + \langle y, \phi\rangle + \langle u, \psi\rangle \qquad (13)$$

The system (11) is homogeneous with respect to the action of an one-parameter group $G = \{g_\lambda\}, \lambda \in \mathbb{R}_+$

$$g_\lambda(y, x, \phi, \psi) \overset{def}{=} \left(\lambda y,\ \lambda^2 x,\ \lambda^3\phi,\ \lambda^4\psi\right)$$

This property was used [9, 10] to find solutions of (11)–(13) in the form of logarithmic spirals

$$\begin{gathered}
x(t) = A_2\,(T-t)^2\, e^{i\alpha ln|T-t|},\ y(t) = A_3\,(T-t)\, e^{i\alpha ln|T-t|},\ u(t) = e^{i\alpha ln|T-t|}, \\
\phi(t) = A_1\,(T-t)^3\, e^{i\alpha ln|T-t|},\ \psi(t) = A_0\,(T-t)^4\, e^{i\alpha ln|T-t|}, \qquad (14)\\
A_0 = 1/126,\ \alpha = \pm\sqrt{5},\ A_{j+1} = -A_j(4 - j + i\alpha),\ j = 1, 2, 3.
\end{gathered}$$

IOP Conf. Series: Journal of Physics: Conf. Series **1163** (2019) 012058 doi:10.1088/1742-6596/1163/1/012058

where T is a time at which solution hits the origin (the hitting time). Here we use the complex notation for vectors in $\mathbb{R}^2$: $re^{i\varphi} = (r\cos\varphi, r\sin\varphi)$.

6. Main Result

The system (2), (5)–(7) does not obey homogeneity. However, we show that in this case there are similar logarithmic spirals. To prove this we use for the Hamiltonian systems (11)–(13) and (8) the procedure of resolution of singularity that was proposed in [9], and then improved in [8]. We use the scheme as in [8], [12,14] and the similar change of coordinates.

Consider the blowing up the singularity for (11)–(13) at the origin by the map

$$B : (x, y, \phi, \psi) \mapsto (\mu, \tilde{x}, \tilde{y}, \tilde{\phi}, \tilde{\psi})$$

$$\mu = \left(|y|^{24} + |x|^{12} + |\phi|^8 + |\psi|^6\right)^{\frac{1}{24}},$$

$$\left(\tilde{y}, \tilde{x}, \tilde{\phi}, \tilde{\psi}\right) = g_{1/\mu}\left(y, x, \phi, \psi\right),$$

Here $\mu \in \mathbb{R}_+$, $A_0 = 1/126$, $A_{j+1} = -A_j(4 - j + i\alpha)$, $\alpha = \sqrt{5}$, and $\tilde{x}$, $\tilde{y}$, $\tilde{\phi}$, $\tilde{\psi} \in \mathbb{R}^2$ lie on the manifold $S = \left\{|\tilde{y}|^{24} + |\tilde{x}|^{12} + \left|\tilde{\phi}\right|^8 + \left|\tilde{\psi}\right|^6 = 1\right\}$.

In the coordinates $(\mu, \tilde{x}, \tilde{y}, \tilde{\phi}, \tilde{\psi})$ and a new time parameter s defined by $ds = \frac{1}{\mu}dt$ the system (11) has the form:

$$\begin{aligned} \mu' &= \mu\mathcal{M}_0(\tilde{x}, \tilde{y}, \tilde{\phi}, \tilde{\psi}), & \tilde{x}' &= \tilde{y} - 2\tilde{x}\mathcal{M}_0, \\ \tilde{\psi}' &= -\tilde{\phi} - 4\tilde{\psi}\mathcal{M}_0, & \tilde{y}' &= u - \tilde{y}\mathcal{M}_0, \\ \tilde{\phi}' &= \tilde{x} - 3\tilde{\phi}\mathcal{M}_0, & u &= \tilde{\psi}/|\tilde{\psi}|. \end{aligned} \tag{15}$$

where $\mathcal{M}_0 = \frac{1}{24}\left(24|\tilde{y}|^{22}\langle\tilde{y}, u\rangle + 12|\tilde{x}|^{10}\langle\tilde{x}, \tilde{y}\rangle + 8|\tilde{\phi}|^6\langle\phi, x\rangle - 6|\tilde{\psi}|^4\langle\psi, \phi\rangle\right)$.

Using similar techniques we apply the blowup for (8) at (q^0, p^0). We get that the blowup for the principal part of the descending system (8) coincides with the system (15). The spirals (14) represent hyperbolic periodic trajectories (circles) in the space $(\tilde{x}, \tilde{y}, \tilde{\phi}, \tilde{\psi})$ and exist in both the Hamiltonian systems. By the invariant manifold theorem for a limit cycle [15] we have the following result.

***Theorem* 2.** In a sufficiently small neighborhood of the origin there exist solutions of (2)–(4) that attain the upper equilibrium position and have the form of logarithmic spirals

$$\begin{aligned} x(t) &= C_x(T-t)^2 e^{i\varrho ln|T-t|}(1 + g_x\left(T-t\right)), \\ y(t) &= C_y(T-t) e^{i\varrho ln|T-t|}(1 + g_y\left(T-t\right)), \\ u(t) &= -C_u e^{i\varrho ln|T-t|}(1 + g_u\left(T-t\right)), \end{aligned}$$

Here $0 < T < \infty$ is a time at which solution hits the origin (the hitting time), $g_{x,y,u}\left(T-t\right) \to 0$ *as* $t \to T$, $\varrho > 0$, $C_{x,y,u} \in \mathbb{C}$.

7. Conclusion

We consider a mathematical model of a spherical inverted pendulum on a movable cart. We generalize the results obtained for the linear model to the nonlinear case. We show that this problem posses a singular point of the second order and there are spiral-similar solution which attains the singular point in finite time. We hope that obtained analytical results can be useful in developing effective algorithms for computer modeling and construction of a stabilizing controller.

Acknowledgments
This research was supported in part by the Russian Foundation for Basic Research under grant no. 17-01-00805.

[1] Liu G, Nešić D and Mareels I 2007 A note on the control of a spherical inverted pendulum *IFAC Proceedings* **40**(12) 11611166
[2] Liu G, Nešić D and Mareels I 2008 Nonlocal stabilization of a spherical inverted pendulum *International Journal of Control* **81**(7) 1035-1053
[3] Yoon M-G 2010 Dynamics and Stabilization of a Spherical Inverted Pendulum on a Wheeled Cart *Intenational Journal of Control, Automation and Systems* **8**(6) 1271-1279
[4] Postelnik L, Guangyu L,Stol K and Swain A 2011 Approximate Output Regulation for a Spherical Inverted Pendulum *Proceedings of the 2011 American Control Conference* 539544
[5] Nelson J and Kraft L G 1994 Real-time control of an inverted pendulum system using complementary neural network and optimal techniques *Proc. American Control Conf* 2553-2554
[6] Prasad L B, Tyagi B and Gupta H O 2014 Optimal Control of Nonlinear Inverted Pendulum System Using PID Controller and LQR: Performance Analysis Without and With Disturbance Input *International Journal of Automation and Computing* **11**(6) 661-670
[7] Varghese E S et al 2017 Optimal control of inverted pendulum system using PID controller, LQR and MPC *IOP Conf. Series: Materials Science and Engineering* **263** 052007
[8] Zelikin M I, Lokutsievskiy LV and Hildebrand R 2017 Typicality of Chaotic Fractal Behavior of Integral Vortices in Hamiltonian Systems with Discontinuous Right Hand Side *Journal of Mathematical Sciences* **221**(1) 1-136
[9] Zelikin M I and Borisov V F 1994 *Theory of Chattering Control with Applications to Astronautics, Robotics, Economics and Engineering (*Birkhäuser, Boston)
[10] Milyutin A A, Iliutovich A E Osmolovkii N P and Chukanov S V 1993 *Optimal Control in Linear Systems* [in Russian] (Nauka, Moscow)
[11] Manita L and Ronzhina M 2017 Optimal control of a spherical inverted pendulum *Lobachevskii Journal of Mathematics* **38**(5) 954957
[12] Lokutsievskiy L V, Manita L A and Ronzhina M I 2017 Optimal solutions in a neighborhood of a singular extremal for a problem with two-dimensional control *Mathematical Theory of Optimal Control Materials of the International Conference dedicated to the 90th birthday of Academician R. V. Gamkrelidze* 94-96
[13] Ronzhina M I and Manita L A 2018 On optimal solutions in a problem with two-dimensional bounded control *Systems Analysis: Modeling and Control. Materials of the International Conference in memory of Academician A.V. Kryazhimskiy* 90-93
[14] Manita L and Ronzhina M 2018 Singular solutions for vibration control problems *Journal of Physics: Conference Series* **955**(1) 1-6
[15] Hartman Ph 1964 *Ordinary differential equations* (J. Wiley & Sons, New York)

IOP Conf. Series: Journal of Physics: Conf. Series **1163** (2019) 012059 doi:10.1088/1742-6596/1163/1/012059

Ab initio calculation of multilayer magnetic structures by VASP on OpenPOWER high performance system

V. Borzilov[1,*], **V. Lozhnikov**[1,†], **P. Prudnikov**[1,‡], **M. Mamonova**[1,§], **A. Mamonov**[1,¶], **A. Sorokin**[2,£], **G. Baksheev**[3,∞]

[1] Department of Theoretical Physics, Omsk State University, Mira prospekt 55-A, Omsk, 644077, Russia
[2] Shared Facility Center "Data Center of FEB RAS", Khabarovsk, Russia
[3] Novosibirsk State University, 2 Pirogova st., Novosibirsk, 630090, Russia

E-mail: *borzilovvo@omsu.ru, †vereskanthem@gmail.com, ‡prudnikp@univer.omsk.su, §mamonova_mv@mail.ru ¶amamonov@gmail.com, £alsor@febras.net, ∞g.baksheev@g.nsu.ru

Abstract. In this work we calculated energy and magnetic characteristics for multilayer ferromagnetic structure Co/Cu[100] by VASP on POWER architecture. We analyzed performance of the VASP package on the OpenPOWER HPC system with several Pascal P100 GPU units in compare with calculation on POWER8 CPU only and on a system with Intel architecture. We revealed that the VASP calculations achieve maximum performance on OpenPOWER System with the GPUs.

1. Introduction

The behavior of multilayer magnetic structures has become of great technological importance due to the applications in magnetic storage devices. The *ab initio* calculations are widely used to calculate some characteristics of solids [1] and multilayer magnetic structures [2]. The main advantage of *ab initio* approach is independence on experimental data. Unlike the case of semi-empirical methods, there is no need for calibration or fitting parameters. Thus, *ab initio* methods can also be used to calculate the characteristics of perspective systems, i.e., for prediction of properties of materials that have not yet been developed. Mainly used packages that can perform *ab initio* calculations are VASP [3–5], Quantum Espresso [6], ABINIT [7]. High performance systems should be used for complex materials simulations with large amount of atoms in supercell. In this paper we calculated energy and magnetic characteristics for multilayer Co/Cu[100] ferromagnetic structure by VASP on POWER architecture. We analyzed performance of the VASP package on OpenPOWER HPC system with several Pascal P100 GPU units in compare with calculation on POWER8 CPU only and on system with Intel architecture. We revealed that the VASP calculations achieve a maximum performance on OpenPOWER system with the GPUs.

2. Vienna Ab initio Simulation Package (VASP)

The central idea of DFT [3] is to consider the electron density $\rho(r)$ instead of the full many-body wave functions $\Psi(r_1, \ldots, r_N)$.

$$\rho(r) = \left\langle \Psi | \sum_{i=1}^{N} \delta(r - r_i) | \Psi \right\rangle, \qquad m(r) = \sum_{v=1}^{N} \Psi_v^*(r) \sigma \Psi_v(r) \tag{1}$$

$$\left\{ -\frac{^-h}{2m}\nabla^2 + V_{\text{eff}} + \sigma \cdot B_{\text{xc}}(r) - \varepsilon_v \right\} \Psi_v(r) = 0, \qquad B_{xc} = \frac{\partial E_{\text{xc}}[\rho(r), m(r)]}{\partial m(r)} \tag{2}$$

In this work we used next methods and approximations:

(i) Projection augmented wave (PAW) method. The number of plane wave components is limited by the Cut-off Energy;

(ii) GGA (generalized gradient approximations);

$$E_{xc}^{GGA}[n_\uparrow(\vec{r}), n_\downarrow(\vec{r})] = \int \varepsilon[n_\uparrow(\vec{r}), n_\downarrow(\vec{r}), \nabla n_\uparrow(\vec{r}), \nabla n_\downarrow(\vec{r})] d\vec{r} \tag{3}$$

(iii) Pseudopotential Perdew–Burke–Ernzerhof (PBE) [8];

(iv) Standard method Monkhorst–Pack grid. A regular grid in k-space [9].

3. VASP compilation

Official support of GPU appear in VASP from version 5.4.1, for our work we used version 5.4.4. VASP have one precompiled configuration file named `makefile.include` with a lot of parameters. Showing all parameters is redundancy and we present main part of it in table 1. We use Intel Parallel Studio XE C/C++ with Intel MKL library to compile the VASP package on X86_64 architecture. The optimization flags were choosed `-O1` and `-O2` because compilation with harder optimization was not complete successfully. It was used Ubuntu 16.04 with 4.4.0-137 kernel. We used XlC 13.1.5 and Xlf 15.1.5 with including ESSL library on CentOS 7 with 3.10.0-514 kernel to compile VASP on IBM Power System S822LC.

Table 1. Compilers options

Compiler	Version	Flags
IBM XL C/C++	13.1.5	-g -q64 -O3 -qarch=pwr8 -qtune=pwr8:st -qfullpath -qsaveopt
IBM XL Fortran	15.1.5	-g -q64 -O3 -qarch=pwr8 -qtune=pwr8:st -qfullpath -qsaveopt -qflag=i:e -qsuppress=cmpmsg
NVIDIA CUDA compilation tools	8.0.61	-DCUDA_GPU -DRPROMU_CPROJ_OVERLAP -DCUFFT_MIN=28 -UscaLAPACK -fPIC -DADD_ -DMAGMA_WITH_MKL -DMAGMA_SETAFFINITY -DGPUSHMEM=300 -DHAVE_CUBLAS
Intel Parallel Studio XE C/C++	2017	-O2 -f_com=no -free -w0
Intel Parallel Studio XE Fortran	2017	-O1 -mkl=sequential -lstdc++

IOP Conf. Series: Journal of Physics: Conf. Series **1163** (2019) 012059 doi:10.1088/1742-6596/1163/1/012059

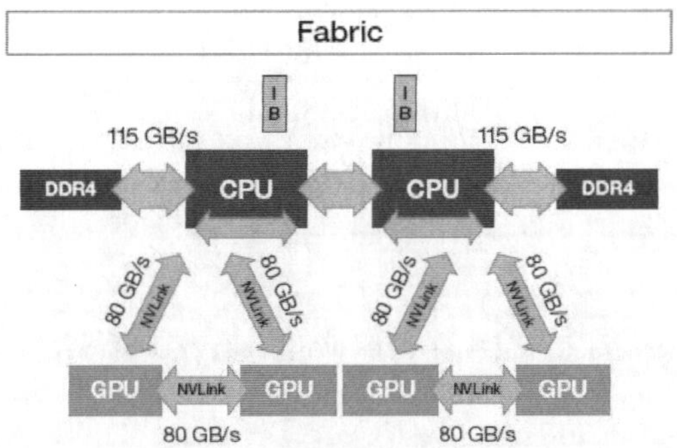

Figure 1. NVLink communications protocol in IBM Power System S822LC

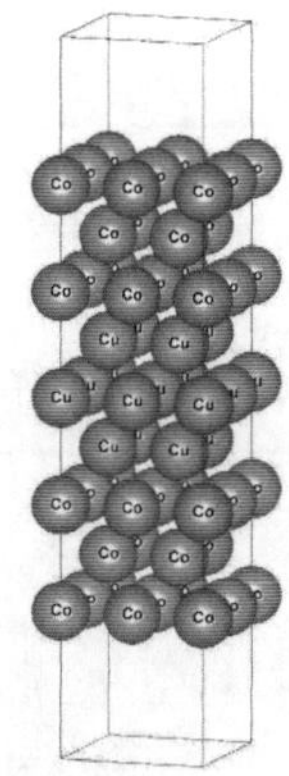

Figure 2. Computed Co/Cu/Co multilayer structure

4. Hardware information

IBM Power System S822LC is two socket HPC system with two POWER8 CPUs with 20 cores running at 4 GHz and interconnected with two NVidia Pascal P100 GPUs with high bandwidth (80GByte in and 80GByte out) NVLink 1.0 interface (figure 1). It is very important for exchange data between multiple GPUs and fast load data from CPU. The major goal of this system is efficiently using GPU units and accelerating calculations. A large part of HPC resources installed during the last decade are based on Intel CPUs. Novel generations of Intel CPUs present a wide spectrum of multicore processors [5]. Intel Core i7 4770 is desktop processor, but it is "tock" model in Intel extensive strategy of microprocessor development – it is mostly complete 22 nm architecture. We compared IBM POWER8 with Intel Haswell because both architectures introduced in one year (2013) and had similar technical process (22 nm).

5. Model and simulation parameters

In this work we calculated energy and magnetic characteristics for multilayer Co/Cu[100] ferromagnetic structure to compare results obtained on CPU and GPU. The computed multilayer structure is shown at figure 2.

Table 2. Simulation parameters

Cutoff energy	500 eV
K-points	12
Number of atoms	36
Thickness of vacuum layers	5 Å

6. Results of performance and accuracy comparison

We used three configurations to compare a VASP calculation times with similar INCAR parameters for IBM POWER8 CPU and Intel Haswell. We set `LREAL=.FALSE.` as described at VASP official documentation and use NCORE=1 with one MPI thread for GPU calculations. We do not use NVIDIA MPS system and do not set NSIM parameter clearly, but we know that it is important for large tasks especially. Core i7 have only 4 real cores and we run VASP with 4 processes only. On POWER8 we run VASP on 8 cores for using most part of one CPU.

IOP Conf. Series: Journal of Physics: Conf. Series **1163** (2019) 012059 doi:10.1088/1742-6596/1163/1/012059

Table 3. Comparison of the accuracy of the calculations performed on the CPU and GPU

Architecture	Total magnetization	Free energy (eV)
Intel Core i7-4770 Haswell	40.646	-202.20538224
IBM POWER8	40.646	-202.20538199
IBM POWER8+Nvidia P100	40.658	-202.04101161

The results of magnetization and free energy calculations (table 3) are well correlated. The calculations with the GPU provide less accuracy but a value of the error is not so sufficient. The times of calculations are different for POWER8 system and Intel Core i7 (table 4). The one POWER8 thread was more efficient than one Intel thread for the VASP calculations. If we use GPU only with one MPI thread we have much better performance (table 4) than Intel or POWER CPUs even without optimizations of VASP parameters.

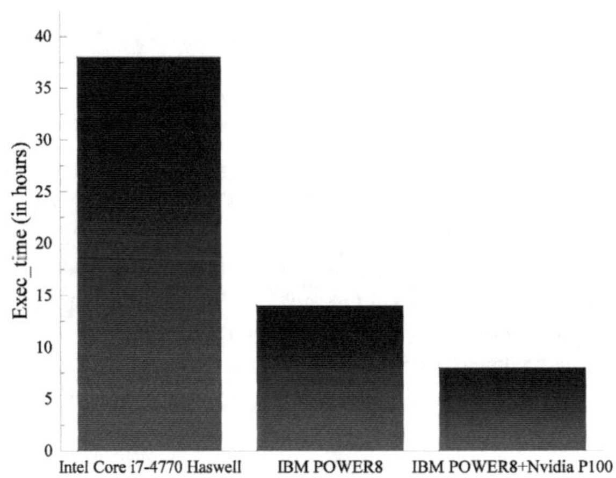

Figure 3. Dependence of execution time on used architecture.

Table 4. Execution times for different computing systems

Architecture	Execution time	Used memory	Number of threads
Intel Core i7-4770 Haswell	38:23:54	6.91 Gb	4
IBM POWER8	14:27:20	6,67 Gb	8
IBM POWER8+Nvidia P100	08:27:08	29,64 Gb	1

7. Conclusion

VASP is widely used by researchers to get characteristics of solids and multilayer magnetic structures. A one of the efficient ways to get results in less time is using of GPU especially in hybrid way with MPS daemon and multiple MPI threads. Therefore, to get maximum performance one need to optimize INCAR parameters for GPU, especially NCORE and NSIM. The values of acquired quantities and accuracy of GPU calculations are in a good agreement with CPU results. In order to use VASP efficiently with GPUs more memory is required in comparing with calculations on CPU.

Acknowledgments

We would like to thank the IBM experts, who help us to optimize VASP package for IBM Power Systems S822LC.

This research was supported by the grants 17-02-00279, 18-42-550003 of Russian Foundation of Basic Research and by the grant MD-6868.2018.2 of the President of the Russian Federation. The simulations were supported by the computational resources of Shared Facility Center "Data Center of FEB RAS" (Khabarovsk) [10]. Computations were performed with the methods and techniques, which were developed under the RFBR scientific project number 18-29-03196.

References

[1] Lejaeghere K, Bihlmayer G, Björkman T, et al 2016 *Science* **351** aad3000
[2] Kondrashov R A, Mamonova M V, Povoroznuk E S, Prudnikov V V 2017 *Lobachevskii J Math* **38** 940
[3] Kresse G and Furthmüller J 1996 *Phys. Rev.* B **54** 11169
[4] Kresse G, Marsman M, Furthmüller J 2015 *VASP THE GUIDE*
[5] Stegailov V and Vecher V (2017) Efficiency Analysis of Intel and AMD x86_64 Architectures for Ab Initio Calculations: A Case Study of VASP. In: Voevodin V, Sobolev S (eds) Supercomputing. RuSCDays 2017. Communications in Computer and Information Science, vol 793. Springer, Cham
[6] Giannozzi P, Baroni S, Bonini N, et al 2009 *Phys. Rev.* B **21** 395502
[7] Gonze X, Amadon B, Anglade P M, et al 2009 *Comput. Phys. Commun.* **180** 2582
[8] Perdew J P, Burke K, Ernzerhof M 1996 *Phys. Rev. Lett.* **77** 3865
[9] Monkhorst H J, Pack J D 1976 *Phys. Rev.* B **13** 5188
[10] Sorokin A A, Makogonov S I, Korolev S P 2017 *Scientific and Technical Information Processing* **4** 302

IOP Conf. Series: Journal of Physics: Conf. Series **1163** (2019) 012060 doi:10.1088/1742-6596/1163/1/012060

On limit laws of multi-dimensional stochastic synchronization models

Anatoly Manita[1,2]

[1] Lomonosov Moscow State University, Faculty of Mechanics and Mathematics, 119991, GSP-1, Moscow, Russia

[2] National Research University Higher School of Economics, Moscow Institute of Electronics and Mathematics, Russia

E-mail: manita@mech.math.msu.su, amanita@hse.ru

Abstract. Lévy stochastic processes and related fine analytic properties of probability distributions such as infinite divisibility play an important role in construction of stochastic models of various distributed networks (e.g., local clock synchronization), of some physical systems (e.g., anomalous diffusions, quantum probability models), of finance etc. Nevertheless, little is known about limit probability laws resulted from the long time behavior of such stochastic systems. In this paper we will focus on the impact of interaction graph topologies on limit laws of multicomponent synchronization models.

Models with synchronizing interaction. We recall a notion stochastic synchronization model (SMM) in a short descriptive manner referring readers to [1] for detailed definitions. There are N components and $x_j \in \mathbb{R}^d$ denotes a state of the component $j \in \{1, \dots, N\}$. The system $x(t) = (x_1(t), \dots, x_N(t))$ evolves in continuous time $t \in \mathbb{R}_+$ and the evolution is composed of two parts: a *free dynamics* and a *synchronizing* pairwise *interaction*. The interaction is defined by the set of nonnegative parameters α_{jk}, $j \neq k$. It is assumed that the component j sends messages to the component k with the rate α_{jk}. Messages contain information on the sender's current state x_j and immediately reach their destinations. We may imagine that there is a network with nodes $1, \dots, N$ and each node shares information with the other nodes by using messages. If $\alpha_{j'k'} = 0$ for some $j' \neq k'$ then it means that there are no messages from j' to k'. For any $j \neq k$ the message flow $T^{j\to k}$ is a Poissonian point process. It is assumed that stochastic flows $T^{j\to k}$ are independent for different pairs (j, k). Thus points of $T = \bigcup_{j\neq k} T^{j\to k}$ form a sequence $T = \{0 < T_1 < T_2 < \cdots\}$ of random epochs when nodes of the network exchange messages and *interact* in the following way. Imagine, for example, that at time T_q the node $j_1^{(q)}$ sends a message to some another node $j_2^{(q)}$. For breavity notation let us write $j_1 = j_1^{(q)}$ and $j_2 = j_2^{(q)}$. After receiving the message from j_1 the node j_2 ajusts its state to the value x_{j_1}: $x_{j_2}(T_q + 0) = x_{j_1}(T_q)$. This is the only jump in the system at the time T_q: $x_j(T_q + 0) = x_j(T_q)$ for all $j \neq j_2$. Between successive epochs of interaction all $x_k(t)$ evolve independently. Namely there is a family of mutually *independent* stochastic processes $x_k^\circ(t)$, $k = \overline{1, N}$, called the *free*

IOP Conf. Series: Journal of Physics: Conf. Series **1163** (2019) 012060 doi:10.1088/1742-6596/1163/1/012060

dynamics of components, such that

$$x_k(t) - x_k(T_{n-1}+0) = x_k^\circ(t) - x_k^\circ(T_{n-1}+0), \qquad T_{n-1} < t \le T_n, \quad n \ge 1, \quad k \in \{1,\dots,N\}.$$

It is very natural to assume that any $x_k^\circ(t)$ is an $\mathbb{R}^d$-valued Lévy process, i.e., a process having independent increments, starting from $x_k^\circ(0) = 0$ and being stochastically continuous [2]. A remarkable feature of Lévy processes is the following property of their increments [2].

$$\mathsf{E}\exp\left(\mathrm{i}\,\lambda_j \cdot (x_j^\circ(t) - x_j^\circ(s))\right) = \exp\left(-(t-s)\,\eta_j(\lambda_j)\right), \qquad \lambda_j \in \mathbb{R}^d, \quad 0 \le s < t. \tag{1}$$

Here the function $\eta_j : \mathbb{R}^d \to \mathbb{C}$ is called the Lévy exponent of $x_k^\circ(t)$. The LHS of (1) is the characteristic function $\psi_{x_j^\circ(t)-x_j^\circ(s)}(\lambda_j)$, i.e., the Fourier transform of the probability distribution of $x_j^\circ(t) - x_j^\circ(s)$. The RSH of the identity (1) provides a very convenient tool for analytical study of Lévy-driven dynamics. It should be noted that the class of Lévy processes is wide enough to cover many important examples such as homogeneous random walks with light-tailed or heavy-tailed jumps, jump-diffusion processes etc. In particular, if $x_j^\circ(t)$ is a classical Brownian motion with diffusion matrix Σ and drift vector $v_j \in \mathbb{R}^d$ then $\eta_j(\lambda_j) = -\mathrm{i}\,v_j \cdot \lambda_j + \frac{1}{2}(\Sigma\Sigma^\top\lambda_j)\cdot\lambda_j$.

To summarize, the SSMs are a special class of continuous time $(\mathbb{R}^d)^N$-valued Markov processes determined by two sets of parameters: the matrix (α_{jk}) and the Lévy exponents $\{\eta_j = \eta_j(\lambda_j)\}$.

Motivations of SSMs. The synchronizing interaction is motivated by many computer science applications with message passing mechanism (parallel computations [3–5], wireless sensor networks [6–8] etc.). SSMs can be also considered as agreement algorithms [9, 10] of a special kind. The idea to use Lévy processes instead of classical random walks is rather popular in physics [11–14], especially in subdomains of social and biological physics.

Known results on the long time behavior. The rate matrix (α_{jk}) determines a message routing graph (MRG) $\mathcal{G} = \mathcal{G}(N, (\alpha_{jk}))$ having vertices $1,\dots,N$ and directed edges (j,k) with weights $\alpha_{jk} > 0$. The presence of the edge (j,k) in $\mathcal{G}$ indicates a possibility of directly sending messages from j to k. We will always assume that the digraph $\mathcal{G}$ is strongly connected. Then the SSM demonstrates a long time synchronization in the following (stochastic) sense [1]: all differences $x_j(t) - x_k(t)$ have limits in distribution as $t \to \infty$,

$$r_{jk}(x(t)) = x_j(t) - x_k(t) \xrightarrow{d} r_{jk}^\infty \tag{2}$$

where $r_{jk}(x) = x_k - x_j$. Moreover, in regular situations the state vector $x(t) = (x_1(t),\dots,x_N(t))$ has no limit in distribution [1, Th. 1]. Denote by ℓ_{jk}^∞ the probability law of the random vector $r_{jk}^\infty \in \mathbb{R}^d$. In [15] some general results for probability laws ℓ_{jk}^∞ were obtained. It was discovered that under particular cases we have the famous Linnik distributions among them.

Infinite divisibility in probability and applications. The probability law of a random vector $\xi \in \mathbb{R}^d$ is called *infinitely divisible* (ID) iff for any n there exist *independent identically distributed* random vectors $\xi_{n,1}, \dots, \xi_{n,n}$ such that $\xi_{n,1} + \dots + \xi_{n,n}$ has the same distribution as ξ. It is a classical result [16] that the ID laws are appearing as limits of sums independent uniformly asymptotically negligible random variables. Since the Central Limit Theorem and the Poisson Limit Theorem are special cases of such summations it is not surprising that the *Gaussian* and *Poisson laws* which we often meet in physics and other application domains are ID. At the same time the uniform distribution in the segment $[0,1]$ is not ID.

The ID property is very desirable when choosing probability laws for constructing new models in statistics [17, 18], physics [19, 20], biology [21, p. 451], actuarial science [18] etc.

IOP Conf. Series: Journal of Physics: Conf. Series **1163** (2019) 012060 doi:10.1088/1742-6596/1163/1/012060

Denoting by $\psi_\xi(\nu) := \mathsf{E}\exp(\mathrm{i}\nu\cdot\xi)$, $\nu\in\mathbb{R}^d$, the characteristic function (CF) of ξ, we can reformulate the definition of ID as follows: ξ is ID iff for any $n\in\mathbb{N}$ there exists a CF $\psi_{(n)}(\nu)$ such that $\psi_\xi(\nu) = \left(\psi_{(n)}(\nu)\right)^n$. Of course, there is no warranty the the n-th root of a CF is again a CF of some other distribution. In general, proofs that concrete probability laws are ID (*or* are not ID) can be extremely complicated. It is well known [2], however, and it is readily seen from (1) that any Lévy process $L(t)$ is ID for any $t>0$.

Infinite divisibility in synchronization models. Since the free dynamics of SSM is ID for any t it is natural to ask whenever this property is inherited by limit laws ℓ^∞_{jk} . This is the main problem discussed in the present paper. The problem is complicated and as we will show soon the answer depends on topology of the graph $\mathcal{G}$. We start from a useful representation for CFs of ℓ^∞_{jk}. Introduce $\boldsymbol{b}_{(2)} = \left(b_{jk},\ j,k=\overline{1,N}\,,\ j\neq k\right)$, a set of variables $b_{jk}\in\mathbb{C}$. Consider CFs $\psi_{r_{jk}(x(t))}(\nu) = \mathsf{E}\,e^{\mathrm{i}\,\nu\cdot(x_k(t)-x_j(t))}$. Theorems 2 and 3 from [15] can be rephrased as follows.

Theorem 1. *Let the graph $\mathcal{G}$ be strongly connected. Then*

(i) for all $j\neq k$ and for all $\nu\in\mathbb{R}^d$ limits $\varphi_{(jk)}(\nu) := \lim\limits_{t\to\infty}\psi_{r_{jk}(x(t))}(\nu)$ exist;

(ii) the functions $\varphi_{(jk)}(\nu)$ are CFs of some probability laws in $\mathbb{R}^d$ and can be represented as composition $\varphi_{(j_1k_1)}(\nu) = (h_{j_1k_1}\circ\boldsymbol{b}_{(2)})(\nu)$ where

- $\boldsymbol{b}_{(2)} : \mathbb{R}^d\to\mathbb{C}^{N(N-1)}$ *such that* $b_{jk} = \eta_j(-\nu)+\eta_k(\nu)$,
- *functions* $h_{j_1k_1} = h_{j_1k_1}(\boldsymbol{b}_{(2)})$ *are solutions of the following system of $(N-1)N$ equations*

$$h_{j_1k_1}\mathcal{Z}_{j_1k_1} = \alpha_{j_1k_1} + \alpha_{k_1j_1} + \sum_{j\notin\{j_1,k_1\}}\alpha_{jj_1}h_{jk_1} + \tag{3}$$

$$+ \sum_{k\notin\{j_1,k_1\}}\alpha_{kk_1}h_{j_1k}, \qquad j_1,k_1=\overline{1,N}\,,\ j_1\neq k_1, \tag{4}$$

and $\mathcal{Z}_{j_1k_1} := b_{j_1k_1} + \sum_{k:\,k\neq k_1}\alpha_{kk_1} + \sum_{j:\,j\neq j_1}\alpha_{jj_1}$.

Item (i) means that $x_j(t)-x_k(t)$ have limits in distribution as $t\to\infty$, i.e., that (2) holds. Hereafter, we will consider the case of *identical components*

$$\eta_j(\lambda) = \eta(\lambda) \qquad \forall j \tag{5}$$

i.e., we assume the probability law of free dynamics $x^\circ_j(t)$ of any component j is the same. Recall that by assumption the processes $\left(x^\circ_j(t),\ t\in\mathbb{R}_+\right)$, $j=\overline{1,N}$, are independent. It follows from (5) that the substitution $b_{jk} = \eta_j(-\nu)+\eta_k(\nu) = \eta(-\nu)+\eta(\nu) = 2\,\mathrm{Re}\,\eta(\nu)$ does not depend on (j,k) so $\varphi_{(j_1k_1)}(\nu) = h_{j_1k_1}(c\mathbf{1})\Big|_{c=2\,\mathrm{Re}\,\eta(\nu)}$ where $\mathbf{1}$ is the vector of all ones.

New probabilistic interpetation for limit laws. Coalescing MCs and first passage distributions. Let X_t, $t\in\mathbb{R}_+$, be a particle walking on the set $\{1,\dots,N\}$ with jump rates $j_1 \overset{\alpha_{j_2j_1}}{\longrightarrow} j_2$. Consider a continuous time Markov chain (MC) $Y_t = (X^{(1)}_t, X^{(2)}_t)$ with state space $\mathcal{S} = \{1,\dots,N\}^2$ defined as follows. $X^{(1)}_t$ and $X^{(2)}_t$ are coordinates of two particles started from different points $k^{(1)}_0$ and $k^{(2)}_0$ and moving according to the rules of X_t. These two particles are walking independently until they meet. Afterwards they are moving together. So we have coalescing MCs. Let σ be the coalescing time $\sigma := \inf\left\{s>0:\ X^{(1)}_s = X^{(2)}_s\right\}$. In terms of the two-dimensional MC Y_t the

IOP Conf. Series: Journal of Physics: Conf. Series **1163** (2019) 012060 doi:10.1088/1742-6596/1163/1/012060

random variable σ is the *first hitting time* of Y_t to the subset $\{(m,m) : m = \overline{1,N}\} \subset \mathcal{S}$. Let $G_{(jk)}(c) := \mathsf{E}\left(\exp(-c\sigma) \,\middle|\, (X_0^{(1)}, X_0^{(2)}) = (j,k)\right)$, $c \geq 0$, denote the Laplace transform (LT) of the probability distribution of σ. It is clear from this definition that $G_{(jk)}(c) = G_{(kj)}(c)$. Consider the SSM $x(t)$ with identical components. The next result states that $h_{jk}(c\mathbf{1}) = G_{(jk)}(c)$.

Theorem 2. *Let the graph $\mathcal{G}$ be strongly connected and assumption (5) holds. Then*

$$\psi_{(jk)}(\nu) = G_{(jk)}(c)\Big|_{c=2\,\mathrm{Re}\,\eta(\nu)} = G_{(jk)}\left(2\,\mathrm{Re}\,\eta(\nu)\right) \qquad \forall j \neq k.$$

Since $G_{(jk)}(c) = G_{(kj)}(c)$ it follows that limit laws ℓ_{jk}^{∞} are symmetric.
Using basic properties of ID laws it is easy to get the following conclusion.

Corollary 3. *If $G_{(jk)}(c)$ is the Laplace transform of an ID law then ℓ_{jk}^{∞} is ID.*

We see that the question about ID of ℓ_{jk}^{∞} is related to the problem of ID of first passage times (FPTs) in very special MCs.

Example 4. The two-component SSM provides the simplest example with $N = 2$ and two rate parameters $\alpha_{12} > 0$ and $\alpha_{21} > 0$. It is easy to see that for the starting point $Y_0 = \left(X_0^{(1)}, X_0^{(2)}\right) = (1,2)$ the coalescing time σ is exponential with rate $\gamma = \alpha_{12} + \alpha_{21}$, so it has probability density $p(t) = \gamma e^{-\gamma t}$, $t \geq 0$, and LT $G_{(12)}(c) = \gamma/(\gamma + c)$. It is well known that the exponential distribution is ID.

Example 5. Let $N \geq 2$ be arbitrary and $\mathcal{G}$ be a complete graph with $\alpha_{jk} = \alpha > 0$, $j \neq k$. This case corresponds to the totally symmetric networks discussed in [15, Sect. 3.2]. Under assumption (5) it follows from (4) that $G_{(jk)}(c) = G_{(12)}(c)$. So (4) reduces to a single equation and we get $G_{(jk)}(c) = 2\alpha/(2\alpha + c)$. These algebraic considerations may be replaced by purely probabilistic arguments. Indeed, if σ_{jk} is distributed as the coalescing time in Y_t with the starting point $Y_0 = (j,k)$ then by symmetry $\sigma_{jk} \overset{d}{\sim} \sigma_{12}$ (i.e., they all have the same distribution). Applying the one step analysis [22] to Y_t it is not hard to derive that distribution of σ_{12} concides with the mixture of an exponential and the distribution of σ_{12} itself,

$$p_{\sigma_{12}}(t)\,dt = \frac{1}{N-1}\,2\alpha\exp\left(-2\alpha t\right)\,dt + \frac{N-2}{N-1}\,p_{\sigma_{12}}(t)\,dt, \qquad t > 0,$$

and, evidently, $p_{\sigma_{12}}(t) = 2\alpha\exp\left(-2\alpha t\right)$. So σ is again exponential and thus it is ID.

The first passage problem is of significant interest for many applied domains such as queueing systems [23], financial models or physics [14, 24]. Despite of the existing progress in modelling many interesting phemomena and in the related asymptotic analysis, results on the ID property of FPTs are far to be exhaustive. There is a huge probabilistic literature on this topic [22, 25–29] but there is no answer which would be applicable to the analysis of a general N-component SSM. Nevertheless, as we will show below, some of these results are useful for particular cases of SSMs.

Computational aspects related to inversion of the LTs of ID laws are also rather complicated [23].

General results about the LT of σ for synchronization models. Analysis of the equations (4) carried out in [15] results in the following conclusions.

Theorem 6. *Let the graph $\mathcal{G}$ be strongly connected. Then*

- *any $G_{(jk)}(c)$, $j < k$, the LT of the first passage time σ, is a rational function $P_{(jk)}(c)/Q_{(jk)}(c)$ with real coefficients;*

- *all poles of* $G_{(jk)}(c)$ *being considered as a function of the complex variable* $c \in \mathbb{C}$ *belong to* $\{c \in \mathbb{C} : \ \mathrm{Re}\, c < 0\}$;
- $\deg P_{(jk)} < \deg Q_{(jk)}$, $\deg P_{(jk)} \le \frac{1}{2}\left(N^2 - N\right) - 1$, $\deg Q_{(jk)} \le \frac{1}{2}\left(N^2 - N\right)$.

There are no known results about ID precisely applicable to this situation. Paper [28] contains a study of rational LTs having negative poles and zeros. The above theorem 6 covers more general LTs with complex (non real) poles and zeros. For many reasons (including computational aspects) it is important to describe digraphs $\mathcal{G}$ for which degrees of $G_{(jk)}$ are less than for the general case. As we will see below this can be done for specific topologies of the MRG $\mathcal{G}$.

Symmetric digraphs. In the case of identical components we assume that N is arbitrary and the digraph $\mathcal{G}$ is symmetric: $\alpha_{jk} = \alpha_{kj}$ for all $j \ne k$. Example 5 is a very particular subcase of this situation. For symmetric digraphs we can slightly improve the general result of Theorem 6. It appears that the rational function $P_{(jk)}/Q_{(jk)}$ has only real negative poles $-\vartheta_M < \cdots < -\vartheta_1 < 0$ and the following representation holds

$$G_{(jk)}(c) = \sum_{m=1}^{M} \frac{d_{jk,m}}{c + \vartheta_m} \tag{6}$$

with some coefficients $d_{jk,m} \in \mathbb{R}$, $d_{jk,1} > 0$, $\sum_m d_{jk,m} = 1$ and $M \le \frac{1}{2}\left(N^2 - N\right)$. The key idea is to connect the system of equations (4) with spectral properties of some symmetric matrix. The proof is rather straighforward and so we omit details.

Many researches were interested in sufficient conditions on parameters $d_{jk,m}$ and ϑ_m for ensuring that the RHS of (6) is a LT of an ID probability law.

If all $d_{jk,m} > 0$ then we have a mixture of exponential laws which is ID (see [25, Sect. XIII.7]). For $d_{jk,m}$ of different signs and ordered ϑ_m $(0 < \vartheta_1 < \vartheta_2 < \cdots)$ Steutel [29] provided a particular answer by proving that finite mixtures with one change of signs are ID.

Unfortunately, neither of these analytical results is applicable to general stochastic synchronization models. In general, coefficients $d_{jk,m}$ in (6) can be positive or negative with two or more changes of signs. It is not clear how to control them in terms of parameters α_{jk}.

Another possible way is to use the probabilistic interpretation of $G_{(jk)}(c)$ as the LT of a first passage time in a special Markov chain.

On ID property of first passage times in continuous time Markov chains. Consider a finite MC with no symmetry assumption. Let n be a number of its states. Miller [27] proved that if $n \le 4$ then the FPT from any state to any other state is ID. By providing a corresponding example he also proved that for $n \ge 5$ the statement is not true.

Ring topology of $\mathcal{G}$. Let N be arbitrary. We say that a digraph $\mathcal{G}$ is a ring iff $\alpha_{jk} = \gamma_{k-j}$ (mod N). This message routing topology depends on $N - 1$ parameters $\gamma_1, \gamma_2, \ldots, \gamma_{N-1}$. In general, $\gamma_m \ne \gamma_{-m}$ hence the digraph is not symmetric. It is easy to see that the LT $G_{(jk)}(c) = H_{k-j}(c)$, i.e., it depends only on $k - j \pmod N$. Such model was studied in [30]. It was proved that $G_{(jk)}(c)$ is a fraction whose nominator and denominator are rational functions of a special form. Using that representation it is easy to see that $G_{(jk)}(c) = P_{(jk)}(c)/Q_{(jk)}(c)$ with $\deg P_{(jk)}(c) \le [N/2] - 1$, $\deg Q_{(jk)}(c) \le [N/2]$. Evidently, the degrees here are much smaller than those in the general case of Theorem 6. Developping the approach of [30] we get a result similar to, but better than, (6).

Theorem 7. *Let a (nonsymmetric) graph be a ring. Then all poles* $G_{(jk)}(c)$ *are strictly negative and the representation (6) holds with some* $M \le [N/2]$.

It was also proved in [30] that for $N \leq 5$ all first passage times LT are ID. It is worth noting that even for $N = 6$ there is an example of $\mathcal{G}$ such that for some pair (j, k) the first passage time LT $G_{(jk)}(c)$ has the form (6) with $M = 3$, $d_{jk,1} > 0$, $d_{jk,2} < 0$, $d_{jk,3} > 0$. This is the case of two changes of signs and the Steutel's theorem [29] does not work here. Nevertheless, applying the state-augmentation method and the Miller's result we can extend results of [30].

Theorem 8. *Let a stochastic synchronization model consist of $N \leq 7$ identical components and its digraph $\mathcal{G}$ is a ring. Then all FPTs σ_{jk} are ID and hence all limit laws ℓ^{∞}_{jk} are ID.*

NNE-ring topology. The so called birth-death processes and skip-free random walks on the one-dimensional lattice are the only classes of continuous time MCs with arbitrary number of states for which many fine probabilistic properties of the first passage times were successfully studied [22, 25, 26, 31]. In these classes, in going from one state to another, the chain must pass all intermediate states at least once. It was proved that the first passage times in such Markov chains are ID. To make use of there result we define the following subclass of ring topologies.

We say that the ring topology satisfies a NNE (nearest neighbors edges) condition iff $\gamma_m = 0$ for $|m| \neq 1 \pmod N$.

Theorem 9. *For arbitrary N in stochastic synchronization models with the NNE-ring topology all Laplace transforms $G_{(jk)}(c)$ are ID.*

The idea of the proof is to construct an auxiliary birth-death process (BDP) having the same FPT distributions as the MC $Y_t = (X_t^{(1)}, X_t^{(2)})$. As in [30] we consider $U_t = X_t^{(1)} - X_t^{(2)}$ as a random walk on $\mathbb{Z}_N$ and reduce the problem to the first passage times to the state $0 \in \mathbb{Z}_N$. Identifying states j and $N - j$ for all $1 \leq j < N/2$ we get a BDP with the finite state space $\{0, 1, \ldots, [N/2]\}$. Now the ID property follows from known results.

Conclusions. We considered a class of stochastic models with synchronizing interaction and studied the question if their limit probability laws inherit the infinite divisibility property from individual dynamics of components. It was shown that this problem is heavily dependent on topology of the message routing graph. An important byproduct of this study is the demonstration of the fact that a complexity of numerical procedures for finding these limit laws may also depend on a chosen network topology. Since Lévy-driven dynamics are popular tools for constructing many physical models we hope that our results will be useful not only in the context of stochastic synchronization models.

References.

[1] Manita A 2016 On behavior of stochastic synchronization models *Journal of Physics: Conference Series* **681**(1) 012024
[2] Applebaum D 2004 *Lévy Processes and Stochastic Calculus* (Cambridge University Press)
[3] Bertsekas D P and Tsitsiklis J N 1989 *Parallel and distributed computation: numerical methods* vol 23 (Prentice hall Englewood Cliffs, NJ)
[4] Fujimoto R 2000 *Parallel and Distributed Simulation Systems* (Wiley)
[5] Shchur L N and Shchur L 2015 Parallel discrete event simulation as a paradigm for large scale modeling experiments. *Data Analytics and Management in Data Intensive Domains* (Obninsk, Russia) pp. 107–113
[6] Sundararaman B, Buy U and Kshemkalyani A D 2005 Clock synchronization for wireless sensor networks: A survey *Ad Hoc Networks (Elsevier)* **3** 281–323
[7] Simeone O, Spagnolini U, Bar-Ness Y and Strogatz S H 2008 Distributed synchronization in wireless networks *IEEE Signal Processing Magazine* **25**(5) 81–97
[8] Manita A 2014 Clock synchronization in symmetric stochastic networks *Queueing Systems* **76**(2) 149–180
[9] Manita A and Manita L 2017 Stochastic time synchronization models based on agreement algorithms *International Conference on Analytical and Computational Methods in Probability Theory* (Springer) pp. 361–375
[10] Manita A and Manita L 2018 Distributed time synchronization algorithms and opinion dynamics *Journal of Physics: Conference Series*, vol 955 (IOP Publishing) vol 955 p. 012038

[11] Zaslavsky G 1995 From Lévy flights to the fractional kinetic equation for dynamical chaos *Lévy Flights and Related Topics in Physics* (Springer) pp. 216–236

[12] Metzler R and Klafter J 2000 The random walk's guide to anomalous diffusion: a fractional dynamics approach *Physics reports* **339**(1) 1–77

[13] de Nigris S, Carletti T and Lambiotte R 2017 Onset of anomalous diffusion from local motion rules *Phys. Rev. E* **95** 022113

[14] Dybiec B, Gudowska-Nowak E and Chechkin A 2016 To hit or to pass it over: remarkable transient behavior of first arrivals and passages for Lévy flights in finite domains *Journal of Physics A: Mathematical and Theoretical* **49**(50) 504001

[15] Manita A 2017 Probabilistic issues in the node synchronization problem for large distributed systems *Lobachevskii Journal of Mathematics* **38**(5) 948–953

[16] Gnedenko B V and Kolmogorov A N 1968 *Limit distributions for sums of independent random variables* (Addison-Wesley)

[17] Katti S 1967 Infinite divisibility of integer-valued random variables *The Annals of Mathematical Statistics* **38**(4) 1306–1308

[18] Steutel F 1979 Infinite divisibility in theory and practice *Scandinavian Journal of Statistics* **6** 57–64

[19] Klauder J R 2007 Infinite divisibility in euclidean quantum mechanics *Contributions in Mathematical Physics* (Springer) pp. 161–176

[20] Uchaikin V V and Zolotarev V M 2011 *Chance and stability: stable distributions and their applications* (Walter de Gruyter)

[21] Holden A 1983 Stochastic processes in neurophysiology: Transformation from point to continuous processes *Bulletin of mathematical biology* **45**(4) 443–465

[22] Keilson J 1981 On the unimodality of passage time densities in birth-death processes *Statistica Neerlandica* **35**(1) 49–55

[23] Abate J, Choudhury G L and Whitt W 2000 An introduction to numerical transform inversion and its application to probability models *Computational probability* (Springer) pp. 257–323

[24] Redner S 2001 *A Guide to First-Passage Processes* (Cambridge University Press)

[25] Feller W 1971 *An Introduction to Probability Theory and Its Applications, Vol. 2* (Wiley)

[26] Keilson J 1963 The first passage time density for homogeneous skip-free walks on the continuum *The Annals of Mathematical Statistics* **34**(3) 1003–1011

[27] Miller H 1979 Infinite divisibility in stochastic processes *The Annals of Probability* **7**(3) 406–417

[28] Sumita U and Masuda Y 1987 Classes of probability density functions having Laplace transforms with negative zeros and poles *Advances in applied probability* **19**(3) 632–651

[29] Steutel F and van Harn K 2003 *Infinite Divisibility of Probability Distributions on the Real Line* (CRC Press)

[30] Manita A 2017 Limit distributions in stochastic networks with message-passing synchronization *International Conference on Analytical and Computational Methods in Probability Theory* (Springer) pp. 347–360

[31] Steutel F 1973 Some recent results in infinite divisibility *Stochastic Processes and their Applications* **1**(2) 125–143

IOP Conf. Series: Journal of Physics: Conf. Series **1163** (2019) 012061 doi:10.1088/1742-6596/1163/1/012061

A new code-based public-key cryptosystem resistant to quantum computer attacks

Elena Egorova[1], **G Kabatiansky**[1], **E Krouk**[2] **and C Tavernier** [3]

[1] Skoltech, Moscow, Russia

This research was supported by RFBR grants 16-01-00716 and 18-07-01427
[2] Higher School of Economics, MIEM, Moscow, Russia

This research was supported by RFBR grant 16-01-00716
[3] ASSYSTEM, France

E-mail: egorovahelene@gmail.com; g.kabatyansky@skoltech.ru; ekrouk@hse.ru; tavernier.cedric@gmail.com

Abstract. We propose a new type of public-key cryptosystems (PKC) which is based on repetition of *different* error-correcting codes. We give a brief analysis of some well known attacks on code-based PKC, including structural ones and show that the scheme could be used as a perspective post-quantum PKC.

1. Introduction

It is well known that with appearance of enough large and stable quantum computer main known public-key cryptosystems (PKC) will be broken by Shor's algorithm [1] and network communications (as well as some other activities) become insecure. PKC which are resistant against known attacks via hypothetical quantum computer are called post-quantum PKC. Among few directions of construction post-quantum PKC like lattice-based cryptography, hash-based cryptography the code-based cryptography is the oldest one and the most elaborated.

First PKC based on linear error-correcting codes was proposed in [2] forty years ago. Since then not many principal improvements were done. Among them a PKC, which is based on syndrome decoding of linear codes, was suggested in [3]. After that many different families of error-correcting codes were investigated as a candidate to replace Goppa codes used in [2]. In this paper we propose a new PKC based on repetition of different error-correcting codes what is a generalization of the construction of [4]. We show that our scheme is resistant to structural attacks of [5], [6] proposed for breaking [4] and we analyse also different types of general decoding attack in our case. We also show how the new PKC can be used for digital signature purpose. Importance of our scheme as well as some other code-based schemes is that there are no known effective attacks on these scheme based on a quantum computer similar to Shor's attack [1] on the famous RSA cryptosystem [7].

2. Main known code-based PKC

Let us start from recalling some basic facts about McEliece scheme [2] and its modifications. McEliece PKC works in the following way - a user A chooses a generator $k \times n$ matrix G_A of some linear (n, k)-code C_A, which has decoding algorithm Φ correcting t errors with polynomial in n complexity. This matrix is a secret, known only to A. The user A also chooses randomly two matrices: $k \times k$ nonsingular matrix S_A and $n \times n$ permutation matrix P_A and then construct *public* matrix $G_{pub} = S_A G_A P_A$, which will be known to all other users. In order to deliver a message m of length k to the user A any other user transmits to A the following vector of length n

$$y = mG_{pub} + e,$$

where e is a vector of weight t which is *randomly* generated by B. The user A after receiving vector y calculates

$$y' = yP_A^{-1} = mG_{pub}P_A^{-1} + eP_A^{-1} = (mS)G_A + e',$$

where $e' = eP_A^{-1}$ and $wt(e') = wt(e) \leq t$ since P is a permutation.Then A applies the decoding algorithm Ψ to vector $y' = m'G + e'$ in order to obtain vector $m' = mS$ and finally A receives $m := m'S^{-1}$. Any other user will deal either with the problem of correcting t errors by looking like arbitrary linear code or with the problem of reconstructing code structure from its public-key matrix (its called structural attack). In the original paper [2] irreducible Goppa codes [8] were chosen as the family of codes for the scheme. In particular, it was suggested to use Goppa code of length $n = 1024$ with $k = 524$ information bits (code dimension) and the minimal code distance $d = 101$.

Later H. Niederreiter proposed a scheme, which is in some sense dual to McEliece scheme, [3]. Saying informally, the user A does the same manipulations but with a parity-check matrix H_A of an (n, k)-code C_A. Namely, the matrix $H_{pub} = \tilde{S}_A H_A P_A$, where $\tilde{S}_A$ is randomly chosen $(n-k) \times (n-k)$ nonsingular matrix, is a public-key matrix. The set of messages is the set of all binary vectors e of length n such that $wt(e) = t$ and the corresponding ciphertext defined as $s = H_{pub}e^T$. These two schemes may have different parameters but they have equivalent security, i.e., if one PKC can be broken than the other also, see [9] or just a few lines prove of this fact in [4]. Because of similarity and equivalent security of these two schemes we consider them below as a single scheme. Note that in [10] it was proposed how to send more information via McEliece scheme by choosing the error-vector e not random but as carrying some information bits while maintaining the same security.

3. Repetition of different codes and a new code-based PKC

Recall that for any error-correcting code C of length n, cardinality M and with the minimal code distance d its L-repetition is the code of length nL, cardinality M and with the minimal code distance dL which consists of the following codewords $v^L = (v||v...||v)$, where $||$ denotes words concatenation.

Sidelnikov proposed how to use this construction in order to build PKC. Namely, he used u-times repetition of binary Reed-Muller codes $RM(s, m)$ of rather small order s, where for different copies of RM-codes different generator (or parity-check) matrices are used. Recall that $RM(s, m)$ of length $n = 2^m$ has the number of information bits equals to $k = \sum_{i=0}^{s} C_m^i$ and distance $d = 2^{m-s}$. The corresponding public generator $k \times un$ matrix of the Sidelnikov scheme has form $G_{pub} = FP$, where P is random $nu \times nu$ permutation matrix and F is $k \times un$ matrix obtained by concatenation of matrices $F_1, \ldots, F_u$, where the parameter u is a positive integer. Each matrix F_i has form $F_i = S_i G_{RM}$, where G_{RM} is the ordinary generator matrix of $RM(r, m)$ code and matrices S_i are random $k \times k$ nonsingular matrices chosen independently. Since matrices S_i are not equal, as it should be for McEliece scheme, the Sidelnikov scheme is (slightly) different from the classical McEliece PKC. Now we generalize the repetition construction to the case of different codes and then generalize Sidelnikov scheme.

Consider two linear codes V_1 and V_2 of the same dimension k but different lengths n_1 and n_2 and let G_1 and G_2 be some generator matrices of these codes. Define a $k \times n$ generator matrix $G_{1,2} = G_1||G_2$ of a new code, which we call pseudo-repetition ("pseudo" because codes are different), as the concatenation of matrices G_1 and G_2, hence $n = n_1 + n_2$. Denote this code as $V_1 \bigsqcup V_2$. Surely this construction can be defined for any multiplicity u of pseudo-repetition, in particular, when $V_1 = V_2 = \ldots = V_u = RM(s, m)$ we have the Sidelnikov scheme.

In order to create a code-based PKC one needs to define which and how errors will be corrected. Let ψ_1 and ψ_2 be decoding algorithms of codes V_1 and V_2 capable to correct t_1 and t_2 errors correspondingly. Then define the following decoding algorithm for code $V_1 \bigsqcup V_2$ capable to correct up to $t = t_1 + t_2$. Let e_1 be the error vector on first n_1 positions and let e_2 be the error vector on second (last) n_2 positions. Then $wt(e_1) \leq t_1$ or $wt(e_2) \leq t_2$ (or both) since $wt(e_1 + e_2) \leq t_1 + t_2$ and hence we can decode the corresponding code and find either the corresponding codeword (or the error-vector if one deals with Niederreiter PKC). It may happen that the other part (with more errors than the corresponding t_i) also can be decoded but incorrectly. Anyway we recover codevectors in both cases and fortunately for the right choice corresponding distance will be at most t and for the wrong one with the probability very close to 1 the corresponding total distance will be larger than t.

Note that the sender (Bob) cannot create (unfortunately) an error vector in such a way that both $wt(e_1) \leq t_1$ and $wt(e_2) \leq t_2$ since separation of the coordinate set on subsets corresponding two codes is unknown (keep secret).

Now let us consider a "toy example" to illustrate our idea. Let V_1 be RM(4,10) code. It has length $n_1 = 1024$, dimension $k = 386$ and $d_1 = 64$. There are different decoding algorithms capable to decode up to $t_1' = 31$ errors (so-called bounded distance decoding), moreover there is a list decoding algorithm capable to decode up to the minimal code distance, see [11], in particular, capable to decode up to $t_1 = 61$ with rather small (polynomial) complexity. The size of the corresponding list is small and to find the codeword at the minimal distance from the received vector is rather easy, see [11] for details. Let V_2 be Goppa code of length $n_2 = 512$, the same dimension $k = 386$ and the constructive distance $d_2 = 29$, see [8]. Hence the standard decoding algorithm of Goppa codes can correct up to $t_2 = 14$ errors. Let us consider two decoding scenarios. In both scenarios we decode Goppa code with correction 14 errors (or less). Difference is only by decoding of the RM code. In the first scenario we decode RM code by bounded distance decoding and it outputs at most one codeword. Then $t = 31 + 14 = 45$ and the sender should generate this errors randomly on the total length $n = 1024 + 512 = 1536$. Therefore the expected number of errors on the positions of RM code will be $45 \times 2/3 = 30$ and the expected number of errors on the positions of Goppa code will be 15, what is very close to the decoding capacity of both algorithms. Hence in this scenario we should try to decode both codes.

In the second case $t = 61 + 14 = 75$. Therefore the expected number of errors on the positions of RM code will be $75 \times 2/3 = 50$ and the expected number of errors on the positions of Goppa code will be 25, what is significantly bigger than the decoding capability of the corresponding algorithm. So in this scenario the receiver (Alice) first will try to decode (by list decoding algorithm) RM(4,10) code, and only in the very rare case when the number of errors on positions of RM code is larger than 61 and its decoding failed then decode Goppa code. Surely the encryption rate is bigger for the second scenario.

4. Analysis of known attacks

The main idea of the new scheme is to hide which positions are positions of RM code and which are - of Goppa code. Let us first consider structural attacks on Sidelnikov scheme proposed in [5, 6]. Note that these attacks work well even for Sidelnikov PKC only in the case of $u = 1$, i.e., without repetition and hence positions of RM code are known.

Now consider the example of the previous section. Both attacks based on the search of minimal codevectors in the dual code. Note that for code $V_1 \bigsqcup V_2$ its dual code contains codes $V_1||0^{n_2}$ and $0^{n_1}||V_2||$, where $||$ denotes concatenation. The dual to RM(4,10) is RM(5,10) and hence the minimal weight is 32. It is known that for binary Reed-Muller code $RM(s, m)$ the number $A_d(s, m)$ of codevectors of weight $d = 2^{m-s}$ equals to

$$A_d(s, m) = 2^s \prod_{i=0}^{m-s-1} \frac{2^{m-i} - 1}{2^{m-s-i} - 1}$$

Hence for the considered case the number A_{32} of codevectors of weight 32 in $RM(5, 10)$ is surely less than $2^5 \prod_{i=0}^{4} 2^6 = 2^{35}$. On the other hand, the total number of binary vectors of weight 32 and length 1536 is

$$\binom{1536}{32} > (3 \times 2^9)^{32}/32^{32} > 2^{48+288-160} = 2^{176}$$

Hence the probability that a randomly chosen vector of weight 32 belongs to the dual to RM(4,10) code is negligibly small,i.e., less that 2^{-176}. Hence this attempt to guess a minimal vector from the dual to RM(4,10)-code failed.

Consider more sophisticated attack. Let H be 1150×1536 pubic parity-check matrix (straightforward calculated from G_{pub}). One may think about a desirable minimal vector v of weight 32 (such that $Hv^T = 0$) as the sum of two vectors of weight 16, i.e., $v = v_L + v_R$ and $Hv_L^T = Hv_R^T$. Let us evaluate the list $\mathbf{L}$ of syndroms for all vectors of length 1536 and weight 16 and then arrange this list, which size is $L = \binom{1536}{16} > (3 \times 2^9)^{16}/16^{16} > 2^{24+144-64} = 2^{104}$, in lexicographical order what takes at least $1150 \times |\mathbf{L}| > 2^{114}$ operations. Then by merging the list with its copy one can find *all* minimal codevectors of weight 32, but the number of operations is infeasible and a *structural* attacks of [5, 6] failed.

Surprisingly the direct attack on the new system is more efficient. Namely, consider the information set decoding algorithm [12] which was proposed as an attack in the original McEliece paper and later on further developed in many papers. In order to decode an $(1536, 386)$-code with correction up to $t = 75$ errors one just chooses randomly 386 positions assuming that they are free from errors with the probability $p = (\frac{1150}{1536})^{75} \approx 2^{-31}$ and then recover the corresponding codevector and check if this vector has (Hamming) distance from the decoded one not more than t. Recovery of a codevector takes $386 \times 1150 \approx 4 \times 10^5$ binary operations, so the total average number of binary operations is $\approx 8 \times 10^{14}$ what is not sufficiently large to consider this example enough secure. To make it resistant to decoding attack one needs to increase code parameters (length, dimension, distance) in a few times.

5. Conclusion

We proposed a new public-key cryptosystem based on (pseudo)repetition of different error-correcting codes (but of the same dimension) which is a generalization of Sidelnikov PKC [4]. Even for rather small values of public key length this system is resistant against structural attacks developed in [5], [6] for breaking Sidelnikov PKC. On the other hand, the "toy example" of the system can be broken via well-known decoding algorithms of arbitrary linear codes, see [12] and its improvement in [13],[14]. To avoid this attack one needs to consider longer codes with bigger error-correction capability or even the same code but to increase its decoding radius by using more complicated (but still of polynomial complexity) decoding algorithms of [16]. Then it becomes possible to correct up to 100 errors just by RM-code (but with small probability of failure) plus 14 errors by Goppa code. Hence even the "toy" example" code can correct totally up to 114 errors and therefore the complexity of the corresponding information set decoding is of order 10^{27} what looks much more reliable than 10^{15}.

Known results shows that currently quantum algorithms speed up decoding of linear codes insignificantly, for instance, the complexity of May-Meurer-Thomae ISD algorithm drops down from $2^{0.05904n+o(n)}$ to $2^{0.05806n+o(n)}$ [17]. Therefore the proposed scheme can be considered as a candidate for post-quantum cryptography. Moreover, one of the main advantage of the new PKC is that it based on decoding algorithms which capable to correct errors beyond so-called bounded decoding, i.e., beyond half of the code minimal distance. Note that almost all previous code-based PKC use only bounded distance decoding but refer to hardness of decoding linear codes what is not correct since NP-hardness is proved only for maximum-likelihood decoding (i.e., minimum distance decoding), see [18],[19]. We use decoding algorithms which are much closer maximum-likelihood decoding what gives an additional reason that the new PKC should be resistant to quantum computer attacks.

One of the direction of further research is to find another, better codes and make the corresponding cryptoanalysis more solid. Another direction of further research is to apply the new scheme of codes pseudo-repetition to constructing code-based digital signature in style of [20] or [21].

References

[1] Shor P W 1997 *SIAM Journal on Computing* **26** 1484–509
[2] McEliece R J 1978 *Jet Propulsion Lab* 114–6
[3] Niederreiter H 1986 *Problems of Control and Information Theory* **15** 159–66
[4] Sidelnikov V M 1994 *Discrete Mathematics* **6** 3–20
[5] Minder L and Shokrollahi 2007 *LNCS* **4515** 347-60
[6] Borodin M A and Chizhov I V 2014 *Discrete Math. Appl.* **4** 191–207
[7] Rivest R, Shamir A and Adleman L 1978 *Communications of the ACM* **21** 120-6
[8] Goppa V D 1970 *Problems of Information Transmission* **6** 207–12
[9] Li X Y, Deng R H and Wang X M 1994 *IEEE Transactions on Information Theory* **40** 271–3
[10] Biswas B and Sendrier N 2008 *PQCrypto 2008* 47–62
[11] Dumer I, Kabatiansky G and Tavernier C 2008 *ACCT 2008* 82–5
[12] Prange E 1962 *IRE Transactions on Information Theory* **8** 5–9
[13] Kruk E 1989 *Problems of Information Transmission* **25** 251-4
[14] Barg A, Krouk E and Tilborg H C A 1999 *IEEE Transactions on Information Theory* **45** 1392–1405
[15] Dumer I 1991 *Proceedings of the 5th Joint Soviet-Swedish International Workshop on Information Theory* 50-52
[16] Dumer I 2004 *IEEE Transactions on Information Theory* **50** 811–23
[17] Kirshanova E 2018 *Lecture Notes in Computer Science* **10786** 507–27
[18] Berlekamp E R, McEliece R J and van Tilborg H C A 1978 *IEEE Transactions on Information Theory* **24** 384–6
[19] Guruswami V and Vardy A 2005 *IEEE Transactions on Information Theory* **51** 2249 – 56
[20] Kabatianskii G, Krouk E and Smeets B 1999 *Lecture Notes in Computer Science* **1355** 161-77
[21] Courtois N T, Finiasz M and Sendrier N 2001 *Lecture Notes in Computer Science* **2248** 157–74

IOP Conf. Series: Journal of Physics: Conf. Series **1163** (2019) 012062 doi:10.1088/1742-6596/1163/1/012062

Multipliers of an Antiphase Solution in a System of Two Coupled Nonlinear Relaxation Oscillators

M M Preobrazhenskaia[1,2]

[1] P. G. Demidov Yaroslavl State University, Sovetskaya str., 14, Yaroslavl, 150000, Russia
[2] Scientific Center in Chernogolovka RAS, Lesnaya str., 9, Chernogolovka, Moscow region, 142432, Russia

E-mail: rita.preo@gmail.com

Abstract. We consider a system of two non-linear differential-difference equations that models a behaviour of two coupled neural oscillators with a synaptic connection of a threshold type. To find a solution of the system means to find normalized membrane potentials of neurons. We prove that there exists an antiphase solution. It is a special relaxation periodic regime such that one membrane potential is a half-period shift of the other membrane potential. We prove that the antiphase solution is stable and construct its asymptotic. We show that there exists a solution with a bursting-effect by choosing a specific connection chain delay. That is, for any natural n there is a connection chain delay such that an antiphase solution has exactly n asymptotically high spikes on a period after a refractive segment.

1. Problem statement

We consider a mathematical model of synaptic coupled neuron type oscillators that is based on an idea of fast threshold modulation:

$$\begin{aligned}\dot{u}_1 &= [\lambda f(u_1(t-1)) + b\,g(u_2(t-h))\ln(u_*/u_1)]u_1,\\ \dot{u}_2 &= [\lambda f(u_2(t-1)) + b\,g(u_1(t-h))\ln(u_*/u_2)]u_2.\end{aligned} \tag{1}$$

This model was suggested in [1] and considered in [2, 3]. Here, $u_1(t), u_2(t) > 0$ are normalized membrane potentials of neurons. Every neuron is modelled by a singularly perturbed differential-difference equation with delay

$$\dot{u} = \lambda f(u(t-1))u. \tag{2}$$

This equation was suggested in [4] (see also [5]). A parameter $\lambda \gg 1$ is large and characterizes the rate of electric processes in the system. A function $f(u) \in C^2(R_+)$ satisfies:

$$f(0) = 1;\;\; f(u) + a,\; uf'(u),\; u^2 f''(u) = O(u^{-1}) \text{ as } u \to +\infty,\; a = \text{const} > 0. \tag{3}$$

The term $bg(u_{j-1}(t-h))\ln(u_*/u_j)u_j$ models a synaptic connection. We consider a connection of a threshold type with a time delay $h > 1$. Parameter $b = \text{const} > 0$, $u_* = \exp(c\lambda)$ is a threshold value, where $c = \text{const} \in R$. A function $g(u) \in C^2(R_+)$ satisfies the following conditions:

$$\forall u > 0\; g(u) > 0,\; g(0) = 0;\;\; g(u) - 1,\; ug'(u),\; u^2 g''(u) = O(u^{-1}) \text{ as } u \to +\infty. \tag{4}$$

A mathematical model of fast threshold modulation was studied in articles [6, 7, 8].

We study the existence and stability of relaxation periodic regimes of system (1). We are interested in a so-called *antiphase* solution (see [7]). In addition, the solution has a special neurodynamic system property — a *bursting-effect* (see [9, 10, 11, 12, 13, 14]).

Definition 1. *A periodic solution of (1) is called antiphase if*

(i) $u_2(t)$ *is a time shift of* $u_1(t)$*:* $u_1(t) = u(t)$, $u_2 = u(t+\Delta)$,

(ii) $\Delta = T/2$, *where* T *is a period of* $u(t)$.

Definition 2. *A bursting-effect is an alternation of several consecutive intensive spikes with refractory period of a membrane potential.*

To define a limit object correctly we rewrite (1) in a suitable form. Using the following exponential substitution

$$u_j = \exp(\lambda x_j), \quad j = 1, 2, \tag{5}$$

we get

$$\begin{aligned} \dot{x}_1 &= F(x_1(t-1), \varepsilon) + b\,(c - x_1)G(x_2(t-h), \varepsilon), \\ \dot{x}_2 &= F(x_2(t-1), \varepsilon) + b\,(c - x_2)G(x_1(t-h), \varepsilon), \end{aligned} \tag{6}$$

where $\varepsilon = 1/\lambda \ll 1$, and

$$F(x, \varepsilon) = f(\exp(x/\varepsilon)), \quad G(x, \varepsilon) = g(\exp(x/\varepsilon)). \tag{7}$$

By (3) and (4), we define the limit functions:

$$R(x) \stackrel{\text{def}}{=} \lim_{\varepsilon\to 0} F(x, \varepsilon) = \begin{cases} 1 & \text{as } x < 0, \\ -a & \text{as } x > 0, \end{cases} \qquad H(x) \stackrel{\text{def}}{=} \lim_{\varepsilon\to 0} G(x, \varepsilon) = \begin{cases} 0 & \text{as } x < 0, \\ 1 & \text{as } x > 0. \end{cases} \tag{8}$$

In order to find an antiphase solution of (1) with a bursting-effect, we need to find a relaxation antiphase regime of (6). More precisely, we are interested in a solution with a given number of relatively short alternating segments of positivity and negativity of the solution which go after a long enough segment where the solution values are negative. By (5), the existence of such a solution of system (6) means that (1) has an antiphase regime with bursting-effect.

2. Analysis of the auxiliary equation

We are looking for a solution of (6) in the following form

$$x_1(t) \equiv x(t), \quad x_2(t) \equiv x(t + \Delta). \tag{9}$$

Such a solution corresponds to discrete travelling waves. This approach was suggested in [15] and considered in several works (for example, [2, 16, 17]). Substituting (9) in system (6), we obtain the following:

$$\dot{x} = F(x(t-1), \varepsilon) + b\,(c - x)G(x(t - h + \Delta), \varepsilon). \tag{10}$$

For a fixed natural n, we are interested in the existence of a relaxation cycle in (10) which contains n disjoint segments with positive values. We suppose that the parameters satisfy the following inequalities

$$(n-1)T_0 + t_0 + 1 < h - \Delta < nT_0, \quad c < -a - \frac{1}{b} - \frac{a+1}{1-\xi}, \tag{11}$$

where

$$\xi \stackrel{\text{def}}{=} \exp(-bt_0), \quad t_0 \stackrel{\text{def}}{=} 1 + 1/a, \quad T_0 \stackrel{\text{def}}{=} 2 + 1/a + a. \tag{12}$$

We explain a meaning of the restrictions (11) below.

Considering a limit case for (10) as $\varepsilon \to 0$ and by (8), we obtain the limit relay equation

$$\dot{x} = R(x(t-1)) + b\,(c-x)H(x(t-h+\Delta)). \tag{13}$$

As in [1, 2, 14, 15], we construct a solution of (13) using step-by-step method. Fix constants $\sigma > 0$, $q_1 \in (0,\sigma)$, $q_2 > \sigma$, and define

$$S \overset{\text{def}}{=} \{\varphi(t) \in C[-h+\Delta-\sigma,-\sigma],\ -q_1 \le \varphi(t) \le -q_2\ \forall\, t \in [-h+\Delta-\sigma,-\sigma],\ \varphi(-\sigma) = -\sigma\}. \tag{14}$$

We use $x_\varphi(t)$ to denote a solution of (13) for $t \ge -\sigma$ with an arbitrary function φ from (14). We prove that there exists a T_*-periodic function $x_*(t)$ such that the solution $x_\varphi(t)$ coincides with it. Let us describe the function $x_*(t)$.

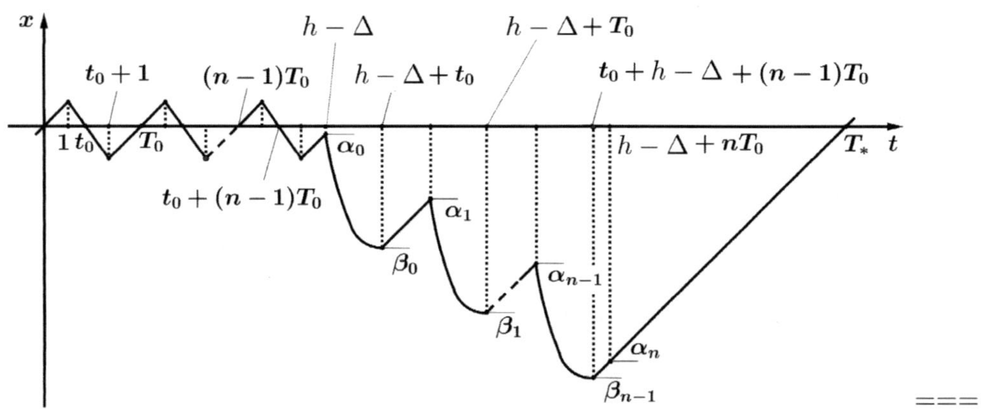

Figure 1. Solution (15) of relay equation (13)

$$x_*(t) \overset{\text{def}}{=} \begin{cases} x_0(t) & \text{as } t \in [0, h-\Delta], \\ x_e(h-\Delta+kT_0, \alpha_k, t) & \text{as } t \in [h-\Delta+kT_0, h-\Delta+t_0+kT_0], \\ t-h+\Delta-t_0-kT_0+\beta_k & \text{as } t \in [h-\Delta+t_0+kT_0, h-\Delta+(k+1)T_0], \\ t-T_* & \text{as } t \in [h-\Delta+nT_0, T_*]. \end{cases} \tag{15}$$

where x_0 is a T_0-periodic solution of an equation $\dot{x} = R(x(t-1))$ (see [4]) defined as follows

$$x_0(t) \overset{\text{def}}{=} \begin{cases} t & \text{as } 0 \le t \le 1, \\ 1-a(t-1) & \text{as } 1 \le t \le t_0+1, \\ t-T_0 & \text{as } t_0+1 \le t \le T_0, \end{cases} \tag{16}$$

and x_e is a solution of an auxiliary problem

$$\dot{x} = 1 + b(c-x), \quad x\big|_{t=\widetilde{t}} = \widetilde{x}; \tag{17}$$

for given constants $\widetilde{t}$, $\widetilde{x}$. Therefore, x_e has the following form

$$x_e(\widetilde{t},\ \widetilde{x}, t) \overset{\text{def}}{=} (\widetilde{x} - 1/b - c)\exp\big(-b(t-\widetilde{t})\big) + 1/b + c. \tag{18}$$

The period T_* is defined by the following identities

$$\alpha_0 \overset{\text{def}}{=} h-\Delta-nT_0, \quad \alpha_k \overset{\text{def}}{=} \xi^k\big(\alpha_0-\eta\big)+\eta,\ k=1,\ldots,n-1, \quad \eta \overset{\text{def}}{=} \frac{1}{b}+c+\frac{a+1}{1-\xi}; \tag{19}$$

$$\beta_k \overset{\text{def}}{=} x_e(h - \Delta + kT_0, \alpha_k, h - \Delta + t_0 + kT_0), \quad k = 0, \ldots, n-1. \tag{20}$$

$$T_* \overset{\text{def}}{=} -\alpha_n + h - \Delta + nT_0. \tag{21}$$

Note that restriction (11) means that the value of the delay $h - \Delta$ lies on a segment of period where the function $x_0(t)$ is negative and increasing.

Sequences $\{\alpha_k\}$, $\{\beta_k\}$ are values of function $x_*(t)$ (see pic. 1). We claim that the sequence $\{\alpha_k\}$ decreases. (As corollary, $\{\beta_k\}$ decreases too.) Indeed, ξ is positive constant smaller than 1 and the coefficient $(\alpha_0 - \eta)$ under ξ^k is negative. The last follows from restriction (11) on c. It also follows from (11) that $\alpha_0 < 0$. This implies that all elements of $\{\alpha_k\}$ are negative. We obtain that $x_*(t)$ has "long" enough segment with negative value.

Let us formulate a statement about relation between periodic solutions of (10) and (13).

Theorem 1. *Fix a natural n. Let the parameters a, b, c, h, Δ meet restrictions (11). Let $x_*(t)$ be a T_*-periodic solution (15) of (13) with exactly n segments on a period where it is positive. Then there exists a sufficiently small $\varepsilon_0 > 0$ such that, for all $\varepsilon \in (0, \varepsilon_0]$, equation (10) with initial function from (14) admits a unique exponentially orbitally stable cycle $x_*^\varepsilon(t)$ of period T_*^ε satisfying the limit relations*

$$\lim_{\varepsilon \to 0} \max_{0 \le t \le T_*^\varepsilon} |x_*^\varepsilon(t) - x_*(t)| = 0, \quad T_*^\varepsilon = T_* + O(\varepsilon). \tag{22}$$

A proof scheme is the same as, for example, in [1]. We give only a short plan here. Let us introduce some notation for its presentation. For an arbitrary function φ from (14), denote by $x^\varepsilon(t, \varphi)$ a solution of (10) with initial condition $\varphi(t)$, $t \in [-h + \Delta - \sigma, -\sigma]$. Note that set (14) is closed, bounded and convex. Suppose that the equation $x^\varepsilon(t - \sigma, \varphi) = -\sigma$ has $2n$ or more positive roots. We denote one with number $2n$ by $T^\varepsilon(\varphi)$. Finally, we define the Poincaré operator $\Pi^\varepsilon : C[-1-\sigma, -\sigma] \to C[-1-\sigma, -\sigma]$ by the formula $\Pi^\varepsilon(\varphi) = x^\varepsilon(t + T^\varepsilon(\varphi), \varphi)$, $-1-\sigma \le t \le -\sigma$.

A proof plan is as follows. We find asymptotic (uniform with respect to φ and t) formulas of the solution $x^\varepsilon(t, \varphi)$ on different time intervals (see similar constructions in [18]). As a corollary, we get that $x^\varepsilon(t, \varphi)$ is close to $x_*(t)$ on the segment $[-\sigma, T^\varepsilon(\varphi)]$. The formulas imply that for an appropriate choice of the parameters σ, q_1, q_2 the operator Π^ε is defined on the set S and takes it into itself. It allows us to use the Schauder principle to prove that there exists a fix point of Π^ε — a periodic solution of (6) with asymptotic $x^\varepsilon(t, \varphi)$. Then we estimate Fréchet derivative of Π^ε on the fix point. From this exponential estimation it follows that the map is compressive and the fix point is unique. Moreover, the estimation gives us an information about multipliers. One of them equals 1, and all other are exponentially small. It means that periodic solution exponentially orbitally stable.

3. Theorem about an existence of an antiphase regime of system (1)

From Theorem 1 it follows that equation (10) admits a cycle $x_*^\varepsilon(t)$ which is close to $x_*(t)$, described by (15). Since a period $T_*^\varepsilon = T_*^\varepsilon(\Delta)$ of the cycle depends on Δ, to prove that antiphase solution of (6) exist we need to prove that the equation

$$T_*^\varepsilon(\Delta) = 2\Delta \tag{23}$$

has a solution.

By (19), (21), (22), we obtain that restriction (23) has the following form:

$$-\xi^n(h - \Delta - nT_0 - \eta) - \eta + h - \Delta + nT_0 = 2\Delta. \tag{24}$$

This implies

$$\Delta = \frac{1}{3 - \xi^n}((h - \eta)(1 - \xi^n) + nT_0(1 + \xi^n)). \tag{25}$$

Put (25) to (11). We get the following restriction on h:

$$t_0 + 1 + (2n-1)T_0 - \frac{1}{2}(1-\xi^n)(\eta + a) < h < 2nT_0 - \frac{1}{2}(1-\xi^n)\eta. \quad (26)$$

From above and Theorem 1 we obtain the following.

Theorem 2. *Fix a natural n. Let the parameters a, b, c, h meet restrictions (11), (26), Δ is described by (25). Then system (6) admits antiphase solution $(x_*^\varepsilon(t), x_*^\varepsilon(t+\Delta))$ with exactly n segments on a period where it is positive.*

Taking into account (5), we obtain that (1) admits an antiphase solution with a bursting.

4. Theorem about a stability of the antiphase regime of system (1)

Investigate stability of the antiphase solution in a phase space $C[-h-\sigma, -\sigma] \times C[-h-\sigma, -\sigma]$ of system (6). Let us proof the following.

Theorem 3. *The antiphase solution $(x_*^\varepsilon(t), x_*^\varepsilon(t+\Delta))$ of (6), satisfying conditions of Theorem 2, is exponentially obitally stable.*

We linearise system (6) on the antiphase solution. We get

$$\begin{aligned}\dot\gamma_1 = -bG(x_*^\varepsilon(t+\Delta-h),\varepsilon)\gamma_1 + A(x_*^\varepsilon(t-1),\varepsilon)\gamma_1(t-1) + \\ +b(c - x_*^\varepsilon(t))B(x_*^\varepsilon(t+\Delta-h),\varepsilon)\gamma_2(t-h),\end{aligned} \quad (27)$$

$$\begin{aligned}\dot\gamma_2 = -bG(x_*^\varepsilon(t-h),\varepsilon)\gamma_2 + A(x_*^\varepsilon(t-\Delta-1),\varepsilon)\gamma_2(t-1) + \\ +b(c - x_*^\varepsilon(t-\Delta))B(x_*^\varepsilon(t-h),\varepsilon)\gamma_1(t-h),\end{aligned} \quad (28)$$

where $G(x,\varepsilon)$ is defened by(7),

$$A(x,\varepsilon) \stackrel{\text{def}}{=} f'(\exp(x/\varepsilon))\exp(x/\varepsilon)/\varepsilon, \quad B(x,\varepsilon) \stackrel{\text{def}}{=} g'(\exp(x/\varepsilon))\exp(x/\varepsilon)/\varepsilon. \quad (29)$$

Since (27), (28) is linear system with periodic coeffitions (of period T_*), we can apply Lyapunov-Floquet theory. To study an antiphase solution stability we estimate its multipliers. To prove Theorem 3 we show that system (27), (28) has unit multiplier and all other its multipliers are less than 1.

Note that if $(\gamma_1(t), \gamma_2(t))$ is a solution of (27), (28) then $(\gamma_2(t+\Delta), \gamma_1(t+\Delta))$ is also a solution. It means that there exists a constant κ such that, for any $t > -\sigma$, we have

$$\kappa\gamma_1(t) = \gamma_2(t+\Delta), \quad \kappa\gamma_1(t) = \gamma_2(t+\Delta). \quad (30)$$

By this and $T_* = 2\Delta$, for any $t > -\sigma$, we get $\kappa^2\gamma_1(t) = \gamma_1(t+T_*)$, $\kappa^2\gamma_2(t) = \gamma_2(t+T_*)$. It means that κ^2 is a multiplier. Thus, we have the following.

Claim 1. *Any κ is corresponded a multiplier $\nu = \kappa^2$ of linearised system (27), (28).*

Furthermore, (30) implies that fuunctions γ_1 and γ_2 have the following form:

$$\gamma_1(t) = \gamma(t), \quad \gamma_2(t) = \kappa\gamma(t-\Delta), \quad (31)$$

where $\gamma(t)$ satisfies the equation

$$\begin{aligned}\dot\gamma = -bG(x_*^\varepsilon(t+\Delta-h),\varepsilon)\gamma + A(x_*^\varepsilon(t-1),\varepsilon)\gamma(t-1) + \\ +\kappa b(c - x_*^\varepsilon(t))B(x_*^\varepsilon(t+\Delta-h),\varepsilon)\gamma(t-\Delta-h).\end{aligned} \quad (32)$$

Let us define multipliers of equation (32). As in [16], we call an eigenvalue of a special monodromy operator by *a multiplier of a differential-difference equation.* To define the operator let us fix real number γ_* and consider the space

$$E = \{\hat{\gamma}(t) \in C[-h-\Delta-\sigma, -\sigma], \quad \hat{\gamma}(-\sigma) = \gamma_*\}.$$

Defenition 3. *A monodromy operator of (32) is an operator $V : E \to E$ such that*

$$V\hat{\gamma} = \gamma(t + T_*^\varepsilon, \hat{\gamma}) \in E, \quad -h-\Delta-\sigma \le t \le -\sigma, \tag{33}$$

where $\gamma(t,\hat{\gamma})$ is a solution of (32) with the initial condition $\hat{\gamma}(t)$, $-h-\Delta-\sigma \le t \le -\sigma$; T_*^ε is a period of antiphase solution.

By (31), we have the following.

Claim 2. *Multipliers of system (27), (28) and multipliers of equation (32) are the same.*

Note that (33) is linear, bounded and depends on κ (since coefficients of (32) depend on κ). Claims 1 and 2 allow us to find multipliers of (27), (28) in a following way. At first, we construct formulas $\mu = \mu(\kappa)$ of multiplier of equation (32) (depending on κ). Secondly, we find κ from the equation $\mu(\kappa) = \kappa^2$. As a result we find desired multipliers.

Consider equation (32). Definition of $A(x,\varepsilon)$ and $B(x,\varepsilon)$ imply that they are close to delta-function. Their values have an order $O(e^{-q/\varepsilon})$ whenever x is not zero. Here and below q is a positive constant (precise value of which is not important). The function γ switch its value as arguments $x_*^\varepsilon(t-1)$, $x_*^\varepsilon(t+\Delta-h)$ of functions A and B equal to zero and $\varepsilon \to 0$. Denote roots of equation $x_*^\varepsilon(t) = 0$ by $t_{k,l}$ and $t_{k,l}^0$, $k = 0, \ldots, n-1$, $l \in \{0\} \cup N$. By (15), (22), we have

$$t_{k,l} \stackrel{\text{def}}{=} kT_0 + lT_* + O(e^{-q/\varepsilon}), \quad t_{k,l}^0 \stackrel{\text{def}}{=} t_0 + kT_0 + lT_* + O(e^{-q/\varepsilon}) \tag{34}$$

Then γ has the following swich point

$$1 + t_{k,l}, \quad 1 + t_{k,l}^0, \quad h - \Delta + t_{k,l}, \quad h - \Delta + t_{k,l}^0. \tag{35}$$

Let us find how γ change value at that points.

Lemma 1.

$$\gamma(1+t_{k,l}+0) = -a\gamma(1+t_{k,l}-0)+O(e^{-q/\varepsilon}), \quad \gamma(1+t_{k,l}^0+0) = -\gamma(1+t_{k,l}^0-0)/a+O(e^{-q/\varepsilon}), \tag{36}$$

$$\gamma(h-\Delta+t_{k,l}+0) = \gamma(h-\Delta+t_{k,l}-0) + \kappa b(c-\alpha_k(h-\Delta))\gamma(t_{k,l}-2\Delta) + O(e^{-q/\varepsilon}), \tag{37}$$

$$\gamma(h-\Delta+t_{k,l}^0+0) = \gamma(h-\Delta+t_{k,l}^0-0) + \kappa b(c-\beta_k(h-\Delta))\gamma(t_{k,l}^0-2\Delta)/a + O(e^{-q/\varepsilon}). \tag{38}$$

The proof of Lemma 1 is technical and use an asyptotic (15), (22) of an antiphase solution. Consider a point set

$$M \stackrel{\text{def}}{=} \{-\sigma, \ t_{k,0} - T_*^\varepsilon, \ t_{k,0}^0 - T_*^\varepsilon : \ k = 0, 1, \ldots, n-1\}, \tag{39}$$

embeded to the segment $[-h-\Delta-\sigma, -\sigma]$. Let us prove the following fact.

Lemma 2. *Let $\hat{\gamma}(t) \in E$ and value of $\hat{\gamma}$ at points of M are known:*

$$\tilde{\gamma} \stackrel{\text{def}}{=} \hat{\gamma}(-\sigma), \quad \tilde{\gamma}_k \stackrel{\text{def}}{=} \hat{\gamma}(t_{k,0} - T_*^\varepsilon), \quad \tilde{\gamma}_k^0 \stackrel{\text{def}}{=} \hat{\gamma}(t_{k,0}^0 - T_*^\varepsilon). \tag{40}$$

Then, across period T_*^ε of $x_*^\varepsilon(t)$, the values maps to numbers

$$\gamma(t_{k,0}) = \tilde\gamma + O(e^{-q/\varepsilon}), \quad \gamma(t_{k,0}^0) = -a\tilde\gamma + O(e^{-q/\varepsilon}), \tag{41}$$

$$\gamma(-\sigma + T_*^\varepsilon) = \xi^n\tilde\gamma + \sum_{k=0}^{n-1} c_k\tilde\gamma_k + \sum_{k=0}^{n-1} c_k^0\tilde\gamma_k^0 + O(e^{-q/\varepsilon}), \tag{42}$$

$$c_k \stackrel{\text{def}}{=} \kappa b(c-\alpha_k)\xi^{n-k}, \quad c_k^0 \stackrel{\text{def}}{=} \kappa b(c-\beta_k)\xi^{n-k-1}/a. \tag{43}$$

The function $G(x_*^\varepsilon(t-h+\Delta))$ is exponentially small on a segment $[-\sigma, t_{0,0}+h-\Delta]$. It implies that the derivative $\dot\gamma(t)$ close to zero on this segment excepting points $1+t_{k,0}$, $1+t_{k,0}^0$. Thus, by (36), we get $\gamma(t) = \tilde\gamma + O(e^{-q/\varepsilon})$ as $t \in [-\sigma, 1+t_{0,0}]$, $[1+t_{k,0}^0, 1+t_{k+1,0}]$, $k \in 0,\dots,n-2$, $[1+t_{n-1,0}^0, h-\Delta+t_{0,0}]$, $\gamma(t) = -a\tilde\gamma + O(e^{-q/\varepsilon})$ as $t \in [1+t_{k,0}, 1+t_{k,0}^0]$, $k = 0,\dots,n-1$. Therefore, (41) is proved.

The function $G(x_*^\varepsilon(t-h+\Delta))$ is exponentially close to 1 on the segments $[h-\Delta+t_{k,0}, h-\Delta+t_{k,0}^0]$, $k = 0,\dots,n-1$. This means that going through the intervals $(h-\Delta+t_{k,0}, h-\Delta+t_{k,0}^0)$, $k = 0,\dots,n-1$, whose length is t_0, the $\gamma(t)$ is multiplied by $\xi = \exp(-bt_0)$. Finally, using Lemma 1, we get (42).

Lemma 2 is proved.

A main result of this section is the following.

Lemma 3. *Monodromy operator (33) has multipliers μ_k, $k \in N$, with the following properties:*
(i) $\mu_1 = 1$,
(ii) $\mu_2, \mu_3 : \ |\mu_2| = |\mu_3| < 1$,
(iii) $\mu_4, \mu_5, \dots : \ |\mu_k| = O(e^{-q/\varepsilon})$.

Let us prove it. For every function $\hat\gamma \in E$, we have $\hat\gamma(t) = \gamma_1(t) + \gamma_0(t)$,

$$\gamma_1(t) \stackrel{\text{def}}{=} \begin{cases} 1 & \text{as } t \in M, \\ 0 & \text{as } t \in [-h-\Delta-\sigma, -\sigma] \setminus M, \end{cases} \qquad \gamma_0(t) \stackrel{\text{def}}{=} \begin{cases} 0 & \text{as } t \in M, \\ \hat\gamma(t) & \text{as } t \in [-h-\Delta-\sigma, -\sigma] \setminus M. \end{cases} \tag{44}$$

Since V is linear, we obtain

$$V\tilde\gamma = V\gamma_1 + V\gamma_0 = \gamma(t,\gamma_1) + \gamma(t,\gamma_0) = \gamma(t,\gamma_1) + O(e^{-q/\varepsilon}).$$

Thus, using Lemma 2, we have that an operator multipliers consist of two groups. The first one contains eigenvalues of $(2n+1)$-dimensional linear operator

$$\begin{pmatrix} \xi^m & c_0 & c_0^0 & \cdots & c_{n-1} & c_{n-1}^0 \\ 1 & 0 & 0 & \cdots & 0 & 0 \\ -a & 0 & 0 & \cdots & 0 & 0 \\ \vdots & \vdots & \vdots & \ddots & \vdots & \vdots \\ 1 & 0 & 0 & \cdots & 0 & 0 \\ -a & 0 & 0 & \cdots & 0 & 0 \end{pmatrix}. \tag{45}$$

The second one contains infinity count of exponentially small multipliers. The characteristic polynomial of matrix (45) has the form

$$\mu^{2n-1}(-\mu^2 + \xi\mu + \kappa(1-\xi^{k+1})) = 0 \tag{46}$$

Solving system (47), (46) we find multipliers:

$$\mu_1 = 1, \quad \mu_{2,3} = -1/2 - \xi^n \pm i\sqrt{3-4\xi^n}/2. \tag{47}$$

Modules of $\mu_{2,3}$ are close to 1, but less than 1.

Lemma 3 is proved. It means that antiphase solution is exponentially orbitally stable.

5. Conclusion

As an example, show (pic. 2) an antiphase solution of (6). We use a program *Tracer3_70* developed by D. S. Glyzin. The function and parameters are as follows:

$$f(u) = a(1-u)/(a+u), \quad g(u) = u/(1+u), \quad h = 16, \quad a = 1.5, \quad b = 2, \quad c = -4.5, \quad \lambda = 100.$$

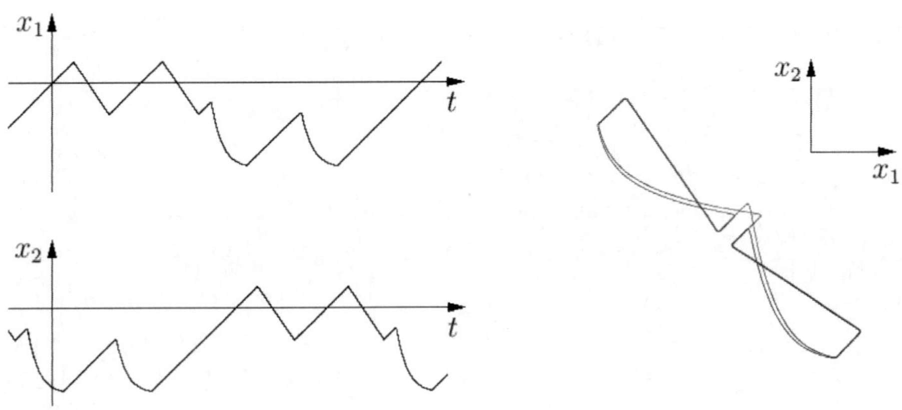

Figure 2. Antiphase solution of (6) that has two segments with positive values.

To sum up, note that the delay h of connection chain can be chosen longer than period in model (2) of alone oscillation. We prove that this allows us to get an antiphase solution of (1) with a bursting-effect.

Acknowledgments
This work was supported by the Russian Science Foundation (project nos. 14-21-00158).

References

[1] Glyzin S D, Kolesov A Yu and Rozov N Kh 2013 *Diff. Eq.* **49** 1227-44.
[2] Glyzin S D, Kolesov A Yu and Rozov N Kh 2015 *Rus. Math. Surveys* **70** 383-452
[3] Preobrazhenskaia M M 2017 *Model. Anal. Inform. Sist.* **24** 186-204
[4] Kolesov A Yu, Mishchenko E F and Rozov N Kh 2010 *Computat. Math. and Math. Physics* **50** 1990-2002
[5] Kashchenko S 2015 *Models of Wave Memory* (Berlin: Springer) p 239
[6] Somers D and Kopell N 1993 *Biol. Cybern.* **68** 393-407
[7] Somers D and Kopell N 1995 *J. Math. Biol.* **33** 261-80
[8] Terman D 2005 *Tutorials in Math. Biosciences I, Lecture Notes in Math.* **1860** 21-68
[9] Chay T R and Rinzel J 1985 *Biophys. J.* **47** 357-66
[10] Ermentrout G B and Kopell N 1986 *SIAM J. Appl. Math.* **46** 233-53
[11] Izhikevich E 2000 *Inter. J. of Bifurc. and Chaos.* **10** 1171-266
[12] Rabinovich M I, Varona P, Selverston A I and Abarbanel H D I 2006 *Rev. Mod. Phys.* **78** 1213-65
[13] Coombes S and Bressloff P C 2005 *Bursting: the Genesis of Rhythm in the Nervous System* (Singapore: World Scientific Publishing Company) p 418
[14] Glyzin S D, Kolesov A Yu and Rozov N Kh 2013 *Math. Notes* **93** 676-90
[15] Glyzin S D, Kolesov A Yu and Rozov N Kh 2011 *Diff. Eq.* **47** 1697-713
[16] Glyzin S D, Kolesov A Yu and Rozov N Kh 2013 *Theor. and Math. Physics* **175** 499-517
[17] Glyzin S D, Kolesov A Yu and Rozov N Kh 2012 *Comput. Math. and Math. Physics* **52** 702-19
[18] Preobrazhenskaia M M 2017 *Model. Anal. Inform. Sist.* **24** 550-66

IOP Conf. Series: Journal of Physics: Conf. Series **1163** (2019) 012063 doi:10.1088/1742-6596/1163/1/012063

Numerical simulation of two- and three-dimensional gravity-capillary waves

A S Dosaev and Yu I Troitskaya

Institute of Applied Physics of the Russian Academy of Sciences, 46 Ul'yanov Street, 603950, Nizhny Novgorod, Russia

E-mail: dosaev@appl.sci-nnov.ru

Abstract. Dynamics of gravity-capillary deep water waves is studied within the framework of potential Euler equations with parameterizations for wind forcing and viscous decay. Conformal mapping based technique for two dimensions and its generalization for weakly three-dimensional waves are employed to model evolution of nonlinear decimeter-range waves and generation of parasitic capillary ripples.

1. Introduction

In nature wind waves of gravity-capillary range are special in many respects. Steep decimeter-range waves develop a characteristic pattern of parasitic capillary ripples on their forward slopes. Parasitic ripples are generated by a narrow pressure distribution associated with sharp crest of the underlying wave [1], and are therefore very sensitive to a change of its parameters. Ripples are drawing energy from the underlying wave and dissipating due to viscousity, making decimeter-range waves subject to nonlinear damping. On the other hand, decimeter waves are also subject to strong wind forcing. Understanding the dynamics of gravity-capillary waves is important for remote sensing, because short ripples are the main scatterers of EM waves in the microwave range, and for studying long waves generation mechanisms, because momentum transfer from atmosphere to ocean occurs mainly due to the interaction of wind with short waves.

We perform numerical simulation of gravity-capillary waves within the framework of fully nonlinear equations of motion (Euler equations) for potential flow with a free surface using parameterizations for wind forcing and viscous decay. For two-dimensional flows a powerful technique of conformal transformations is available, and numerical methods based on it have high accuracy and computational efficiency [2]. Another common approach to modeling potential waves, applicable to both two- and three-dimensional motions, is to use boundary integral equation methods. They, however, have high computational costs. Here to take an account of three-dimensional effects we employ a quasi-three-dimensional model put forward by Ruban [3]. It is based on the method of conformal transformations and allows an efficient implementation using a Fast Fourier Transform. The model assumes narrow directional distribution of waves while not imposing any limitations on their steepness. We model gravity-capillary waves dynamics in both two- and three-dimensional setups and study evolution of horizontal asymmetry of wave profiles.

IOP Conf. Series: Journal of Physics: Conf. Series **1163** (2019) 012063 doi:10.1088/1742-6596/1163/1/012063

2. Basic equations

Within our approach we assume water to be an ideal incompressible fluid and the wave motion to be potential. In the Cartesian coordinate frame with the x axis coinciding with the unperturbed water surface and the y axis directed upward, governing equations for the free surface $y = \eta$ and velocity potential φ are:

$$\Delta\varphi = 0, \quad (1)$$

$$\eta_t + \nabla_\perp \eta \cdot \nabla_\perp \varphi = \varphi_y, \quad (2)$$

$$\varphi_t + \frac{1}{2}|\nabla\varphi|^2 + p + T\varkappa + g\eta = 0 \quad (3)$$

where $\nabla_\perp$ denotes horizontal component of gradient, T is surface tension coefficient and $\varkappa$ is free surface curvature. At infinity the deep water boundary condition holds:

$$|\nabla\varphi| \to 0 \quad \text{as} \quad y \to -\infty$$

Other effects, such as viscous damping and wind forcing, are incorporated into the pressure term p. Viscous dissipation is modeled in the spirit of [4] by the term

$$p_{visc} = -4\nu\Delta_s\varphi,$$

where Δ_s denotes the Laplacian operator on the free surface, ν is kinematic viscousity. Wind input is required in order to balance energy losses due to dissipation; here we parameterize it as

$$p_{wind} = 0.04u_*^2\frac{\partial\eta}{\partial x}$$

where u_*^2 is wind friction velocity, so that it corresponds to the Miles' generation for the small slopes [5].

2.1. Numerical model for 2D waves

To solve the system (1)-(3) we employ a method based on the use of conformal mapping. The flow domain is mapped into the lower half-plane of complex variable $w = u + iv$:

$$x + iy = z(u + iv, t), \quad (4)$$

so that the function $z(w)$ is analytic in the flow domain. In the new independent variables (u, v) the Laplace equation (1) for the potential φ retains its form, while the free surface becomes a straight line, being mapped onto the real axis $v = 0$. That makes solving the Laplace equation trivial. "Complex potential" $\theta = \varphi + i\psi$, where ψ is the stream function and harmonic conjugate of φ, becomes an analytic function in the half-plane $v < 0$. On the real axis $\psi(u) = \hat{H}\varphi(u)$, where $\hat{H}$ is Hilbert transform

Evolution equations for the mapping z and potential φ can be derived from (1)-(3):

$$z_t = z_u(\hat{H} - i)\frac{\hat{H}\varphi_u}{|z_u|^2}, \quad (5)$$

$$\varphi_t = -\frac{\varphi_u^2 - (\hat{H}\varphi_u)^2}{2|z_u|^2} + \varphi_u\hat{H}\left(\frac{\hat{H}\varphi_u}{|z_u|^2}\right) - p - T\varkappa - gy$$

The equations that we have beed actually solving numericaly can be obtained by a further change of variables

$$R = \frac{1}{z'}, \qquad V = \frac{i\theta'}{z'},$$

IOP Conf. Series: Journal of Physics: Conf. Series **1163** (2019) 012063 doi:10.1088/1742-6596/1163/1/012063

which leads to a system with improved numerical stability (Dyachenko equations [6]):

$$\begin{aligned} R_t &= i(U'R - UR'), \\ V_t &= i(UV' - R\mathcal{P}'(|V|^2 + 2p + 2T\varkappa)) + g(R-1), \end{aligned} \tag{6}$$

where prime denotes differentiation with respect to w,

$$U = \mathcal{P}\left(RV^* + R^*V\right), \quad \mathcal{P} = \frac{1}{2}(1 + i\hat{H}),$$

asterisk denotes complex conjugate. The system (6) was integrated numerically using Dormand-Prince method of Runge-Kutta family with adaptive step size.

2.2. Numerical model for 3D waves

As was shown by Ruban [3], method of conformal transformations can be applied to a problem of weakly three-dimensional waves' evolution, and the corresponding approximate equations of motion retain some of the good computational properties of the two-dimensional system (5). The method is based on the assumption of narrow directional distribution of waves, while no limitation is imposed onto waves' steepness.

A new horizontal axis q pointed perpendicular to the "main" horizontal direction x is introduced, so that conformal mapping z and potential φ become functions of q:

$$x + iy = z(u + iv, q, t), \qquad \varphi = \varphi(u, v, q, t)$$

The characteristic scale of change along the q direction is assumed to be large: $\epsilon^2 = l_x/l_q \ll 1$. The case of $\epsilon = 0$ corresponds to a purely two-dimensional motion. From a perturbation theory with regard to small parameter ϵ the first order correction to (5) is obtained in [3], [7] together with some possible regularizations for the resulting equations. The resulting formulas are somewhat too cumbersome to be quoted here, but the three-dimensional corrections to the RHS of (5) can still be computed using Fast Fourier Transforms and local operations in either coordinate- or wavenumber representation.

It remains an open question to what extent those approximate equations of motion can describe the dynamics of waves of gravity-capillary or capillary ranges.

3. Evolution of gravity-capillary waves

In our numerical experiments we prescribe the initial conditions as a pure Stokes wave and then observe its evolution under the action of wind forcing. In a strictly periodic setup, and if the wind forcing is not too strong so the dominant wave does not grow to the point of breaking, the evolution leads to a stationary configuration where the wind input is balanced by the nonlinear damping due to parasitic ripples. An example of such process is shown on figure 1.

Very short gravity-capillary waves represent an interesting special case, when the resulting stationary configuration depends not only on the magnitude of wind forcing, but also on the way the wind friction velocity changes in time before arriving at the final value. Two possible stationary configurations for $\lambda = 5$ cm waves can be seen on figure 4. The difference vanishes for longer waves.

An important property of these stationary wave profiles is that they are asymmetric with regard to the vertical axis, that is, their front slope is steeper than the rear. This feature may be important to remote sensing, because it affects the inclination of the array of scatterers (that is, capillary ripples) to the horizon. The front-back asymmetry of wave profile η can be quantified as [8]

$$A = \frac{\langle (\hat{H}\eta)^3 \rangle}{\sqrt{\langle \eta^2 \rangle}^3},$$

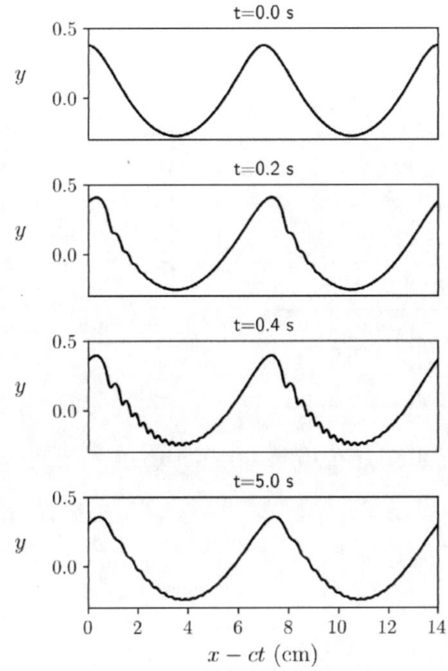

Figure 1. Evolution of 2d wave profile under the action of wind: $\lambda = 7$ cm, $u_* = 10$ cm/s

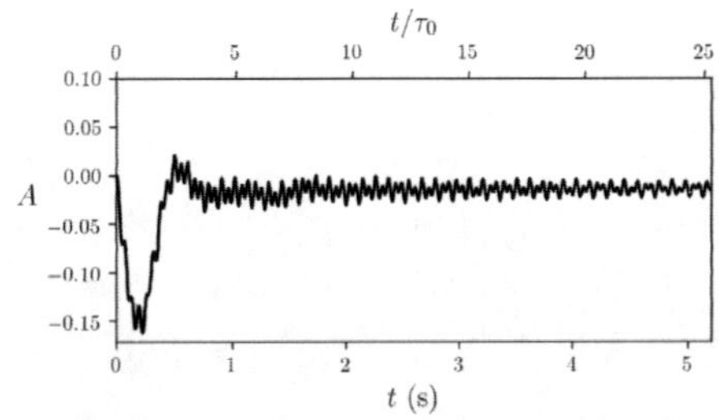

Figure 2. Evolution of profile asymmetry for the example on figure 1

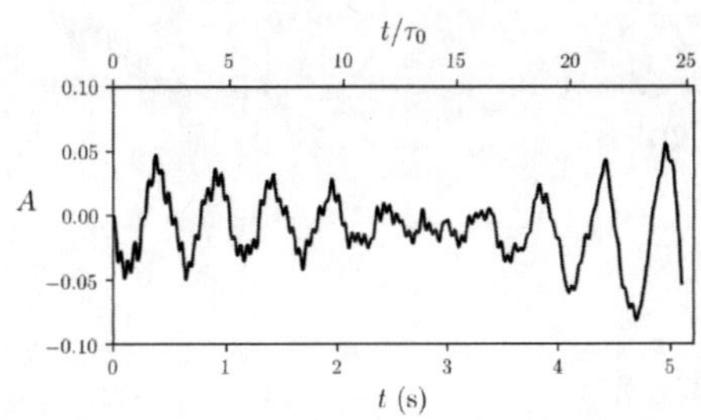

Figure 3. Asymmetry of 3d wave profile along the line $q = 0$; $\lambda = 7$ cm, $u_* = 10$ cm/s

where $\hat{H}$ is Hilbert transform. Steeper front slope corresponds to negative values of A. As the figure 2 shows, after a few large oscillations (during which asymmetry can be positive or negative) the field settles at a moderate value of negative asymmetry.

Evolution of individual waves in a modulated wavetrain can differ substantially from the periodic scenario. Figure 5 shows an example of a three-dimensional simulation, where the initial conditions were prescribed as a Stokes wave modulated in both longitudinal and transversal directions. Corresponding evolution of profile asymmetry (in the $q = 0$ plane) is shown on figure 3. It can be seen that the waves in a modulated wavetrain oscillate in a similar manner as in the strictly periodic two-dimensional setup, but the modulation prevents their arrival at a stationary configuration.

Acknowledgements
This research was supported by RFBR grant No. 18-35-00658.

References

[1] Longuet-Higgins M 1995 *Journal of Fluid Mechanics* **301** 79–107
[2] Chalikov D and Sheinin D 2005 *Journal of Computational Physics* **210** 247–273
[3] Ruban V P 2005 *Physical Review E* **71** 055303
[4] Ruvinsky K D, Feldstein F I and Freidman G I 1991 *Journal of Fluid Mechanics* **230** 339–353
[5] Plant W J 1982 *Journal of Geophysical Research: Oceans* **87** 1961–1967
[6] Zakharov V E, Dyachenko A I and Vasilyev O A 2002 *European Journal of Mechanics-B/Fluids* **21** 283–291
[7] Ruban V P 2010 *The European Physical Journal Special Topics* **185** 17–33
[8] Elgar S 1987 *IEEE transactions on acoustics, speech, and signal processing* **35** 1725–1726

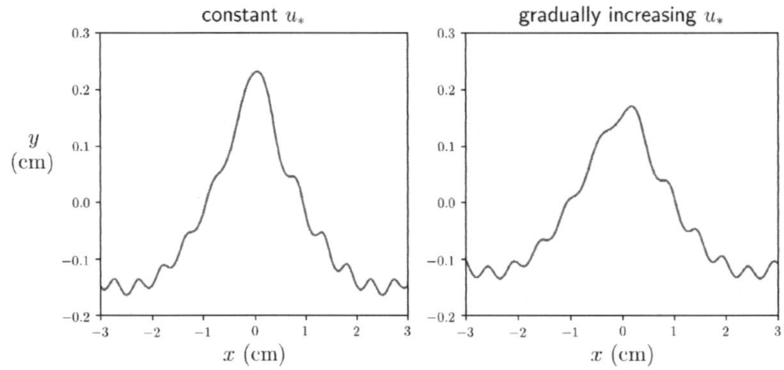

Figure 4. Stationary gravity-capillary wave profiles obtained in different setups; $\lambda = 5$ cm, $u_* = 13.5$ cm/s

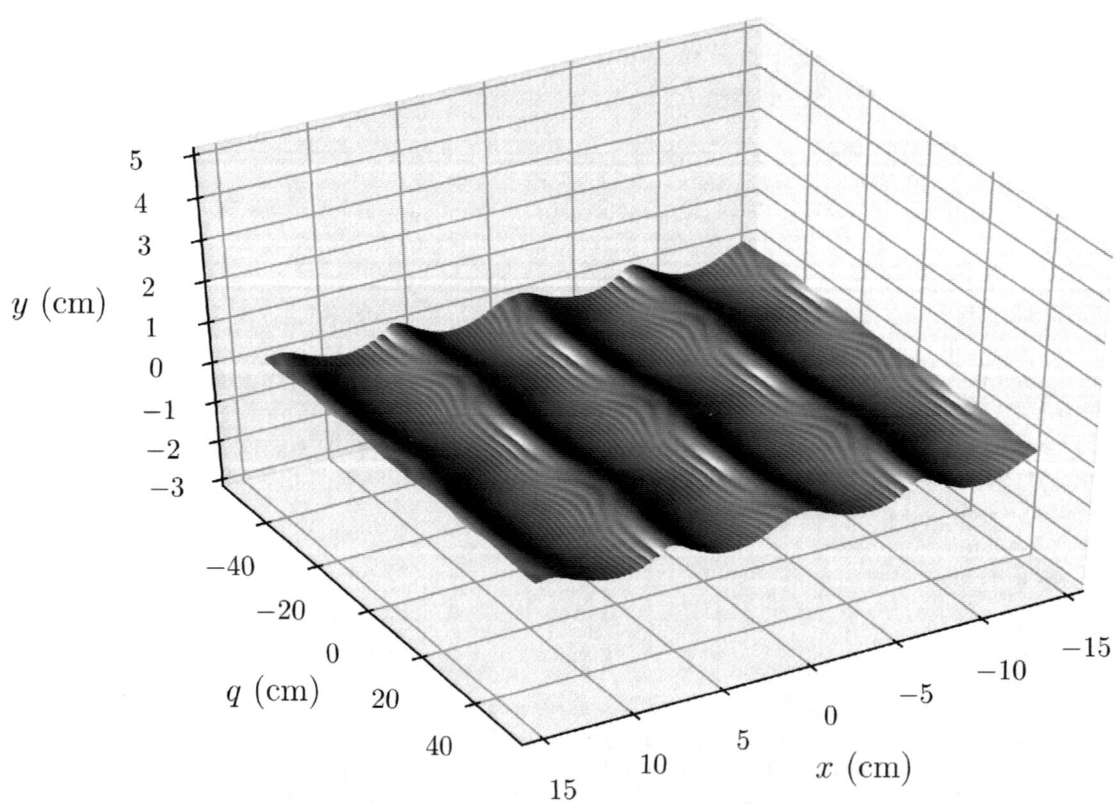

Figure 5. Numerical simulation of a modulated three-dimensional gravity-capillary wave, $\lambda = 7$ cm, $ak \sim 0.25$

IOP Conf. Series: Journal of Physics: Conf. Series **1163** (2019) 012064 doi:10.1088/1742-6596/1163/1/012064

Hegselmann-Krause model of opinions dynamics of interacting agents with the random noises

E S Volkova[1] **, L A Manita**[1]**, A D Manita**[1, 2]

[1] National Research University Higher School of Economics, Moscow Institute of Electronics and Mathematics, Russia
[2] Lomonosov Moscow State University, Faculty of Mechanics and Mathematics, 119991, GSP-1, Moscow, Russia
E-mail: esvolkova_2@edu.hse.ru, lmanita@hse.ru, amanita@hse.ru

Abstract. We consider the Hegselmann-Krause bounded confidence model of opinion dynamics. We assume that the opinion of an agent is influenced not only by other agents, but also by external random noises. The case of independent normally distributed external noises is considered. We perform computer modeling of deterministic and stochastic models. The properties of the models were analyzed and the difference in their behavior was revealed. We study the dependence of the number of a confidence clusters on the parameters of the problem such as the initial profile of opinions, the level of confidence, the variance of noise.

1. Introduction

During recent decades models of opinion dynamics in groups of interacting agents have become widespread [1]-[7]. The main issues that are studied are reaching consensus (agreement) in the group, clustering opinions [1]-[3], conditions for the existence of consensus [4]-[7], the rate of convergence to consensus [8]-[9]. The opinion dynamics depends on the consensus algorithm, the topology structure of the group of agents and so on. In this paper we study the Hegselmann – Krause bounded confidence model (H-K model). We introduce a stochastic Hegselmann – Krause model, analyze properties of the corresponding stochastic process and compare its behavior with the deterministic H-K model.

2.Hegselmann -- Krause bounded confidence model [2]-[4]

Consider a group of m agents. Each agent has an opinion which is represented by a real number. Let $x_i(t)$ be an opinion of the agent i at the moment $t \in T = \{0,1,2,\dots\}$. We assume that $x_i(t) \in [0,1]$ for all t and $i = 1,\dots m$. Denote $x(t) = (x_1(t),\dots,x_m(t))$. The vector $x(t)$ is called an opinion profile at time t. The agent i interacts only with those agents whose opinions differ from x_i not more than a certain confidence level ε. For each agent i denote a confidence set by $D(i,x)$:

$$D(i,x) = \{1 \le j \le m, |x_i - x_j| \le \varepsilon\} \tag{1}$$

On the next time step the agent i replaces its current opinion by the average of the agent opinions that are in $D(i,t)$:

$$x_i(t+1) = \left|D\big(i, x(t)\big)\right|^{-1} \sum_{j \in D(i,x(t))} x_j(t)\,,\ t \in T \tag{2}$$

Here $|D(i,x)|$ is the number of elements of $D(i,x)$.

An opinion profile x is a consensus state if $x_i = \tilde{x}$ for all $i = 1, \ldots, m$. We will call $\tilde{x}$ the consensus value. We say that the model reaches the consensus $\tilde{x}$ if $x_i(t) \to \tilde{x}$ for all i. The main properties of H-K model are the following.

1. The dynamics does not change the order of opinions.
2. If the split between agents i and j occurs at the moment t_0, then this split remains for all $t \geq t_0$. (If $|x_i - x_j| > \epsilon$ then we say that there is a split between agents i and j.)
3. If agents reach a consensus, then it is reached at the finite time.

3.Stochastic Hegselmann – Krause model

Suppose that the opinions of agents can be influenced not only by other agents, but also by external random factors. Then the dynamics of $x(t)$ is described by

$$x_i(t+1) = g\big(x_i^*(t+1)\big) \tag{3}$$

$$x_i^*(t+1) = \left|D\big(i, x(t)\big)\right|^{-1} \sum_{j \in D(i,x(t))} x_j(t) + \sigma\delta_i(t), \mathrm{t} \in \mathrm{T} \tag{4}$$

$$g(y) = \begin{cases} 0, & y < 0 \\ y, & 0 \leq y \leq 1 \\ 1, & y > 1 \end{cases} \tag{5}$$

We assume that $\delta_i(t), i = 1, \ldots, m$ are independent identically distributed (i.i.d.) random variables, $\delta_i(t) \sim N(0,1)$. We use the function g(y) since opinion values cannot be outside the interval [0; 1].

We are interested which properties of the deterministic H-K model are preserved for the stochastic model. Since $x(t)$ is a stochastic process, the above items $(1) - (3)$ do not hold. Therefore, we introduce the notion of a confidence cluster.

Definition. The set of agents $I = \{i_1, \ldots, i_s\}$ form a confidence cluster at time t, if the following conditions are satisfied:

1. $\forall i_k \in I\ \exists i_l \in I, i_l \neq i_k: \left|x_{i_k}(t) - x_{i_l}(t)\right| < \varepsilon$
2. $\forall i_k \in I, \forall j \in \mathrm{N} \backslash I: \left|x_{i_k}(t) - x_j(t)\right| > \varepsilon$

4.Simulations

We performed computer simulations using the Python language to reveal the properties of the stochastic H-K model and to compare the deterministic model with the stochastic model. We simulate the stochastic and deterministic H-K models for $m = 30$. Random initial opinions are taken uniformly from [0,1]. Simulations allow us to formulate the following properties of the stochastic H-K model:

1. Opinions of agents can change their orders; therefore, agents who were in the same cluster at the initial moment may occur in different clusters (see Figure 1).
2. Clusters can exchange agents.

IOP Conf. Series: Journal of Physics: Conf. Series **1163** (2019) 012064 doi:10.1088/1742-6596/1163/1/012064

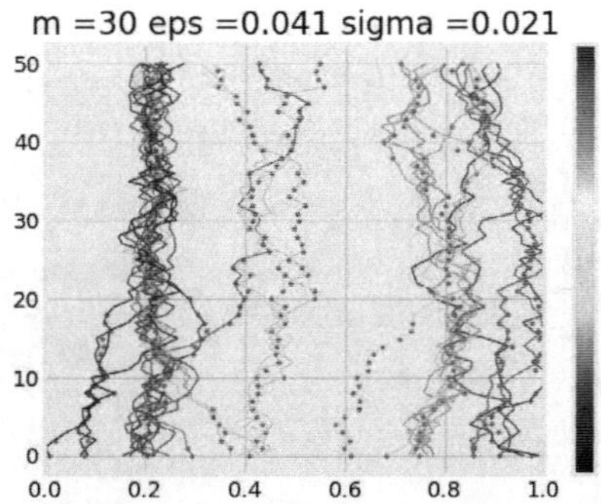

Figure 1. Evolutions of clusters. The x- axis represents the opinion space [0,1], the $y-$ axis represents the time.

3. The number of clusters depends on the choice of initial opinions. Figure 2 shows that for different initial opinions we obtain the different effects. The number of clusters in the first case equals 5, in the second case equals 4.
4. The number of clusters depends on the choice of initial opinions. Figure 2 shows that for different initial opinions we obtain the different effects. The number of clusters in the first case equals 5, in the second case equals 4.
5. The number of clusters varies during the evolution (Fig.2)

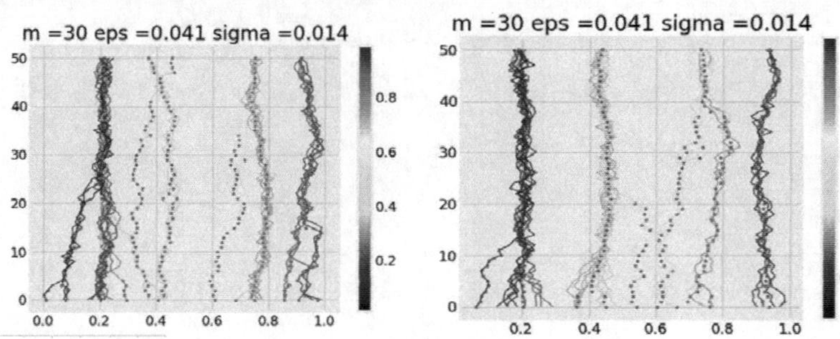

Figure 2. Number of clusters in the model with noise for different initial opinion profiles.

6. The number of clusters depends on the choice of the concrete realization of the random noise. Figure 3 shows two different realizations for $m = 30$, the fixed initial opinion profile, $\varepsilon = 0.041$ and $\sigma = 0.014$.

(a)

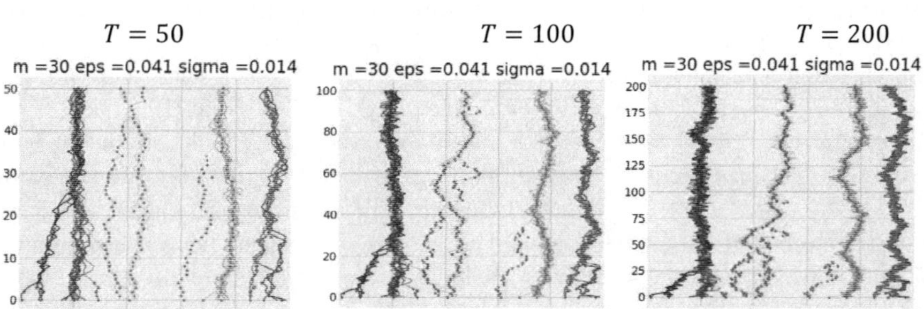

(b)

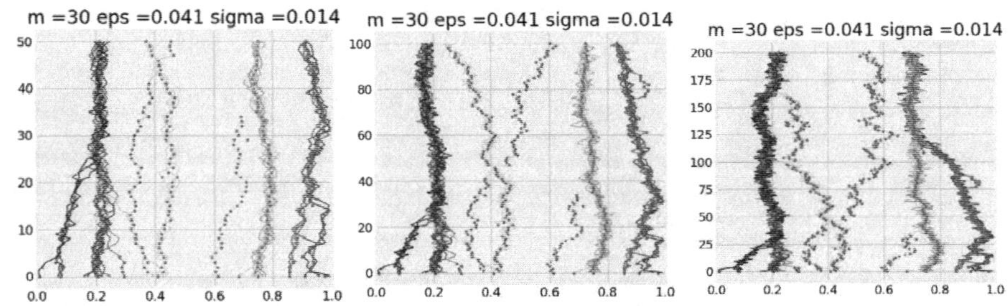

Figure 3. The number of clusters depends on the choice of the concrete realization of the random noise. (a) and (b) represent two different realizations of the noises

In the stochastic model agents don`t reach a consensus, but their opinions can be divided into the confidence clusters. We compare the number of groups of opinions in the model without noise with the number of confidence clusters in the noisy model.

The simulations show that noise can affect the opinions of agents in a such way, that the number of clusters (for some parameter σ at a certain time) will decrease compared to the number of opinions groups in the model without noise (for a fixed parameter ε, array $\hat{\delta}$ and a fixed initial opinions $x_i(0)$). As an example, consider the initial opinions uniformly distributed in the interval $[0; 1]$. Fix the value $\varepsilon = 0{,}1$, $\sigma = 0.03, T = 50$. Figure 4 shows the fragmentations for the model without and with noise:

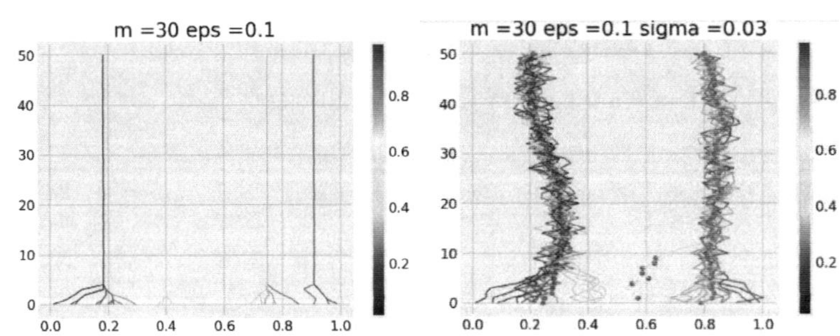

Figure 4. (Left) opinion clustering in the model without noise and (right) opinion clustering in the model with noise

We have four opinion subgroups in the deterministic H-K models and two confidence clusters in the model with the noise. The number of clusters is less than the groups of opinions in the model without noise.

For a fixed set of initial opinions of m agents $x_0 = [{x_0}^0, {x_0}^1, \dots, {x_0}^{m-1}]$ and a set of random noises $\hat{\delta}$ in the Hegselmann-Krause model it is possible to choose the parameters T, σ, ε in such a way, that the number of clusters in the model with random noises will be less than the number agents of groups in the model without random noises, and vice versa.

We change the parameter ε from 0.001 to 0.191 with a step of 0.1, and calculate the number of opinion groups for H-K model without noise (Fig.5 blue line). For stochastic H-K model for each ε we change the parameter σ from 0 to 0.1 with a step 0.001, and look for the minimum number of a confidence clusters by the parameter σ. The algorithm was done several times for different realizations of noises - $\widehat{\delta_1}$, $\widehat{\delta_2}$, $\widehat{\delta_3}$, which were set randomly, following the formula (4) (on Fig.5 different realizations are marked with colored lines – red, grey and yellow). Figure 5 shows the dependence of the number of confidence clusters (in the case of a model with the noise of the minimum number of confidence clusters by σ) on the parameter ε .

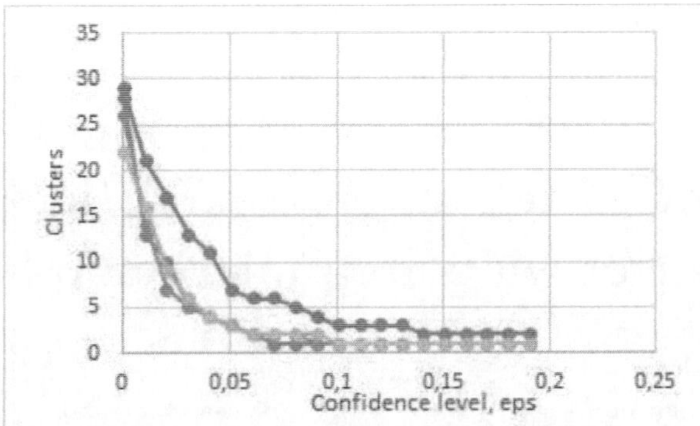

Figure 5. the dependence of the number of a confidence clusters (in the case of a model with the noise of the minimum number of confidence clusters by σ) on the parameter; blue line – simulation without noise, red line - with the set of noise $\widehat{\delta_1}$, grey line - with the set of noise $\widehat{\delta_2}$, yellow line - with the set of noise $\widehat{\delta_3}$.

5.Conclusions

We consider the Hegselmann-Krause bounded confidence model with noise. Using computer modeling we analyze the dependence of the opinion dynamics on the parameters of the problem such as the initial profile of opinions, the level of confidence, the variance of noise.

References

[1] Friedkin N and Johnsen E 1999 Social influence networks and opinion change *Advances in group processes* 16 1-29

[2] Krause U 2000 A discrete nonlinear and non-autonomous model of consensus formation *Communications in Difference Equations, Amsterdam: Gordon and Breach Publ.* 227 – 236

[3] Dittmer J C 2001 Consensus formation under bounded confidence *Nonlinear Analysis* 47 615 – 4621

[4] Hegselmann R and Krause U 2002 Opinion dynamics and bounded confidence models, analysis, and simulation *Journal of Artifical Societies and Social Simulation (JASSS)* 5(3) 1-33

[5] Hegselmann R and Krause U 2005 Opinion dynamics driven by various ways of averaging *Computation Economics* 25 381-405

[6] Wei Su, Ge Chen, Yiguang Hong 2017 Noise leads to quasi-consensus of Hegselmann-Krause opinion dynamics *Automatica* 85 448-454

[7] Mathias J-D, Huet S, Deffuant G 2016 Bounded Confidence Model with Fixed Uncertainties and Extremists: The Opinions Can Keep Fluctuating Indefinitely *Journal of Artificial Societies and Social Simulation* 19 (1)

[8] A Manita and L Manita 2018 Distributed Time Synchronization Algorithms and Opinion Dynamics *Journal of Physics: Conference Series* 955 (1), 012038

[9] Manita L and Manita A 2017 Stochastic Time Synchronization Models Based on Agreement Algorithms In: Rykov V., Singpurwalla N., Zubkov A. (eds) *Analytical and Computational Methods in Probability Theory ACMPT 2017.* Lecture Notes in Computer Science, vol 10684. Springer, Cham

Configuration of vortex-antivortex lattices at output mirror of wide-area microchip laser

A.Yu.Okulov
Russian Academy of Sciences, 119991, Moscow, Russian Federation.
E-mail: alexey.okulov@gmail.com

Abstract. Square vortex lattices predicted theoretically and experimentally observed in diode-pumped solid state microchip lasers are shown to have a remarkable symmetry. These lattices are formed by counter-rotating vortices. The interaction of vortices with nonlinear gain medium leads to precession of vortices and collective excitations with dynamics identical to acoustical and optical vibrations of atomic lattices.

1. Introduction

Pattern formation outside the equilibrium demonstrates a rich variety of emerging structures [1, 2]. The vortex lattices in high Fresnel number solid-state microchip lasers arising within a broad range of experimental parameters [3] were predicted theoretically in [4]. The subsequent numerical investigation [5] have shown the deep similarity of optical vortex lattices in lasers with spatial patterns that spontaneously occur in Faraday instability [6]. Trajectories of individual particles of ensemble in Faraday convection are chaotic but the averaged mass flow in Faraday square cells manifests itself as vortex lattice. The elementary cell consists from a four vortices. Each pair of whirls located at the corners of the cell are co-rotating and each pair of adjacent whirls at the side of the cell are counter-rotating. The corresponding field of velocities averaged over ensemble of particles is continuous though each given particle in ensemble moves on complex trajectory [6].

The same patterns were observed in wide area diode-pumped solid state microchip lasers [7]. The Fresnel number N_f in these experiments had been varied from 10 to 1000, the number of synchronously counter-rotating vortices along the side of the square was about $N_L = 30$. The spatially periodic vortex-antivortex patterns recorded experimentally in a broad range of system parameters are in remarkable difference from conventional sets of Gaussian-Hermite (HG) or Gaussian-Laguerre (LG) eigenmodes. The square pattern occurs due to phase-locking of HG or LG modes via nonlinearity inherent to gain medium [8]. The theoretical study in [4] and subsequent experimental confirmation [3] overlooked the phase structure of optical field $\theta(\vec{r}) = arg[E_n(\vec{r})]$. The investigation of spatial phase distribution of optical field and associated probability flow with appropriate field of velocities $\vec{v}(\vec{r}) = \hbar\nabla\theta(\vec{r})/m$ given by Madelung transform have shown the existence of regular array of whirls with opposite circulations [9]. Current survey is aimed to study nonstationary regimes of vortex-antivortex lattices formation when stationary solutions are perturbed by noise sources due to spontaneous emission $\delta E_n(\vec{r})$ and fluctuations of optical pump $\delta N_n(\vec{r})$.

2. Master equation and observed dynamics

The master equation for the square lattice pattern formation is of Ginzburg-Landau type [10]. The other model relevant to laser cavity under consideration is equivalent to Fox-Lee integral with nonlinear gain included in explicit form [11]:

$$E_{n+1}(\vec{r}) = \int\int_{-\infty}^{\infty} K(\vec{r}-\vec{r'})f(E_n(\vec{r'}))d^2\vec{r'}, \tag{1}$$

$$K(\vec{r}-\vec{r'}) = \frac{ikD(\vec{r'})}{2\pi L_r}\, exp\,[ik(\vec{r}-\vec{r'})^2/2L_r]. \tag{2}$$

$$f(E_n(\vec{r})) = \frac{\sigma L_a N_n(\vec{r})E_n(\vec{r})(1+i\delta\omega T_2)}{2} + E_n(\vec{r}) + \delta E_n(\vec{r}), \tag{3}$$

inversion dynamics is given by:

$$N_{n+1}(\vec{r}) = N_n(\vec{r}) + [\frac{N_0(\vec{r}) - N_n(\vec{r}) + \delta N_n(\vec{r}))}{T_1} - \sigma N_n(\vec{r})\, c\, \epsilon_0\, |E_n|^2/{}^{-}h\omega]\frac{2L_r}{c}, \tag{4}$$

where $E_n(\vec{r})$ is electric field envelope at output mirror, f is nonlinear gain transfer function, L_r is cavity length, $L_a << L_r$ is a thickness of amplifying medium relevant to experimental situation [12], R is reflectivity of output mirror, $D(\vec{r}) = R\exp[-|\vec{r}|^2/D_0^2]$ is diaphragm function corresponding to transversely inhomogeneous linear losses, key parameter for comparison with experiments [3] was $D_0 \sim 50-2000\mu m$, $G_n(\vec{r}) = \sigma N_n(\vec{r})L_a$ is transversely inhomogeneous linear gain, σ is simulated emission cross-section, $N_n(\vec{r})$ is density of resonant ions per unit volume, $\delta\omega$ is detuning from gain line center, T_2 and T_1 are transverse and longitudinal relaxation times correspondingly, $\delta E(\vec{r})$ is random noise due to spontaneous emission term emulated as multimode random process [13], $\delta N(\vec{r})$ is fluctuating part of optical pump also defined as spatially smooth multimode random process, $N_f = D_0^2/(\lambda L_r)$ is Fresnel number, λ is lasing wavelength. The convergence rate to equilibrium solutions (stationary eigenmodes) for dicrete time step $\delta t = 2L_r n/c$ proved to be 20 - 100 iterates [14].

For modeling of nonstationary behavior approximately the same amount of transient iterates had been required [15] to attain a realistic oscillation regime with smooth spatial field profile $E_n(\vec{r})$. The most reliable convergence to realistic solutions had been observed for relaxation oscillations (fig.1,2). For low Fresnel number $1 < N_f < 10$ the transient behavior of numerical scheme had been pretty good realistic from the very first iterates started from completely random transverse field $E_n(\vec{r})$ taken as multimode random process with N_m spatial harmonics with random phases $\delta\phi_n$ [13]:

$$\delta E_n(\vec{r}) = \tilde{N}_m^{-1}\sum_m^{\tilde{N}_m} |A_1 + A_2 + ... + A_m + ...|^2 \;\; ; A_m = \delta A_0 \exp[i\vec{k}_m \cdot \vec{r} + \delta\phi_n], \tag{5}$$

Numerical arrays $E_n(\vec{r})$ (complex) and $N_n(\vec{r})$ (real) were composed of [128,128] and [256,256] points to facilitate the usage of fast Fourier transform [15] optimized for standard Fortran compiler. In most realizations at $1 < N_f < 10$ the transverse field distribution $E_n(\vec{r})$ had been almost identical to fundamental TEM_{00} gaussian mode $E_n(\vec{r}) \sim \exp[-|\vec{r}|^2/D_0^2]$ with slightly different D_0.

Increase of Fresnel number have led to perpetual increase of complexity of output pattern. In most cases for $10 < N_f < 50$ the stationary output resulted in selection of Hermite- Gaussian and Laguerre-Gaussian functions. All these low-order eigenfunctions of **2D** of harmonic oscillator [16, 17] were also subjected to relaxation oscillations (fig.3) due to intermediate value of cavity lifetime τ_c:

$$T_1 < \tau_c < T_2, \;\; \tau_c = 2L_r n/(c(1-R)), \tag{6}$$

IOP Conf. Series: Journal of Physics: Conf. Series **1163** (2019) 012065 doi:10.1088/1742-6596/1163/1/012065

The characteristic frequency ω_{rel} for observed relaxation oscillations was close to:

$$\omega_{rel} \sim \sqrt{T_1 \tau_c} \sqrt{G-1}; \quad G = \sigma N_0(\vec{r}=0). \tag{7}$$

In a parameter region between $8 < N_f < 50$ and $1.5 < G-1 < 3$ a certain realizations on initial conditions $\delta E_n(\vec{r})$ have led to localized excitations close to dissipative solitons [11, 18] of $2D$ hyperbolic secant profile [19, 20].

3. Singular lattices and their vibrations

The stationary patterns at higher Fresnel numbers $10^2 < N_f < 10^3$ demonstrated a remarkable phase locking phenomena. The synchronized modes of empty resonator form an almost perfect periodic lattice of dark spots [9]. Each dark spot is surrounded by a nearly circular flow resembling isolated optical vortex alike Laguerre-Gaussian beam [21, 22]. The phase of optical field is a precise indicator of electromagnetic energy circulation [23]. When phase is increased from 0 to 2π along right-handed circular path around amplitude $E_n(\vec{r})$ zeros, the azimuthal flow of probability to find photon in a given volume is accompanied by right-handed azimuthal component of Pointing vector $\vec{S} = \epsilon_0 c^2 \vec{E} \times \vec{H}$ [9]. The swirling of electromagnetic energy is counter-clockwise in opposite case of left-handed probability flow(fig.5). A part of numerically obtained patterns is composed from zeros of amplitudes equispaced in a checkerboard geometry with alternating circulations as is shown in [8]. The macroscopic photons wavefunction $\Psi \sim E_n(\vec{r})$ [21] which is visualized by electric field amplitude $E_n(\vec{r})$, demonstrates the regular lattices of field zeros with counter-rotating whirls around [fig.6]. In contrast to optical speckle patterns [24] where vortex-antivortex pairs are distributed chaotically in space the lattice of whirls in wide-aperture lasers is periodic in space [fig.5]. The far field pattern consists of four lobes whose separation is defined by wavelength λ and period p of the near field pattern [5].

The other set of numerical realizations appears to be identical to Abrikosov lattice obtained analytically $\Psi \sim \vartheta_3[\sqrt{2\pi i\kappa}(x+iy);1]$ [25],where κ is ratio of coherence length to penetration depth, ϑ_3 is Jacobi theta function (fig.4). Exactly as in the early models of type-II superconductors placed in external magnetic field [25] the vortices at zeros of amplitudes have co-directed angular momenta but the field of velocities is continuous as well due to rotation around bright spots. The energy flow around maxima of intensity is opposite to flow around zeros. Both types of vortex lattices obtained numerically have remarkable square symmetry found in [25] yet. Noteworthy the experimentally observed lattices in superconductors of the second type have triangular symmetry.

The noise sources due to spontaneous emission $\delta E_n(\vec{r})$ and fluctuations of optical pump $\delta N_n(\vec{r})$ have led to vibrations of vortex lattices. To detect vibrations the power spectrum $I(\Omega) = FFT[|E_n(\vec{r})|^2]$ had been recorded at several points $\vec{r}$ of transverse section. In the stationary regime with well formed vortex lattice the spectrum $I(\Omega)$ proved to be independent of location $\vec{r}$. The large maximum near frequency of relaxation oscillations had been recorded in all oscillation regimes and for all transverse spatial distributions. The remarkable feature of obtained patterns are small maxima of power spectrum beyond relaxation frequency. The early numerical study reported this spikes at power spectrum as collective vibrations of vortex lattice [4]. The mechanism responsible for lattices vibrations had been reported as vortex precession around equilibrium position. In our numerical model the observed vortex lattices of 4x4, 7x7 and 9x9 whirls were practically unmovable at a long discrete time n intervals. Nevertheless the sufficiently large vortex lattice composed of 9x9 whirls produced numerical noise spectrum with a small but clearly visible collective excitation spikes (fig.3). As is known from lattice vibration theory there exist two basic types of collective lattice vibrations with a different dispersion laws. The acoustic mode corresponds to running waves whose group velocity tends to zero at the edge of the first Brillouin zone. The optical mode of vibrations possesses the higher frequencies with significantly smaller group velocities. The discrete time n to record this this spectrum $I(\Omega)$ was about 500 time steps (iterates of maps(1-4)). The runtime took for about a one minute on *dual core* 1.86Ghz processor per each FFT accelerated iteration of discrete master equations (1-4).

4. Conclusion

The computational model for eigenmodes $E_n(\vec{r})$ of microchip laser resonator is composed of convolution integral with nonlinear kernel evaluated via Fast-Fourier transform [11, 14] and the relaxation oscillator $N_n(\vec{r})$ for gain medium coupled with wavefunction $E_n(\vec{r})$. The regular square lattices (fig.4,5) appear in a well defined range of key system parameters namely laser gain and Fresnel number. The inherent feature of these nonstationary vortex patterns are vibrations of observed lattices which are known as optical and acoustical ones (fig.3).

References

[1] Cross M. C. and Hohenberg P. C. 1993 Pattern formation outside of equilibrium *Rev. Mod. Phys.*, **65**, 851
[2] Okulov A.Yu. and Oraevsky A.N. 1986 Spatio-temporal behavior of a light pulse in nonlinear nondispersive media *J.Opt.Soc.Am* B **3** 741
[3] Chen Y. F. and Lan Y. P. 2001 Transverse pattern formation of optical vortices in a microchip laser with a large Fresnel number *Phys. Rev.* A **65**, 013802
[4] K.Staliunas and C.O.Weiss 1995 Nonstationary vortex lattices in large-aperture class B lasers *J.Opt.Soc.Am* B **12**, 1142
[5] Okulov A.Yu. 2008 3D-vortex labyrinths in the near field of solid-state microchip laser *J.Mod.Opt.* **55** , 241-259
[6] Francois N., Xia H., Punzmann H., Ramsden S. and Shats M., 2014 Three-dimensional fluid motion in Faraday waves: creation of vorticity and generation of two-dimensional turbulence *Phys. Rev.* X **4**, 021021
[7] Chen Y. F. and Lan Y. P. 2003 Observation of transverse patterns in an isotropic microchip laser *Phys. Rev.* A, **67(4)**, 043814
[8] Okulov A.Yu. 2004 3D-configuration of the vortex lattices in microchip laser cavity, QCMC-2004, Glasgow UK, *AIP Conference Proceedings* **734**, p.366
[9] Okulov A.Yu. 2009 Vortex-antivortex wavefunction of a degenerate quantum gas *Laser Physics* **19**, 1796-1803
[10] Burlak G. and Malomed B.A. 2018 Interactions of three-dimensional solitons in the cubic-quintic model *Chaos* **28**, 063121
[11] Okulov A.Yu. and Oraevsky A.N.1988 Spatiotemporal dynamics of a wave packet in nonlinear medium and discrete maps, *Proceedings Lebedev Physics Institute (in Russian) N.G.Basov ed., Nauka, Moscow* **187**, 202-222
[12] Vadimova O.L., Mukhin I.B., Kuznetsov I.I.,Palashov O.V.,Perevezentsev E.A. and Khazanov E. A. 2013 Calculation of the gain coefficient in cryogenically cooled Yb : YAG disks at high heat generation rates *Quantum Electronics* **43**, n.3, 201-206
[13] Okulov A.Yu.1991 The effect of roughness of optical elements on the transverse structure of a light field in a nonlinear Talbot cavity *J.Mod.Opt.* **38**, n.10, 1887
[14] Okulov A.Yu.1994 On correlations between cavity mode and inversion profile in a solid-state microchip laser *Optics and Spectroscopy* **77**, n.6, 888-892
[15] Okulov A.Yu. 1993 Scaling of diode-array-pumped solid-state lasers via self-imaging *Opt.Comm.* **99**, p.350-354
[16] Chen Y. F.and Lan Y. P., 2002 Observation of laser transverse modes analogous to a SU(2) wave packet of a quantum harmonic oscillator *Phys. Rev.* A**66(5)**, 053812
[17] Chen Y. F. , Hsieh Y. H., and Huang K. F. 2018 Originating an integral formula and using the quantum Fourier transform to decompose the Hermite-Laguerre-Gaussian modes into elliptical orbital modes *OSA Continuum*, **1**, Issue 2, pp. 744-754
[18] Okulov A.Yu. 2000 Spatial soliton laser: geometry and stability *Optics and Spectroscopy* **89**, 145-147
[19] Okulov A.Yu. 2004 Exact theory of localized and periodic structures in an active optical resonator *Bulletin of Lebedev Physical institute*, n.10
[20] Malomed B.A. and Towers I. 2002 Stable (2+ 1)-dimensional solitons in a layered medium with sign-alternating Kerr nonlinearity *J.Opt.Soc.Am* B **19** 537-543
[21] Okulov A.Yu. 2012 Rotational Doppler shift of the phase-conjugated photons, *J.Opt.Soc.Am* B **29**, 714-718
[22] Rsheed A. Al, Lyras A., Lembessis V. E., Aldossary O. M. 2016 Guiding of atoms in helical optical potential structures *J. Phys.* B **49** 125002
[23] Rsheed A. Al, Lyras A., Lembessis V. E., Alqarni A., Alshamari S., Siddig A., Aldossary O. M. 2016 Rotating optical tubes for vertical transport of atoms *Phys. Rev.* A **94**, 063423

[24] Okulov A.Yu. 2009 Twisted speckle entities inside wavefront reversal mirrors *Phys. Rev.* A **80**, 013837
[25] Abrikosov A.A. 1957 On the magnetic properties of superconductors of the second group *JETP* **5** 1174

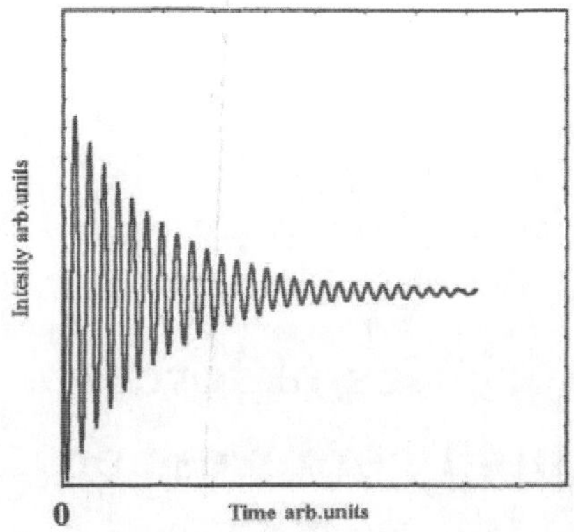

Figure 1. Temporal evolution of intensity $|E_n(\vec{r})|^2$ under stepwise gain switching.

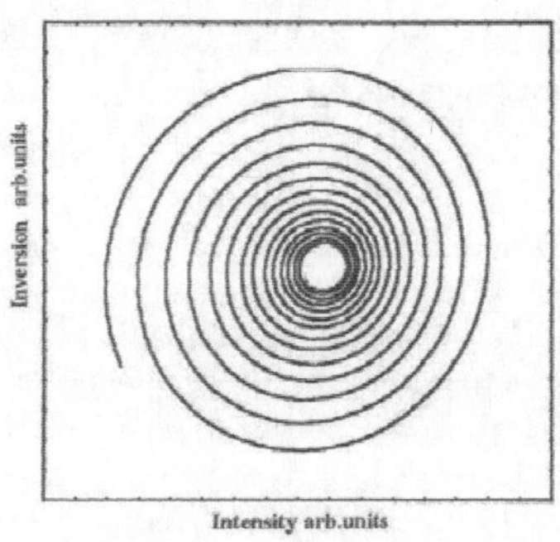

Figure 2. Temporal evolution in phase-space of intensity $|E_n(\vec{r})|^2$ and inversion N_n under stepwise gain switching.

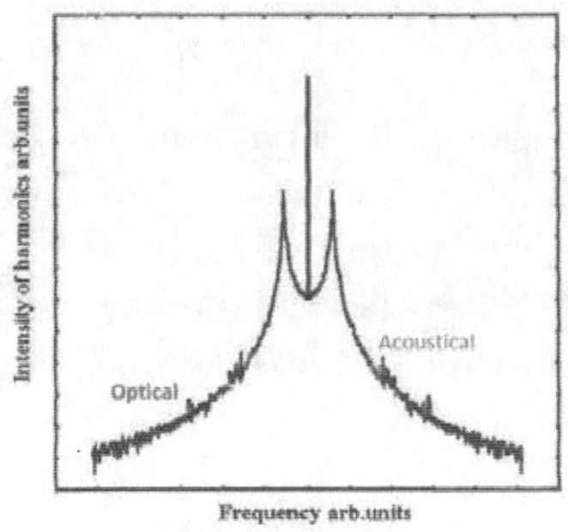

Figure 3. The power spectrum of wide area (high Fresnel number $100 < N_F < 1000$) stationary emission pertubed by spontaneous noise added at each round trip n. Noteworthy the collective excitations spikes interpreted as acoustical and optical vibrations of vortex lattice of 9x9 whirls .

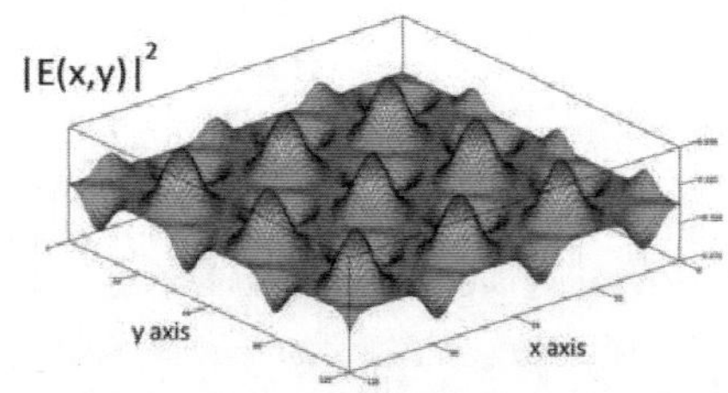

Figure 4. The stationary transverse intensity distribution $|E_n(\vec{r})|^2$ of wide area laser stationary emission with high Fresnel number $100 < N_F < 1000$.

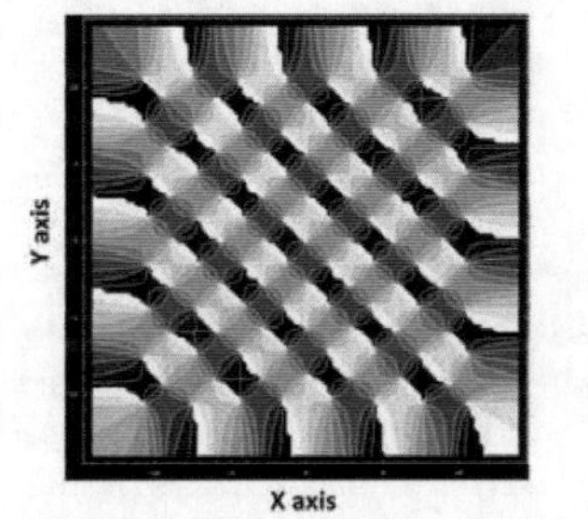

Figure 5. The grey scale stationary transverse phase distribution $arg[E_n(\vec{r})|]$ of wide area (high Fresnel number $100 < N_f < 1000$) laser . Black color corresponds to phase $arg[E_n(\vec{r})|] = 0$ which gradually increases to $arg[E_n(\vec{r})] = 2\pi$ in circumvention around vortex cores.

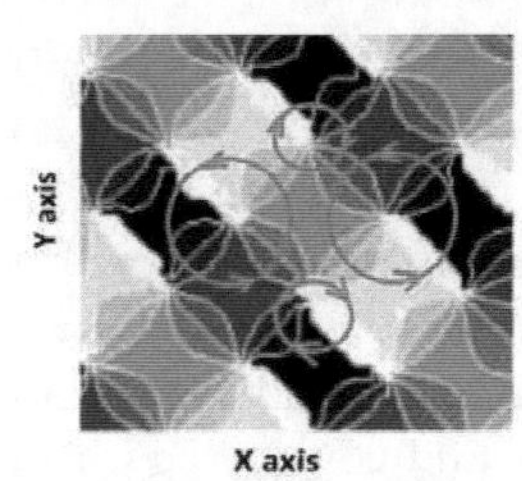

Figure 6. The enlarged elementary cell grey scale transverse phase distribution $arg[E_n(\vec{r})]$ of wide area laser nonstationary emission.

IOP Conf. Series: Journal of Physics: Conf. Series **1163** (2019) 012066 doi:10.1088/1742-6596/1163/1/012066

Calculation of the invariant characteristics of forced oscillations of a beam with longitudinal compression and it's phenomenological model

Glyzin Sergey[1,2], **Lokhanin Mikhail**[1] **and Sirotin Dmitriy**[1,2]

[1] *P.G. Demidov Yaroslavl State University, Russian Federation, Yaroslavl, Sovetskaya str., 14*
[2] *Scientific Center in Chernogolovka RAS, Russian Federation, Chernogolovka, Lesnaya str., 9*

E-mail: glyzin@gmail.com, lokhaninmv@mail.ru, hhiks@yandex.ru

Abstract. This paper is devoted to the comparison of the invariant characteristics of experimental data and solutions of the phenomenological model. Set of experimental data was obtained in the process of registering oscillations of a beam with longitudinal compression. In the course of the experiment excitation of oscillations is performed by exposure of alternating magnetic field on a magnet placed on the loose end. Depending on the frequency of external stimulus both chaotic and periodic oscillations were obtained. Based on the analysis of experimental arrays a phenomenological model was developed. To verify it's adequacy, a number of invariant numerical characteristics were calculated, including the frequency spectrum, correlation dimension, Lyapunov exponents, and β-statentropy. Based on the comparison of the characteristics obtained for the experimental data and for the mathematical model, it is concluded that it is qualitatively corresponds with experiment.

1. Introduction

Mechanical and electro-mechanical systems with chaotic behavior have long been in the sphere of keen interest of both theorists and creators of experimental devices with chaotic behavior. In addition to that beam oscillatory systems with several minima of potential under the influence of external periodic forces have been actively studied recently. Rather considerable literature is devoted to magnetoelastic beams with Euler instability, we refer only to works [1, 2, 3] and the bibliography in them.

2. Description of the experiment

2.1. Beam structure

The beam under consideration was implemented on the basis of a pair of steel strips connected at the loose ends and compressed in the longitudinal direction with a strained kevlar thread. At the loose end there is an aluminium foil shutter and a small SmCo5 magnet fixed on it, which is affected by an alternating magnetic field, their mass can be neglected, since it is rather small. The whole structure is bolted to a massive (≈ 5 kg) steel plate, which fixes the entire installation (see fig. 1, a more detailed description is contained in [4]).

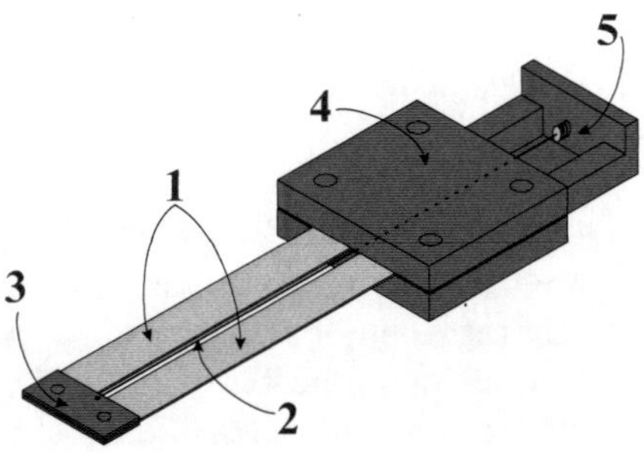

Figure 1. 1 - Elastic steel strips. 2 - Kevlar string. 3 - Clamping plates. 4 - Structure fixing plates. 5 - Spring tightener. Aluminium foil curtain (the main element of the deflection sensor) and a small magnet are attached on the loose end.

2.2. Phase portraits of oscillations

Depending on the frequency of external stimulus, a large set of data was obtained. This set contains both ordered periodic and chaotic oscillations. It can be noted that the transition to chaos is carried out by period doubling. To reconstruct the attractor $(x_n, x_n + \tau)$, a set of phase portraits was built, some of which are shown in Fig. 2, where τ is equal to the index corresponding to the first zero or the first value of the autocorrelation function nearest to zero.

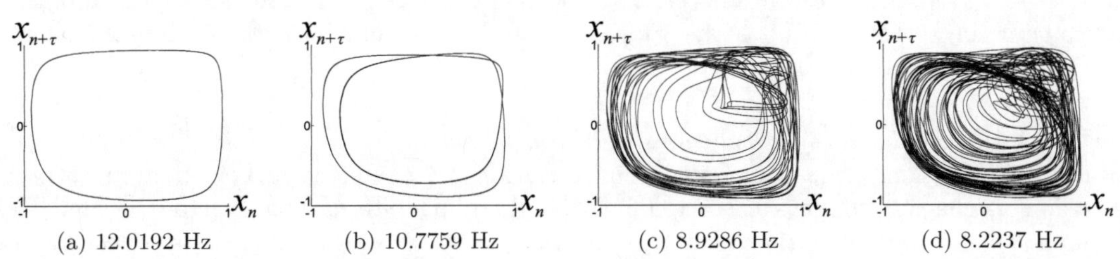

Figure 2. Phase portraits of the reconstructed from experimental data attractor

3. Phenomenological model

3.1. Construction of the mathematical model of the experiment

To construct the phenomenological model of the experiment described, several assumptions were made (see also descriptions of the construction of similar models in [5, 6, 7]). First of all, the movement of the material point in the force field was considered, and not the movement of the beam. Thus, the potential function must have the following properties: it must have a maximum at zero ($x = 0$) deviation and two minima at $x = \pm x_0$. This means that the equilibrium position at $x = 0$ is unstable, and the equilibrium at $x = \pm x_0$ is locally stable. The simplest realization of such potential is given by a fourth-degree polynomial

$$U(x) = Ax^4 - Bx^2. \tag{1}$$

Putting down the equation of the balance of forces, we have the equation

$$m\ddot{x} = F = F_{pot} + F_{diss} + F_{ext} = -\frac{\mathrm{d}U}{\mathrm{d}x} - \eta\dot{x} + F_{ext}, \tag{2}$$

where m is the mass, F_{pot} is the potential force, F_{diss} is the friction force and F_{ext} is the external force. Given the type of potential (1), we have

$$F_{pot} = -4Ax^3 + 2Bx = -4Ax(x^2 - x_0^2),$$

and $x = \pm x_0 = \pm\sqrt{B/(2A)}$ are points of stable equilibrium. Whereas we assume that the external force is given by the formula $F_{ext} = F_0 \cos(\omega t)$, thereby, the equation (2) is written as

IOP Conf. Series: Journal of Physics: Conf. Series **1163** (2019) 012066 doi:10.1088/1742-6596/1163/1/012066

$$\ddot{x} = -\eta m^{-1}\dot{x} - 4Am^{-1}(x^2 - x_0^2) + F_0 m^{-1}\cos(\omega t). \tag{3}$$

For the variable x we select a new scale $y = xx_0^{-1}$, then

$$\ddot{y} = -\eta m^{-1}\dot{y} - 4Ax_0^2 m^{-1} y(y^2 - 1) + F_0 m^{-1} x_0^{-1}\cos(\omega t). \tag{4}$$

Consider oscillations near one of the points $y = \pm 1$. By replacing $y = 1+\xi$ or $y = -1+\xi$ and assuming external forces to be absent, we linearize the resulting equation at zero. As a result, the equation of small free oscillations near the equilibrium points will look like

$$\ddot{\xi} = -\eta m^{-1}\dot{\xi} - 8Ax_0^2 m^{-1}\xi.$$

In this case the oscillation frequency is equal to $\omega_0^2 = 8Ax_0^2 m^{-1}$. This allows us to normalize the time $\tau = \omega_0 t$, as the result we obtain the Duffing equation with external influence:

$$\ddot{y} + \alpha\dot{y} + \frac{1}{2}y(y^2 - 1) = f\cos(\Omega\tau), \tag{5}$$

where $\alpha = \eta\omega_0 m^{-1}$ is the loss parameter, $f = F_0(mx_0\omega_0^2)^{-1}$ is the parameter of the intensity of external influence, $\Omega = \omega\omega_0^{-1}$ is the frequency of the external impact, normalized on the frequency of small free oscillations.

3.2. Selection of parameters of the phenomenological model

It is known that, when the parameters in the equation (5) change, the Feigenbaum scenario of transition to chaos through a cascade of period doubling bifurcations can be realized (see for example [5]). This being the case, the following method was proposed for selecting the parameters of the problem (5). For fixed values of the parameters α and f of the equation (5) such that when Ω changes, there is a Feigenbaum scenario of transition to chaos, the frequency Ω, at which the first period doubling occurred, and the interval Ω, in which (5) has a chaotic attractor is determined. Further, the parameters α and f are chosen in such a way that the ratio of the length of the chaotic interval to the first doubling is close to the experimental data. It appeared that, the most similar pattern to the experiment is observed at the values of the parameters $\alpha = 0.988$ and $f = 0.745$. With these values, the first doubling of the period occurs approximately at $\Omega \approx 0.71$, and chaotic oscillations occur in the range from $\Omega \approx 0.76$ to $\Omega \approx 0.854$ (see fig. 3). It is convenient to trail the doubling bifurcations and chaotic oscillations of the (5) model by the values of the Lyapunov indices (Fig. 4) calculated on the attractor (an algorithm from [8, 9] is used).

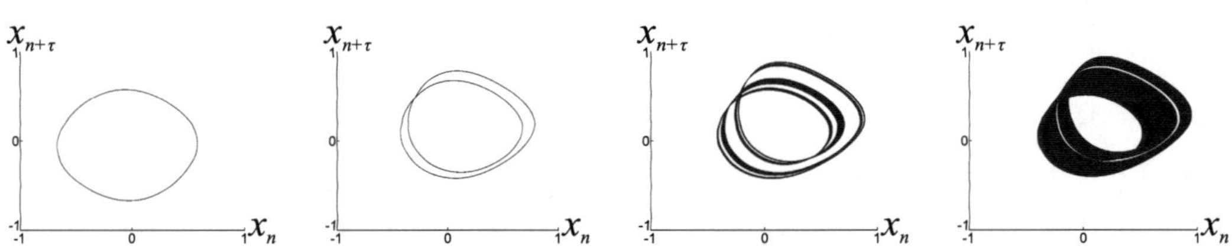

(a) $\Omega = 0.6315$ (b) $\Omega = 0.717$ (c) $\Omega = 0.76$ (d) $\Omega = 0.804$

Figure 3. Phase portraits of the attractor of the phenomenological model (5)

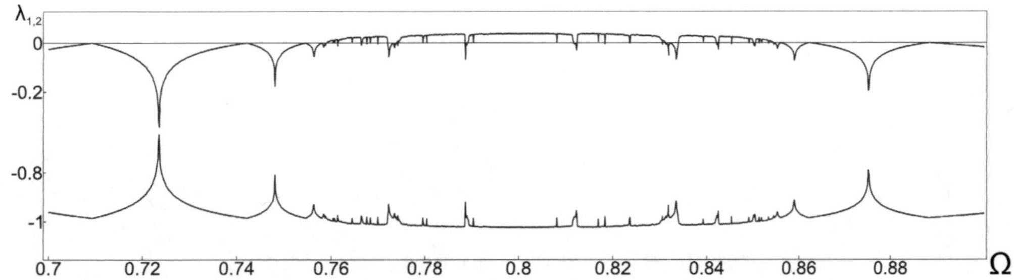

Figure 4. Dependence of the first and second Lyapunov exponents on Ω

4. Comparison of the invariant characteristics of the experimental system and the phenomenological model

To effectively compare the qualitative behavior of the solutions of the phenomenological model (5) and experimental data, it is required to find some invariant characteristics, which could be found in both cases. Comparing these indicators, it is possible to choose the parameters of the problem (5) so that they are close to the corresponding characteristics from the experiment. It is important to note that a direct comparison of the chaotic solution of the equation (5) and experimental data cannot give close results due to the fact that the trajectories of the chaotic regime (5) in the case of the positivity of the highest Lyapunov exponent exponentially scatter.

The following parameters were chosen as invariant characteristics by which the behavior of the experimental and model systems will be compared: the spectral density function, the correlation integral and the correlation dimension, and finally, the β-statentropy.

4.1. Calculation of spectral density functions

One of the standard methods for assessing the characteristics of signals is the spectral analysis of the received data sets. For data corresponding to different frequencies of external influence in the experiment, the frequency spectra were calculated on the basis of the fast Fourier transform. Also for the equation (5) with the above values of the parameters the data sets were formed and the frequency spectra were also calculated for them. The resulting spectral density functions are shown in Fig. 5 and 6. Comparison of these functions shows that they correspond to each other, but the presence in the experimental data of the random component does not make it possible to use this method to the full extent.

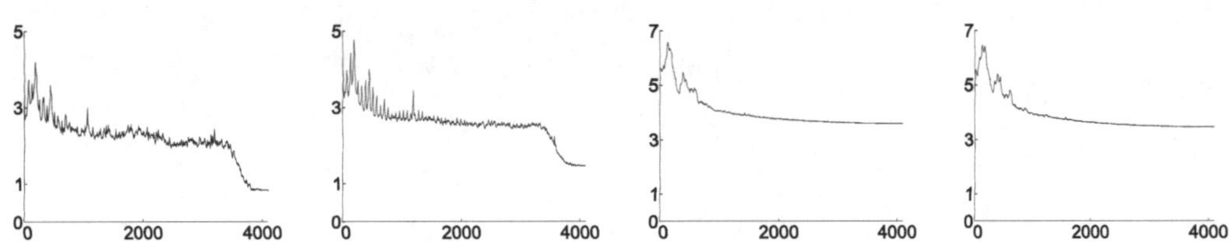

(a) $12.0192Hz$ (b) $10.7759Hz$ (c) $8.9286Hz$ (d) $8.2237Hz$

Figure 5. Frequency spectrum of experimental data

4.2. Calculation of the correlation dimension

Another important invariant characteristic of an attractor is its correlation dimension D_c. The definition of this characteristics is based on the calculation of the correlation integral (see [10])

$$C(\delta) = \frac{1}{n^2} \sum_{i,j} \theta(\delta - |y_i - y_j|), \quad \text{where} \quad \theta(x) = \begin{cases} 0, x \leq 0 \\ 1, x > 0 \end{cases}, \tag{6}$$

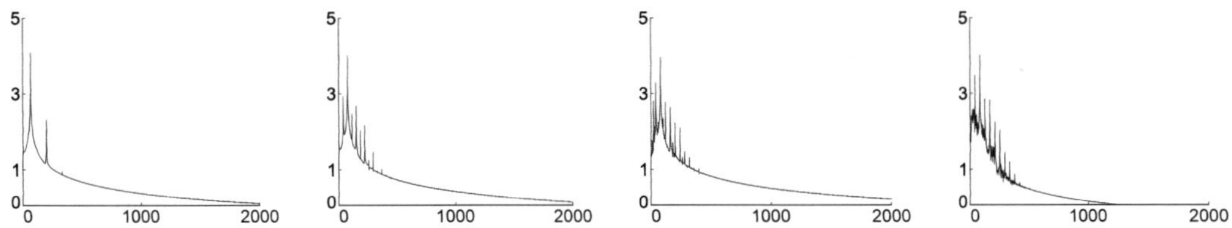

(a) $\Omega = 0.6315$ (b) $\Omega = 0.717$ (c) $\Omega = 0.76$ (d) $\Omega = 0.804$

Figure 6. Frequency spectrum of model data

n is the number of points in the data set, and y_j is specially prepared data. The value of $C(\delta)$ depends on δ in a power-law manner and has the form $C(\delta) = \delta^{D_c}$. To estimate the dimension D_c, the dependence of $\log C(\delta)$ on $\log \delta$ for several values of δ is calculated and the slope from the most linear segment that defines D_c is evaluated. Fig. 7 shows the dependence of D_c on the frequency of external influence for the experiment and on the Ω parameter for the model equation (5). It is easy to see that the graphs 7 (a) and 7 (b) are close to each other.

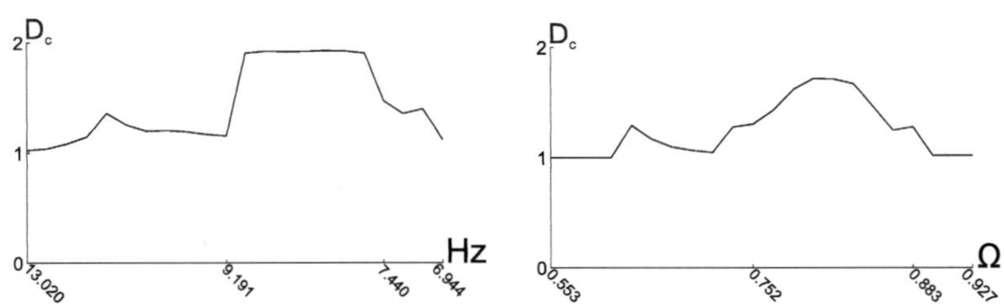

(a) Experimental data (b) Model data

Figure 7. Dependence of the correlation dimension on the frequency of external influence

4.3. Calculation of the β-statentropy

Numerical analysis of the experimental data and the model equation showed that the most preferable invariant characteristic of the attractor, which is reliably calculated in both cases is the β-statentropy. In accordance with the algorithm described in [11, 12], the β-statentropy is calculated using the following formulas:

$$\eta_n^{(k)}(\beta) = \begin{cases} \frac{\beta r_n^{(k)}(\beta)}{k(r_n^{(k+1)}(\beta) - r_n^{(k)}(\beta))}, & \beta \neq 0 \\ \frac{1}{k(r_n^{(k)}(0) - r_n^{(k+1)}(0))}, & \beta = 0 \end{cases}, \tag{7}$$

where

$$r_n^{(k)}(\beta) = \frac{1}{n} \sum_{j=1}^{n} \gamma\Big(\overset{(k)}{\min_{i:i\neq j}} \rho(\xi_i, \xi_j)\Big), \quad \gamma(t) = \begin{cases} t^\beta, & \beta \neq 0 \\ -\ln t, & \beta = 0 \end{cases}.$$

Here ξ_j, $j = 1, \dots, n$ are specially processed elements of the data array, $\overset{(k)}{\min_{i:i\neq j}} \rho(\xi_i, \xi_j)$ is the k-th minimum distance between points ξ_j and ξ_i. The most illustrative results were obtained using the formula (7) with $\beta = 0$, $k = 0$. Table 1 summarizes the results for selected experimental data arrays and their corresponding solutions of the (5) equation. Chaotic oscillations correspond to positive values of η, and ordered (in this case, periodic) values correspond to zero. It is easy to see that the entropy estimation for experimental and model data is in coherence with each other.

IOP Conf. Series: Journal of Physics: Conf. Series **1163** (2019) 012066 doi:10.1088/1742-6596/1163/1/012066

Table 1. Results of β-statentropy calculation.

Frequency (Hz)	η	Ω	η
8.9286	0.1387	0.7600	0.1054
8.6806	0.1379	0.7750	0.1248
8.4459	0.1371	0.7900	0.1365
8.2237	0.137	0.8040	0.139
8.0128	0.1349	0.8180	0.1343
7.8125	0.1327	0.8530	0.1146
7.6220	0.1073	0.8602	0.0828

5. Conclusion

In conclusion, note that a large amount of data obtained in the course of the full scale experiment allowed to propose a phenomenological model of the process. Comparison of the invariant numerical characteristics of the experimental data and the solution of the phenomenological model showed that it adequately reflects the basic properties of the experimental system. The most successful research tool is the calculation of the β-statentropy, which gives most fitting results for the experiment and the model.

Acknowledgments

This work was supported by the Russian Science Foundation (project nos. 14-21-00158)

References

[1] Moon F C, Holmes P J 1979 A magnetoelastic strange attractor *Journal of Sound and Vibration* **65**:2 275 - 296

[2] Tam J I, Holmes P 2014 Revisiting a magneto-elastic strange attractor *Journal of Sound and Vibration* **333**:6 1767 - 1780

[3] Kumar K A, Ali S F, Arockiarajan A 2017 Magneto-elastic oscillator: Modeling and analysis with nonlinear magnetic interaction *Journal of Sound and Vibration* **397** 265 - 284

[4] Glyzin S D, Lokhanin M V, Sirotin D M Invariant Characteristics of Forced Oscillations of a Beam with Longitudinal Compression *Modeling and Analysis of Information Systems* **25**:1 54 - 62. (In Russian)

[5] Ueda Y 1978 Randomly transitional phenomena in the system governed by Duffings equation, *Nagoya University Institute of Plasma Physics*, Japan, 1978, IPPJ - 341.

[6] Kapitanov D V, Ovchinnikov V F, Smirnov L V The dynamics of an axially loaded elastic bar after loss of stability *Problems of strength and plasticity*, **76**:3 205 - 216 (in Russian)

[7] Saranin V A About chaotic behaviour of an electrostatic pendulum at parametrical influence *Bulletin of Perm University. Series: Physics* **27-28** 18 - 23 (in Russian)

[8] Wolf A, Swift J B, Swinney H L, Vastano J A 1985 Determining Lyapunov exponents from a time series *Physica D: Nonlinear Phenomena* **16** 285 - 317

[9] Glyzin D S, Glyzin S D, Kolesov A Yu and Rozov N Kh 2005. The Dynamic Renormalization Method for Finding the Maximum Lyapunov Exponent of a Chaotic Attractor *Differential Equations* **41**:2 284 - 289

[10] Grassberger P, and Procaccia I 1983 Measuring the strangeness of strange attractors *Physica D: Nonlinear Phenomena* **9(1)** 189 - 208.

[11] Timofeev E A 2005 A Consistent Estimator of the Entropy of Measures and Dynamical Systems *Math. Notes* **77**:6 831 - 842

[12] Timofeev E A 2006 Invariants of measures admiting statistical estimates *St. Petersburg Math. J.* **17**:3 527-551.

IOP Conf. Series: Journal of Physics: Conf. Series **1163** (2019) 012067 doi:10.1088/1742-6596/1163/1/012067

Architecture of a quantum computing platform

Marat Ismagilov, Rustam Sayfutdinov, Alexander Vasiliev
Kazan Federal University, Russia
E-mail: vav.kpfu@gmail.com

Abstract.
We present an architecture of a cloud-based quantum computing platform. It is based on the original model of a nanophotonic quantum processor with integrated quantum memory and a quantum transistor. At the moment the platform works with an integrated quantum simulator that implements the behavior of the underlying model.

1. Introduction

Nowadays there is an interdisciplinary effort worldwide toward scalable quantum computational devices and quantum communication systems, that promise new possibilities for solving computing-intensive or previously not efficiently solvable problems. However, quantum processors can still be used only in restrictive laboratory conditions. Thus, it was a natural decision to propose a cloud-based access to the quantum computing devices and integrate them with a high-performance computing platform. This led to the new area of quantum programming and a rapidly increasing number of companies (including IBM, Google, Microsoft) developing their own quantum computing platforms. The new quantum programming community is seeking for the optimal programming languages and compiler design for realistic (already created or future) quantum hardware [1].

In this research we develop an architecture of a quantum computing platform with the cloud interface, that is based on the original model of a nanophotonic quantum processor with integrated quantum memory and a quantum transistor [2]. At the moment the platform works with an integrated quantum simulator that implements the behavior of the underlying model [3]. As shown earlier this model is universal, i.e. it is capable of performing arbitrary quantum computations. Additionally, it allows to speed up the implementation of controlled operations, that are heavily used in most multi-qubit algorithms.

The platform has two main components: the server-side framework and the client SDK, which gives the possibility of coding the mixture of quantum and classical algorithms in a conventional environment.

2. Preliminaries

Let us recall the computational model given in [3].

A Quantum Processing Unit (QPU) is a device consisting of 4 modules controlled by a classical device as illustrated in Fig. 1. They include:

- An s-qubit *Quantum Memory* module, that allows to store quantum states. The contents of the memory is described by the current state $|\psi\rangle \in \mathcal{H}^{2^s}$. The quantum memory module

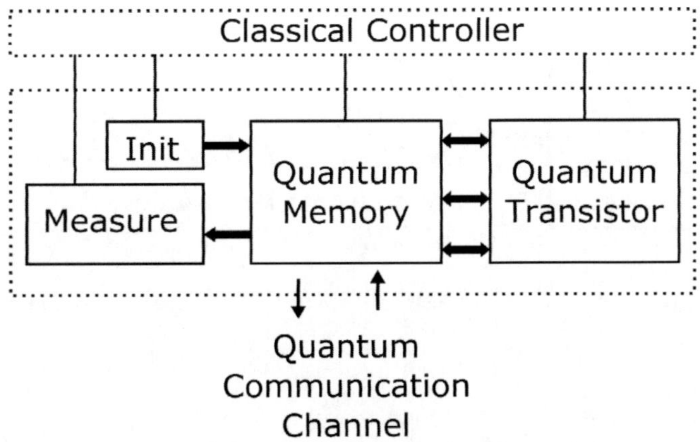

Figure 1. A model of a Quantum Processing Unit (QPU).

is capable of exchanging qubits with quantum transistor and quantum communication channel.

- A *Quantum Transistor* is a 3-qubit device capable of performing quantum transformations. There are three types of operations described in a physical architecture: QET, PHASE and CQET [5]. Generally, we can consider an array of quantum transistors working in parallel, but here for simplicity we leave just one module.
- *Init* is an initialization module that can create a quantum state given its classical description and save it to the quantum memory. Without loss of generality we can consider a single-qubit initialization in state $|0\rangle$, since any other state can be obtained via universal set of operations.
- *Measure* is a measurement module that implements a photon detecting procedure and gives a classical bit out of qubit being measured.

The proposed model is capable of performing arbitrary quantum computation, since the available set of operations provides a possibility to implement CNOT and any single-qubit operation.

A quantum algorithm in this model can be described by a (t, s) quantum program, which is the sequence of t instructions of several types over s-qubit quantum memory. These parameters correspond to the key complexity measures: t is the time of computation and s is the required space.

3. Cloud Architecture for QPU

Based on the proposed QPU model we have constructed a cloud-based quantum computing platform. The platform has two main components: the server-side framework and the client-side software development kit.

3.1. Client SDK

We have created a Software Development Kit (SDK) that communicates with the server-side framework and provides the means for quantum programming at the client side.

The Python-based SDK hides most of the complex details and gives abstractions that make coding quantum algorithms as easy as shown in Listing 1.

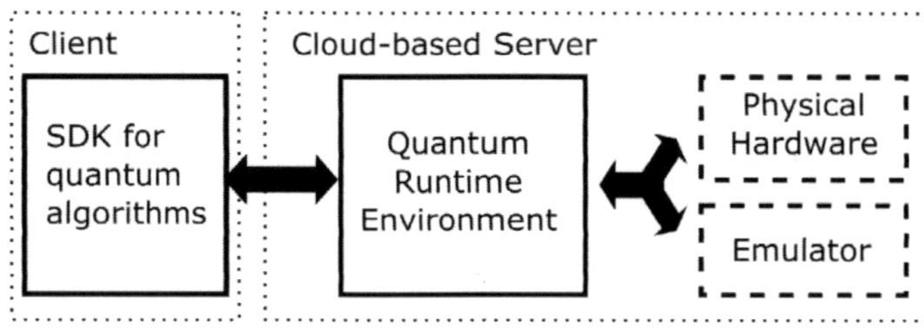

Figure 2. A cloud-based quantum computing platform.

Listing 1. Usage example

```
import kmqc

from kmqc import gates, Program


p = Program(
        gates.X(1),
        gates.H(0),
        gates.H(1),
        gates.CNOT(0, 1),
        gates.H(0))
endpoint = 'endpoint'
user_id, api_key = 'user_id', 'api_key'
conn = kmqc.connect(endpoint, user_id, api_key)
r = conn.execute(p)
print(r)
```

3.2. Server-side Framework

The server-side Framework provides an application programming interface (API) for the remote clients of the quantum computing platform. Any incoming quantum program is translated to a lower level of abstraction transforming logical operations and logically encoded qubits into physical ones. The translation is performed based on the formal grammar that uses basic operations of the QPU model as the terminal symbols and the conventional quantum operators as the nonterminal ones.

As mentioned earlier, there are three types of computational operations given by the physical architecture [3]: QET, CQET and PHASE. These operations constitute a universal basis and allow to perform arbitrary quantum computation. In particular, any single-qubit operation can be described using following operators [4]:

$$U_1(\mu) = \begin{pmatrix} 1 & 0 \\ 0 & e^{i\mu} \end{pmatrix} \sim R_z(\mu), \tag{1}$$

$$U_2(\varphi, \mu) = R_z\left(\varphi + \frac{\pi}{2}\right) R_x\left(\frac{\pi}{2}\right) R_z\left(\mu - \frac{\pi}{2}\right), \tag{2}$$

$$U_3(\theta, \varphi, \mu) = R_z(\varphi + 3\pi) R_x(\pi) R_z(\theta + \pi) R_x(\pi) R_z(\mu), \tag{3}$$

where $R_x(\theta)$ can be implemented by QET $(-\theta)$ and $R_z(\theta)$ is equivalent (in a logical encoding of qubits $|0\rangle_L = |01\rangle$, $|1\rangle_L = |10\rangle$) to PHASE (θ) [5].

An entangling operation is given by the CQET [3], which can be used to implement the widely used CNOT operation.

Altogether, the server-side framework maintains access to the following logical gates:

- R_x, R_z, CNOT;
- U_1, U_2, U_3;
- R_y, H, S, $S^\dagger$, T, $T^\dagger$, CCNOT, X, Y, Z.

4. Conclusion

Thus, we have proposed a quantum computing platform with the cloud interface based on the mathematical model of the Quantum Processing Unit (QPU). This platform contributes to the area of quantum programming providing API to the unique model of a nanophotonic quantum processor with integrated quantum memory and a quantum transistor, that has certain advantages for implementation of controlled transformations.

Acknowledgments

The research was funded by the subsidy allocated to Kazan Federal University for the state assignment in the sphere of scientific activities, project # 1.12878.2018/12.1.

References

[1] Chong F T, Franklin D and Martonosi M 2017 *Nature* **549** 180–7
[2] Moiseev S A and Andrianov S N 2016 *Optics and Spectroscopy* **121** 886–96
[3] Ablayev F, Andrianov S, Fetisov D, Moiseev S, Terentyev A, Urmanchev A and Vasiliev A 2018 *AIP Conference Proceedings* **1936** 020014
[4] Cross A W, Bishop L S, Smolin J A and Gambetta J M 2017 *Preprint* quant-ph/1707.03429
[5] Ablayev F M, Andrianov S N, Moiseev S A and Vasiliev A V 2013 *Lobachevskii Journal of Mathematics* **34** 291–303

IOP Conf. Series: Journal of Physics: Conf. Series **1163** (2019) 012068 doi:10.1088/1742-6596/1163/1/012068

Nonequilibrium critical dynamics of low-dimensional frustrated magnets and multilayer structures

P V Prudnikov†, V V Prudnikov, V O Borzilov, M M Firstova, and A A Samoshilova

Department of Theoretical Physics, Omsk State University, Mira prospekt 55-A, Omsk, 644077, Russia

E-mail: †prudnikp@univer.omsk.su

Abstract. A Monte Carlo simulation of the non-equilibrium behavior of multilayer magnetic nanostructure is carried out. Equilibrium properties, critical temperatures T_c and hysteresis effects were studied for various thicknesses of ferromagnetic layers. It is shown that, in contrast to bulk magnetic systems, the aging effects in nanostructure arise not only at the ferromagnetic ordering temperature T_c but also within a wide temperature range at $T \leq T_c$. Simulation of transport properties in multilayer structure permitted to calculate non-equilibrium values of magnetoresistance.

1. Introduction

A significant interest has been recently focused on non-equilibrium processes in magnetic low-dimensional materials. The reduction of the dimension of magnets is accompanied by an increase in fluctuations of the spin density and the manifestation of the effects of critical slowing down and "aging" in the non-equilibrium behavior of low dimensional magnetic systems [1]. Thin films and low-dimensional magnets demonstrate the slow critical evolution from a non-equilibrium initial state. Aging, coarsening and memory effects are nontrivial features in the non-equilibrium behavior of such systems with slow dynamics [2]. The magnetic properties of multilayer magnetic systems have been widely investigated over the past years, since they are widely used in magnetic storage devices [3]. The antiferromagnetic coupling was crucial for the discovery of the giant magnetoresistance (GMR). It kickstarted the field of nanomagnetism and spintronics [4]. The use of synthetic antiferromagnets in magnetic random access memory (MRAM) [5] can reduce the critical current [6] and the time for switching [7].

The nanoscale periodicity in magnetic multilayer structures gives rise to the mesoscopic effects of the strong spatial spin correlation with the slow relaxation dynamics of magnetization accompanying the quenching of the system in the non-equilibrium state. In contrast to the bulk magnetic systems, where the slow dynamics and aging effects manifest themselves near the critical point [2], magnetic superstructures with nanoscale periodicity allow increasing the relaxation time owing to the effects related to the larger characteristic spin-spin correlation length. That is why the aging and nonergodicity effects can be experimentally observed in the multilayer Co/Cr magnetic structure [1] within a wider temperature range as compared to that for the bulk magnetic systems. We have performed in the paper [8] a numerical Monte Carlo simulation of the non-equilibrium behavior of the multilayer Co/Cr/Co magnetic structure with the thickness of Co films equals to $N = 3$ monolayers. Calculations of the two-time correlation functions and the staggered magnetization allow us to reveal the aging effects, which are characterized by slowing down of correlation and relaxation processes with an increase in the system's "age" t_w as the time between a sample preparation and the beginning of measurement of its characteristics. The revealed aging effects for our model of multilayer structure are in good agreement with results of experimental observations in the paper [1]. It was shown that aging in the non-equilibrium critical behavior of multilayer magnetic structure is occurred not only at T_c but also within a wide range of temperatures at $T \leq T_c$. Therefore, the existence of these non-equilibrium features should surely be taken into account in any applications of the multilayer magnetic structures for spintronic devices based on the giant magnetoresistance phenomena.

2. Equilibrium properties of the magnetic structure and hysteresis effects

Magnetic order in the multilayers is complex due to a strong influence of the shape and the magnetocrystalline anisotropy of the sample. The presence of anisotropy leads to dimensionality crossover in Heisenberg films [9, 10]. Low-dimensional magnets [11] and multilayers based on anisotropic Heisenberg films [8, 12, 13] demonstrate a dependence on initial states of the non-equilibrium critical evolution. For the statistical Monte Carlo description of ferromagnetic films contacting with nonmagnetic film, we apply the anisotropic Heisenberg model with the Hamiltonian of the spin system in the form

$$H = -J_1 \sum_{i,j} \left[\boldsymbol{S}_i \boldsymbol{S}_j - \Delta(N) S_i^z S_j^z \right] - A \sum_i (S_i^z)^2 - \boldsymbol{h} \sum_i \boldsymbol{S}_i \qquad (1)$$

where $\vec{S}_i = (S_i^x, S_i^y, S_i^z)$ is a three-dimensional unit vector, fixed in cite i of the fcc-lattice, $J_1 > 0$ is the exchange integral describing the interaction between the neighboring spins in film. The temperature T of the system is measured in units of the exchange integral J_1/k_b. A characterizes a single-ion anisotropy. The simulations were performed for films with linear sizes $L \times L \times N$ with applied periodic boundary conditions in the plane of the film. $L \times L$ gives the number of spins in each layer, and N is the number of monolayers in the thin film. $\Delta(N)$ is the parameter characterizing the effective influence of anisotropy generated by the crystal field of the substrate on magnetic properties of film subject to its thickness N in terms of monoatomic

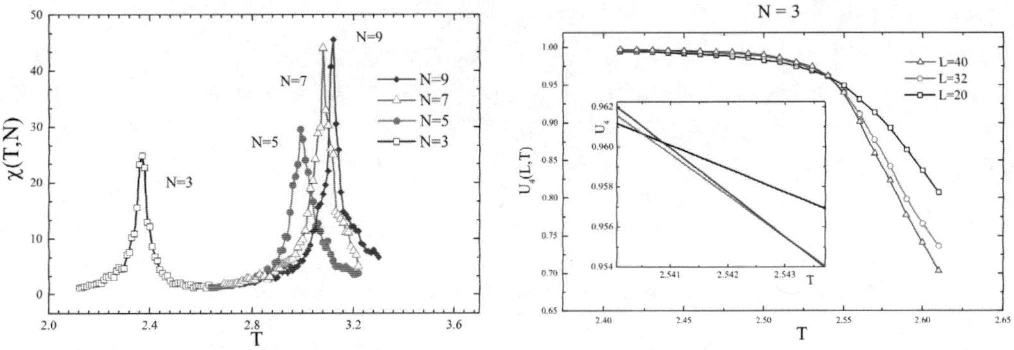

Figure 1. Temperature dependence of the staggered susceptibility $\chi_{stg}(T, N)$ (left) and Binder cumulant $U_4(T, L)$ for $N = 3$. Approximation of cumulant values for different film linear sizes L is given in insertion. Triangle from crossing lines determines a value of the critical temperature T_c.

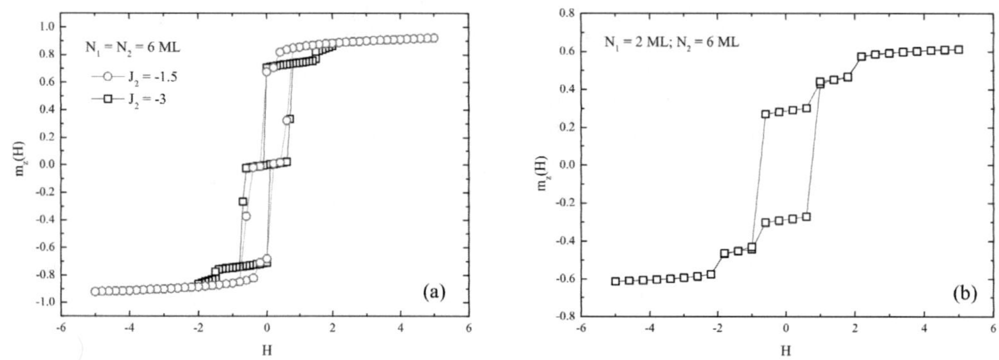

Figure 2. Magnetic hysteresis loops at $T = 0.8$ for $A = 1$ and $\Delta = 0$: $N_1 = N_2 = 6$ ML symmetric structure (a) $N_1 = N_2 = 6$ ML non-symmetric structure (b)

layers (ML). $\Delta = 0$ corresponds to the isotropic Heisenberg model, $\Delta = 1$ to the XY-model. The Hamiltonian (1) describes magnetic properties of systems with anisotropy of "easy" magnetic plane type with magnetization oriented in XY plane. The presence of anisotropy leads to the appearance of long-range magnetic order in two-dimensional Heisenberg systems with the critical temperature $T_c \neq 0$.

In this work we consider the multilayer magnetic structure consisting of two ferromagnetic films separated by the nonmagnetic film. The thickness of this nonmagnetic metal layer is selected for obtaining GMR effects in such a way that the long-range and oscillatory RKKY interlayer exchange interaction between spins of the ferromagnetic layers has an effective antiferromagnetic character. Through this interaction, the magnetizations of the adjacent ferromagnetic layers are oriented opposite to each other. When this structure is placed in an external magnetic field, the magnetizations of layers begin to orient parallel that leads to a significant change in the electrical resistance.

We calculated the equilibrium characteristics of the multilayer magnetic structure with the aim to determine the critical temperatures $T_c(N)$ of the ferromagnetic phase transition in films with different thicknesses N. For a more accurate determination of the critical temperatures, we considered the structures with films of different linear sizes $L = 20$, 32, and 40. We calculated such characteristics as the staggered magnetization $\mathbf{m}_{stg} = \mathbf{m}_1 - \mathbf{m}_2$, where $\mathbf{m}_1$ and $\mathbf{m}_2$ are the magnetizations of the films

$$m_n = |\mathbf{m}_n| = \frac{1}{N_s}\left[\sum_{\alpha\in\{x,y,z\}}\left(\sum_i^{N_s} S_i^\alpha\right)^2\right]^{1/2}, \quad n = 1, 2, \tag{2}$$

the staggered susceptibility (Fig. 1)

$$\chi_{stg} = \frac{1}{T}\left(\langle m_{\text{stg}}^2\rangle - \langle m_{\text{stg}}\rangle^2\right), \tag{3}$$

where $N_s = NL^2$ is the total number of spins in the film and angular brackets denote the statistical averaging.

The most accurate values of the critical temperatures $T_c(N)$ for structures with different film thicknesses N were obtained by the method of intersection of curves for temperature dependencies of the Binder cumulant $U_4(N, T, L)$ for systems with different linear sizes L:

$$U_4(N, T, L) = 1/2\left(3 - \frac{\langle m_{\text{stg}}^4\rangle}{\langle m_{\text{stg}}^2\rangle^2}\right). \tag{4}$$

IOP Conf. Series: Journal of Physics: Conf. Series **1163** (2019) 012068 doi:10.1088/1742-6596/1163/1/012068

The scaling dependence of the cumulant

$$U_4(N, T, L) = u(L^{1/\nu}(T - T_c)) \tag{5}$$

makes it possible to determine the critical temperature $T_c(N)$ from the coordinate of the intersection points of the curves specifying the temperature dependence $U_4(N, T, L)$ for different L (Figs. 1). Consequently, we determined for magnetic structures with film thicknesses $N = 3, 5, 7, 9$ ML the following values of magnetic ordering temperatures: $T_c(N = 3) = 2.5413(8)$, $T_c(N = 5) = 2.9033(12)$, $T_c(N = 7) = 3.0356(6)$, $T_c(N = 9) = 3.1014(11)$.

The magnetic hysteresis loops for the multilayer system at $T = 0.8$ with $A = 1$ and $\Delta = 0$ are presented in Figure 2. It is interesting to see that the hysteresis loops are changed with increasing of exchange interlayer interaction. The presence of new additional steps on hysteresis loop is connected with additional magnetic states (Figure 2a). The shape of the hysteresis loop for non-symmetric structure (Figure 2b) is formed by three subsequent transitions between ferro-antiferro-ferromagnetic states.

3. Non-equilibrium behavior of the magnetic structure and aging effects

As it was shown in [1,8], the effects of slow dynamics appear in magnetic superstructures within a wide range of temperatures at $T \leq T_c$.

One of important features arisen in non-equilibrium behavior of systems with slow dynamics [2] is dependence on an initial states. The non-equilibrium behavior of a system is realized during transition at the starting instant $t = 0$ from the initial state at temperature T_0 to the state with temperature T_s differing from T_0. The accompanying equilibration process is characterized by relaxation time $t_{\rm rel}(T_s)$, and equilibrium corresponding to temperature T_s is reached in times $t \gg t_{\rm rel}(T_s)$, while the system dynamics prove stationary and invariant with respect to time reversal. However, in times $t \ll t_{\rm rel}(T_s)$, the evolution of the system depends on its initial state. In this connection, the non-equilibrium behavior of the system depends on whether it evolves from a high-temperature $T_0 > T_s$ or a low-temperature $T_0 < T_s$ initial state. The high-temperature initial state for magnetic systems is created at $T_0 > T_c$ and characterized by initial magnetization $m_0 = 0$, while the low-temperature initial states with $T_0 < T_c$ are characterized by $m_0 \neq 0$. The non-equilibrium behavior of a systems with slow-dynamics demonstrates the breakdown of translational invariance in time due to the long-time influence of non-equilibrium initial states. It manifests itself through two-time characteristics of the system, such as the autocorrelation function

$$C(t, t_w) = \frac{1}{V} \int d^d x \left[\langle S(x, t) S(0, t_w) \rangle - \langle S(x, t) \rangle \langle S(0, t_w) \rangle \right], \tag{6}$$

where the waiting time t_w is the time between a sample preparation and the beginning of measurement of its characteristics and $t - t_w$ is the time of observation with $t - t_w, t_w \ll t_{\rm rel}$.

Description of the non-equilibrium critical behavior of a number of model statistical systems such as three-dimensional Ising model, the two-dimensional XY model, and multilayer magnetic structure [2] shows that the two-time dependent quantities demonstrate so-called aging. This phenomenon is characterized by both translation symmetry breaking in time and a slow-down of relaxation and correlation processes with increasing "age" t_w of the sample. The aging is displayed most pronouncedly in two-time dependence of the autocorrelation function (6) during evolution of system from the high-temperature initial state. The aging effects in investigated magnetic structure dependence on ferromagnetic film thicknesses N in a wide quenching temperature range $T_s \leq T_c$, we concentrate our efforts uppermost on study of autocorrelation function two-time behavior with evolution from the high-temperature initial state with $T_0 > T_c$ and $m_0 = 0$.

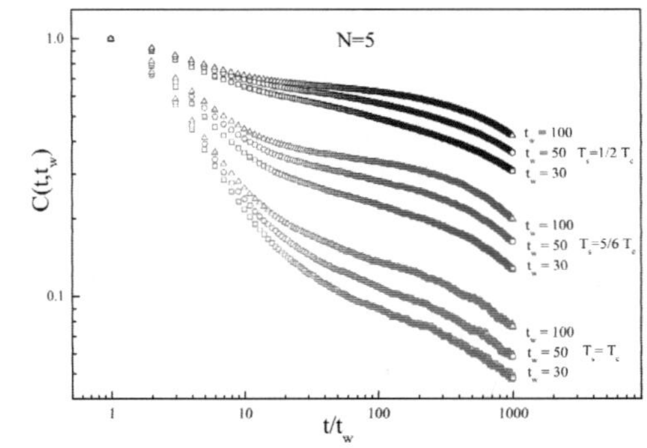

Figure 3. Autocorrelation function versus the observation time $t - t_w$ for structure with films thickness $N = 5$ ML at quenching temperatures $T_s = T_c = 2.9033$, $(5/6)T_c$, and $(1/2)T_c$.

We computed the two time-dependent autocorrelation function taken in the form

$$C(t, t_w) = \left\langle \frac{1}{N_s} \sum_{i=1}^{N_s} \mathbf{S}_i(t)\mathbf{S}_i(t_w) \right\rangle - \left\langle \frac{1}{N_s} \sum_{i=1}^{N_s} \mathbf{S}_i(t) \right\rangle \left\langle \frac{1}{N_s} \sum_{i=1}^{N_s} \mathbf{S}_i(t_w) \right\rangle . \quad (7)$$

The simulated structures were characterized by the linear size of the films $L = 64$ and thicknesses $N = 3$, 5, 7, 9 ML. The quenching temperatures $T_s(N)$ were taken equal and also fractional to critical temperatures, i.e., $T_s = T_c(N)$, $(5/6)T_c(N)$, $(2/3)T_c(N)$, and $T_c(N)/2$. We used the waiting times $t_w = 10$, 30, 50, 100 MCS/s in the study of the two-time dependence of $C(t, t_w)$. Statistical averaging of the autocorrelation function $C(t, t_w)$ was carried out on 500 MC runs for every t_w.

As an example, the obtained time dependence of the autocorrelation function $C(t, t_w)$ from observation time $t - t_w$ for waiting time values $t_w = 30, 50, 100$ MCS/s is presented in Fig. 3 for structure with films thickness $N = 5$ ML at quenching temperatures $T_s = T_c = 2.9033$, $(5/6)T_c$, and $(1/2)T_c$.

The curves of $C(t, t_w)$ clearly demonstrate the aging effects in this magnetic structure, i.e., the slowing down of time correlations with increasing system age t_w. We must note that the aging in the multilayer structures arises not only at $T_s = T_c$ as in bulk systems, but also in the low-temperature phase at $T_s < T_c$. Furthermore, the comparison of curves in Fig. 3 at temperatures $T_s = T_c$, $(5/6)T_c$, and $(1/2)T_c$ shows that the aging effects becomes stronger with decrease of quenching temperature relative to the critical temperature, i.e. the falling of time correlations at $T_s < T_c$ with the same waiting times t_w become more slow than at the critical temperature. Reason for this behavior is connected with XY-type anisotropy, which is realized in Co/Cu(100)/Co structures, and with extremely slow dynamics in two-dimensional XY-model characterized by aging not only near the temperature of the Berezinskii-Kosterlitz-Thouless phase transition T_{BKT}, but also in the entire range of the existence of the low-temperature phase [15–18].

At the next stage of this work, we study of influence of non-equilibrium behavior of the multilayer magnetic structure on its magnetoresitance with realization of evolution from both high-temperature and low-temperature initial states with the films magnetization $m_0 = 0$ and $m_0 = 1$, correspondingly. For magnetoresistance calculations we used methods, which were developed in [19, 20].

We calculate two-time dependence of the magnetoresistance $\delta(t, t_w)$ on observation time $t - t_w$ and waiting time t_w. The waiting time t_w characterizes the time between a sample preparation

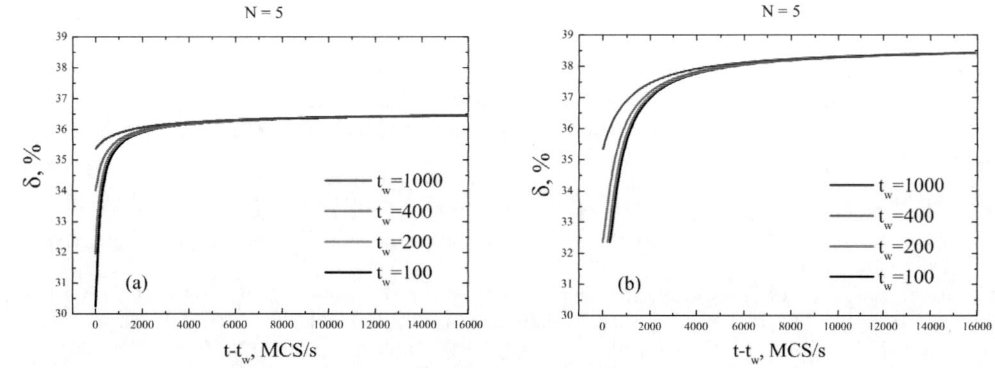

Figure 4. Time dependence of the magnetoresistance δ for the $N = 5$ at temperatures $T_s = T_c(N)/4$ for different waiting times $t_w = 100$, 200, 400, and 1000 MCS/s with evolution from (a) the high-temperature and (b) low-temperature initial states.

in non-equilibrium initial state and the beginning of measurement of its magnetoresistance. During the time period $t - t_w \ll t_{\rm rel}$, where $t_{\rm rel}$ is the system's relaxation time, the temporal behavior of the system is influenced by its initial state. As an example, we present in Fig. 4 calculated time dependence of the magnetoresistance $\delta(t, t_w)$ for the film thicknesses $N = 5$ on observation time $t - t_w$ with the duration at 10 000 MCS/s for waiting times $t_w = 100$, 200, 400, 1000 MCS/s with evolution from the high-temperature initial state (Fig. 4a) and the low-temperature initial state (Fig. 4b) at temperatures $T_s = T_c(N)/4$ (for $N = 5$ $T_s \simeq 231.5$ K). Values of the magnetoresistance $\delta(t, t_w)$ given in Fig. 4 were averaged over 250 runs.

4. Conclusions

In the present paper a Monte Carlo simulation of the non-equilibrium behavior of multilayer nanostructure is carried out with consideration of different thicknesses N of the ferromagnetic films and variation of temperature in wide range with $T \leq T_c(N)$. Two-time dependencies of the autocorrelation function demonstrate aging effects in magnetic structure for all temperatures in low-temperature phase. This phenomenon is characterized by slowing-down of correlation processes with increasing the waiting time t_w, and it is connected with increasing correlation length for transverse spin-spin correlations when the temperature is decreased that leads to increasing correlation and relaxation times in these systems.

We have realized after that the study of influence of non-equilibrium behavior in Co/Cu(100)/Co structure on its magnetoresistance. It is revealed aging effects in time dependence of the magnetoresistance $\delta(t, t_w)$, which is characterized by dependence of its values on waiting time t_w as the beginning of magnetoresistance measurement. The existence of these non-equilibrium effects should surely be taken into account in any applications of the multilayer magnetic structures for spintronic devices based on the giant magnetoresistance effect.

Acknowledgments

This research was supported by the grants 17-02-00279, 18-42-55003 of Russian Foundation of Basic Research and by the grant MD-6868.2018.2 of the Council of the President of the Russian Federation.

The simulations were supported in through computational resources provided by the Shared Facility Center "Data Center of FEB RAS" (Khabarovsk), by the Supercomputing Center of Lomonosov Moscow State University, by Moscow Joint Supercomputer Center and by St.

Petersburg Supercomputer Center of the Russian Academy of Sciences.

References

[1] Mukherjee T, Pleimling M, and Binek Ch 2010 Phys. Rev. B **82** 134425
[2] Prudnikov V V, Prudnikov P V, and Mamonova M V 2017 Phys. Usp. **60** 762
[3] Chen Y, Song D, Qiu J, et al. 2010 IEEE Trans. Magn. **46** 697
[4] Duine R A, Lee K-J, Parkin S S P, Stiles M D 2017 arxiv.org:1705.10526
[5] Apalkov D, Dieny B, Slaughter J M 2016 Proceedings of the IEEE **104** 1796
[6] Hayakawa J, Ikeda S, Lee Y M, Sasaki R, Meguro T, Matsukura F, Takahashi H, and Ohno H 2006 Jpn. J. Appl. Phys. **45** L1057
[7] Bergman A, Skubic B, Hellsvik J, Nordström L, Delin A, and Eriksson O 2011 Phys. Rev. B **83** 224429
[8] Prudnikov V V, Prudnikov P V, Purtov A N, Mamonova M V 2016 JETP Lett. **104** 776
[9] Prudnikov P V, Prudnikov V V, Menshikova M A, and Piskunova N I 2015 J. Magn. Magn. Mater. **387** 77
[10] Prudnikov P V, Prundikov V V, Medvedeva M A 2014 JETP Lett. **100** 446
[11] Popov I S, Prudnikov P V, Ignatenko A S, Katanin A A 2017 Phys. Rev. B **95** 134437
[12] Prudnikov V V, Prudnikov P V, Purtov A N, Mamonova M V, Piskunova N I 2018 J. Magn. Magn. Mater. doi:10.1016/j.jmmm.2017.11.084
[13] Prudnikov V V, Prudnikov P V, Mamonova M V 2018 JETP **127** 731
[14] Huang F, Kief M T, Mankey G J, and Willis R F 1994 Phys. Rev. B **49** 3962
[15] Berthier L, Holdsworth P C W, Sellitto M 2001 J. Phys. A **34** 1805
[16] Prudnikov V V, Prudnikov P V, Alekseev S V, Popov I S 2014 Phys. Met. Metall. **115** 1186
[17] Prudnikov V V, Prudnikov P V, Popov I S 2015 JETP Lett. **101** 539
[18] Prudnikov V V, Prudnikov P V, Popov I S 2018 JETP **126** 369
[19] Prudnikov V V, Prudnikov P V, and Romanovskiy D E 2016 J. Phys. D: Appl. Phys. **49** 235002
[20] Prudnikov V V, Prudnikov P V, Romanovskii D E 2015 JETP Lett. **102** 668

IOP Conf. Series: Journal of Physics: Conf. Series **1163** (2019) 012069 doi:10.1088/1742-6596/1163/1/012069

Application of 3D basic operators method extension to astrophysical problems

I A Kondratyev[1,2], **S G Moiseenko**[1]

[1] Space Research Institute RAS, Profsoyuznaya 84-32, 117997, Moscow, Russia

[2] National Research Nuclear University MEPhI, Kashirskoe highway, 31, 115409, Moscow, Russia

E-mail: mrkondratyev95@gmail.com, moiseenko@iki.rssi.ru

Abstract. Basic (support) operators method (Samarskii's method, operator-difference method) has proven itself well in 2D numerical simulations of the astrophysical problems. An idea of the operator approach consists of the inclusion of boundary conditions in a finite difference form into the grid analogue of solving problem and formulation of the finite difference problem as an operator equation. The finite difference operators are constructed in the way to fulfill corresponding relations between continuous operators (for instance, div(rot)=0, div is conjugated to -grad; -div(grad) is self-conjugated etc.). The approach allows obtaining completely conservative finite difference schemes. The matrix which corresponds to the self-conjugated operator is symmetrical and can be inversed efficiently by modern iteration methods. In the paper a 3D generalization of 2D grid analogues for continuous differential operators using a cell-node approximation on triangular grid was realized for tetrahedral mesh. For testing we calculated a 3D Newtonian gravitational potential, and a stationary heat transfer equation in spherical layer with the first type boundary condition on the inner surface and the third type boundary condition on the outer one was calculated. The method was applied to simulation of anisotropic heat transfer in magnetized neutron star crust.

1. Introduction

Numerical simulations of physical problems become more and more important nowadays. The operator approach in numerical methods [1] was proposed by Soviet and Russian mathematician Alexander Andreyevich Samarskii. His idea consists of the inclusion of boundary conditions into a finite-difference form and formulation a finite-difference problem as an operator equation. It allows to study operators properties and use corresponding efficient methods for the solution of the numerical problem. The idea of the basic (support) operator method consists of the grid analogue construction of one of the continuous operator (e.g. div), which approximates corresponding continuous operator. The grid analogue of the relative conjugate operator (e.g. -grad) is constructed using a corresponding integral relation. In this way operator-difference equations will have the same properties as the differential problem. For example, self-conjugated continuous operator remains symmetrical in the grid form, and sparse and symmetrical matrix corresponds to it. Also basic operators method allows to obtain completely conservative finite-difference schemes. After obtaining the grid formulation of the solving problem, the sparse and symmetrical matrixes can be inversed efficiently by modern iterative or direct methods. Two dimensional operator-difference completely conservative Lagrangian method for the magnetohydrodynamical equations was built by Ardeljan et al. in the series of papers [2, 3, 4, 5] for triangular grid of a variable structure. The method was successfully applied for the different astrophysical problems [6, 7, 8]. The mesh remapping procedure was developed in [2], the criteria for a mesh rarefaction and contraction were suggested in [8]. In astrophysical simulations we often have huge gradients or many orders variations of physical values and the mesh remapping procedure allows to refine the mesh in regions with strongly changing parameters and rarefy the grid in the regions with smooth flow. In this paper the 3D grid analogues for the differential operators in Cartesian coordinates were described. Following the operator formalism from [2, 3, 4, 5], the 3D grid analogues of basic differential operators were suggested and the operator formulations of some boundary problems were written.

2. Grid analogue of the gradient operator (nodes to cells)

We use the grid which consists of the tetrahedra. So-called cell-node approximation is used: node grid functions are defined in the grid nodes, and cell grid functions are defined to be constant in the grid cells and the boundary nodes. Following [2, 3, 4], let's introduce linear spaces of cell grid functions, node functions and boundary node functions. This formal approach allows to study the approximation and stability of the numerical schemes using efficient modern methods.

First we introduce the grid gradient operator, which converts the scalar node function into the cell vector function. For definition of the grid analogue of the differential gradient we have used the following invariant definition of the *grad* operator:

$$\nabla p = \lim_{V \to 0} \frac{1}{V} \int_S p * d\vec{S}. \tag{1}$$

Let p is a scalar node grid function. Inside each cell we additionally define it as a linear interpolation of node values of the cell's nodes. After integration of this linear interpolation, taking into account mean value theorem, we get the definition of operator *grad* grid analogue of:

$$(\nabla_{\triangle} p)_i = \frac{1}{V_i} \sum_{k=1}^{4} (\bar{p}_k S_k \vec{n}_k)_i. \tag{2}$$

Here $\bar{p}_k = \frac{(p_1+p_2+p_3)_k}{3}$ - is an average interpolated value of p in the k-th tetrahedron face, indexes 1,2 and 3 correspond to p node values on the k-th face; $\vec{n}_k$ is a unit external normal to the k-th face, S_k is a square of the k-th face and V_i is a volume of the cell with the index i. Expression (2) is the first order approximation for the differential operator *grad*. The grid analogues of other (nodes to cells) operators (such as *div*(*vector*), *grad*(*vector*), *rot*(*vector*) etc.) can be derived in a similar way.

3. Grid analogue of the divergence operator (cells to nodes)

Let us introduce the scalar product in the linear grid spaces $(p, g)_\alpha = \sum_s U_s p_s g_s$, α corresponds to node and cell spaces. For the cells $U_s = V_i$, for nodes $U_s = W_j$, here $W_j = \frac{1}{4}\sum_{k=1}^{K_j} V_k$ is a "node volume" (see [2] and references in it).

The grid analogue for divergence operator is constructed to be conjugated to the $-grad$ grid operator from (2). The Green formula and its grid analogue was used (for the case when all functions vanish on the domain boundary or the domain is infinite):

$$\int p\nabla \cdot \vec{v} dV + \int \vec{v} \cdot \nabla p dV = 0 \tag{3}$$

IOP Conf. Series: Journal of Physics: Conf. Series **1163** (2019) 012069 doi:10.1088/1742-6596/1163/1/012069

$$(\vec{v}, \nabla_{\triangle} p) + (\nabla_{\times} \cdot \vec{v}, p) = 0; \quad \sum_{l=1}^{N_l} \nabla_{\times} \cdot \vec{v}_l p_l W_l = -\sum_{k=1}^{K_j} \vec{v}_k \nabla_{\triangle} p_k V_k. \tag{4}$$

Grid analogue of the Green formula (4) is written in terms of a scalar product in grid spaces. After substitution the $\nabla_{\triangle}$ from (2) and rearrangement of terms for the j-th node the expression for the grid operator *div* can be written in the form:

$$(\nabla_{\times} \cdot \vec{v})_j = -\frac{1}{3W_j} \sum_{k=1}^{K_j} \tilde{\vec{v}}_k \cdot (\vec{n}_1 S_1 + \vec{n}_2 S_2 + \vec{n}_3 S_3)_k, \tag{5}$$

here K_j is a number of adjacent cells for j node, $\tilde{\vec{v}}_k$ - value of the cell function $\vec{v}$ in the k-th cell. Indexes $1, 2$ and 3 correspond only to those cell faces that contain the j-th node. Summation is done over all cells adjacent to the j-th node.

4. Boundary grid operators

For the numerical simulation of boundary value problems we need to formulate grid analogues for the different boundary conditions. Following [2], a boundary operator Φ is introduced on the tetrahedral grid. The Φ operator acts on elements from boundary nodes space and results into boundary nodes space. It is constructed in the way to fulfill the Green grid analogue formula including boundary:

$$(p, \nabla \cdot \vec{v}) + (\nabla p, \vec{v}) = \oint p\vec{v}d\vec{S}; \quad \sum_{l=1}^{N_l} \nabla_{\times} p_l \vec{v} W_l + \sum_{k=1}^{K_j} p_k \nabla_{\triangle} \cdot \vec{v}_k V_k = \sum_{q=1}^{K_\gamma} \Phi \cdot \vec{v}_q p_q W_q, \tag{6}$$

here K_γ is a number of nodes neighboring to the boundary node γ, W_q - is a "volume" of the boundary node [2].

The boundary operator Φ can be introduced in 3 ways. The first one is a simple numerical integration by the cell surfaces near the corresponding boundary node

$$(\Phi \cdot \vec{v})_\gamma = \frac{\vec{v}_\gamma}{3W_\gamma} \cdot \sum_{q=1}^{K_\gamma} \vec{n}_q S_q. \tag{7}$$

The second way is the integration with taking into account linear interpolation of boundary function

$$(\Phi \cdot \vec{v})_\gamma = \frac{1}{9W_\gamma} \sum_{q=1}^{K_\gamma} (\vec{v}_1 + \vec{v}_2 + \vec{v}_\gamma)_q \cdot \vec{n}_q S_q. \tag{8}$$

In (8) $\vec{v}_1$ and $\vec{v}_2$ are the values of function $\vec{v}$ in the boundary nodes of the q-th boundary cell face, adjacent to the node γ.

To obtain the third formula, fictitious cells introduced outside the computational domain were used, following [5].

Finally we get the expression for the boundary operator Φ:

$$(\Phi \cdot \vec{v})_\gamma = \frac{1}{12W_\gamma} \left(\sum_{q=1}^{K_\gamma} (\vec{v}_1 + \vec{v}_2 + \vec{v}_\gamma)_q \cdot \vec{n}_q S_q + \vec{v}_\gamma \cdot \sum_{q=1}^{K_\gamma} \vec{n}_q S_q \right). \tag{9}$$

Grid approximation of the *div* operator including the boundary can be written in the form ($(\nabla \cdot \vec{v})_\gamma = \nabla_{\times} \cdot \vec{v} + \Phi \cdot \vec{v} + O(h)$, h is a characteristic cell size). 3D boundary operators for other grid analogues of cell to node operators can be constructed in a similar way.

IOP Conf. Series: Journal of Physics: Conf. Series **1163** (2019) 012069 doi:10.1088/1742-6596/1163/1/012069

5. Test calculations

To test the described operators we solved several 3D problems numerically. The first one is a Dirichlet problem for Poisson equation. The physical interpretation of the solution is a Newtonian gravitational potential generated by the spatial distribution of the matter. Differential formulation is (gravitational constant $G = 1$)

$$\begin{cases} \Delta\Psi = 4\pi\rho, \\ \Psi \mid_\Gamma = \Psi_\gamma. \end{cases} \tag{10}$$

We consider $\rho = \dfrac{z}{4\pi\sqrt{x^2+y^2+z^2}}$ further.

The general solution of the (10) is given by the volume potentials formula

$$\Psi(\vec{r}) = -\iiint\limits_V \frac{d\vec{r}^*\rho(\vec{r}^*)}{|\vec{r}^* - \vec{r}|}. \tag{11}$$

For the numerical solution of the problem (10) we need to know the boundary values of the gravitational potential. The approximation of the formula (11) on the tetrahedral grid gives the expression for the calculation of the boundary values Ψ_γ.

$$\Psi_\gamma(\vec{r_q}) = -\sum_{k=1}^{K_j} \frac{V_k\rho(\vec{r}_\Delta)}{|\vec{r}_\Delta - \vec{r}_q|}, \tag{12}$$

where $\vec{r}_\Delta$ is an average cell value of $\vec{r}$, K_j is a total number of grid nodes.

Similarly to the approach developed in [2] we include the grid boundary conditions into the grid operator form of the problem. We write the Poisson equation in the boundary knots: $\delta\nabla_\times \cdot \nabla_\triangle\Psi_\gamma = \delta 4\pi\rho$, and distinguish the boundary in that equation:$\nabla_\times \cdot \nabla_\triangle(I-\delta)\Psi + \nabla_\times \cdot \nabla_\triangle\delta\Psi = 4\pi\rho$. Further, we subtract one from another and get:

$$(I-\delta)\nabla_\times \cdot \nabla_\triangle(I-\delta)\Psi = 4\pi(I-\delta)\rho - (I-\delta)\nabla_\times \cdot \nabla_\triangle\Psi_\gamma \tag{13}$$

operator δ works as follows: it equals to 0 in the internal nodes and 1 on the boundary, I is a unit operator.

It can be seen that grid formulation of the problem gives us one operator equation which includes the boundary condition. $(I-\delta)\nabla_\times \cdot \nabla_\triangle(I-\delta)$ is a grid Laplace operator, which is self conjugate as a continuous Laplace operator.

The solution of the operator-difference equation (13) and analytical gravitational potential are represented on the fig. 1. The maximal relative numerical error on the 3D grid of 2000 knots was $\leq 5\%$.

The second test problem is stationary heat transfer equation in the spherical layer with the Dirichet boundary condition on the inner bound and the 3rd type boundary condition on the outer bound. In the general form the problem can be written as:

$$\begin{cases} \nabla \cdot A(T,\vec{r})\nabla T = f(\vec{r}), \\ T \mid_{inn} = T_0, \\ (\vec{n}, A(T,\vec{r})\nabla T) + g(T,\vec{r}) \mid_{out} = 0. \end{cases} \tag{14}$$

here $A(T,\vec{r})$ is a thermal conductivity, it can be scalar or tensor function, $g(T,\vec{r})$ in the boundary condition corresponds to physical processes on the boundary surface (convection, radiation, etc.).

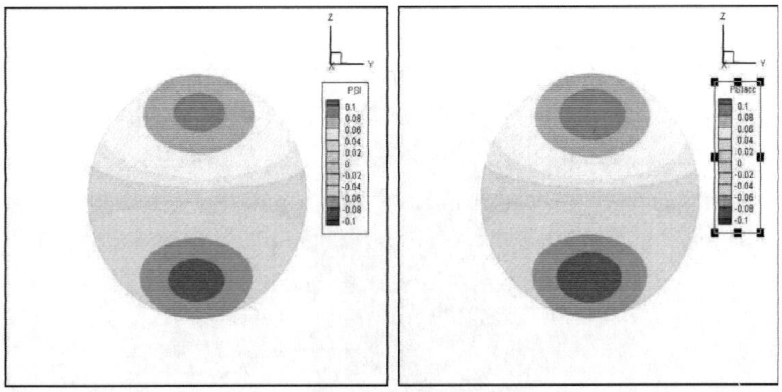

Figure 1. The gravitational potential Y-Z cross section, left - numerical, right - analytical solution

We represent the problem (14) as one grid equation including boundary conditions following the approach from [2].

$$(I-\delta_1)\nabla_\times \cdot A(P_\triangle T)\nabla_\triangle (I-\delta_1)T + (I-\delta_1)\nabla_\times \cdot A(P_\triangle T)\nabla_\triangle T_0 - \delta_2\Phi \cdot \vec{n}g(T) = (I-\delta_1)f. \quad (15)$$

$P_\triangle T$ is an interpolation of the node function T into the cells (heat fluxes are defined in the cells and boundary nodes), indexes 1 and 2 in the δ operators correspond to inner and outer boundary respectively.

6. Heat transfer simulation in neutron star crust

Neutron star (NS) is a final stage of stellar evolution of massive stars with masses more than $\sim 8M_\odot$ and less than $25-30M_\odot$ (upper limit depends on particular theoretical model; $M_\odot$ is a solar mass). It is a product of core-collapse supernova explosion, bright and high-energy astrophysical phenomenon. NS is a very compact object with linear sizes approximately 10-20 kilometres. The density in the central parts of the NS can reach $10^{14-15}g/cm^3$, the magnetic fields are $\sim 10^{12}$G on the NS surface. The aim of this work is to calculate the temperature distribution on the surface of the neutron star by the solution of the heat transfer equation for crust in the presence of the strong magnetic field. The anisotropy of the temperature distribution is connected with the dependence of nonlinear thermal conductivity tensor function on the magnetic field configuration.

The problem was simulated in 2D (axial symmetry) in [9, 10].

Thermal conductivity tensor was derived in the paper of Bisnovatyi-Kogan and Glushikhina [11] for the plasma with arbitrary degenerate electrons and nondegenerate ions in the presence of magnetic field. It was obtained with the Chapman-Enskog method from the Boltzmann equation. The thermal conductivity tensor takes into account heat fluxes along and across the magnetic field and the Hall effect. It is written for the strongly degenerate electrons as follows

$$\begin{cases} A_{ij} = \frac{k^2 T n_e}{m_e}\tau_d\big(\lambda^{(1)}\delta_{ij} + \lambda^{(2)}\varepsilon_{ijk}\frac{B_k}{B} + \lambda^{(3)}\frac{B_i B_j}{B^2}\big), \\ \lambda^{(1)} = \frac{5\pi^2}{6}\big(\frac{1}{1+(\omega\tau_d)^2} - \frac{6}{5}\frac{(\omega\tau_d)^2}{(1+(\omega\tau_d)^2)^2}\big), \\ \lambda^{(2)} = -\omega\tau_d\frac{4\pi^2}{3}\big(\frac{1}{1+(\omega\tau_d)^2} - \frac{3}{4}\frac{(\omega\tau_d)^2}{(1+(\omega\tau_d)^2)^2}\big), \\ \lambda^{(3)} = (\omega\tau_d)^2\frac{5\pi^2}{6}\big(\frac{1}{1+(\omega\tau_d)^2} - \frac{6}{5}\frac{1}{(1+(\omega\tau_d)^2)^2}\big), \end{cases} \quad (16)$$

where k is a Boltzmann constant, n_e is a concentration of electrons, m_e is an electron mass, ω is a cyclotron frequency, τ_d is an average time between electron-ion collisions in the absence of

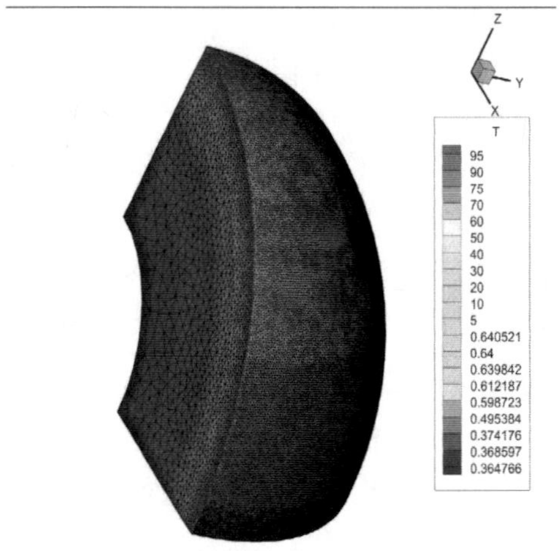

Figure 2. Temperature distribution in the neutron star crust (in units 10^5K), $r_{inn} = 4km, r_{out} = 10km$

magnetic field. $\omega\tau_d$ is a magnetization parameter, which corresponds to the dependence of a spatial temperature anisotropy on the magnetic field.

On the inner boundary of the crust in our simulations we assume constant temperature T_0, on the outer boundary we assume radiation with the Stephan-Boltzmann law, $g(T, \vec{r}) = \sigma T^4$ in (14). Also we assume the absence of heat sources in the interior of the crust ($f = 0$).

After the approximation of the continuous problem (14) using the developed 3D operators and the inclusion of the boundary conditions into the operator equation we get nonlinear set of algebraic equations which is solved using the Newton method.

On the fig. 2 some preliminary numerical results for octant are shown for the constant magnetic field, $B_z = 5 \cdot 10^{11}G, n_e = 10^{29} \div 10^{31}cm^{-3}$. The temperature distribution on the outer crust surface is anisotropic: at the magnetic pole it is hotter ($T = 6.5 \cdot 10^4K$), than at the equator ($T = 3.5 \cdot 10^4K$).

We plan to simulate this problem in the full 3D formulation with various magnetic field configurations and more realistic physical parameters. Results will be published elsewhere.

Acknowledgments

We thank G.S.Bisnovatyi-Kogan and M.V. Glushikhina for useful comments. This research was supported by Russian Science Foundation grant No.18-12-00378.

References

[1] Samarskii A A 1989 *Finite difference schemes theory* Moscow: Nauka
[2] Ardeljan N V and Kosmachevskii K V 1995 *Comput. Math. Modelling* **6** 209
[3] Ardeljan N V and Kosmachevskii K V 1987 *Problems of Construction and Research of Conservative Difference Schemes for Magneto-gas-dynamics* MSU, Moscow
[4] Ardelyan N V and Gushin I S 1982 *Vestnik MSU* **15**, 3 3
[5] Ardelyan N V 1983 *Comp. Math. Math. Phys.* **23** 1168
[6] Ardeljan N V, Bisnovatyi-Kogan G S and Moiseenko S G 2000 *Astron. Arstopys.* **355** 1181
[7] Ardeljan N V, Bisnovatyi-Kogan G S and Moiseenko S G 2005 *Mon. Not. R. Astron. Soc.* **359** 333
[8] Ardeljan N V, Bisnovatyi-Kogan G S, Kosmachevskii K V and Moiseenko S G 1996 *Astronomy and Astrophysics Supplement* **115** 573
[9] Geppert U, Kuker D and Page D 2004 *Astron. Astrophys.* **426** 267
[10] Perez-Azorin J F, Miralles J A and Pons J A 2006 *Astron. Astrophys.* **451** 1009
[11] Bisnovatyi-Kogan G S and Glushikhina M V 2018 *Plasma Physics Reports* **44** 4 355

IOP Conf. Series: Journal of Physics: Conf. Series **1163** (2019) 012070 doi:10.1088/1742-6596/1163/1/012070

Quasi-Stable Solutions of the Genetic Networks Models

S D Glyzin[1,2], A Yu Kolesov[1] and N Kh Rozov[3]

[1] P.G. Demidov Yarostavl State University, Sovetskaya str., 14, Yaroslavl, 150000, Russia
[2] Scientific Center in Chernogolovka RAS, Lesnaya str., 9, Chernogolovka, Moscow region, 142432, Russia
[3] Lomonosov Moscow State University, Leninskiye Gory, Main Building, Moscow, 119991, Russia

E-mail: glyzin@uniyar.ac.ru; kolesov@uniyar.ac.ru; fpo.mgu@mail.ru

Abstract.
We consider dynamic properties of solutions of the simplest genetic network called repressilator. This network contains three elements unidirectionally coupled into ring. More specifically, the first of them inhibits the synthesis of the second, the second inhibits the synthesis of the third, and the third, which closes the cycle, inhibits the synthesis of the first one. The interaction of the protein concentrations and of mRNA (message RNA) concentration is surprisingly similar to the interaction of six ecological populations – three predators and three preys. This new mathematical model is represented by a system of unidirectionally coupled ordinary differential equations. The existence and stability of special periodic motions (traveling waves) for this system are studied. It is shown that, with a suitable choice of parameters and an increasing number m of equations in the system, the number of coexisting traveling waves increases indefinitely, but all of them (except for a single stable periodic solution for odd m) are quasi-stable. The quasi-stability of a cycle means that some of its multipliers are asymptotically close to the unit, while the other multipliers (except for a simple unit one) are less than unity in absolute value.

1. Introduction

Consider the model system of ordinary differential equations

$$\dot{p}_j = -p_j + \frac{\alpha}{1 + u_{j-1}^{\gamma}} + \alpha_0, \quad \dot{u}_j = \beta(p_j - u_j), \quad j = 1, 2, 3, \quad u_0 = u_3, \tag{1}$$

describing the simplest genetic network (see [1]). Here m_j and p_j $j = 1, 2, 3$ are concentrations of matrix ribonucleic acid (mRNA) and protein of the genetic oscillator, α, α_0, γ, β are positive parameters.

Suppose that in (1) the parameters β and α_0 are small. Then, replacing $\beta t \to t$ and discarding the parameter α_0, one obtains a singularly perturbed system, to which the well-known Tikhonov reduction principle [2] is applied. After reduction procedure we have the following system:

$$\dot{u}_j = -u_j + \frac{\alpha}{1 + u_{j-1}^{\gamma}}, \quad j = 1, 2, 3, \quad u_0 = u_3. \tag{2}$$

IOP Conf. Series: Journal of Physics: Conf. Series **1163** (2019) 012070 doi:10.1088/1742-6596/1163/1/012070

The problem of self-excited oscillations in (2) and similar systems arising in the modeling of gene networks has been extensively investigated (see, e.g., [3, 4, 5, 6, 7, 8, 9, 10]). It is known that, for $\gamma > 2$ and a suitable increase of α, the model (2) has a stable cycle that is self-symmetric (i.e., invariant under cyclic permutations of the coordinates). The self-symmetry property implies that this cycle can be represented in the form

$$(u_1, u_2, u_3) = (u(t), u(t+\Delta), u(t+2\Delta)), \tag{3}$$

where $\Delta > 0$ is the phase shift. Here $u(t)$ is a 3Δ-periodic function.

Phenomenological model of genetic oscillator can be derived also from population dynamics systems. The interaction of the concentrations u_j and p_j described above is similar to the interaction of six ecological populations – three predators and three preys. Indeed, suppose that u_j, $j = 1, 2, 3$ and p_j, $j = 1, 2, 3$ are the population densities of the predators and preys, respectively. The gene network can be modeled using Yu. S. Kolesovs approach [11]. In the case of an arbitrary number of elements this approach yields the system

$$\dot{p}_j = \frac{r_1}{1+a}[1 + a(1 - u_{j-1}) - p_j]p_j + \alpha, \quad \dot{u}_j = r_2[p_j - u_j]u_j, \quad j = 1, 2, \ldots, m, \ u_0 = u_m, \tag{4}$$

where r_1, r_2, a, and α are positive constants.

As well as the system (1), the new mathematical model of repressilator (4) can be simplified. At first assume that $r_2 \gg 1$ and $r_1 = r \sim 1$. Then, according to the reduction principle [2], as $r_2 \to +\infty$, we have $p_j = u_j$, $j = 1, 2, \ldots, m$. For the components u_j, we obtain the system

$$\dot{u}_j = \frac{r}{1+a}[1 + a(1 - u_{j-1}) - u_j]u_j + \alpha, \quad j = 1, 2, \ldots, m, \quad u_0 = u_m,$$

which, after the normalizations $u_j/(1+a) \to u_j$ and $\alpha/(1+a) \to \alpha$, becomes

$$\dot{u}_j = r[1 - u_j - a\,u_{j-1}]u_j + \alpha, \quad j = 1, 2, \ldots, m, \quad u_0 = u_m. \tag{5}$$

We consider special periodic solutions of the system (5) that can be represented in the form

$$u_j = u(t + (j-1)\Delta), \quad j = 1, 2, \ldots, m, \quad \Delta = \text{const} > 0. \tag{6}$$

This solution we call a traveling wave of (5).

Below, the existence and stability of such solutions are analyzed in the case where $r \gg 1$, $\alpha \ll 1$, and the parameter a is of the order of unity. More precisely, we assume throughout that

$$a = \text{const} > 1, \quad \alpha = r\exp(-b\,r), \quad r \gg 1, \quad b = \text{const} > 0. \tag{7}$$

It will be shown that, under conditions (7), the number of coexisting periodic solutions (6) of the system (5) increases indefinitely as $r \to +\infty$ and $m \to +\infty$ consistently. However, all of them (except for a single stable solution for odd m) are quasi-stable. Namely, the stability spectrum of each of these periodic solutions contains a nonempty group of multipliers $\nu \in \mathbb{C}$, $\nu \neq 1$, lying at a distance of order $\exp(-c\,r)$, $c = \text{const} > 0$ from the unit circle.

The quasi-stability phenomenon of solutions of the model of gene network is the main subject of the paper.

IOP Conf. Series: Journal of Physics: Conf. Series **1163** (2019) 012070 doi:10.1088/1742-6596/1163/1/012070

2. General research scheme

To simplify the subsequent analysis, in system (5), we make the exponential changes of variables $u_j = \exp(x_j/\varepsilon)$, $j = 1, 2, \ldots, m$, where $\varepsilon = 1/r \ll 1$. As a result, in view of relations (7), the system becomes

$$\dot{x}_j = 1 - \exp\left(\frac{x_j}{\varepsilon}\right) - a\exp\left(\frac{x_{j-1}}{\varepsilon}\right) + \exp\left(-\frac{b + x_j}{\varepsilon}\right), \quad j = 1, 2, \ldots, m, \tag{8}$$

where $x_0 = x_m$. According to (6), we are interested in periodic solutions of system (8) that can be represented in the form

$$x_j = x(t + (j-1)\Delta, \varepsilon), \quad j = 1, 2, \ldots, m, \tag{9}$$

where $\Delta > 0$ and $x(t, \varepsilon)$ is a T-periodic solution of the auxiliary delay equation

$$\dot{x} = 1 - \exp\left(\frac{x}{\varepsilon}\right) - a\exp\left(\frac{x(t-\Delta)}{\varepsilon}\right) + \exp\left(-\frac{b + x}{\varepsilon}\right). \tag{10}$$

Direct verification shows that components (9) satisfy system (8) if and only if $T = m\Delta/k$, $k \in \mathbb{N}$.

Given a positive integer k, assume that Eq. (10) has the required periodic solution $x(t, \varepsilon)$ of period $T = m\Delta/k$. Then the stability analysis of the corresponding cycle (9) is reduced to analyzing the location of multipliers of the linear system

$$\dot{g}_j = A(t + (j-1)\Delta, \varepsilon)g_j + B(t + (j-1)\Delta, \varepsilon)g_{j-1}, \quad j = 1, 2, \ldots, m, \tag{11}$$

where $g_0 = g_m$ and the coefficients $A(t, \varepsilon)$ and $B(t, \varepsilon)$ are given by

$$A(t,\varepsilon) = -\frac{1}{\varepsilon}\left(\exp\left(\frac{x(t,\varepsilon)}{\varepsilon}\right) + \exp\left(-\frac{b + x(t,\varepsilon)}{\varepsilon}\right)\right), \; B(t,\varepsilon) = -\frac{a}{\varepsilon}\exp\left(\frac{x(t-\Delta,\varepsilon)}{\varepsilon}\right). \tag{12}$$

With (11), in what follows, we will need the auxiliary linear delay equation

$$\dot{g} = A(t, \varepsilon)g + \varkappa B(t, \varepsilon)g(t - \Delta), \tag{13}$$

where $g(t)$ is a scalar complex-valued function and $\varkappa$ is an arbitrary complex parameter. More precisely, we will be interested in its multipliers $\nu_l(\varkappa)$, $l = 1, 2, \ldots$, arranged in decreasing order of moduli.

Let us explain the meaning of a multiplier as applied to Eq. (13). For a fixed number $\sigma_0 > 0$, consider the space $E = C[-\Delta - \sigma_0, -\sigma_0]$ of continuous complex-valued functions $g(t)$ for $-\Delta - \sigma_0 \le t \le -\sigma_0$ with the norm $||g||_E = \max\limits_{-\Delta-\sigma_0 \le t \le -\sigma_0} |g(t)|$. The monodromy operator of Eq. (13) is a bounded linear operator $V : E \to E$ acting on an arbitrary function $g_0(t) \in E$ according to the rule

$$Vg_0 = g(t + m\Delta/k, \varkappa, \varepsilon), \quad -\Delta - \sigma_0 \le t \le -\sigma_0, \tag{14}$$

where $g(t, \varkappa, \varepsilon)$ is the solution of Eq. (13) on the time interval $-\sigma_0 \le t \le m\Delta/k - \sigma_0$ with an initial function $g_0(t)$, $-\Delta - \sigma_0 \le t \le -\sigma_0$. The spectrum of this operator is always discrete, since some power of V is compact (for $m/k > 1$, V is compact itself). By the multipliers of Eq. (13), by analogy with ordinary differential equations, we mean the eigenvalues of operator (14).

To study the relation between the multipliers of system (11) and Eq. (13), the so-called tuning method with respect to the parameter $\varkappa$ was proposed in [12]. According to this method, we consider the family of equations

$$\left[\nu_l(\varkappa)\right]^k = \varkappa^m, \quad l \in \mathbb{N}. \tag{15}$$

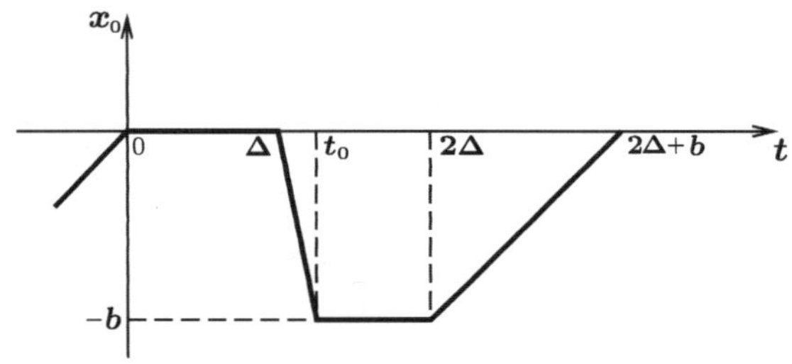

Figure 1. Periodic function $x_0(t)$

It turns out that there is a correspondence between the nonzero roots of these equations and the multipliers of system (11). More precisely, the following assertion holds (see [12, 13, 14]).

Lemma 1. *For each multiplier ν of system (11), there is a positive integer l_0 that such*

$$\nu = \nu_{l_0}(\varkappa_0), \tag{16}$$

where $\varkappa_0$ is a root of Eq. (15) at $l = l_0$. Conversely, given some $l = l_0$, if Eq. (15) has a nonzero root $\varkappa = \varkappa_0$, then the original system (11) has a multiplier of form (16).

Thus, the analysis of the existence of cycles (9) in system (8) is reduced to the search for periodic solutions of the auxiliary scalar equation (10) with periods $T = m\Delta/k$. The stability of traveling waves is analyzed separately and, by Lemma 1, is reduced to the asymptotic computation of the roots of Eqs. (15). Below, both these issues are studied for positive integer m and k satisfying the conditions

$$m \geq 3, \quad 2 < m/k < a + 1. \tag{17}$$

3. Analysis of the auxiliary nonlinear equation

Our nearest goal is to show that, for any fixed values of the parameters a, b, Δ satisfying the inequalities

$$a > 1, \quad b > 0, \quad \Delta(a-1) > b, \tag{18}$$

and for all $0 < \varepsilon \ll 1$, the auxiliary equation (10) has a nontrivial periodic solution.

For a formulation of corresponding result we consider periodic function

$$x_0(t) = \begin{cases} 0 & 0 \leq t \leq \Delta, \\ -(a-1)(t-\Delta) & \Delta \leq t \leq t_0, \\ -b & t_0 \leq t \leq 2\Delta, \\ t - 2\Delta - b & 2\Delta \leq t \leq T_0, \end{cases} \qquad x_0(t + T_0) \equiv x_0(t), \tag{19}$$

where $t_0 = \Delta + b/(a-1)$ and $T_0 = 2\Delta + b$ (this function is plotted in Fig. 1).

Lemma 2. *Under conditions (18), for all sufficiently small $\varepsilon > 0$, Eq. (10) has a cycle $x = x_*(t, \varepsilon)$ of period $T_*(\varepsilon)$ that satisfies, as $\varepsilon \to 0$, the asymptotic equalities*

$$\max_{0 \leq t \leq T_*(\varepsilon)} |x_*(t,\varepsilon) - x_0(t)| = O(\varepsilon), \quad T_*(\varepsilon) = T_0 + O(\varepsilon). \tag{20}$$

We describe the general scheme for proving this lemma. Let σ_0 be a fixed constant satisfying the conditions

$$0 < \sigma_0 < \min\left(\frac{b}{2}, \frac{\Delta}{2}, \frac{b}{2(a-1)}, \frac{1}{2}\left(\Delta - \frac{b}{a-1}\right), \frac{\Delta}{a-1}\right). \tag{21}$$

The set $S \subset C[-\Delta - \sigma_0, -\sigma_0]$ of initial functions $\varphi(t)$ is defined as

$$S = \{\varphi(t) : \ -q_1 \le \varphi(t) \le -q_2 \quad -\Delta - \sigma_0 \le t \le -\sigma_0, \ \varphi(-\sigma_0) = -\sigma_0\}, \tag{22}$$

where $q_1 > q_2 > 0$ are universal (independent of t, ε, φ) constants, which will be specified later. Below, we are interested in the solution $x = x_\varphi(t, \varepsilon)$, $t \ge -\sigma_0$ of Eq. (10) with an arbitrary initial value $\varphi(t) \in S$, $t \in [-\Delta - \sigma_0, -\sigma_0]$.

For each function $\varphi(t) \in S$, let $t = T_\varphi(\varepsilon)$ denote the second positive root of the equation

$$x_\varphi(t - \sigma_0, \varepsilon) = -\sigma_0 \tag{23}$$

(if it exists). The operator Π from S to $C[-\Delta - \sigma_0, -\sigma_0]$ is defined as

$$\Pi(\varphi) = x_\varphi(t + T_\varphi(\varepsilon), \varepsilon), \quad -\Delta - \sigma_0 \le t \le -\sigma_0. \tag{24}$$

We proof that for a suitable choice of the parameters q_1, q_2 operator (24) is defined on set (22) and, moreover, $\Pi(S) \subset S$, $T_\varphi(\varepsilon) > \Delta \ \forall \varphi \in S$. Furthermore, since S is closed, bounded, and convex and the operator Π is compact by virtue of the inequality $T_\varphi(\varepsilon) > \Delta$, the Schauder principle implies that Π has at least one fixed point $\varphi = \widetilde{\varphi}(t, \varepsilon)$ in S. It is also clear that the solution of Eq. (10) is periodic with period $T_*(\varepsilon) = T_\varphi|_{\varphi = \widetilde{\varphi}}$. This solution is the desired one, since it possesses asymptotic properties (20).

To implement the above-described scheme, we need to know a uniform (with respect to $\varphi \in S$) asymptotic representation of the solution $x_\varphi(t, \varepsilon)$ on various intervals of t. The corresponding constructions are divided into eight stages. The proof of Lemma 2 due to the bulkiness is not given (see [15]).

4. Analysis of the auxiliary linear equation

In this section, we study the asymptotic behavior of multipliers of a linear equation similar to (13), namely,

$$\dot{g} = A_*(t, \varepsilon) g + \varkappa B_*(t, \varepsilon) g(t - \Delta), \tag{25}$$

with coefficients given by (12) at $x(t, \varepsilon) = x_*(t, \varepsilon)$.

Consider the set of initial functions

$$B_0 = \{g_0(t) \in E : \ g_0(-\sigma_0) = 0, \ ||g_0|| \le 2\}, \tag{26}$$

where, as in Section 2, E is the space $C[-\Delta - \sigma_0, -\sigma_0]$ over the field of complex numbers and $|| * ||$ is the norm in E (defined in a usual manner). Let $g_1(t, g_0, \varkappa, \varepsilon)$ denote the solution of Eq. (25) with an arbitrary initial function $g_0(t)$ from set (26), and let $g_2(t, \varkappa, \varepsilon)$ be the solution of this equation with the initial function $g_2 \equiv 1$, $t \in [-\Delta - \sigma_0, -\sigma_0]$. Integrating Eq. (25) by the method of steps, we see that, on the interval $t \in [-\sigma_0, T_*(\varepsilon) - \sigma_0]$, the functions are power functions of $\varkappa$, i.e.,

$$g_1(t, g_0, \varkappa, \varepsilon) = \sum_{n=1}^{n_0} g_{1,n}(t, g_0, \varepsilon) \varkappa^n, \quad g_2(t, \varkappa, \varepsilon) = \sum_{n=0}^{n_0} g_{2,n}(t, \varepsilon) \varkappa^n, \tag{27}$$

where

$$n_0 = \begin{cases} T_*(\varepsilon)/\Delta & \text{if } T_*(\varepsilon)/\Delta \text{ is an integer,} \\ \lfloor T_*(\varepsilon)/\Delta \rfloor + 1 & \text{otherwise,} \end{cases} \tag{28}$$

and $\lfloor * \rfloor$ is the integer part. Note also that

$$g_{1,n}(t, g_0, \varepsilon) \equiv g_{2,n}(t, \varepsilon) \equiv 0 \quad t \in [-\sigma_0, (n-1)\Delta - \sigma_0], \ \ n \ge 2. \tag{29}$$

For an asymptotic analysis of the multipliers of Eq. (25), we need the following result.

Lemma 3. *There exists a sufficiently small* $\varepsilon_0 > 0$ *such that, for all* $0 < \varepsilon \leq \varepsilon_0$, $\varkappa \in \Lambda_0 \overset{\text{def}}{=} \{\varkappa \in \mathbb{C} : \ |\varkappa| \leq 1\}$, $g_0 \in B_0$ *on the interval* $-\sigma_0 \leq t \leq T_*(\varepsilon) - \sigma_0$, *the estimates*

$$\sum_{n=0}^{n_0} \left| \frac{\partial^n g_1}{\partial \varkappa^n}(t, g_0, \varkappa, \varepsilon) \right| \leq \exp\left(-q/\varepsilon\right), \quad \sum_{n=0}^{n_0} \left| \frac{\partial^n g_2}{\partial \varkappa^n}(t, \varkappa, \varepsilon) \right| \leq M \tag{30}$$

hold with constants $q, M > 0$ *independent of* $t, \varepsilon, \varkappa, g_0$. *Moreover, as* $\varepsilon \to 0$, *the asymptotic representations*

$$\frac{\partial^n}{\partial \varkappa^n} g_2(T_*(\varepsilon) - \sigma_0, \varkappa, \varepsilon) = [\varkappa^2]^{(n)} + O\left(\exp\left(-q/\varepsilon\right)\right), \quad n = 0, 1, \ldots, n_0 \tag{31}$$

hold uniformly in $\varkappa$ *(here and below,* $[*]^{(n)}$ *denotes the nth derivative with respect to* $\varkappa$*).*

Proof. The first inequality in (30) is proved by induction. For this purpose, the time interval $[-\sigma_0, T_*(\varepsilon) - \sigma_0]$ is divided into subintervals $[(n-1)\Delta - \sigma_0, n\Delta - \sigma_0]$, $n = 1, 2, \ldots, n_0 - 1$, and $[(n_0 - 1)\Delta - \sigma_0, T_*(\varepsilon) - \sigma_0]$, where n_0 is given by (28). The proof of Lemma 3, like Lemma 2, is divided into 8 stages. This technical result is not given here(see [15]).

Now we pass to the asymptotic computation of the multipliers of Eq. (25). For this purpose, we consider the monodromy operator $V_*(\varkappa, \varepsilon)$ of this equation acting on the space E according to the rule (similar to (14))

$$V_*(\varkappa, \varepsilon) g_0 = g(t + T_*(\varepsilon), \varkappa, \varepsilon), \quad -\Delta - \sigma_0 \leq t \leq -\sigma_0, \tag{32}$$

where $g(t, \varkappa, \varepsilon)$, $-\sigma_0 \leq t \leq T_*(\varepsilon) - \sigma_0$, is the solution of Eq. (25) with an initial function $g_0(t) \in E$. Let $\nu_s(\varkappa, \varepsilon)$, $s \in \mathbb{N}$, denote the eigenvalues of operator (32) arranged in decreasing order of moduli. The following assertion holds.

Lemma 4. *For any* $R > 0$, *there are* $\varepsilon_0 = \varepsilon_0(R) > 0$, $q = q(R) > 0$, *and* $\delta = \delta(R) > 0$ *such that, for all* $0 < \varepsilon \leq \varepsilon_0$, $\varkappa \in \Lambda_{\delta,R} \overset{\text{def}}{=} \{\varkappa \in \mathbb{C} : \ \exp(-\delta/\varepsilon) \leq |\varkappa| \leq R\}$

$$\sup_{s \geq 2} |\nu_s(\varkappa, \varepsilon)| \leq \exp\left(-q/\varepsilon\right). \tag{33}$$

The multiplier $\nu_1(\varkappa, \varepsilon)$ *is simple, depends analytically on* $\varkappa \in \Lambda_{\delta,R}$, *and, as* $\varepsilon \to 0$, *has the asymptotic representations (uniform in* $\varkappa$*)*

$$\nu_1(\varkappa, \varepsilon) = \varkappa^2 + O\left(\exp\left(-q/\varepsilon\right)\right), \quad \frac{\partial \nu_1}{\partial \varkappa}(\varkappa, \varepsilon) = 2\varkappa + O\left(\exp\left(-q/\varepsilon\right)\right). \tag{34}$$

5. Main results

Recall that the analysis of the existence of traveling waves (9) for system (8) is reduced to the search for periodic solutions of the auxiliary equation (10) with the periods T given by the equalities $T = m\Delta/k$, $k \in \mathbb{N}$. In this context, in what follows, the periodic solution of Eq. (10) provided by Lemma 2 and its period are denoted by $x_*(t, \varepsilon, \Delta)$ and $T_*(\varepsilon, \Delta)$, respectively, in order to emphasize that these functions depend on Δ. Let m and k be fixed positive integers related by inequalities (17). These inequalities imply that the quantity $\widehat{\Delta}_{(k)} = b/(m/k - 2)$ satisfies the condition $\widehat{\Delta}_{(k)} > b/(a-1)$. In what follows, we assume that the parameter Δ in (10) ranges over some interval $K \subset (b/(a-1), +\infty)$ for which $\Delta = \widehat{\Delta}_{(k)}$ is an interior point. Then we have the asymptotic representation (uniform in $\Delta \in K$)

$$T_*(\varepsilon, \Delta) = 2\Delta + b + O(\varepsilon), \quad \varepsilon \to 0. \tag{35}$$

In view of formula (35), it is easy to see that the equation

$$T_*(\varepsilon, \Delta) = m\Delta/k \tag{36}$$

IOP Conf. Series: Journal of Physics: Conf. Series **1163** (2019) 012070 doi:10.1088/1742-6596/1163/1/012070

has at least one root $\Delta = \widehat{\Delta}_{(k)}(\varepsilon)$ such that

$$\widehat{\Delta}_{(k)}(\varepsilon) = \widehat{\Delta}_{(k)} + O(\varepsilon), \quad \varepsilon \to 0. \tag{37}$$

Therefore, the following assertion holds.

Theorem 1. *Let $a > 1$, $b > 0$, and m, k be positive integers related by inequalities (17). Then there is a sufficiently small $\varepsilon_0 > 0$ such that, for all $0 < \varepsilon \le \varepsilon_0$, system (8) has a cycle (traveling wave)*

$$C_k : \quad x_j = \widehat{x}_{(k)}(t + (j-1)\widehat{\Delta}_{(k)}(\varepsilon), \varepsilon), \quad j = 1, 2, \ldots, m, \tag{38}$$

where $\widehat{x}_{(k)}(t,\varepsilon) = x_*(t,\varepsilon,\Delta)|_{\Delta=\widehat{\Delta}_{(k)}(\varepsilon)}$, and $\widehat{\Delta}_{(k)}(\varepsilon)$ is root (37) of Eq. (36).

Let us analyze the stability of cycle (38).

Theorem 2. *Cycle (38) is exponentially orbitally stable if $k = (m-1)/2$ and quasi-stable otherwise.*

Proof sketch. For cycle (38), we consider a variational system similar to (11), namely,

$$\dot{g}_j = \widehat{A}_*(t + (j-1)\Delta, \varepsilon) g_j + \widehat{B}_*(t + (j-1)\Delta, \varepsilon) g_{j-1}, \quad j = 1, 2, \ldots, m, \quad g_0 = g_m, \tag{39}$$

where $\Delta = \widehat{\Delta}_{(k)}(\varepsilon)$, and $\widehat{A}_*(t,\varepsilon), \widehat{B}_*(t,\varepsilon)$ are coefficients (12) with $x(t,\varepsilon) = \widehat{x}_{(k)}(t,\varepsilon)$ and $\Delta = \widehat{\Delta}_{(k)}(\varepsilon)$. Let $\mathscr{V}(\varepsilon)$ denote the shift operator along the solutions of this system over the time from $t = -\sigma_0$ to $t = m\Delta/k - \sigma_0$. Obviously, the stability properties of cycle (38) can be determined via the asymptotic computation of the spectrum of $\mathscr{V}(\varepsilon)$. The corresponding analysis is based on Lemma 1, according to which any eigenvalue ν of this operator is given by the equality $\nu = \widehat{\nu}_{l_0}(\varkappa_0, \varepsilon)$, where $\widehat{\nu}_l(\varkappa, \varepsilon)$ are the multipliers of the auxiliary equation (25) for $\Delta = \widehat{\Delta}_{(k)}(\varepsilon)$ and $\varkappa_0$ is a nonzero root of the equation

$$[\widehat{\nu}_l(\varkappa, \varepsilon)]^k = \varkappa^m, \quad l \in \mathbb{N} \tag{40}$$

for $l = l_0$. Thus, the proof of Theorem 2 is reduced to analyzing of the roots of Eqs. (40).

First, we determine the values of $\varkappa$ for which it Eqs. (40) make sense. Specifically, we show that these equations have no roots in the set

$$\{\varkappa \in \mathbb{C} : \ |\varkappa| > R\} \tag{41}$$

for sufficiently large fixed $R > 0$. For this purpose, we need the following result.

Lemma 5. *It is true that*

$$||\mathscr{V}(\varepsilon)||_{\mathbb{R}^m \to \mathbb{R}^m} \le M\varepsilon^{-n_0(m-1)}, \quad M = \text{const} > 0, \tag{42}$$

where n_0 is a positive integer given by the equality (similar to (28))

$$n_0 = \begin{cases} m/k & \text{if } m/k \text{ is an integer,} \\ \lfloor m/k \rfloor + 1 & \text{otherwise.} \end{cases} \tag{43}$$

The proof of Lemma 5 is carried out similarly to Lemma 2.

The values of the parameter $\varkappa$ can be tentatively localized by applying this lemma. Specifically, combining estimate (42) with relations (15) and (16), which hold for any multiplier of system (39), we conclude that

$$|\nu| = |\widehat{\nu}_{l_0}(\varkappa_0, \varepsilon)| = |\varkappa_0|^{m/k} \le ||\mathscr{V}(\varepsilon)||_{\mathbb{R}^m \to \mathbb{R}^m} \le M\varepsilon^{-n_0(m-1)}.$$

IOP Conf. Series: Journal of Physics: Conf. Series **1163** (2019) 012070 doi:10.1088/1742-6596/1163/1/012070

Thus, all possible roots $\varkappa_0$ of Eqs. (40) lies to the disk

$$\left\{\varkappa \in \mathbb{C}: \ |\varkappa| \leq M^{k/m}\varepsilon^{-\alpha_0}\right\}, \tag{44}$$

where $\alpha_0 = n_0(m-1)k/m$ and M is the constant from(42).

To reduce the set (44) of admissible values of $\varkappa$, we consider the operator obtained from (32) at $\Delta = \widehat{\Delta}_{(k)}(\varepsilon)$. Note that relations (27) – (29) and estimates (30) imply the representation

$$\widehat{V}_*(\varkappa,\varepsilon) = \sum_{n=0}^{n_0} \widehat{V}_{*,n}(\varepsilon)\varkappa^n, \tag{45}$$

where $\widehat{V}_{*,n}(\varepsilon) : E \to E$ are bounded linear operators satisfying the estimates

$$\begin{aligned} ||\widehat{V}_{*,n}(\varepsilon)||_{E\to E} \leq M_n, \quad M_n = \text{const} > 0, \quad n = 0,1,2, \\ ||\widehat{V}_{*,n}(\varepsilon)||_{E\to E} \leq \exp\left(-q/\varepsilon\right), \quad n = 3,\dots,n_0. \end{aligned} \tag{46}$$

It follows from (45) and (46) that

$$\sup_{l\geq 1} |\widehat{\nu}_l(\varkappa,\varepsilon)| \leq \sum_{n=0}^{n_0} |\varkappa|^n ||\widehat{V}_{*,n}(\varepsilon)||_{E\to E} \leq M_0 + M_1|\varkappa| + M_2|\varkappa|^2 + \exp\left(-q/\varepsilon\right)\sum_{n=3}^{n_0}|\varkappa|^n. \tag{47}$$

Finally, it should be noted that the inequality

$$M_0 + M_1|\varkappa| + M_2|\varkappa|^2 + \exp\left(-\frac{q}{\varepsilon}\right)\sum_{n=3}^{n_0}|\varkappa|^n < |\varkappa|^{m/k}. \tag{48}$$

holds on the subset $\{\varkappa \in \mathbb{C}: \ R \leq |\varkappa| \leq M^{k/m}/\varepsilon^{\alpha_0}\}$ of set (44) for sufficiently large fixed $R > 0$ by virtue of the condition $m/k > 2$ (see (2.10)).

Estimates (47) and (48) imply that, with a suitable choice of $R > 0$, Eqs. (40) have no roots in set (41). In what follows, we assume that the constant R is suitably chosen.

The above analysis implies that the consideration of Eqs. (40) can be restricted to the values of $\varkappa$ from the set $\Lambda_{\delta,R}$, where the constant δ is determined by R according to Lemma 4 at $\Delta = \widehat{\Delta}_{(k)}(\varepsilon)$. Note also that, by virtue of (34), Eq. (40) with $l = 1$ has exactly $m-2k$ simple roots in the indicated set. These roots include the unit one, since, for $\varkappa = 1$ and $\Delta = \widehat{\Delta}_{(k)}(\varepsilon)$, Eq. (25) is the linearization of Eq. (10) around the cycle $x = \widehat{x}_{(k)}(t,\varepsilon)$ and, hence, admits a unit multiplier. The other $m-2k-1$ roots of the equation have, as $\varepsilon \to 0$, the asymptotic representation

$$\varkappa_s = \theta_s + O\left(\exp\left(-q/\varepsilon\right)\right), \quad \theta_s = \exp\left(i2\pi s(m-2k)^{-1}\right), \quad s = 1,2,\dots,m-2k-1. \tag{49}$$

Summarizing, we note that all multipliers $\nu \in \mathbb{C}$ of cycle (38), except for the simple unit one, are divided into two groups. Indeed, according to (15), (16), (34), and (49), there is a group of so-called critical multipliers that are exponentially close to the unit circle. More precisely, as $\varepsilon \to 0$, they satisfy the asymptotic equalities

$$\nu_s = \theta_s^2 + O\left(\exp\left(-q/\varepsilon\right)\right), \quad s = 1,2,\dots,m-2k-1. \tag{50}$$

The other $2k$ multipliers correspond to the roots of Eqs. (40) lying in the disk $\{\varkappa \in \mathbb{C} : |\varkappa| \leq \exp(-\delta_0/\varepsilon)\}$, where $\delta_0 = \text{const} > 0$. Moreover, since $\nu = \varkappa^{m/k}$, these multipliers are exponentially small in absolute value.

It remains to note that, in the case $k = (m-1)/2$, the group of critical multipliers (50) is empty; therefore, cycle (38) is exponentially orbitally stable. In the case $k \neq (m-1)/2$, this cycle is quasi-stable. Theorem 5.2 is proved.

IOP Conf. Series: Journal of Physics: Conf. Series **1163** (2019) 012070 doi:10.1088/1742-6596/1163/1/012070

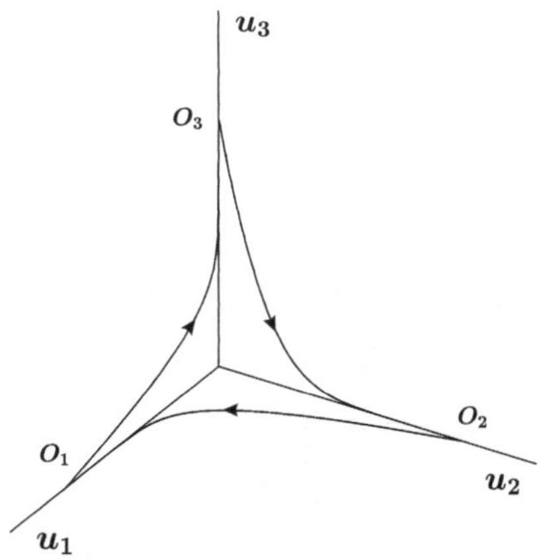

Figure 2. Stable homoclinic triangle of the system (1) for $m = 3$, $r = 1$, $a = 10$, $\alpha = 0$

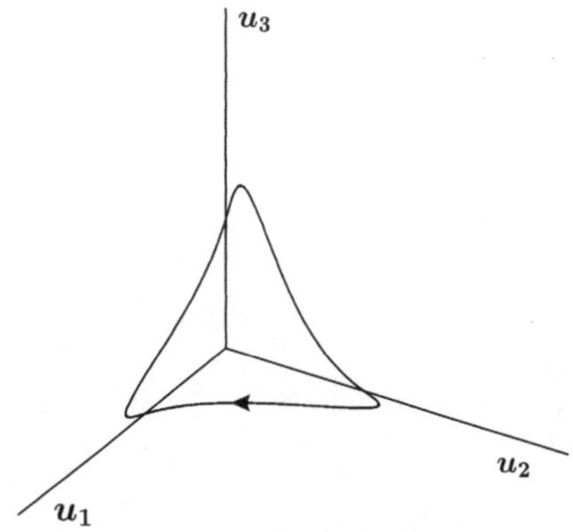

Figure 3. Stable cycle of the system (1) for $m = 3$, $r = 1$, $a = 10$, $\alpha = 0.01$

Note that, for fixed $a > 1$ as $m \to +\infty$, the number of indices k satisfying conditions (17) increases indefinitely. From this and Theorems 1 and 2, it follows that the number of coexisting traveling waves (38) also increases indefinitely as $\varepsilon \to 0$ and $m \to +\infty$ consistently. However, all of them (except for a single stable periodic motion for $k = (m-1)/2$) are quasi-stable. Thus, a quasi-buffer phenomenon occurs in this case. In contrast to the usual buffer phenomenon, which is associated with the unlimited accumulation of coexisting attractors, in this case, quasi-stable structures are accumulated unlimitedly with growing m.

6. Conclusions

First, we discuss the limits of the applicability of model (5). Note that, in contrast to system (1) we cannot set $\alpha = 0$. This is clear even for $m = 3$. Indeed, in the three-dimensional case, for $a > 2$ and $\alpha = 0$, system (5) has a stable homoclinic triangle formed by the saddles $O_1 = (1, 0, 0)$, $O_2 = (0, 1, 0)$, $O_3 = (0, 0, 1)$ and the corresponding separatrices (see Fig. 2, where this triangle is shown for $m = 3$, $r = 1$, $a = 10$, $\alpha = 0$). It is also clear that this steady state is not biologically reasonable, since it corresponds to the extinction of one of the genes. For $\alpha > 0$, the stable homoclinic triangle transforms into a stable cycle lying in the cone $\mathbb{R}^3_+ = \{(u_1, u_2, u_3) : u_j > 0, \ j = 1, 2, 3\}$. In other words, the self-excited oscillations are properly regularized. For $m = 3$, $r = 1$, $a = 10$, and $\alpha = 0.01$, the above cycle has the form shown in Fig. 3. Thus, the case of (7) corresponds to the limit of applicability of model (5). That is why it exhibits a quasi-buffer phenomenon. Due to this phenomenon, under conditions (7), the phase point of system (7) can stay in a neighborhood of the quasi-stable cycle (6) over a time interval on the order of $\exp(cr)$, $c = \text{const} > 0$. Thus, we deal with an effect similar to the well-known Arnold diffusion.

Note that quasi-stable regimes are biologically implementable for two reasons. First, although the time of their existence is finite, it can be comparable with the lifetime of the system. Second, from a biological point of view, it is the small values of α that are of greatest interest.

It should be noted that, for $\alpha \sim 1$, quasi-stable structures collapse. A numerical analysis shows that, in this case for odd m, the only attractor of system (5) is a traveling-wave cycle, which is the continuation of cycle (38) with respect to the parameter α for $k = (m-1)/2$.

For $m = 9$, $r = 1$, $a = 10$, and $\alpha = 0.01$, the component $u = u_1(t)$ of this cycle in the plane (t, u) is plotted in Fig. 4. In the case $m = 2m_0$, where $m_0 \geq 1$, system (5) represents a genetic trigger [16]: for $a > 1$, and $\widetilde{\alpha} < (a-1)/4$, where $\widetilde{\alpha} = \alpha/r$, it admits two stable equilibria

$$O_1 = (u_1^0,\ u_2^0,\ u_1^0,\ u_2^0,\ \ldots,\ u_1^0,\ u_2^0), \quad O_2 = (u_2^0,\ u_1^0,\ u_2^0,\ u_1^0,\ \ldots,\ u_2^0,\ u_1^0)$$

with components

$$u_1^0 = \frac{\widetilde{\alpha}}{(a-1)u_2^0}, \quad u_2^0 = \frac{1}{2}\left(1 + \sqrt{1 - \frac{4\widetilde{\alpha}}{a-1}}\right).$$

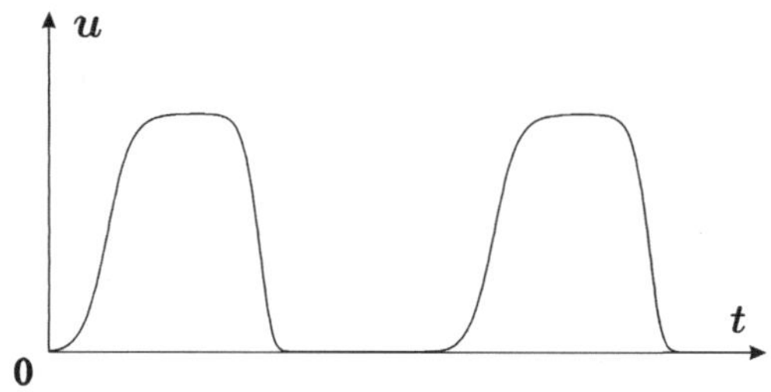

Figure 4. The component $u = u_1(t)$ of traveling-wave cycle for $m = 9$, $r = 1$, $a = 10$, $\alpha = 0.01$

Acknowledgments
This work is supported by the Russian Science Foundation (project no. 14-21-00158).

References

[1] Elowitz M B and Leibler S 2000 A synthetic oscillatory network of transcriptional regulators *Nature* **403** 335 - 338

[2] Tikhonov A N 1952 Systems of differential equations containing small parameters in the derivatives *Mat. Sb.* **31**(3) 575 - 586

[3] Volokitin E P 2004 On limit cycles in the simplest model of a hypothetical genetic network *Sib. Zh. Ind. Mat.* **7**(3) 57 - 65

[4] Buse O, Kuznetsov A and Pérez R A 2009 Existence of limit cycles in the repressilator equations *Int. J. Bifurcation Chaos* **19**(12) 4097 - 4106

[5] Buse O, Pérez A, and Kuznetsov A 2010 Dynamical properties of the repressilator model *Phys. Rev. E* **81**(066206) 066206-1066206-7

[6] Likhoshvai V A, Matushkin Yu G, and Fadeev S I 2003 Problems in the theory of the functioning genetic networks *Sib. Zh. Ind. Mat.* **6**(2) 64 - 80

[7] Likhoshvai V A, Demidenko G V, Fadeev S I, Matushkin Yu G, and Kolchanov N A 2004 Mathematical simulation of regulatory circuits of gene networks *Comput. Math. Math. Phys.* **44**(12) 2166 - 2183

[8] Fadeev S I and Likhoshvai V A 2003 On hypothetic genetic networks *Sib. Zh. Ind. Mat.* **6**(3) 134 - 153

[9] Kolesov A Yu, Rozov N Kh and Sadovnichii V A 2016 Periodic solutions of travelling-wave type in circular gene networks *Izv. Math.* **80** 523 - 548

[10] Glyzin S D, Kolesov A Yu, and Rozov N Kh 2017 Existence and stability of the relaxation cycle in a mathematical repressilator model *Math. Notes* **101** 71 - 86

[11] Kolesov A Yu, and Kolesov Yu S 1997 Relaxation Oscillations in Mathematical Models of Ecology (Am. Math. Soc., Providence, 1997).

[12] Glyzin S D, Kolesov A Yu, and Rozov N Kh 2013 Relaxation self-oscillations in Hopfield networks with delay *Izv. Math.* **77** 271 - 312

[13] Glyzin S D, Kolesov A Yu, and Rozov N Kh 2013 Periodic traveling-wave-type solutions in circular chains of unidirectionally coupled equations *Theor. Math. Phys.* **175** 499 - 517

[14] Glyzin S D, Kolesov A Yu, and Rozov N Kh 2014 The buffer phenomenon in ring-like chains of unidirectionally connected generators *Izv. Math.* **78** 708 - 743

[15] Glyzin S D, Kolesov A Yu, and Rozov N Kh 2018 Quasi-Stable Structures in Circular Gene Networks *Comput. Math. and Math. Phys.* **58** 659 - 679

[16] Romanovskii Yu M, Stepanova N V, and Chernavskii D S 2003 Mathematical Modeling in Biophysics (Inst. Kompyut. Issled., Moscow, 2003) [in Russian].

IOP Conf. Series: Journal of Physics: Conf. Series **1163** (2019) 012071 doi:10.1088/1742-6596/1163/1/012071

The application of the CLOPE algorithm for clustering a set of the network packets

A V Belov [1] T D Novikova [2]

1. School of The Applied Mathematics of the National Research University Higher School of Economics, Myasnitskaya ulitsa 20, 101000, Russian Federation
2. Department of Systems and Software Engineering, NUST MISIS, Moscow, Leninskiy prospekt 4 Moscow, 119049, Russian Federation

Abstract - For the modern software that uses network communication protocols the problem of the ensuring reliability is acute. To solve such an important problem the stress testing is used. This type of the testing involves the generation of a large number of the test data, including sets of network packets. Reducing the stored data after the testing process is the main task. This task can be solved by clustering the set of received packets. To solve this problem it is proposed to use the clustering algorithm for categorical data of CLOPE. This algorithm allows to cluster datasets without the information about the source clusters. It has the low computational complexity and ease of implementation. The article describes the preparation and results of sets of the network packets processing experiments. The article shows that the CLOPE algorithm can be effectively used for the clustering network packets received during stress testing. The results of the research extend the toolkit for the SW stress testing process.

1. Introduction

For testing software which used network messaging one of the most popular methods is the stress testing. Tests are developed using the black box method. Test scenarios are selected according to the functional profile and usually using a pseudo-random selection algorithm [1]. This algorithm can be used for the testing non-functional attributes such as SW reliability and SW performance.

However the negative consequence of using the random testing is that the obtained test results, and the program execution paths do not show the reasons why they are performed and do not allow us to understand which parts of the input data are the cause of this execution.

Another important feature is a large amount of data sets generated during testing. This feature of testing leads to the "path explosion problem" [2] when there are many semantically similar traces that can not be resolved to one branch of the SW application code.

Thus, it becomes necessary to select groups depending on certain signs in the set of paths obtained which will reduce the number of data stored in the test results and highlight those parts of the data, changing which changes the behavior of the application. In the case of using selective testing in applications that use network interaction, it becomes expedient to replace clustering with a set of traces to clustering a set of packages, each of which would in fact be an unique identifier of the path.

2. Choosing an approach to the clustering

To improve the efficiency of the testing process, it is proposed to use the network packet clustering. It is required to construct an algorithm $\alpha: X \rightarrow Y$ where X – the test network packets and Y – a set of clusters (initially nothing is known about Y, even the number of clusters can be unknown) [3]. Each cluster consists of close packages and packages from the different clusters are the significantly different.

IOP Conf. Series: Journal of Physics: Conf. Series **1163** (2019) 012071 doi:10.1088/1742-6596/1163/1/012071

The clustering algorithm will allow:

- Identify the structure of the unknown network protocol for researcher.
- Evaluate the test coverage of the software being investigated using coverage metrics based on the input data and metrics based on the control flow, since the input network packets may contain data used in the program environment.
- Map selected clusters of the packets that lead to a failures with the types of errors.
- Select packet masks that lead to errors.

When an approach to clustering was choosing it was necessary to specify a measure of the proximity of objects. For string sequences the Hamming metric is most often used:

$$d_{ij} = \sum_{k=1}^{p} |x_{ik} - x_{jk}| \quad (1)$$

Where is d_{ij} – distance between two objects,

x_{ik} – value of k-coordinate i-object,

x_{jk}– value of k-coordinate j-object.

To solve the problem of the clustering a set of network packets it is necessary that the final form of clusters has to be an arbitrary form. Since the estimated number of clusters at the initial stage of the study is not known it is necessary to use an algorithm that does not use this input parameter.

To solve this problem is proposed the CLOPE Algorithm (Clustering with sLOPE), a non-hierarchical iterative cluster analysis method for the processing the large sets of the categorical data.

3. CLOPE algorithm for the stress testing process

The original CLOPE algorithm is a transaction data clustering algorithm (a transaction is some arbitrary set of the objects of finite length). The main idea of this method is to use a global optimization criterion based on the maximization of the cost function as applied to clustering problems [4].

During the execution of the algorithm, a relatively the small amount of information about each cluster is required to be stored in RAM and the minimum number of passes is made on the data set. When one uses the CLOPE algorithm the number of clusters is selected automatically and depends on the repulsion coefficient - the parameter that determines the level of the similarity of transactions within the cluster. The repulsion factor is set by the user: The larger the repulsion coefficient the lower the level of transaction similarity and as a result more clusters will be created.

Transaction database consisting of a set of transactions. Each transaction is a set of objects. A set of clusters is a partition of a set such that each element of C is called a cluster, n is the number of transactions, A is the length of the transaction, k is the number of clusters. For the task of clustering network protocols, A is the maximum packet length.

In this case, the parameter r is set – the repulsion coefficient.

Cj – cluster can be characterized by the following parameters:

D (Cj) – the set of objects in the Cj cluster.

Occ (Cj) – the number of occurrences of the object I in the cluster Cj

IOP Conf. Series: Journal of Physics: Conf. Series **1163** (2019) 012071 doi:10.1088/1742-6596/1163/1/012071

$S(C_j) = \sum_{i \in D(C_j)} Occ(i, C_j) = \sum_{t_i \in C_j} |t_i|$ – is the total number of elements in the cluster Cj.

W (Cj) = | *D (Cj)* | – cluster width.

H(Cj) = *S(Cj)/W(Cj)* – is the height of the cluster.

CLOPE algorithm includes 2 main phases (Fig.1):

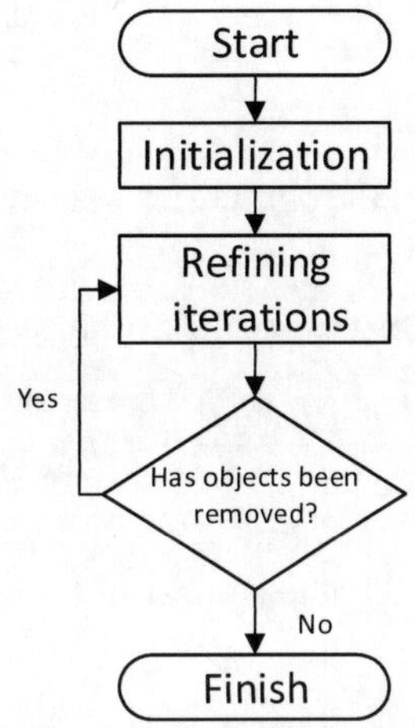

Figure 1. CLOPE flowchart.

At the initialization phase the clustering cost is calculated for the each packet in the set and then the refinement occurs where the packets are moved between the clusters to maximize the clustering cost function – profit.

$$Profit(C) = \frac{\sum_{i=1}^{k} H(C_i) \times |C_i|}{\sum_{i=1}^{k} |C_i|} == \frac{\sum_{i=1}^{k} \frac{S(C_i)}{W(C_i)^r} \times |C_i|}{\sum_{i=1}^{k} |C_i|} \quad (2)$$

After that refinement iterations take place in order to improve the existing distribution. As part of the refinement iterations the entire set of transactions is re-iterated. For each the cost of deletion from the current cluster is calculated:

$$Cost^{-}(C_i, t_n) = \frac{\left(S(C_i) - length(t_n)\right) * (|C_i| - 1)}{W(C_i/t_n)^r} - \frac{S(C_i) * |C_i|}{W(C_i)^r} \quad (3)$$

And also the cost of adding to the another cluster is calculated:

$$Cost^{+}(C_i, t_n) = \frac{\left(S(C_i) + length(t_n)\right) * (|C_i| + 1)}{W(C_i \cup t_n)^r} - \frac{S(C_i) * |C_i|}{W(C_i)^r} \quad (4)$$

Based on the calculated values it is decided to move the transaction to another cluster or this transaction remains in the current cluster. As with initialization it is done in such a way that the total cost for the deleting from the current cluster and adding to another cluster is the highest (if for all movements the total cost is a negative then the transaction remains in place)

IOP Conf. Series: Journal of Physics: Conf. Series **1163** (2019) 012071 doi:10.1088/1742-6596/1163/1/012071

4. Quality metrics

For the CLOPE algorithm the quality parameters are as follows [2]:

- The average similarity of elements

$$\overline{M_{similarity}} = \frac{1}{E}\sum_{i=0}^{E}\left(\frac{1}{O(C_i)}\sum_{j=1}^{O(C_j)}\frac{|\{L_m|L_m \in D_j, L_m \in C_i\}, m = 1, \dots, W(D_j)|}{W(D_j)}\right) * 100 \quad (5)$$

Where $O(C_i)$ is the number of packets in the cluster, L_m – cluster package, D_j – the first cluster, C_j – second cluster, E - number of clusters.

- The average distance between clusters, as the average value of the " Manhattan" distance

$$\overline{M_{manh}} = \frac{\sum_{i=1}^{E}\sum_{j=1}^{E}|Occ(L_n, C_i') - Occ(L_n, C_j')|}{(E-1)^2}, i \neq j, n = 1, \dots, W(C_i') = 1, \dots, W(C_j') \quad (6)$$

Where C'_j and C'_i are clusters. $Occ\ (L_n, C_j)$ - the number of the occurrences of the object L_n in the cluster C_j.

- The minimum distance between clusters

$$M_{min} = \min\left|Occ(L_n, C_i') - Occ\left(L_n, C_j'\right)\right|, i = 1, \dots, E, j = 1, \dots, E, \\ i \neq j, n = 1, \dots, W(C_i') = 1, \dots, W(C_j') \quad (7)$$

- The maximum distance between clusters

$$M_{max} = \max\left|Occ(L_n, C_i') - Occ\left(L_n, C_j'\right)\right|, i = 1, \dots, E, j = 1, \dots, E, \\ i \neq j, n = 1, \dots, W(C_i') = 1, \dots, W(C_j') \quad (8)$$

- The number of iterations of the algorithm

5. Computational experiments

Computational experiments were conducted on a test bench which is a network server that uses the HTTP protocol for the message transfer. Clustering was performed with the value of the repulsion coefficient r: 2.7. This repulsion coefficient was used for a specific set of data generated using the Peach utility.

Table 1. Example of input data

Cluster	Packet
Cluster 1 Number of packets: 121	POST HTTP/1.1 Host: 127.0.0.1\r\n Content-Type: text/html; charset=utf-8\r\n Content-Length: 30 Number:455 Value1: AAAAAAAAAAAAAAAAAAAAAAAAAAAAAA AaaaaaAAAAAAAAAAAAAAAAAAAA...

IOP Conf. Series: Journal of Physics: Conf. Series **1163** (2019) 012071 doi:10.1088/1742-6596/1163/1/012071

	POST HTTP/1.1 Host: 127.0.0.1\r\n Content-Type: text/html; charset=utf-8\r\n Content-Length: 30 Number:455 Value1: AAAAAAAAAAAaaaaAAAAAAAAAAAAAAAAAAA AaaaaaAAAAAAAAAAAAAAAAAAAAAA…
Cluster 2 Number of packets: 101	POST HTTP/1.1 Host: 127.0.0.1\r\n Content-Type: text/html; charset=utf-8\r\n Content-Length: 8 Number:460 Value1: 01 03 05 07 08 09 00 POST HTTP/1.1 Host: 127.0.0.1\r\n Content-Type: text/html; charset=utf-8\r\n Content-Length: 8 Number:460 Value1: 01 03 05 07 08 09 10 11
Cluster 3 Number of packets: 235	POST HTTP/1.1 Host: 127.0.0.1\r\n Content-Type: text/html; charset=utf-8\r\n Content-Length:50 Number:416 Value1:JJHAJJJJJJJJJJJJJJJJJJJJJJHSBBBBBBBBBBBBBSHTTBBAGGSDN POST HTTP/1.1 Host: 127.0.0.1\r\n Content-Type: text/html; charset=utf-8\r\n Content-Length:50 Number:416 Value1:JJHAJJJJJJJJJJJJJJJJJJJHSBBBBBBBBBBBBBSHTTBBAGGSDNTTT

The main scheme of the computational experiment is on the Figure 2.

IOP Conf. Series: Journal of Physics: Conf. Series **1163** (2019) 012071 doi:10.1088/1742-6596/1163/1/012071

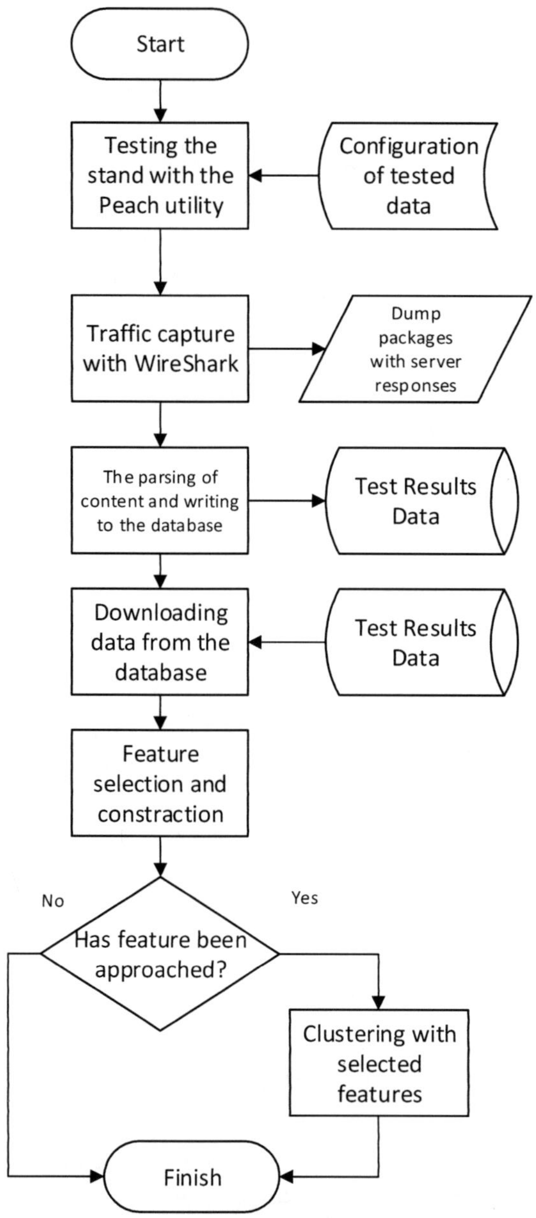

Figure 2. The computational experiment flowchart

The results of the computational experiment are presented in the Table 2.

Table 2. Test results on a test bench

Parametres	Value
The average similarity of the elements	62.33%
Average distance between clusters	74.35%

Minimum distance between clusters	27%
Maximum distance between clusters	95%
Number of iterations of the algorithm	17

If we draw conclusions from the obtained results we can see that in the clustering kit generated on the test bench the generated packets are similar to each other (the average similarity is more than 50%). The obtained similarity can be explained by the nature of the generated data since the transfer via the HTTP protocol implies the preservation of identical field names in the packet structure.

6. Conclusion

According to the results of successful testing of the clustering algorithm it was noted that the existing solution may improve the implementation of the mechanism for selecting the optimal parameters of the algorithm. Clusters have been identified that reflect the behavior of the software under investigation. This model allows clustering data only after performing the stress testing and isolating the types of software behavior of the interest to the researcher based on the data of the tested network packets. However, in the further investigations, it is possible to improve the result obtained by solving the problem of the selecting the parameters of the algorithm for each set of the packages individually.

Using the CLOPE algorithm for the clustering network packets will achieve the following goals:

- To identify the structure of the network protocol used in the software.
- To evaluate the software test coverage and to reduce the time for analyzing the test results.
- To reduce the volume of the stored data after testing replacing the set of the elements included in one cluster with one typical representative which will save memory for the database.

7. .References

[1] Richard Hamlet (1994). "Random Testing". In John J. Marciniak. Encyclopedia of Software Engineering (1st ed.). John Wiley and Sons. ISBN 0471540021.
[2] Yan Shoshitaishvili, Ruoyu Wang, Christopher Salls, Nick Stephens, Mario Polino, Andrew Dutcher, John Grosen, Siji Feng, Christophe Hauser, Christopher Kruegel, and Giovanni Vigna. SoK: (State of) The Art of War: Offensive Techniques in Binary Analysis. In IEEE Symposium on Security and Privacy, 2016.
[3] Igor Mandel. Cluster analysis - M: Finance and Statistics– 1988. – 176 p.
[4] Katerina Novokhatska, Oleksii Kungurtsev Application of clustering algorithm CLOPE to the query grouping problem in the field of materialized view maintenance // J. Comput. Inf. Technol. 2016. Vol. 24, № 1. P. 79–89.

IOP Conf. Series: Journal of Physics: Conf. Series **1163** (2019) 012072 doi:10.1088/1742-6596/1163/1/012072

Dissipative Dynamics of a Single Polymer in Solution: A Lowe-Andersen Approach

Suman Majumder, Henrik Christiansen and Wolfhard Janke
Institut für Theoretische Physik, Universität Leipzig, Postfach 100 920, 04009 Leipzig, Germany

E-mail: suman.majumder@itp.uni-leipzig.de
henrik.christiansen@itp.uni-leipzig.de
wolfhard.janke@itp.uni-leipzig.de

Abstract. We study the equilibrium dynamics of a single polymer chain under good solvent condition. Special emphasis is laid on varying the drag force experienced by the chain while it moves. To this end we model the solvent in a mesoscopic manner by employing the Lowe-Andersen approach of dissipative particle dynamics which is known to reproduce hydrodynamic effects. Our approach captures the correct static behavior in equilibrium. Regarding the dynamics, we investigate the scaling of the self-diffusion coefficient D with respect to the length of the polymer N, yielding results that are compatible with the Zimm scaling $D \sim N^{-3/5}$.

1. Introduction

Dynamics of a polymer chain in a dilute solution, although being extensively studied, is still a topic of utmost importance [1, 2]. In particular, this topic serves as a benchmark for establishing a coarse-grained or mesoscopic approach to understand more realistic problems on larger time and length scales. The dynamics of a single chain, generally, is characterized by the self-diffusion coefficient D which scales with chain length N as

$$D \sim N^{-x}. \quad (1)$$

In the free-draining limit where hydrodynamic effects are absent or screened, one has $x = 1$, whereas in the non-draining limit where hydrodynamic interactions are important, one expects $x = 3/5$. The former is referred to as Rouse scaling [3] and the latter as Zimm scaling [4].

Over the years these results have been verified both experimentally [5] as well in computer simulations [6]. Before the advent of present-day sophisticated experimental setups, when it was difficult to keep track of a single polymer movement, numerical simulations were considered to be the only way to verify the available theoretical understanding. Monte Carlo simulations cannot capture the real dynamics due to the absence of hydrodynamic effects. Even when doing Molecular Dynamics (MD) simulations, one has to be careful about the choice of the thermostat. The effect of hydrodynamics is achieved via the preservation of local linear and angular momentum during the entire simulation.

In the context of polymer dynamics another major issue is the consideration of explicit solvents. The introduction of dissipative particle dynamics (DPD) has eased this task [7]. There one has the luxury of considering the solvent explicitly via a mesoscopic approach which allows to access larger time and length scales [8, 9]. In addition, DPD also allows to tune the Schmidt number, i.e., the ratio of the kinematic viscosity and the self-diffusion coefficient. This makes consideration of solvents having viscosities comparable to real fluids quite plausible. However, from a technical point of view, one has to be cautious when integrating the equations of motion using a Verlet-type algorithm in the DPD formalism as it often disturbs the detailed balance and thereby does not produce the correct equilibrium properties unless sufficiently small time steps are used [10, 11]. There have been few attempts to use more advanced integration schemes that can overcome this difficulty. In this context, Lowe instead of aiming at improving the integration scheme, modified the DPD approach in the spirit of the Andersen thermostat, of course, with the effect of hydrodynamic interactions being intact [12]. With Lowe's approach, which also goes by the name Lowe-Andersen (LA) thermostat, one is not only allowed to use relatively larger time steps [13] but also has the option to tune the dissipation of velocities of particles, thereby gaining access to fluids with Schmidt numbers as high as observed in real fluids [12]. Motivated by these advantages, here, we construct a model polymer with explicit solvent particles and perform MD simulations for a wide range of effective solvent viscosities, i.e., frictional-drag experienced by the particles. Our results show that the dynamics in good solvent produces the theoretically expected Zimm scaling valid in the presence of hydrodynamics.

The paper is organized as follows. In the next section we present the details of our model and the method of simulation. Following that, we present results concerning static and dynamic properties of our model. In the final section we present a summary of our results and an outlook to future work.

2. Model and Method

We consider a bead-spring model of a flexible homopolymer in three spatial dimensions. The bonds between successive monomers are maintained via the standard finitely extensible non-linear elastic (FENE) potential

$$E_{\text{FENE}}(r_{ii+1}) = -\frac{K}{2}R^2 \ln\left[1 - \left(\frac{r_{ii+1} - r_0}{R}\right)^2\right], \tag{2}$$

with $K = 40$, $r_0 = 0.7$ and $R = 0.3$. The monomers and the solvent molecules both are considered to be spherical beads of mass $m = 1$ and diameter σ. All nonbonded interactions, i.e., solvent-solvent, solvent-monomer, and monomer-monomer interactions are modeled by

$$E_{\text{nb}}(r_{ij}) = E_{\text{LJ}}\left[\min(r_{ij}, r_c)\right] - E_{\text{LJ}}(r_c), \tag{3}$$

where $E_{\text{LJ}}(r)$ is the standard Lennard-Jones (LJ) potential given as

$$E_{\text{LJ}}(r) = 4\epsilon\left[\left(\frac{\sigma}{r}\right)^{12} - \left(\frac{\sigma}{r}\right)^6\right] \tag{4}$$

with $\sigma = r_0/2^{1/6}$ as the diameter of the beads, $\epsilon(=1)$ as the interaction strength and $r_c = 2^{1/6}\sigma$ as cut-off radius that ensures a purely repulsive interaction.

As already mentioned, we simulate our system via MD simulations at constant temperature using the LA thermostat. In this approach, one updates the position $\vec{r}_i$ and velocity $\vec{v}_i$ of the i-th bead using Newton's equations as follows,

$$\frac{d\vec{r}_i}{dt} = \vec{v}_i, \frac{d\vec{v}_i}{dt} = \vec{f}_i, \tag{5}$$

where $\vec{f}_i$ is the conservative force (originating from the bonded and nonbonded interactions) acting on the particle. This part of the simulation is the usual microcanonical MD. For

IOP Conf. Series: Journal of Physics: Conf. Series **1163** (2019) 012072 doi:10.1088/1742-6596/1163/1/012072

controlling the temperature with the LA thermostat, one considers a pair of particles within a certain distance R_T. Then, with a probability $\Delta t\Gamma$, a bath collision is executed following which the pair gets a new relative velocity from the Maxwellian distribution. Here, Δt is the width of the time step chosen for the updates in Eq. (5) and Γ determines the collision frequency. The exchange of relative velocities with the bath is only done on its component parallel to the line joining the centers of the pair of particles, thus conserving the angular momentum. Additionally, the new velocities are distributed to the chosen pair in such a way that the linear momentum is also conserved. In summary, the work flow for our LA approach has the following form:

(i) Update $\vec{v}_i \leftarrow \vec{v}_i + \frac{1}{2m}\vec{f}_i\Delta t$ and $\vec{r}_i \leftarrow \vec{r}_i + \vec{v}_i\Delta t$.

(ii) Calculate $\vec{f}_i$ and then $\vec{v}_i \leftarrow \vec{v}_i + \frac{1}{2m}\vec{f}_i\Delta t$.

(iii) Choose all pairs of particles with $r_{ij} < R_T$ and with probability $\Delta t\Gamma$ do the following:

- (a) Draw $\vec{v\prime}_{ij}\cdot\vec{n}_{ij}$ from the distribution $\xi_{ij}\sqrt{2k_BT/m}$ where $\vec{v\prime}_{ij}$ is the new relative velocity of particles i and j, $\vec{n}_{ij}$ is a unit vector, ξ_{ij} is a Gaussian white noise, and T is the desired temperature.
- (b) Calculate the change $2\vec{\Delta}_{ij} = \vec{n}_{ij}(\vec{v\prime}_{ij} - \vec{v}_{ij})\cdot\vec{n}_{ij}$.
- (c) Distribute the change as $\vec{v}_i \leftarrow \vec{v}_i + \vec{\Delta}_{ij}$ and $\vec{v}_j \leftarrow \vec{v}_j - \vec{\Delta}_{ij}$.

(iv) Calculate physical quantities and go to (i).

It is to be noted that the LA approach is an alternative approach to DPD, however, with the liberty to use large Δt. The other advantage of the method is that by varying R_T and Γ one can tune the bath collision frequency, i.e., effectively controlling the frictional drag or in other word the solvent viscosity. In this work, we restrict ourselves to the case where $R_T = r_c$ and vary Γ within the range $[0.1, 250]$ with the goal to cover solvents with diverse viscosity. We do our simulations using LAMMPS [14] which we modified to implement the LA thermostat.

We first generate a random walk of length N on a simple-cubic lattice and then put this walk or chain in a box of size $L = 1.25N^{3/5}$. Subsequently, we insert solvent particles keeping the density fixed to $\rho = 0.7$ and make sure that the solvent particles do not overlap with the monomers of the polymer chain. Then we run our MD simulation with LA thermostat at temperature $T = 1.0$ for 10^7 MD steps with $\Delta t = 0.005$ and allow the system to equilibrate. In our simulations, the unit of temperature is ϵ/k_B (where we chose $k_B = 1$) and the unit of time is the standard LJ time unit $\tau = (m\sigma^2/\epsilon)^{1/2}$. Once the system is equilibrated, we let it run for another period of $10^4\tau$ and simultaneously start measuring various physical quantities that will be presented subsequently. We have used polymers of chain length $N \in [16, 512]$. All results presented are averaged over 100 different independent runs for $N < 512$ and 50 runs for $N = 512$.

3. Results

As a first step to benchmark our proposed framework we calculate the radius of gyration as

$$R_g = \sqrt{\sum_{i,j}\frac{(\vec{r}_i - \vec{r}_j)^2}{2N^2}}, \tag{6}$$

which is a measure for the size of the polymer. Under good solvent condition, this scales with the chain length N as $R_g \sim N^\nu$ where the critical exponent $\nu \approx 3/5$. In Fig. 1(a) we show the plots of R_g as function of N for three different Γ values of the solvent. A fitting of the data using the form $R_g = R_0N^\nu$ provides $\nu \in [0.58, 0.61]$. Fixing $\nu = 3/5$ in the fitting, also works quite well as shown by the continuous lines in Fig. 1(a). This implies that our framework reproduces the correct equilibrium static behavior of a polymer in good solvent, irrespective of the value of Γ.

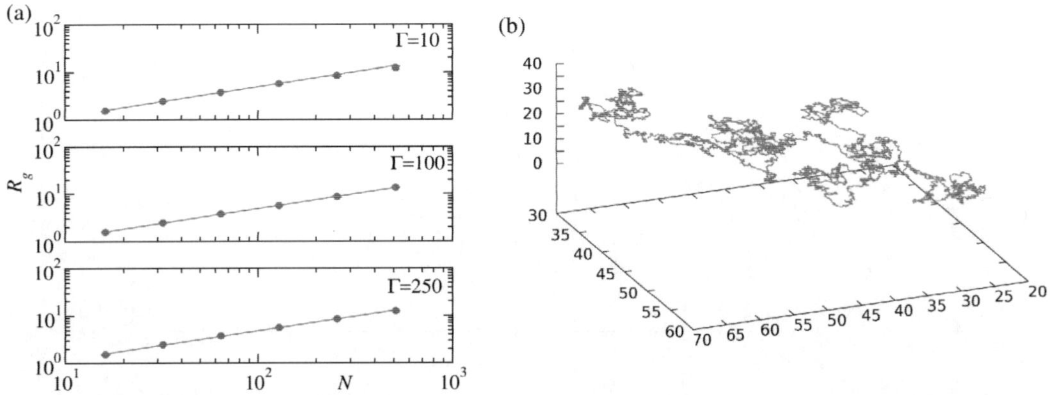

Figure 1. (a) Radius of gyration R_g as function of the chain length N in good solvents for three different Γ. The solid line in each case is a fit using the form $R_g = R_0 N^{3/5}$, the expected scaling behavior. (b) The trajectory of the center of mass of a polymer of length $N = 512$, over a time period of 200τ in equilibrium at $T = 1.0$.

Next we move to the dynamic properties. In Fig. 1(b), we show the equilibrium trajectory of the center of mass (cm) of a polymer of length $N = 512$ over a time interval of 200τ. Tracking the motion of the cm of a polymer is crucial when one wants to measure its diffusion in the solvent. The trajectory seems to be stochastic in nature and hence is indicative of a Brownian motion. To check the nature of the motion we calculate the mean squared displacement of the cm of a polymer given as

$$MSD = \langle [\vec{R}_{\text{cm}}(t) - \vec{R}_{\text{cm}}(t_0)]^2 \rangle, \tag{7}$$

where $\vec{R}_{\text{cm}}(t)$ is the position vector of the cm of the polymer at time t, and t_0 is the time when the measurement starts. From Einstein's equation it is known that for Brownian motion in the long-time limit [1, 2]

$$MSD = A + 6Dt, \tag{8}$$

where D is the self-diffusion coefficient of the polymer and A is a constant. In Fig. 2(a) we show the plots of MSD as function of time t for different chain lengths N in a solvent with $\Gamma = 10$. In the large t limit, for all N the data is consistent with $\sim t$ behavior, whereas at early times for a brief period it is expectedly ballistic in nature, i.e., $\sim t^2$. A similar behavior is observed in Fig. 2(b) where we show plots of MSD for polymers having fixed chain length $N = 128$ in solvents having different Γ.

To have a more comprehensive understanding of the effect of variation of the solvent, i.e., Γ, we aim to calculate the self-diffusion coefficient D of the polymer. In this regard, one can calculate the velocity (of the cm of the polymer) autocorrelation function $C(t) = \langle \vec{v}(t_0) \cdot \vec{v}(t) \rangle$ which is related to D via the Green-Kubo relation as

$$D = \frac{1}{d} \int_{t_0}^{t} C(t) dt, \tag{9}$$

where d is the spatial dimension. However, here, we calculate D using the Einstein relation (8) as follows. We pick two times t_1 and t_2 in the long-time limit. From Fig. 2, one can easily observe that for $t > 100\tau$ for all N and Γ, the mean squared displacement MSD is consistent with the linear behavior. Thus, we chose t_1 and t_2 to be such that $t_2 > t_1$ and $t_1 \geq 100\tau$. Then

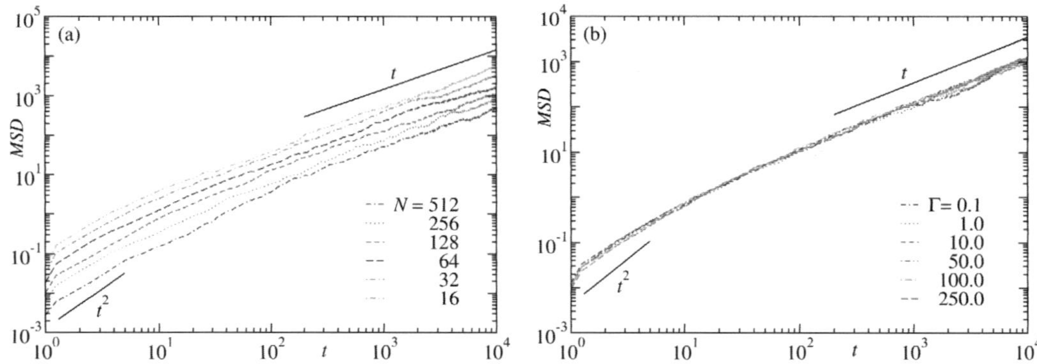

Figure 2. (a) Time dependence of mean squared displacement MSD of the center of mass of a polymer for different chain lengths N in a solvent having $\Gamma = 10$ at $T = 1.0$. The solid line at small t corresponds to ballistic motion $\sim t^2$ and the one at large t corresponds to Brownian motion $\sim t$. (b) Same as (a) but for a polymer of fixed length $N = 128$ in solvents having different Γ. The solid lines there have the same meaning as in (a).

using (8) one can write down

$$D = \frac{1}{6}\left[\frac{MSD(t_2) - MSD(t_1)}{t_2 - t_1}\right]. \tag{10}$$

Equation (10) provides a set of values of D for different choices of the pair (t_1, t_2). This allows one to have appropriate error bars, independent of the usual fitting exercise using the form (8).

Figure 3(a) demonstrates the scaling of the self-diffusion coefficient D with chain length N for different solvents as indicated by the Γ values therein. The dashed line there corresponds to the Rouse scaling with the exponent $x = 1$ in the form (1) whereas the solid line represents the Zimm scaling with $x = 3/5$. Consistency of our data with the solid line indicates the validity of the Zimm scaling, expected in the presence of hydrodynamic effects. Fits using the form $D = D_0 N^{-x}$ yield x within $[0.6, 0.68]$, slightly higher than $x = 3/5$ expected for the Zimm scaling. However, a fit using the same form by fixing $x = 3/5$ also yields reasonably acceptable $\chi_r^2 (< 2.5)$ values, where χ_r^2 is the goodness of fit parameter χ^2 divided by the degrees of freedom.

Finally, in Fig. 3(b) we show the dependence of the self-diffusion coefficient D on the collision frequency Γ that controls the effective viscosity of the solvent. The data show no signature of strong dependence, rather in a broader sense D seems to be pretty constant for different Γ. This behavior is similar to the conclusion drawn in Ref. [13] where, for an ideal gas, they did not observe any strong dependence of D on Γ. Nevertheless, the data presented here are for a relatively high temperature $T = 1.0$, hence, the influence of viscosity is not severely pronounced. At low temperatures, the diffusion might get suppressed due to high effective viscosity [15].

4. Conclusion

Motivated by dissipative particle dynamics, here, we have constructed an explicit solvent model for a polymer. Instead of using the standard approach of doing it, we rely on the Lowe-Andersen approach of the bath collision. This allows us to control the drag force applied by the solvent on the polymer, i.e., the effective viscosity. Via the scaling of the radius of gyration R_g with the chain length N as $R_g \sim N^{3/5}$ we confirm that our approach yields the known static critical exponent. The method conserves both the linear and angular momenta locally, thereby preserving the hydrodynamics. The scaling of the self-diffusion coefficient D with chain length N indicates a much faster dynamics than the Rouse dynamics, and in fact is pretty consistent

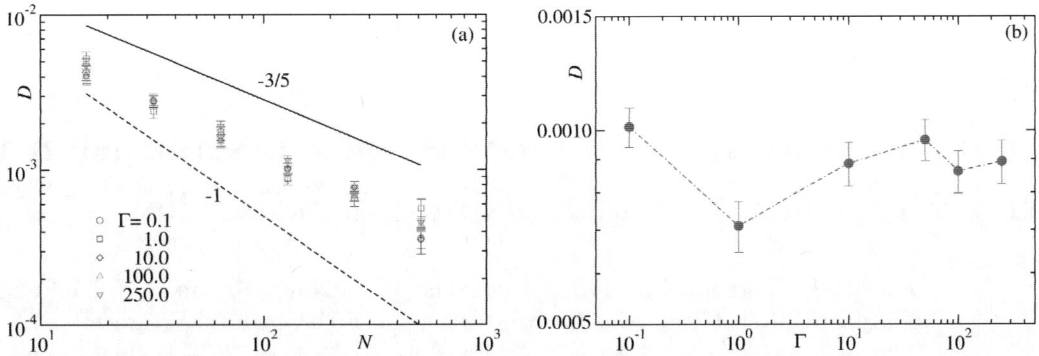

Figure 3. (a) Plots showing the dependence of the self-diffusion coefficient D on the chain length N of a polymer for different solvents with Γ as indicated. The solid line and dashed lines represent respectively the Zimm and Rouse scaling. (b) Variation of D as function of Γ for a polymer of length $N = 128$ at $T = 1.0$.

with Zimm scaling $D \sim N^{-3/5}$ valid in the presence of hydrodynamic effects. The successful application of this method opens up opportunities to explore other aspects of polymer dynamics including nonequilbrium scenarios, e.g., during collapse of a polymer [16, 17, 18, 19]. There, the effect of varying the collision frequency Γ has a stronger influence on the structure formation, and in principle such phenomena could well be tuned by the degree of dissipation [20].

Acknowledgments

The work was funded by the Deutsche Forschungsgemeinschaft (DFG) under Grant Nos. JA 483/33-1 and SFB/TRR 102 (project B04), and further supported by the Deutsch-Französische Hochschule (DFH-UFA) through the Doctoral College "$\mathbb{L}^4$" under Grant No. CDFA-02-07, the EU Marie Curie IRSES network DIONICOS under Contract No. PIRSES-GA-2013-612707, and the Leipzig Graduate School of Natural Sciences "BuildMoNa".

References

[1] de Gennes P-G 1985 *Scaling Concepts in Polymer Physics* (Cornell University Press, Ithaca)
[2] Doi M and Edwards S F 1986 *The Theory of Polymer Dynamics* (Clarendon, Oxford)
[3] Rouse P E 1953 *J. Chem. Phys.* **21**, 1272
[4] Zimm B H 1956 *J. Chem. Phys.* **24**, 269
[5] Smith D E, Perkins T T and Chu S 1996 *Macromol.* **29** 1372
[6] B Dünweg and Kremer K *Phys. Rev. Lett.* **66** 2996
[7] Hoogerbrugge P J and Koelman J M V A 1992 *Europhys. Lett.* **19**, 155
[8] Espanol P and Warren P B 1995 *Europhys. Lett.* **30**, 191
[9] Groot R D and Warren P B 1997 *J. Chem. Phys.* **107**, 4423
[10] Marsh C A and Yeomans J M 1997 *Europhys. Lett.* **37**, 511
[11] Hafskjold B, Liew C C and Shinoda W 2004 *Mol. Simul.* **30**, 879
[12] Lowe C P 1999 *Europhys. Lett.* **47**, 145
[13] Koopman E A and Lowe C P 2006 *J. Chem. Phys.* **124**, 204103
[14] Plimpton S 1995 *J. Comput. Phys.* **117**, 1; `http://lammps.sandia.gov`
[15] Majumder S, Christiansen H and Janke W 2018 work in progress
[16] de Gennes P-G 1985 *J. Phys. (France) Lett.* **46** 639
[17] Majumder S and Janke W 2015 *Europhys. Lett.* **110** 58001
[18] Majumder S, Zierenberg J and Janke W 2017 *Soft Matter* **13** 1276
[19] Christiansen H, Majumder S and Janke W 2017 *J. Chem. Phys.* **147** 094902
[20] Majumder S, Christiansen H and Janke W 2018 in preparation

IOP Conf. Series: Journal of Physics: Conf. Series **1163** (2019) 012073 doi:10.1088/1742-6596/1163/1/012073

2D melting and motility induced phase separation in Active Brownian Hard Disks and Dumbbells

Pasquale Digregorio[1], Demian Levis[2,3], Antonio Suma[4,5], Leticia F. Cugliandolo[6], Giuseppe Gonnella[1], Ignacio Pagonabarraga[2,3]

[1] Dipartimento di Fisica, Università degli Studi di Bari and INFN, Sezione di Bari, via Amendola 173, Bari, I-70126, Italy

[2] CECAM Centre Européen de Calcul Atomique et Moléculaire, Ecole Polytechnique Fédérale de Lausanne, Batochimie, Avenue Forel 2, 1015 Lausanne, Switzerland

[2] UBICS University of Barcelona Institute of Complex Systems, Martí i Franquès 1, E08028 Barcelona, Spain

[4] SISSA - Scuola Internazionale Superiore di Studi Avanzati, Via Bonomea 265, 34136 Trieste, Italy

[5] Institute for Computational Molecular Science, Temple University, Philadelphia, PA 19122, USA

[6] Sorbonne Université, Laboratoire de Physique Théorique et Hautes Energies, CNRS UMR 7589, 4 Place Jussieu, 75252 Paris Cedex 05, France

Abstract. Recently, we characterized the complete phase transition diagram in the ϕ-Pe parameter space for two models of active brownian particles in two dimensions. These models are composed of hard disks and dumbbells, respectively, the former being isotropic and the latter anisotropic. Here, we want to outline all the most significant features of these two paradigmatic models and compare them.

Remarkably, the phase diagrams of the two models are affected differently by the introduction of activity. Disks present a two-stage melting scenario from Pe=0 to about Pe=3, with a first order phase transition between liquid and hexatic and a Berezinskii-Kosterlitz-Thouless transition between hexatic and solid. At higher activities, the three phases are still observed, but the transition between liquid and hexatic becomes a BKT transitions without a distinguishable coexistence region. Dumbbells, instead, present a macroscopic coexistence between hexatically ordered regions and disordered ones, over a finite interval of packing fractions, for all activities, included Pe=0, without any observable discontinuity in the behavior upon increasing Pe.

1. Introduction

Active materials evolve out of thermal equilibrium because their constituents are able to extract energy from the environment and inject it into the system, breaking detail balance [1]. Among others, Active Brownian Particles (ABP) model constitutes a standard paradigmatic model to study the impact of activity on soft matter [2, 3, 4]. For many of the active systems cited above, self-propulsion is able to trigger a motility-induced phase separation (MIPS) between a low-density gas-like phase and dense stable aggregates [5], reminiscent of the equilibrium liquid-gas transition but in the absence of cohesive forces and without a thermodynamic framework to support it [6, 7].

It is worth saying that the most intriguing phenomena induced by activity, as well as the most promising experimental applications, concern systems confined in 2D layers. Being interested in further exploring aggregation phenomena and thermodynamic phases of passive and active two-dimensional systems at any finite density [8, 9, 10] and trying to establish a reference phase diagram for active matter, we naturally faced the long debated problem, still not fully understood, of the melting transition in two dimensions. .

While phase transitions in 2D systems with short-range interactions do not involve any spontaneous symmetry breaking mechanism [11], also ordered crystal-like phases have long been known to exist [12]. A general scenario was proposed to explain solidification in two dimensions by Halperin, Nelson and Young [13, 14, 15], according to which the ordering transition occurs in two steps: one from the isotropic liquid to a hexatic phase, with short-range translational order and quasi-long-range orientational order, and a second one from the hexatic to the solid, characterized by quasi-long-range translational order and a true long-range hexatic one. Both of this two transition are Berezinskii-Kosterlitz-Thouless (BKT) transitions, mediated by the unbinding of topological defects. Very recently, numerical simulations [16, 17, 18] and experimental results [19] reached a consistent understanding of the full melting transition for hard disks, which is still a two-step transition, but with a first-order liquid-hexatic transition and a BKT hexatic-solid one.

2. Methods

2.1. Active brownian disks

Active brownian disks constitute the simplest model of self-propelled brownian isotropic particles. The stochastic equations of motion for N repulsively interacting hard disks of mass m, self-propelled by an active force $\mathbf{F}_{\text{act}}$ constant in modulus and directed along $\mathbf{n}_i =$

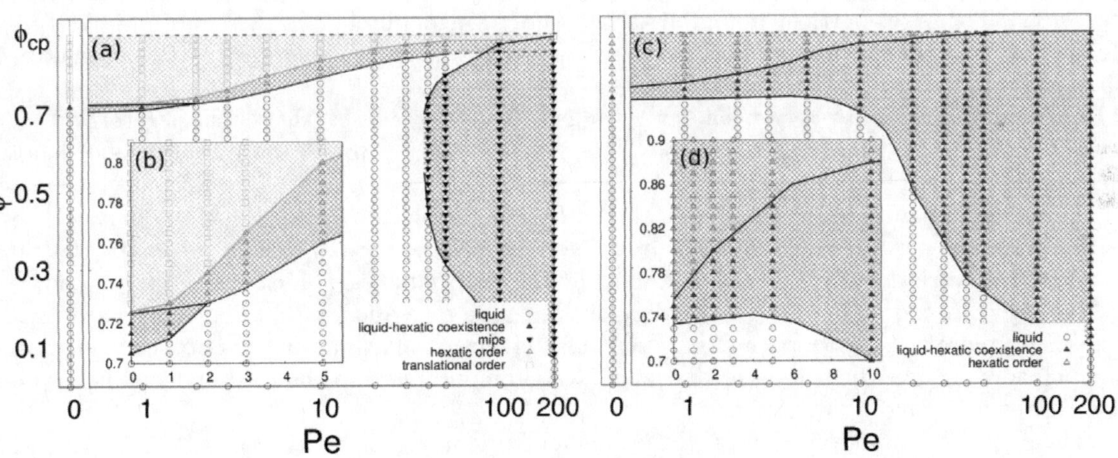

Figure 1. (a) Phase diagram of active brownian disks evolving according to eq. 1. Figure adapted from [20]. Symbols are used to point out the simulations performed. Black empty circles are shown in the liquid phase, blue empty triangles in the hexatic one, orange empty squares in the solid one. Filled symbols are used for the coexistence regions, circles for the low-Pe liquid-hexatic coexistence, triangles for the MIPS. (b) Enlargement of the small activity region across the coexistence. (c) Phase diagram of active brownian dumbbells (see eq. 2), adapted from [21]. Filled blue triangles are used here in the whole connected coexistence region. (d) Zoom of the coexistence region at low Pes.

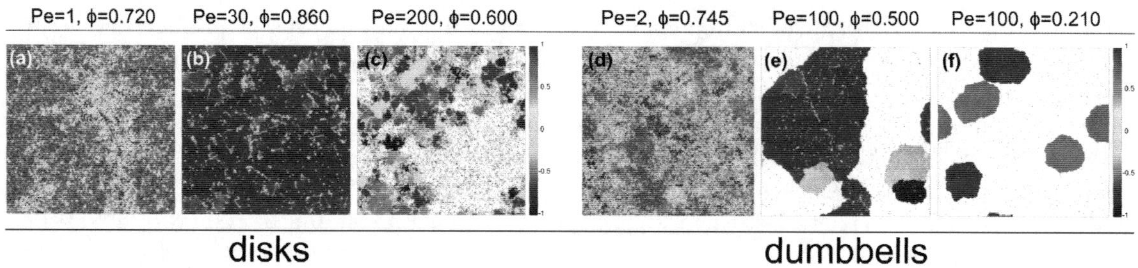

Figure 2. Color representation of the local hexatic parameter ψ_{6j} for disks (left) and dumbbells (right) and for cases given in the labels above the pictures. See the text for details about the color code. The following regions of the phase diagram are shown for disks: (a) liquid-hexatic coexistence at low Pe; (b) hexatic phase at intermediate Pe; (c) MIPS region at high Pe; and for dumbbells: (d) coexistence region at low Pe; (e) MIPS at high Pe, in the middle of the coexistence curve; (f) MIPS at high Pe, close to the lower limit of the coexistence region.

$(\cos\theta_i(t), \sin\theta_i(t))$, are:

$$m\ddot{\mathbf{r}}_i = -\gamma\dot{\mathbf{r}}_i + F_{\text{act}}\mathbf{n}_i - \boldsymbol{\nabla}_i \sum_{j(\neq i)}^{N} U(r_{ij}) + \boldsymbol{\xi}_i \;, \qquad \dot{\theta}_i = \eta_i \;, \tag{1}$$

with $\mathbf{r}_i$ the position of the center of the i-th particle, $r_{ij} = |\mathbf{r}_i - \mathbf{r}_j|$ the inter-particle distance. The inter-particle potential is the short-range purely repulsive potential $U(r) = 4\varepsilon[(\sigma/r)^{64} - (\sigma/r)^{32}] + \varepsilon$ if $r < \sigma_d = 2^{1/32}\sigma$, and 0 otherwise, where σ_d is the diameter of the disks.
The terms $\boldsymbol{\xi}$ and η are zero-mean independent Gaussian noises that verify $\langle \boldsymbol{\xi}_i(t)\boldsymbol{\xi}_j(t')\rangle = 2\gamma k_B T\delta_{ij}\delta(t-t')\mathbf{1}$, $\langle \eta_i(t)\,\eta_j(t')\rangle = 2D_\theta\delta_{ij}\delta(t-t')$, with $D_\theta = 3k_BT/\gamma\sigma_d^2$. The units of length, mass and energy are given by σ_d, m and ε, and are typically set to one. Two relevant parameter of the model are the surface fraction occupied by the particles $\phi = \pi\sigma_d^2 N/(4V)$, where $V = L^2$ is the system's surface, and the Péclet number $\text{Pe} = F_{\text{act}}\sigma_d/(k_BT)$, which measures the ratio between the work done by the active force and the thermal energy K_BT. We tune L and F_{act} at fixed $\gamma = 10$ and $k_BT = 0.05$, according to the value of ϕ and Pe that we want to obtain, in order to explore the whole space parameter. This choice of the values for friction coefficient and temperature ensures the overdamped regime, as commonly used in the field and justified by the low-Re regime of most experiments.

The equation of motions are integrated numerically, using the software LAMMPS [22], with a velocity Verlet algorithm and an additional Langevin-type thermostat.

2.2. Active dumbbells

Dumbbells are diatomic molecules consisting of two identical disks of diameter σ_d. The equations of motion for $2N$ disks are:

$$m\ddot{\mathbf{r}}_i = -\gamma\dot{\mathbf{r}}_i + \mathbf{F}_{\text{act}} - \boldsymbol{\nabla}_i \sum_{j(\neq i)}^{2N} U(r_{ij}) + \boldsymbol{\chi}_i \;. \tag{2}$$

The potential is the same as the one defined above, while $\boldsymbol{\chi}_i$ are uncorrelated Gaussian noises with zero mean and variance $2\gamma k_BT$. The active force has constant magnitude F_{act} and is kept along the tail-to-head direction. The bond between the disks of each dumbbell are kept rigid at

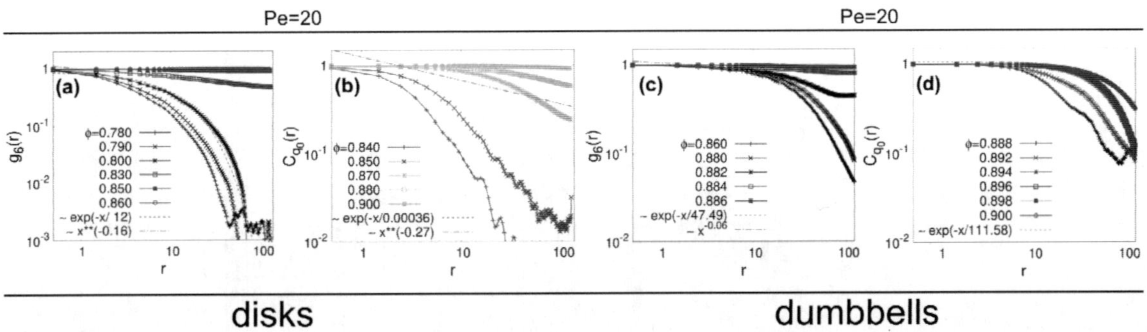

Figure 3. Correlation functions at fixed Pe= 20 and different ϕs as defined in the keys. (a) Hexatic correlations functions for disks across the liquid-hexatic transition; (b) translational correlation functions for disks across the hexatic-solid transition; (c) hexatic correlation functions for dumbbells across the liquid-hexatic transition; (d) translational correlation functions for dumbbells up to the close-packing limit.

a distance σ_d using a RATTLE scheme [23]. The surface fraction is $\phi = \pi\sigma_d^2(2N)/(4V)$, while the Péclet number is $\mathrm{Pe} = 2F_{\mathrm{act}}\sigma_d/(k_BT)$.
Note that the dumbbell model differ from colloids in the way the rotational diffusion is implemented. While the latter is unaffected by the surrounding particles, because the equation of motion for θ is independent from the atoms positions, dumbbells rotate due to the combination of the noise of the dimers. Thus, when dumbbells start to cluster together, they will not be able to rotate.
The equation of motions are integrated in the same way as the ones of colloids.

3. Phase diagram of the two models

The phase diagram for isotropic particles (Fig. 1(a,b)) has been established according to a systematic evaluation of several quantities. As described in details in [20], we used local surface fraction pdfs and virial pressure for self-propelled disks defined as in [6] to provide the limits of the coexistence regions throughout the parameters space. Spatial correlation functions $g_6(r)$ for the two-point correlation of the local hexatic parameter $\psi_6(\mathbf{r}_j) = N_j^{-1}\sum_{k=1}^{N_j} e^{i6\theta_{jk}}$ and finite size analysis of its 4-th order cumulant allowed us to distinguish between liquid structure and hexatic one, and consequently to locate the liquid-hexatic transition for all values of Pe. Besides, correlation functions $C_{\mathbf{q_0}}(r)$ of local translational order parameter $e^{i\mathbf{q_0}\cdot\mathbf{r_i}}$, being $\mathbf{q_0}$ the maximum of the diffraction peak in the reciprocal space, have been used to identify hexatic-solid transition.

Both the two peaks structure in density pdfs and the double loop shape in the equation of state consistently demonstrate that the liquid-hexatic transition in the passive limit is a first order phase transition, as expected from repulsive disk with a $\sim r^{-64}$ repulsive interaction [18], and that, notably, the first-order nature of this phase transition is preserved after bringing the system out of equilibrium by adding non-zero small activity, up to Pe$\sim$ 3. For larger Pes liquid-to-hexatic transition undergoes a crossover to a continuous transition. On the other hand, upon increasing activity from the passive limit, the BKT nature of hexatic-solid transition is maintained at all finite activities. Activity has the effect to destabilize the two ordered phases and, as a consequence, hexatic and solid are pushed towards higher surface fraction. Nevertheless, remarkably, both hexatic and solid phases are demonstrated to exist as stable thermodynamic phases up to very high activities. This can be directly observed from the hexatic and translational correlation functions shown, for fixed Pe= 20 in Fig. 3(a,b). Note that above $\phi \gtrsim 0.800$ the hexatic correlation length diverges as expected in the hexatic phase

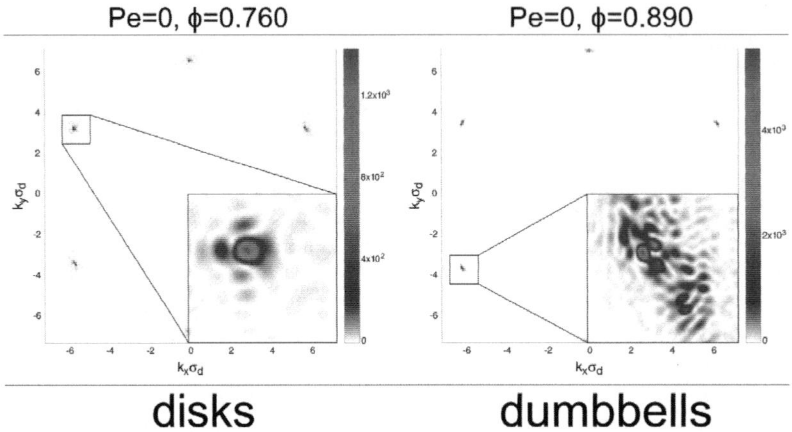

Figure 4. Static structure factor $S(q) = N^{-1} \sum_{i,j} e^{\mathbf{iq}\cdot(\mathbf{r}_i - \mathbf{r}_j)}$ for ordered configurations of disks and dumbbells, with an enlargement of one of the six peaks.

from Halperin-Nelson theory. Above $\phi \gtrsim 0.870$ also the positional correlation length diverges indicating the emergence of quasi-long-range translational order.

Hexatic order for self-propelled disks can also be inferred qualitatively by providing a color map for the local hexatic parameter defined before. We include some examples of this map in Fig. 2 for few representative cases at different points in the Pe-ϕ space reported within the figure. The color scale refers to the projection of the local hexatic parameter along the direction in the complex plane of the sample average of the parameter itself. The specific nature of the configurations shown in Fig. 2 is provided within the caption.

For what concerns dumbbells, we studied the phase diagram in great details in [21], focusing on the emerging relations between the two-dimensional phase transition scheme for passive dumbbells and the motility-induced phase separation. As clearly shown in Fig. 1(c,d), we found a strongly different, but still much interesting scenario, never seen with disks, demonstrating at least that MIPS is a strong system-dependent phenomenon.

We used the same quantities described above for the disks in order to explore extensively the parameter space. In the limit of passive dumbbells we observed that, similarly to isotropic particles, our molecules undergo a first order phase transition between isotropic liquid and a hexatic phase for the mutual orientation of the centers of mass of the single beads. This represents a notable extension to molecular systems of the most recent results obtained for hard disks [17, 19] and is a novelty in the understanding of the nature of 2d melting.

The most remarkable point we found is that, turning on activity and increasing Pe from the passive limit we did not find any finite critical value of activity which triggers the motility-induced phase separation, since we observe macroscopic phase separation for each probed Pe value. As shown in Fig. 3(c), increasing the surface fraction and crossing the upper limit of the coexistence region, quasi-long-range orientational order arises in the system, revealing that coexistence is between a liquid-gas phase and a hexatic phase everywhere in the spanned activity range. Moreover, the finite range of packing fractions where liquid and hexatic coexist is continuously connected throughout the phase diagram, suggesting a crucial interplay between the liquid-hexatic first order phase transition and the motility-induced phase separation.

From both the phase diagram of Fig. 1(a,b) and translational correlation functions for dumbbells shown in Fig. 3(d) is worth noting that we did not find any certain evidence of the existence of a solid phase for dumbbells. Neither for passive molecules, nor with activity, we observe quasi-long-range positional order for the centers of mass of the beads. As explained in [24], the reason for that comes from the shape of the objects in the system. Indeed, since each head-to-tail distance is fixed to the same value σ_d for all values of the global packing fraction, the beads cannot arrange on a perfect triangular lattice, as they are free to do in the disk geometry, at any global $\phi < \phi_{cp}(\sim 0.900)$. More quantitatively, the lack of positional order for dumbbells can be deduced from the the structure of the diffractions peaks in the reciprocal space for the ordered phase. In Fig. 4 the structure factor for passive disks in the solid phase and the one for passive dumbbells at very high packing fraction, close to the maximum packing value, shown the six-fold structure expected for a hexagonal packing. In the same figure is also shown an enlargement of the structure of the peaks used, as described before, for calculating the translational correlation function. While for the disks this peaks has a regular symmetric structure, as the power-law singularity expected in a 2D solid [14], for the dumbbells it is much more broadened. Clearly, the structure of the six peaks representing the triangular order for the beads merges with the internal structure of the molecules, strongly indicating the absence of quasi-long-range positional order.

4. Conclusions

This wide analysis of two of the most simple and representative models of active systems has revealed many interesting features both of 2D melting scenario for passive anisotropic objects and activity-driven aggregation. Furthermore, it inspires fundamental questions about the nature of the ordered phases in presence of activity and about the strong observed dependence of motility-induced phase separation on the shape of the self-propelled constituents. First, we aim to explore further the profound differences between active disks and dumbbells and establish stronger relationship between them. Moreover, would be of great relevance to investigate the role of topological defects in the active melting transitions, comparing the results with the general framework for defects of the KTHNY theory.

References

[1] Marchetti M C, Joanny J F, Ramaswamy S, Liverpool T B, Prost J, Rao M and Simha R A 2013 *Rev. Mod. Phys.* **85** 1143
[2] ten Hagen B, van Teeffelen S and Löwen H 2011 *J. Phys. C: Cond. Matt.* **23** 194119
[3] Romanczuk P, Bär M, Ebeling W, Lindner B and Schimansky-Geier L 2012 *Eur. Phys. J. Spec. Topics* **202** 1–162
[4] Cates M and Tailleur J 2013 *EPL* **101** 20010
[5] Cates M E and Tailleur J 2015 *Annu. Rev. Cond. Matt. Phys.* **6** 219
[6] Levis D, Codina J and Pagonabarraga I 2017 *Soft Matter* **13** 8113–8119
[7] Solon A P, Stenhammar J, Cates M E, Kafri Y and Tailleur J 2018 *Phys. Rev. E* **97** 020602
[8] Bialké J, Speck T and Löwen H 2012 *Phys. Rev. Lett.* **108** 168301
[9] Briand G and Dauchot O 2016 *Phys. Rev. Lett.* **117** 098004
[10] Klamser J U, Kapfer S C and Krauth W 2018 *arXiv:1802.10021*
[11] Mermin N D and Wagner H 1966 *Phys. Rev. Lett.* **17** 1133
[12] Alder B J and Wainwright T E 1962 *Phys. Rev.* **127** 359
[13] Kosterlitz J M and Thouless D J 1973 *J. Phys. C: Solid State Physics* **6** 1181
[14] Halperin B I and Nelson D R 1978 *Phys. Rev. Lett.* **41** 121
[15] Young A 1979 *Phys. Rev. B* **19** 1855
[16] Bernard E P, Krauth W and Wilson D B 2009 *Phys. Rev. E* **80** 056704
[17] Bernard E P and Krauth W 2011 *Phys. Rev. Lett.* **107** 155704
[18] Kapfer S and Krauth W 2015 *Phys. Rev. Lett.* **114** 035702
[19] Thorneywork A L, Abbott J L, Aarts D G A L and Dullens R P A 2017 *Phys. Rev. Lett.* **118** 158001
[20] Digregorio P, Levis D, Suma A, Cugliandolo L F, Gonnella G and Pagonabarraga I 2018 *Phys. Rev. Lett.* **121**(9) 098003
[21] Cugliandolo L F, Digregorio P, Gonnella G and Suma A 2017 *Phys. Rev. Lett.* **119** 268002
[22] Plimpton S 1995 *J. Comp. Phys.* **117** 1–19
[23] Allen M P and Tildesley D J 1989 *Computer simulation of liquids* (Oxford University Press)
[24] Petrelli I, Digregorio P, Cugliandolo L F, Gonnella G and Suma A 2018 *arXiv:1805.06683*

IOP Conf. Series: Journal of Physics: Conf. Series **1163** (2019) 012074 doi:10.1088/1742-6596/1163/1/012074

Population annealing molecular dynamics with adaptive temperature steps

Henrik Christiansen[1], **Martin Weigel**[2] **and Wolfhard Janke**[1]

[1] Institut für Theoretische Physik, Universität Leipzig, Postfach 100 920, 04009 Leipzig, Germany

[2] Applied Mathematics Research Centre, Coventry University, Coventry CV1 5FB, England

E-mail:
henrik.christiansen@itp.uni-leipzig.de
martin.weigel@complexity-coventry.org
wolfhard.janke@itp.uni-leipzig.de

Abstract. Population annealing is a novel generalized-ensemble simulation scheme used in large-scale parallel Monte Carlo simulations of disordered spin systems and similar problems. In a recent publication we proposed a generalization of this method to molecular dynamics simulations of biopolymers. In the present article we review this work and introduce a scheme for automatically choosing the temperature steps based on the observed distribution of potential energies in the population of simulated replica.

1. Introduction

Due to a rugged free-energy landscape, studies of the folding behavior of peptides are among the computationally most difficult problems in computer simulations. In such problems, the systems get trapped in local minima that are separated by high barriers in free energy that are increasingly difficult to overcome by activated dynamics as the temperature is gradually lowered. Several numerical approaches have been developed to overcome such trapping problems and the ensuing loss of ergodicity in simulations of biopolymers. Parallel tempering (PT), sometimes also known as replica exchange, has previously been shown to have the potential of successfully sampling the full configuration space when combined with molecular dynamics (MD) simulations of peptides [1, 2]. In PT, one considers a number of replicas ($\leq 10^2$) of the same system, which are, a priori, simulated independently at different temperatures, ranging from unphysiologically high temperatures where configurations equilibrate quickly, forgetting their previous history, down to the physiologically relevant temperatures, where the peptide is in a folded state. The simulations running at neighboring temperatures are subsequently permitted to exchange configurations according to an acceptance probability taking into account the Boltzmann weight at the two temperatures. This setup results in a diffusive motion of replicas in temperature space, thus allowing the replicated system to escape from local free-energy minima by heating copies up to the highest simulated temperatures before they cool down again, typically arriving in a different local minimum from the one occupied previously. Although PT can easily be implemented in parallel, the number of processors which can effectively be used is limited, as a setup with too many closely spaced temperatures does not improve the equilibration and even slows down the random walk in temperature space.

IOP Conf. Series: Journal of Physics: Conf. Series **1163** (2019) 012074 doi:10.1088/1742-6596/1163/1/012074

In contrast, population annealing (PA) [3, 4, 5, 6] is a simulation scheme introduced in the context of Monte Carlo (MC) simulations that can make use of very large numbers of parallel workers and hence physical cores and massively parallel accelerators such as GPUs [7] without ensuing problems of bad parallel scaling or decreasing simulational efficiency. In this method one considers a multitude (typically at least 10^4) of independent replicas which are initially equilibrated at a high (or potentially even infinite) temperature to get an ensemble of configurations that is perfectly or nearly perfectly uncorrelated. The temperature is then lowered in a sequence of steps, similar to what is done in simulated annealing for a single replica [8]. However, in contrast to simulated annealing, the population is additionally resampled to the new temperature after a number of simulation steps. This resampling leads to a significant speed-up in equilibration and eliminates energetically stuck configurations (reminiscent of the "go with the winners" strategy of Ref. [9]). Observables in PA are calculated using an average over the ensemble, whereas in PT averages are taken over the time series at a fixed temperature. Population annealing has been shown to perform well, e.g., in the simulation of spin glasses [4, 6]. We recently showed how PA can be combined with MD (PAMD) to efficiently simulate macromolecular systems such as the folding of the peptide met-enkephalin [10]. In the present work we demonstrate how the schedule of temperature steps in this scheme does not need to be pre-determined, but it is much preferable to decide about the next temperature to include as one goes along by utilizing the information about the distribution of potential energies resulting from the parallel simulation of a population of replicas.

2. Model and method

The algorithm for PAMD as described in Ref. [10] comprises the following steps:

(i) Set up an equilibrium ensemble of R independent copies of the system at some high temperature T_0.

(ii) Choose the next temperature T_i from a pre-defined sequence.

(iii) Resample the ensemble of systems to the new temperature $T_i < T_{i-1}$ by replicating each copy a number of times proportional to the relative Boltzmann weight $\tau_j \propto e^{-(1/k_B T_i - 1/k_B T_{i-1})E_j}$, where E_j is the potential energy of the jth replica. Similarly to PT, the momenta are easily adjusted to the new temperature by rescaling $p_k \to \sqrt{\frac{T_i}{T_{i-1}}}p_k$.

(iv) Update each copy with θ simulation steps of the underlying MD algorithm.

(v) Calculate observables $\mathcal{O}$ at temperature T_i as population averages.

(vi) Goto step (ii) until T_i reaches or falls below the target temperature T_N.

In our previous work [10], studying the folding of the penta-peptide met-enkephalin with the amino-acid sequence Tyr-Gly-Gly-Phe-Met, we adapted a temperature set (T_i = 700, 585, 489, 409, 342, 286, 239, 200 K) from a PT simulation of the same peptide [2]. However, finding an appropriate temperature set in PT simulations is a non-trivial task, for which a number of solutions are being discussed [11, 12, 13]. It is clear, for instance, that too large temperature steps in PT lead to low acceptance rates of swap moves, whereas too many small steps lead to a slowing down resulting from the scaling properties of the temperature random walk. Similar difficulties arise in PAMD with a fixed temperature protocol. In population annealing, however, it is possible to exploit the fact that one has access to the energy histogram at temperature T_{i-1} from the population of replicas at the point of having to decide about the next step, T_i, in the temperature scheme. To benefit from this additional information, we apply single-histogram reweighing [14] to get an estimate of the energy histogram at T_i. We then require that the histograms of potential energy at T_i and T_{i-1} have a certain prescribed overlap $\alpha^\star$ at each step. The actual overlap can be estimated before taking the real temperature step from the following

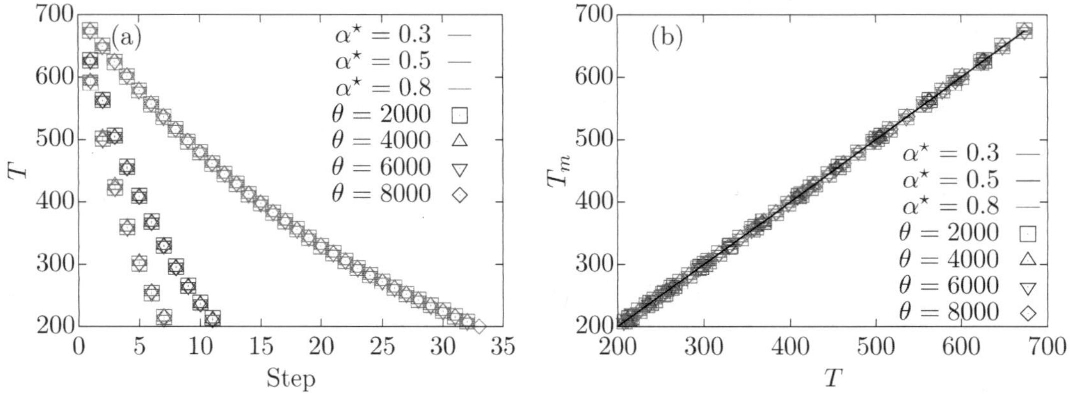

Figure 1. (a) Temperatures found by the PAMD simulation with adaptive temperature stepping are shown for different values of $\alpha^\star$ and number θ of MD steps. The symbols encode the MD steps θ and the colors relate to the values of $\alpha^\star$. (b) The observed temperature T_m (measured just before the resampling step) during the simulations is shown against the heatbath temperature T for the same values of θ and $\alpha^\star$ as in (a).

expression [7],

$$\alpha(T_{i-1}, T_i) = \frac{1}{R}\sum_{j=1}^{R} \min\left(1, \frac{\exp[-(1/k_BT_i - 1/k_BT_{i-1})E_j]}{Q(T_{i-1}, T_i)}\right). \tag{1}$$

Here E_j is the energy of the jth replica and $Q = \sum_{j=1}^{R} \exp[-(1/k_BT_i - 1/k_BT_{i-1})E_j]$ serves as a normalization. Before the simulation one then only has to choose a fixed value of $\alpha = \alpha^\star$, and any given value of $\alpha^\star$ results in a certain temperature sequence. In the above algorithm, we hence propose to replace step (ii) by the following adaptive variant:

(ii') Find the next temperature T_i in such a way, that $\alpha(T_{i-1}, T_i) = \alpha^\star$.

The question of what are reasonable values of $\alpha^\star$ is investigated in the following section.

In the simulations reported below we used OpenMM [15] to perform the MD, relying on the Langevin thermostat and a velocity-Verlet like integrator. The interactions of the peptide are modeled via the AMBER force-field ff94 [16]. It is worthwhile to point out that for PAMD simulations to work properly the thermostat has to be stochastic as otherwise all of the resampled copies of a population member will follow exactly the same trajectory in further evolving the system, thus defying the decorrelating purpose of the resampling step.

3. Results

We fixed the relevant PAMD parameters at $R = 10\,000$ replicas, a time step of $\mathrm{d}t = 0.5$ fs, and a friction coefficient $\gamma = 1$/ps. For the constant energy overlap simulations, we chose $T_0 = 700$ K as the initial high temperature and set $T_N = 200$ K as the target temperature. The start configurations at T_0 were taken from a canonical simulation of length 25 ns, followed by short simulations for each replica to further decorrelate the population. To find a suitable set of overlaps $\alpha^\star$ and corresponding number θ of MD simulation steps per temperature, we tested $\alpha^\star = 0.3$, 0.5, 0.8 and $\theta = 2000$, 4000, 6000, 8000. In Fig. 1(a) we show the temperatures realized by the constant energy overlap PAMD runs by choosing $\alpha^\star$ and θ as indicated above.

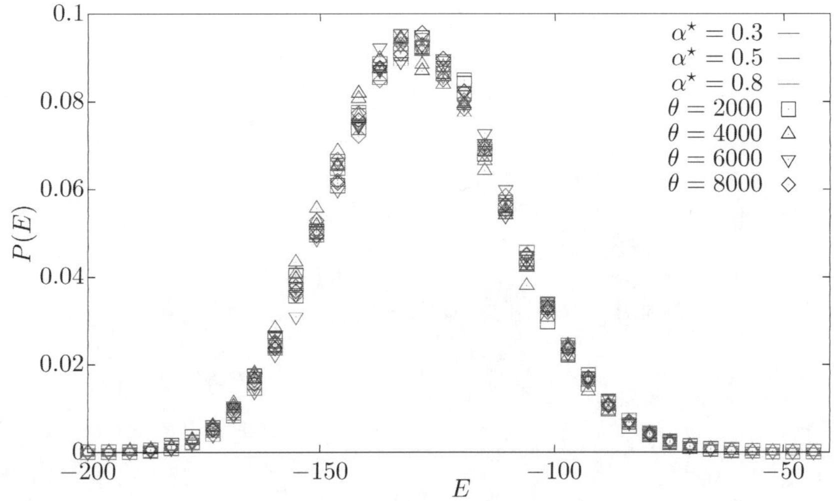

Figure 2. The normalized energy histogram from the simulations with different energy histogram overlaps $\alpha^\star$ and MD simulation steps θ, reweighted to $T = 200$ K using WHAM from the PAMD simulations with adaptive temperature steps. The symbols and colors have the same meaning as in Figs. 1(a) and (b).

As is evident from Fig. 1(a), $\alpha^\star = 0.3$ leads to a temperature set which is very close to the one used in Refs. [2, 10]. Thus the simulation with $\alpha^\star = 0.3$ and $\theta = 4000$ is close to equivalent to our previous study in [10], and it is hence used as a benchmark. In the regime of sufficiently large θ to ensure equilibration in between temperature steps, the adaptive stepping algorithm is insensitive against the chosen θ and mainly depends on $\alpha^\star$. This is also reflected in Fig. 1(b), where we plot the measured temperature T_m against the externally imposed (thermostat) temperature T. Also plotted is a solid line, corresponding to the expected behavior in properly equilibrated simulations, namely $T_m = T$. Note that this type of plot is mostly used in MD simulations, where the intrinsic temperature is readily defined in terms of the kinetic energy, while in MC simulations proper equilibration is mostly ascertained by checking the stationarity of time series of suitably chosen observables.

In Fig. 2 the histogram of potential energy at $T = 200$ K is presented for the different values of $\alpha^\star$ and θ used in Figs. 1(a) and (b). The histograms were obtained by means of the weighted histogram analysis method (WHAM) [17]. One advantage of PAMD simulations is, that one can easily seed the WHAM with good estimates of the partition functions $\mathcal{Z}$, because [5, 7]

$$Q(T, T') = \frac{\mathcal{Z}(T)}{\mathcal{Z}(T')}. \tag{2}$$

This allows one to only perform very few iteration steps in the WHAM while still getting very good estimates of the partition functions. From Fig. 2 it is clear that all considered simulation protocols produce an energy histogram that is well equilibrated.

The quality of sampling of configuration space can be more readily judged from Ramachandran plots, showing the dihedral angles ϕ and ψ along the backbone of the peptide. Such plots are shown in Fig. 3 for the center amino acid GLY3. Panel (a) shows the result of a PT simulation with the same parameters as used in Ref. [10] for reference, panels (b) and (c) correspond to PAMD simulations with adaptive temperature steps using $\alpha^\star = 0.5$, $\theta = 4000$

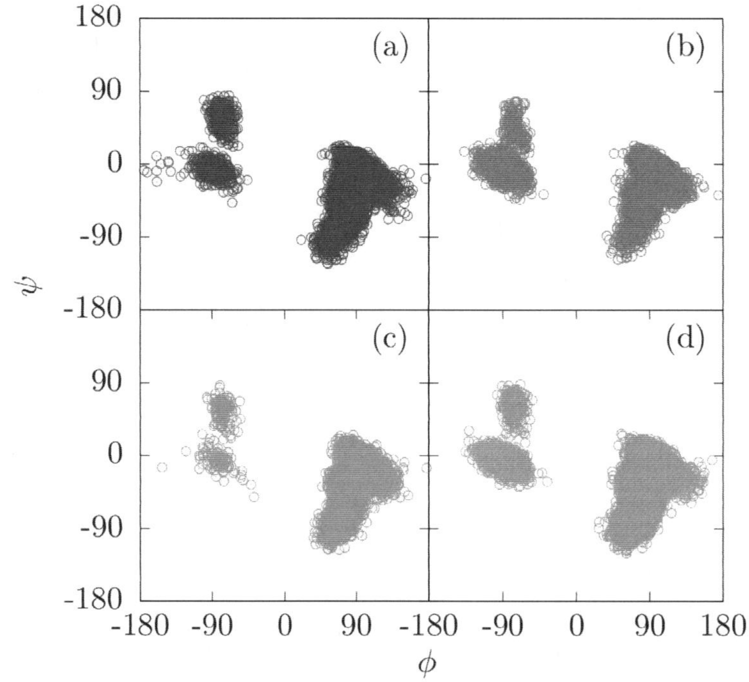

Figure 3. Ramachandran plots showing the realized values of backbone dihedral angles ϕ and ψ for GLY3. (a) The PT simulation with a total runtime of 200 ns already discussed in Ref. [10]. (b) PAMD simulation with adaptive temperature stepping, $\alpha^{\star} = 0.5$, $\theta = 4000$ and $R = 10\,000$. (c) Same as for (b) but for $\alpha^{\star} = 0.8$ and $\theta = 2000$. (d) PAMD simulation using the given temperature set from Ref. [2] and $\theta = 4375$, however, with $R = 20\,000$ replicas.

(b) and $\alpha^{\star} = 0.8$, $\theta = 2000$ (c), respectively. All three simulations show broadly comparable sampling of the configuration space. It appears from a comparison of (b) and (c), that it is favorable to rather choose a larger number of steps θ than increasing the overlap $\alpha^{\star}$. A similar improvement as in (b), however, is achieved by increasing the population size R, see Fig. 3(d). However, this comparison does not take into account the computational resources used. Relative to the PT simulation shown in (a), the adaptive PAMD simulations in (b) and (c) used 1.525 and 1.875 times more computing cycles, respectively, whereas the non-adaptive but larger PAMD simulation shown in (d) also used 1.875 times the cycles of (a). The sampling of (b) and (d) is in both cases slightly superior to (c). One should keep in mind that the wall-clock time in PAMD with increasing number of replicas R can be kept very low when sufficient parallel resources are available. The minimum wallclock time in PA is predominantly limited only by the short simulation time t_s needed for a single replica

$$t_s = \frac{t}{R}, \tag{3}$$

where t is the total simulation time.

IOP Conf. Series: Journal of Physics: Conf. Series **1163** (2019) 012074 doi:10.1088/1742-6596/1163/1/012074

4. Conclusion and Outlook

We have performed population annealing molecular dynamics (PAMD) simulations with adaptive temperature steps for the folding of the penta-peptide met-enkephalin. It was shown that using the criterion of a constant energy overlap significantly simplifies the task of choosing a proper temperature set, a notorious problem also encountered in parallel tempering simulations. We find reliable results in a wide range of possible overlap values $\alpha^\star$ with similar computational efforts if the number θ of MD steps is correspondingly adapted. In the regime where θ is large enough, the precise value of θ has little influence on the actual schedule proposed. If additional parallel computational resources are available, it seems to be preferable to use those to simulate more replicas in the population rather than increasing θ or $\alpha^\star$, but the precise optimum in this multi-dimensional parameter space will depend on the system under consideration.

Acknowledgments

This project was funded by the Deutsche Forschungsgemeinschaft (DFG) under Grant No. SFB/TRR 102 (project B04), and further supported by the Deutsch-Französische Hochschule (DFH-UFA) through the Doctoral College "$\mathbb{L}^4$" under Grant No. CDFA-02-07, the EU Marie Curie IRSES network DIONICOS under Grant No. PIRSES-GA-2013-612707, and the Leipzig Graduate School of Natural Sciences "BuildMoNa".

References

[1] Hansmann U H E 1997 *Chem. Phys. Lett.* **281** 140
[2] Sugita Y and Okamoto Y 1999 *Chem. Phys. Lett.* **314** 141
[3] Iba Y 2001 *Trans. Jpn. Soc. Artif. Intell.* **16** 279
[4] Hukushima K and Iba Y 2003 *AIP Conf. Proc.* **690** 200
[5] Machta J 2010 *Phys. Rev. E* **82** 026704
[6] Wang W, Machta J and Katzgraber H G 2015 *Phys. Rev. E* **92** 063307
[7] Barash L Y, Weigel M, Borovský M, Janke W and Shchur L N 2017 *Comput. Phys. Commun.* **220** 341
[8] Kirkpatrick S, Gelatt C D and Vecchi M P 1983 *Science* **220** 671
[9] Grassberger P 2002 *Comput. Phys. Commun.* **147** 64
[10] Christiansen H, Weigel M and Janke W 2018 *preprint arXiv:1806.06016*
[11] Katzgraber H G, Trebst S, Huse D A and Troyer M 2006 *J. Stat. Mech.: Theory and Exp.* **2006** P03018
[12] Bittner E, Nußbaumer A and Janke W 2008 *Phys. Rev. Lett.* **101** 130603
[13] Vogel T and Perez D 2015 *Phys. Rev. Lett.* **115** 190602
[14] Ferrenberg A M and Swendsen R H 1988 *Phys. Rev. Lett.* **61** 2635
[15] Eastman P, Swails J, Chodera J D, McGibbon R T, Zhao Y, Beauchamp K A, Wang L P, Simmonett A C, Harrigan M P, Stern C D, Wiewiora R P, Brooks B R and Pande V S 2017 *PLoS Comput. Biol.* **13**
[16] Cornell W D, Cieplak P, Bayly C I, Gould I R, Merz K M, Ferguson D M, Spellmeyer D C, Fox T, Caldwell J W and Kollman P A 1995 *J. Am. Chem. Soc.* **117** 5179
[17] Kumar S, Rosenberg J M, Bouzida D, Swendsen R H and Kollman P A 1992 *J. Comput. Chem.* **13** 1011

IOP Conf. Series: Journal of Physics: Conf. Series **1163** (2019) 012075 doi:10.1088/1742-6596/1163/1/012075

Quasi-energies of coupled qubits: Magnus-Floquet states and their probing by weak signal

M. V. Bastrakova[1], N.V. Klenov [2,3] and A. M. Satanin[3,4]

[1] Lobachevsky State University of Nizhny Novgorod, Nizhny Novgorod 603950, Russia
[2] Faculty of Physics, Lomonosov Moscow State University, Moscow 119991, Russia
[3] Dukhov All-Russia Research Institute of Automatics (VNIIA), Moscow 101000, Russia
[4] Russia National Research University Higher School of Economics, 101000 Moscow, Russia
Email: sarkady@mail.ru

Abstract. The behaviour of two coupled qubits in a periodic electromagnetic field of arbitrary amplitude was investigated. Quasienergy levels of the system as a function of the driving field parameters and the qubit coupling energy were found on the basis of the Floquet theory and the Magnus approximation. It is shown that the quasienergy levels may intersect depending on the control parameters. This fact can manifest itself as appearance of the novel resonances and interesting nonlinear phenomena.

1. Introduction

Recently, the influence of the alternating field on quantum systems (atoms, molecules) has been actively studied due to the creation of high-power lasers and the development of spectroscopy of fast processes. This made it possible to detect some interesting nonlinear effects: (i) dynamic Stark effect (modification of the spectrum in the monochromatic driving field); (ii) the Autler-Townes effect (the splitting of the frequency into a periodic driving field); (iii) induced transparency effect; (iv) laser cooling, etc. (see articles [1-3] for more details). Similar effects can also be observed when the microwave driving field is applied to manipulate the states of artificial atoms – superconducting qubits. Moreover, since the qubits have micron dimensions, the observation of these nonlinear effects may be realized at lower amplitudes of the alternating field [4].

This work is devoted to the development of a technique for calculating quasi-energy levels of two coupled qubits in a strong alternating field. Our approach is based on approximate calculations of the evolution operator (propagator) for the quantum system on a field period by using of the Magnus method [5-7] and Floquet theory (outside the perturbation theory) [8-11] (see for review [12, 13]). The resulting expression for the propagator is further used to calculate the qubit quasienergies depending on the system parameters: the amplitude of the driving field and the qubit coupling parameter. Note that the obtained dependences are important for development of the microwave spectroscopy of qubits. The technique developed in the present work allows us to consider the scenario when the main signal (pump) may be send on the strip line simultaneously with a weak test signal (probe), which allows to probe the quasienergy levels.

2. Quasienergy

Let us remind briefly the main ideas underlying the concept of quasi-energy [8-10]. If the Hamiltonian of a multilevel quantum system is periodic in time $H(t)=H(t+T)$, then the solution of the Schrödinger equation $i\hbar|\dot{\psi}(t)\rangle = H(t)|\psi(t)\rangle$ according to Floquet's theorem can be decomposed into a complete set of Floquet functions $|\Phi_k(t)\rangle$

$$|\psi_k(t)\rangle = |\Phi_k(t)\rangle e^{-iQ_k t/\hbar}, \tag{1}$$

where Q_k are time-independent quasienergies [11]. They can be found from the solution of the equation:

$$U(T)|\Phi_k(0)\rangle = e^{-iQ_k T/\hbar}|\Phi_k(0)\rangle, \tag{2}$$

where $U(T)=\hat{P}e^{-\frac{i}{\hbar}\int_0^T H(t)dt}$ is an evolution operator on a period, $\hat{P}$ is the chronological ordering. Since quasienergies are not uniquely defined, $Q_k^{'} = Q_k + n\hbar\omega$, we will translate them in the first "Brillouin zone" ($0 \le Q_k \le \hbar\omega$). The evolution operator is determined by solving the equation:

$$i\hbar\frac{\partial U}{\partial t} = H(t)U \ . \tag{3}$$

In the theory of magnetic resonance (when describing the spin motion in an alternating high-frequency field), an approximate solution of Eq. (3) and the expression for quasi-energies have been obtained in the framework of the resonance perturbation theory (Rabi approximation) [14]. In this case, they are weakly dependent on the field amplitude. It is also known that the Rabi approximation is not applicable in the case of periodic signals of large amplitude, when the level shifts in the field are comparable to the distances between the levels. In this case, the influence of the driving field cannot be considered in the framework of standard perturbation theory and should be applied an approach beyond of the Rabi approximation.

In this work, to calculate the evolution operator on the period $U(T)$, the Magnus expansion will be used as an approximation. We consider the propagator on the field period in the form:

$$U(T) = e^{-iM(T)}, \tag{4}$$

where M is a hermitian operator, $M^+ = M$. According to [5-7], the operator M is generally defined by an infinite series

$$M(T) = \sum_{j=1}^{N} M_j(T), \tag{5}$$

where the terms of the Magnus operator are proportional to the corresponding powers of the Hamilton operator $H(t)$. The advantage of the Magnus approximation method is that the solution of the evolutionary problem "step by step" on the field period is not required to find quasienergies Q_k. We need to find and diagonalize the evolution operator Eq. (4) only at the moment of time $t=T$. Since the M operator commutes with the Hamiltonian, the problem reduces to the diagonalization of the Magnus operator.

Note that an important issue for the use of the Magnus approximation approach is the convergence of the series Eq. (5). We will not deal with mathematical subtleties in this paper, and we only will refer to the discussion of this question in terms of the eigenvalues of the operator M in [15]. In practice, this criterion is reduced to the evaluation of the norm of the Hamiltonian of the system according to:

$$\int_0^T \|H(t)\|_2 \, dt < \pi \,, \tag{6}$$

where $\|\cdot\|_2$ is the Euclidean norm of the Hamiltonian $H(t)$.

3. Quasienergies of two interacting qubits in a strong field

The main features of the behavior of two superconducting flux coupled qubits studied in experiments [16, 17] can be described by the Hamiltonian:

$$H(t) = -\frac{1}{2}\begin{pmatrix} \varepsilon_1(t)+\varepsilon_2(t)+J & \Delta_2 & \Delta_1 & 0 \\ \Delta_2 & \varepsilon_1(t)-\varepsilon_2(t)-J & 0 & \Delta_1 \\ \Delta_1 & 0 & -\varepsilon_1(t)+\varepsilon_2(t)-J & \Delta_2 \\ 0 & \Delta_1 & \Delta_2 & -\varepsilon_1(t)-\varepsilon_2(t)+J \end{pmatrix}, \tag{7}$$

where $\varepsilon_i(t)$ is the control parameter of qubit i (i = 1, 2), Δ_i is the corresponding tunneling matrix element, and the parameter J quantifies the strength of the interaction between the qubits. For simplicity, we will discuss the case when the qubits are affected by the monochromatic driving field ($T = 2\pi/\omega$), which will be described by functions:

$$\varepsilon_{1,2}(t) = \varepsilon_{1,2}^{dc} + \varepsilon_{1,2}^{ac}\cos(\omega t), \tag{8}$$

where ε_i^{dc} and ε_i^{ac} are the amplitude of the dc- and ac-fields. It is assumed that a sequence of synchronized pulses of a driving field acts on the qubit system. The pulse duration is much longer than the field period. In the absence of the ac field ($\varepsilon_i^{ac} = 0$) dispersion curves depending on the constant field are found from the solution of the equation $H|\psi_k\rangle = E_k|\psi_k\rangle$ which is shown in the figure 1 (a).

We investigate the behavior of quasienergy levels at arbitrary amplitude of the external driving field on the basis of direct numerical solution of the Eqs. (3) and (2) of the Floquet theory, as well as on the basis of the Magnus approximation from Eq. (4). For two coupled qubits we take into account three terms in the expansion (5). For the case $\varepsilon^{dc} = \varepsilon_2^{dc} = k\varepsilon_1^{dc}, \varepsilon^{ac} = \varepsilon_2^{ac} = k\varepsilon_1^{ac}$, $\omega = 1$ GHz, the Magnus operator Eq. (5) on the field period may be written as:

$$M(T) = \begin{pmatrix} -\varepsilon^{dc}(1+k)\pi - J\pi + \varepsilon^{ac}\pi(\Delta_1^2 + k\Delta_2^2) & \frac{\pi\Delta_2}{4}\left(-4+\left(\varepsilon^{ac}\right)^2 k^2 - 4\varepsilon^{ac}k(J+k\varepsilon^{dc})\right) & \frac{\pi\Delta_1}{4}\left(-4+\left(\varepsilon^{ac}\right)^2 - 4\varepsilon^{ac}(J+\varepsilon^{dc})\right) & 0 \\ \frac{\pi\Delta_2}{4}\left(-4+\left(\varepsilon^{ac}\right)^2 k^2 - 4\varepsilon^{ac}k(J+k\varepsilon^{dc})\right) & -\varepsilon^{dc}(1+k)\pi + J\pi + \varepsilon^{ac}\pi(\Delta_1^2 - k\Delta_2^2) & 0 & \frac{\pi\Delta_1}{4}\left(-4+\left(\varepsilon^{ac}\right)^2 + 4\varepsilon^{ac}(J-\varepsilon^{dc})\right) \\ \frac{\pi\Delta_1}{4}\left(-4+\left(\varepsilon^{ac}\right)^2 - 4\varepsilon^{ac}(J+\varepsilon^{dc})\right) & 0 & \varepsilon^{dc}(1-k)\pi + J\pi - \varepsilon^{ac}\pi(\Delta_1^2 - k\Delta_2^2) & \frac{\pi\Delta_2}{4}\left(-4+\left(\varepsilon^{ac}\right)^2 k^2 + 4\varepsilon^{ac}k(J-k\varepsilon^{dc})\right) \\ 0 & \frac{\pi\Delta_1}{4}\left(-4+\left(\varepsilon^{ac}\right)^2 + 4\varepsilon^{ac}(J-\varepsilon^{dc})\right) & \frac{\pi\Delta_2}{4}\left(-4+\left(\varepsilon^{ac}\right)^2 k^2 + 4\varepsilon^{ac}k(J-k\varepsilon^{dc})\right) & \varepsilon^{dc}(1-k)\pi - J\pi - \varepsilon^{ac}\pi(\Delta_1^2 + k\Delta_2^2) \end{pmatrix}$$

We present the behavior of the dispersion curves $E_k(\varepsilon^{dc})$ in figure 1 (a) for two coupled qubits. Let us mentally draw on dependencies $E_k(\varepsilon^{dc})$ a set of lines parallel to the vertical axis at distances $n\hbar\omega$ from each other and then move the fragments of dispersion curves from each line to the first Brillouin zone $-\hbar\omega/2 \le Q_k \le \hbar\omega/2$ (dashed gray lines in figure 1 (a)). The obtained picture will approximately correspond to the pictures shown in figure 1 (b), which were calculated using the numerical solution of Eq. (2) or the approximation decomposition of Magnus Eq.(4). As shown in figure 1 (b) the dependence of quasienergies on the parameter ε^{dc} is very simple at $\varepsilon^{dc} \gg \Delta_i$: the quasienergies behave in accordance with the almost-linear laws of dispersion of the uncoupled qubits. Note that when $\varepsilon^{dc} \sim \Delta_i$, the curvature of the qubit dispersion plays an important role in the formation of the quasienergy.

IOP Conf. Series: Journal of Physics: Conf. Series **1163** (2019) 012075 doi:10.1088/1742-6596/1163/1/012075

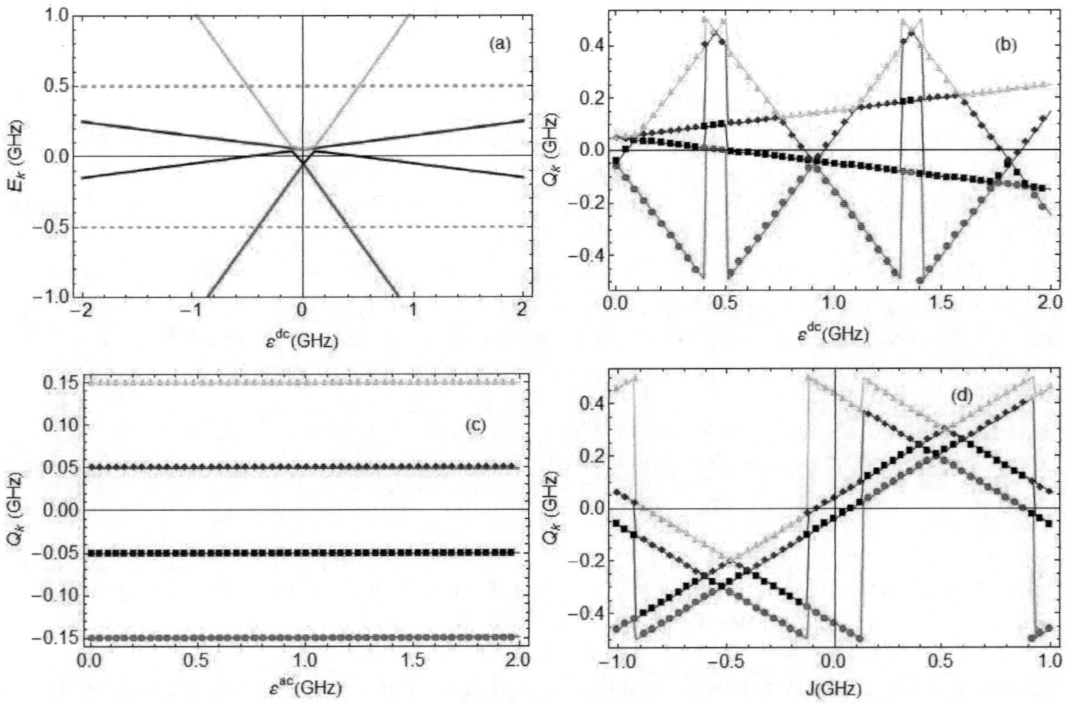

Figure 1. (a) Dispersion curves E_k of two coupled qubits as a function of the amplitude dc-field ε^{dc}. Quasienergies Q_k (b, c, d) as a function of the amplitude of the amplitude dc-field ε^{dc} (b), ac-field ε^{dc} (c) and the interaction parameter of qubits J (d). The solid line represents the numerical calculation, the dots show the solution based on the expansion of Magnus. The following parameters of the qubit and fields were used: $\omega = 1\,\text{GHz}$, $\Delta_2 = 2\Delta_1 = 0.02$ GHz, $k = 1.2$, $\varepsilon^{dc} = 1\,\text{GHz}$ (c), $\varepsilon^{dc} = 0.4\,\text{GHz}$ (d), $\varepsilon^{ac} = 1.2\,\text{GHz}$ (b, d), $J = 0.1\,\text{GHz}$ (d, c). The colors on the figures establish the correspondence between the levels (a) and the quasi-levels (b,c, and d).

Figure 1 (b, c, d) demonstrates the quasienergy behavior of two coupled qubits when the field parameters $\varepsilon^{dc}, \varepsilon^{ac}$ and the interaction parameter of the qubits J are changed. It is found experimentally that the qubits coupling parameter can be varied in a wide range, for example, by means of an additional superconducting circuit placed between the qubits [18]. In addition, the so-called programmable qubits are being actively developed, in which changing the parameters of the system it is possible to organize ferromagnetic ($J > 0$) or antiferomagnetic ($J < 0$) interaction in pairs between different qubits [19].

The results of the Magnus approximation (denoted by points in figure 1 (b, c, d)) describes the behavior of quasi-levels calculated numerically on the basis of Floquet theory (solid lines in figure 1 (b, c, d)) with high accuracy. Application of Magnus approximation for calculation $\Delta_i T << 1$ that corresponds to the chosen parameters of modeling and according to the analysis of experimental works on research of superconducting flux qubits. Note that in some cases tunnel level splitting Δ_i (i = 1, 2) is small in comparison with other energy parameters of the system [16, 17].

4. Conclusion

For the observation of atomic spectra it is necessary to excite atoms by a weak variable field and observe the emitted radiation by changing the frequency of the probe signal. For example, we can use a weak magnetic field for scanning when a system is irradiated by a monochromatic field. According to the Bohr's postulates transitions between levels take place when the distance between them is compared with the energy of the electromagnetic field quantum.

In the case of a driving field of large amplitude, we can only talk about the transitions between the quasilevels of the qubits, which are shown in figure 1 (b, c, d). As one can see, when the control parameters of the qubits are changed, there is an intersection of quasilevels, which should be accompanied by the appearance of resonance of the response function. If we add a weak probe signal to the Hamiltonian of qubits Eq. (7), it is easy to obtain an analogue of the formula for the transition probability with the usual perturbation theory [20]. In this case, peaks of response function appear if the differences of quasienergies coincide with the energy of the absorbed quantum. Note that the above ambiguity of quasienergies ($Q_k^{'} = Q_k + n\hbar\omega$) is insignificant when considering (neglecting the transfer processes), if the splitting of levels is small compared to $\hbar\omega$.

Thus, the behavior of qubits in a strong driving field demonstrates new possibilities of nonlinear spectroscopy, which carries additional information about the qubit parameters.

Acknowledgments

This work was supported by the RNF No. 18-72-00158 and the RFBR (grants No. 18-07-01206 and 16-07-01012).

References

[1] Scully M O, Zubairy M S 1997 *Quantum optics* (Cambridge: Cambridge University Press).
[2] Harris S E, Field J E, and Imamoglu A 1990 *Phys. Rev. Lett.* **64** 1107.
[3] Fleischhauer M, Imamoglu A, and Marangos J P 2005 *Rev. Mod. Phys* **77** 633.
[4] Abdumalikov A A, Astafiev O, Zagoskin A M, Pashkin Yu A, Nakamura Y, Tsai J-S 2010 *Phys. Rev. Lett.* **104** 193601
[5] Magnus W 1954 *Commun. on Pure and Appl. Math.* **7** 649.
[6] Pechukas P, Light J L 1966 *Journ. of Chem. Phys.* **44** 3897.
[7] Denisenko M V, Klenov N V, and Satanin A M 2018 *JETP* **155** (1).
[8] Zeldovich Ya B 1967 *Sov. Phys. JETP* **24** 1006.
[9] Ritus V I 1967 *Sov. Phys. JETP* **24** 1041;
[10] Grifoni M and Hanngi P H 1998 *Phys. Rep.* **304** 219.
[11] Shirley J H 1965 *Phys. Rev.* **138** B979.
[12] Blanes S, Casas F, Oteo J A and Ros J 2009 *Physics Reports*, **470**, 151.
[13] Mananga E S, Charpentier T 2016 *Phys. Rep.* **609** 1.
[14] Rabi I 1927 *Phys. Rev.* **29** 174.
[15] Moan Ch and Niesen J 2008 *Found. Comput. Math.* **8** 291
[16] Grajcar M, Izmalkov A, van der Ploeg S H W, Linzen S, Il'ichev E, Wagner Th, Hübner U, Meyer H-G, van den Brink A M, Uchaikin S, and Zagoskin A M 2005 *Phys. Rev. B* **72** 020503 (R).
[17] Il'ichev E, Shevchenko S N, van der Ploeg S H W, Grajcar M, Temchenko E A, Omelyanchouk A N, and Meyer H-G 2010 *Phys. Rev. B* **81** 012506.
[18] van der Ploeg S H, Izmalkov A, van den Brink A M, Hübner U, Grajcar M, Il'ichev E, Meyer H G, Zagoskin A M 2007 *Phys. Rev. Lett.* **98** 057004.
[19] Johnson M W, Amin M H, Gildert S, Lanting T, Hamze F, Dickson N, Harris R, Berkley A J, Johansson J, Bunyk P, Chapple E M, Enderud C, Hilton J P, Karimi K, Ladizinsky E, Ladizinsky N, Oh T, Perminov I, Rich C, Thom M C, Tolkacheva E, Truncik C J, Uchaikin S, Wang J, Wilson B, Rose G 2011 *Nature* **473** 194.
[20] Landau L D and Lifshitz E M 1965 *Quantum Mechanics* (*A Course of Theoretical Physics* vol 3).

IOP Conf. Series: Journal of Physics: Conf. Series **1163** (2019) 012076 doi:10.1088/1742-6596/1163/1/012076

A new method of simulations for the propagators of multi-qubit registers

M. V. Bastrakova[1], N.V. Klenov [2,3] and A. M. Satanin[3,4]

[1] Lobachevsky State University of Nizhny Novgorod, Nizhny Novgorod 603950, Russia
[2] Faculty of Physics, Lomonosov Moscow State University, Moscow 119991, Russia
[3] Dukhov All-Russia Research Institute of Automatics (VNIIA), Moscow 101000,
[4] Russia National Research University Higher School of Economics, 101000 Moscow, Russia
Email: sarkady@mail.ru

Abstract. A numerical method of simulations for the evolution operators (propagators) of coupled qubits was developed on the basis of Magnus representation. The propagator of a multi-qubit system can be expressed by a $N \times N$ matrix whose size increases as $N = 2^n$ with the number of qubits n. Therefore, the standard approach to the propagator calculation requires the involvement of numerically expensive procedures. We have shown that the calculations of the propagator at the current time can be reduced to the inversion of the Vandermonde matrix, which occurs when constructing interpolation polynomials. In this case, due to the recurrence relations, the number of required flops increases only as $\sim O(N^2)$ (in opposite to $\sim O(N^3)$ flops for the matrix of the general form). This fact allows us to speed up the calculation of the propagator for multi-qubit systems.

1. Introduction

Recently the problem of calculating the propagator for multi-qubit systems has been considered as high topic of interest in the area of quantum communications, since its implementation is a basis of the control systems for quantum registers. We define here for our aim the quantum register as an array of (coupled) two-level systems – qubits. These systems can be coupled into clusters by a direct interaction and may be controlled by a set of gate operations determined by the time-dependent Hamiltonian of the system $H(t)$. If the qubits do not interact initially, then the wave function $\psi(0)$ is as a direct product of two-component one-qubit functions, for instance, prepared in the ground states. We are interested here in the evolution of the full function $\psi(t) = U(t)\psi(0)$ under the action of gate operations. The dynamics is determined by the unitary operator $U(t) = \hat{T}\exp\left(-i\int_0^t H(t')dt'\right)$, where $\hat{T}$ is the operator of chronological ordering [1]. It is known that the calculation of the propagator for a multi-qubit system is a numerically expensive procedure, since with the increase in the number of qubits n the size of the propagator is determined by the matrix $N \times N$, with the size increasing as $N = 2^n$.

In this present work, we developed a numerical method for propagator simulations based on the Magnus representation in the exponential form. After then we applied the Cayley-Hamilton theorem to rewrite the propagator as a matrix polynomial. We have shown than that the simulations of the propagator at the current time can be reduced to the solution of the standard procedure of inversion of the Vandermonde matrix, which occurs when constructing interpolation polynomials. In this case, due to the recurrence relations, the number of required flops increases only as $\sim O(N^2)$ (a in opposite to $\sim O(N^3)$ flops for the matrix of the general form). This fact allows us to expedite the calculation of the propagator for multi-qubit systems. The method described below is a numerical implementation of the general algorithm for calculating the propagator (see article [2]).

2. Method of solution

It is known that the evolution operator $U(t)$ obeys the equation

$$i\hbar\frac{\partial U(t)}{\partial t} = H(t)U(t), \tag{1}$$

where $H(t)$ is the Hamiltonian of a system which may be considered as a matrix $N \times N$ for the n qubits in the cluster ($N = 2^n$). It was shown in [2] that the Magnus expansion [3, 4] in the form of an exponent from some operator $M(t)$ (the "Magnus"operator) can be used to calculate the propagator $U(t)$ of the n-qubit system. Using the substitution $U(t) = e^{-iM(t)}$ in the solution of Eq. (1), we obtain a recurrent formula for the calculation $M(t)$ in the form of a series of degrees of the Hamiltonian with any required precision [5]:

$$M = \sum_{i=1}^{\infty} M_i, \quad M = \frac{1}{\hbar}\int_0^t dt_1\left(H(t_1) - \frac{1}{2}[M_{n-1}(t_1), H(t_1)] - \frac{i}{12}[M_{n-1}(t_1), [M_{n-1}(t_1), H(t_1)]] + \cdots \right). \tag{2}$$

Obviously, the dimension of the matrix $M(t)$ is also will be equal to $N \times N$. Further, according to the well-known Cayley-Hamilton theorem [6], the propagator $M(t)$ can be rewritten as a matrix polynomial and the maximum degree of the resulting polynomial is not higher $N-1$:

$$U(t) = \sum_{j=0}^{N-1} u_j M^j(t). \tag{3}$$

The expansion coefficients u_j can be written explicitly through the eigenvalues of the matrix $M(t)$. The matrix $M(t)$ can be diagonalzed by an unitary transformation: $M(t) = S^{\dagger}(t)\mu(t)S(t)$, where S is the unitary matrix and obeys the property: $S^{\dagger}(t)S(t) = I$, and where the elements $\mu_j(t)$ ($j = 1, 2, \cdots, N$) are eigenvalues of the matrix $M(t)$. Note that the matrix has no trace ($Tr(M(t)) = 0$), or $\sum_{j=1}^{N} \mu_j(t) = 0$.

Applying this unitary rotation to both sides of Eq. (3), we obtain:

$$\sum_{j=1}^{N} \mu_k^{j-1} u_{j-1} = e^{-i\mu_k}, \quad k = 1, \ldots, N \tag{4}$$

where on the left side there is appeared a Vandermonde matrix, the elements of which are the powers of the eigenvalues of the matrix $M(t)$. By inverting the Vandermonde matrix, one can find all the coefficients of the expansion of the evolution operator Eq. (3) through the eigenvalues of the matrix $M(t)$. It is obvious that up to $N \le 4$ expansion coefficients are found analytically, and at $N > 4$ by numerical methods.

Thus, the practical value of the developed technique is that the vector of states of qubits at the current time is determined according to Eq. (3) the "instant" value of the propagator, the elements of which are calculated according to the above scheme.

3. Results and discussion

In this paper we consider only the small registers of interacting superconducting qubits consisting of three ($n=3$) and four ($n=4$) coupled qubits [7]. We compare the results for populations of levels obtained by the proposed approximation method of Magnus Eq. (3) and on the basis of numerical solution of Eq. (1).

For practical purposes, it is useful to simulate the control of qubits by ultra-short unipolar pulses (fluxons in Josephson transmission lines, JTLs, see Fig 1 (a) for example). These pulses propagate along the nonlinear waveguide lines which are coupled magnetically with the qubits. Implementation of such control scheme based on Josephson lines with appropriate parameters is described in [8-10]. The Hamiltonian of such registers can be represented as:

$$H(t)=-\frac{1}{2}\sum_{j=1}^{n}\left(\Delta_j\sigma_x^{(j)}+\varepsilon_j(t)\sigma_z^{(j)}\right)+\frac{1}{2}\sum_{k=1}^{n}\sum_{l=k+1}^{n}J_{kl}\sigma_z^{(k)}\sigma_z^{(l)}, \tag{5}$$

where Δ_j is the tunnel energies of each of the qubits in the register, $\varepsilon_j(t)=\varepsilon_{0,j}+\varepsilon_{m,j}e^{-\frac{(t-t_{0,j})^2}{2\tau_j^2}}$ – is the driving unipolar pulses of Gaussian shape with amplitudes $\varepsilon_{m,j}$, durations τ_j and bias $\varepsilon_{0,j}$. Interaction in the register of qubits is carried out in pairs, and the coupling constant is defined by J_{kl}. The expression (5) contains Pauli matrices $\sigma_x^{(j)}=\begin{pmatrix}0&1\\1&0\end{pmatrix}$ and $\sigma_z^{(j)}=\begin{pmatrix}1&0\\0&-1\end{pmatrix}$ for each qubit $j=1...n$.

We believe that the system is initialized at the initial time $\varepsilon_j(0)=\varepsilon_{0,j}$, populated only the ground state $W_1=|\langle\alpha_1|\alpha_1\rangle|^2=1$, where $H|\alpha_k\rangle=E_k|\alpha_k\rangle$, $k=1...n^2$. In the field of unipolar pulses, we will observe the dynamics of the population levels of qubits $W_k(t)=|\langle\alpha_k|\psi(t)\rangle|^2$, where the wave function of the system is defined at the current time: $\psi(t)=U(t)\psi(0)$. To compare with the calculations by the interpolation scheme, in parallel the evolutionary dynamics is studied numerically on the basis of the direct solution of Eq. (1) on the basis of the Cayley scheme. The unitary rotation at the short time δt may be calculated as:

$$U(\delta t)=\frac{I-iH\delta t/2\hbar}{I+iH\delta t/2\hbar}, \tag{6}$$

where I is the identity $N\times N$ matrix. Another way is to calculate the evolutionary dynamics by using the Magnus decomposition according of the Eq. (3). It is easy to show that the numerical solution Eq. (6) and finding time dependent vector $|\psi(t)\rangle=U(t)|\alpha_1\rangle$ it is necessary to employ $\sim O(2^{3n})$ computational operations, in contrast to realize the Magnus method $|\psi_M(t)\rangle=U_M(t)|\alpha_1\rangle$, in which to invert the Vandermonde matrix it is necessary $\sim O(2^{2n})$ operations, where n is the number of qubits in the register.

Figure 1 (a) schematically demonstrates a method for controlling the states of three coupled qubits ($n=3$ in Eq. (5)) by a sequence of unipolar pulses propagating along the Josephson transmission line. The levels of the qubits shown in the inset figure 1 (b), which are obtained by the solution $H|\psi_k\rangle=E_k|\psi_k\rangle$ depending on the bias. There is a redistribution of population levels in the field of pulses. We can attain the necessary population levels, if we change the parameters of the control fields in a certain way. For example, figure 1 (b) shows the process of rapidly reading the states of a three qubit system. That is the state of the system after the pulse action has not changed. As shown in figure 1 (b) an approximate submission of the evolution operator according to Magnus (points in figure 1 (b)) and the numerical solution of the Schrödinger equation (solid curves in figure 1 (b)) agree well with each other under the chosen parameters.

IOP Conf. Series: Journal of Physics: Conf. Series **1163** (2019) 012076 doi:10.1088/1742-6596/1163/1/012076

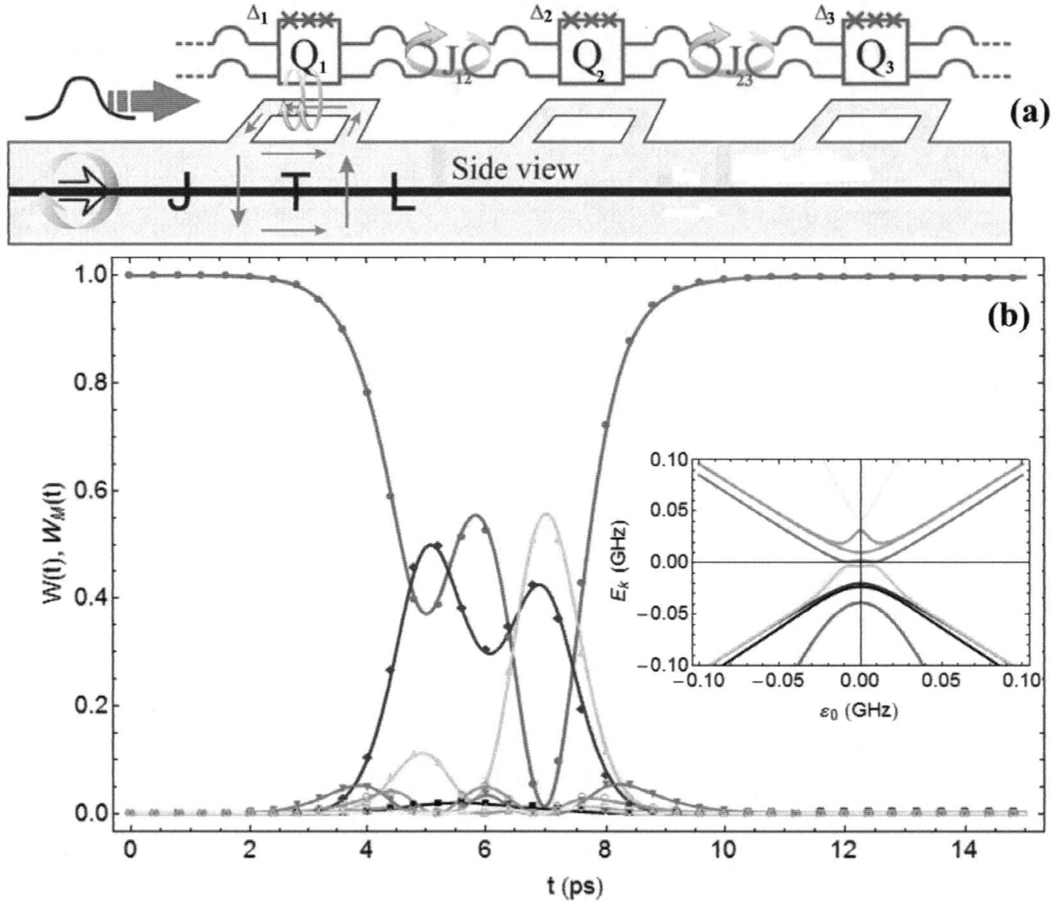

Figure 1 (a) Scheme of three coupled Josephson qubits controlled by unipolar pulses. (b) Dynamics of population levels of qubits ($\Delta_1 = \Delta_3 = 0.01$ GHz, $\Delta_2 = 0.015$ GHz, $J_{12} = J_{23} = 0.01$ GHz $J_{13} = 0.005$ GHz), where solid curves $W_k(t)$ represent the results of numerical simulation, and the points showing analytical solution $W_{k,M}(t)$. Here the colors correspond to the following levels: k = 1 – red, 2 – black, 3 – blue, 4 – green, 5 – gray, 6 – orange, 7 – pink, 8 – yellow. These are parameters of unipolar pulses: $\varepsilon_{0,1} = \varepsilon_{0,2} = \varepsilon_{0,3} = 0, \varepsilon_{m,1} = \varepsilon_{m,3}$ =2.5 GHz, $\varepsilon_{m,2} = 1.5\varepsilon_{m,1}$, t_{01} =5 ps, t_{02} = 6 ps, t_{03} =7 ps, $\tau_1 = \tau_3$ = 1 ps, $\tau_2 = 2\tau_1$.

Similar calculations were performed for a register consisting of four interacting qubits ($n = 4$ in Eq. (5)). The energy levels of the four qubits as a function of the bias parameter $\varepsilon_0 = \varepsilon_{0,j}$ $(j = 1...n)$ are shown in the figure 2 (a). As shown in figure 2, we can control the populations of multi-level system ($N = 2^4$) under the influence of four pulse sequence. We select amplitudes and durations of unipolar pulses for realization of the required gate operations, by analogy with one and two qubits [11]. Probabilities of the level populations calculated on the basis of Magnus theory (points in figure 2) are in good correspondence (solid curves in figure 2) with direct numerical calculation according to the Eq. (6) within the applicability of the Magnus decomposition approach [3], at $\Delta_j \tau_j \ll 1$ ($j = 1...n$). For demonstration in figure 2 (a) shows the evolution of the populations of the first five levels. On the figure 2 (b) we present a histogram of the population distribution of all 16 levels at the time of $t = 15$ ps, when unipolar impact is over. As can be seen from the figure 2 (b), by choosing the appropriate parameters of the control pulses, it is possible to create the required population of a certain group of levels, which can be useful for design of devices for initialization of quantum states.

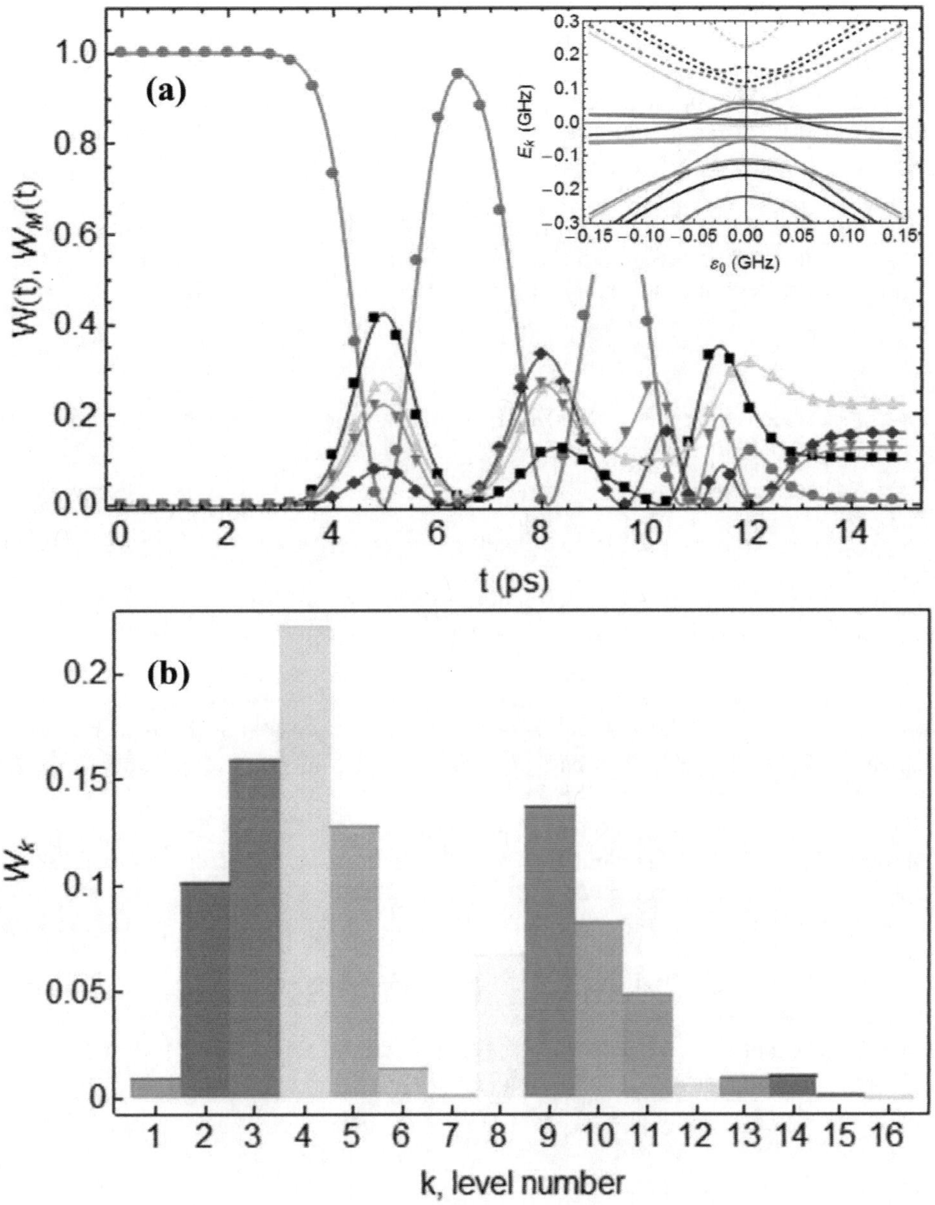

Figure 2 (a) Time dependent population dynamics of the first five levels of four coupled qubits ($\Delta_1 = \Delta_4 = 0.01$ GHz, $\Delta_2 = \Delta_3 = 0.015$ GHz, $J_{12} = J_{23} = J_{34} = 0.01$ ГГц, $J_{13} = J_{24} = 0.005$ GHz), where solid curves $W_k(t)$ represent the results of numerical simulation, and points showing analytical solutions $W_{k,M}(t)$. The colors are similar to those shown in Fig. 1. (b) This is a histogram of the populations of the levels of four coupled qubits at time t = 15 ps. Here are the parameters of unipolar pulses: $\varepsilon_{0,j} = 0$, $\varepsilon_{m,1} = 2.6$ GHz, $\varepsilon_{m,2} = 2.3$ GHz, $\varepsilon_{m,3} = 3.6$ GHz, $\varepsilon_{m,4} = 2$ GHz, $t_{01} = 5$ ps, $t_{02} = 8$ ps, $t_{03} = t_{04} = 11$ ps, $\tau_j = 1$ ps, $j = 1...4$.

4. Conclusion

The proposed approach allows speeding up the processes of numerical calculations of multi-qubit systems. Since the Magnus method is based on the interpolation procedures it is not tied to choose the number of steps in the discrete schemes. The developed technique can be generalized to the case of a large number of qubits in the register. We can also study the behavior of the response functions of the system from the field parameters after unipolar action, when the state of the system will be steady-state. At the same time, this approach allows one to determine quickly the state of the system at the time of interest (no need to realize the "step by step" procedure). This significantly reduces the time spent on numerical simulation.

From a practical point of view, the obtained results have demonstrated the effectiveness of the Magnus methods for the analysis of control procedures of multi-qubit systems with picosecond pulses realized in superconducting electronic circuits [9-12]. Note that this technique allows us to go beyond of the perturbation theory and to consider the pulses of large amplitude to obtain noticeable populations.

Acknowledgments
This work was supported by the RNF No. 18-72-00158 and the RFBR (grants No. 18-07-01206, 16-29-09515, and 16-07-01012).

References

[1] Landau L D and Lifshitz E M 1965 *Quantum Electrodynamics* (A Course of Theoretical Physics vol 4).
[2] Denisenko M V, Klenov N V, and Satanin A M 2018 *JETP* **155** (1).
[3] Magnus W 1954 *Commun. on Pure and Appl. Math.* **7** 649.
[4] Pechukas P, Light J L 1966 *Journ. of Chem. Phys.* **44** 3897.
[5] Prato D and Lamberti P W 1997 *The Journal of Chemical Physics* **106** 4640.
[6] Gantmacher F R 1959 *The Theory of Matrices*. (Chelsea Publidhing Company, New York)
[7] Niskanen A O, Harrabi K, Yoshihara F, Nakamura Y, and Tsai J S 2006 *Phys. Rev. B* **74**, 220503(R).
[8] Ustinov A V, Cirillo M, and Malomed B A 1993 *Phys. Rev. B* **47**, 8357(R).
[9] McDermott R, Vavilov M G, Plourde B L T, Wilhelm F K, Liebermann P J, Mukhanov O A, and Ohki T A 2018 *Quantum science and technology* **3** 024004.
[10] Soloviev I I, Klenov N V, Bakurskiy S V, Pankratov A L, and Kuzmin L S 2014 *Appl. Phys. Lett.* **105** 202602.
[11] Soloviev I I, Klenov N V, Bakurskiy S V, Kupriyanov M Y, Gudkov A L, Sidorenko A S 2017 *Beilstein J. Nanotech.* **8** 2689.
[12] Klenov N V, Kuznetsov A V, Soloviev I I, Bakurskiy S V, Denisenko M V, and Satanin A M 2017 *Low Temperature Physics* **43**(7) 789.

IOP Conf. Series: Journal of Physics: Conf. Series **1163** (2019) 012077 doi:10.1088/1742-6596/1163/1/012077

Inhomogeneous magnetization of a thin film of a ferromagnetic semiconductor in an electric field

V M Chetverikov

Department of Applied Mathematics of Moscow Institute of Applied Mathematics, National Research University Higher School of Economics, Moscow, Russia

E-mail: vchetverikov@hse.ru

Abstract. A theoretical model describing the spontaneous magnetization of a ferromagnetic semiconductor (InMn)As film in the presence of an external electric field directed across the film is considered. It is assumed that the ions of a manganese impurity with spin 5/2 are acceptors, have a uniform spatial distribution inside the semiconductor, and do not change their position under the action of an external field. The motion of holes with spin ½ changes their spatial distribution under the action of the field. The exchange interaction between manganese ions and holes allows the formation of magnetization that is non-uniform across the film thickness.

In particular, the existence of a piecewise continuous solution describing the presence of a phase transition boundary for magnetization inside a ferromagnetic semiconductor film is shown.

1. Introduction

It is known that semiconductors and ferromagnetic materials play complementary roles in the technology of storage and the current processing of information [1]. That is why the study of phenomena occurring in a film of ferromagnetic metals on the surface of a semiconductor was the most attractive [2]-[3].

The discovery in 1992 of a hole-like ferromagnetic order in (InMn)As motivated studies on the Mn-doped GaAs semiconductor and other III-V group materials [4]-[7]. The ferromagnetic transition, which was found in GaAs at temperatures above 100 K, gave rise to the hope that a new magnetic medium would appear which could open new paths for the technology of information processing and storing [8]. Report [9] on the detection of ferromagnetism in (GaMn)N at room temperature (which was predicted in [10]) increased the interest in this class of materials.

We assume that the Mn atom acts as an acceptor upon substitution on the cation site of the III-V semiconductor, leaving the negatively charged manganese ion half-filled d shell with angular momentum L = 0 and spin S = 5/2 [8]. There are also experimental evidences [7], [11], [12] that, in these materials, the ferromagnetism occurs because of the interaction between the local moments of different Mn ions with the help of holes arising in the semiconductor valence band. To control the magnetization of the above-mentioned connections, an external electric field can be used. Detection of many new properties of semiconductor heterostructures [13]-[17] allows one to explore the physics of previously unavailable combinations of quantum structure and magnetism in semiconductors.

2. Derivation of the system of equations

In this paper, we consider a mathematical model [18] describing the magnetization of a thin (of the order of 1-10 nm) ferromagnetic semiconductor film $(Ga_{0.95}Mn_{0.05})As$ in the presence of a constant electric field perpendicular to the surface. It is assumed that the longitudinal dimensions of the film are much larger than the thickness but smaller than the size of the ferromagnetic domain.

Therefore, we here do not take into account the magnetic interactions, which are much smaller than the exchange interactions. The problem becomes one-dimensional, depending on one x coordinate perpendicular to the film surface. The value of L is the thickness of the film. We consider the free energy F per unit surface area of the film

$$F = E_{ex} + F_1 + F_2 + F_3 + F_4 \tag{1}$$

where E_{ex} is the energy of exchange interaction of mobile holes and fixed ions of manganese, F_1 is the free energy associated with the entropy of the distribution of spins and over the ions of manganese, F_2 is the free energy of holes, F_3 is the energy of the electrostatic field, and F_4 is the energy of interaction of mobile holes with an electrostatic field.

We introduce the density of manganese ions with the direction of spins 5/2 and -5/2 along the easy magnetization axis $n_-^{5/2}(x), n_-^{-5/2}(x)$ and the density of holes with different projections on this axis $n_+^{1/2}(x), n_+^{-1/2}(x)$. The superscript here indicates the spin projection, and the subscript indicates the sign of the charge.

$$E_{ex} = -\varepsilon_{ex} \cdot \int_0^L \left(n_+^{1/2}(x) - n_+^{-1/2}(x)\right) \cdot \left(n_-^{5/2}(x) - n_-^{-5/2}(x)\right) dx, \quad \varepsilon_{ex} > 0 \tag{2}$$

$$F_1 = -\Theta \cdot \int_0^L \left(n_-^{5/2}(x) \cdot ln\left(n_- / n_-^{5/2}(x)\right) + n_-^{-5/2}(x) \cdot ln\left(n_- / n_-^{-5/2}(x)\right)\right) dx,$$
$$n_- = n_-^{5/2}(x) + n_-^{-5/2}(x) \tag{3}$$

The value Θ in (3) is equal to the product $k \cdot T$, where T is the temperature on the Kelvin scale and k is Boltzmann's constant. The second equation in (3) corresponds to the immobility of manganese ions whose total density is given by a quantity n_- that does not depend on the coordinate

$$F_2 = \int_0^L \left(\Omega(\mu^+(x), \Theta) + \mu^+(x) \cdot n_+^{1/2}(x)\right) dx + \int_0^L \left(\Omega(\mu^-(x), \Theta) + \mu^-(x) \cdot n_+^{-1/2}(x)\right) dx \tag{4}$$

The thermodynamic potential F_2 of holes of mass m determines the relationship between the chemical potential μ and the density of holes n_+. The first integral in (4) corresponds to holes, with the direction of the spin along the axis of easy magnetization, the second integral, to holes with the opposite direction of the spin. The following equations hold for the thermodynamic potential [19]:

$$\Omega(\mu, \Theta) = \left(\frac{m \cdot \Theta}{2\pi}\right)^{3/2} \cdot \Theta \cdot Li_{5/2}\left(-\exp(\mu / \Theta)\right), \quad Li_j(z) = \sum_{k=1}^{\infty} \frac{z^k}{k^j},$$
$$n_+^{1/2}(x) = -\frac{\partial}{\partial \mu} \Omega(\mu, \Theta)\Big|_{\mu = \mu^+(x)} = -\left(\frac{m \cdot \Theta}{2\pi}\right)^{3/2} \cdot Li_{3/2}\left(-\exp(\mu^+(x) / \Theta)\right), \tag{5}$$
$$n_+^{-1/2}(x) = -\frac{\partial}{\partial \mu} \Omega(\mu, \Theta)\Big|_{\mu = \mu^-(x)} = -\left(\frac{m \cdot \Theta}{2\pi}\right)^{3/2} \cdot Li_{3/2}\left(-\exp(\mu^-(x) / \Theta)\right).$$

The values of the parameters are given in atomic units, in which the unit of length is $\hbar^2 / (m_e \cdot e^2) \approx 0.529 \cdot 10^{-10} m$ and the unit of energy is $m_e \cdot e^4 / \hbar^2 \approx 27{,}21\, eV$.

When writing formulas (5), we use the notation $Li_j(z)$ for a special function — polylogarithm. There is another notation $PolyLog[j, z]$ in the Mathematica package. In the calculations related to formulas (5), an asymptotic formula for polylogarithms was used,

$$Li_j(-e^M) \underset{M \to \infty}{\to} -M^j / \Gamma(j+1), \Gamma(z) = \int_0^{\infty} e^{-t} \cdot t^{z-1} dt, \tag{6}$$

which ensured the 1% accuracy of calculations with $M = \mu / \Theta > 10$. Using asymptotic formula (6), we write the last two equations in (5) in the form

$$n_+^{\pm 1/2}(x) = \frac{n_-}{2}\left(\frac{\mu^{\pm}(x)}{\varepsilon_F}\right)^{3/2}, \quad \varepsilon_F = \left(3\pi^2 \cdot n_-\right)^{2/3} \cdot (2m)^{-1}. \tag{7}$$

For mobile holes, the electroneutrality condition is true:

$$\int_0^L \left(n_+^{1/2}(x)+n_+^{-1/2}(x)-n_-\right)dx=0. \tag{8}$$

The energy of the electrostatic field F_3 and the interaction energy F_4 of mobile holes with this field are determined by the following formulas in which χ is the dielectric constant of the semiconductor and E is the strength of the external electrical field:

$$F_3=\frac{\chi}{8\pi}\int_0^L\left(\frac{d\varphi}{dx}\right)^2 dx \quad , \tag{9}$$

$$\frac{d\varphi}{dx}\Big|_{x=0}=\frac{d\varphi}{dx}\Big|_{x=L}=-\frac{E}{\chi}, \tag{10}$$

$$F_4=\int_0^L\left(n_+^{1/2}(x)+n_+^{-1/2}(x)-n_-\right)\cdot\varphi(x)dx. \tag{11}$$

It is assumed that a thin ferromagnet film is separated from the metal electrodes by sufficiently thick layers of an insulator, so that the electric field does not cause a current.

As a result of minimizing the total free energy F with respect to unknown functions $\varphi(x), n_-^{5/2}(x), n_+^{1/2}(x), n_+^{-1/2}(x)$, we obtain four equations that allow us to find the density of the magnetic moment as a function of the coordinate x and the temperature T:

$$M(x,T)=g\cdot\left(\frac{5}{2}\left(n_-^{5/2}(x)-n_-^{-5/2}(x)\right)+\frac{1}{2}\left(n_+^{1/2}(x)-n_+^{-1/2}(x)\right)\right). \tag{12}$$

In formula (12), it is assumed that the gyromagnetic ratio g is the same for a hole and an Mn ion.

A variation in the free energy (1) with respect to the potential together with the boundary conditions (10) leads to the obvious boundary-value problem:

$$\frac{d^2\varphi}{dx^2}=-\frac{4\pi}{\chi}\cdot\left(n_+^{1/2}(x)+n_+^{-1/2}(x)-n_-\right). \tag{13}$$

A variation in the free energy (1) with respect to the density $n_-^{5/2}(x)$ gives a local connection between the differences in manganese densities with different projections of spins on the axis of easy magnetization with the same difference in the hole densities:

$$n_-^{5/2}(x)-n_-^{-5/2}(x)=n_-\cdot\mathrm{Tanh}\left(\frac{\varepsilon_{ex}}{\Theta}\cdot\left(n_+^{1/2}(x)-n_+^{-1/2}(x)\right)\right). \tag{14}$$

A variation in the free energy (1) with respect to the densities $n_+^{1/2}(x), n_+^{-1/2}(x)$, with the electroneutrality condition (8) taken into account, leads to a local connection between the chemical potentials, the electric potential, and the difference in the hole densities with different spin directions:

$$\mu^{\pm}(x)=\varepsilon_F-\varphi(x)\pm\varepsilon_{ex}\cdot n_-\cdot\mathrm{Tanh}\left(\frac{\varepsilon_{ex}}{\Theta}\cdot\left(n_+^{1/2}(x)-n_+^{-1/2}(x)\right)\right)=$$

$$=\varepsilon_F\cdot\left(1-\psi(x)\pm\delta\cdot\mathrm{Tanh}\left(\frac{\varepsilon_{ex}}{\Theta}\cdot\left(n_+^{1/2}(x)-n_+^{-1/2}(x)\right)\right)\right), \tag{15}$$

$$\psi(x)=\varphi(x)\cdot\left(\varepsilon_F\right)^{-1},\quad \delta=\varepsilon_{ex}\cdot n_-\cdot\left(\varepsilon_F\right)^{-1}.$$

The constant ε_F (Fermi energy) in (15) appears from the condition $\mu^{\pm}(x)=\varepsilon_F$ in the absence of an electric field and under the condition $n_+^{1/2}(x)-n_+^{-1/2}(x)=0$.

3. System of 5 equations for three functions

Finally, we can write a system of five equalities for three functions: dimensionless potential $\psi(x)$ of the self-consistent electric field, density of the spatial distribution of positive hole charge $q(x)$ and the spatial distribution of the difference $\Delta(x)$ in the densities of holes with different spin directions.

The transition from unknown hole densities $n_+^{1/2}(x), n_+^{-1/2}(x)$ to the functions $q(x)$ and $\Delta(x)$ is determined by two equalities

$$n_+^{1/2}(x)+n_+^{-1/2}(x)=q(x)\cdot n_-,\quad n_+^{1/2}(x)-n_+^{-1/2}(x)=\Delta(x)\cdot n_-,$$
$$n_+^{1/2}(x)\ge 0,\quad n_+^{-1/2}(x)\ge 0,\quad n_->0,\quad \Delta(x)\in[-1,1],\quad x\in[0,L]. \tag{16}$$

We note that the density of magnetization (12) that we are ultimately interested in is completely determined by the function $\Delta(x)$.With the help of the functions $q(x)$, $\Delta(x)$, the above five-equation system has the following form:

$$\frac{d^2\psi}{dx^2}=-\gamma\cdot\left(q(x)-1\right),\quad x\in(0,L),\qquad \gamma=4\pi\cdot n_-\cdot\left(\chi\cdot\varepsilon_F\right)^{-1}>0, \tag{17}$$

$$\frac{d\psi}{dx}\Big|_{x=0}=\frac{d\psi}{dx}\Big|_{x=L}=-A,\quad A=E\cdot\left(\chi\cdot\varepsilon_F\right)^{-1}>0,\ L>0\quad , \tag{18}$$

$$q(x)=(1/2)\cdot\left(1-\psi(x)+a(x)\right)^{3/2}+(1/2)\cdot\left(1-\psi(x)-a(x)\right)^{3/2},$$
$$a(x)=\delta\cdot\mathrm{Tanh}[(b/T)\cdot\Delta(x)],\quad b=\varepsilon_{ex}\cdot n_-/k>0. \tag{19}$$

$$\Delta(x)=(1/2)\cdot\left(1-\psi(x)+a(x)\right)^{3/2}-(1/2)\cdot\left(1-\psi(x)-a(x)\right)^{3/2}. \tag{20}$$

Equations (17), (18) are simply equations (13) and (10) written by using the dimensionless potential $\psi(x)$. Equations (19)-(20) directly follow from equations (15), (16), in which the densities $n_+^{1/2}(x), n_+^{-1/2}(x)$ are expressed through the self-consistent potential $\psi(x)$ and the difference in the densities $\Delta(x)$. In this case, equation (20) can be represented in a slightly different form:

$$\Delta(x)=\Lambda\left(\psi(x),\Delta(x)\right)\cdot\mathrm{Tanh}[(b/T)\cdot\Delta(x)]/q(x),$$
$$\Lambda\left(\psi(x),\Delta(x)\right)=\delta\cdot\{((3/2)\cdot(1-\psi(x))^2+(1/2)\cdot(\delta\cdot\mathrm{Tanh}[(b/T)\cdot\Delta(x)])^2\}, \tag{21}$$

and precisely this form will be used to solve the joint system of equations (17) - (19) and (21) to find the functions $\psi(x)$ and $\Delta(x)$.

In the absence of an electric field, all the functions appearing in (17) - (21) are spatially homogeneous: $\psi(x)=\psi_{01}=const, q(x)=1, \Delta(x)=\Delta$. The values ψ_{01}, Δ depend on the temperature T and are determined by solving the system of equations:

$$(1-\psi_{01}+a_0(\Delta))^{3/2}+(1-\psi_{01}-a_0(\Delta))^{3/2}=2,$$
$$a_0(\Delta)=\delta\cdot\mathrm{Tanh}[(b/T)\cdot\Delta], \tag{22}$$
$$\Delta=\delta\cdot\left((3/2)\cdot(1-\psi_{01})^2+(1/2)\cdot(a_0(\Delta))^2\right)\cdot\mathrm{Tanh}[(b/T)\cdot\Delta].$$

The system of equations (22) is a special case of system (19), (21) and always has trivial solution $\psi_{01}=0, \Delta=0$. This solution corresponds to the absence of magnetization (12). Note that if system (22) has solution $\psi_{01}\ne 0, \Delta>0$, then it also has solution $\psi_{01}\ne 0, \Delta<0$.

In the region $\delta\in(0,0.25)$ and in the temperature region $T\in[0,T_C)$, there is a solution of equations (22) with the one percent accuracy: $\psi_{01}=0.25\cdot(a_0(\Delta))^2, \Delta>0$. Here T_C is determined by the formula

$$T_C=(3/2)\cdot b\cdot\delta. \tag{23}$$

Obviously, Δ tends to zero as T tends to the Curie temperature T_C. This follows from the structure of the third equation in (22).

Checking the fulfillment of the applicability condition for the asymptotic expression for polylogarithms $\mu/\Theta > 10$ for chemical potential $\mu = \mu_0^- = \varepsilon_F \cdot (1 - 0.25 \cdot \delta^2 - \delta)$ in the temperature range $T \in [0, T_C)$ shows that this condition is satisfied only for $\delta < 0.225$. If we take into account the increase in the chemical potential up to $\mu = \mu_C^- = \varepsilon_F$ at $T = T_C$, then the limit on the value δ can be relaxed up to $\delta < 0.258$. Since $\delta \sim \varepsilon_{ex} \cdot (n_-)^{1/3}$, the possibility to use asymptotic formulas (6) with the specified accuracy imposes upper limits on the magnitude of the exchange interaction constant ε_{ex} and the manganese impurity density n_-.

In a spatially homogeneous case (without an external electric field), the Curie temperature is determined by relation (23) at any point of the film under consideration. However, in the presence of an electric field, some non-uniformity of spontaneous magnetization occurs in the film thickness: in each section $x_0 \in [0, L]$ of thickness, there is a local critical temperature,

$$T_{cr}(x_0) = (3/2) \cdot b \cdot \delta \cdot (1 - \psi(x_0)) \quad . \tag{24}$$

Formula (24) has the following meaning: at a temperature $T = T_C(x_0)$ in the region of the film $x \in [0, x_0]$, $\Delta(x) = 0$, but $\Delta(x) > 0$ in the region $x \in (x_0, L]$. The function $T_{cr}(x_0)$ is an increasing function of the coordinate x_0, and $T_{cr}(L)$ is the Curie temperature throughout the film thickness. Formula (24) is a consequence of equation (21) which always has trivial solution $\Delta = 0$, but there may still exist the piecewise continuous solution described above. The non-negative solution $\Delta(x) > 0$ in the area $x \in (x_0, L]$ should have the property: $\Delta(x) \to 0$ as $x \to x_0 + 0$ at a certain temperature $T = T_{cr}(x_0)$.

In the absence of an external electric field, the problem becomes spatially homogeneous and

$$T_{cr}(L) = (3/2) \cdot b \cdot \delta \cdot (1 - \psi(L))^{1/2} = (3/2) \cdot b \cdot \delta \cdot (1 - \psi_{01})^{1/2} . \tag{25}$$

Since $\psi_{01} = 0.25 \cdot (a_0(\Delta))^2$, for $\Delta = 0$ the value ψ_{01} is zero according to the second relation in (22). Therefore, formula (25) becomes formula (23) when the external field is turned off.

4. Conclusions

The considered model allows us to estimate the spatial magnetization distribution of a thin film of a ferromagnetic semiconductor in the presence of an external electric field.

It would be much more difficult to analyze the model without detailed numerical calculations if we did not use asymptotic formulas for polylogarithms. Precisely this permits writing equations with quite observable properties. In particular, one can obtain simple approximate formulas for the dependence of the Curie temperature on the spatial coordinate across the film.

The proposed formulas well approximate the results of numerical calculations. In addition, a simple verification criterion is proposed to control the accuracy of the approximation used for polylogarithms.

It is shown that, in the model under study, there are piecewise continuous solutions that describe the presence of the phase transition boundary for magnetization inside the film of a ferromagnetic semiconductor. The location of this phase transition boundary depends on the external electric field and the temperature.

Acknowledgments

In conclusion, the author is deeply grateful to professors Kozhushner M.A., Posvaynskii V.S., Trakhtenberg L.I. for the opportunity to get acquainted with their article [18] before it publication.

References

[1] Jungwirth T, Atkinson W A, Lee B H and MacDonald A H 1999 Interlayer coupling in ferromagnetic semiconductor superlattices *Physical Review B* **59**, No 15, 9818-9821.

[2] Matsukura F, Tokura Y and Ohno Hideo 2015 Control of magnetism by electric fields. *Nature nanotechnology*, vol **10** 209-219

[3] Ohno H, Chiba D , Matsukura F, Omiya T, Abe E, Dietl T, Ohno Y & Ohtani K 2000. Electric-field control of ferromagnetism *Nature* Vol **408** 21/28 944-946

[4] Ohno H, Munekata H, Penney T, von MolnaÂr S & Chang L L 1992 Magnetotransport properties of p-type (In,Mn)As diluted magnetic III-V semiconductor *Phys. Rev. Lett.* **68**, 2664-2667

[5] Munekata H, Zaslavsky A, Fumagalli P & Gambino R J 1993 Preparation of (In,Mn)As/(Ga,Al)Sb magnetic semiconductor heterostructures and their ferromagnetic characteristics *Appl. Phys. Lett.* **63**, 2929-2931

[6] Ohno H. 1998 Making nonmagnetic semiconductors ferromagnetic *Science* 281, 951-956

[7] Ohno H. 1999 Properties of ferromagnetic III-V semiconductors. *J. Mag. Magn. Mater.* **200**, 110-129

[8] Jungwirth T, Konig Jurgen, Sinova Jairo, Kucera J and MacDonald A H 2002 Curie temperature trends in III.Mn.V ferromagnetic semiconductors, *Physical review B* **66**, 012402

[9] Reed M L et al 2001 Room temperature ferromagnetic properties of (Ga,Mn)N *Appl. Phys. Lett.* **79**, 3473-3475

[10] Dietl T, Ohno H and Matsukura F 2001 Hole-mediated ferromagnetism in tetrahedrally coordinated semiconductors *Physical Review B*, Volume **63**, 195205-1—195205-21.

[11] Linnarsson M, Janzén E, Monemar B, Kleverman M and Thilderkvist A 1997 Electronic structure of the (GaMn)As center *Phys. Rev. B* **55**, 6938-6944

[12] Okabayashi J. et al 1998 Core-level photoemission study of $Ga_{1-x}Mn_xAs$ *Phys. Rev. B* **58**, R4211-R4214

[13] Ye P D et al 1995 Electrons in a Periodic Magnetic Field Induced by a Regular Array of Micromagnets *Phys. Rev. Lett.* **74** 3013-3016

[14] Carmona H et al 1995 Two Dimensional Electrons in a Lateral Magnetic Superlattice *Phys. Rev. Lett.* **74**, 3009-3012

[15] Smorchkova P, Samarth N, Kikkawa J M and Awschalom D D 1997 Spin Transport and Localization in a Magnetic Two-Dimensional Electron Gas *Phys. Rev. Lett.* **78**, 3571-3574

[16] Koshihara S et al 1997 Ferromagnetic Order Induced by Photogenerated Carriers in Magnetic III-V Semiconductor Heterostructures of (In,Mn)As/GaSb, *Phys. Rev. Lett.* **78**, 4617-4619

[17] Haury A et al 1997 Observation of a Ferromagnetic Transition Induced by Two-Dimensional Hole Gas in Modulation-Doped CdMnTe Quantum Wells *Phys. Rev. Lett.* **79**, 511-514

[18] Kozhushner M A, Lidskii B V, Posvaynskii VS, Trakhtenberg L I 2016 Effect of electric field on the magnetic characteristics of a ferromagnetic nanosemiconductor *JETP* v 123, No. 6, 1227-1232

[19] Landau L D, Lifshitz E M 1976 *Statistical Physics,* v.5, Part 1, (Moscow: Nauka)

IOP Conf. Series: Journal of Physics: Conf. Series **1163** (2019) 012078 doi:10.1088/1742-6596/1163/1/012078

Hole distribution in a film of a ferromagnetic semiconductor in the presence of an external electric field

A E Aminova, **V M Chetverikov**

Department of Applied Mathematics of Moscow Institute of Applied Mathematics, National Research University Higher School of Economics, Moscow, Russia

E-mail: vchetverikov@hse.ru

Abstract. The results of numerical calculations for the mathematical model proposed for describing the magnetization in a thin film of a ferromagnetic semiconductor at temperatures below the Curie temperature in the presence of an external electric field are presented. The theoretical prediction of the existence of a piecewise continuous solution, which describes the presence of the phase transition boundary for magnetization inside the film, is confirmed. The location of this phase transition boundary depends on the external electric field and temperature.

1. Introduction

The possibility of changing the spontaneous magnetization of a ferromagnetic semiconductor by applying an electric field was demonstrated in [1]. A thin film of a ferromagnetic semiconductor $(In_{0.97}Mn_{0.03})As$ 5 nm thick was placed on a substrate of different semiconductors, and the electric field, perpendicular to the film, changed its magnetization. These studies continue the earlier works [2] - [5], where a hole-like ferromagnetic order in (InMn)As and other materials of group III-V was studied.

In such a system, ferromagnetism is provided by the exchange interaction of holes with manganese ions by the Zener mechanism, i.e. indirect ferromagnetic interaction between ions is provided due to the sd -exchange of Vonsovsky-Ziner between holes and ions. The dependence of the density of states of the hole gas in the impurity band of energy, which noticeably overlaps with the valence band, is rather complicated. And for simplicity in the model calculation, it was assumed that this dependence is the same as in a gas of some effective mass. Theoretical assessments of these phenomena were presented in [6].

The quantum-statistical theory of influence of an electric field on the distribution of magnetization in a current-isolated ferromagnetic semiconductor was proposed in [7]. It was assumed that the manganese ions with spin 5/2 have a given spatially uniform distribution in the film, but may have a spatially non-uniform distribution of the spin direction on the axis of the easy magnetization, i.e.

$$n_- = n_-^{5/2}(x) + n_-^{-5/2}(x)\,. \tag{1}$$

The distribution density of mobile holes, which had spin ½, was not assumed to be spatially uniform $n_+(x) = n_+^{1/2}(x) + n_+^{-1/2}(x)$, but the electroneutrality condition was satisfied for the system of ions and holes:

$$\int_0^L \left(n_+^{1/2}(x) + n_+^{-1/2}(x) - n_-\right) dx = 0\,, \tag{2}$$

where L was the thickness of the film. The magnetization density of the film was determined as

$$M(x,T) = g \cdot \left(\frac{5}{2}\left(n_-^{5/2}(x) - n_-^{-5/2}(x)\right) + \frac{1}{2}\left(n_+^{1/2}(x) - n_+^{-1/2}(x)\right)\right), \tag{3}$$

depending on the coordinates and the temperature. In formula (3), it was assumed that the gyromagnetic ratio g was the same for a hole and an Mn ion. The energy of the exchange interaction between Mn ions and holes was described by the formula

$$E_{ex} = -\varepsilon_{ex} \cdot \int_0^L \left(n_+^{1/2}(x) - n_+^{-1/2}(x)\right) \cdot \left(n_-^{5/2}(x) - n_-^{-5/2}(x)\right) dx, \quad \varepsilon_{ex} > 0. \tag{4}$$

The relationships between the spatial distributions of ions and holes and the expression for the chemical potential of holes with different spin directions in the state of thermodynamic equilibrium had the forms

$$n_-^{5/2}(x) - n_-^{-5/2}(x) = n_- \cdot \mathrm{Tanh}\left(\frac{\varepsilon_{ex}}{\Theta} \cdot \left(n_+^{1/2}(x) - n_+^{-1/2}(x)\right)\right), \tag{5}$$

$$\mu^{\pm}(x) = \varepsilon_F - \varphi(x) \pm \varepsilon_{ex} \cdot n_- \cdot \mathrm{Tanh}\left(\frac{\varepsilon_{ex}}{\Theta} \cdot \left(n_+^{1/2}(x) - n_+^{-1/2}(x)\right)\right),$$
$$\varepsilon_F = \left(3\pi^2 \cdot n_-\right)^{2/3} \cdot (2m)^{-1}. \tag{6}$$

The value Θ in (5), (6) is equal to the product $k \cdot T$, where T is the temperature on the Kelvin scale and k is Boltzmann's constant. The value of ε_F corresponds to the Fermi energy for holes of mass m in the absence of an electric field, and $\varphi(x)$, to the potential of a self-consistent electric field in formula (6). The values of the parameters are given in atomic units in which the unit of length is $\hbar^2 / (m_e \cdot e^2) \approx 0.529 \cdot 10^{-10} m$, and the unit of energy is $m_e \cdot e^4 / \hbar^2 \approx 27,21\, eV$.

2. System of 5 equations for three functions

In this paper, we propose to use asymptotic relations for polylogarithms:

$$Li_j(-e^M) \underset{M\to\infty}{\to} -M^j/\Gamma(j+1), \Gamma(z) = \int_0^\infty e^{-t} \cdot t^{z-1} dt,$$

which allow us to obtain an algebraic relationship between the distribution densities of holes and their chemical potentials. Polylogarithms arise when writing Fermi-Dirac integrals. Replacing them with an asymptotic expression introduces an error in calculations that is less than one percent if the inequality $M = \mu / \Theta > 10$ is true. The fulfillment of this inequality should be verified for all solutions obtained in the temperature range not exceeding the Curie temperature.

This replacement allows us to write the ratio

$$n_+^{\pm 1/2}(x) = 2^{-1}(\varepsilon_F)^{-3/2} \cdot n_- \cdot (\mu^{\pm}(x))^{3/2}, \tag{7}$$

allowing us significantly to simplify the original task with the help of three new functions: the dimensionless potential $\psi(x) = \varphi(x) \cdot (\varepsilon_F)^{-1}$ of the self-consistent electric field, the density of the spatial distribution of positive hole charge $q(x) = (n_-)^{-1} \cdot \left(n_+^{1/2}(x) + n_+^{-1/2}(x)\right)$, and the spatial distribution of the difference $\Delta(x) = (n_-)^{-1} \cdot \left(n_+^{1/2}(x) - n_+^{-1/2}(x)\right)$ in the densities of holes with different spin directions.

The function $\Delta(x)$ determines the density of magnetization by formulas (3), (5), but it itself must satisfy the equation

$$\Delta(x) = \Lambda\left(\psi(x), \Delta(x)\right) \cdot \mathrm{Tanh}[(b/T) \cdot \Delta(x)] / q(x),$$
$$\Lambda\left(\psi(x), \Delta(x)\right) = \delta \cdot \{((3/2) \cdot (1 - \psi(x))^2 + (1/2) \cdot (\delta \cdot \mathrm{Tanh}[(b/T) \cdot \Delta(x)])^2\}, \tag{8}$$

where $q(x)$ is a function of $\Delta(x)$ and $\psi(x)$ according to the definition and formulas (6), (7):

$$q(x) = (1/2) \cdot \left(1 - \psi(x) + a(x)\right)^{3/2} + (1/2) \cdot \left(1 - \psi(x) - a(x)\right)^{3/2},$$
$$a(x) = \delta \cdot \mathrm{Tanh}[(b/T) \cdot \Delta(x)], \quad b = \varepsilon_{ex} \cdot n_- / k > 0. \tag{9}$$

Three more equalities represent the nonlinear Poisson equation for the potential $\psi(x)$ and the boundary conditions at the lower and upper boundaries of the film:

$$\frac{d^2\psi}{d\,x^2} = -\gamma\cdot\left(q(x)-1\right), \quad x\in(0,L), \quad \gamma = 4\pi\cdot n_-\cdot\left(\chi\cdot\varepsilon_F\right)^{-1} > 0, \tag{10}$$

$$\frac{d\psi}{d\,x}\Big|_{x=0} = \frac{d\psi}{d\,x}\Big|_{x=L} = -A, \; A = E\cdot\left(\chi\cdot\varepsilon_F\right)^{-1} > 0, \; L>0 \quad . \tag{11}$$

The value χ in (10), (11) is the dielectric constant of the semiconductor.

3. Results of analysis of the system of equations (8) - (11)

At the temperature $T=0$, the solution $\psi_0(x)$ of the boundary value problem (10) - (11) can easily be found numerically. Then from equation (9) one can easy find the charge density of holes

$$q_0(x) = (1/2)\cdot\left(1-\psi_0(x)+\delta\right)^{3/2} + (1/2)\cdot\left(1-\psi_0(x)-\delta\right)^{3/2} \tag{12}$$

and the magnitude of the difference $\Delta_0(x)$ from equation (8):

$$\Delta_0(x) = \delta\cdot\{((3/2)\cdot(1-\psi_0(x))^2 + (1/2)\cdot\delta^2\}/\,q_0(x)\,. \tag{13}$$

It is easy to see that the function $-\Delta_0(x)$ is also a solution to the problem in question. This is a manifestation of the general property of system (8) - (11): if any pair of functions $\psi(x), \Delta(x)$ is a solution, then the pair $\psi(x), -\Delta(x)$ is also a solution. Therefore, in the future we will limit ourselves to finding only positive functions $\Delta(x)$.

At very high temperatures (at temperatures above the Curie temperature) (formally for $T=\infty$) the solution $\psi_\infty(x)$ of the boundary value problem (10) - (11) can also be easily found numerically. Then from (9) it is easy to find the charge density of holes

$$q_\infty(x) = \left(1-\psi_\infty(x)\right)^{3/2} \tag{14}$$

and the relation $\Delta_\infty(x)=0$ from equation (8). Note that this solution exists at any temperature, but it corresponds to zero spontaneous magnetization.

Figures 1-5 show the results of calculations of the potential, the hole charge density and the hole density differences with opposite spin directions for the model case [7]

$$m=2,\; \varepsilon_{ex}=5,\; \chi=5,\; g=2,\; n_-=3\cdot10^{-4},\; E=0.01 \tag{15}$$

by formulas (8) - (14). The density of manganese ions given in (15) in atomic units corresponds to the density $2\cdot10^{21}\;cm^{-3}$ in ordinary units. The thickness of the film $L=100$ in atomic units corresponds to the thickness of 5.29 nm.

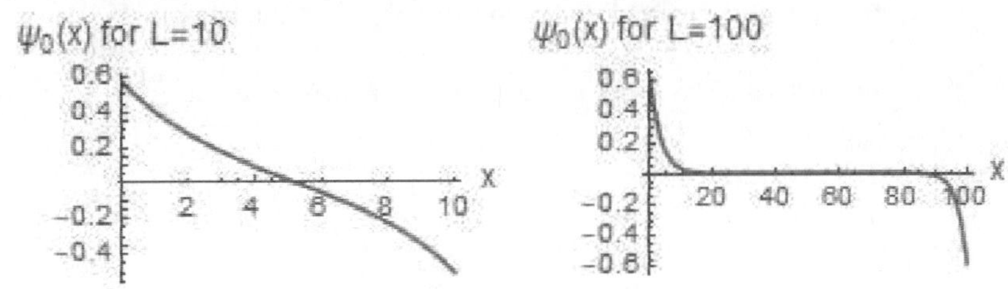

Figure 1. Dependence of the potential of a self-consistent electric field on the transverse film coordinate x at $T=0$ for different film thicknesses L.

Comparison of figures 1 and 2 shows that the potential of the self-consistent electric field at temperatures above T_C differs from the potential at $T=0$ very small.

IOP Conf. Series: Journal of Physics: Conf. Series **1163** (2019) 012078 doi:10.1088/1742-6596/1163/1/012078

The same situation for the spatial density of holes (regardless of the spin direction) is shown by comparison of figures 3 and 4.

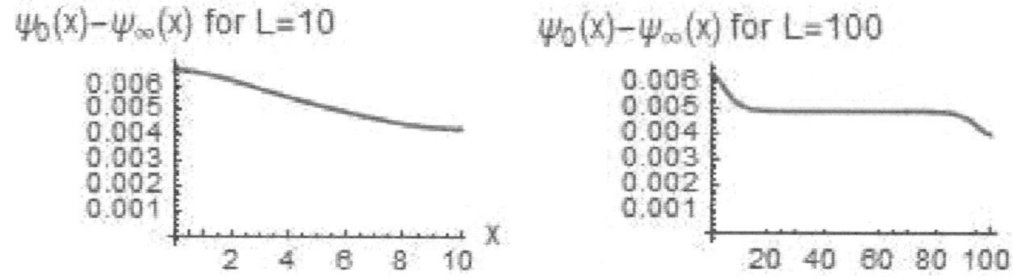

Figure 2. Comparison of the potentials of a self-consistent electric field with $T=0$ and $T=\infty$ for different film thicknesses L.

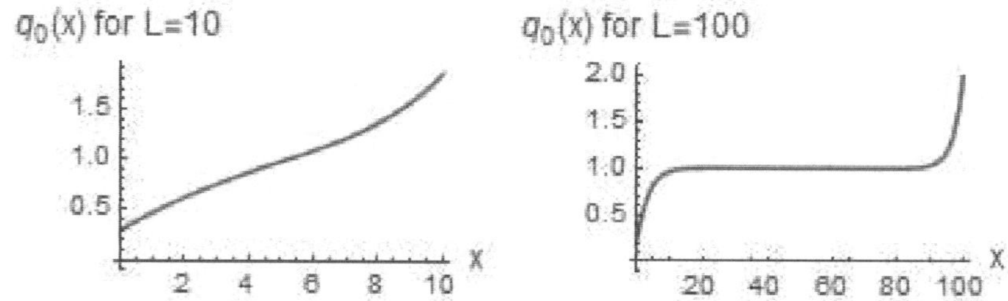

Figure 3. Dependence of the dimensionless density (charge) of all holes, regardless of the spin direction, on the transverse film coordinate x at $T=0$ for different film thicknesses L.

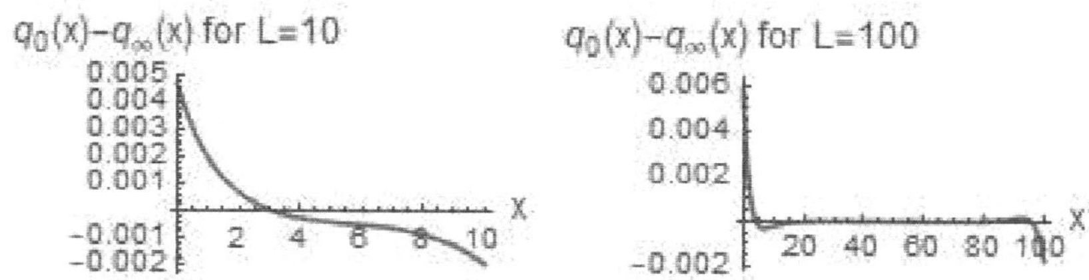

Figure 4. Comparison of the densities (charge) of all holes, regardless of the spin direction, with $T=0$ and $T=\infty$ for different film thicknesses L.

However, the value $\Delta(x)$ shows a strong change with increasing temperature from $\Delta_0(x)$ at $T=0$ (see figure 5) to the value $\Delta_\infty(x)=0$ for $x\in[0,L]$ at $T>T_C$.

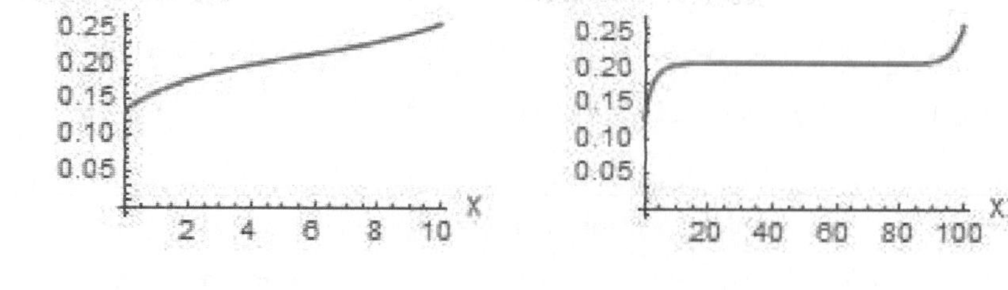

Figure 5. Dependence of the difference of the dimensionless density of holes with opposite spin orientation on the transverse film coordinate x at $T = 0$ for different film thickness L

The calculations by formulas (8) - (14) with $T = 0$ and with $T = \infty$ are rather simple, but the calculations at temperatures from zero to the Curie temperature require an iteration procedure for each value of T.

Numerical results for the function $\Delta(x)$ at four different temperatures are shown in figure 6. When $T = 80$, the distributions $\Delta(x)$ are piecewise continuous: $\Delta(x) = 0$ for $x \in [0,1.8]$ at $T = 80$ and $x \in [0,87]$ at $T = 100$.

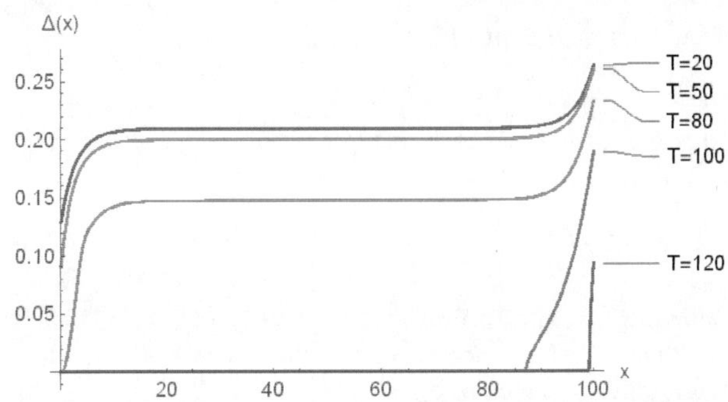

Figure 6. Distribution of the magnitude $\Delta(x)$ at different temperatures. Distributions at $T = 0$ and $T = 20$ are hardly distinguishable.

If we define the local Curie temperature $T_{cr}(x_0)$ in the cross section x_0 of a ferromagnetic semiconductor by the relation

$$T_{cr}(x_0) = (3/2) \cdot b \cdot \delta \cdot (1 - \psi(x_0))^{1/2}, \tag{16}$$

then it will exactly correspond to the data shown in figure 6.

Calculations show that $T_{cr}(1.8) = 80$ and $T_{cr}(87) = 100$ for the film thickness $L = 100$. Formula (16) is a consequence of equation (8) and allows determining the Curie temperature for the entire film thickness L as a whole by the formula

$$T_C = T_{cr}(L) = (3/2) \cdot b \cdot \delta \cdot (1 - \psi(L))^{1/2}. \tag{17}$$

To ensure that the condition of applicability of the asymptotic formula for polylogarithms at temperatures $T \le T_C = (3/2) \cdot \delta \cdot b \cdot \left(1 - \psi(L)\right)^{1/2}$, it is sufficient to satisfy the condition

$$(2/3) \cdot (1 - \psi(0)) > 10 \cdot \delta^2 \cdot \left(1 - \psi(L)\right)^{1/2}. \tag{18}$$

For the model parameters given in (15), condition (18) is satisfied.

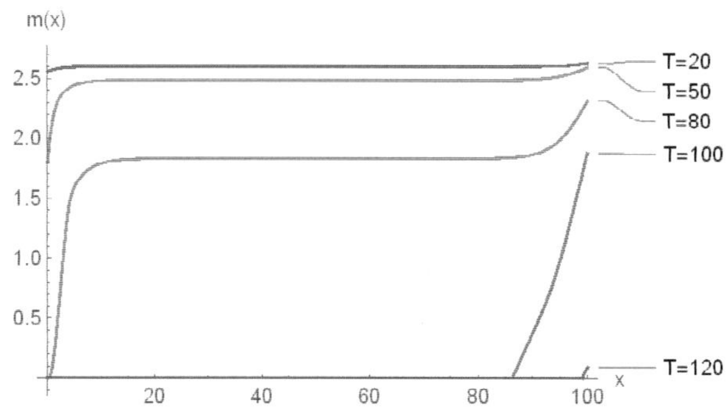

Figure 7. Distribution of magnetization $m(x)$ at different temperatures.

The density of magnetization of a film defined by formula (15) can be represented as

$$M(x,T) = g \cdot n_{-} \cdot m(x)$$

$$m(x) = (1/2)\left(\Delta(x) + 5 \cdot \mathrm{Tanh}[\left(b/T\right) \cdot \Delta(x)]\right).$$

As the comparison of figures 6 and 7 shows, the difference in magnitudes is explained by the large contribution of manganese ions to the magnetization.

4. Conclusions

It is the presence of an electric field that allows one to see the appearance of the phase transition boundary for magnetization inside a film of a ferromagnetic semiconductor. The proposed theoretical formulas for connecting the local Curie temperature for the boundary with the coordinates of the boundary itself correspond to numerical calculations.

References

[1] Ohno H, Chiba D , Matsukura F, Omiya T, Abe E, Dietl T, Ohno Y & Ohtani K 2000. Electric-field control of ferromagnetism *Nature* Vol **408** 21/28 944-946

[2] Ohno H, Munekata H, Penney T, von MolnaÂr S & Chang L L 1992 Magnetotransport properties of p-type (In,Mn)As diluted magnetic III-V semiconductor *Phys. Rev. Lett.* **68**, 2664-2667

[3] Munekata H, Zaslavsky A, Fumagalli P & Gambino R J 1993 Preparation of (In,Mn)As/(Ga,Al)Sb magnetic semiconductor heterostructures and their ferromagnetic characteristics *Appl. Phys. Lett.* **63**, 2929-2931

[4] Ohno H. 1998 Making nonmagnetic semiconductors ferromagnetic *Science* 281, 951-956

[5] Ohno H. 1999 Properties of ferromagnetic III-V semiconductors. *J. Mag. Magn. Mater.* **200**, 110-129

[6] Jungwirth T, Konig Jurgen, Sinova Jairo, Kucera J and MacDonald A H 2002 Curie temperature trends in III.Mn.V ferromagnetic semiconductors, *Physical review B* **66**, 012402

[7] Kozhushner M A, Lidskii B V, Posvaynskii VS, Trakhtenberg L I 2016 Effect of electric field on the magnetic characteristics of a ferromagnetic nanosemiconductor *JETP* v 123, No. 6, 1227-1232

IOP Conf. Series: Journal of Physics: Conf. Series **1163** (2019) 012079 doi:10.1088/1742-6596/1163/1/012079

Matrix multiplication and universal scalability of the time on the Intel Scalable processors

Alexander Russkov[1], **Lev Shchur**[2,3]

[1] Science Center in Chernogolovka, 142432 Chernogolovka, Russia
[2] National Research University Higher School of Economics, 101000 Moscow, Russia
[3] Landau Institute for Theoretical Physics, 142432 Chernogolovka, Russia

Abstract. Matrix multiplication is one of the core operations in many areas of scientific computing. We present the results of the experiments with the matrix multiplication of the big size comparable with the big size of the onboard memory, which is 1.5 terabyte in our case. We run experiments on the computing board with two sockets and with two Intel Xeon Platinum 8164 processors, each with 26 cores and with multi-threading. The most interesting result of our study is the observation of the perfect scalability law of the matrix multiplication, and of the universality of this law.

1. Introduction

At the fall of 2017 the new architecture of the computing nodes from Intel, combining the Intel(R) Xeon Scalable [1] architecture of the processor boards and Intel(R) Omni-Path Architecture [2] of the interconnections comes to the open market. It is a common thought, that these architectures will be one of the main basis for the high-performance computing in the nearest years. The new architecture becomes subject for the computer experiments with the goal to understand it's practical features, and some experiments are already reported. Application of the combination of Intel-OPA with previous architecture Intel(R) Xeon Phi discussed in [3]. The experiments with the High-Performance LINPACK (HPL) on the cluster equipped with Intel(R) Xeon Platinum 8160 processors demonstrate some interesting features of the memory organization, which shows, in particular, the increased DRAM traffic and increased L2 cache miss rates [4]. The technical details on the performance of the particular processors from the Intel Scalable series were published in the review [5].

In this communication, we present preliminary results of the experiments with the multiplication of the big matrices which requires a huge RAM storage of hundreds of gigabytes (GB). The main question is how the DRAM capacity boundaries and the number of cores in the processor influence the computing time.

Experiments have been done on a single node, the rack S2600WF with two Intel(R) Xeon Platinum 8164 2.0 GHz processors with 2×768 GB 2666MHz DDR4 onboard memory. The software packages used are CentOS Linux release 7.5.2804, gcc compiler release 4.7.1, Intel C++ with libraries release 2018.3.

We use Intel(R) Math Kernel Libraries (MKL) [6] which automatically supports features of the Intel Scalable(R) architecture, in particular, the AVX-512 extension, multi-threading, and NUMA.

2. Number of cores influence

The Xeon Platinum 8164 processor has 26 cores and supports 52 threads. Figure 1 demonstrates the boundary of 52 with the multiplication of the two matrices with the linear size 32000 with the single floating point precision and of the linear size 64000 with the double floating point precision. There are two facts on the processor performance which can be extracted from the figure. Firstly, the performance is saturated for the number of threads larger than 52, clearly showing the available number of threads per one processor. Secondly, performance is limited mainly by the memory exchange, as it is factor 8 slower at the peak performance for the matrix size 64000 compared with the matrix size 32000. This factor is due to the 8 times larger memory used for the matrix multiplication of linear size 64000, factor 4 comes from the number of matrix elements and factor two from the number representation, `real*8` instead of `real*4` for the linear size 32000. We will provide more clear evidence for this observation in the following analysis.

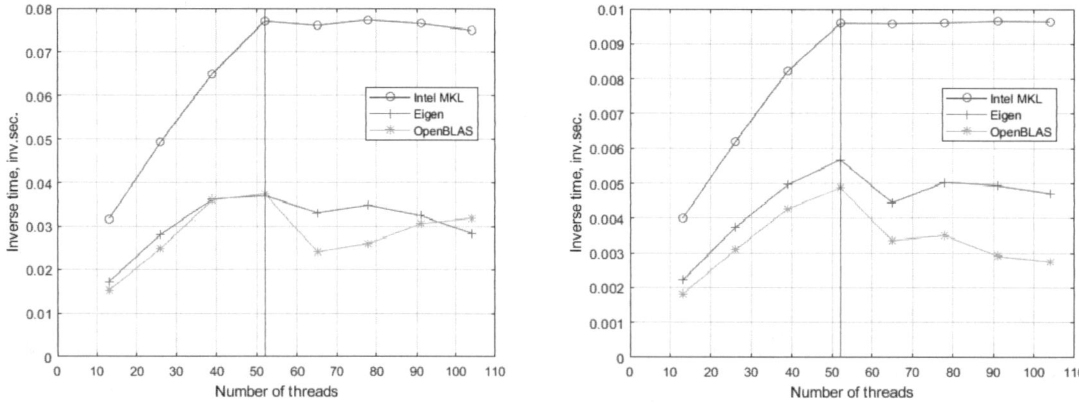

Figure 1. Inverse time for the square matrix multiplication with the linear sizes 32000 (left panel) and 64000 (right panel) as function of the number of threads, for three libraries - Intel MKL [6], Eigen [7] and OpenBLAS [8].

As seen from both panel in Fig. 1 the efficiency growth almost linear with the number of threads up to 52, and demonstrate almost perfect scalability.

We plot also results of matrix multiplication using Eigen library [7] and OpenBLAS library [8], which demonstrates that these libraries are not using all new features of Scalable architecture yet: firstly, the scalability of performance is not linear, the efficiency grows slower than in the case of MKL, and it is saturated early for both libraries, Eigen and OpenBLAS, and even goes a bit lower for the number of threads larger than 52.

3. The RAM size influence

Figure 2 shows the time of multiplication of two square matrices of the linear size 128000. The memory necessary for matrix multiplication with the data type `real*4` of elements is about 196 billion bytes (three matrices of the size 65 billion bytes), and it is $393{\cdot}10^9$ bytes for the `real*8` elements. The tasks are fit within the memory accessible by a single processor, which is 768GB, accordingly Figure 2 does not show any influence of the memory size, and the ratio of the corresponding bars of the same color is about factor of 2, which is just the ratio of the number of bytes for two data types. At the same time, it is visible, that multiplication with the Eigen library becomes faster than with the OpenBLAS library for the larger system sizes, the trend which was already visible in the right panel of Figure 1.

IOP Conf. Series: Journal of Physics: Conf. Series **1163** (2019) 012079 doi:10.1088/1742-6596/1163/1/012079

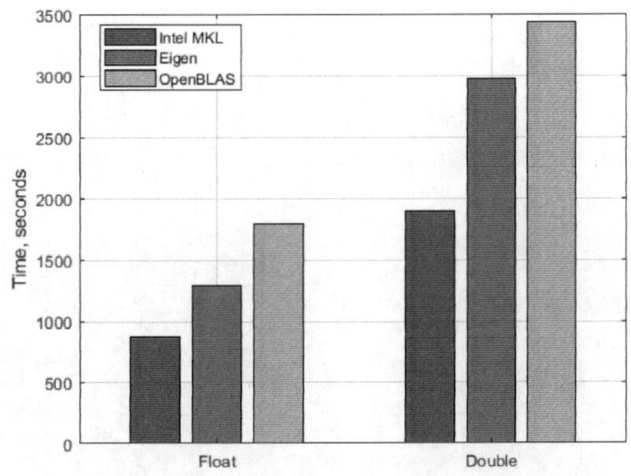

Figure 2. Comparison of the times for the multiplication of the two square matrices of the linear size 128000 using the single precision (left group) and the double precision (right group) for three libraries - Intel MKL [6], Eigen [7] and OpenBLAS [8]. The number of threads is 52.

In the next set of the experiments, we check how the DRAM size accessible for the single processor does influent the efficiency of the simulations. For that reason, in the first experiment, we perform in parallel two sets of matrix multiplication. We choose the size of the matrices in a way, they just fill the whole RAM. It is 6 square matrices of the linear size 128000 with `real*8` elements. In total it is 786 billion bytes which just fits in the DRAM accessible by the single processor. The left panel of the Figure 3 shows the time variation of the two parallel processes as a function of the number of threads. Data for the time can be fit with the expression

$$T = T_{\infty} + A/N_t, \quad (1)$$

i.e. inversely proportional to the number of threads N_t. The resulting fit is shown in the left panel of Figure 3 as the solid line, and fitting parameters are $T_{\infty} \approx 2335$ and $A \approx 41100$. The time of the multiplication achieves the asymptotic value already at the cores threshold of 26.

Next experiment has been performed with the four sets of the same matrix multiplication running in parallel, therefore, the total memory for the matrices exceeds two times the memory directly reachable by the single processor and still fits the onboard memory of 1.5TB. The right panel of the Figure 3 shows the running time of the four processes (marked by circles, stars, rhombs, and squares) as function of the number of threads, and it is seen that first saturation is at the number which is half of the number of cores per processor, and the second saturation is double of the number of cores per processor, which is the maximum number of threads per processor. Probably, the saturation at the value of 13 threads is due to the memory effect but it is not clear. We plan to conduct more experiments to clear up the effect.

It is interesting that data in both panels of the Figure 3 can be collapsed on one curve. For that reason, we divide the number of threads at the left panel by factor 2 and multiply the time at the left panel by factor 2. Resulting data are shown in the Figure 4 by solid stars and squares, while keeping the data of the right panel of Figure 3 unchanged. Figure 4 shows excellent coincidence of the data sets. One can say that the expression (1) is the universal function for the scalability of the matrix multiplication on the Intel Scalable architecture, probably with some restrictions which are the subject of the future work.

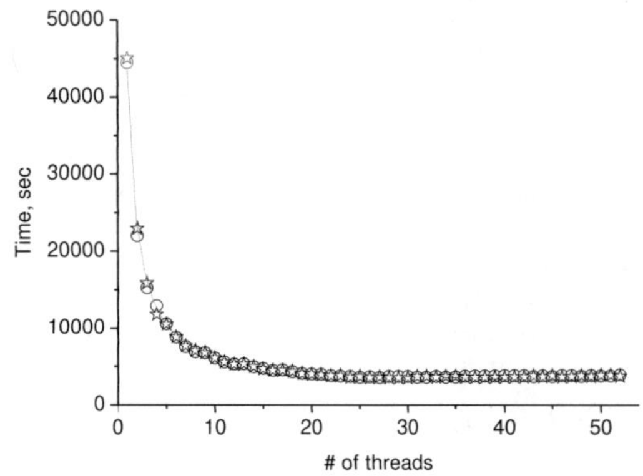

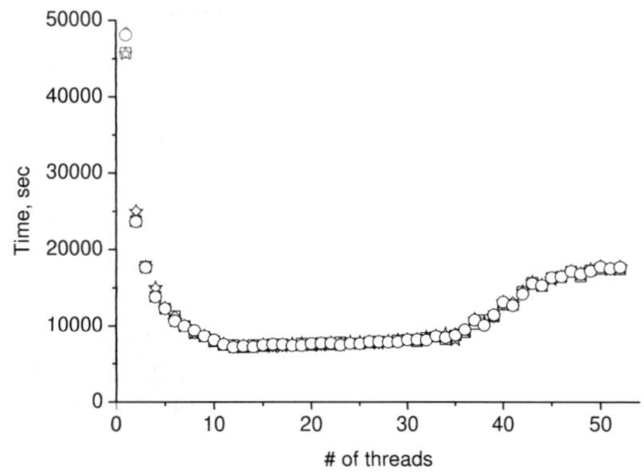

Figure 3. Left panel: comparison of the times for the multiplication of the two pairs (circles and stars) of the square matrices of the linear size 128000 using the double precision. The line is the fit to the data (see the text). Right panel: comparison of the times for the multiplication of the four pairs (circles, stars, rhombs, and squares) of the square matrices of the linear size 128000 using the double precision.

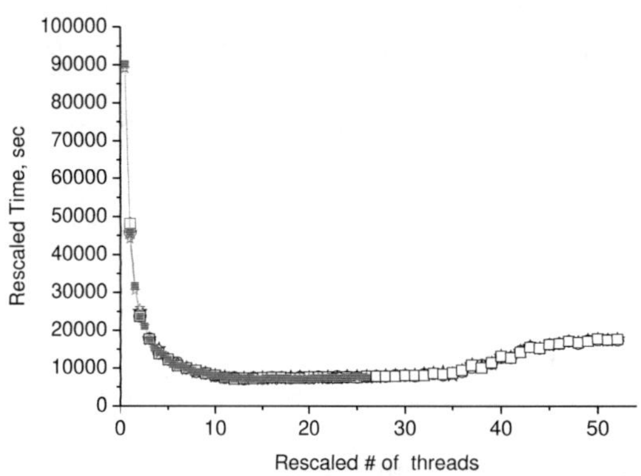

Figure 4. The data collapse as the combination of the data sets from the left and right panels of the Fig 3. See text for explanation.

4. Discussion

Presented experiments with the matrix multiplication of the big size on the Intel Xeon Platinum 8164 with 26 cores and 52 threads (the total onboard number of cores 52 and threads 104) and the big memory of 768GB per processor (total onboard memory is 1.5TB)) shows interesting features of the Intel Scalable Architecture. We found the clear saturation of the efficiency with the number of threads larger than maximum value per processor, which is 52 in our case. We found that libraries Eigen [7] and OpenBLAS [8] is less effective at the moment on the Intel Scalable Architecture comparing with the Intel MKL.

The most interesting results of the communication are the perfect scalability of the matrix multiplication time (see expression (1)) and the universal behavior of the scalability as shown in the Figure 4.

Future experiments can be done to understand better the performance of the Intel Scalable Architecture in computing using the matrix multiplication which is one of the cores of any scientific simulations.

Acknowledgments

The research is done within the framework of the research plan 0236-2014-0002.

References

[1] https://ark.intel.com/products/series/125191/Intel-Xeon-Scalable-Processors
[2] https://www.intel.com/content/www/us/en/high-performance-computing-fabrics/omni-path-driving-exascale-computing.html
[3] P. Boyle, M. Chuvelev, G. Cossu, C. Kelly, C. Lehner, L. Meadows, *Accelerating HPC codes on Intel(R) Omni-Path Architecture networks: From particle physics to Machine Learning*, arXiv:1711.04883.
[4] J. D. McCalpin, HPL and DGEMM Performance Variability on the Xeon Platinum 8160 Processor, 2018 SC18: The International Conference for High Performance Computing, Networking, Storage, and Analysis (SC), Dallas, TX, US, , pp. 225-237.
[5] https://colfaxresearch.com/xeon-2017/
[6] https://software.intel.com/en-us/mkl
[7] https://eigen.tuxfamily.org/, branch 3.3, update 21.09.2018
[8] https://www.openblas.net/, release 0.3.4

AUTHOR INDEX

Institute of Physics
Dirac House, Temple Back
Bristol BS1 6BE UK

ISSN: 1742-6588
ISBN 978-1-5108-8490-8

9 781510 884908

CURRAN ASSOCIATES INC.
proceedings.com